Kevin Tian Yang

Personal Collection

6/22/2016

THE SOVIET UNION AND COMMUNIST CHINA 1945–1950

THE SOVIET UNION AND COMMUNIST CHINA 1945–1950

THE ARDUOUS ROAD TO THE ALLIANCE

DIETER HEINZIG

AN EAST GATE BOOK

M.E. Sharpe
Armonk, New York
London, England

An East Gate Book

Library of Congress Cataloging-in-Publication Data

Heinzig, Dieter.
[Die Sowjetunion und das kommunistische China 1945–1950. English]
The Soviet Union and Communist China 1945–1950 : the arduous road to the alliance / Dieter Heinzig.
p. cm.
Includes bibliographical references and index.
ISBN 0-7656-0785-9 (alk. paper)
1. Soviet Union—Foreign relations—China. 2. China—Foreign relations—Soviet Union. 3. Soviet Union—Foreign relations—1945–1991. 4. China—Foreign relations—1912–1949. I. Title.

DK68.7.C6 H4513 2003
327.47051′09′044—dc21 2003014563

Printed in the United States of America

The paper used in this publication meets the minimum requirements of the
American National Standard for Information Sciences
Permanence of Paper for Printed Library Materials

BM (c) 10 9 8 7 6 5 4 3 2 1

Long live comrade Stalin, teacher of world revolution and best friend of the Chinese people!

—Mao Zedong in Moscow on February 17, 1950, before his return to China

I will write a book about Stalin's mistakes and crimes myself. But it will be so terrifying that I will not allow it to be published for 10,000 years.

—Mao Zedong to Aleksei Kosygin in February 1965

Contents

Abbreviations

CC	Central Committee
CCP	Chinese Communist Party
CCR	Chinese Changchun Railroad
CER	Chinese Eastern Railroad
Cominform	Information Bureau of the Communist and Workers' Parties
Comintern	Communist International
CPSU	Communist Party of the Soviet Union
CPSU(B)	Communist Party of the Soviet Union (Bolshevists)
ECCI	Executive Committee of the Communist International
GRU	Glavnoe Razvedyvatel'noe Upravlenie (Chief Department of Reconnaissance)
KMT	Kuomintang
MGB	Ministerstvo Gosudarstvennoi Bezopasnosti (Ministry of State Security)
MID	Ministerstvo Inostrannykh Del (Ministry of Foreign Affairs)
MPR	Mongolian People's Republic
MVD	Ministerstvo Vnutrennikh Del (Ministry of the Interior)
NKVD	Narodnyi Komissariat Vnutrennikh Del (People's Commissariat of the Interior)
OSS	Office of Strategic Services
PCC	Political Consultative Conference
PLA	People's Liberation Army
PRC	People's Republic of China
RCP(B)	Russian Communist Party (Bolshevists)
RKKA	Raboche-krest'ianskaia Krasnaia Armiia (Red Workers' and Peasants' Army)
UNRRA	United Nations Relief and Rehabilitation Administration

Note on Transliteration

Russian words, with the exception of Mikoyan, Yeltsin, Moscow, Yalta, Crimea, and Kronstadt, are transcribed according to the modified Library of Congress system. Chinese words are basically transcribed in *hanyu pinyin.* A few well-known proper names constitute exceptions to this rule. They will be presented in the written form that was customary internationally at the time of the events described: Sun Yat-sen, Chiang Kai-shek, Chiang Ching-kuo, T.V. Soong, Chang Tso-lin, Chang Hsueh-liang, Peking, Peiping, Mukden, Harbin, Kirin, Port Arthur, Dairen, Jehol, Chahar, Manchukuo, Tientsin, Tsingtao, Nanking, Sian, Yenan, Sinkiang, Urumchi, Kulja, Yangtze, Chungking, Canton, Whampoa, Hong Kong, Kuomintang, and Yenching University. The Pinyin equivalent will be provided in parentheses.

Preface

The idea for this book took shape during the 1980s as the contemporary literature from the People's Republic of China (PRC) began—at first sporadically and then ever-more frequently—to cite from documents in the Party archives that dealt with the history of the relationship between the Communist International (Comintern) and the Chinese Communist Party (CCP). As time passed, albeit very slowly, this literature began to include the relations between Moscow and the CCP during the first years after the dissolution of the Comintern in 1943. Archival documents on Party history dealing with the relationship with the Soviet Union soon followed and important contemporary witnesses made published statements. This new development was a result of the intellectual liberalization that followed Mao Zedong's death.

Following the collapse of Communism in the Soviet Union, the doors to the Moscow archives began to open—if not completely, then at least enough to leave a sizable crack. Here, too, contemporary witnesses began to present their memoirs to the public.

These events have meant an almost dramatic improvement in the availability of sources. Western research had previously been dependent on publications from the USSR and the People's Republic of China, in which censorship had ensured that the historiographical boundaries the Party had established were observed strictly. These publications seldom cited archival materials.

Until the late 1950s, the literature in the People's Republic of China and the Soviet Union gave the impression that practically no problems had ever arisen between Moscow or the Comintern and the Chinese Communist Party. It presented a deterministic story in which the October Revolution had triumphed in Russia in 1917, and the proletarian revolution in China in 1949, in keeping with the iron laws of historical materialism; and it further claimed that these two great socialist states had established a perpetual alliance in February 1950, again in keeping with historical laws. During the 1950s, the historiography on both sides never tired of following the official propaganda line and praised the harmony of the eternal brotherhood with tremendous emotion.

During the Sino-Soviet conflict of the 1960s and 1970s, the leaderships of both states—with the Party historians in their wake—for the first time admitted that differences had repeatedly emerged in the relationship between Moscow and the CCP ever since the 1920s. To the extent that they introduced any new historical facts at all, they did so to put the other side in the wrong, to accuse it of deviating from the true path of Marxism-Leninism, and to attribute to it the sole blame for disrupting bilateral relations. Western historians often found themselves unable to separate fact from fiction as here, too, neither side generally made any attempt to lend credibility to their polemical attacks by citing archival documents.

The new types of sources that have become available since the 1980s make it possible to gain insight into hitherto unknown events that are of central importance to the development of the relationship between Moscow and Communist China. We can now partially or completely fill in numerous blanks. We can understand more fully the conflicts that emerged during the Comintern period and confidently answer questions that had hitherto remained controversial, such as whether the Chinese Communist Party ever requested weapons—in vain—from Moscow during the Sino-Japanese War and whether Stalin really restrained the Chinese comrades during the revolution in 1945. Stalin's double game in Manchuria after the Second World War appears in a new light. It is now possible to provide convincing documentation of Anastas Mikoyan's secret meeting with the Chinese Party leadership and Liu Shaoqi's secret meeting with Stalin in 1949, which Western research did not even know had taken place. And we can now for the first time reliably reconstruct Mao Zedong's visit to Moscow between December 1949 and February 1950, and the negotiations for the treaty of alliance and accompanying agreements.

The author had hoped that this new openness with respect to access to archival documents would be extended to the whole period of relations between the Soviet Union and Communist China, so that it would be possible to reassess the entire history of these relations. But this hope has unfortunately not materialized. The conclusion of the Sino-Soviet treaty complex in early 1950 proved to be a temporal break, after which the flood of new sources practically dried up. Exceptions such as the Soviet documentation of the correspondence between Moscow, Pyongyang, and Peking with respect to the Korean War and individual documents from the 1950s and 1960s merely prove the rule.

The author therefore decided to place the period between the two Sino-Soviet treaties of alliance—the one the USSR concluded with the Nationalist Government on August 14, 1945, and the one it signed with the government of the People's Republic of China on February 14, 1950—at the center of his study. There are several reasons for this choice. The most important is that voluminous new archival source material and statements from contemporary witnesses from both sides are only available on the Soviet and the Chinese side for this period. It is also a crucial period, during which the strategic coalition between the Soviet Union and Communist China gradually took shape. We will discuss the period from 1921 to 1945 in an introductory chapter and developments from 1950 to the present in a final chapter. The book thus addresses—albeit to differing degrees of intensity—the entire period of relations between the Soviet Union and Communist China between 1921 and the present.

Some of the new sources have already been used in Western monographs.[1] A considerable number of these only became available after they were published, however, and are assessed for the first time in the present study. Many of these sources continue to be available only in unpublished form. These include documents in the Russian presidential archives (AP RF), the Russian Foreign Ministry archives (AVP RF), and the successor institutions (RTsKhIDNI and TsKhSD) to the Central Party archives of the Communist Party of the Soviet Union (CPSU). The author also interviewed important contemporary witnesses, such as people on both sides who interpreted during summit meetings and high-ranking Soviet diplomats who served in China at the time. The study focuses on processing and analyzing new sources. Particular care is taken to criticize the sources, and familiar sources will be taken into account for this purpose.

My colleagues Joachim Glaubitz, Klaus Sagaster, Wolfgang Lippert, Lev Deliusin, Chen Jian, Andrei Ledovskii, Konstantin Shevelev, Peter M. Kuhfus, Zhu Ruizhen, Dietrich Frenzke, Liparit Kiuzadzhian, Eva-Maria Stolberg, Vasilii Sidikhmenov, Shi Yinhong, Zheng Yifan, Chen Zhihong, Gao Fang, Zhang Wenwu, Wang Jifei, Gu Junli, Yang Yunruo, Wang Jin, Yang Kuisong, Li Haiwen, Zhou Wenqi, Xue Xiantian, Xing Shugang, Wang Qi, Shen Jinfu, and Liu Keming have given me incalculable assistance by exchanging ideas, providing assistance, and being there to help with materials.

Some contemporaries who witnessed events described in the book firsthand agreed to be interviewed: Lev Deliusin, Rishat Kudashev, Andrei Ledovskii, Sergei Tikhvinskii, Yan Minfu, Zhu Ruizhen, Li Yueran, Shi Zhe, Rong Zhi, Sergo Mikoyan, Nikolai Fedorenko, and Zhao Yan.

The Federal Institute for East European and International Studies, Cologne, gave me valuable assistance. Its director, Heinrich Vogel, provided constant support and encouragement as the book evolved. Anna Eckner assisted me tirelessly with the translation of Chinese texts and read proofs. Johann Matern energetically procured Russian-language materials. And Barbara Langer, the institute's publication officer, provided assistance in publishing the work.

I thank everyone named above from the bottom of my heart. I, of course, am alone responsible for the views expressed in the book.

I would like to express my gratitude to the Deutsche Forschungsgemeinschaft (DFG) for financing my research activities in Moscow and Peking and to the Goethe-Institut Inter Nationes e.V. for helping to fund the present English-language edition. I would also like to extend my appreciation to my old friend David J.S. King, who did the translation, and Patricia Loo, who is in charge of M.E. Sharpe's Asian Studies Program, for their great cooperation.

I dedicate this book to my wife, Ellen, and my son, Henning. Without my wife's patience and thoughtfulness, I would never have been able to write it.

Dieter Heinzig
Erftstadt, June 2002

THE SOVIET UNION AND COMMUNIST CHINA 1945–1950

1

Background

The Emancipation of the Chinese Communist Party from Moscow

As the Second World War drew to a close, leaders in the Soviet Union and the Chinese Communist Party (CCP) had every reason to look back with mixed feelings on the evolution of their mutual relations to that point. On the one hand, they continued to feel the bond of their common Marxist-Leninist ideology and eschatological belief in world revolution. On the other, both sides had constantly pursued their own policy interests, and these had repeatedly led to conflicting goals and irritations.

The first tensions had already emerged during the 1920s. Mao Zedong's rise to power as Party leader, which began in 1935, increased the number of conflicts considerably and initiated the process by which the CCP became emancipated from Moscow—that was completed by the end of the war. The period from 1935 to 1945 is thus considerably more significant to the topic of this book than the one that preceded it. In this chapter, we will therefore focus our historical review on this period, and the following section will provide only a cursory sketch of developments between 1921 and 1935.[1]

The CCP on the Comintern Leash (1921–35)

We can only understand the early relations between Moscow and the Chinese Communists if we view them in the context not only of Soviet policy toward China but also of the USSR's overall foreign policy. The latter shifted during the first half of the 1920s from an internationalist to a nationalist Russian course. In 1927, Stalin reduced the result of this process casually and pedantically to the slogan that "he is revolutionary and internationalist who is prepared to protect the Soviet Union unreservedly and unconditionally."[2]

The Communist International (Comintern), founded in 1919, became the leash by which Moscow communicated its policies to the other Communist parties. From 1920, the resolutions of the congresses and the Executive Committee of the Comintern (ECCI) were binding on all Communist parties as "sections" of the Comintern.[3] The Comintern was entitled by statute to intervene in the affairs of individual parties at any time. All important Comintern resolutions were formulated not by the Comintern itself, however, but in the Politburo of the Central Committee (CC) of the Russian Communist Party.[4] After its Fourth World Congress in 1922, the Comintern advocated Soviet interests ever more clearly. Nikolai Bukharin demanded that the proletariat of other countries defend the Soviet state.

At the Fifth World Congress (1925), the sections were required to take the experiences of the Russian CP extensively into account. Between 1925 and 1930, the Comintern finally became a tool Stalin used to discipline foreign Communist parties in the pursuit of Soviet interests. On a visit to Moscow in 1928, the Chinese Communist Zhang Guotao received the impression that the Comintern was no longer "the general headquarters for the world revolution" but had become "Stalin's plaything for bullying the Communists of various nations."[5]

Moscow had already begun to pursue nationalist Russian interests associated with the traditions of czarist expansionism in China in mid-1919. One clear indication of this is the fact that the Soviet government rescinded the offer it had already made to Peiping to return the Chinese Eastern Railroad (CER) without compensation.[6] It became clear that the Soviet Union wished to preserve or restore Russia's traditional influence in Manchuria and Outer Mongolia. Three years later, Chang Tso-lin, then the ruler of Manchuria, told Sun Yat-sen what he thought about Soviet behavior to China. He said that Moscow was pursuing an imperialist policy in China, because, despite all its protestations of friendship, it was not returning the Chinese Eastern Railroad or Mongolia.[7] In subsequent years, the Soviet leadership's attempts to follow the legacy of czarist Russia in its policy toward China offended all political forces in China—from the warlords through the Kuomintang (KMT) to the Communists, whose leaders considered themselves Chinese patriots in addition to accepting the dogma of world revolution. As will be shown in detail below, this was particularly true of Mao Zedong.[8]

Since 1922, the reins of China policy in the Soviet leadership had clearly all come together with Stalin.[9] In 1928, Otto Kuusinen, the ECCI secretary, confirmed that Stalin was the supreme authority on all matters that involved China.[10] It is therefore legitimate to use the terms "Soviet Union," "Soviet leadership," "Comintern," "Moscow," and "Stalin" interchangeably with regard to Soviet policy toward China, as we have done in the present book.

The Chinese CP was founded in 1921 under the auspices of the Comintern. Differences had already emerged between Moscow and the Chinese Communists during the 1920s. They were primarily due to the fact that Stalin chose to deal with the KMT as the USSR's preferred partner in China—a strategy that he essentially maintained until the end of the 1940s.[11] His idea was that the KMT, as the strongest progressive political force in the country, would best be able to defeat the reactionary warlords, unify China, remove it from any (non-Soviet) foreign influence, and thus prepare it, if unintentionally, for the Communist seizure of power, which the CCP would carry through under Soviet direction. The Chinese Communists, whose primary goal was to seize power in the struggle with the KMT, initially played no significant role in Moscow's calculations because they were too weak militarily and politically. This lasted until Stalin realized in late 1948 that they were about to be victorious in the civil war.

The first tensions emerged when Moscow instructed the CCP through the Comintern to form a bloc with the KMT in 1922. This constituted the basis of Soviet military aid between 1923 and 1927, which for the most part benefited only the KMT. Despite the fact that Chiang Kai-shek had severely restricted the influence of the Communists and Soviet military advisers in the KMT in a kind of coup d'état on March 20, 1926, the Comintern continued to cling to the alliance with Chiang.

After the signs of the imminent massacre of unionized workers Chiang arranged in Shanghai on April 12, 1927, had become clear, the Comintern advised the Communists and the workers to avoid open conflict and thus was in great part responsible for the fact that they were headed for disaster. Even after that, Comintern delegates in Wuhan continued to seek to maintain the united front with Chiang. Only several days later did the ECCI bring itself to condemn Chiang Kai-shek as a "traitor to the revolution." Chiang's right wing in the KMT, which Stalin wanted to "utilize to the end, squeeze out like a lemon, and then fling away,"[12] made the break with the Chinese Communists and the Comintern in April 1927. Stalin then instructed the CCP leadership through Mikhail Borodin to support the left-wing KMT government in Wuhan, to which the Communists should cling like to a lifesaver.[13] When the Wuhan KMT also expelled the Communists in July, the Soviet military advisers had no alternative but to leave China. Stalin cynically blamed the CCP leadership under Chen Duxiu for clinging so long to the alliance with the KMT: Chen was made the scapegoat, and Qu Qiubai replaced him as secretary-general.[14]

In the summer of 1927, Moscow and the Chinese CP resorted to organizing armed uprisings in four provinces, all of which failed. This was apparently done on the initiative of Stalin, who was trying in this way to use the dispute with the Trotskyite opposition to cover over the errors in his China policy to that point. The Comintern agent Besso Lominadze and the chief Soviet military adviser, Vasilii Bliukher, took part in the consultations with the CCP leadership. The first revolts took place on August 1 in Nanchang. The "autumn harvest uprisings" followed in rural areas of Hunan, Hubei, Jiangxi, and Guangdong provinces and in Changsha city. An armed revolt in Canton in December, which the Comintern agent Heinz Neumann had played a significant role in planning and implementing, also failed.[15] These setbacks resulted in bloody losses to the Chinese Communists, while Neumann escaped. Despite this, it was the CCP leadership—and not the Comintern—that had to exercise self-criticism in connection with the failed uprisings and for clinging so long to its cooperation with the KMT at the Sixth Comintern Congress in July 1928: the Party leadership had, it confessed, made "serious opportunistic mistakes." And at the Seventh Party Conference of the CCP in June–July 1928 in Moscow, which the Comintern kept on a tight leash, the CCP leadership under Qu Qiubai, but not Stalin, bore the full blame for the failure of the revolts during the second half of 1927: it had allegedly followed a "putschist line." The congress took place in the shadows of the conflict between Stalin and Bukharin. The CCP leadership had already criticized Mao previously for the failure of the "autumn harvest uprisings" he had planned in Hunan.[16]

Moscow's tactic of infiltrating Chinese Communists into the KMT came to nothing. In December 1927, the Nationalist Government of the reunited KMT in Nanking broke off diplomatic relations with the USSR. Soviet policy toward China had thus failed across the board. Chiang Kai-shek conquered the warlords in North China in 1928 and began a campaign of annihilation against the Communists in the south.

It was obvious that the directives from Moscow had resulted in a debacle for the Chinese Communists. Only scattered cells in a few large cities, the remains of rebel troop units, and numerous peasant troops were left to the Party. Independently of the

Party leadership, which remained obedient to the Comintern and was seeking to reconstruct Communist organizations in the cities, Mao united his peasant troops with troops under Zhu De in the mountains of the border area between Hunan and Jiangxi in spring 1928, and formed a Soviet region. Soviet regions also arose in other provinces such as Hunan, Hubei, Anhui, and Fujian. They were conceived as fixed territorial bases from which the revolution would be carried to the cities. Mao had drawn the conclusion from the failure of the city revolts, which the Comintern had recommended, that the revolution could only be carried out successfully from the villages. A new trend in Chinese Communism that was largely independent of Moscow and of the CCP leadership of the time thus began that would ultimately be successful.[17]

Stalin viewed Mao's and Zhu's actions with skepticism—above all because he had not approved them in advance. Moscow had after all also learned from the mistakes of the past with respect to revolutionary tactics. In February 1928, the ECCI recommended to the CCP leadership to avoid coup attempts in the cities and peasant uprisings for the time being. They should instead carry out land reform in Soviet regions and begin a guerrilla war from there—an idea not unlike Mao's. These recommendations also influenced the resolutions of the Sixth CCP Party Congress.[18] From February 1930, the CCP leadership under Li Lisan criticized Mao Zedong and Zhu De—albeit from Shanghai—for clinging to the "old-fashioned ideas" of "evasive distraction." Their troops, it argued, should move against important cities and major lines of communication instead. The concept of surrounding the cities from the countryside was supposedly a great mistake. In August, Li developed a general plan for conquering Changsha, Nanchang, and other major cities.[19] To carry it out, he requested the support of a military intervention by Moscow in North and East China. When the Comintern refused to provide this, Li accused it of violating the principle of internationalism and threatened to terminate his Party's loyalty to Moscow. The "Li Lisan line," which was not based on instructions from Moscow, ended in fiasco when the conquest of Changsha failed in August and September.[20]

The Comintern then intervened directly in its Chinese section's leadership structure. It ordered Li Lisan to come to Moscow, where members of the ECCI cross-examined him in November, ordered him to exercise self-criticism, and prohibited him from returning to China. At the same time, the ECCI sent Pavel Mif (Mikhail Fortus), deputy head of the Eastern Secretariat of the ECCI, to China, where he arranged to replace the CCP leadership with a group of Chinese comrades loyal to the Comintern and trained in Moscow in January 1931. Their opponents in the Party gave them nicknames such as "the twenty-eight Bolsheviks [*ershiba ge buershiweike*]" and "the Chinese Stalin group" [*Zhongguo Si-da-lin pai*]." They were led by Wang Ming (Chen Shaoyu), who had been trained in Moscow at the Sun Yat-sen University in the mid-1920s and was distinguished by his unquestioning loyalty to the Soviet—or to what he took to be the Soviet—line until 1974, when he died in exile in Moscow. He returned to Moscow in 1932, where he represented the Chinese CP in the Comintern until 1937. In this capacity, he supported the Party leadership that remained in Shanghai until 1933 in its attempts to gain control of the various rural Soviet regions, and of the Central Soviet Region under Mao Zedong in Jiangxi in particular.[21]

For various reasons, the Comintern's influence diminished somewhat after 1931. The CCP leadership apparently heard from a courier in late 1935 that in response to the resolutions of the Seventh World Congress of the Comintern, the ECCI would in future no longer intervene in the Chinese Communists' organizational affairs and would only continue to supervise their policies.[22] It became clear one year later that this supervision continued to be designed to serve Soviet and not Chinese interests when the Comintern interpreted Japan's aggression against China as actually directed against the Soviet Union, called on all Communist parties "to protect the USSR,"[23] and made no mention of protecting China.

Thus concluded the first chapter in the relations between Moscow and the Chinese Communists, which had for the most part been characterized by a high degree of Soviet influence over the Chinese CP. For the most part, it did not end because of the restrictions the Comintern imposed on itself in late 1935. Instead, developments within the Communist Party that Moscow could not influence largely brought about the change.

Mao Zedong's Rise to Party Leader: Prerequisite for Emancipation

The new development, which began in 1935, led step-by-step to the CCP's emancipation from the Soviet Union by 1945. The most important step in this development can be seen to have begun when Mao Zedong was able to assert his authority against the faction subservient to Moscow that had predominated to that point in an Enlarged Session of the Politburo in Zunyi (Guizhou Province) during the legendary Long March in January 1935. Mao had already realized that directives from Moscow revealed little familiarity with the situation in China and were unlikely to lead the revolution to victory and had begun to act independently from as early as 1927. He appears to have assumed military leadership de facto after the Zunyi Conference and to have thus laid the foundations for his rise to Party leader.

Secretary-general Bo Gu (Qin Bangxian) and the German Comintern military adviser Otto Braun (Li De; Hua Fu), who led the Stalin group in the Party leadership at that time and held the military leadership in their hands along with Zhou Enlai, were relieved of this responsibility at the Zunyi Conference and could exert practically no further influence on military decisions after the beginning of 1936. At Zunyi, Mao moved up into the Standing Committee of the Politburo and Bo Gu was removed from this body. In the two following months, Bo Gu also lost his position as "first officer of the Central Committee [*zhongyang fuzeren* or *zong fuzeren* (a sort of secretary-general)]" and was replaced by Luo Fu (Zhang Wentian). At the same time, Mao moved into the military leadership and finally assumed control of the "military group of three [*sanren junshi xiaozu*]," which also included Zhou Enlai and Wang Jiaxiang. This group of three constituted the supreme command of the military during the remainder of the Long March. Mao was able to attain such a prominent position in this way primarily because the main problems of that time involved the military, and the Party as such had withdrawn to the background.[24] By the end of the Long March in October 1935, Mao was apparently in charge of the three most important governing bodies in northern Shaanxi: the Party leadership (represented by Luo Fu), the Military Commission of the Central Committee (represented by Zhou Enlai), and the Soviet government (represented by Bo Gu).[25]

Moscow could not have influenced these events, even if it had wished to do so, as the radio connection with the leadership of the Chinese CP had been interrupted in the late summer or fall of 1934 and could only be resumed in mid-1936.[26] This fact is emphasized in the historiography of the People's Republic of China, where the Zunyi Conference is celebrated as the breakthrough on the road toward emancipation from Moscow: "The conference took place while the connection between the CCP and the Comintern had been interrupted. For the first time, the CCP chose its own road reflecting conditions in the country in an independent and sovereign way. . . . This was an existential turning point in the history of the CCP."[27] Researchers debate about how Stalin and the Comintern received the news of Mao's increased power; this cannot yet be documented.[28] There are indications, however, that Wang Ming and Kang Sheng of the Chinese Comintern section in Moscow at least were dissatisfied with it.[29]

Mao expressed this new sense of self-confidence toward Moscow remarkably clearly in the summer of 1936:

> In the USSR, the Communist Party is in power, yet even there the Third International does not rule nor does it have any direct political power over the people at all. Similarly, it can be said that although the Communist Party of China is a member of the Comintern, still this in no sense means that Soviet China is ruled by Moscow or by the Comintern. We are certainly not fighting for an emancipated China in order to turn the country over to Moscow![30]

Wang Ming returned to China at the end of 1937 with the first Soviet aircraft to land in Yenan. By playing intensively on the Comintern's authority, he was able to get his way at least formally against Mao—for instance, on the issue of Mao's refusal to cooperate closely with the KMT against Japan in particular—at a Politburo session in December 1937. Georgi Dimitrov, secretary-general of the Comintern between 1935 and 1943, supposedly advised Wang Ming not to seek the position Luo Fu held as secretary-general of the CCP. If Wang Ming did seek the position anyway, for which there is evidence, he was unsuccessful; the position was abolished.[31]

Relations between Moscow and the Chinese leadership were damaged further when Wang Ming insisted at the December session that the CCP had to fight Trotskyism just as ruthlessly as Stalin was fighting it in the Soviet Union. The application for readmission into the Party of the former Party leader, Chen Duxiu, who had been expelled from the Party for Trotskyism at the beginning of 1929, could never be accepted. It did not matter whether Chen really was a Japanese spy—he had to be one, Wang Ming argued. If Chen was readmitted anyway, there would be "unthinkable consequences" if Stalin found out about it.

The Party leadership resolved to reject Chen's application. It appears, however, that many members of the leadership were uncomfortable with Wang Ming's fanatical insistence that they take measures to persecute Trotskyism. Mao was one of them. He had previously intervened in favor of Chen's readmission and stated in a private conversation that Stalin was making a serious mistake in "suppressing counterrevolutionaries." Wang Ming had had five long-serving Party functionaries tortured and executed as Trotskyites in Sinkiang. As Zhang Guotao had worked closely with them, he felt threatened himself and left the Party in April 1938. Regardless of whether it is true that Wang's attacks on

Chen were actually aimed at Mao, as has occasionally been claimed, the decision of Zhang Guotao, who had previously been a weighty opponent to Mao Zedong, to leave the Party was a significant victory on Mao's road to become the undisputed Party leader.[32]

Wang Ming's star began to wane when his strategy for the defense of Wuhan, which he carried out despite instructions from Yenan, proved a failure in the summer and fall of 1938 while Mao's strategy in the north led to the reinforcement of the Communist bases. They could not help but see this even in Moscow. Furthermore, Ren Bishi, the new CCP representative in the Comintern, was able to convince Dimitrov that spring that Mao and not Wang Ming should be seen as the Chinese Communists' actual leader. *Pravda* published a photo of Mao (and Zhu De) for the first time on July 6 and described him as "Chairman of the government of the special region of the Republic of China."[33] In August 1938, Dimitrov acknowledged Mao's leadership role clearly to Ren Bishi and Wang Jiaxiang and distanced himself from Wang Ming.[34] The plenary session of the CCP Central Committee that met between September and November 1938 in Yenan revealed that Mao had again gained ground against Wang Ming, although he had not actually made the breakthrough, and that Mao and Stalin had reached a compromise. Mao bowed—at least pro forma—to pressure from Moscow to allow the CCP to cooperate with the KMT against the Japanese and recognized the KMT as the "party in power" and Chiang Kai-shek as "leader" of the "Chinese people." Moscow in turn accepted Mao's demand for the independence of the CCP in the alliance with the KMT—also at least pro forma—and approved Mao's tactic of "continuous guerrilla warfare" in the countryside with the goal of extending the Communist bases even if this would lead to tensions with the KMT, and came to terms with Mao's position as de facto Party leader.[35]

After disputes with the Comintern around New Year's 1941, Mao began to discredit his critics around Wang Ming and Bo Gu, who were loyal to Moscow and described themselves as "one hundred percent Bolsheviks [*bai fen zhi bai de buershiweike*]" and who at first apparently also included Zhou Enlai and Luo Fu, as "subjectivists" and "dogmatists."[36] In October 1941, he finally succeeded in forcing Wang Ming, who was representing Moscow's line in an intense debate, into a corner.[37] Wang supposedly found it necessary to use the radio of the Soviet representatives in Yenan to communicate with Moscow—without Mao's knowledge.[38] Years later, Mao counted the situation Stalin created by supporting the "Wang Ming line" in retrospect as one of the three main areas of conflict in the relations between Moscow and the CCP.[39] Wang Ming, Mao said, had been "Stalin's descendant [*Si-da-lin de houdai*]."

From 1942 to 1944, Mao carried out a "rectification movement [*zhengfeng yundong*]" in Yenan with the help of security chief Kang Sheng to force all Party members to be loyal to him personally and to his "Sino-Communist" line. The campaign was clearly designed above all—in Soviet jargon—to eliminate all Wang Ming's "internationally" oriented supporters, which Mao dismissed as the "Moscow group." Party members who remained in the Soviet Union were labeled "dogmatists" and at times also "right-wing opportunists." In December 1943—that is, after the Comintern had been dissolved—Dimitrov, who had since become responsible for contacts with the Communist parties in the Central Committee apparatus of the Communist Party of the Soviet Union (CPSU), accused the CCP leadership of mistreating members of the Moscow group, naming Zhou Enlai and Wang Ming specifically, and criticized the role that Kang Sheng played in this.[40] Mao's campaign created an atmosphere of fear. Methods of psychological terror also known from other states under Communist rule were employed. Numerous Party members were confronted with absurd accusations and forced to accuse themselves and denounce others. Numerous people were arrested and imprisoned for years; some were clearly driven to suicide. There are supposed to have also been cases of poisoning and executions. Wang Ming was destroyed politically, but he was allowed—clearly out of consideration for Moscow—to remain a member of the Central Committee.[41]

The historiography of the People's Republic of China admits that Kang Sheng forced people to make confessions. But at the same time it claims that he had learned this method from the purges in the Comintern and Soviet Union. Furthermore, Mao Zedong and the Party leadership had already corrected these "mistakes" long before Dimitrov made his criticism.[42]

When the rectification movement was called off, Mao had definitively defeated his opponents. The dissolution of the Comintern on June 8, 1943, helped Mao to eliminate the Moscow group. When this became known in Yenan, Mao could not disguise his satisfaction.[43]

Mao's approach in eliminating the Moscow group was skillful in that he constantly praised the Comintern, the Soviet Union, and Stalin to the skies on every appropriate occasion while at the same time he gradually isolated his pro-Soviet opponents. On March 20, 1943, he formally assumed the office of chairman of the Politburo and Secretariat. The Seventh Party Congress (April–June 1945) formally put an end to this development: Mao was elected chairman of the Central Committee on June 19, 1945. At the same time, a grandiose Mao cult was initiated. On the eve of the Party Congress, a "Resolution on certain questions of the history of our Party" was published, which falsified Party historiography, constituted a form of hagiography of Mao, and settled accounts with his most prominent opponents in the Party, such as Chen Duxiu, Li Lisan, and Zhang Guotao. Out of consideration for relations with Moscow, Wang Ming's name was not mentioned. The quality of the distortion of history in the resolution can easily be compared to Stalin's notorious "short textbook [*kratkii kurs*]" on the history of the CPSU, which it may even have been emulating.[44]

The "Sinification of Marxism" as Ideological Challenge to Moscow

Following the rectification movement, the "Sinification of Marxism," or adaptation of Marxist theory to social reality in China as Mao understood it, was completed. This was seen at the time as an ideological challenge to the Soviet Union, whose leadership prided itself on possessing the monopoly on the only valid and globally conceived interpretation of Marxism-Leninism.

The Sinification of Marxism was closely linked to a Sino-centric and at the same time Asia-centric worldview, of which Mao had already given an indication in the summer of 1936. At that time, Mao claimed for the first time that the Chinese revolution was a key factor in the world situation. When it developed its full strength, the masses of many colonial countries would follow China's example, he claimed.[45]

During the second half of 1937, Mao began to develop a philosophical basis for his adaptation of Marxism to concrete social conditions in China.[46] He employed the formula Sinification of Marxism for the first time at the Sixth Plenary Session of the Central Committee in October 1938, although it was not included—clearly out of deference to Moscow—in the final resolution. With this formula, Mao established his claim to a special form of Marxism valid only for China: "Consequently, the Sinification of Marxism [*Ma-ke-si zhuyi de Zhongguo hua*]—that is to say, making certain that in all of its manifestations it [Marxism] is imbued with Chinese peculiarities, that is, that it is practiced according to these peculiarities—becomes a problem that must be understood and solved by the whole Party without delay."[47]

Mao also sought in this way to lay the ideological foundations for any differences between the actions of the Chinese Communists and Moscow's directives that were designed primarily to preserve Soviet interests, and to set the revolution in China and the CP's seizure of power as the primary goal.[48]

In January 1940, Mao distanced himself clearly from the "Soviet road." The "Chinese democratic republic that we desire to establish now" could only be a "new-democratic republic," he stated. In it, the proletariat (as the leading force), the peasantry, the intelligentsia, and other sections of the petty bourgeoisie would constitute the basic forces determining China's fate. The new democratic republic would differ both from the old European-American style capitalist republics and also from "the socialist republic of the Soviet type." Mao went even further: he exalted the "New Democracy" as a model for the revolutions in "all colonial and semi-colonial countries." In this context, Mao distinguished three types of state system: the type represented by the "old democracy" in many of the capitalist countries that had since degenerated into a "bloody military dictatorship of the bourgeoisie," or was on the way to becoming one; the type represented by the Soviet Union under the dictatorship of the proletariat, in which the "old democracies" would become transformed and which for a certain period would be the dominant form throughout the world; and, finally, the "transitional form of state" of the New Democracy, which was "to be adopted in the revolutions of the colonial and semi-colonial countries" as currently existed in China—that is, as a united front against Japan made up of the "joint dictatorship of several anti-imperialist classes."[49]

At the Seventh Party Congress in the spring of 1945, Mao self-confidently allowed Liu Shaoqi, then secretary of the Central Committee and deputy chairman of the Central Committee Military Commission of the CCP, to praise him as a genius, one who had already begun the work of sinifying Marxism: "Comrade Mao Zedong . . . is a creative and talented Marxist, combining as he does the universal truth of Marxism—the highest ideology of mankind—with the concrete practice of the Chinese revolution. . . . The thought of Mao Zedong is . . . Communism and Marxism applied to China . . . is the best expression of Marxism applied to a given nation."[50]

In the same context, Liu went yet another step further by claiming that the "Thought of Mao Zedong" would make significant contributions to the struggle for the emancipation of the peoples of all countries in general, "and of the peoples in the East in particular."[51] The Chinese Party leadership's claim to have given Marxism under Mao Zedong's leadership an independent form that could serve as a model for Asian and other colonial and semicolonial countries was raised repeatedly in the following years until 1951.[52] Again it was Liu Shaoqi, since elected deputy Politburo chairman, who claimed in the spring of 1946 that Mao as a Chinese had transformed Marxism "from a European to an Asiatic form."[53] Marx and Lenin had been Europeans, he argued, who wrote in European languages about European histories and problems and only seldom discussed Asia or China. As conditions in other Southeast Asian countries were similar to those in China, the Chinese road would influence all these countries.

This clearly constituted a challenge to the Soviet Union's position as "center of the world revolution" with universal ideological claims. Moscow observed Mao's behavior uneasily.[54] The same was true of China's implicit attempts to place the Thought of Mao Zedong on the same level with Marxism and Leninism. Moscow passed over statements to this effect in silence.

We can assume that in the years 1937–38, Mao was interested in using the introduction of the formula Sinification of Marxism primarily to distance himself from Moscow and emphasize his right to pursue his own road in the ideological area. He did so at a very theoretical and abstract level then. It would be another two decades before this road assumed a concrete form that drastically affected society. Mao's disastrous concept of the Great Leap Forward and of people's communes in the late 1950s, which had not been arranged with the Soviet leadership, had the effect of a huge and directed provocation in Moscow and played a significant role in the development of the Sino-Soviet conflict.

The CCP and the Soviet Union During the Sino-Japanese War: Allegiance with Reservations

The Soviet leadership never successfully regained total control of the Chinese CP after Mao Zedong began his rise to Party leadership in Zunyi. Emancipation from Moscow in this sense meant that the Chinese Party leadership often but by no means always followed Soviet instructions, which were presented in the form of recommendations after the dissolution of the Comintern in 1943.

The change took the form among other things of a certain ambivalence in Mao's behavior toward Moscow in which statements of ideological loyalty were interspersed with nationalist outbursts. Two quotations from the succeeding period provide examples of this.

In the fall of 1935, Mao stressed at a Party conference that "we are not an independent communist party; we are a branch of the Comintern. Our Chinese revolution is a part of the world revolution."[55] On the other hand, as mentioned above, he also said to an American journalist in mid-1936 that although the CCP was a member of the Comintern, Communist China was by no means ruled from Moscow, and the CCP would not hand China over to Moscow.[56]

During the Second World War, the CCP leadership followed—at least outwardly—the shifts in the changing course of Soviet policy toward the "imperialist states." It did not always accept Moscow's suggestions regarding the Party's policy in the Chinese civil war or the struggle against the Japanese aggressors. Here they followed the internally accepted slogan that they did have to listen to the Soviets, but also that they could not listen to them completely (*bu ke bu ting, you bu ke jin ting*).[57] The differences between them involved the Soviet attempt to make the CCP fight decisively and in coordination with the KMT against Japan. The greater the Chinese resistance, the more Japan would be tied down in China, and the threat of a Japanese attack on the Soviet Union would be reduced correspondingly. Stalin's calculations conflicted with Mao's line, which was designed to preserve Chinese strength in the war against Japan so that it would be ready for the final struggle against the KMT. Another important source of conflict was Moscow's refusal to deliver weapons and other military equipment to the Chinese comrades out of consideration for the KMT. We will provide further details about both points of contention below.

Current historiography in the People's Republic of China also reflects the CCP leadership's ambivalence at that time. There is a balance between recognition and criticism of Moscow's policy toward China during the war against Japan. There is praise for the Soviet contribution in creating the united front between the KMT and the CCP against the Japanese; for the USSR as "China's main support" in the war against Japan; and for the military destruction of the Japanese Guandong army by the Red Army in Manchuria. The historiography condemns the fact that Moscow had respected only "Chiang Kai-shek and his army of two million" and underestimated the military strength of the CCP. After the dissolution of the Comintern, the CPSU had presumed to take a leadership role toward the Chinese Communists, had intervened in their internal affairs with false decisions, and had sought to restrain them from the revolution. And it had damaged China's sovereignty very severely in the Yalta accords.[58]

The Soviet Union, the CCP, the Kuomintang, and the United Front Against Japan: Origin of a Concept

There are debates in the literature over whether—and if so, in what form—there was dissension between Moscow and the CCP leadership in the period between August 1935 and August 1937 over the creation of a united front between the CCP and the KMT to fight the Japanese.[59] This is not the place to present in detail the complex arguments of the opposing sides, whose differing opinions were due to the fact that they use different sources or interpreted the same sources differently. Instead, we will examine the plausibility of the core of these arguments and introduce new documents to filter out an approximation of the sequence of events.

The Seventh Congress of the Comintern in July–August 1935 announced the dissolution of the international antifascist united front. With reference to China, the Comintern also recommended in this context the creation of a united front against the Japanese. It did not, however, explicitly call for the KMT—and specifically a KMT led by Chiang Kai-shek—to be included in such a united front. Even the concept that Wang Ming, and the Chinese section of the Comintern he led, developed during the Congress and the following months did not make this demand explicitly. Since the fall of 1933, the section had occasionally used slogans such as "unite all enemies of Japanese imperialism [*lianhe yiqie Riben diguozhuyi de diren*]"[60] and considered a united front with Chiang Kai-shek to be possible "if he really ends the struggle against the Red Army and turns his weapons against the Japanese imperialists."[61] It seems, however, that the Comintern, Wang Ming's section, and the CCP leadership officially placed Japan and the KMT (in Party jargon, "Japanese imperialism and its bandits who are lackeys and traitors to the fatherland [*Riben diguo zhuyi ji qi zougou maiguo zei*]")[62] on more or less the same level—at least until December 1935—and essentially pursued the goal "resist Japan and fight Chiang Kai-shek [*kang Ri fan Jiang*]" or "resist Japan and punish Chiang [*kang Ri tao Jiang*]," even if this line had begun to soften gradually in the CCP leadership due the impression that the KMT military was becoming increasingly superior toward the end of the year.[63]

As radio connections had been interrupted, the CCP leadership did not hear about the Seventh Comintern Congress's resolutions immediately. Recent Chinese sources indicate that it was only informed of the situation at the end of November 1935, when Lin Yuying (Zhang Hao), Communist trade union representative and member of the Chinese Comintern section in Moscow, arrived in North Shaanxi as an ECCI courier. Lin delivered the message that the Chinese CP should form an anti-Japanese united front with all available forces and at the same time intensify the struggle against Chiang Kai-shek—that is, follow the *kang Ri fan Jiang* line. At the end of December 1935 in Wayaobao, the Politburo accordingly adopted the strategic line "to mobilize, unite, and organize all revolutionary forces all over China, of the entire nation to fight against the current main enemies: Japanese imperialism and the traitor to the fatherland Chiang Kai-shek."[64] The civil war was to be combined with the national war.[65]

Zhang Guotao also mentions Lin's report, in which the Seventh Comintern Congress was supposed to have proclaimed the "new line of the anti-Japanese national united front." But Zhang then continues by saying that according to Lin's presentation the Comintern had then taken the position that this front should take the place of the Soviet movement. The previous slogan of *kang Ri fan Jiang* was to be replaced by "*lian Jiang kang Ri* [unite with Chiang and resist Japan]." Both the leadership in northern Shaanxi and Zhang's headquarters in Xikang supposedly accepted the new line.[66]

It is currently impossible to clarify this contradiction beyond any doubt. The best guess would be that Zhang Guotao was confusing his dates. He may have mistakenly moved the transmission of the new Comintern line, which was resolved later, and the CCP's approval of this line back in time and connected it with Lin Yuying's mission. It is only possible to document the change in the Comintern's *kang Ri fan Jiang* line in March 1936 from other sources. News of the new ECCI resolution did not reach the CCP leadership immediately due to the continuing interruption of the radio connection. After it had been reestablished, Dimitrov still criticized the CCP leadership for continuing to cling to the slogan *kang Ri fan Jiang* at the end of July and demanded the creation of an anti-Japanese united front with the KMT.[67]

It is not yet possible to document the reasons for the change of the Comintern's opinion in the sources. One can imagine that information that Moscow had received in late 1935 about German and Japanese preparations for an anti-Comintern pact provided significant impetus to it.[68] Since this time, Stalin had been tortured by fear of a coordinated military advance by Germany and Japan against the Soviet Union.[69] The plan that Japanese foreign minister Hirota Koki submitted to the National Government in early October 1935 and published in January 1936[70] probably also played a role. It included the suggestion that Nanking should develop close economic ties to Manchukuo and pursue "the annihilation of Communism" alongside Japan. As the "sources of the Communist threat came from a country in the north," China would have to collaborate with Japan to defend its northern border areas. It seems plausible to assume that fears of cooperation between Japan and China against the USSR drove Stalin to do everything in his power to prevent an alliance between Tokyo and Nanking. The creation of an anti-Japanese united front in China that was as comprehensive as possible and in which the Communists would also participate seemed to be a means to achieve this goal.

The Party's leaders apparently adopted the new Comintern line of an anti-Japanese united front only on August 10, 1936, and proclaimed the slogan "compel Chiang to resist Japan [*bi Jiang kang Ri*]."[71] It cannot be documented whether this reflected their own opinion or pressure due to the change of policy in Moscow at the end of July—the Comintern was again able to communicate with the Chinese comrades by radio from mid-May. Mao Zedong later maintained that the CCP accepted the KMT's conditions for establishing the united front in 1936–37 because the foreign menace from Japan was threatening the country and the Communists were, first of all, Chinese.[72] The ECCI compromised with the CCP leadership in a directive on August 15. It made it clear that the creation of a united front with the KMT did not mean that the Red Army and the CCP would be integrated into the KMT army. On the contrary: the Party and the army had to preserve their political and organizational independence and increase their power. This instruction followed the military and strategic line that had long been agreed upon by Moscow and the CCP leadership.[73]

On August 25, 1936, the same day they asked Moscow for heavy weapons for use in the struggle against the KMT, the Chinese Communists, who had initiated a campaign against the Japanese in early 1936, for the first time officially and publicly proposed to the KMT leadership that they establish a united front against the Japanese.[74] The harsh conditions Nanking imposed on the Communists, whereby the Soviet regions were to be dissolved, the Red Army[75] cut to 3,000 men, and all senior officers sent abroad, strengthened the CCP leadership in its intentions to connect with units opposed to Chiang within and outside the KMT to force Chiang Kai-shek to undertake their common struggle against the Japanese enemy, in keeping with the slogan coined at the beginning of August, *bi Jiang kang Ri*.[76]

The anti-Japanese united front between the Kuomintang and the CCP was not established on the CCP's initiative, however, but in the wake of the Sian incident and the war against Japan that began during the summer of 1937.

On December 12, 1936, Chang Hsueh-liang, Chiang's agent in the fight against the Communists in Shaanxi, had Chiang Kai-shek captured in Sian without the CCP leadership's knowledge and confronted him with the demand to suspend the civil war and include all parties and groups—that is, also the Communists—in the government to carry out the "struggle for the salvation of the fatherland" against Japan together. The CCP leadership was enthusiastic about this development and understood it as an opportunity to achieve the plan it had long been pursuing to eliminate Chiang Kai-shek and create an anti-Japanese united front with Chang's troops and also with other KMT units that would include the USSR and the Mongolian People's Republic (MPR). Zhu De advocated asking Chang to kill Chiang Kai-shek.[77] Mao Zedong agreed, but for tactical reasons recommended that they avoid the word "kill." Chang Hsueh-liang's reaction to the telegram the CCP leadership sent, which also included a plan for combating an offensive of groups loyal to Nanking, was hesitant, and he insisted that the Communists first demonstrate whether the Soviet Union would continue to support a joint northwestern army. In the following days, the CCP leadership sent numerous telegrams to Moscow in which, among other things, they expressed the opinion that Chiang Kai-shek should be condemned before a People's Court, his subordinates convinced to form an alliance, and a "government of revolutionary defense" established in Sian as a counterweight to Nanking. In a telegram to the Nanking government, it demanded that Chiang Kai-shek be removed from office and tried in court.[78]

Stalin took a different position. In his view, only Chiang Kai-shek was in a position to resist Japanese aggression effectively in China; he also took the views of Washington and London into consideration.[79] Immediately after the incident, *Pravda* published an adventurous interpretation: the "Japanese agent" Wang Jingwei, "spurred on by his foreign masters," had instigated Chang Hsueh-liang's revolt "against the central government" in order to create further chaos in the country and thus make things easier for the aggressor.[80] The Comintern replied to the CCP leadership in very similar terms in two telegrams that arrived on December 16: regardless of his intentions, which may have been very different, Chang's behavior assisted Japanese aggression. The Comintern instructed the CCP leadership to seek a peaceful solution as long as Chiang Kai-shek was prepared to include representatives of the anti-Japanese movement in his government, guarantee democratic rights for the Chinese people, terminate the fight against the Red Army, collaborate with Communist armed forces in the fight against Japan, and cooperate with states that sympathized with China's resistance to Japanese imperialism. The telegrams from Moscow struck the CCP leadership like lightning from the bright blue sky. Mao Zedong was furious and feared that Chang Hsueh-liang would accuse the CCP of plotting if they followed the Moscow line. In the end, however, the leadership complied in the belief that it would have to sacrifice its own interests in this case, for better or for worse, for those of the Comintern and thus of the Soviet Union.[81]

Chinese public opinion on the whole was indignant about Chiang Kai-shek's arrest and Chang Hsueh-liang's attempt to deal with the Communists. From a military perspective, it was unclear how the rebellion would fare as the Nanking government set significant numbers of troops in motion toward Sian. Zhou Enlai suggested to the CCP leadership that it should guarantee Chiang Kai-shek's safety for tactical reasons, but link this guarantee to the condition that Nanking not send any troops against Sian. Zhou and Ye Jianying were also among those who urged Chiang Kai-shek in Sian to agree to form an anti-Japanese united front. On December 18, Chiang Kai-shek made verbal concessions and claimed that he would abandon his plans to annihilate the Communists and fight against Japan. The CCP Central Committee then informed the KMT leadership that Chiang Kai-shek would be set free if he changed his policy of civil war into one of fighting against Japan. The message also marked a turning point in the policies of the Chinese CP, which took the form of the slogan *lian Jiang kang Ri*. Pressure from Moscow probably played no small role in this decision.[82]

Chiang Kai-shek's release on December 25, 1936, cleared the way for negotiations for an anti-Japanese united front between the KMT and CCP. Moscow was relieved: the danger that Japan would attack the Soviet Union seemed to be removed for the time being because it was somewhat more tied down militarily in China. Mao Zedong at first feared that Chiang Kai-shek would undertake a campaign against the CCP to avenge its secret cooperation with Chang Hsueh-liang.[83] Even after the anti-Japanese alliance was formed, both parties in the civil war continued to have considerable reservations toward the new alliance, which they saw more from a tactical than a strategic perspective, despite the unity they portrayed publicly. The CCP leadership carried out its policies toward the KMT in close consultation with Moscow. Differences frequently emerged in the negotiations with Nanking during the following months. These involved primarily the Communists' demand that their troops remain independent. And, under pressure from the Comintern, the Chinese CP also accepted compromises involving terminology. In February 1937, it offered the KMT to replace "the Soviet government" with the "government of the Special Region of the Republic of China" and the "Red Army" with the "National Revolutionary Army."[84]

The Japanese attack on Chinese units in the vicinity of the Marco Polo Bridge not far from Peiping on July 7, 1937, the subsequent major Japanese offensive in the area of Peiping and Tientsin, the beginning of the shelling of Shanghai by the Japanese navy on August 13, and the conclusion of the Sino-Soviet nonaggression pact on August 21 led both to the beginning of nationalist military resistance to Japanese aggression and to a breakthrough in the negotiations between the KMT and the CCP. After the Japanese attack, the Communist leadership informed Chiang Kai-shek of the desire "of all officers and units in the Red Army" "to fight against the enemy invasion under the leadership of the Generalissimo." On August 22, the Nanking government agreed to incorporate what had been the Red Army as the "Eighth Route Army [*ba lu jun*]" under the command of Zhu De and Peng Dehuai into the government army. This was essentially a formal and not really a practical process, however. Internal documents reveal that Mao considered the united front with the KMT in the fight against Japan a "minimal program"—that is, a temporary tactic—and the "Chinese revolution," on the other hand, as the "maximal program." Zhou Enlai had already linked the two programs together causally earlier and put the result in a prophetic form: "The first day of the anti-Japanese war will mean the beginning of the end for Chiang Kai-shek."[85]

Soviet Go-Betweens in Yenan: Comrades or Enemies?

Three Russians arrived in Yenan at an unknown time after Otto Braun had returned to the Soviet Union in late summer 1939. They presented themselves as TASS representatives, maintained a radio connection with Moscow, and remained until October 1943. According to Soviet accounts, the CCP leaders treated them in an increasingly unfriendly manner in the second half of 1941 and were almost hostile after December. The Chinese leaders avoided meeting them and had them watched.[86]

Three more individuals joined the group in May 1942. Among them was Petr Vlasov (cover name [a kind of alias]: Petr Vladimirov; Chinese name: Sun Ping), who operated as as contact man between the Comintern and the CCP Central Committee and presented himself as a TASS military correspondent. It would appear that he really worked for the Soviet Army secret service (Glavnoe Razvedyvatel'noe Upravlenie, or GRU). Vlasov took over leadership of the group. His main task was apparently to inform Moscow regularly of important events in Yenan and on Japanese troop movements in Manchukuo and North China. He was apparently also supposed to increase Soviet pressure on the CCP leadership, which had already existed since 1940, to ensure that the Chinese Communists did not cause trouble with the KMT and fought Japan at its side as decisively as possible. Vlasov maintained close contact with the few pro-Soviet members of the Chinese Party leadership, such as Wang Ming and Bo Gu, who, he reported, were belittled by Mao as the "Moscow group" or "the Muscovites," while Vlasov praised them for their "loyalty to internationalism and friendship with the Soviet Union." Clearly, Vlasov and Mao Zedong each had access to a radio set with its own code and used it to maintain a connection with Moscow. Neither generally knew what the other radioed to Moscow or what information he received. For unknown reasons, Mao occasionally used Vlasov's set. We do not know how Moscow interpreted Vlasov's and Mao's radio messages; they were certainly often different.[87]

In the context of Petr Vlasov's role in Yenan, it would seem appropriate to discuss the reliability of a book published in Moscow in 1973 as a serious source. It appeared in the form of a diary that Vlasov had supposedly kept during his time in Yenan between 1942 and 1945.[88] The question of authenticity is important because the information it contains —if accurate—would allow us unique insights into the events in Yenan at that time, including the CCP's relations with Moscow and Washington, from the perspective of an outsider, and in this respect would be even better than Edgar Snow's *Red Star over China.* The question is also important because some Western authors cite the book uncritically as a source, although they could have known the true story of how it was written.[89]

We have known since 1990 that this publication was not a diary but a fictitious account written by Petr Vlasov's son Iurii in the form of a diary. The younger Vlasov has since admitted that the CPSU Central Committee Secretariat commissioned him to write the book "in the context of worsening relations with China" and in view of an imminent Sino-Soviet military confrontation following the Ussuri conflict (a shooting incident between Soviet and Chinese border guards on an island in the Ussuri River) in 1969. Publication of the book was probably handled by Oleg Rakhmanin, deputy head of the Central Committee department responsible for relations with ruling Communist parties, who at that time coordinated all anti-Chinese propaganda. Iurii Vlasov said that the manuscript was created on the basis of the following information: coded radio messages his father sent to Moscow, other Soviet secret service material, and information from two of his father's colleagues, who were with him for part of the time in Yenan. He also relied on memories of conversations that he said he had had with his father between 1948 and his father's death in 1953.[90]

Iurii Vlasov admits that his book had been subjected to censorship "at all levels of power through the highest."[91] The diary is permeated from beginning to end with an abysmal mistrust of Mao Zedong and his political line, which was supposedly designed solely to exploit the Soviet Union rather than to behave in solidarity with it. We can assume that some sections, and those sections in particular that seem stylistically polished in their ideological abstraction, were invented or corrected on Rakhmanin's instructions. One occasionally also finds inaccurate descriptions of facts that involve the Yenan environment.[92] The information Iurii Vlasov later published about his father included inconsistencies.[93] On the one hand, Mao is supposed to have hated Petr Vlasov so much that he sought to have him removed physically. On the other, he made him one of his confidants, had his wife visit him in a Moscow hospital, toasted him officially during the festivities in connection with the establishment of the People's Republic of China on October 1, 1949, and subsequently gave him a valuable clock from the possessions of the Guangxu emperor with words to the effect that no one had helped the Chinese Communists as much as he had.

The bulk of the book appears to reproduce contemporary observations that Petr Vlasov and his Soviet comrades made in Yenan.[94] This can be seen by comparing the text with indications from other sources and by examining the plausibility of the statements in the overall historical context. A Russian sinologist, who had access to some of Petr Vlasov's radio messages in the CPSU archives, concluded that they were reproduced accurately in the "diary."[95] This, he reported, was true, for example, of the statement that Mao and Liu Shaoqi were of the opinion that the Comintern was a "foreign agency" that wanted to impose its will on the CCP, had a hostile opinion of the peculiarities of the Chinese revolution, and had damaged it more than once.[96] The indignant statement in October 1942 that the Eighth Route Army was not fighting against the Japanese troops is also factually correct.[97] Only people as close to the CCP leadership as Petr Vlasov, who clearly received some Party secrets from members of the Moscow group, would have known about many of the actions in the leadership that the diary addresses and presents credibly.

Until it can be determined which portions are genuine by comparing them with archival documents and reports by other contemporary witnesses, one must employ caution and only use the book selectively as a source—that is, when statements can be confirmed by other evidence or indications or appear plausible in the general context. This can, for example, hardly be assumed of a far-reaching claim like the one that Mao Zedong and the majority of Chinese CP leaders allegedly already viewed the Soviet Union with "organic animosity [*organicheskaia nepriiazn*"]" during the war.[98] It is more difficult to assess statements like the one according to which Mao and those in the Party leadership who thought like him allegedly said that they wanted to use the USSR primarily as a tool in the conflict with the KMT and that they were willing to accept the possibility of a war between the USSR and the United States to achieve this.[99] The literature from the People's Republic of China, however, confirms that the CCP leadership had planned in 1936 to conclude an anti-Japanese treaty of alliance with the USSR and the Mongolian People's Republic after establishing a joint government with Chang Hsueh-liang.[100] Otto Braun also claimed at that time that Mao had been planning since 1935 to draw the Soviet Union into the conflicts, first with KMT China and then with Japan.[101]

It is conceivable that, while the German army advanced as far as Moscow at the end of 1941, Mao actually did make the derisive comments about Stalin that are cited literally in the diary: "And Stalin babbled on about how they would go no further. Let's see how this leader babbles now in Moscow. . . ."[102] Something similar is true of another statement Mao is supposed to have made about Stalin: "He does not know China and also cannot know it, and yet he presumes to be able to make judgments about everything. All his so-called theses about our revolution are meaningless babble. And they babble exactly the same stuff in the Comintern."[103] It is conceivable that Mao did let out something similar to his closest confidants and that Wang Ming, Bo Gu, or some other member of the Moscow group heard about it and fed the information to Vlasov.

And yet, even if one assumes that large parts of the book are based on Petr Vlasov's contemporary reports, one should still employ them carefully. After all, what Iurii Vlasov tells us about his father reveals that he was a fanatical adherent of the Comintern, which he perceived from a Stalinist perspective, and even the slightest deviation from Moscow's policies—or from what he saw as Soviet interests—seemed like treason to him. He watched the actions of the CCP leadership with a distrust that seems almost pathological. In this, Petr Vlasov is comparable to Otto Braun, who admitted when looking back on his mission in China: "I always took the Soviet Union's position as the touchstone for every Communist, regardless of nationality or situation."[104]

In the following, the author will cite Vlasov's diary only when—and then without comment—he considers the particular statements to be genuine and also accurate in the aforementioned sense.

Petr Vlasov returned to the Soviet Union with his group in November 1945. According to his son's account, some of his Soviet comrades who had been recalled from Yenan in October 1943 had denounced him to Dimitrov. They reported that he had overstepped his authority in Yenan and proved himself politically blind by supporting Wang Ming unilaterally. Vlasov's apartment was searched. But Dimitrov trusted Vlasov. When he returned from China, his son reports, he was awarded high honors and ordered to report to Stalin. One of the comrades returning with Vlasov, the doctor and secret service agent Andrei Orlov (cover name: Terebin), had acted as Mao Zedong's and Jiang Qing's personal physician. Orlov worked for the GRU and sent his radio messages to the Soviet General Staff in Moscow, which forwarded them to Stalin.[105] When asked by Mao, who appreciated his medical skills, Stalin sent him to Yenan for a second time in 1946. He again became active for the secret service there and maintained radio contact with Moscow using the equipment Vlasov had left behind. The Soviet physician Mel'nikov, who worked for the Ministerstvo Vnutrennikh Del (MVD) in Yenan, later denounced Orlov for allegedly spying for Washington—according to Shi Zhe in the fall of 1949—and Orlov was recalled to Moscow. Shi Zhe had allegedly also criticized him to the MVD shortly before. In the notorious MVD prison Lefortovo, Orlov was forced to make false confessions under torture in which he also incriminated Petr Vlasov. On Stalin's instructions, Orlov was released, rehabilitated, and died shortly afterward in an airplane accident. Iurii Vlasov assumed that he had been killed for political reasons. Petr Vlasov worked as consul in Shanghai from 1949 to 1951 and as ambassador to Burma from 1952. His son is convinced that he was killed with an "anti-cancer vaccination" on Lavrentii Beriia's orders in 1953.[106]

We will probably have to answer the question we asked in the title of this section as follows: Stalin's go-betweens in Yenan tended to be more enemies than comrades to Mao Zedong.

Moscow's One-Sided China Policy During the Sino-Japanese War

As mentioned above, Stalin, in his fear of a two-front war with Japan and Germany, assumed that it would seriously tax Japan militarily—and perhaps even bleed it dry—to become bogged down in the war with China and make it difficult for Tokyo to begin a war against the Soviet Union. It was therefore important to support China in the war against Japan and to bring together all the anti-Japanese forces in China, including the alliance between the KMT and the CCP. An exaggerated and at the same time cynical way to describe this policy would be to say that Stalin was firmly resolved to fight against Japan to the last Chinese.[107] The CCP leadership's policy was very different: it was designed above all to preserve its own strength during the war against Japan and to prepare for the final struggle with the KMT.

During the war, it was not solely Moscow's but also Washington's policy in China to convince both parties in the civil war to put their struggle aside in order to resist Japanese aggression together. With the exception of these efforts, the Soviet Union concentrated its entire involvement in the war primarily on Europe.

Speaking to the future Soviet chief military adviser to the KMT government in Chungking, Vasilii Chuikov, in the fall of 1940, Stalin reduced his policy goals in China to the following formula: "The most important thing is to unite all the forces in China to beat back the aggressor. . . . Only when the Japanese aggressor's hands have been tied can we avoid a two-front war, should German aggressors attack our country."[108]

Moscow apparently based its decisions primarily on two considerations in determining who should receive material assistance. On the one hand, the USSR had recognized the KMT government diplomatically and had obligated itself in the Sino-Soviet nonaggression pact of August 21, 1937, not to support any "third state" or any "third power" (Chinese: *di san guo;* Russian*: tret'ia derhzava*) that acted aggressively toward the KMT government.[109] Although there could be questions about the interpretation of this clause under international law, Nanking could conceivably have extended it to cover domestic opponents as well. The other calculation, which had arisen in the context of the relative military strengths of the parties in the Chinese civil war and constituted the basis for the nonaggression pact, was more important: at the beginning of the war against Japan, the KMT army was about 2 million strong, while the Red Army included approximately 100,000 soldiers.[110] Moscow decided to ally itself with the stronger party and not to counteract this strategy by assisting its rival in the civil war.[111]

Before he left for China, Stalin mentioned both factors to Chuikov, who had wondered why he had been assigned to Chiang Kai-shek and not the Chinese Red Army. Stalin argued quite openly as follows:

> It might seem that the Chinese Communists are closer to us than Chiang Kai-shek. It might seem that we should also give them the most aid. . . . But this aid would look like an export of the revolution to a country with which we maintain diplomatic relations. The CCP and working class are still too weak to become the leaders in the struggle against the aggressor. . . . Relevant treaties were concluded with Chiang Kai-shek's government. Familiarize yourself with all these documents. You will act in strict accordance with them.[112]

Stalin's go-between in Yenan, Petr Vlasov, himself used the same argumentation between 1942 and 1945:

> The Soviet Union is not in a position to help the special region [dominated by the CCP] with equipment and weapons, and it is just this that they accuse us of and this is probably their main grounds for discontent.
>
> Any delivery of weapons from the USSR to the special region would be in violation of the treaty with the central government in Chungking and mean the collapse of the anti-Japanese united front. Such military deliveries would create a gulf between Chungking and Moscow. . . . From this point of view, this would provoke armed struggle between the Kuomintang and the Chinese CP.
>
> Such a development would be a wonderful prospect for the Japanese fascists. . . .
>
> But they were counting on military support from the USSR here [in Yenan] and not at all, it appears to me, in order to reinforce the resistance against the Japanese occupation forces. . . . I think they are thinking about a civil war here.[113]

It can be demonstrated in very general terms that Stalin paid less attention to East Asia after the conclusion of the neutrality pact with Japan in April and the German attack on the Soviet Union in June 1941. This only changed again after the Western Allies opened the second front in Europe in June 1944.[114]

American diplomats also became convinced that Stalin preferred to deal with China through a KMT led by Chiang Kai-shek. In a memorandum to President Roosevelt at the Cairo Conference, Special Envoy William Averell Harriman explained in November 1943 that the Chinese government wanted peace and a strong central government in China in the postwar era. They could only achieve this goal, he argued, with Chiang Kai-shek's help. Stalin "desires some solution of the Chinese Communist problem," he said, by getting Chiang either to accept the CCP as an independent political party or to include them in his government.[115] Stalin's assessment reflected his tendency, which had been clear since the 1920s, to consider the KMT to be more powerful politically and militarily than the Chinese CP. It cannot yet be demonstrated whether Moscow really wanted a united China ruled from a strong center or would perhaps have preferred a divided, weak China. The People's Republic of China in any case has repeatedly accused the USSR of seeking the second alternative.

In June 1944, Stalin repeated to Harriman that he thought that Chiang Kai-shek was "the best of the lot" and the one to undertake the unification of China; the Communist leaders were not as good and would not able to bring about the unification.[116] In April 1945, Stalin had positive things to say about Chiang to Ambassador Patrick J. Hurley. He described him as "selfless" and "a patriot."[117] Shortly afterward, he assured Harry L. Hopkins explicitly that the Soviet Union would do everything it could to promote the unification of China under Chiang's leadership; no Communist leader, he said, was strong enough to unite China.[118]

To American diplomats, Soviet leaders distanced themselves clearly from the Chinese Communists. In June 1944, Stalin told Harriman disparagingly that these were "not real Communists, but 'margarine' Communists," though they were "real patriots and want to fight the Japs."[119] Stalin used the term "margarine Communists" to Hurley, too.[120] Some impoverished Chinese described themselves as Communists, Viacheslav Molotov explained, "but were related to Communism in no way at all."[121] They were merely expressing dissatisfaction with their economic conditions in this way, he said, and would forget this political inclination when their economic conditions improved. The Soviet government should not be associated with these "communist elements" and also could not be blamed for this situation.

It can be assumed that Stalin's and Molotov's decision to distance themselves from Yenan reflected a tendency to think little of the Chinese Communists that really existed in Moscow. Stalin is supposed to have derided the Chinese comrades as "radish Communists" (red outside and white inside) at a Politburo session, and he apparently did use the nickname "Pugachev"—the leader of an eighteenth-century popular uprising in Russia, who had ultimately been executed—to refer to Mao in the circle of his closest colleagues.[122] Nonetheless, the statements we have cited should not be taken completely at face value. They were also intended, one should assume, to arouse trust in the American allies, on whose undiminished military commitment against Germany the Soviet Union urgently depended. It is therefore plausible to assume that Stalin and Molotov were seeking to the best of their abilities to use their behavior to combat the conception of a Communist world conspiracy—which also involved China—that was still common in Washington and Chungking. It cannot be ruled out entirely that Stalin informed the CCP in Yenan of this deception and asked for it to understand its posture toward Washington. This could be what a directive of the CCP leadership is referring to in late November 1945 when it said that the USSR had no choice but "to distance itself from the Chinese Communists officially."[123] On the other hand, it is not very probable that Stalin and Molotov were seeking through their statements to convince the American leadership to deliver weapons to the CCP as well as the KMT in order to reinforce its military effectiveness against Japan.[124]

However that may be, the Soviet messages were not without effect in Washington. Harriman got the impression in 1944 and 1945—which was not actually entirely incorrect—that the Chinese Communists were fighting their way to power essentially on their own and disregarding Stalin's advice.[125] Under Ambassador Hurley's influence, President Truman became convinced that the Chinese Communists operated independently of Moscow, and George C. Marshall was convinced that it did not receive any support from foreign Communists.[126] These assessments, too, were in large part accurate.

No Soviet Weapons for the Chinese Comrades

The Soviet Union refused to deliver weapons to the Chinese CP during the Sino-Japanese War and thereby significantly damaged relations with the Chinese comrades. Only two exceptions to Stalin's rule have become known. They did not involve Chinese but rather Soviet interests, however: in the context of his fear of a Japanese attack on the USSR, Stalin asked Mao Zedong to assign units to the border with the Mongolian People's Republic to hold off Japanese troops and offered in this context to supply them with modern military equipment.[127]

Stalin had behaved differently before the war began. The question of the possibility of arms deliveries from the USSR to the CCP apparently first arose in 1935 and was closely linked to the establishment of the regions under Communist rule in North and Northwest China as a result of the Long March. In the summer of 1935, the Party leadership in Northwest Sichuan for the first time debated where they should go next. Zhang Guotao insisted on the creation of a new base region in Sichuan, Gansu, and Xikang, while Mao Zedong argued for continuing the march to the north. No agreement was reached during the Maoerkai session of the Politburo in early August. Mao Zedong and the other members of the Party leadership elected in Zunyi marched with their troops toward Gansu, Ningxia, and Shaanxi and arrived in northern Shaanxi in October. Zhang Guotao and Zhu De pushed on with their units toward the southwest to Xikang. In the summer of 1936, Zhang moved toward Gansu and Shaanxi, but then marched toward Sinkiang and finally had to break off his advance in Gansu.[128]

The decision in the summer of 1935 to march toward the north or northwest involved advancing into regions close to the USSR geographically. This plan was intended to make it possible to receive material assistance from the Soviet Union, which they desperately needed to compensate for recent military losses. The plan to move closer to the "home of the world proletariat, the Soviet Union, and the Mongolian People's Republic" was apparently first discussed in early August.[129]

In mid-September 1935, Mao Zedong recommended that they move up to the Soviet Union geographically in order to receive political and material assistance, and "military aid involving airplanes and cannon." The CCP was not an independent Party but a branch of the Comintern, he argued. The Chinese revolution was a part of the world revolution. "We will first establish a base on the Soviet border and then spread out to the east." The Party leadership adopted Mao's suggestions and resolved that they should "use guerrilla warfare to fight their way through to the Soviet border [*jingguo youji zhanzheng da dao Sulian bianjie qu*]," "establish a connection with the Comintern [*datong guoji lianxi*]," "receive instructions and assistance from the Comintern [*dedao guoji zhidao yu bangzhu*]," reorganize the troops, give them an opportunity to rest, and increase their troop strength.[130]

In November, Lin Yuying, who had been sent by the ECCI to Shaanxi, informed the Party leadership that Stalin had no objection if the Communist troops approached the Soviet Union and Outer Mongolia. Lin reportedly informed Zhang Guotao that Stalin agreed with Zhang's plan to establish a base in Sinkiang and West Gansu and that the Soviet Union would provide the necessary weapons for this and train the Red Army, thus developing it into an effective military force.[131] Luo Fu then advised Mao Zedong in letters to move forward via Ningxia to Outer Mongolia to receive technical assistance there.[132] Mao declared that he was "in full agreement [*wanquan tongyi*]." He suggested, however, that they first extend the Soviet region to the Mongolian border, then "conquer an area [*yi pian*] contiguous with the Soviet Union and Mongolian Republic," and "with armed force keep Japanese imperialism from attacking North China, swallowing up the entire country, and then attacking the Soviet Union."[133] The strategy of establishing a territorial connection with the USSR and Mongolia was called *datong guoji luxian* (advance to the international line, or follow the Comintern line) or *datong Sulian* (advance to the Soviet Union).[134]

The question that has thus far remained unanswerable arises here: How is Stalin supposed to have heard about his Chinese comrades' plans and desires? After all, the radio connection between the CCP leadership and Moscow was interrupted during this period.

Lin Yuying clearly delivered the news that the Comintern had corrected its earlier position that the CCP should first conquer key cities or a few provincial capitals. This was now viewed as a Trotskyite idea in Moscow, he said. In the meantime, the Comintern was of the opinion that the Chinese Communists had already achieved an initial victory in a few provinces.[135]

The *datong Sulian* strategy was confirmed and worked out at the end of December 1935. At the Wayaobao Conference, the Politburo resolved to consolidate the Soviet regions in Shaanxi and Gansu and advance from there in an "Eastern Expedition [*dongzheng*]" via Shanxi to Suiyuan to establish a connection with Outer Mongolia. In this way, they sought "to unite the Soviet Red Army and the Chinese Red Army on the basis of the fight against the common enemy, Japanese imperialism [*ba Sulian hongjun he Zhongguo hongjun . . . jiehe qilai*]" and above all to receive equipment and materials from Moscow.[136] In any case, the Red Army's primary goal in 1936 was to annihilate the troops of the "robber and traitor to the fatherland"[137]—that is, the KMT. There were varying opinions on all these questions at the Wayaobao Conference. Otto Braun expressed concern that the advance to Shanxi and Suiyuan might provoke a Japanese attack on the Soviet Union.[138]

Braun opposed such plans at a session of the Party leadership that took place in early January 1936 and subsequently in a letter to Mao Zedong. He expressed suspicions that Mao wanted to entangle the Soviet Union in conflicts first with the KMT and then with Japan to exploit it for his own purposes. The advance to the Mongolian border could serve Tokyo as an excuse to attack the Mongolian People's Republic, which in turn would force the USSR to intervene militarily. Since 1934, Mao had stated repeatedly that the main contradiction in the world was the conflict between China and Japan, so that the Chinese people's struggle against Japanese aggression and the KMT was the deciding factor in the revolutionary struggle on the world scale. Consistent with Stalin's line, Braun believed instead that the defense of the Soviet Union, the cradle of socialism against potential enemies, was the key issue and saw in the Chinese comrades' war against the Japanese a way to relieve the pressure on the Soviet Union of a potential threat from Japan. In his view, this was the Chinese comrades' clear "internationalist" duty. Mao in turn also expected the Comintern's and the Soviet Union's assistance in the civil war against the KMT and the fight against the Japanese as part of their "internationalist" duty. The conflict between Braun and Mao at the beginning of 1936 thus reflects a conflict of goals between the two Communist parties that was due to conflicts between their nationalistic, and not their ideological, interests.[139]

What is not quite clear in this picture is how Stalin, as mentioned above, is supposed to have approved the CCP's plan to advance to the border with the Soviet Union—which Otto Braun apparently did not know. If this is true, Stalin, unlike Braun, with whom he could not communicate at this time, had not understood the extent of this strategy or was not as mistrustful as his man in China.

The CCP appears in fact to have been following two lines at the same time. On the one hand, it wanted—as its main goal—as always to link the civil war with the war against Japan in keeping with the slogan "follow the anti-Japanese line [*datong kang Ri luxian*]." On the other, it sought—in the longer term—to extend the Soviet region toward the Soviet or Mongolian border in order to obtain Soviet assistance and, if possible, to draw the Soviet Army into the fight against the KMT and Japan: *datong guoji luxian* or *datong Sulian.*[140]

The Eastern Expedition that began in February 1936 ended in failure in May. KMT troops beat the Red Army in Shanxi and drove them back across the Huanghe to Shaanxi.[141] Now the CCP leadership began to define the *kang Ri fan Jiang* strategy to mean the alliance of Chang Hsueh-liang's Northeastern Army with Yang Hucheng's Northwestern Army against Chiang Kai-shek and Japan. They hoped thereby to establish a regional center with a common government in Ningxia-Gansu, create a territorial link from there via Outer Mongolia or Sinkiang to the USSR and thus to receive Soviet assistance, and finally to conclude an anti-Japanese treaty of alliance with Moscow and Ulan Bator. In connection with increasing difficulties with finances and provisions, the CCP leadership requested Soviet military assistance, including airplanes and artillery, through the Comintern in late August on the grounds that it was necessary for the conquest of western Gansu, Ningxia, and Suiyuan. It is possible that this was done through Deng Fa, who was sent to Moscow for that purpose.[142] The Soviet leadership agreed in mid-September to deliver assistance—among other things, 200 tons of military material (*junshi wuzi*)—if the Red Army could conquer Ningxia and receive the assistance there. This attempt to build a territorial bridge to the Soviet Union also failed in October–November, however. Moscow's attempts to send aid deliveries via Sinkiang were also unsuccessful due to transport problems. The CCP leadership finally began to consider the *datong Sulian* plan a long-term goal.[143]

One wonders why Stalin, who had already decided that spring to support a united front against Japan that included the KMT in China, was prepared at the same time to support with weapons the Communists, who were still fighting against the KMT army, and thus to run the risk of a counterproductive effect on the government in Nanking. We lack sources that would allow us to answer this question. The most plausible hypothesis appears to be that he was counting on the deliveries' remaining secret, and also on the fact that they dealt with a power base bordering the Soviet sphere of influence and dominated by his Chinese comrades and wanted to have a second iron in the fire should the anti-Japanese united front fail and an anti-Soviet axis be created between Nanking and Tokyo.

Soviet willingness to deliver weapons to the CCP ended abruptly for the aforementioned reasons with the beginning of the Sino-Japanese War and the conclusion of a nonaggression pact between Moscow and Nanking. Such requests to Stalin from Yenan subsequently fell on deaf ears.

As soon as he arrived in Yenan in late 1937, Wang Ming told his comrades the truth: the Soviet Union would indeed not be in a position to transport weapons and ammunition to Yenan by air. This was due to an agreement with the KMT that flights by the Soviet air force over Chinese territory had to be approved by the KMT government. Wang's flight to Yenan had thus been secret and illegal. Mao, who was supposedly particularly interested in the arms deliveries, protested, sobbing: "If so much can be given to Chiang Kai-shek, why can't we get a small share?"[144]

Mao probably was not the only one to hold this opinion. Many Communists in Yenan were clearly disappointed that almost all the civilian and military assistance from the USSR went to the KMT while the Chinese comrades received only medicine, medical equipment, and political literature, including the works of Lenin and Stalin. Sarcastic protests could be heard in Yenan, such as "for the bourgeoisie, weapons; for the proletariat, books."[145] It appears that the CCP leadership did not protest to the Comintern openly at the time about this one-sided support for the KMT for reasons of discipline and because they lacked any other allies.

The Soviet chief military adviser to the KMT in 1938 and 1939 states that the USSR sent "significant assistance" to the Chinese Communists by land via Sinkiang, but provided no concrete information about what this supposedly included.[146] As if to justify the one-sided character of the assistance, the Soviet ambassador to the Nationalist Government in 1938 claimed that Soviet aid to the KMT had contributed to Chiang Kai-shek's willingness to form a united front with the CCP and had saved the Chinese Communists from annihilation by the KMT; it was clear to the KMT leadership that they could not at the same time receive Soviet assistance and fight the Communists.[147] In July 1943, for example, when the KMT leadership was planning a major attack on the Chinese border region, Moscow made it clear that the Soviet weapons were not intended for use in the civil war but solely in the war against Japan.[148]

The Chinese Communists thus had to accept the fact during the war that practically all Soviet aid between 1937 and 1941 went to the Nationalist Government. The only case of a weapons shipment from the USSR to the CCP that has become known involved 6 antiaircraft guns and 120 machine guns that were delivered in early 1943 through the KMT to Communist units in the Manchurian-Mongolian border area. This exceptional Soviet behavior was probably motivated by the fact that Moscow was particularly interested in strengthening Chinese resistance against Japan and in keeping Japanese troops tied down in China during the critical period around the siege of Stalingrad.[149]

On the other hand, in the wake of the Sino-Soviet nonaggression pact concluded on August 21, 1937, in Moscow, the KMT received three major credits for a total of US$250 million to be used primarily for the purchase of Soviet weapons. The weapons deliveries included among other things more than 1,000 artillery pieces, more than 900 airplanes, 82 tanks, 2 million artillery shells, almost 32,000 bombs, nearly 10,000 machine guns, and 50,000 rifles. The aid deliveries were usually transported by truck via Sinkiang. More than 1,500 Soviet military advisers and about 2,000 members of the air force fought on the Chinese side against Japan. The chief military advisers with the KMT army were first Mikhail Dratvin; then Aleksandr Cherepanov in 1938–39; and then Chuikov, from December 1940 to February 1942. In any case, Moscow did not provide this aid free of charge. In exchange, the USSR received valuable goods such as raw leather, tungsten, and other rare metals, and also large quantities of food. The last shipment of military goods was clearly delivered in August 1941. In October 1941, the Soviet Union declared that it was no longer in a position to make further aid deliveries, because of its own war against Germany. In the same month, American military assistance—at first very modest—began.[150]

In a publication in January 1940, Mao Zedong commented with reference to Chiang Kai-shek's alleged rejection of further Soviet aid in 1927: "In particular, Soviet assistance is absolutely indispensable for China's final victory in the War of Resistance. Refuse Soviet assistance, and the revolution will fail."[151] This observation sounds particularly strange because the problem at that time (1927) was not so much that Soviet support was being rejected, as that Moscow was no longer prepared to provide any such assistance at all.

In January 1941, Mao Zedong requested through Chuikov that the Soviet Union supply weapons and send out technical personnel. He argued that without heavy weapons—including airplanes—the Red Army could not capture any cities such as Lanzhou. Stalin refused indirectly by claiming that the Communists did not have to attack any fortified cities but needed instead to isolate Japanese bases with their superior Red Army and force the enemy to waste its resources. He then made the frivolous suggestion that the Chinese comrades ask the KMT to give them some of the Soviet weapons aid—knowing full well that Chiang Kai-shek would never dream of fulfilling such a request.[152]

After the German attack on the Soviet Union in the summer of 1941, Stalin again and increasingly began to fear a two-front war with Japan and Germany. This was apparently why he reversed himself and threw out as bait to the CCP that he would deliver military equipment—on the condition that Communist troops marched toward the Mongolian People's Republic to receive it there. Stalin to all appearances wanted in this way to motivate the hesitant Mao Zedong to disrupt the Japanese attack on the USSR that he feared. Mao refused, as it seemed too risky to him in this situation to expose his border region militarily in any way.[153]

Conflict Between Moscow and Yenan over Policy Toward the KMT and Japan During the War

When the war against Japan began on July 7, 1937, the Red Army was incorporated into the KMT army as the Eighth Route Army. A few days later, Mao Zedong clearly observed at a Politburo session (the Luochuan Conference) that the KMT would not change its reactionary nature and therefore could not be victorious against Japan. It would splinter. One part would capitulate and the other would be annihilated. Although formally subject to the supreme command of Chiang Kai-shek, the Eighth Route Army had to remain independent in fact and subject to the command of the CCP leadership, Mao said. The Communists could not let themselves be carried away with patriotism. Their troops had to increase their fighting strength, avoid direct confrontation with the Japanese, limit themselves to guerrilla warfare in the enemy hinterland, and construct bases there. The fight against Japan played an important role in resolving domestic political contradictions. Secretary-General Luo Fu apparently formulated this position even more clearly by returning to the line that had originally been arranged between the Comintern and the CCP: the Chinese Communists had to fight against both Japan and the KMT. The defeat of the one also meant the defeat of the other. Both would lead to the victory of the Chinese Communists. Mao and Luo's conception is reminiscent of the Chinese stratagem "Sitting on the mountain, watch the tigers fight [*zuo shan guan hu dou*]"—the expectation was that the tigers would tear each other to pieces and leave the valley to the observer.[154]

Zhang Guotao and Zhou Enlai objected. They apparently advocated fighting the Japanese aggressor alongside the KMT to the best of their abilities and without any major reservations; that is, they wanted to take the "anti-Japanese united front" seriously. They could not count on Chiang to capitulate, they argued. They had committed themselves to the KMT to form the united front and had to keep their word. To avoid direct confrontation with the Japanese would create the fatal impression among the people that the Communists were not doing their best in the war of resistance. In the end, they agreed on a compromise that only barely hid the differences.

In late 1937, Wang Ming brought the Comintern's instructions back to Yenan from Moscow. As was to be expected, the Comintern stated that the struggle against Japan could not be won by fighting the KMT but only by collaborating closely with it militarily; the two armies were to be combined organizationally and were to operate under the supreme command of the KMT. In mid-December, Wang insisted at a Politburo session that the CCP should cooperate seriously with the KMT against Japan. This was of great international significance, he argued: if China provided effective resistance, Japan would be unable to attack the Soviet Union, and this would be good for the international proletarian revolution. Wang Ming apparently appeared before the session like an imperial envoy delivering an edict from the ruler. Mao and Luo avoided challenging the Comintern's authority openly. Unlike at the Luochuan Conference in August, they did not oppose Wang Ming's line, which was approved unanimously, though it did not become a formal resolution.[155]

Following the session, the CCP declared solemnly in numerous publications that it was sincere in its desire to stand by the KMT to save the nation in the war of resistance.[156] It seems, however, that by then Mao's conversion to Moscow's line was already essentially formal and tactical. In the fall of 1937, the Eighth Route Army received the order to expend 70 percent of its military efforts on the expansion of the Soviet regions, 20 percent against the KMT, and 10 percent against Japan.[157] Developments during the following years show that Mao clung to his principle that they had to ally themselves with the KMT, but should also fight it.[158] He may have felt encouraged in his double-dealing when Wang Ming made it public at the December session that Stalin had stated that the CCP had no choice but to depend on itself.[159] But Chiang Kai-shek was also just as interested in obstructing the expansion of the region controlled by the Communists as in blocking the advance of the Japanese troops. He deployed 150,000 to 200,000 of his best troops to blockade the Communists within the "special region."[160]

The anti-Japanese united front therefore really existed more in theory than in practice. The Nationalist Government's power to command the Communist troops existed only pro forma.[161] Nonetheless, they did work together somewhat satisfactorily at first before increasing tensions arose in late 1938 as a result of, among other things, the KMT's dissolution of Communist mass organizations in Wuhan.[162]

The constant Soviet pressure on the CCP leadership to attack Japan alongside the KMT in the greatest possible military style in order to tie Japan down in China and restrain it from attacking the Soviet Union resulted in significant differences between Moscow and Yenan from late 1940.[163]

The Soviet Union began very early on to tell the Chinese Red Army to employ trench warfare tactics to stop the Japanese advance, while the CCP leadership preferred guerrilla warfare.[164] There are indications that there was Soviet pressure behind a large-scale but unsuccessful offensive by the Communist Eighth Route Army in the second half of 1940, the "battle of the hundred regiments [*bai tuan dazhan*]," in which over 400,000 soldiers participated on the Communist side and which was in fact not carried out as a guerrilla campaign.[165] This was the only major Communist offensive during the entire war.[166] According to Chuikov, Soviet chief military adviser in Chungking in December 1940, Mao Zedong was of the opinion that it had not served the interests of China or the CCP but merely "to hinder Japanese aggression against the Soviet Union." Thereafter, in Chuikov's opinion, the Chinese Communists had essentially ended the war against Japan.[167]

From mid-October, the KMT exploited the weakening of the Communist troops to attack them militarily. The CCP leadership began to believe at that time that Chiang Kai-shek was planning to capitulate to Japan and become a Pétain. In early November 1940, Mao therefore toyed for a while with the idea of following the slogan "go on the offensive politically and go on the offensive militarily as well [*zhengzhi shang jingong, junshi shang ye shi jingong*]," terminating the anti-Japanese united front with Chungking, and attacking the KMT with 150,000 men.[168] The Soviet leadership, on the other hand, continued to be interested in maintaining the united front and took the position that they had to preserve Chiang Kai-shek from being taken in by the pro-Japanese clique around He Yingqin.[169] Moscow criticized the CCP's hard line against the KMT as "leftist."[170]

Mao at first toed the Moscow line and issued the word that they should protect Chiang Kai-shek from the influence of "the hidden pro-Japanese clique" and be friendly to him.[171] Under pressure from the Comintern, he also altered his previous position that it would be equally damaging regardless of whether Chiang turned to Japan or to the United States and England.[172] He now took the position that it would be negative if Chiang joined the German, Italian, and Japanese camp, while it would be beneficial if he worked with Washington.[173] He went so far as to advocate a unification of the forces of the CCP, the Chinese people, the Soviet Union, and the United States and England (in that order).

In early 1941, conflict emerged again between Moscow and Yenan over the Soviet attitude toward the KMT. In keeping with the agreement between the KMT and the CCP of June 1940, most of the Communist groups stationed south of the Yangtze under Chen Yi had marched to northern Jiangsu and had forced out the KMT troops that were there. The position toward the Communists hardened in Chungking. In late December 1940, a part of the Communist New Fourth Army, the only regular Red Army unit to remain south of the Yangtze, did not cross the Yangtze and march from southern Anhui (Wannan) to the north as Chungking had ordered, but instead moved south toward Jiangxi. Kuomintang troops then surrounded the Communist units, attacked them, and wore them down in a battle involving heavy losses on both sides (the Wannan incident); the New Fourth Army lost 7,000 men. The CCP leadership resolved to prepare militarily to beat back the attack.[174] Communist units attacked KMT lines of communication. The leadership of the Eighth Route Army was instructed not to send any further reports to Chiang Kai-shek. The situation worsened dramatically. The united front now existed practically only on paper; that is, in the sense of a truce that was not always respected.[175]

When he decided to attack the New Fourth Army, Chiang Kai-shek was concerned about a negative reaction from Moscow. He even anticipated that Soviet military aid would be broken off and consoled himself with the hope that the United States would support him that much more. In fact, Ambassador Aleksandr Paniushkin did criticize Chiang Kai-shek for the attack on the New Fourth Army. The Soviet Ministry of Foreign Affairs urged the KMT ambassador, Shao Lizi, to maintain the united front. The Soviet press expressed concern but did not clearly take its Chinese comrades' side. Moscow continued weapons deliveries to Chungking via Sinkiang and Gansu, although Yenan had repeatedly asked that they be stopped. The USSR concluded a commercial agreement with the KMT government. Chief military adviser Chuikov advised the CCP leadership to accept the instructions from the KMT and have the rest of the New Fourth Army march north. Dimitrov insisted that the CCP had to follow its political guidelines strictly and avoid conflict with Chiang Kai-shek and thus took a position designed essentially to placate and appease.[176]

Yenan was disappointed. To leading comrades, Mao Zedong reacted negatively to Dimitrov's intervention. He observed that current relations with the KMT were advantageous neither to the CCP nor to the revolution. Chiang Kai-shek had begun to break with the CCP, he argued. It would take a while and "is very useful to us [*dui women shen you li*]." Moscow's policies (*yuanfang de zhengce;* literally, "the policies from afar") had not agreed with the CCP's expectations for three months, and things were going back and forth. They should secretly prepare a counterattack on the KMT, he argued, that could begin in four or six months. Until Chiang actually made the break, however, they had to delay things—that distant Chuikov was right. Mao concluded, "It would be very bad if the Soviet Union continued to help Chungking with weapons [*Sulian ru jieji Chongqing wuqi, shen wei bu hao*]." They should consider how they could stop the weapons shipments.[177]

Two historians from the People's Republic of China characterized the resulting situation accurately as follows: "At that time, the differences between the CCP and the CPSU and Comintern were very clear to everyone. . . . The former advocated the fundamental interests of the Chinese revolution and the latter the practical interests of the socialist Soviet Union."[178]

When Chiang Kai-shek initiated an anti-Japanese offensive in late January in Henan and signaled that he would be willing to come to terms with the CCP, Yenan also began to employ calmer tones and considered civil war now to be avoidable.[179] In mid-February, however, Chuikov reported that it was now about time to force Chiang Kai-shek, who had been on the defensive since the Wannan incident, to make concessions and not the opposite; never before had Chiang Kai-shek been subjected to so much criticism from within and without, and never before had the Communist Party received such broad support in Chinese public opinion (literally, among the masses).[180] Indeed, public opinion in China, stirred up with anti-Japanese emotions, was revolted by Chungking's actions against the New Fourth Army, and in the wake of the battle of the hundred regiments the impression predominated that the Communists were fighting against Japan with more determination than the KMT. Washington also warned Chiang Kai-shek not to put the united front with the CCP at risk. The Chinese Communists' prestige had grown. Despite this, Chiang Kai-shek observed in early March that a break between the KMT and the CCP was unavoidable. They could deal with the Communists in three months, he said, militarily if not politically.[181]

On April 13, 1941, the Soviet Union and Japan signed a five-year neutrality pact. On this occasion, Stalin showered the Japanese foreign minister, Matsuoka Yosuke, with compliments in Moscow. Japan was interested in gaining a free hand for its plans to expand in China and Southeast Asia, and the Soviet Union wanted to avoid a two-front war. The pact, and a supplementary declaration in which Moscow and Tokyo agreed to respect each other's spheres of interest in Outer Mongolia and Manchukuo, provoked a protest from the Chinese Nationalist Government.[182]

The incident, on the other hand, clearly pleased Mao Zedong. Internally, he welcomed the fact that signature of the two documents completely dismissed the danger that the Soviet Union might be attacked by Japan. The danger that the KMT would capitulate to the Japanese and move against the Chinese Communists had diminished, he said, and England and the United States had been seriously embarrassed. It did not seem to upset Mao at that time that Moscow sanctioned Japanese rule in Manchuria. The anticipated weakening of the KMT regime was more important to him: the pact "dealt" Chiang Kai-shek a "severe blow [*geile yige yanzhong daji*],"[183] he said. After the treaty was concluded, Communist troops fought even less actively against the Japanese army than before.[184]

The leadership also defended the neutrality pact as a magnificent victory of Soviet foreign policy in propaganda intended for Party members and in public. The pact supposedly established peace in the eastern regions of the USSR and gave Moscow a greater voice in the West and the East. Soviet assistance for China in its war of resistance against Japan would, it said, not be affected.[185]

The peace the USSR had achieved with Japan with the neutrality pact stood in stark contrast to the persistent calls to continue the fight against Japanese occupation that Moscow directed at the KMT and the CCP.[186] In any case, the Soviet Union ordered CCP units fighting in the ranks of the Soviet Far East Army not to send any further troops to Manchuria, as this could provoke Tokyo.[187] These units constituted the remnants of the Communist United Anti-Japanese Army (*kang Ri lianjun*) of Manchuria that had been able to escape to the Soviet Union after its defeat by Japanese troops in 1940 and 1941.[188] Zhou Baozhong, its commander, protested against the Soviet demand.[189] Guerrilla warfare in Manchuria was part of the Chinese revolution, he argued. The Soviet Union should take this into account and not simply look after its own needs. The Chinese units' battles "objectively protected the Soviet Union [*keguan shang jiu shi baowei Sulian*]." Zhou concluded by observing that although he agreed not to send the main armed forces to Manchuria, he would nonetheless set smaller units in motion in that direction.

In May 1941, Chuikov ordered Zhou Enlai and Dong Biwu to stage an offensive along with KMT troops stationed in southern Shanxi against the Japanese army to make it more difficult for Japan to attack the Soviet Union.[190] The CCP leadership reacted evasively. An internal document reveals the tactical considerations involved here clearly: They had to fight Japan in order not to provoke the scorn of the KMT, but should not fight too hard (*bu da de tai xiong*), as the Japanese would otherwise carry out reprisals and the KMT would exploit the situation to attack the regions under Communist control. "It is enough to make sure that the KMT believes we are really fighting [*shi guomindang juede women zheng zai da jiu hao le*]."[191]

When Chuikov went yet another step further and proposed that the KMT leadership should "punish [*chufa*]" all troops unwilling to cooperate—whereby he clearly meant primarily the Communist units—he unleashed a vehement protest from the Chinese comrades.[192] Mao instructed Zhou Enlai to inform Chuikov that he "should not babble on at random [*bu yao suibian luanshuo*]" and that he should instead analyze the KMT precisely. He would then realize that it was not fighting so much against Japan as against the CCP.[193] The CCP was prepared to fight the Japanese together with the KMT, but would not allow the KMT to goad it into rushing forward mindlessly and would advance regularly after examining the situation precisely. The Red Army, unlike the KMT troops, had already done enough in the fight against Japan since the defeat at Wuhan (Mao may have been referring to the "battle of the hundred regiments").[194] Chuikov would hopefully not fall for the KMT propaganda that suggested the CCP "moved troops without attacking" and "made concessions to Japan."[195] Zhou and Dong refused to discuss the CCP's military plans with Chuikov.[196]

It was at this time, as mentioned above, that the CCP leadership understood relations with Moscow in the following very significant way: "We should listen to the Soviets, but we should not listen to them completely [*bu ke bu ting, you bu ke jin ting*]."[197]

It is not clear to what extent the Chinese Communists accepted Moscow's changing assessment of the international situation during the late 1930s and early 1940s. In the spring of 1939, the CCP leadership had identified—at least to outsiders—with Stalin's new appraisal of international relations, whereby the Western democratic and not the fascist states were the Soviet Union's most dangerous enemies, and the CCP leaders had reached the conclusion that Great Britain was the "most reactionary country in the world."[198] Moscow had thus abandoned the line of the "international antifascist united front" resolved by the Seventh Comintern Congress. After the Ribbentrop-Molotov pact became known and the Second World War broke out, Mao did, however, play down the differences between "capitalist" and "fascist" states in an original fashion: "No matter if it is Germany, Italy, Japan, Britain, the U.S.—they are all imperialist countries taking part in the war directly or indirectly for only one aim, which is to carve up the world and rob the peoples of the world."[199]

Following the German attack on the USSR on June 22, 1941, the Comintern and the Soviet government called on all foreign parties to return to the international antifascist united front. Yenan declared its solidarity immediately. There were differences, however, in the way it formulated the call. Whereas Moscow mentioned only the struggle between the USSR and Germany, the CCP leadership determined that all actions directed against Germany, *Italy, and Japan,* and all that supported the Soviet Union *and China,* were to be considered "good, useful, and just."[200] Mao Zedong stated in an internal instruction that under the new conditions the Party had to stand by the united front with the KMT, drive the Japanese imperialists out of China, and thus help the Soviet Union, and that it should ally itself with antifascist forces in the United States and England against the common enemy.[201] Historians from the People's Republic of China observe that such a formulation could not have satisfied Moscow. The instructions from Moscow had focused even more than before on the USSR's interests and had thus highlighted the contradictions between the Comintern and the CCP.[202]

Following the attack of the German army, the fear of a two-front war flared up more fiercely in Moscow—and not entirely without reason. Ribbentrop was pressing his colleague, Matsuoka, who was interested in a war against the USSR, for Japan to occupy Vladivostok and march westward to unite with the German army before the onset of winter.[203] When the Soviet-Japanese neutrality pact was concluded, Matsuoka had told Ribbentrop that the pact would immediately become invalid in the case of a war between Germany and the Soviet Union.[204]

Between the summer of 1941 and the beginning of 1943—that is, until the turning point in the war, the battle for Stalingrad—Stalin asked Mao Zedong five times to send troops to the vicinity of the Great Wall or the border with the Mongolian People's Republic to tie down Japanese forces in the case of a potential Japanese attack on the USSR and thus to relieve the Soviet Army. In two cases, this request was tied to an offer to supply the troops involved with modern Soviet military equipment. Mao refused or avoided the question.[205]

Stalin asked Mao for help for the first time in late June 1941, after learning of supposedly extensive Japanese troop movements toward the Soviet border.[206] He asked the CCP leadership to attack the Japanese units and destroy their supply lines, and in particular to interrupt the railroad lines to Peiping, Zhangjiakou (Kalgan), and Baotou.[207] In addition, the Eighth Route Army should initiate an offensive alongside KMT troops in southern Shanxi to decrease Japanese pressure on the eastern part of the USSR.[208] Stalin was apparently so obsessed with the idea of a Japanese attack that he did not believe reports from a Soviet spy in Tokyo, Richard Sorge, which did not support the idea.[209]

According to Soviet statements, Zhu De reported to Moscow that if Japan attacked the USSR, the Eighth Route Army would deploy all forces at its disposal against the Japanese and provide sufficient support to the Soviet Union.[210] Mao confirmed this promise on July 3. But the CCP leadership refused to attack the Japanese as Moscow requested. It did reportedly have a plan drawn up to sabotage Japanese transportation routes, and it began to undertake the relevant action and provide espionage support to the Soviet side.[211] It also stated internally that Communist forces were prepared "to fight alongside the Soviet Army [*peihe Sujun zuozhan*]" in the case of a Japanese attack on the USSR, but, on the other hand, that it did not expect such an attack to happen at the moment.[212] When Chuikov asked in mid-July what military action the CCP was preparing to support the USSR, he received the following answer:[213] The Eighth Route Army had already begun to destroy transportation routes and collect information. They could not agree to undertake a large-scale attack on the Japanese, however, as they were far inferior to the enemy in terms of troop strength and equipment. They had too little ammunition, too few machine guns and cannon, and absolutely no explosives. The CCP could therefore do no more than secure their bases behind enemy lines and carry out a broad and long-term guerrilla war. They could not "bet everything on one card [*zhuazhu yizhi*]." The CCP leadership rejected the Soviet suggestion that the Eighth Route Army should advance to the communications routes to Outer Mongolia, at which time they would receive further equipment from the Soviet Union, on the grounds that this was too dangerous.[214]

Mao's reaction is understandable. He needed his troops at this time to resist the increasing pressure of the KMT blockade on the Communist border region of Shaanxi,

Gansu, and Ningxia, as well as the increasingly intense Japanese offensives against the region.[215] Yenan withdrew its troops from the Japanese front in the summer of 1941 and avoided larger-scale confrontations with the Japanese in order to concentrate on securing its main base. Regardless of the situation in which the Soviet brother state found itself, Mao was in any case not in a position to sacrifice the Chinese Communists' own interests for the sake of the Soviet Union.[216]

According to Soviet accounts, Moscow again asked Yenan on September 3 what the CCP would do to support the Soviet Union in the case of a Japanese attack. Mao Zedong responded with "confused and evasive statements," expressed reservations, and made impossible demands. When the Soviet interlocutor insisted on a clear answer, Mao accused him of "lacking dialectical thinking" and simply broke off the discussion.[217]

Moscow was dissatisfied with its Chinese comrade's evasive reaction. On October 7, Dimitrov sent a list of fifteen questions in an inquisitorial style to discover what military measures the CCP was planning alongside the KMT to tie Japan down in China and save the Soviet Union from becoming embroiled in a two-front war. The telegram's style revealed doubts about the Chinese comrades' solidarity with the USSR.[218]

The CCP leadership apparently had not expected this. To placate Moscow, it now gave the impression of being more willing to compromise with the KMT. It dropped all of the prerequisites for participating in and sending a representative to a consultative conference that it had insisted on to the KMT following the Wannan incident and had already reduced following the outbreak of the German-Soviet War.[219]

But Moscow was expecting something more along the lines of a large-scale strategic offensive from Yenan to be carried out jointly with the KMT against the Japanese army to tie it down and prevent a Japanese attack on the Soviet Union.[220] The CCP leaders were not prepared to do this, however, unless Japan actually attacked the USSR. They stuck to their previous position that they would carry out guerrilla warfare for the time being to "conserve and maintain its forces and await the appropriate time."[221]

Dimitrov's fifteen-point list provoked vehement debates in the CCP leadership. The instruction from Moscow again improved Wang Ming's position. He made himself spokesman of the Soviet position and maintained at a session of the Central Committee Secretariat on October 8 that the Party's policies were too far to the left and obstructed the creation of a united front with the KMT. They violated "the principles of internationalism." In plain words, they did not represent the interests of the Soviet Union decisively enough. He argued that "the most important thing they had to do today was to fight together against Japan [*jintian zhuyao shi gongtong da Riben*]," and they had to involve all population groups—even big landlords and big capitalists—to this end. They should cultivate relations with Chiang Kai-shek and the KMT. The fight between the New Democracy and feudalism, which remained the ultimate goal, had to take second place to this. With this, Wang clearly opposed the basic positions that Mao represented.[222]

Mao feared that Dimitrov's telegram and Wang Ming's presentation could seriously damage his prestige. Several other top leaders—allegedly including Wang Jiaxiang, Chen Yun, Ren Bishi, and Kang Sheng—may have been irritated by the intervention from Moscow. Many Party comrades were accustomed to view Stalin and the Soviet Union uncritically—from Mao's perspective, "dogmatically." On the other hand, the call for cooperation with Chungking had lost its innocence since the Wannan incident. Nonetheless, the situation following Dimitrov's and Wang Ming's interventions, according to a historian from the People's Republic of China, involved a "potential danger" for Mao, who was more determined with each passing day "to decide the practical questions involved in the Chinese revolution independently [of Moscow], and in a sovereign way."[223]

Mao Zedong therefore felt it necessary to settle things generally with Wang Ming. At the session of the Secretariat on October 8, he accused him of violations of the Party's general line, including inadequate determination in the conflict with the KMT, inadequate support for the policy of guerrilla warfare, and abuse of office.[224] Mao was able to prevail over Wang Ming in circumstances that are not known in detail. Those present at the session of the Secretariat condemned Wang's position "almost unanimously"[225] and resolved to convene a Politburo session at which Wang should answer for "his mistakes in the period before the Sixth Plenary Session in Wuhan and for his position on current policy questions."[226] It appears that Wang thus lost his last battle against Mao and "Mao Zedong's line achieved a decisive victory."[227] In the judgment of a historian from the People's Republic of China, the CCP's relations with the Comintern and the Soviet Union, which in previous years had been characterized by the principle that "the lower level obeys the upper level [*xia ji fucong shang ji*]," began to change at that point.[228] It was in fact a process that had already begun previously, but now reached its climax.

Wang Ming reported himself ill after the session of the Secretariat and did not appear at the Politburo session. He later claimed that Mao considered his behavior to be so dangerous that he had begun immediately afterward to have poison administered to him to kill him. His health had been so damaged by this, he said, that he had to remain in bed for four years. The last attempt to poison Wang was allegedly made in 1962, when he had already been living in exile in Moscow for many years.[229]

Nothing is known about any answer by the CCP leaders to Dimitrov's fifteen-point list. In practice, at least, the Party stuck to its line: to "carry out broad-based guerrilla warfare" and "conserve and maintain its own forces [*jiesheng yu baocun ziji de shili*]" and await an appropriate time."[230]

In late October 1941, the CCP leadership convened an "antifascist conference" in which representatives from ten East and Southeast Asian countries participated.[231] Mao called for the creation of an "antifascist alliance of all the peoples of the East" with headquarters in Yenan. There is no evidence that this initiative had been discussed with the Comintern. In that case, Moscow probably viewed it as an unacceptable unauthorized action by the Chinese comrades.

The CCP leadership clearly continued the line it had begun—defensive action toward the Japanese troops and defense of the Communist border region—in 1942, too. It did this despite the fact that the Soviet Army's military defeats, Japan's military successes, and an increased Japanese presence in Manchuria had further increased the fear in Moscow that Japan would enter the war against the USSR. Without calling it by name, a Comintern resolution criticized the CCP in late February 1942 for fighting primarily against the KMT rather than turning against the Japanese militarily and thus reducing Japan's pressure on the USSR; this, it said, was a sign of "nationalism."[232]

According to Soviet sources, the CCP troops' contribution in the fight against Japan decreased constantly from 1941 to 1942 in keeping with the tactical consideration that Mao Zedong is supposed to have used: "It is better that we conserve our forces, defeat the KMT, seize power in China, and then liberate China with the help of the USSR, Great Britain, and America from the Japanese conquerors. . . ."[233]

There was cooperation but possibly also conflict with respect to Sinkiang between Moscow and Yenan at this time. There are indications that they worked together in April 1942 to prepare to overthrow Sheng Shicai, who had up to then been the pro-Soviet governor-general of Sinkiang, to prevent him from removing Sinkiang from the Soviet sphere of influence and subordinating it to the Nationalist Government in Chungking. The attempted coup, which began with the murder of Sheng's younger brother Shiqi, failed. Mao Zedong's younger brother Zemin was one of the eighty-eight conspirators who were later executed. Despite Soviet threats to Sheng Shicai, Chiang Kai-shek declared in the summer of 1942 that Sinkiang was subject to the KMT government. At the end of the year, Sheng expelled the Soviet advisers and their military staff. The last units of the Soviet Army withdrew from Sinkiang in October 1942. One suspects that Sheng Shicai changed sides because he assumed the USSR would be defeated in the war against Germany in 1942 and therefore did not expect any further Soviet aid for Sinkiang.[234]

The historiography of the People's Republic of China presents a different version. The conflict according to this account had escalated when Sheng Shicai himself had his brother, who had been educated in Moscow and become a Communist, murdered when he sought to turn against his elder brother's "anti-Soviet and anti-communist intrigues"—referring clearly to the planned orientation toward Chungking.[235] As Sheng Shicai wished to employ the murder to prepare a conspiracy against the Chinese Communists active in Sinkiang, they tried to return to Yenan via the Soviet Union. Moscow agreed, but Sheng stopped them from leaving the country.[236] The Soviet military adviser in Urumchi did not attempt to assist the imprisoned CCP members by informing Yenan of their desperate plight. Moscow exploited the situation to curry favor in Chungking by suggesting to the KMT leadership that the USSR had repeatedly urged Sheng Shicai between 1934 and 1941 to be loyal to the Nationalist Government. Chiang Kai-shek supposedly succumbed to the Soviet blandishments but at the same time decided to block Moscow's actions in Sinkiang and to bring Sinkiang under the control of the Nationalist Government with Sheng's help. In this way, Sinkiang reportedly became an "anti-Communist bulwark" from the spring of 1943. To the extent that they passed through this important corridor, contacts between Moscow and the CCP were significantly limited.[237]

Sheng Shicai claimed that Stalin obstructed his planned membership in the Chinese Party leadership, which was already approved by the CCP in 1937, and gently pressured him to become a member of the CPSU. If this is true, this would be an early example of rivalry between Moscow and the Chinese Communists with respect to Sinkiang.[238]

In May 1942, the Soviet Defense Ministry asked the CCP leadership to send troops to southern Manchuria in the event of a Japanese attack on the USSR to fight Japan alongside the Soviet Army there. Yenan ordered Peng Dehuai and Nie Rongzhen to see what would be possible.[239]

On the basis of information Petr Vlasov radioed from Yenan to Moscow, Dimitrov admonished Mao Zedong in mid-June that the CCP leadership should do "everything in its power to improve relations with Chiang Kai-shek and strengthen the Chinese united front against Japan." The KMT was stronger militarily than the CCP and therefore had to take the lead. Dimitrov accused the CCP representatives in Chungking—above all, Zhou Enlai—of giving Chiang Kai-shek an opportunity to turn the mood in the KMT against the Communists by their thoughtless behavior. He ordered Mao to instruct his representatives in Chungking to adhere strictly to the policy of improving relations between the KMT and CCP.[240]

According to Vlasov, Mao Zedong informed him in the summer and fall of 1943 that he still was not prepared to fight against Japan. He only proposed to do so in the event of a Soviet-Japanese war, but even then it would only be in the form of "small operations," as he wanted to conserve his own forces. All Moscow's requests that they "disrupt the Japanese in any way in the preparation of the war against the USSR" remained unsuccessful, he said.[241]

Mao's behavior is comprehensible. The KMT had further reinforced its blockade of the border regions under Communist rule, in the meantime, and, following Sheng Shicai's change of sides to Chungking, were seeking to interrupt Yenan's supply lines to Sinkiang. Yenan had to deploy its troops primarily to avert this danger. It would have been almost suicidal to undertake a major military engagement with the Japanese army. On top of that, Mao apparently calculated generally that the fight against the Japanese aggressor should be left primarily to Chungking's troops, which would be weakened in this way. The Red Army, which was inferior to the KMT, should instead be protected for the final revolutionary conflict with the KMT that would begin after the end of the war against Japan.[242]

It cannot be shown whether Chinese Communists ever went so far as to use this calculation to justify secretly cooperating with the Japanese against the KMT. Both Chiang Kai-shek and Vlasov's diary make this claim. Chiang states that Yenan had transmitted KMT military deployment and operational plans to the Japanese in 1943 and 1944.[243] The Communists supposedly hoped to lure the Japanese troops deep into Southwest China in order to expand the areas they controlled in Shaanxi to the northwest unmolested by the KMT army, which would be preoccupied with the Japanese. They were supposedly even prepared to undertake pincer attacks with the Japanese on the KMT government's main bases in Sichuan and Guizhou. This last is improbable. Vlasov puts it more vaguely and merely observes that he had heard in the summer of 1945 that a permanent connection had existed "for a long time" between the Japanese headquarters in Nanking and the CCP leadership through an agent sent by Mao Zedong.[244]

When the ECCI proclaimed its intention to dissolve the Communist International in May 1943, the Chinese Party leadership greeted the idea with relief, because this would free the CCP "of the obligations resulting from the statutes and resolutions of the Congresses of the Communist International."[245] The Party had received a lot of assistance from the Comintern, it said. Nonetheless, China's Communists had long since been capable "of independently determining their own political direction."[246] The leadership cadres of the Communist parties of the various countries, Mao said, had already reached adulthood and achieved political maturity; the Chinese CP was an example of this.[247] All Communist parties would now become "more national," a road the CCP had already taken some time before.

One cannot help but see the self-confidence in these words. There was allegedly "practically a festival atmosphere" at the Politburo session in which Mao announced the resolution to dissolve the Comintern; Mao himself had not disguised his "triumphant glee."[248] The dissolution assured him that his "internationalist" opponents in the Party leadership would never again be able to operate against him with Comintern directives from Moscow. On the other hand, he continued to depend on the Soviet Union. He hoped—now perhaps more than ever—that Moscow—and Washington also—would support him by restraining the KMT from fighting the CCP before Japan capitulated. Since the battle for Stalingrad, they could anticipate a Soviet victory against Hitler, which would put Moscow in a position to play a stronger role again in China and East Asia.[249]

As it had previously, Moscow endeavored to assist the Chinese comrades in this way. When the KMT army threatened the Communist special region with a major attack from the south in July 1943, the Soviet Union intervened with Chiang Kai-shek in response to a request from Yenan and made it clear that Soviet military assistance was only intended for use in the fight against Japan and not in a civil war against the Chinese Communists.[250]

The Comintern was dissolved on June 8, 1943. In a speech to the Seventh Party Congress of the CCP in the spring of 1945, Mao recognized—not very convincingly—the Comintern's "outstanding contribution" to China and stated, with a deep bow toward Moscow, that the CPSU had become the commander in chief (*zong siling*) that all the Communist parties and the Chinese CP as well listened to since its dissolution.[251]

The Beginnings of a CCP Foreign Policy: Flirting with the United States at the End of the War

The CCP leaders did not, however, seek assistance only from the Soviet Union in the conflict with the KMT and Japan; they also carried out such actions toward Washington. It is only from this period that we can speak of the first attempts at an independent foreign policy of the Chinese Communists.

Yenan's attempts to make use of the United States in the context of a united front reached an initial high point in 1944. This was due to the Communists' expectation that Washington would be particularly interested in cooperating with them militarily after the disastrous defeat that a massive Japanese spring offensive had dealt to the KMT army—just as Great Britain had recently opted to support the guerrilla movement instead of the exile government in Greece. The situation also seemed favorable as—against Chiang Kai-shek's initial resistance—a first group of U.S. Army observers, the Dixie Mission, arrived in Yenan in late July and a second group one month later to hold talks with top leaders in the CCP like Mao Zedong, Zhou Enlai, Lin Biao, Ye Jianying, and Zhu De. They were supposed to determine the Communists' military potential and how they could convince them to increase their contributions to the war effort. The Americans' visit initiated a period of intensive flirtation between Yenan and Washington.[252]

In light of the new situation, the CCP Central Committee released a directive to all functionaries involved in contacts with the foreigners on August 18, 1944. It stated that a visit like that of the Dixie Mission should not be seen as an ordinary affair but instead as the "development of our united front in the international sphere" and the "beginning of our work in the field of foreign policy [*women waijiao gongzuo de kaishi*]."[253] It meant "the beginning of practical contact, after the first step had been taken to get to know our new democratic China." This new diplomacy was still only "semiautonomous," however, as the Chinese (including the CCP) and the Allies still recognized the KMT government as the central government and the KMT rejected the CCP's autonomous diplomacy. And yet the policy of the "international united front [*guoji tongzhan*]"—like the "domestic united front [*guonei tongzhan*]" eight years earlier—would further strengthen the CCP and thus help to improve prospects that the revolution would be victorious. To do this, the tactical principles of the domestic united front (a tactical and temporary alliance with a weaker enemy against the main enemy), had also to be applied to the international united front. The previous cooperation with the United States could gradually be transformed into "cooperation and assistance in the fight against the enemy"—by which the Central Committee meant both the Japanese, who were the only enemy to be named explicitly, and the KMT, to which the directive for understandable reasons only referred cautiously. On the basis of this military cooperation, it would be possible also to collaborate culturally and later politically and economically. As early as the welcome dinner, Mao Zedong asked one of the members of the Dixie Mission whether the State Department could set up a consulate at Yenan.[254]

As can be seen from the directive on August 18, the CCP leadership hoped for great things from the Dixie Mission. This document marked the beginning of Communist China's foreign policy. It expressed the expectation that the "allied states"—meaning primarily the United States—would open "diplomatic institutions [*waijiao jiguan*]" in the regions under Communist rule, send troops into these regions for the common fight against Japan, and help the Red Army with "weapons, ammunition, supplies, medicine, and technical assistance." The Party would allow these states to open press agencies in these regions and send priests and parsons to exercise their religion and would return churches and other buildings to them for this purpose. It invited states like America, England, and Canada to provide assistance in the form of medicine, tools, and money. It welcomed foreign investments and international technical cooperation.[255]

The Party leadership's main goal appears to have been to get Washington to recognize and assist the Chinese Communists as well as the KMT government. It hoped in this way also to weaken the alliance between Washington and Nanking.[256]

The aforementioned directive contains tactical instructions about how to deal with the American visitors in Yenan to earn their sympathy and achieve the goals that had been set. The following is a sample:

> In all things that we let the foreigners know and in which we wish to let them participate, we can arrange this on our own initiative and according to plan. Even if we want something, we can make them come to us with it of their own accord, as, for example, in the case of assistance with arms and ammunition. The KMT tries to talk them into doing things every day, they want this and that. We don't raise the question at first and achieve [in this way] instead their respect, and they will raise the question with us. . . . With respect to state and Party secrets, one should refuse to answer or pass them on. If it is not good to answer, one should evade the question and not speak. . . .[257]

The Communists conversing with the Americans in Yenan presented the KMT as corrupt, dictatorial, and militarily incompetent. Mao Zedong said that Chiang Kai-shek was stubborn and that "fundamentally he is a gangster," and asked if Washington really wanted to continue to recognize and support the ineffective KMT government, which lacked popular support.[258] Mao abused Chiang by calling him a "turtle's egg [*wangba dan*]" and said he would curse him openly to his face if he were there.[259]

The CP leaders presented themselves as the best Chinese partner in the war against Japan, asked the members of the Dixie Mission to stop Chiang Kai-shek from unleashing a civil war, and requested American support in the fight against the Japanese and the reconstruction of China after the war. The army's commander in chief, Zhu De, suggested placing part of his forces under American command. Mao Zedong explained that Chinese and American interests fit together "economically and politically": "We must cooperate and we must have American help."[260] He asked Washington in careful terms to deliver weapons to all the forces fighting against Japan, "which will include the Communists," and advocated an American landing in China.[261] Mao suggested that the Chinese Communists were not dependent on Moscow and expected little assistance from the USSR, which had its own problems, at that time or after the war.[262]

The Americans' Chinese interlocutors in Yenan took pains to give the impression that the CCP maintained practically no relations with Moscow. The same was true of the Soviet go-betweens present in Yenan—of whom John Service, one of the Americans, met three, including Petr Vlasov and Andrei Orlov.[263] They introduced themselves to the Americans with only their Chinese names and suggested falsely that they send out their news only through Chungking and the KMT censorship. The Chinese Communists complained that they knew "less than anyone else" what the Soviet Union "is likely to do." In this way, the Americans received the mistaken impression that no Soviet aircraft had landed in Yenan since November 1942 and that the Soviet go-betweens could barely speak Chinese, were not China experts, and were not treated as important people by the CCP leaders. The Chinese Communists constantly denied that they had any relations with the Soviet government—which was formally true because contacts with Moscow took place at the Party level. It cannot, however, be documented that there were any arrangements between Moscow and Yenan to mislead Washington about the true extent of the cooperation between the Soviet and Chinese Communists.[264]

Vlasov, the leader of the Soviet contact group in Yenan, commented sarcastically on the visit of the Dixie Mission as follows:

> It does not bother Mao that the Americans would not allow the national war of liberation to assume the character of a revolution. It was important to increase [his] strength with their help and drive out old Chiang by frightening the White House with the decay and inevitable collapse of the Chungking system, with the unreliability of this system in terms of its economic prospects. Mao behaved like a merchant proving that his product is best.[265]

The CCP leaders' perfectly staged efforts to produce a favorable impression on their American guests paid off. The members of the Dixie Mission were enthusiastic about what they experienced in Yenan. In the reports that they sent back to Washington, they praised the discipline and fighting strength of the Communist troops, highlighted their military successes against Japan, and pointed out the efficiency and integrity of their leaders. These characteristics were not present to the same degree in the KMT army, they suggested. The observers recommended supporting the Communists with better weapons. Their reports produced for the first time a degree of sympathy for the regime in Yenan in parts of American public opinion and the government. The KMT received the reports with considerable mistrust.[266]

The American visitors informed the Communist leaders that Washington was considering assisting in the creation of a coalition government made up of the KMT and the CCP. The CCP leadership felt encouraged by this to attempt major advances through various channels to the chief of staff to Chiang Kai-shek, General Joseph W. Stilwell. In a letter to Stilwell in September 1944, Zhou Enlai criticized the KMT for its "defeatism" toward Japan, described the Communists' military achievements as outstanding, and complained about the KMT army's constant pressure on the Communist regions, which could escalate into civil war at any time. In the end, Zhou Enlai requested at least half of the ammunition and goods destined for China.[267]

At the same time, the CCP leaders assured Stilwell that their troops were prepared to fight under his command but not under one imposed by Chiang Kai-shek—an offer that Zhou Enlai had already made to Stilwell in 1942.[268] In October, Peng Dehuai said he was ready to come to the assistance of the U.S. Army with a million regular soldiers and at least the same number of People's Militia if it should decide to land between Shanghai and Shandong.[269] In November, Zhou Enlai invited the American treasury secretary, Henry M. Morgenthau Jr., to come to Yenan. At the same time, an American journalist got the impression after talks with CCP leaders in Yenan that the United States was now the Chinese Communists' main partner in matters involving foreign policy. The Communist leadership would do everything to win the friendship of the United States.[270]

On November 7, 1944, Patrick Hurley, the first high-ranking representative from Washington, arrived in Yenan. In President Roosevelt's name, he was supposed to bring together the Chinese military forces to defeat Japan with American support. He presented the Communist leadership with a five-point plan, which, among other things, included a demand that all armed units be subject to the Nationalist Government, which Yenan could not accept. The CCP leaders convinced the well-intentioned but naive Hurley to co-sign a five-point counterproposal that included the Communist suggestion to form a coalition government, which Yenan knew full well Chiang Kai-shek would never accept. The Americans allegedly allowed themselves to be lured into a hole with Hurley's signature, as Mao Zedong reported with great pleasure within the Party. Mao argued further that they would thus have no alternative but to exert pressure on Chiang Kai-shek. They had naively sought concessions that no one wanted to make, he said. Chiang Kai-shek wrote in his diary that he had thought Hurley to be too experienced and intelligent to allow the Communists to deceive him; but that this had not been the case. The KMT government rejected the five-point plan that Hurley had presented. It proposed a three-point plan that included the demand that all Communist troops be subjected to the government in Chungking. In a talk with Hurley's envoy, Mao rejected the plan and heaped vile invective on Chiang Kai-shek. On December 8, Zhou Enlai warned Hurley that

Yenan would publish the five-point plan he had co-signed. Hurley, who had been nominated as ambassador to China at about this time, now finally felt that the wool had been pulled over his eyes and cursed Zhou angrily—in his absence—as a "motherfucker." His attempt at mediation had failed.[271]

In mid-November, Roosevelt expressed the wish to Chiang Kai-shek that the KMT and the CCP form a military cooperation; this, he suggested, was also the wish of the Soviet leadership.[272] In early December, Mao again expressed his interest in a landing by the U.S. Army to an American officer, promising that it would receive military support and assuring him that the Communist troops would "serve with all our hearts under an American general, with no strings or conditions attached."[273]

The American side had proposed military cooperation to Yenan for the first time in October 1944 in the context of plans for landing U.S. troops on the Chinese coast, which the Communist leaders would help to prepare.[274] In the second half of December, General Albert C. Wedemeyer, commander in chief of U.S. armed forces in China and the China branch of the Office of Strategic Services (OSS), precursor to the Central Intelligence Agency (CIA), made an offer to the CCP leaders to deliver weapons and technical assistance without Ambassador Hurley's knowledge. This was to be achieved in the context of sabotage missions that American technicians would carry out alongside Communist units behind Japanese lines. Another proposal called for the deployment of American paratroopers on the coast of Shandong, who would be supported by Communist units until American troops landed from the sea. In making his offers, Wedemeyer did not take into account Hurley's condition for military cooperation with the Communists: that they unite with Chungking politically.[275]

The Communist leadership eagerly accepted the American proposals. It could appreciate the advantages of a deployment of the U.S. Army in its area of operations that could help them gain access to weapons and ammunition. It also recognized the opportunities for manipulation that the lack of coordination between Hurley and Wedemeyer or the OSS gave them and took the chance of attempting to exploit this opportunity fully. On January 9, 1945, Yenan suggested in a telegram to Wedemeyer that Roosevelt should receive Mao Zedong and Zhou Enlai in Washington as "leaders of a primary Chinese political party." Zhou requested that the telegram be shown to Wedemeyer but not to Hurley, whom he did not trust to be discreet. In a second telegram on the following day, he offered to present evidence to Wedemeyer that the KMT leadership was negotiating with Japan for the "sellout of American interests in China."[276]

Wedemeyer seems to have ignored Zhou's request and made the telegram accessible to Hurley. When Hurley also learned on January 13 from Chiang Kai-shek that Wedemeyer had negotiated behind his back with Yenan to provide military cooperation, he "exploded in all directions." His dislike of the Chinese Communists, which had begun at about the time of his nomination as ambassador, increased considerably. In his protest in Washington on January 14, 1945, he explained the Communists' more standoffish position on cooperation with the KMT by suggesting that Yenan had achieved an important goal in the negotiations with the promise of assistance in the form of weapons and had become more confident. Wedemeyer's proposals had given the Communists everything they wanted: recognition, material support, and the destruction of the Nationalist Government. Hurley advised Roosevelt to receive Chiang and Mao, if necessary, together—on the condition that they had first reached an agreement.[277]

Washington agreed with Hurley and praised him. The administration apparently did not respond to the proposal for a visit by Mao and Zhou. Both officers who had offered to cooperate with Yenan were transferred. Wedemeyer found himself forced to instruct his staff to provide assistance to third parties only with Chiang Kai-shek's approval, and that the same applied to negotiations.[278]

The CCP leaders' attempt to hitch the White House to their wagon by circumventing Hurley had failed. For a time, Mao Zedong apparently considered freezing all diplomatic relations with the United States and the KMT until further notice. Petr Vlasov, Moscow's man in Yenan, observed at this time that the CCP leadership had realized "that its chances of succeeding in negotiations with the Americans were gradually diminishing."[279] In late January, however, Mao did declare his readiness to give in to Hurley's request and send Zhou Enlai back to the negotiating table in Chungking after a six-week interruption. The negotiations ended in mid-February without result.[280]

The reasons mentioned above were not the only ones that seemed to lead the CCP leaders to review their improved relations with the United States from late 1944. For one thing, the Soviet representatives had expressed Moscow's concern in late 1944 about the increasingly close relations between the CCP and the United States. Furthermore, military successes south of the Yangtze had so renewed the confidence of the leadership of the CCP, and the increasing sovietization of Eastern Europe had so strengthened its revolutionary ambitions, that Mao Zedong announced his determination in December 1944 to seize power in China while the war was still in progress.[281]

Yenan now assumed a more self-confident stance toward the Americans. In late December 1944, the CCP leadership abandoned the position it had previously held on military cooperation with the U.S. Army in internal consultations and established new conditions for military cooperation with the United States: the American troops would have, among other things, to "obey the political instructions of the Chinese CP [*fucong zhonggong de zhengce faling*]" and equip twenty Communist brigades with weapons.[282] One month later, Mao Zedong confirmed this policy. He rejected a proposal from Hurley that an officer from the U.S. Army should assume supreme command of the Communist armed forces. This would mean placing the Red Army under foreign command and turning its soldiers into colonial troops.[283] On the other hand, at about the same time, Yenan presented the Americans in Chungking alleged proof of contact between the KMT and Japan and—unsuccessfully—requested aid in the amount of US$20 million. In mid-February, Hurley and Wedemeyer left Chungking to return to Washington.[284] During their absence, the Communists were not admitted either to the American embassy or to U.S. Army headquarters.

It appears that Yenan now saw ever-less sense in continuing to beg for American assistance. The fact that Stalin apparently informed Mao Zedong of the upcoming Yalta Conference in early February appears to have played a significant role in this decision. Stalin suggested in doing so that the Soviet Union would again turn its attention increasingly to East Asia after the victory over Germany—a change that had already become clear since the Western Allies landed in Normandy in June 1944.[285] In addition, the Communist armed forces' military successes south of the Yangtze in late 1944 continued into the new year. Both facts could have further increased Mao's self-confidence, and he may have expected the United States and the KMT to be interested now in seeking a political compromise with the CCP. The increased emphasis on ideological positions the aforementioned factors provoked was expressed clearly at a Central Committee meeting in March 1945 when Mao Zedong demanded that a fundamental distinction be made between the Soviet Union on the one hand and the United States, Great Britain, and other Western countries on the other, and that the CCP should see it as its first foreign policy priority to ally itself with the Soviet Union.[286] Those who opposed continuing to flirt with Washington and advocated armed struggle were clearly establishing their position step-by-step in the Party leadership at this time.[287]

By March, the CCP leaders apparently had little hope of being able to use the United States as originally planned. Petr Vlasov observed in mid-March that "the political and military alliance with the Americans" had clearly failed when the CCP leadership got the impression that it would not receive any weapons from the United States to be used against the KMT. In Yenan, "the earlier optimistic mood quickly turned sour." They were to hold the door open for negotiations, but at the same time they let the Americans know that Communist China was not on its own, because it had an ally in the Soviet Union.[288]

The situation was not yet entirely clear at that time, however. Despite the factors mentioned above, Mao Zedong continued in principle to pursue the apparently friendly united front line toward the United States for a while.[289] At the Central Committee session in March 1945, he recommended that the CCP should attempt to come to terms with Great Britain, the United States, and other states "prudently and carefully."[290] At the same time, he sought to seduce one American diplomat by saying that the United States and China "complement each other economically" and will not compete. The United States was thus not only "the most suitable country to assist this economic development in China," it was also "the only country fully able to participate."[291]

The question of military cooperation was also put back on the table. The Communist press announced in late March that Yenan's 3 million regular troops and militia could, once armed, occupy a portion of the coast until the Americans landed: "The connection between the rich resources of the Allied nations and our inexhaustible human potential spells victory."[292] On April 1, Mao insisted in the talk with the aforementioned American diplomat that the Chinese Communists would continue to offer and practice cooperation with the Americans "in any manner possible to them"—regardless of "whether or not they receive a single gun or bullet."[293] It would continue to seek "America's friendship and understanding" because China would need it during the period of postwar development. Mao repeated earlier offers that the U.S. Army could count on the effective support of the People's Liberation Army (PLA) in the event of ground activity in China. This overture to the United States on April 1 is the last that has become known.

On the following day, Hurley announced at a press conference that the United States would in the future only support the KMT and distanced himself from "armed political parties"—a clear reference to the CCP—and warlords, which were obstacles in the path toward Chinese unification.[294] Yenan was deeply depressed and considered its hopes for American assistance to have failed definitively.[295] In an internal assessment, the CCP leadership described Hurley's statement as a "serious political blow to the positions of the Communist Party within and outside of China." In private, Mao was annoyed with "the USA's new policy on the China question in general and toward the Communist Party in particular." The CCP leaders at first wanted to react with a sharp public counterstatement, but then decided not to make one. In their first published reaction, they guaranteed that they would give their allies "any possible help if they desire it and provided that they fight against the Japanese."[296]

Mao Zedong's strikingly cautious treatment of the United States in March and the beginning of April was very probably motivated by the fact that at this time it was still unclear whether—and if so, when—the Soviet Union would enter the war against Japan. Only when this occurred, according to Yenan's reasoning, could the CCP begin a civil war against the KMT, which the Soviet Union would hopefully support. As long as this was uncertain, however, they should not offend Washington and Chungking completely.[297]

On April 5, the Soviet Union provided a signal that pointed in the direction Yenan desired. It terminated the April 13, 1941, neutrality pact with Japan early with an implicit reference to the clause *rebus sic stantibus:* Japan had since allied itself with Germany and supported that country in the war against the Soviet Union while it fought the United States and Great Britain, which were allied with the USSR.[298] Stalin's man in Yenan observed that there had been "sincere jubilation" there.[299] Mao had been "happy and excited" all day, Vlasov reported, and the Chinese in Yenan had sung hymns of praise to the Soviet Union and Stalin. Everyone, from simple Party member up to highest Party functionary, had taken it for granted that the Soviet Army would not only defeat the Japanese units in Manchuria and hand their weapons over to the CCP troops, but that it would liberate Inner Mongolia for the Chinese comrades as well. Mao now allegedly counted on "making the special region into a huge territory that the Soviet Union would conquer for him." In any case, Yenan did expect the USSR to enter the war against Japan soon and initiated preparations at the regional level to cooperate militarily with the Soviet Army to reconquer Manchuria.[300]

The Communists were careful not to let this assessment of the situation leak to the public, however. Their treatment of the United States in published materials remained tolerant and even friendly until mid-May. When Roosevelt died on April 12, Yenan honored him with great sympathy. The CCP leaders stated about his successor, Harry S. Truman, that they "expected much" from and had faith in him. This could be an indication that following the initial storms of enthusiasm about the termination of the Soviet-Japanese neutrality pact, Yenan had become sober and perhaps even skeptical about whether Moscow was really prepared to deploy the Soviet Army to support a Communist revolution in China. Furthermore, Moscow still had not declared war on Tokyo even after the German capitulation on May 8. The fact that the leadership at that time still hoped to break through to the Yangtze delta and South China coast and cooperate with the American troops who were still expected to land there clearly also played a significant and perhaps even cardinal role in Yenan's continued conciliatory treatment of the United States.[301]

In late May, the pendulum swung back. The CCP leadership recognized that it had lost the race to the coast to the KMT army. News also made its way to Yenan that

Tokyo was extending peace feelers toward Washington. The leadership feared that in this case the Japanese army would capitulate to the American armed forces that landed in China. The U.S. Army would then take control of the relevant areas and ultimately hand them over to the KMT and if necessary support them militarily in the struggle against the Communists. Yenan now viewed the landing of the American troops it had been expecting impatiently in an entirely different light: as an intervention of "American imperialism."[302]

A newspaper article on June 1 reflected this new view of the situation.[303] It claimed that "profascist elements" in the United States and Great Britain had since thrown dark shadows on the Asian war's originally emancipatory goals "with their calculations of a war for colonies." The "nauseous attitude of certain Allied people" was reminiscent of the tragic history of 1918—the reference was to the Allied intervention in Siberia. The Communist leadership was strengthened in its convictions when the FBI arrested American diplomat John Service—who, along with other members of the Dixie Mission, had advocated supporting the Chinese Communists in early June on the charge of betraying secrets.[304]

The change also affected the Seventh Party Congress of the CCP (April 23–June 11, 1945), which was in its final stages at that time. Mao Zedong predicted in a concluding speech in late May that Washington would adopt a policy of the containing the Soviet Union, opposing Communism, establishing absolute hegemony in East Asia, and transforming China into a "semi-colony of America." The policy was paralleled by the fact that the Greek government was suppressing Communist rebels in Greece with British aid; it could get worse for the CP in China. This was a clear call, Mao said, for an outburst of ideological confrontation between international "revolutionaries" and "reactionaries"; the latter had to be opposed under the slogan "Down with the Imperialists!" and with preparations for the struggle after the war in China. The CCP's practice of cooperating with the United States that had existed during the war would quickly come to an end, Mao said, and be replaced by a new period of "harsh attacks [*da ma*]."[305]

Mao Zedong's statements during this period reflected not only a readiness to fight, but also concern about the growing tensions between the Soviet Union and the Western powers that could be observed following the capitulation of Germany. An escalation could weaken Soviet influence in Asia and increase the likelihood of American assistance for the KMT in its struggle against the Communists. Mao sought to reassure himself and his listeners at the Party Congress behind closed doors by indicating that Stalin had spoken of an upcoming period of peace in Europe. The initiative lay in Moscow while Great Britain, the United States, and China remained passive, Mao said. They should place their trust in the Soviet Union's strength. During the final weeks of the Party Congress, Mao was clearly leaning again toward a more flexible foreign policy. He feared that Chiang Kai-shek might lash out militarily at any time to force a decision in the civil war.[306]

In mid-June, following the end of the Seventh Party Congress, on the other hand, it appeared for a time that Yenan wanted to set the stage for a break with the KMT and United States. The CCP leaders called publicly for a counterattack on the reactionaries and announced that they would not participate in the People's Political Conference called for July 7 in Chungking, which the KMT had planned as the preliminary to a national assembly to be convened on November 12.[307] Instead, a preparatory conference to a Chinese Liberated Regions People's Representative Congress was to be held in Yenan to parallel the Chungking conference.[308] They also convened the congress itself—apparently as a kind of antiparliament—on November 12. The first such tendencies, against both American and Soviet interests, could already be seen in late February. At that time, the Communist Party leadership announced its intention to establish an Association of the Trade Unions of all Liberated Soviet Regions, which would in turn promote a United Administrative Center of all Soviet Regions in China to overturn "the regime of Chiang Kai-shek's one-party dictatorship."[309] Attempts in late April and early May to establish a Women's Organization of all Liberated Regions and a Youth Organization of the Liberated Regions were part of the same policy.[310] Stalin's man in Yenan already assessed these efforts in February as Mao Zedong's call "to unleash a civil war," as Mao did not expect serious support from either Washington or Moscow.[311] In mid-May, there was allegedly no more urgent problem for Mao than to establish—as a step toward civil war—a separate government that he wanted to call either the United Committee of all Liberated Regions, or the United Committee for the Liberation of the Chinese People—an idea that was supposedly controversial in the Party leadership.[312]

In late June, shortly before negotiations for a treaty of alliance between the Soviet Union and KMT China began in Moscow, Yenan increased the intensity of its anti-American propaganda. For the first time since the Americans had entered the war in the Far East in December 1941, the CCP leadership addressed the American people directly and threatened that "imperialists like Hurley" would be taught a lesson if they continued to oppose the Chinese people.[313] Mao cursed Hurley publicly as an imperialist and predicted that the American government would sink "into the stinking and deep latrine [*you chou you shen de fenkeng*] of Chinese reaction" if they continued to follow Hurley's policies.[314] Shortly afterward, Mao made similarly sharp comments to Chinese visitors: "Since I was able to fight Japan with these few rusty rifles, I can fight the Americans too."[315]

The Chinese Communists' eleven-month flirtation with Washington had come to an end. Yenan still did not go so far as to break off relations completely. Instead, it waited cautiously to see how the United States relations with the USSR and the KMT would develop in future.

Interim Conclusions

From the very beginning, Stalin pursued a policy toward the Chinese CP that focused on Soviet interests—embellished ideologically by the Comintern. As Mao rose to become Party leader, he increasingly confronted Moscow with his own interests, which focused primarily on a Communist seizure of power in China.

With the exception of Li Lisan, who Moscow wanted to overtake on the "left" in 1930, the CCP leadership—steered by the Comintern—proved itself on the whole a pliable Soviet policy instrument between 1922 and 1935. Moscow's strategy of forming a bloc out of the KMT and the CCP failed in 1927, and the policy of armed revolts arranged with Moscow also ended in fiasco.

The strategy that Mao Zedong and Zhu De pursued from 1928 in the border region between Hunan and Jiangxi—that is, of driving the revolution from the villages—was not arranged either with the Comintern or with the Shanghai headquarters of the CCP. Moscow—unlike Shanghai—had actually come to conclusions similar to those of Mao and Zhu at the time, but nonetheless considered Mao's and Zhu De's actions with suspicion, as they had been carried out independently of the Comintern's instructions. When the Party leadership under Li Lisan also proved insubordinate in 1930 by preparing, against Moscow's will, armed revolts in the cities that failed again and from the very beginning, Stalin unabashedly intervened in the leadership structure of the Chinese Comintern branch and established a new Party leadership that was obedient to Moscow.

Mao's rise to Party leader, which took off in early 1935, introduced a new period in relations with Moscow. The transition to greater independence that reflected Mao's will was facilitated at first by the fact that for two years there was no radio contact between Moscow and the CCP. Stalin's hesitation to recognize Mao as Party leader reinforced the feeling of indifference in the relations between the two sides. The last obstacles in Mao's path to undisputed Party leader were set aside when Mao was confirmed as leader by Moscow in the summer of 1938 and when he definitively eliminated Wang Ming and the other pro-Soviet forces in the Party in October 1941. Mao had already been working on the Sinification of Marxism without consulting first with Stalin.

From this time, the slogan within the Chinese Party leadership was that one had to listen to the Soviets about the policy to be pursued in China, but that one should "not listen to them completely." Mao did as a general rule recognize verbally whatever line had been promulgated from Moscow. This was in principle true when issues of international policy were involved. He often resisted, however, when the particular Comintern line involved domestic Chinese policy. This was the case with respect to the pressure that Stalin exerted to convince the Chinese Communists to cooperate more intensively with the KMT in the fight against Japan. Mao also refused to comply when Moscow ordered him to undertake military offensives against the Japanese in the Sino-Soviet border area to disrupt suspected Japanese plans to move against the Soviet Union without taking into account the security of the Communist regions.

Mao had recognized that the primary goal of Soviet China policy during the war was to combine the entire military potential of the KMT and the CCP to tie Japan down in China and keep it from overrunning the USSR. The CCP's primary interest, on the other hand, was to preserve its strength for the final struggle for power in China, which would begin at the end of the war. The reluctance of the Chinese Communists was further provoked by Stalin's refusal to deliver weapons to them, whereas Moscow generously equipped the KMT government as the incomparably stronger military force. He either refused requests for the delivery of weapons from the CCP directly or made them dependent on fulfilling conditions that would serve the USSR's security but that Yenan could not accept without endangering the security of its own territory. It could be only slight consolation that Moscow repeatedly warned Chungking that Soviet weapons were not to be employed against the Chinese Communists.

It was primarily the CCP's economic and military weakness that convinced the leadership in Yenan to attempt to obtain the assistance that the Soviet Union denied from the United States as soon as an opportunity to do so seemed to present itself in the summer of 1944. This effort failed when Washington imposed the condition on its assistance that the Communists cooperate closely with and subordinate themselves to the KMT government—which the CCP leaders refused to do. Yenan's attempted tactical rapprochement with Washington in 1944–45, which clearly had not been arranged with Moscow and constituted the CCP's first steps toward an independent foreign policy, again demonstrated the extent of the Chinese Communists' emancipation from Moscow.

2

Moscow's Two-Faced Policy Toward China Between 1945 and 1948

The Soviet Union, the United States, the Kuomintang, and the CCP at the End of the Second World War

By the end of the Second World War, China was becoming involved in a complex diplomatic game in which the primary actors—the USSR, the United States, the KMT, and the CCP—all sought to achieve the strongest possible starting position from which to influence the postwar order. In pursuing this goal, they watched one another suspiciously, and each sought to use the others for their own purposes. This game continued after the war was over—violently in the case of the two Chinese rivals—until it finally led to the establishment of an alliance between the Soviet Union and Communist China in 1949–50 and the unambiguous fronts of the Korean War.

At this point, it is essential to analyze relations between Soviet and Chinese Communists in the context of this quadrilateral structure and not solely in terms of their bilateral relations. The evolution of these bilateral relations would otherwise remain largely incomprehensible.

It is possible to reconstruct the policies of the United States and the KMT somewhat plausibly because the most important documents and numerous contemporary accounts are accessible. The sources also allow us satisfactory insights into the relations between these two actors and the Soviet Union and the Chinese Communists. On the other hand, the condition of the sources made it impossible up to now to assess the development of relations between the Soviet Union and the Chinese Communist Party. As already mentioned, recent access to Soviet and Chinese archival documents and contemporary witness reports has significantly improved this situation.

At the end of the war and in the immediate postwar period, both Washington and Moscow continued, as before, to be interested in unifying China under the leadership of the KMT but with the participation of the Communists as a kind of junior partner.[1] Roosevelt, Stalin, and Churchill came to an understanding on February 11, 1945, in Yalta that the Soviet Union would recognize the KMT government as the legitimate government of China without any consideration of the CCP's interests as the price of concessions from Chungking with respect to Soviet influence in Outer Mongolia and Manchuria. American and Soviet China policy became uncertain in the months following the Japanese capitulation. This was due primarily to the fact that shadows had been falling on the Soviet-American military alliance since the spring of 1945 in response to their differing conceptions of Europe's future, which provided a foretaste of the upcoming Cold War.

Both Washington and Moscow had a foreboding of the future rivalry between the United States and the Soviet Union in world politics, but leaders in both capitals still tended to seek to delimit their spheres of interest by peaceful means in 1945. They sought to determine how far they could go without provoking the use of force by the other side.

The United States was interested in having the Soviet Union enter the war against Japan as soon as possible and in Soviet recognition of and support for the Kuomintang government. In return, Washington was prepared to exercise its influence with the KMT in keeping with the Yalta agreement to convince it to recognize the de facto Soviet protectorate over Outer Mongolia and Soviet control of strategically important territories in Manchuria. After it became involved in the European theater of the war in June 1944 in particular, the United States had encouraged China to intensify the war effort against Japan in order to relieve pressure on its own forces on the Japanese front. The United States counted primarily on its traditional ally, the KMT, in this and supported this government as much as possible. The Americans were, however, determined not to involve their own troops in a civil war.[2]

It remained in the American interest to avoid a civil war between Chinese Nationalists and Communists. This was due in part to Washington's desire for the two parties to fight together against Japan. On the other hand, Washington did not want to end up in the awkward position of having to support the KMT party unilaterally in a civil war and thus come into conflict with Moscow. Both motives also applied mutatis mutandis to the Soviet Union's position.[3] Furthermore, Roosevelt wanted to prevent the Soviet Union from exerting a strong influence in postwar China, which he perceived as a future great power. In his view, the East Asian Pacific area should constitute a primarily American-Chinese sphere of influence. The Soviet Union was thus to be contained by including it along with the United States, China, and Great Britain in a group of states in which China and Great Britain were to be subordinated to the United States. Roosevelt had already spoken to Molotov of the "four policemen" that would guarantee the peace after the war in the spring of 1942.[4]

Soviet policy toward China should be seen in the context of Moscow's goals in Asia—as in eastern Central Europe—after the war. They wished to create a security barrier, a buffer zone against Japan in particular but also against the United States and an increasingly powerful China. In this way, Stalin also wished to create the conditions for projecting Soviet power into East Asia and the Pacific and strengthen the Soviet Union's global geopolitical position. By acting and planning in this way, he was essentially continuing the policies of the czars. To compensate for the losses that Russia had suffered in its defeat in the war against Japan in 1904–5, Stalin sought to incorporate southern Sakhalin and the Kuril Islands into the Soviet Union. He wanted to establish a foothold in Japan, on the Korean Peninsula, and in Manchuria; maintain a degree of influence in Sinkiang; have the de facto protectorate over Outer Mongolia confirmed; and obstruct a military presence of the United States on the mainland.[5]

At first, the Soviet leadership seemed to have accepted in principle that the United States could consider China essentially to belong to its sphere of interest. In August 1944, Molotov even expressed the hope to Patrick Hurley, Roosevelt's personal representative in China, that the United States would take "the lead economically, politically, and militarily in Chinese affairs."[6] At the end of 1945, the Soviet Foreign Ministry continued to assume that "if before the war the English and in part the Japanese were masters in China, the United States of America will now be masters in China."[7] By this model, the ministry continued, Chiang Kai-shek wanted to exploit "the Soviet Union and the USA and the conflicts between them in particular" to strengthen the Nationalist Government. Stalin thus achieved a kind of tacit agreement with Roosevelt and Churchill in Yalta to the effect that Outer Mongolia and Manchuria should be treated as a Soviet sphere of influence and the rest of China south of the Great Wall as an American sphere of interest as the meaning of the Open Door policy. This did not, however, make the Soviet leadership trust the United States completely with respect to Manchuria: "The United States claim the right to advance both into north China and also into Manchuria. This is one of the most important problems for the Soviet Union. We have rid ourselves of the Japanese neighbors on our borders, and we cannot allow Manchuria to become a showcase for the economic and political influence of another great power."[8]

Nevertheless, Stalin wanted to avoid anything that could lead to open conflict or even war with the United States. This included civil war in China, which would place moral pressure on the Soviet Union to side with the CCP. Like Washington, Moscow considered Chiang Kai-shek to be the strong man in China, in comparison with whom the Chinese Communists seemed little more than a *quantité négligeable*. Stalin always sided with military strength. He had to know that the CCP only had at most a million poorly trained and poorly equipped soldiers to set up against a well-trained, 4.5-million-strong KMT army equipped with modern weapons.[9] Both the Soviet and the American leadership sought to promote cooperation between the KMT and CCP armies in the fight against Japan. On the other hand, Stalin considered the Chinese CP a useful vehicle for projecting Soviet power into China—and Manchuria in particular. Given these interests, Soviet double-dealing was only logical.

A powerful anti-Communist attitude underlay Chiang Kai-shek's policy. Deeply suspicious, he saw the hand of the Communist world conspiracy at work everywhere. He stared at his primary domestic political opposition, the Chinese Communists, as a rabbit stares at a snake, and perceived the Soviet Union as their patron and secret alliance partner. He sought the support of the United States primarily as the most important foreign guarantor of his power in China, but also did not pass up any opportunity to take advantage of the mistrust that predominated in relations between Washington and Moscow.[10]

In any case, he distrusted the Chinese Communists significantly more strongly than he distrusted the Soviet Union, with which he sought to reach a compromise and which he repeatedly urged to renounce its support for the CCP. He was interested in seeing the Soviet Union enter the war against Japan as soon as possible because he feared the anti-Japanese war might otherwise drag on for a long time. He hoped that the postwar order would see the restoration of China in its "historical borders" and reconquest of the "lost territories." This involved the institution of full sovereignty over the border regions of Sinkiang, Outer Mongolia, and Manchuria and the rejection of the Open Door policy the United States favored. In addition, the idea that he had to be armed for the final conflict with the Chinese Communists after the Second World War obsessed Chiang, and he considered everything, even the war against the Japanese invaders, less important than this perception. In this, he did not differ significantly from the CCP, which sought to preserve its army during the anti-Japanese war for the decisive conflict with the KMT after Japan's defeat.

The Chinese Communists continued to be interested primarily in leading their revolution to victory and in seizing power in China through the conflict with the KMT. Oriented for ideological reasons toward the Soviet Union but abandoned by it to their own resources with respect to arms deliveries, they did not hesitate to ask Washington for military aid in their search for allies in 1944–45 or to seek to isolate Chiang Kai-shek by cooperating closely with the United States. As stated explicitly in a Central Committee resolution in early 1946, the Communists sought to employ the classic stratagem "using barbarians to control barbarians [*yiyi zhi yi*]" in dealing with relations between Moscow and Washington.[11]

The Chinese Communist Party also sought to reestablish China within its historical borders. In the summer of 1936, Mao said that the immediate task was to "regain all our territories we have lost. . . . This means that Manchuria must be regained."[12] At about the same time, he pointed out that "when the people's revolution has been victorious in China, [the] Outer Mongolian Republic will automatically become a part of the Chinese federation, of their own will."[13]

What Stalin's man in Yenan observed in December 1944 appears essentially accurate: "Whatever Mao does, his entire policy is based on the assumption of assistance from the Soviet Union. In every critical situation, this policy speculates on the unwavering support of the Soviet Union."[14] Nonetheless, the CCP leadership had to observe that the Soviet Union had agreed in the treaty of alliance concluded with Chungking on August 14, 1945, "to distance itself officially from the Chinese Communists" in return for privileges in Manchuria.[15]

A comparison of their goals shows that the KMT and the CCP were more or less in agreement in three areas by the end of the war. Both rejected the American-Soviet consensus on the establishment of spheres of influence, the American Open Door policy, and the creation of a coalition Chinese government.

The Sino-Soviet Treaty of Alliance of August 14, 1945: Moscow's Grab for Czarist Privileges

There is considerable evidence that the CCP leaders began to pursue the Soviet option again more seriously as relations with the United States continued to worsen. Mao Zedong stated at the Seventh Party Congress in mid-May 1945 that the Soviet Union was the CCP's only true friend; it was the model and "commander in chief" of the world revolution. The fact that Yenan had received no help from Moscow to that point was a result of the military situation. As tensions intensified between the Soviet Union and the United States and the socialist camp became stronger, as could be anticipated after the Second World War, "international help would certainly come" to Yenan. China's Communists should be in a position after the war to continue the struggle against the KMT with the help of the Soviet Union—by means of the long-planned march to Manchuria in particular. The CCP should receive material and technical support from Moscow and establish a permanent base for the Chinese revolution in Manchuria.[16]

Moscow did little to reciprocate Yenan's advances at that time. At the beginning of June, the news trickled through to Chungking that Stalin had renewed his assurances to Washington that he would work with the Americans to establish a unified China under Chiang Kai-shek. Published information from Moscow in mid-July showed that the American-Soviet military alliance continued to function well with respect to China, which must have irritated the Chinese Communists. It stated—not entirely accurately—that negotiations between Stalin and the KMT foreign minister, T.V. Soong (Song Ziwen), for a Sino-Soviet treaty had taken place "in a friendly atmosphere" and revealed "broad mutual understanding."[17]

The real shock hit the Communists on August 14, 1945, however.[18] On this date, after three weeks of negotiations with the KMT government, the Soviet Union signed the Treaty of Friendship and Alliance, which was valid for thirty years. It obligated both parties to provide mutual assistance in the war against Japan, not to enter into separate negotiations with Japan, and not to conclude an armistice or peace treaty with Japan without the consent of the other party. In it, Moscow implicitly recognized the Nationalist Government as the only Chinese government. The parties to the treaty obligated themselves to "non-interference in the internal affairs" of the other party (Article 5). They assured each other "every possible economic assistance in the postwar period" (Article 6). In an exchange of notes, Moscow assumed the obligation "to give to China its moral support as well as aid in military supplies and other material resources, such support and aid to be entirely [Chinese text: *wanquan;* Russian text: *polnost'iu*] given to the Nationalist Government as the Central Government of China." The Soviet government agreed further to respect "China's full sovereignty" over Manchuria—and thus the Nationalist Government's authority in Manchuria—and not to interfere in Sinkiang. China in turn undertook to recognize "the independence of Outer Mongolia" as long as a plebiscite to be held after the Japanese capitulation confirmed the Mongolian people's desire for independence.[19]

Three thirty-year agreements were concluded along with the Sino-Soviet treaty of August 14.[20] They called for joint Sino-Soviet operation of the Manchurian railroads in joint use of the harbor of Port Arthur (Lüshun) as a marine base, with defense of the harbor transferred to the Soviet Union, and the lease of half of the harbor facilities in Dairen (Dalian) to the Soviet Union free of charge, in theory leaving Dairen a free port accessible to all states. The details of the cooperation between the Soviet Army and the Nationalist Government in Manchuria were established in a further agreement, and it was agreed that the Nationalist Government should take over the administration of the territories liberated from Japanese rule.[21] Stalin had already committed himself to withdraw Soviet troops from Manchuria three months at the latest after the Japanese capitulation.[22]

The Yalta Agreement of February 11, 1945 (I)

As mentioned above, the United States, the USSR, and England agreed in principle to establish the agreements mentioned above, accept the status quo in Outer Mongolia, and allow the conclusion of a treaty of friendship and alliance between the USSR and China in a secret agreement (henceforth referred to as the Yalta agreement) at the Yalta Conference on February 11, 1945.[23] Roosevelt had long called for the pro-American Nationalist Chinese government to find a compromise with Moscow to keep the Soviet Union from intervening in China at the end of the war.[24] The agreement in principle to reestablish the special privileges Russia had once enjoyed in Manchuria but had lost after the Russian defeat in the war against Japan in 1905—in particular, the lease of Port Arthur to the Soviet Union and joint administration of the Chinese Changchun Railroad[25] (henceforth CCR) were the price Roosevelt and Churchill had to pay for the Soviet Union's willingness to enter the war against Japan and for its support of the Nationalist Government in the fight against Japan, at the latest three months after the capitulation of Germany, which was also agreed upon in the secret agreement of February 11. The agreement also included confirmation of the Soviet Union's de facto protectorate over Outer Mongolia. Roosevelt made concessions to Stalin at China's expense in order to avoid American losses in the final stages of the war.[26] He was reacting to Stalin's unmistakable question: Why should Russia enter the war against Japan if she were not to get anything for it?[27] In any case, it was clearly generally assumed at the time the Yalta arrangement was made that the war against Japan would last at least a further eighteen months after the capitulation of Germany.[28]

Stalin also exploited the psychologically favorable and deep-seated anti-Japanese resentment that linked him with Roosevelt and Churchill at the time that they prepared together for the final conflict against Japan. In justifying the need to be granted privileges in Manchuria, Stalin referred to the ancient injustice that Japan had supposedly inflicted on Russia forty years earlier. In the final version of the Yalta agreement therefore, the introduction to the section presenting the Soviet Union's special rights in Manchuria states: "The former rights of Russia violated by the treacherous attack of Japan in 1904 shall be restored. . . ."

Background: Russian Privileges in Manchuria

Stalin was referring to Russia's defeat in the Russo-Japanese War of 1904–5, which had arisen in response to rivalries for influence in Manchuria and on the Korean Peninsula. Russia was compelled by the Treaty of Portsmouth in 1905 to lease to Japan the Russian leasehold property on the Liaodong Peninsula, including Port Arthur and Dairen, and to transfer to Japan the Southern Manchurian Railroad between Changchun and Dairen, which was administered jointly with China. Russia continued to hold the existing rights to the Chinese Eastern Railroad (CER).[29]

Most of the special privileges the czar had obtained between 1896 and 1898 by exploiting China's weakness following the 1894–95 war with Japan were thus lost. In the secret Russian-Chinese treaty of alliance on May 22, 1896,[30] which was directed against Japan, China approved construction of a railroad (the CER) from the Russian border across Manchuria to Vladivostok. It was constructed almost exclusively with Russian capital borrowed from France. In this way, Russia shortened the connection with Vladivostok considerably and saved 35 million rubles that it would have had to spend to construct the rest of the line across Russian territory. It was agreed that China would be able to purchase the line after thirty-six years,[31] and that it would be transferred to China free of charge after eighty years (i.e., in 1976). Using pressure and bribery, the czar was also able to convince the Empress Dowager in 1898 to lease the Liaodong Peninsula to Russia,[32] whereby Russia gained among other things its first ice-free naval base on the Pacific, and to approve construction of a further rail line (the Southern Manchurian Railroad) that would connect Port Arthur with the CER. At about the same time, Russia convinced China to exclude foreign activity generally in Manchuria.

The development of Manchuria into a Russian sphere of influence was thereby concluded. Aleksandr Izvol'skii, Russian minister of foreign affairs from 1906 to 1910, described the Manchurian spheres of influence as an "important colonial possession" of Russia.[33] Although the agreement with China read differently, Russian officials were practically the sole authority in the area of the railroad. In 1899, Russia and Great Britain undertook to respect mutually their spheres of influence in Manchuria and Central China. Stalin could not have stated any more plainly that he was following in the footsteps of Russian imperialism than by demanding the restoration of the Russian sphere of influence, lost in 1905, at Yalta in 1945. By agreeing to it, Roosevelt essentially betrayed the American tradition of anticolonialism that had been initiated in China in 1899. At that time, Washington had called on all great powers to follow an Open Door policy—that is, to grant equal trading privileges to all states and give the Chinese government the sole right to collect customs duties, among other things. But in 1945, the American interest in seeing the USSR enter the Pacific war as soon as possible was so overwhelming that Roosevelt repressed such scruples, if they even arose.

This is not the place to recapitulate the complicated history of the consequences of the Russian privileges in Manchuria in detail. Let the following summary suffice. In 1905, Russia and Japan divided Manchuria in a secret agreement into a Russian sphere of influence in the north and a southern Japanese sphere of influence in the south. Following the October revolution, Soviet Russia at first offered to transfer the CER to China without compensation, but withdrew the offer again shortly afterward.[34] China was granted the right to purchase the CER in the Sino-Soviet agreement of May 31, 1924, in which Moscow renounced Russian privileges, concessions, and consular jurisdiction and recognized China's sovereignty in Outer Mongolia.[35] In a supplementary agreement, the two parties agreed to administer the railroad jointly until China would purchase it.[36]

Sino-Soviet relations became tense as the Chinese Communists sought to infiltrate the KMT beginning in 1927. This, and the increasingly nationalist self-confidence of the Nanking government after the capture of Peking by KMT troops in the summer of 1928, led in 1929 to the attempt to transfer the CER to Chinese ownership, or at least an attempt to take full control of the railroad, which was supposed de jure to be administered jointly and yet was under Soviet administration in fact. Chang Hsueh-liang, the ruler in Manchuria, who had since become linked to the KMT government, had the Soviet consulate general in Mukden and the Soviet trade missions in Manchuria closed, confiscated the CER telegraph system, expelled several Soviet railroad employees including the director general, and arrested numerous others. Moscow then sent troops into Manchuria and inflicted bloody casualties on the Chinese army. The Soviet Army's military superiority and international diplomatic pressure forced the Chinese government to concede. In December 1929, it signed a Sino-Soviet protocol in which it agreed to reestablish the status quo ante.[37]

Beginning in September 1931, the Japanese army occupied Manchuria in a few months. After it had transformed the country into the Japanese puppet state Manchukuo in February 1932, Tokyo increasingly exerted pressure on the Soviet position in the CER. It confiscated Soviet equipment, interrupted the connection with the Trans-Siberian Railroad, and arrested large numbers of Soviet staff, holding them captive for long periods and in individual cases even killing them. Moscow gave in to Japanese pressure and sold the CER to Japan in March 1935 for 140 million yen—a small amount given the 625 million yen Moscow originally requested. Having already ceded Port Arthur, Dairen, and the Southern Manchurian Railroad to Japan in 1905, the Soviet Union now lost its last base in Manchuria with the CER.[38]

The Yalta Agreement of February 11, 1945 (II)

Let us now return to the Yalta agreement of February 11, 1945. China was not represented at the conference at Yalta. Chiang Kai-shek, who was anxiously seeking American and Soviet assistance against the Communists, was in a weak position. Roosevelt nonetheless did consider it proper out of consideration for his Chinese ally to take care to include a sentence in the agreement whereby the agreement on Outer Mongolia, the Manchurian harbors, and the CCR "will require concurrence of Generalissimo Chiang Kai-shek." This was nothing more than a pro forma gesture, however. It was contradicted by a phrase further below in the agreement that definitively and unmistakably stated that the conferees had agreed that the Soviet claims "shall be unquestionably fulfilled after Japan has been defeated." Two clauses in the agreement also weakened China's position in the planned treaty negotiations with the USSR. They stated that "the preeminent interests of the Soviet Union shall be safeguarded" with respect to Dairen and the CCR and that the American president would take steps to obtain Chiang Kai-shek's approval on the regulation of the questions involving Port Arthur, Dairen, the CCR, and Outer Mongolia. Roosevelt and Churchill did not consult Chiang prior to or during the negotiations for the agreement, as British foreign minister Anthony Eden registered with concern.[39]

Stalin did not consider it necessary to keep his Chinese comrades informed, either.[40] In retrospect, Mao Zedong commented bitterly that Roosevelt and Churchill had resolved to hand over all of China to the United States.[41] At that time, Mao said, Stalin "did not support our Communist Party either materially or morally, and in particular not morally, but Chiang Kai-shek."

The February 11 agreement was kept so secret that Roosevelt for a while denied to the U.S. ambassador in China, Patrick Hurley, that it even existed. When Chiang Kai-shek heard about suggestions that Stalin had made about the Far East at the Yalta Conference through his ambassador in Washington in mid-March,[42] he allegedly could not believe the news and wondered whether China had "really been sold out at Yalta."[43] Roosevelt regretted concluding the agreement a few days before his death (on April 12) and sought a remedy—apparently because in retrospect he considered the surrender of Chinese sovereign rights without the participation of the Chinese government to have been improper.[44] The new president, Harry S. Truman, only instructed Hurley to present the contents of the agreement officially and confidentially to the Chinese president on May 9, which was done on May 15.[45] Its full text was finally made available to Chiang on June 15.[46] The governments of the three signatory states published it only on the first anniversary of the signing, February 11, 1946, and in doing so provoked a storm of indignation in the broader Chinese public—with respect to the Soviet designs on Outer Mongolia in particular.[47] The fact that China had recognized the independence of Outer Mongolia one month earlier contributed significantly to this reaction.

Stalin respected the commitment he had made in the February 11 agreement to enter the war against Japan on the side of the Allies at most three months after the capitulation of Germany (May 8–9, 1945): on August 8, he declared war on Tokyo. On the following day, the Soviet Army[48] occupied Manchuria, although Moscow had not informed its Chinese comrades in Yenan of this in advance.[49] It continued to advance despite the Japanese offer to surrender on August 10, which the Allies—and Moscow—received the following day. When the USSR finally declared a cease-fire on August 23–24, it had stationed approximately 300,000 soldiers in Manchuria under the command of Marshal Rodion Malinovskii.[50]

By placing its signature under the August 14, 1945, treaty with the USSR, the KMT government implicitly recognized the agreements that had been dictated to it in Yalta. Washington and Chungking hoped that this treaty would make it definitively clear to the Chinese Communists that the Soviet Union supported the Nationalist Government and not the CCP. This bitter realization, it was hoped, would force the CCP to settle with and enter into a coalition led by the KMT. The U.S. ambassador in China, Hurley, was sure that Moscow would take part in the entire plan. In July 1945, he observed accurately in a report:

> The Chinese Communists do not believe that Stalin has agreed or will agree to support the National Government of China under the leadership of Chiang Kai-shek. The Chinese Communists still fully expect the Soviet [sic!] to support the Chinese Communists against the National Government. . . . Nothing will change their opinion . . . until a treaty has been signed between the Soviet and China in which the Soviet agrees to support the National Government. When the Chinese Communists are convinced that the Soviet is not supporting them, they will settle with the National Government if the National Government is realistic enough to make generous political settlements.[51]

It was presumably this view of the situation, among other things, that moved Chiang Kai-shek to ask the USSR for assistance in the conquest of Manchuria in early June and to accept the use of the Manchurian railroads, Dairen, and Port Arthur by the Soviet Army. Chiang preferred to see the Soviet Army rather than CCP troops occupy Manchuria.

Negotiations for the Treaty of Alliance

The negotiations for the August 14, 1945, Sino-Soviet Treaty of Friendship and Alliance took place in two rounds—from June 30 to July 13 and from August 7 to 14—in Moscow.[52] On the Chinese side, the negotiations were led by Prime Minister (more precisely, President of the Executive Yuan) T.V. Soong, who also held the position of foreign minister during the first round. The Chinese delegation included Chiang Kai-shek's oldest son, Chiang Ching-kuo, who had lived in the USSR for a long time and had learned to speak Russian fluently.[53] During the second round, Wang Shijie, who had been appointed foreign minister in the meantime, accompanied Soong. Stalin, assisted by Minister of Foreign Affairs Vyacheslav Molotov and Ambassador Appolon Petrov, headed the Soviet delegation. During the negotiations, T.V. Soong maintained close contact with the American ambassador in Moscow, W. Averell Harriman, who in turn kept President Truman up to date.[54] In early July, Truman informed Harriman that the United States did not want "to to act as interpreter on any point of the Yalta agreement during the present discussions in Moscow."

Molotov, Petrov, and important ministers greeted the Chinese delegation cordially at Moscow airport on June 30—in an incomparably greater show of friendship than was shown to Mao Zedong on his arrival in Moscow in December 1949.[55] That afternoon, Stalin, Molotov, and Petrov received the Chinese group. Soong presented his credentials to Stalin, in which Chiang Kai-shek expressed the hope that the negotiations would establish a solid basis for a friendly, close, and long-lasting cooperation. Referring to Sun Yat-sen's legacy, Soong also expressed the hope that they could establish cordial relations between the USSR and China. According to the Soviet protocol, Stalin replied: "The czarist government previously worked for a division of China. New people are now in power in Russia, who respect China's integrity and sovereignty. I think that the two sides can reach a mutual understanding."[56] According to the Chinese protocol, Stalin formulated his comments as follows: "There are new people in power in Russia. In the past, Russia wanted an alliance with Japan in order to break up China. Now, we want an alliance with China to curb Japan."[57] The talk apparently took place in a friendly atmosphere.[58]

In the first substantive talk on July 2, Stalin raised the topic that was most important to him, as he considered it to constitute the basis for the negotiations that had just begun: the Yalta agreement. He asked whether Soong was aware of the three-power agreement that had been reached at the Crimean conference with respect to the Far East.[59] Soong said that he was familiar with it; the United States had informed his government of the contents of the agreement. He also said that he was prepared to discuss all the points in the agreement.[60]

The sharpest disputes during the negotiations involved the status of Outer Mongolia. Although this region had achieved independence de jure following the 1911 Chinese revolution, it had in fact been transformed into a Soviet protectorate in 1921. This had constituted the primary obstacle in the negotiations for the Soviet-Chinese treaty of May 31, 1924. Although Moscow had ultimately recognized Chinese sovereignty over Outer Mongolia in the final version of the treaty, nothing had changed in its actual status as a result by 1945.[61]

On July 2, Soong was the first to raise the sensitive Mongolian question. He referred to a discussion that Chiang had had with Ambassador Petrov in Chungking: "With respect to the Mongolian People's Republic, Chiang Kai-shek said that the question of Outer Mongolia cannot be resolved at present. We are of the opinion . . . that it is necessary to leave this question open and do not intend to raise it at all at present."[62] Before the beginning of the Moscow negotiations, Chiang had correctly pointed out to Petrov that Moscow had after all recognized China's sovereignty over Outer Mongolia in the Sino-Soviet treaty of May 31, 1924.[63] China was prepared, however, Chiang added, to grant Outer Mongolia a high degree of autonomy. Furthermore, if China made concessions to Moscow on the Mongolian question, London would demand similar concessions with respect to Tibet.

Stalin contradicted Soong on July 2 and insisted firmly that China had to recognize the "independence" of Outer Mongolia. He presented several arguments to justify his position. He was dealing with Soviet security interests in the Far East with reference to Japanese aggression. Even if Japan should suffer a defeat, it could recover from it in fifteen to seventeen years, he said, as Germany had done after the First World War. The USSR had to retain the right, he argued, to protect the territory of Outer Mongolia with its troops. And this, he said, was in both the Mongolian and also the Chinese interest. Stalin argued sophistically: "When we are strong, China will also be strong." The Mongolian People's Republic did not want to belong to China, he said. It was seeking "an independent existence and also national independence." With remarkable chutzpah, Stalin finally brought out, with the hint of a threat, the argument that it would be "better for China to cut Outer Mongolia off [*otrubit'*], and that if it did not do so, Outer Mongolia would be a factor that would unify all the Mongols, which would be disadvantageous to China."[64]

Stalin then explained the Soviet motives in seeking the Yalta agreement on the Far East. With respect to Port Arthur, the Chinese Changchun Railroad, southern Sakhalin, and Outer Mongolia, the USSR first had to strengthen its strategic position against Japan in order to defeat Japan alongside China. The Soviet Union was not seeking any profit from the Mongolian People's Republic, Port Arthur, or the railroad. Furthermore, the Soviet government had to justify its decision to attack Japan to the war-weary Soviet people.[65]

The Chinese side provided the greatest resistance on the Mongolian question from the very beginning. Chiang Kai-shek felt he had support from Roosevelt and Truman's interpretation that the formulation used in the Yalta agreement—that the status quo in Outer Mongolia should be preserved—meant that China should retain its sovereignty legally.[66] For a time, the negotiations risked failure over the Mongolian issue. The Mongolias had been an inseparable part of China in Chinese historical consciousness since the Yuan dynasty in the thirteenth century. Chiang Kai-shek had to fear that to surrender Outer Mongolia would raise doubts in Chinese public opinion about his legitimacy as the moving force in Chinese national consciousness. Outer Mongolia's strategic significance also played a role in the calculations on both sides. Stalin apparently feared that if it became Chinese, either Japan, the United States, or a future strong China could threaten the Trans-Siberian Railroad from it, while Chiang saw the danger that if the Soviets continued to occupy it, they would be able to make a more rapid military attack on Peiping.[67]

On July 2, Soong stated the Chinese position forcefully. Stalin, he said, did not understand China's position on the Mongolian question. The Chinese government could not declare "that it was surrendering a part of its territory" under any circumstances. It would not even raise this question and would let "sleeping dogs lie." If the Chinese government surrendered Outer Mongolia, which the USSR had previously recognized as an integral part of China, its position "would be shaken in the eyes of the people." Stalin had himself said, Soong added, that he wanted to see stability in China. It would be best not to create additional difficulties for China. To avoid doing this, "we are not addressing the question of Outer Mongolia now."[68]

The dispute continued at the following session, which took place on July 7. Both sides referred to the single sentence in the February 11 Yalta agreement that dealt with the Mongolian question. This read: "The *status quo* in Outer Mongolia (The Mongolian People's Republic) shall be preserved."[69] The two sides had conflicting interpretations of the term "status quo." Soong interpreted it legally—that is, in the sense that China "cannot recognize the independence of Outer Mongolia."[70] In view of the Sino-Soviet treaty of May 31, 1924, he was correct in this. Stalin, on the other hand, argued that the status quo should be understood as the actual situation and said accordingly that they had to understand it "in the sense of recognizing the independence of the Mongolian People's Republic."[71] Soong now made a further concession to the Soviet side by stating that his government had no objection to the continued presence of Soviet troops in Outer Mongolia. When he then implored Stalin to "understand for one moment that no Chinese government could recognize the independence of Outer Mongolia and remain in power," the latter, unmoved, replied that the Soviet government had remained in power despite the fact that it had recognized the independence of Finland.[72] Soong rejected this historical parallel. Stalin then threatened that "nothing would come of the talks" if the Chinese side stuck to its position. Soong asked him to consider that even the most liberal representative of public opinion in China would not support any government that surrendered Outer Mongolia. Even the Communists would use this fact to criticize the government. The session ended without results and in a tense atmosphere after barely an hour.

Chiang Kai-shek now made an unusual attempt to break the ice. He instructed his son to hold a talk with Ambassador Petrov on the following day. Chiang Ching-kuo explained to Petrov that he had come as an "unofficial representative" of his father to discuss the situation, which had become complicated, in the spirit of mutual understanding. He was concerned that the negotiations had "come up against an impasse." As a China expert, Petrov had to realize that public opinion in China would not allow Chiang Kai-shek to recognize the independence of Outer Mongolia at that time. Petrov remained firm. Outer Mongolia was "de facto an independent state," and acceptance of the statement the Soviets proposed involved "only the recognition of the actual state of affairs." Chiang declared that everything would be fine if there were no Mongolian question as Soong agreed in principle with the Soviet drafts of the treaty and agreements. Unmoved, Petrov replied "that the negotiations would be unlikely to get around the impasse [*edva li vyidut iz tupika*]" if the Chinese delegation did not make any statement about the independence of the Mongolian People's Republic.[73]

This could possibly be a talk that Chiang Ching-kuo later claimed—without indicating any date—to have held with Stalin.[74] If this hypothesis is correct, it is probable that Chiang was seeking to increase his own prestige by raising the rank of his interlocutor. It is true, however, that the contents of the talk that he reports differ from those described by Petrov. Chiang indicates that Stalin had referred openly to China's supplicant position while talking with him: "But you must understand that today you need my help and I do not need yours." He needed Outer Mongolia for military and strategic reasons, he said. If an army attacked the Soviet Union from Mongolia and interrupted the Trans-Siberian Railroad, he said, "then we are finished." Stalin mentioned Japan, China, and the United States as potential attackers. He would never be satisfied, Stalin reportedly told him with a serious face, until he received Mongolia. Regardless of whether Chiang Ching-kuo's talk with Stalin actually took place, the arguments Chiang attributes to Stalin probably come very close to the Soviet leader's actual reasoning on Outer Mongolia.

The breakthrough on the Mongolian question came at the July 9 session. Stalin's resolute attitude and the realization that Roosevelt had not been prepared to intervene on China's behalf in Moscow apparently moved Chiang Kai-shek to change his position. Chiang convinced himself to give way on the Mongolian question as long as Stalin accepted his position on Manchuria, Sinkiang, and the Chinese Communists. Soong read a telegram from Chiang containing these requests at the session. They essentially included the following:[75] The Soviet Union should, first of all, recognize China's sovereignty and territorial integrity in Manchuria. China was prepared, it said, to use Port Arthur as a naval base alongside the USSR and to make Dairen a free port for twenty years. But China would have to administer the harbors of Port Arthur and Dairen, "as the sovereignty and integrity of Manchuria can only be preserved under these conditions." Similarly, although the two parties could use the CCR together for twenty years, the profits should be divided equally and the administration had to remain in Chinese hands. Second, the Soviet Union should help China to suppress the uprising (literally, "with removing the disorder [*v ustranenii besporiadkov*]") in Sinkiang and resume relations and trade. Third, and finally, Moscow should not support the Chinese Communists in their struggle against the KMT government and should give the aid intended for China exclusively to the KMT government.

Because the question of Outer Mongolia had proved an obstacle in the negotiations, Chiang stated at the end of his telegram that the Chinese government was prepared "to grant Mongolian independence" after Japan had been defeated and the three aforementioned conditions had been fulfilled (literally, "after achieving the goals described in the three aforementioned points") "in the interests of a lasting common peace." A plebiscite would have to be held there, however. Chiang concluded his telegram with grandiose words: the Chinese government hoped that "the Soviet Union could appreciate China's colossal sacrifice and also the sincerity with which it made this sacrifice to secure the peace and a lasting cooperation between the two countries" (author's translation from the Russian protocol).

If Chiang had hoped that Stalin would react with spontaneous gratitude and relief to the Chinese concession on the Mongolian question, he was mistaken. Stalin first criticized Soong's earlier observation that he, Soong, had not recognized the independence of Manchuria in 1933 despite Japanese pressure. Stalin asked sullenly whether Soong was drawing a parallel between Manchuria and Mongolia. When Soong denied this, Molotov responded that a parallel of this sort was in fact being drawn. Stalin then repeated his interpretation of the term "status quo" in the Yalta agreement, and Soong countered it with his own. Stalin then observed bluntly that Chinese sovereignty had in fact already ceased to exist twenty-four years earlier. He wanted China to state its renunciation of Outer Mongolia immediately and not wait until after the victory over Japan and a plebiscite that would not benefit China, anyway.[76]

Stalin addressed only the three conditions Chiang Kai-shek raised after this cool reaction on the Mongolian question. He first commented on the third condition, involving Moscow's relations with the Chinese CP. Given their significance for the main theme of the present book, let us reproduce these statements literally. Stalin said:

> With regard to the Chinese Communists, we are not supporting them and are not thinking of doing so. China has only one government and if there is a second government on its territory that is an internal Chinese problem. As is known, we have provided aid to Chiang Kai-shek before. This is how it will be in future as well. When it becomes necessary to help China, this assistance will be made available to Chiang Kai-shek's government.[77]

In an apparent attempt to convince the Chinese side of the sincerity of these claims, Stalin informed Soong that Sun Yat-sen's son, Sun Ke (Sun Fo), a high-ranking KMT functionary, was secretly supplying important information to the Chinese Communists.[78]

When Soong inquired in a different context what Stalin thought of the Chinese Communists, the following dialogue ensued:

> *Comrade Stalin* asked what the Chinese government wants from the Soviet Union. They apparently want the Soviet Union not to arm the Communists and for all assistance to be provided directly to Chiang Kai-shek.
>
> *Soong* replied in the affirmative.
>
> *Comrade Stalin* said he agreed to that and asked if the Chinese government did not want the Soviet Union to begin disarming the Chinese Communist troops.[79]
>
> *Soong* said it would be impossible to comply with such a request and pointed out that the Chinese government was seeking to resolve the Communist problem by political means.
>
> *Comrade Stalin* pointed out that it would be very good if they could find the political means to resolve the problem of the mutual relations between the CCP and Kuomintang. The Chinese Communists were good patriots, he said, who were fighting against the Japanese successfully.
>
> *Soong* said the central government was seeking to merge the armies and create a united force in the country.
>
> *Comrade Stalin* replied this was a completely legitimate desire as in a state there should be one army and one government.[80]

Stalin was following the February 11 Yalta agreement in these statements. It obligated the USSR in the final paragraph to "render assistance" to China (represented by the Nationalist Government) "with its armed forces for the purpose of liberating China from the Japanese yoke." Stalin's claim that the Soviet Union had not previously supported the CCP materially was essentially true. He did not seriously mean what he said at the same time about his intention not to do so in the future—at least in Manchuria, as will be discussed in detail in this chapter ("Moscow's Double Game in Manchuria")—and Stalin had probably already planned to break the August 14 Sino-Soviet treaty at the time of the negotiations. In this respect, the dialogue reproduced above already reflected the duplicitous game that would characterize future Soviet policy toward China.

With reference to Chiang Kai-shek's second demand, regarding Sinkiang, Stalin asked whether he should send troops. When Soong did not address this question and instead answered that the USSR should stop the smuggling of weapons across the Soviet border to Sinkiang in the future, Stalin, according to the Soviet protocol, denied such smuggling existed and responded evasively to other Chinese requests.[81] According to the Chinese protocol, he said that he would put an end to the smuggling.[82]

Stalin made it clear that he did not fully agree with Chiang Kai-shek's demands with respect to Manchuria. He declared his willingness to surrender the Russian right to participate in the administration of the neutral zone bordering on Port Arthur as originally established in the convention of April 25, 1898. But he stated imperatively that the "administration of the harbor at Port Arthur has to be Soviet." The Yalta agreement even provided some legal grounds for this demand as it had called for the reestablishment of Russian rights as they had existed before 1904 and stipulated concretely the restoration of the "lease of Port Arthur as a naval base of the USSR." With respect to the CCR, Stalin said that he did "not at all agree to transfer these rail lines immediately to Chinese possession. . . . These rail lines were constructed by Russians with Russian money." Soong insisted on letting the past be past; furthermore, the USSR had sold the railroad to the Japanese. Stalin responded sharply that Chiang Kai-shek's telegram gave the impression that China was doing the USSR a favor (*okazyvaet milost'*) in allowing it to use the railroad. He said he insisted on a joint administration.[83]

Although they could not reach full agreement on the Mongolian question on July 9, they did achieve a consensus in principle; the rest involved primarily questions of procedure and timing. Soong said to Molotov: "We have now . . . reached an agreement on the substance, we now only have to deal with the wording."[84]

This consensus opened the door to tough negotiation on almost all treaty questions under discussion. It began with a meeting of Soong and Molotov on July 10—drafts of the treaty and agreements had been exchanged the evening before—and ended with a regular session on July 12. On July 11, Stalin accepted that China would only recognize the independence of the Mongolian People's Republic after the victory over Japan and a plebiscite—in return for Soong's concession that the signatories point out at the beginning of the relevant joint statement that the Mongolian people had already repeatedly expressed their desire for independence.[85] Among other things, the conflict was over the borders of Outer Mongolia, ownership of the CCR, stationing of Soviet troops along the railroad, the question of whether such troops could be transported to Port Arthur by rail, the Soviet request to station troops in Dairen, and the general interpretation of the term "preeminent interests of the Soviet Union" in Dairen and on CCR property, as it was used in the Yalta agreement. Difficulties arose not least because the terse Yalta agreement, which consisted of only eleven sentences, was very susceptible to interpretation.[86]

The negotiations were interrupted on July 13 because Stalin had to leave Moscow to attend the Potsdam Conference (July 17–August 2). An attempt by Chiang Kai-shek to get Truman to exert pressure on Stalin in the Chinese interest during the conference failed because Stalin and Molotov made clear that the schedule for the USSR's entrance into the Pacific war depended on a successful conclusion to the Sino-Soviet negotiations, and Truman was interested in seeing it enter the war soon. As a result, Truman urged

Chiang "to continue your efforts to reach complete understanding"[87] with Stalin. Soong, who took a harder line than Chiang—on the Mongolian question in particular—then resigned his position as foreign minister and was replaced by Wang Shijie. He apparently feared that it would damage his image to yield. Hurley, the American ambassador in Chungking, attacked the Chinese in the rear by telling his Soviet colleague Petrov about Truman's pressure on Chiang and advised the Soviet side to recommend to Chiang, through his son, that he recall Soong, that "big reactionary," lest he "obstruct the negotiations" in his "ambition." Petrov refused and commented that this would provoke a big scandal. On August 6, Chiang sent Soong back to lead the negotiations in Moscow, although Soong had asked to be relieved of this position as well.[88]

When the second round of Sino-Soviet negotiations began in Moscow on August 7, American interests with respect to the Soviet Union had changed. An atom bomb had been tested successfully for the first time in a New Mexico desert on July 16. This event, along with the dropping of an atom bomb over Hiroshima on August 6, made it probable that the war against Japan would end soon, and the American interest in having the Soviet Union enter the war against Japan sank to zero.[89] Stalin, who was interested in establishing a Soviet presence in Manchuria that could only be achieved before the end of the war, on the other hand, began to hurry for the same reasons and had the Soviet Army occupy Manchuria on August 9. This in turn increased the Chinese side's interest in concluding a treaty as soon as possible—especially when Tokyo accepted the Allied demand for an unconditional surrender on August 10. Truman ceased to exercise restraint with respect to the Moscow negotiations and asked Stalin on August 8 not to press China for further concessions—in particular, not for any involving the inclusion of Dairen in a Soviet military zone.[90]

At the August 10 session,[91] the Mongolian question was still the point that aroused the greatest emotions. This can be seen from the following passage in the protocol:

> *Soong* said that the Chinese delegation had made many concessions on the question of the MPR, the question of Port Arthur, etc.
>
> *Comrade Stalin* replied that there was no point in talking about the MPR, as Mongolia had already been lost to China for a long time.
>
> *Soong* replied that Outer Mongolia made up $^1/_7$ of Chinese territory.
>
> To which *Comrade Stalin* observed that Outer Mongolia was a desert.[92]

According to the Chinese protocol, Stalin added to this final statement: "A small piece of land in Guangdong Province is twenty times more valuable."[93]

Truman's pressure on Stalin and the increased Chinese interest in concluding the treaty quickly also forced the two negotiators to make concessions on various aspects of the Port Arthur and Dairen questions. Stalin recognized China's unlimited sovereignty in Sinkiang: "Sinkiang is a Chinese province and the Chinese can do what they want there. The Soviet Union does not plan to intervene in the internal affairs of the province."[94] This was an important guarantee for China because the USSR had intervened massively in the border region during the 1930s and 1940s and had provided military support for the Kazakh and Uzbek uprisings in the Altai and Ili regions belonging to Sinkiang as late as 1943–44.[95] The withdrawal of Soviet troops from Manchuria, Stalin continued, would be completed in at most three months—whereby he fulfilled a Chinese request. At the end of the session, the two parties disagreed only on details of the administration and use of Port Arthur and Dairen, the administration of the CCR, and the question of whether China should recognize the Mongolian People's Republic within its existing borders (the Soviet position) or within newly established borders (the Chinese position).

Soong apparently feared that the Soviets would consider a section of the Altai region that had been part of Outer Mongolia before 1911 and had gone to Sinkiang after the founding of the MPR to belong to the MPR if something else were not stated explicitly.[96] Stalin (in fact or seemingly) suspected something else. He said that "if the Chinese government was looking for an excuse" to withdraw the agreement it had already given to recognize the MPR with the demand that they limit the borders, "it should say this directly and openly."[97] He also sought to exert pressure by again presenting his original demand for the immediate recognition of the MPR and pointed to the alleged desire "of the Mongols" to incorporate Inner Mongolia. They should be warned not to seek this, and he therefore recommended that the Chinese drop their demand for a revision of the borders.

At the August 10 session, the two parties again discussed their relations with the CCP, and Stalin warned against combating the Communists in Manchuria. In the context of a sentence about national unity and democratization contained in the Soviet draft of a declaration about Manchuria and Sinkiang, which the Chinese side wanted to delete because it was an internal Chinese affair, Stalin said that if KMT forces attacked the Chinese Communists in future, the Chinese government "could hardly expect our support, as it would be difficult to grant this support." The Soviet Union did not want to intervene, and yet in the context of conflicts between Communist and KMT troops "it would be very difficult for the Soviet Union to provide moral support to the Nationalist Government."[98]

Stalin finally agreed to delete the controversial sentence. He pointed out, however, that the Soviet side was making numerous concessions and that "the Chinese Communists would probably criticize the Soviet government for approving the aforementioned points that involved support for the Nationalist Government." He was referring to the passage in the draft statement in which the Soviet government agreed to grant China moral, military, and other material support, "such support and aid [to] be entirely given to the Nationalist Government as the Central Government of China."[99] Soong replied that the Nationalist Government was sincerely attempting to collaborate with the Communists and did not want a civil war. It was not true that they had recently initiated military actions against the Communists. It had been Communist troops, he said, who had attacked first and provoked the conflict. The Communists, Soong said, did not at all want the treaty or agreements to be signed. Stalin answered that it would be sad if that were the case.[100]

At the end of a talk between the delegations that took place on August 12 without Stalin, it was established that there were still six problem areas under discussion: the civil administration of Port Arthur and Dairen, the Joint Military Commission in Port Arthur, the ownership of buildings and equipment in Dairen, the administration of the CCR, customs control of goods transported from the USSR and back on the CCR, and the status of the Dairen harbor commander. Molotov demanded that the administration of Port Arthur and Dairen be subordinate to the Soviet Union; Soong wanted to grant the Soviet side only the right to provide advice. Molotov advocated a Soviet majority in the membership of the Joint Military Commission in Port Arthur and a Soviet chairman; Soong demanded parity in the membership of the commission and a Chinese chairman. Molotov insisted that all buildings and equipment in Dairen were Soviet property; Soong claimed them for China and was prepared to lease some of them to the USSR. Molotov took the position that the manager of the two rail lines that had been combined to form the CCR had to be a Soviet citizen and his deputy a Chinese. Soong demanded that the arrangement Molotov proposed could be adopted on one of the two lines, while the opposite would be true of the other. Molotov insisted that goods transported from the USSR to Port Arthur and Dairen should not be subjected to customs or examined by customs officials; Soong wanted to accept only the latter. The Soviets took the position that the Dairen harbor commander should be a Soviet citizen and the Chinese argued that he should be Chinese.[101]

A dramatic development now took place on the Chinese side with regard to the Mongolian question, which had been excluded from the August 12 talk with Molotov. In a telegram to the Chinese delegation on the same day, Chiang Kai-shek insisted that the borders of Outer Mongolia be defined before its independence could be recognized. Soong and Wang felt that it would be impossible for the Soviet side to concede on this point, wanted to sign as soon as possible, and ultimately ignored Chiang's instructions.[102]

At the final session that began at about midnight on August 13–14, Soong declared that Chiang agreed to recognize the independence of the MPR "within its current borders." The "Mongolian obstacle," the most important impediment in the negotiations, was thus finally cleared out of the way.[103]

The two parties agreed to accept a compromise proposed by Stalin on the CCR: membership of the administration would consist of equal numbers, the Chinese chairman would have two votes; and the manager would be a Soviet citizen. The parties also reached a compromise solution on customs control, whereby only military goods would not be controlled. The question of ownership in Dairen was resolved as follows: the harbor was considered Chinese property, but some of the buildings were leased free of charge to the USSR. Soong agreed to Molotov's proposal that no buildings be leased to third-party states, but asked that this regulation not be established in writing. Stalin agreed. He added, however, that he hoped the Chinese government "was not trying to deceive its alliance partner," and would not violate this oral agreement. Soong said that this would be a crime.

The compromise on the question of the Dairen harbor commander was as follows. The harbor commander would be a Soviet citizen, nominated by the manager of the CCR with the approval of the Chinese mayor of Dairen, and his deputy would be a Chinese. On the question of the military administration of Port Arthur, the parties agreed that the Soviet side would nominate the majority of the members of the military commission and its chairman. The civil administration would be a Chinese problem, in which, however, Soviet interests had to be taken into account. The leadership positions in the civil administration were to be filled in consultation with Soviet military command.

The session ended with a conflict over the financing of the Soviet units that had occupied Manchuria, which Stalin, looking forward to the end of the war, numbered at approximately 1 million. Stalin demanded that the Chinese side cover the maintenance of these troops. Soong objected that China had been completely destroyed after eight years of war and the government could not even feed its own army. The United States financed U.S. troops itself, Soong added, and also helped China out through the lend-lease program. Stalin argued that the United States was a very rich country. Molotov threatened that the Soviet troops in Manchuria would have to take food away from the population, which would be significantly worse. The two sides could not reach an agreement and resolved to work up several different financial proposals and implement the most acceptable draft.

No decision was made during the negotiations about "war booty," that is, possession of the Japanese investments in Manchuria.[104] Truman learned from Harriman that Stalin had raised the question of "war trophies" when the negotiations were resumed on August 7 and had suggested that the USSR laid claim to some of Japanese property, but had avoided T.V. Soong's questions on the matter. Harriman was alarmed and warned Truman that if the Soviets defined war trophies as they had in Germany, they would be able "to strip Manchuria of certain of its industries and to obtain permanently complete industrial domination of Manchuria." Harriman would be proven right: the Soviet Army later used the unclarified legal situation to dismantle most of the Manchurian industrial equipment and transport it to the USSR.[105]

On August 14, the day the entire set of agreements was signed, Soong met with Molotov one last time to sign a secret protocol.[106] It established a special jurisdiction for the Soviet military staff in more or less the following terms: "Russian military courts and authorities will have exclusive jurisdiction over all members of their forces."[107] The protocol was in all probability kept secret because Chiang Kai-shek had to fear massive protests in Chinese public opinion if it should become known. The two sides would also have had to deal with interventions from the United States and Great Britain, who had relinquished all their extraterritorial rights in 1943.[108]

Results of the Negotiations

If we compare the two sides' starting positions with the results of the negotiations, we reach the following conclusion: Neither Stalin nor Chiang Kai-shek was able to enforce his demand that the administration of Port Arthur be exclusively in Soviet or Chinese hands, respectively. The result called for "joint use": a Chinese civil administration with a degree of Soviet influence and a Soviet military administration with considerable Chinese participation. The Chinese side was able in this way to circumvent the term "lease," which was employed in the Yalta agreement with respect to the use of Port Arthur as a naval base by the USSR, which it considered discriminatory. The two sides reached a similar compromise with respect to Dairen, but here the Chinese side was able get its own way on the property question. With respect to the CCR, for which Chiang originally requested a fully Chinese administration, the negotiations resulted in an equal division of profits and joint management. On the question of customs controls on Soviet goods transported to and from Port Arthur and Dairen, they also met in the middle. If Stalin had clung inflexibly and firmly to the Yalta agreement, which had called for the restoration of Soviet rights in Manchuria as they had existed before 1904 and had sought to safeguard "the preeminent interests of the Soviet Union" with respect to Dairen and the CCR, he would have been able to impose his maximum demands.

This record reveals that it is not true that Stalin had all his wishes fulfilled, as is occasionally claimed in the secondary literature.[109] Chiang Kai-shek, however, was able to get Stalin to make the compromises described above only after he had shown that he would accept his demands on the Mongolian question. On this point—as he received no help from Truman—he found himself forced, after putting up bitter resistance, to give in across the board, in exchange for improvements in his position in other controversial areas. The fact that he got his way when he insisted on recognizing the independence of the MPR only after the victory over Japan and not immediately was merely a small token victory. The same was true of the agreement that the recognition should be made dependent on the outcome of a plebiscite. As he should have known, the pro-Soviet result of the plebiscite was certain in advance.

According to the Mongolian government commission's protocol on the implementation of the October 20, 1945, plebiscite, 100 percent of the participants in fact voted for the independence of the Mongolian People's Republic; dissenting votes were not registered.[110] We can assume that this spectacular result was achieved by means of the tested Soviet mixture of propaganda, pressure, and forgery. The protocol reveals that the popular vote was not secret; votes had to be signed or, in the case of illiteracy, sealed with a thumbprint.[111] A Soviet spy reported that some Mongols believed the plebiscite was being implemented on Moscow's initiative "with the goal of subsequently conquering Mongolia and incorporating it into the USSR."[112] We suspect, nonetheless, that a majority, if not 100 percent, of the population would still have preferred "independence"—that is, a Soviet protectorate rather than a return to China—in a free vote, given the deep-seated resentment of Chinese domination. However that may be, China recognized the Mongolian People's Republic with reference to the plebiscite on January 5, 1946.[113]

Stalin could be satisfied with the results of the negotiations. The USSR had, with few sacrifices—that is, a participation in the war against Japan lasting only ten days—created the preconditions for legal recognition of its de facto domination of Outer Mongolia and reacquired most of the former czarist privileges in Manchuria. It had considerably improved its strategic position in East Asia and toward the Pacific. The reestablishment of its rights in the CCR shortened the railroad connection between the European part of the USSR and the Soviet Far East. The use of the ice-free ports at Port Arthur and Dairen improved the access of the Soviet navy and merchant marine to the Pacific and to the oceans of the world. Molotov reported how Stalin gloated when he described the USSR's gains in the east, using a wall map in his dacha at the end of the war: "The Kuril Islands are now ours, Sakhalin is completely ours and, look, how good! And Port Arthur is ours, and Dairen is ours—Stalin drew a circle around China with his pipe—and the CCR is ours. China, Mongolia, everything is fine. . . ."[114]

But given its weak international position, the Nationalist Government was not dissatisfied overall with the set of treaties, either. Moscow had recognized it as the only legal government of China and described it as the exclusive recipient of Soviet aid, which implicitly ruled out the feared support to the CCP. In Manchuria, it had reestablished sovereignty over a territory that had only been nominally under its control from its seizure of power in 1927–28 through 1931 and subsequently was not under its control at all. After repeated interventions during the 1930s and 1940s, Moscow had recognized Chinese sovereignty in Sinkiang and committed itself not to intervene. The bitterest pill that Chiang had to swallow involved the legal seal on the loss of Outer Mongolia—albeit sweetened by the fact that the Soviet Union had already ruled this area de facto for almost a quarter century. The establishment of special Soviet rights in Manchuria must have been unpleasant for Chiang, but it must have seemed tolerable given the aforementioned gains in other areas. Chiang assumed that the Soviet troops would transfer power in Manchuria to the Nationalist Government and not to the Communists, because Stalin did not want to risk losing the advantages he had gained at Yalta.[115]

It cannot be surprising that KMT newspapers such as the *Zhongyang ribao* celebrated the treaty as a great success and an "epoch-making document for changing the history of all mankind."[116] Other journalists were also full of praise, even if some also occasionally expressed criticism of the surrender of Outer Mongolia.[117] In a secret report for Chiang Kai-shek, the dean of the Law Faculty at Yenching University in Peiping criticized the surrender of Outer Mongolia and the granting of Soviet privileges in Manchuria as too high a price for the creation of an anti-Japanese alliance with the USSR.[118] Given their geographical proximity, he explained, both regions were of key significance to China's "stability" (he presumably meant "security"). As a countermeasure against the threat represented by the Soviet presence in the MPR and Manchuria, he recommended transferring the capital from Nanking to Peiping as soon as possible. Three days after the treaty was concluded, T.V. Soong took the position to Truman that Stalin would only feel bound by the treaty as long as China was strong.[119] Many old KMT members apparently were not satisfied with the treaty, but they did not say anything in public.[120]

Despite these critical voices, the Central Executive Committee of the KMT and the Supreme National Defense Council approved the set of treaties unanimously on August 24. After the treaty had been signed, Chiang told Ambassador Hurley that he was in general satisfied with the treaty.[121] Foreign Minister Wang Shijie said something similar to Hurley, as did Madame Chiang to Truman, in the next few days. Wang Shijie noted with satisfaction in his diary that the treaty he had signed in Moscow would drastically limit the CCP's freedom of action.[122] Minister of War He Yingqin gloated that the treaty would be "a great disappointment to the Communists."[123] The secondary historical literature of the People's Republic of China generally recognizes that the KMT delegation fought valiantly during the treaty negotiations.[124]

The Sino-Soviet Treaty of Alliance of August 14, 1945: Moscow's Betrayal of the Chinese Communists

The treaty of August 14 came as a shock to the Chinese Communists. When they heard about it the day after the signing, they gave foreign interlocutors the impression of feeling "hurt and bewildered," "startled and confused"; of having been sold out to Chiang Kai-shek.[125] Stalin neither consulted nor informed their leadership during consultations for either the February 11 agreement at Yalta or the Sino-Soviet treaty negotiations.[126] Zhou Enlai told the head of the American observer mission at the time that the CCP expected KMT troops to attack during the following days.[127] In late 1945, Stalin also confirmed that the Chinese Communists had "not expected [*ne poschitalis'*] this document."[128]

Mao Zedong and the other Communist leaders considered it probable that the American-Soviet confrontation would constitute the main conflict in the world at the end of the war in Europe.[129] The struggle between the CCP and the KMT, they believed, was subordinate to the confrontation between the two superpowers. When they learned about the conclusion of the Yalta agreement, they realized to their amazement that there were still areas of cooperation and compromise between Moscow and Washington, in addition to areas of confrontation. The conclusion of the Sino-Soviet treaty confused their worldview. They had to agree with the American ambassador, Hurley, when he observed that the treaty "demonstrated conclusively" that the Soviets supported Chiang Kai-shek.[130] They were shocked to learn that Moscow had acquired privileges in Manchuria.[131] They had until recently still been convinced that the Soviet Union would not demand any "concessions or special rights" in Manchuria for sending in troops because the days of Russian imperialism were over.[132]

On August 8, the USSR declared war on Japan. The following day, the Soviet Army marched into China without previously informing the comrades in Yenan.[133] It advanced along two lines, one through northern Manchuria toward Harbin and the other into Jehol and Chahar provinces. Despite the Japanese offer to surrender on August 10, which the Allies—including the USSR—accepted the following day, it continued to advance. After the cease-fire, which the Soviets proclaimed on August 23–24, Moscow ultimately had approximately 300,000 soldiers in Manchuria.[134]

The CCP leadership responded enthusiastically to the Soviet arrival. It apparently considered it a signal to begin the final conflict with the KMT, during which the Soviet Union would stand at its side. In a congratulatory telegram to Stalin on August 9, Mao wrote that "a hundred million people in China's liberated areas" were prepared to cooperate with the Soviet Army.[135] The CCP leadership observed in instructions to its subordinate organizations the following day that this "major and sudden historical change [*weida lishi tubian*]" put the Party in a position to extend the liberated regions quickly and capture cities and important communication links.[136] The civil war, it said, was about to begin.[137]

On August 10 and 11, Zhu De, commander in chief of the Communist troops, issued orders to units in various parts of the country to disarm the Japanese and the Chinese troops that collaborated with them, occupy major cities, advance into Manchuria and Inner Mongolia, and take control of important rail lines all over China. The primary goal was to control North and Central China. Troops were set in motion to capture cities like Peiping, Shanghai, Nanking, Tientsin, Shanhaiguan, Tangshan, Jinan, and Tsingtao. They planned to collaborate (*peihe*) with Soviet Army fighting there when they advanced into Manchuria.[138]

The CCP leadership's dream since 1935 of a joint struggle of the Chinese Communists and the Soviet Union against Japan (and for the revolution in China) finally appeared to be fulfilled, as stated in a Politburo resolution: "the Soviet Union is preparing to attack the barbarous Japanese imperialists. This gives the Chinese revolution . . . and the Soviet Union's struggle against aggression a common objective, a common basis: the defeat of Japanese imperialism. The Soviet Union is the most powerful assistant of the Chinese revolution."[139]

After the Japanese surrender, the Communists' initial enthusiasm was subdued by the fact that Japanese officers followed the instruction issued by Chungking and supported by the United States to surrender only to the KMT army. Chiang Kai-shek instructed the Communist troops to remain where they were and to await orders. Washington recognized Chiang as the commander in chief of the Chinese theater of war and offered to assist in transporting KMT units into areas occupied by the Japanese. Yenan reacted to this with anti-American attacks. With few exceptions, the Communist troops were unable to capture larger cities or communication centers.[140]

The treaty of August 14 and the Soviet behavior that followed finally shattered the CCP leadership's grandiose plans and its hope that the Soviet Army would intervene in its favor. Telegrams that Stalin sent to Yenan dating from August 16 indicated that the USSR had reached an understanding with the United States. Under Soviet pressure, Washington would not allow Chiang Kai-shek to wage a civil war and would support negotiations between the KMT and the CCP in Chungking. In return for this concession, the USSR had had to guarantee that it would hand over the areas occupied by the Soviet Army exclusively to the KMT.[141]

In telegrams on August 20 and 22, Stalin advised the Chinese comrades not to begin the revolution at that time. He informed the CCP leadership that it could not wage a civil war since this would jeopardize the existence of the Chinese nation. Because the international community and the Chinese people wanted peace, Mao should begin peace negotiations with Chiang Kai-shek in Chungking—a request that Stalin made at the request of the foreign minister of the KMT government.[142] Moscow and Washington would guarantee Mao's security in Chungking. The Soviet Union could not support offensive actions by Communist troops that were designed to annihilate the KMT domination all over China and advised the Communists to act peacefully for the time being so they could gather their forces.[143]

Mao Zedong was apparently annoyed and rejected the idea in smaller circles: "If the people struggle for the overthrow, the nation can be destroyed [*renmin weile fanshen douzheng, minzu jiu hui miewang*]!"[144] At least one case has become known in which the comrades were told behind closed doors that the Soviet Union's "concrete policy" "cannot be understood."[145] They could only proceed from hope.

Relations between the CCP leadership and Moscow had now reached their lowest point in the period prior to the establishment of the People's Republic of China. Everything would suggest that the members of the leadership considered the conclusion of the Sino-Soviet treaty of August 14 and Moscow's subsequent recommendations a betrayal of the Chinese revolution. In late 1945, Stalin said that the Chinese Communists had turned to Moscow at that time to ask for advice.[146] They had not expected him to advise them to travel to Chungking for negotiations, he said. He had thus provoked "great confusion [*perepolokh*]" among them.

Mao's unambiguous criticisms during the 1950s and 1960s confirm this assessment. Talking with Soviet ambassador Pavel Iudin more than ten years later, he accused Stalin of having

> assessed the situation in China and the possibility of the development of the revolution incorrectly [after the war]. He continued to have more faith in the strength of the KMT than of the CCP. In 1945, he insisted on peace with the Chiang Kai-shek people, on the united front with the KMT, and on creating a "democratic republic." In particular, the CC of the CCP, for some reason, received a secret telegram in the name of the "CPSU(B)" (in fact from Stalin) that insisted that Mao Zedong make a trip to Chungking, for talks with Chiang Kai-shek. The CC of the CCP was against this trip, because it expected provocations from Chiang Kai-shek. Nonetheless . . . I was forced to go, because Stalin insisted on it.[147]

Two years later, Mao again complained to Iudin about Stalin with a clear reference to the situation in August 1945:

> At the most decisive turning point in the revolution [*Zhongguo de zui jinyao guantou*], he did not allow [*bu zhu*] us to carry out the revolution, he was against the revolution. He thus committed a mistake as great as the one committed by Zinov'ev. . . . The Third International had already been dissolved but it was still issuing orders and saying that if we did not negotiate peace with Chiang Kai-shek but waged a civil war, the Chinese nation risked destruction [*Zhongguo minzu you miewang de weixian*]. But we were not at all destroyed.[148]

In this context, Mao considered Stalin's behavior at that time and the CCP leadership's reaction one of the three most difficult problem areas in the relations between the two parties during the Stalin era. The other two areas were Moscow's support of Wang Ming and Stalin's arrogant attitude during the negotiations with Mao between December 1949 and February 1950 in Moscow.[149]

In 1962, Mao complained in an internal speech:

> They [the Soviets] simply would not permit the Chinese revolution. That was in 1945. At that time, Stalin placed obstacles in the path of the Chinese revolution and announced that no civil war could be instigated, that we should collaborate with Chiang Kai-shek, as the Chinese nation would otherwise be annihilated. At the time, we did not obey and the revolution was victorious.[150]

In 1948, Stalin said that he advised the CCP leadership against the revolution after the war and admitted—a rare use of self-criticism by the leader, who was extremely pigheaded in his old age—that he had been mistaken in his assessment of the balance of power in China at that time. He told Yugoslav and Bulgarian Communists:

> It is true, we also have made mistakes. For instance, after the war we invited the Chinese comrades to come to Moscow and we discussed the situation in China. We told them bluntly that we considered the development of the uprising in China had no prospects, that the Chinese comrades should join the Chiang Kai-shek government and dissolve their army. The Chinese comrades agreed here in Moscow with the views of the Soviet comrades, but went back to China and acted quite otherwise. They mustered their forces, organized their armies and now, as we see, they are beating Chiang Kai-shek's army. Now, in the case of China, we admit we were wrong. It has proved that the Chinese comrades and not the Soviet comrades were right.[151]

Milovan Djilas, who was present at this talk, confirmed the basic content of Stalin's statement:

> True, we, too, can make a mistake. Here, when the war with Japan ended, we invited the Chinese comrades to agree on a means of reaching a modus vivendi with Chiang Kai-shek. They agreed with us in word, but in deed they did it their own way when they got home: they musted their forces and struck. It has been shown that they were right, and we were not.[152]

In August 1945, the CCP leadership clenched its teeth, so to speak, and bowed to Stalin's recommendations.[153] It knew very well that it would annoy or even lose its only potential ally if it rebelled against Moscow's wishes in the conflict with the KMT. It also did not feel itself to be strong enough to begin the final conflict with its rival in the civil war without Soviet aid. Stalin's telegrams of August 20 and 22 apparently provoked a discussion in the Party leadership about whether a civil war could damage the Soviet Union or even unleash a Third World War.[154] The leadership did immediately draw the practical consequences and altered its strategy. This involved primarily the decision not to try to capture major cities. On August 21, the leadership abandoned the plan to take Shanghai that had been confirmed one day before.[155] Immediately after it received the August 22 telegram, the Central Committee issued the following directive on the same day:

> 1. The Soviet Union cannot support us because of the limitations imposed by the Soviet-Chinese treaty and in the interests of preserving peace in the Far East. Chiang Kai-shek is exploiting his legal status to accept the capitulation of the enemy troops. The enemy and its marionettes can hand over the major cities and important communication links only to Chiang Kai-shek. In view of this situation, our army must alter course [*yinggai bian fangzhen*]. . . . We must place the emphasis on occupying small cities with the necessary troop strength, conquering extensive rural areas, extending and consolidating the liberated areas, and mobilizing the masses for the struggle. We must take care to organize the training of the troops, adjust to the new situation in time, and plan for the long term. . . .
> 2. . . . The USA, like the USSR, disapproves of a civil war in China. Our Party is preparing itself under the three great slogans "Peace, Democracy, Solidarity [*heping, minzu, tuanjie*]" to negotiate with the KMT to struggle for conditions that are advantageous to our Party and for the people. . . .
> 3. In the major cities, activists should continue to mobilize the masses . . . organize subversive actions, and seize a position for our Party in the cities.[156]

Petr Vlasov's "diary" is strikingly silent about the Sino-Soviet treaty, Stalin's telegrams, and the CCP leadership's reactions on the decisive days between August 14 and August 22.[157] Only in an entry dated August 26 does it say that Moscow had informed Yenan of its intention to observe the treaty concluded with Chungking and not to intervene in China's internal affairs in order to avoid conflict with the United States and being drawn into another world war; this, it said, was "a hard blow" to Mao.[158] The diary's silence at the time of these dramatic events may mean that contacts between Stalin and Mao went around Vlasov, or at least that the corresponding entries fell victim to Soviet censors.

The CCP leadership sought to put a brave face on a bad situation in discussions within the Party. It sought to justify the Soviet policy and make the situation look better to surprised and probably also annoyed Party members. Instructions from headquarters said that the Soviet Union sincerely supported the Chinese revolution and would allow the Chinese CP to follow its own path as long as it did not violate Soviet treaty obligations directly. One regional Party leadership argued as follows: The USSR signed the treaty of friendship with China, it said, to maintain and consolidate peace in the Far East.[159] This was of great advantage to the Chinese people and disadvantageous to Japan. But the Soviet Union could not very well also directly help the Chinese Communists at the same time because of these restrictions.[160]

Mao Zedong did not let the Party membership see that he was upset with Moscow, either, and to this end adopted the Soviet arguments—at least pro forma—as his own. In a talk with comrades in late August,[161] he argued that "confined by the needs of maintaining international peace as well as the Chinese-Soviet treaty, the Soviet Union was not in a position to freely support us." As in Europe, where the USSR would interfere only in Bulgaria but not in Greece, because the latter country was part of the British sphere of influence, the Soviets could not become too involved in China, because it was under the protection of the Americans. "If the Soviet Army assisted us, the United States would certainly support Chiang, and, as a result, the cause of international peace would suffer and a world war might follow." On the same day, Mao declared at a meeting of the Politburo[162] that the stage dominated by the anti-Japanese war was over and that they were moving into a period of peaceful development. Peace was achievable, he said: the Soviet Union, the United States, England, and the Chinese people needed it; even the KMT could not bring itself to fight a civil war. The new slogan was "Peace, democracy, solidarity." At about the same time, he stated within the Party: "We must now, with regard to the country as a whole, essentially go the way of France, which means accept a government led by the capitalist class in which the proletarian class participates."[163] Shortly afterward, he justified his willingness to negotiate with Chiang Kai-shek to Communists in Chungking with the argument that this was a suggestion from the Soviet Union, which wanted them to establish a coalition government with the KMT.[164] A government like that "is not in full accordance with Soviet ideals, but good compared to the outbreak of a violent civil war."

We do not know whether and, if so, to what extent Mao really allowed Stalin's arguments to influence him. He had in any case understood that his expectation that the cooperation between the allies in the Second World War had now ended was premature.

On August 15—just one day after the conclusion of the Sino-Soviet treaty of alliance—the CCP ended its hostile propaganda toward the United States and the KMT and again adopted more friendly tones.[165] Yenan ordered its troops to collaborate with the U.S. Army. In late August, the CCP Central Committee approved the August 14 treaty in public statements—with the exception of the clause on Outer Mongolia—and praised it as the "first treaty based on equal rights with a foreign state in the history of our country."[166] There can be no doubt that such an assessment cost the Chinese Communists a significant degree of self-denial.

On August 24, at Stalin's insistence, Mao Zedong accepted the invitation that Chiang Kai-shek had already expressed three times; he arrived along with Ambassador Harriman in Chungking on August 28.[167] On his arrival, Mao looked to one observer like someone going to his own execution.[168]

Hurley's prediction that only a Soviet treaty with Chungking would convince the Communists to negotiate with the KMT had been confirmed. Elated, he flew to Yenan with the KMT general Zhang Zhizhong to pick up Mao Zedong.[169] The Soviet ambassador in Chungking, Petrov, had declined an invitation from Hurley to accompany him—presumably in order to allow the CCP leadership to save face and not draw unnecessary attention to the close collusion between the United States and the Soviet Union in public.[170] Before Mao left, he had requested and received Soviet guarantees for his personal security.[171]

On a visit to the Soviet embassy in Chungking, Mao Zedong declared to Ambassador Petrov that the CCP was willing to collaborate with the KMT for a long period of time, but he also complained about Chiang Kai-shek's intention to maintain the "one-party dictatorship" and weaken if not annihilate the regions under Communist rule. Petrov declared that he was satisfied that the two parties to the civil war were willing to negotiate seriously and pointed out the advantages of maintaining the peace and of an evolutionary development toward "democratization."[172]

The Chungking negotiations were interrupted on October 10. Both sides announced that they wanted to convene a Political Consultative Conference to implement in practice the resolutions on which they had been able to reach agreement.[173] Mao had agreed officially to reduce the Communist armed forces to sixteen to twenty divisions, but he was determined, as he secretly informed Ambassador Petrov, to circumvent this agreement by increasing the strength of the divisions and thus to retain all of his troops.[174] The two sides in the negotiations had been unable to reach agreement on the key question of who would nominate the governors in the regions under Communist control—that is, the question of who had political power in these regions.[175] On October 11, Mao flew back to Yenan.

Criticism was apparently voiced in Yenan, by Zhu De in particular, about Mao Zedong's decision to negotiate in Chungking at all or the possibility that he had conceded too much.[176] It seems improbable that the negotiations had been nothing more than a diversionary tactic from the beginning, a smoke screen behind which CCP troops would arrange with Moscow to move into Manchuria.[177] At least there is no evidence of this.

In 1945, Stalin sacrificed the Chinese comrades' revolutionary hopes on the altar of the Soviet Union's interests. He perceived these interests in two ways. On the one hand, he sought to get approval from Washington and London for the maintenance or restoration of traditional Russian privileges in Outer Mongolia and Manchuria. On the other, he wanted to appease the United States and convince it to cooperate in the East Asian sphere. The Chinese partner of the two Western powers with whom Stalin had to work to achieve this, however, was not the CCP but the Nationalist Government of the KMT.

Moscow's Double Game in Manchuria

What the Chinese Communists probably did not yet know in August 1945 was that Stalin planned to circumvent the August 14 treaty concluded with the Nationalist Government on one point: with respect to Manchuria. He wanted to do more there than merely restore Russian privileges from the czarist era, which he had already been able to do de jure with the February 11 Yalta agreement and the August 14 treaty with China. He also wanted to preserve these privileges indefinitely and develop Manchuria into a buffer state between the Soviet Union and the rest of China that would in future be subject to the control of the American-dominated KMT and a security barrier between the Soviet Union and Japan, as it recovered its strength. The best guarantee for this appeared to be a Manchuria controlled by its Chinese fraternal Party. Stalin had therefore to all appearances been firmly resolved since the Soviet Army's invasion of Manchuria to bring the CCP, which he saw as the more reliable partner, to power.[178]

The Dilemma Between Legal Obligations and Political Goals

Stalin felt compelled to play a duplicitous game to achieve these goals. To the outside world, he had to give the impression that he was fulfilling the conditions of the set of treaties of August 14, in which the USSR had recognized the Nationalist Government as the only Chinese government and China's full sovereignty in Manchuria.[179] One of the agreements on cooperation in Manchuria during the war signed on August 14 dealt essentially with the transfer of Manchurian territories liberated from the Japanese to the Nationalist Government.[180] The agreement was not clear about military force. On the one hand, it said in Point 1 that the commander in chief of the Soviet armed forces held the "supreme authority and responsibility in all matters relating to the prosecution of the war." It then established in Point 5 that the Nationalist Government would assume full authority "in civil affairs" (Russian version: *po linii grazhdanskikh del*), or "in the direction of public affairs" (Chinese version: *guanli gongwu*), whenever a part of Manchuria ceased to be an immediate battleground. The Chinese version thus included military power and the Russian did not. It cannot be demonstrated to what extent the different wording in the Russian formulation involved an intent that the Chinese side did not see through. The Soviet Army apparently sought to exploit the confused legal situation to help the Chinese Communists ultimately to seize power. They could only refer to this clause legitimately de jure until the cease-fire was proclaimed on August 23–24, however—unless they were going to claim that fighting continued despite the cease-fire.

It will come as no surprise that such a complicated game led to some irritations both among Soviet commanders and officers of the Chinese Red Army. It was also occasionally necessary for the Soviet side to change direction temporarily and for tactical purposes to save face in Chungking and Washington. The Chinese comrades and military in particular did not always understand this. The Communist seizure of power in Manchuria described in this section thus followed tortuous paths.

In the first weeks following the Soviet invasion of Manchuria, which took place on August 9, 1945, Moscow did not behave unambiguously there toward the Chinese comrades. There are examples both of support and disavowal. In mid-September, the Soviet Army agreed with Yenan and helped CCP units to establish themselves in Manchuria secretly. It altered its policy during November. For a four-month period, the Soviet Army supported KMT troops in taking control of key positions in Manchuria and expelled the Chinese comrades. From mid-March 1946, as the Soviet Army began to withdraw finally, Stalin encouraged the CCP leadership to take the offensive and help Communist troops to move into the evacuated regions of northern Manchuria. The Soviet Army provided CCP troops with weapons and ammunition from the former Japanese arsenals throughout the entire period of the Soviet occupation of Manchuria, which ended on May 3, 1946.

This section will examine the reasons that underlay the changing Soviet policy in Manchuria in detail. Let us simply note here that Soviet regard for the treaty of August 14, 1945, and for the U.S. Army's military support of the KMT played an important role.

While Stalin initially tended to recognize China generally as a part of the American sphere of interest at the end of the war, given the Soviet Union's weakness in East Asia, he paid special attention to Outer Mongolia and Manchuria. As we presented in detail above,[181] he had the existing de facto Soviet protectorate of Outer Mongolia confirmed by the United States and Great Britain in the Yalta agreement and by China in the treaty of alliance. The same was true—at least in part—of his wish to have the inclusion of Manchuria within the Soviet sphere of interest respected, and the military occupation represented an important additional trump for him. Regardless of the considerable disruptions in relations between Moscow and the Chinese Communists in other areas, both sides were tacitly united at that time in the attempt to prevent the Nationalist Government from taking over power in Manchuria and including it within the American sphere of interest.

The Manchurian card had several different aspects for Stalin. He acquired military and economic privileges in Manchuria in the treaty of August 14 and in this way put a foot in the door to China, so to speak. The use of Port Arthur as a naval port for use defensively or offensively put the Soviet Union in a position to reinforce its security position in Northeast Asia and create an additional counterweight to the overwhelming U.S. Pacific Fleet. The economic influence that joint ownership of the Manchurian railroads, with the associated real estate, companies, and privileges in Dairen guaranteed, put the Soviet Union in a favorable position to increase its control of the Manchurian economy in the future. Stalin attempted to use the Soviet military presence in Manchuria as a China policy trump card in the conflict with Washington and the West and trade Soviet withdrawal from Manchuria for an American withdrawal from North China. This is the context in which we should see Stalin's Manchurian policy toward the CCP. Stalin's deep-seated fear of a third world war also led to a firm determination to avoid a military confrontation between the Soviet Army and CCP troops and the U.S. Army at all costs.[182]

The Chinese Communists ascribed a key role to Manchuria in their strategic plans during the period following the war. They planned to conquer Manchuria, develop it into a permanent base, and receive Soviet aid through the territorial connection to the Soviet Union and the Mongolian People's Republic. Mao Zedong said: "At present, our bases are not yet secure; we have no base. But when we have Manchuria, we will have a base."[183] During the months of April and May 1945, Mao had already planned to pull together 150,000 to 200,000 soldiers from various military regions and deploy them "along with the Soviet Union to reconquer Manchuria [*peihe Sulian shoufu Dongbei*]."[184]

We can find no evidence of any military cooperation between the Soviet Army and Communist units subject to Yenan in Manchuria in the period prior to the conclusion of the Sino-Soviet treaty of August 14. By the beginning of June, however, Moscow began to plan to deploy the remains of the Communist-led United Anti-Japanese Army, which had been driven from Manchuria into the Soviet Union in 1940–41. In August 1942, these units were reorganized with the new designation Eighty-eighth International Brigade and placed under the command of Zhou Baozhong, who had been trained in the Soviet Union. They were to play a role in coordinating the Soviet Army and CCP troops in the "future great changes" in Manchuria, as they were then called. Japanese resistance shattered the attempts undertaken from 1944 to infiltrate CCP units from the south into Jehol and southern Liaoning. Zhou Baozhong was nominated to be an adviser to the Soviet Far East Headquarters in early June 1945 and regularly participated in its meetings. In mid-June, China sent some 250 Communist officers to Siberia for a special training course where they were prepared to carry out special tasks during the imminent Soviet advance into Manchuria. Zhou subjected himself only reluctantly to Soviet command, however. In August, he and a small unit slipped back into Manchuria along with the Soviet Army, and established himself firmly on the Soviet border in the north.[185]

Immediately after the Soviet Army marched into Manchuria on August 9, headquarters in Yenan issued orders for several units to advance into North China and cooperate militarily with Soviet troops.[186] The order was published ostentatiously along with six other orders, however, and—like the others—was designed in part to deceive public opinion. It was conceived only "for external propaganda" "in order to achieve as quickly as possible a public starting position domestically and abroad," as the Central Committee told the relevant troop units at the same time in secret instructions.[187] One can speculate about what was behind this formulation, however. It is possible that Yenan hoped to influence the treaty negotiations that were coming to an end in Moscow between the USSR and the KMT government and to subject the Soviet leadership to moral pressure. In any case, only one single unit in fact marched to Manchuria; the rest advanced into the area of Chahar, Jehol, and Hebei provinces outside Manchuria to receive the Soviet troops and coordinate with them there without at first mentioning any plans to conquer Manchuria. Chinese Communist units in fact met up with Soviet-Mongolian troops at Zhangjiakou (Kalgan) in Chahar on August 17. The latter refused to capture Zhangjiakou along with the CCP units, but did not stop the Chinese Communists from conquering the city themselves.[188]

Stalin was also correct, formally, when he told a distrustful Harriman in late August that there had not been any contacts up to then between the Soviet Army and Chinese Communist troops in Manchuria, as Zhangjiakou was indeed outside the borders of Manchuria proper.[189]

Telegrams from Moscow that arrived in Yenan on August 14 contained the bad news that the Soviet leadership could not support a civil war in China and that it had to hand the territories the Soviet Army conquered in Manchuria over to the KMT in keeping with the treaty of alliance concluded with Chungking and an agreement with Washington.[190] Shortly afterward, the CCP leadership accepted the new position and reacted with the corresponding reservations. The military leadership of the CCP did put five regiments together to march into Manchuria on August 20, but given the confused situation had them take up positions in Jehol and wait on the border of Manchuria.[191]

Internal instructions from the end of August reveal the CCP leadership's uncertainty.

It sent 1,000 civilian cadres to Manchuria, as it saw no problem in doing so. Yenan generally had the idea at that time that although it should recognize the KMT's authority in Manchuria formally, it should in fact exercise real control through the Communist cadres. Although "it cannot not be decided at the present time" whether the Party leadership would be able to send troops there, it said, that step could not be ruled out altogether. The units—under cover names such as "Manchurian Army [*Dongbei jun*]" and "volunteer army [*yiyong jun*]"—that had already been selected for deployment in Manchuria should therefore advance from Shanxi, Chahar, Hebei, and Shandong toward Manchuria. They should "await further orders" in Jehol, which—like Chahar—was part of Manchukuo and not part of the area covered by the Sino-Soviet treaty of August 14. If possible, they should march further; if not, they should develop their positions in Jehol into a strong base. There were still no KMT troops in Jehol and Chahar at that time; the Communists held military superiority in Shandong, Hebei, Shanxi, and Suiyuan.[192]

Preoccupied by Stalin's telegrams of August 20 and 22, the Chinese Party leadership thus initially acted cautiously. A Central Committee instruction in late August stated that the Soviet Union had to hand Manchuria over to the KMT government in order to preserve the peace in the Far East and fulfill its obligations under the August 14 treaty. The Soviet Army would not be able to establish official contact with or support the CCP or troop units that had advanced to the north. The Soviet Army would only assume a laissez-faire posture and show sympathy to the Communists if their actions did not affect Soviet treaty obligations. They could only march into Manchuria and establish unofficial contact with the Soviet Army if it raised no objections. This should not be done too openly (*bu yao sheng zhang*), it said, or published in the newspapers. They should not try to force the Soviet troops to make official contact, request any support from them, or create diplomatic difficulties for them. There would be a favorable opportunity to conquer Manchuria and Jehol and Chahar when Soviet troops withdrew three months later, however.[193]

At about this time—that is, the end of August and beginning of September—the Soviet Army apparently signaled its willingness to cooperate in the northeast, including Manchuria, for the first time to the CCP. It handed several places in Inner Mongolia and Jehol over to Communist troops. They conquered Shanhaiguan, the "gateway to Manchuria," in a joint attack. Communist units were able to take over significant quantities of Japanese weapons and supplies by mid-September. With Soviet tolerance or assistance, Communist units pushed forward through Chahar and Jehol and along the Changchun Railroad into Manchuria. The Soviet Army also helped the Chinese comrades to establish local administrations in many Manchurian areas.[194]

On September 11, the Party leadership resolved to send 25,000 to 30,000 troops from Shandong to Manchuria "without causing a stir" via "a decentralized route by sea" with designations such as the "Manchurian Volunteer Army [*Dongbei yiyong jun*]" or other local Manchurian names. The following orders were given to the troops: They should not capture any major cities at first, but only rural areas, and small and middle-sized cities that were not occupied by the Soviet Army and lines of communication. They should mobilize the masses, establish a local administrative authority, bring enemy troops under control, collect weapons, and support the Soviet Army in creating a democratic order. They should do so secretly: they should not use either the name of the Party or the correct designation of the military unit (Eighth Route Army) in official—as opposed to unofficial—negotiations with the Soviet Army.[195]

The Soviet Union's policy toward the CCP on Manchuria and the bordering regions occupied by the Soviet Army proved inconsistent during the first three weeks after the conclusion of the treaty of August 14, 1945, however, and led to differing results in practice. The Soviet units had apparently received no general or binding orders from Moscow to help the Chinese comrades to establish themselves. The behavior of the Soviet commanders varied accordingly. The CCP troops usually received assistance, but occasionally also rejection. Military considerations and the personal sympathies of local commanders for the Chinese Communists often clearly played a role when Soviet commanders cooperated with the CCP. The picture thus was not uniform.

In Dairen, Soviet authorities did not allow local Communists to operate openly in the city in late August and preferred to work with the Manchukuo authorities that were still there. Relations became so tense that the Chinese Communists even refused to reveal their Party membership to the Soviets. The brutal behavior of members of the Soviet Army toward civilians throughout the Soviet occupation created a problem for the CCP's prestige among the Manchurian population. Robbery, murder, rape, and looting were the order of the day. The garrison of the Eighth Route Army in Mukden protested to the Political Department of Soviet headquarters and called on it to restore discipline. The Communists were particularly bitter when a Soviet soldier who had been caught looting shot a senior officer from their ranks. There continued to be resentment in Manchuria against the Soviet Union and against the CCP as collaborators with the USSR for a long time. In late 1949, handwritten slogans could still be seen in Harbin such as "Down with Red Imperialism!" and "Down with the traitor to the fatherland Mao Zedong!"[196]

In early September, friction arose between the Soviet commander, A.I. Kovtun-Stankevich, and Zeng Kelin, the commander of the Communist garrison. Zeng was able to circumvent the Soviet commander's resistance with a trick and march into the city. When he then attempted to take over the city administration, in which he wanted to collaborate closely with the Soviets, Kovtun-Stankevich ordered him to withdraw from the city and referred to instructions from Moscow and Soviet obligations under the treaty of August 14. Zeng refused to do so without first consulting the Party leadership.[197]

The Soviet Army delayed and practically refused to hand over power to Chungking during the first months. It did not allow KMT troops to advance into Manchuria—until they invaded in force in early November.

Designing a Dual Strategy: The Secret Meeting in Yenan and Its Consequences

The attitude Moscow had adopted toward the Chinese comrades since mid-August—as a general rule benevolent, at times indifferent, and occasionally negative—was secretly replaced by a joint strategy in mid-September.

On September 14, Lieutenant Colonel Dmitrii Belorussov landed in a Soviet military aircraft in Yenan as Malinovskii's representative to discuss future actions in Manchuria with the Party leadership. This was the first official visit by a representative of the government of the USSR to the headquarters of the CCP. Zeng Kelin was also in the aircraft.

He asked Belorussov to be taken along because he wanted to consult in person with the Party leadership about the tense relations with the Soviet Army. Mao Zedong was negotiating in Chungking at that time. Belorussov's primary interlocutors were Zhu De and Liu Shaoqi.[198]

Newly accessible information allows us to reconstruct the course of the talks—if not exactly, at least in their broad outline. Belorussov delivered a message from Malinovskii, which probably was in writing, that said that neither KMT nor CCP troops could move into Manchuria before the Soviet Army withdrew. Kuomintang units that had moved in without Soviet permission had already been disarmed. The Eighth Route Army should thus not move into Manchuria openly before the Soviet Army withdrew. Chinese Communist units should pull back from areas in which they had already occupied major cities such as Mukden, Dairen, and Changchun; or, according to another version, from areas where they were stationed that were occupied by Soviet troops. The Soviet Army, it said, would withdraw shortly (*bu jiu*). "China should then decide for itself how Chinese troops should move into Manchuria." The Soviet Union "does not intervene in China's domestic politics [*bu ganshe Zhongguo neizheng*]."

According to a Chinese account that was apparently based on documents from the CCP archives, there were also "unofficial talks [*fei zhengshi tanhua*]" between Belorussov and Party leaders.[199] This was clearly a reference to discussions held after an emergency session of the Politburo, which was convened by Liu Shaoqi and had taken place in connection with Belorussov's report. It was agreed during this session that although the Party leaders could not ignore Soviet wishes, they should continue the strategy of developing bases in Manchuria. They should not recall the troops that had already been sent to Manchuria from Shandong. They should not withdraw from the Eighth Route Army's long time battleground in Jehol and Chahar. Out of consideration for relations between Moscow and the Nationalist Government, they should have the troops use cover names in Manchuria, rather than operate as the Eighth Route Army. They should adopt a double strategy in Mukden, Dairen, and Changchun: "withdraw publicly and secretly return [*gongkai che, mimi you qu*]." Most of the troops should remain in Mukden while a small part withdrew into the countryside. It was then resolved to found the Northeast Bureau of the CCP, which would include Peng Zhen, Chen Yun, Cheng Zihua, Lin Feng, and Wu Xiuquan and be under the leadership of Peng Zhen.[200]

Belorussov was apparently informed of the results of the Politburo session in the "unofficial talks." He conceded to the Chinese comrades by dropping his insistence that CCP troops withdraw immediately from areas occupied by the Soviet Army; still, Communist Party functionaries would have to follow Soviet orders and withdraw when told to do so.[201] Zhu De agreed to withdraw Communist troops from larger cities like Changchun, Mukden, Dairen, and Pingqun.[202] At Zhu's request, the Soviet side said it would hand over the entire province of Jehol, which had been part of the former Manchukuo, and parts of West Liaoning, where Communist bases had been established since the beginning of the Sino-Japanese War, to the Eighth Route Army.[203]

The two sides agreed that the CCP should send troops to Manchuria secretly, but only to areas outside the major cities. The Eighth Route Army should not appear in its own name and should employ other designations—such as, for example, "local security forces [*difang baoan dui*]."[204] Belorussov promised support to Communist armed forces operating under such cover names and asked Yenan to send representatives to Manchuria to negotiate on the spot and coordinate the actions of the Soviet Army with those of the CCP.[205]

Belorussov urged the Chinese comrades to be careful:

> In Manchuria, one can only act, not speak [*Zai Dongbei zhi neng zuo, bu neng shuo*]. Under no circumstances can the "self-administration" [a pseudonym at the time for the civil and military organizations of the CCP] be addressed in Manchuria nor can be mentioned that the KMT is not allowed to send any troops to Manchuria, which would offer America and Chiang a pretext and place the Soviet Union in a difficult position.[206]

So much for the Chinese secondary literature. Belorussov's interpreter, Vasilii Sidikhmenov, claims unconvincingly that Belorussov did not carry out any negotiations in Yenan and merely handed over to the CCP leadership a letter that Malinovskii had drawn up on orders from Stalin. According to Sidikhmenov, Malinovskii recommended in his letter that the CCP leadership withdraw its troops from the cities and move them into rural areas. This recommendation was justified, Sidikhmenov argues, because the American secret service had sent information about Soviet aid for the Chinese Communists—such as the surrender of entire Japanese weapons depots and permission to establish themselves militarily in the cities—to Washington, where it had encountered sharp criticism. Molotov put it as follows in a telephone conversation with Malinovskii: "The so-called Communist troops are pushing us into a conflict with America." But this is precisely what Stalin wanted to avoid. Stalin therefore felt compelled, according to Sidikhmenov, to employ a more cautious "policy of maneuvering [*politika lavirovaniia*]" that would extensively avoid giving the impression that the Soviets were supporting the CCP in Manchuria. The recommendation in Malinovskii's letter that the Communist troops should pull back into rural areas, Sidikhmenov states, was based on the conviction that the U.S. intelligence services were present only in the cities.[207]

A comparison of the Chinese publications with Sidikhmenov's account gives the impression that the Soviet leadership was interested primarily in having the units of the Eighth Route Army disappear from cities occupied by the Soviet Army for cosmetic reasons: to avoid giving the impression of cooperation between the Soviet Army and CCP troops in order to maintain relations with the United States and the Nationalist Government. On the other hand, it had no objection to the Eighth Route Army taking up positions in rural areas.

Liu Shaoqi asked Belorussov to take Politburo members Peng Zhen and Chen Yun along on the return flight to establish the apparatus of the Northeast Bureau of the CCP in Mukden and coordinate the actions of the Soviet Army and the CCP.[208] On September 17,[209] six or seven Chinese Communists boarded the Soviet aircraft, including the members of the Northeast Bureau, Peng Zhen, Chen Yun, and Wu Xiuquan.[210] After a dangerous emergency landing in Shanhaiguan, in which some of them were injured, they traveled on to Mukden. Approximately a quarter of the Central Committee members were subsequently transferred gradually to Manchuria.[211] These included such significant leaders as Gao Gang, Lin Biao, and Li Fuchun.

An author from the People's Republic of China, who has access to Party archives, convincingly interprets what we now know about Belorussov's "official" and "unofficial" statements as follows: Following the Soviet occupation of Manchuria, he argues, Stalin had, on the one hand, stated publicly (*gongkai*) that the Soviet Army did not maintain any contacts with the CCP there and also did not plan to maintain any, that he hoped the KMT would send personnel as soon as possible to take over the garrisons, that a treaty would be reached between the KMT and the CCP. On the other hand, he secretly (*mimi*) instructed the Soviet Army to support the rapid dissemination of Communist forces in Manchuria secretly (*anzhong*) and coordinate its actions with the CCP. The Soviet Union, he argued, formally (*zai xingshi shang*) strictly prohibited the CCP troops from entering Manchuria "for foreign policy reasons." In secret (*anzhong*), however, it supported the CCP's efforts to conquer Manchuria. This had no doubt encouraged the CCP leaders in their decision to conquer Manchuria.[212]

As mentioned before, Soviet behavior toward the Chinese Communists had not been uniform up to then in practice. It cannot be established unambiguously how the policy was changed in favor of a consistent—even if secret—cooperation with the CCP. One contemporary witness in the military field indicates that Stalin had not originally planned to hand Manchuria over to the Chinese Communists, but had ultimately decided to help them when they arrived there.[213] Another contemporary witness, who was a Soviet diplomat in China at the time, claims that Stalin sought to establish Manchuria as a buffer state under the control of the CCP from the very beginning, because he believed it would be easier to establish a kind of Soviet protectorate with the Communists than with the KMT—which was certainly true. The second interpretation seems more plausible.[214]

The meeting between Belorussov and the CCP leadership in Yenan provided the initial spark for the arrangement of the joint strategy toward Manchuria that subsequently took effect between Moscow and Yenan. New sources allow us to reconstruct convincingly the process by which this happened. We will outline the most important stages in the following.

On the basis of Belorussov's "unofficial" assurances, which may have gone even further than is clear from the sources accessible to us so far, the CCP Politburo resolved on August 15 "to shift the strategic focus to Manchuria [*ba zhanlüe zhongdian fang zai Dongbei*]" and to send the 100,000 soldiers and 20,000 civilian cadres that were originally to be deployed in South China there. At the session, Zhu De said: "The Central Committee should quickly send people to Manchuria, should prepare 400,000 to 600,000 soldiers." The Central Committee came to this conclusion: "Manchuria must be conquered by our forces, Jehol and Chahar must come completely under our control." There was currently a "unique opportunity [*qian zai yi shi zhi ji*]." If they brought Jehol, Chahar, and eastern Hebei under their control and stationed several thousand soldiers on the coast of Manchuria, it might be possible to stop Chiang Kai-shek from advancing into Manchuria. "If we conquer Manchuria, we will be able to shore up North and Central China, and this will change our Party's position." This, he said, was the "decisive part."[215]

On September 16, the Soviet-Mongolian headquarters of the Transbaikal Military Region responsible for Manchuria went even further than the agreements reached in Yenan. It informed the CCP leadership that "the main forces of the Eighth Route Army should be urgently ordered to move northward immediately" to take over the regions the Soviet Army occupied in Inner Mongolia and Manchuria. The goal, it said, was "to secure control of the north and of the Inner Mongolian regions and thus to maintain the connection to Outer Mongolia and the Soviet Union better." Moscow could assist the CCP units with weapons. If absolutely necessary, they could also withdraw into Outer Mongolia, but they should not turn toward the south. They should "no longer allow the enemy to interrupt the connection between the two sides." The Soviet ambassador in Chungking also "secretly [*sixia*]" informed the CPP leaders that it was necessary "to secure the Zhangjiakou-Gubeikou-Shanhaiguan line [i.e., the land access to Manchuria] to keep Chiang Kai-shek from attacking." The Chinese Communists should do everything in their power to conquer Manchuria.[216]

According to an earlier Western hypothesis, the Soviet representative at the negotiations in Yenan prohibited the Eighth Route Army's entrance into Manchuria and offered as an alternative to support a "Manchurian popular movement" in the hope that it would favor separatism more or be more friendly to the Soviet Union than Yenan.[217] This assumption can by now be considered to have been refuted. Speculations about Soviet attempts to employ other individuals such as Zhou Baozhong, Zhang Xuesi (Chang Hsueh-liang's brother), or Li Lisan for separatist purposes do not seem very convincing.[218]

On September 17, Zhu De and Liu Shaoqi suggested the strategic line "advance to the north, defend ourselves to the south [*xiang bei fazhan, xiang nan fangyu*]" to Mao Zedong and Zhou Enlai, who were negotiating in Chungking. Specifically, they recommended conquering Manchuria with all their forces and bringing Jehol and Chahar under control. On the order of 100,000 to 150,000 troops would have to be moved both to Manchuria and eastern Hebei and Jehol, and the main forces of the New Fourth Army would also have to be set in motion in that direction from south of the Yangtze.[219]

After Mao and Zhou had agreed, the Party leadership instructed all Central Committee bureaus by telegraph on September 19: "The countrywide strategic guideline is to advance to the north and defend oneself from the south. If only we can get Manchuria and the provinces of Jehol and Chahar under control, and there is a coordinated struggle from the liberated areas of the entire country and the people of the entire country, we can guarantee the victory of the Chinese people." Units stationed south of the Yangtze were to be moved to the north, 80,000 men in the New Fourth Army had to be set in motion from East China toward Shandong and eastern Hebei, and the main forces from Shandong and the bulk of the civilian cadres should be sent to eastern Hebei and Manchuria. Everything was to be done to take control of Manchuria, Jehol, Chahar, and eastern Hebei.[220]

In the discussion that took place in the context of this resolution, Zhu De stressed the priority of advancing into Manchuria and the need to take into account the possibility of setbacks in the south to achieve this: "If we are successful only in the north, if we are unable to consolidate the south or even lose some areas, that is the way it has to be. . . . We have to be prepared to create the conditions for an exchange, we must show initiative, must create a situation in which the north belongs to us."[221]

A Chinese historian with access to the CCP archives correctly describes the dramatic decision-making process that took place between September 14 and 19, 1945, as follows: "It is not difficult to recognize that the Soviet Union surrendered to foreign policy constraints at that time and formally held the armed forces of the CCP strictly back from marching into Manchuria, but secretly fully supported the conquest of Manchuria by the Chinese Communists. This played a significant role in the great decision by the CC of the CCP to conquer Manchuria."[222] The battle for Manchuria, the CC instructions said, was not merely a conflict between the two parties to the Chinese civil war; it was part of the "serious struggle for the world."[223]

The CCP leadership's September 19 resolutions were implemented rapidly. The Communist troop presence in Manchuria was reinforced significantly. Reports of the numbers do not agree entirely, however. One author with access to the CCP archives reports that 80,000 men were stationed in "all Manchuria" at the beginning of October.[224] By the end of the year, troop strength was supposedly increased to 280,000 men.[225] According to still other reports, some 200,000 Communist soldiers moved into Manchuria by November and more than 300,000 by December.[226] The differences could be—at least partially—explained by different understandings of the term "Manchuria" respectively "the Northeast [*dongbei*]," whereby in one case the term was used in the narrow sense of Liaoning, Kirin, and Heilongjiang provinces and in the other in the broad sense, which would include Jehol, Chahar, and northern Hebei provinces.

The headquarters of the "Autonomous People's Army [*renmin zizhi jun*]"—a cover name arranged with Moscow—were established in Manchuria with Lin Biao as commander in chief and Peng Zhen as political commissar. Regional and local "coalition governments" were formed that as a general rule were dominated by Communists.[227]

In late September, the new line agreed upon between Moscow and Yenan had already had the effect that the Soviet Army increasingly tolerated administrations dominated by Communists in Manchuria and allowed Chinese units fighting in their ranks to unite with regular Communist troops. They worked harmoniously together in Harbin in early October. Soviet policy was not always consistent, however. From mid-September, for example, the Soviet commanders of most of the former Japanese arsenals refused to distribute weapons to Communist units. In some cases, military equipment was handed over, however.[228] In Mukden, the Soviet commanding officer in late September was "cold and detached" toward the Northeast Bureau of the CCP established there and treated its members as "uninvited guests." Numerous local Communist functionaries criticized Soviet behavior. The central Party leadership did not make their criticism its own, however—at least not in public.[229]

The reasons for this diverse treatment by the Soviet Army are obvious. In view of the tense conflict between the public obligation to turn over the administration in Manchuria to the KMT government on the one hand and the secret agreements with the Chinese Communists on the other, the Kremlin leadership was forced to perform a difficult balancing act. From mid-September, it repeatedly demanded that Communist troops be withdrawn from Mukden and other major cities in keeping with its obligation to Chungking—clearly primarily because their presence there, unlike in rural areas, could hardly be kept secret. On the other hand, it sought to create the conditions for a Communist seizure of power in Manchuria in keeping with the Yenan agreements.[230]

Moscow reacted nervously when U.S. warships carrying over 50,000 soldiers, tanks, and aircraft landed in North China on September 30. The American armed forces took control of strategically important ports on the coast of Shandong and Hebei and occupied Tientsin, Peiping, and other important places in eastern Hebei along with KMT units. In late October, some 300,000 Soviet soldiers in Manchuria faced some 110,000 U.S. troops in the rest of China. By the same time, the Army air corps had transported two KMT armies with 56,000 men from Central to North China. President Truman and Secretary of State James Francis Byrnes intended for the American troops primarily to assist the KMT in disarming the Japanese army. The War and Navy Ministries, which were clearly convinced that Moscow wanted to transform Manchuria and North China into a separate state under Soviet domination, advised that Chiang Kai-shek should be provided with assistance to occupy this territory militarily. Moscow and Washington each suspected more than before that the other side wanted to bring all or part of China under its control.[231]

On October 1, the Soviet ambassador in Chungking, Petrov, surprised the KMT government with the information that the Soviet withdrawal from Manchuria would begin immediately and be completed by December 1.[232] The USSR therefore requested that the KMT army send representatives to Changchun to discuss the methods of the withdrawal. The CCP leadership was also informed of this schedule.[233] By an agreement reached between Stalin and Foreign Minister T.V. Soong on July 11, the Soviet withdrawal was to begin three weeks after the Japanese capitulation and be completed "within three months after the defeat of Japan," or mid-November—a deadline that by now was getting closer.[234]

Stalin apparently assumed that the landing of the U.S. troops was a reaction to the continuing Soviet military presence in Manchuria. He felt he was being put under pressure. In an attempt to avoid direct confrontation with the United States, he began to act more conciliatory toward the KMT by seeking to show his commitment to the treaty. He intended the announcement of the withdrawal from Manchuria by December 1 at the latest as an appropriate signal. The Soviet leadership subsequently pushed the deadline for withdrawal back another four times—in the two first cases with the agreement of the KMT government—and finally to the end of April 1946.[235] The last Soviet soldier left Manchuria on May 3.[236]

On the other hand, Stalin increasingly felt that the KMT army's advance into North China forced him to negotiate to create favorable conditions for the USSR after the Soviet Army's withdrawal from Manchuria. If the Soviet Army now felt it to be necessary to leave Manchuria soon, it should—secretly—do everything possible to help the Chinese comrades keep the KMT out of Manchuria and seize power themselves. In this way, it was hoped, a buffer state would emerge that was friendly to the Soviet Union. In early October, Stalin urged Yenan to transfer 250,000 to 300,000 soldiers to the area of Shanhaiguan and the vicinity of Mukden to block access to Manchuria. At the same time, the Soviet Army provided 300,000 rifles, 100 machine guns, and 15 artillery pieces to the Chinese Communists—apparently from Japanese stockpiles. Soviet headquarters transmitted Stalin's words of praise to the CCP: it was "courageous and mature" and enjoyed his "great confidence." At the same time, the military leadership confirmed its intention to transfer all of Manchuria to the CCP despite its treaty obligations and ordered it to take over governmental power in the provinces and cities quickly. It encouraged the Chinese comrades to move the main body of their armed forces—at least 500,000 men—to Manchuria, as "industry in Manchuria is developed; the borders to the east, west, and north are secure [a reference to the borders with North Korea and the Mongolian People's Republic, controlled by the Soviet Union, and the actual Soviet border]; Manchuria can function completely independently; and if one holds Manchuria, one can conquer all of China."[237]

The Yenan leadership was not prepared for such a rapid development. In late September, it was still following the plan of "focusing on the key cities and villages that were supported in the rear by the Soviet Union, Outer Mongolia, Korea, and Jehol" in their advance; and it only wished to advance along the coast of southern Manchuria later.[238] There were only 80,000 Communist soldiers in all of northeastern China, and most of these were also poorly trained. In response to the Soviet arguments, it now considered sending 50,000 to 80,000 soldiers to Manchuria by the end of October and another 100,000 by the beginning of November. It considered the Soviet request to move 300,000 men there immediately to be impossible, however. According to another source, Mao ordered 80 percent of all available troops to move toward Manchuria and other areas in northeastern China in mid-October.[239]

The pressure on Moscow increased when Chiang Kai-shek informed the Soviet side on October 1 of his intention to send thirteen KMT divisions to Manchuria, which U.S. Navy transport ships would land in Dairen alongside American personnel. Stalin refused on October 6 on the pretext that Dairen could not be used as a military port according to the August 14 agreement.[240]

It would be legitimate to interpret the agreement in this way if one interpreted the clause on Dairen that declared the city "a free port, open to the trade and shipping of all nations"[241] *e contrario*, which does not seem conclusive. At the time, the Soviet argument sounded like a pretext. Moscow also rejected Chungking's request to provide Soviet ships to transport the KMT units.[242] Chiang Kai-shek then informed Ambassador Petrov in the middle of the month that he would have to turn to Washington to transport the troops; at the same time, he again requested—apparently unsuccessfully—to be allowed to land in Dairen and also Port Arthur.[243]

In contrast to its initially accommodating attitude on October 1, Moscow now reacted harshly to Chungking's insistence. In the second half of October, Soviet military police searched KMT offices in Changchun, arrested employees of this office temporarily, sealed the office space of the Manchurian headquarters of the KMT army, and broke off negotiations.[244] The Soviet Army handed the garrisons of the Manchurian port cities of Yingkou and Huludao over to CCP troops. It allowed them to take up military positions in the vicinity of Changchun airport to obstruct the arrival of KMT personnel by aircraft, which also happened.[245] It generally encouraged the Chinese comrades to show greater determination. It informed the Manchurian Bureau of the CCP: "If we previously thought that you should be a bit cautious, you should now feel like masters and act more freely."[246] The regional Manchurian CCP leadership got the impression that the Soviet Army would fight alongside the Communist troops "if the stubborn side attacks" and prepared to seize power.[247]

When the commander of the Manchurian headquarters of the KMT troops, General

Xiong Shihui, arrived in Changchun along with Chiang Kai-shek's son Ching-kuo to negotiate the withdrawal of the Soviet troops with Malinovskii three days later, he presented a plan to send two KMT armies into Manchuria from Hebei: one by sea via Dairen, and the other by land via Shanhaiguan. Malinovskii did not comment on the landing in Dairen, but he refused to allow the KMT to transport troops by air to Changchun or the main body of the KMT forces to advance into Manchuria before the Soviet withdrawal was complete. On October 21, he again refused to allow a landing in Dairen, but he let it be understood that he had nothing against the landing at the ports of Huludao and Yingkou under consideration as an alternative by the KMT government. Chiang Kai-shek accepted this *nolens volens* and announced three days later that the landings would begin on October 29. CCP units fired on an advance guard on October 27 in Huludao, however. On November 2, similar units also obstructed a landing in Yingkou. In both cases, the Chinese Communists cooperated closely with the Soviet Army, according to contemporary American information. When the Nationalist Government protested, the Soviet Union reacted with the hypocritical comment that it did not want to interfere in domestic Chinese conflicts.[248]

Moscow also revealed a degree of firmness to Chungking and Washington during October as it sought to make the withdrawal of the Soviet Army dependent on a U.S. withdrawal. It is worth noting that Secretary of State Byrnes proposed to President Truman at about the same time to offer Stalin the withdrawal of U.S. troops in return for the withdrawal of the Soviet units, but the proposal was rejected.[249]

In late October and early November, Stalin again assumed a cooperative attitude toward the KMT. His fear of a confrontation with the United States appears to have played a central role in this. This is stressed by historians from the People's Republic of China who have access to Party archives, and is confirmed by the sources: The Soviet Union reportedly encouraged the CCP to confront the KMT, but at the same time "repeatedly warned it not to fight" where the U.S. Army was located. Under absolutely no circumstances, the sources stated, should the Americans be brought into this war.[250]

This very danger threatened now to become acute. On the one hand, the KMT government increasingly insisted on disembarking with logistical support from the U.S. Army in the Manchurian ports of Yingkou and Huludao. Stalin intervened and had Marshal Malinovskii inform General Xiong Shihui on October 29 in Changchun that Moscow would now permit the KMT units to land on the Manchurian coast. Shortly afterward, Malinovskii assured General Xiong in a private meeting that the Soviet government did not wish to allow the actions of the Chinese Communists to disrupt the good relations between Moscow and Chungking.[251]

Moscow apparently also saw the danger of a confrontation with the United States—perhaps especially—in the military advance Chang had ordered by land via Shanhaiguan. After Communist resistance had obstructed an attempt to land an advance guard in Huludao on October 27, KMT headquarters directed the American ships to northeastern Hebei, where they landed the well-armed Thirteenth KMT Army at Qinhuangdao on October 30. Following the failed landing attempt at Yingkou, another KMT army was landed at the same place. Qinhuangdao is located about 100 kilometers south of Shanhaiguan, the gateway to Manchuria, where the Great Wall reaches the Pacific. The

KMT troops moved along the Tientsin-Mukden railroad line to the north toward the Manchurian border.[252]

As they advanced, they received support from the U.S. Army, which, however, would not become involved in future struggles.[253] The leadership in Yenan calculated that twenty American officers were attached to each KMT regiment.[254] There were approximately 100,000 Communist soldiers in the entire region, but they were exhausted and in some cases poorly trained; the KMT had 80,000 men. The battle for the strategically important border city of Shanhaiguan began on November 3 against Communist defenders who, with approximately 30,000 soldiers, were greatly outnumbered. Thus began the actual military conflict for Manchuria. On November 16, KMT troops conquered Shanhaiguan with the support of American military advisers.[255]

An event that might also have involved the United States affected Soviet willingness to cooperate with the KMT. As the Soviet Army began to withdraw from Changchun on November 12, CCP troops moved in. A Communist took over the position of mayor, and Communist units surrounded KMT headquarters and switched off the water and the power.[256] As they expected the Soviets to withdraw from all areas south of Harbin by November 20, they could expect the same thing to happen there. Chiang Kai-shek decided on November 17 to move the Manchurian headquarters of the KMT army from Changchun to the recently conquered Shanhaiguan as a protest against Moscow's sabotage of the surrender of Manchuria agreed upon in the treaty. Chiang informed the Soviet ambassador of this on November 15. At the same time, he protested to President Truman that Moscow was maliciously violating the August 14 treaty with the intention of hindering the takeover of power in Manchuria by the Nationalist Government and supporting the Communist troops in their conquest. The Soviet Union's policy of obstruction endangered the unity of China and peace in East Asia. Chiang requested that Truman act together with him. Truman agreed. On November 17, General Xiong Shihui left Changchun in protest by aircraft.[257]

Moscow reacted promptly. Stalin must have feared a loss of face and significant tensions with Washington that would affect the Conference of Foreign Ministers in Moscow (December 16–26) in particular. On the same day, Ambassador Petrov assured the Nationalist Government that the headquarters of the Soviet troops in Manchuria had provided no assistance of any kind to the Chinese Communists. Despite a request from Mao Zedong to delay the introduction of KMT troops for as long as possible, he repeated an offer he had made on November 13 to allow Chinese government troops to land in Changchun and Mukden immediately and give them support from the Soviet Army. An agreement was published on November 27 that obligated both sides to a smooth transfer of power in Manchuria. Moscow said that it would arrange for the CCP troops to withdraw from Mukden and Changchun and to allow the KMT units to land over an air bridge. Chungking agreed in return to send its representatives back to Changchun. The Soviets had already ordered the CCP troops to withdraw from the city on November 19.[258]

Chiang Kai-shek's request to Ambassador Petrov in secret talks to extend the deadline for the withdrawal of the Soviet Army to ensure an orderly transfer of the administration to the KMT units in view of the advance of Communist troops contributed further to the reduction of tension between Moscow and Chungking. Stalin accepted the proposal because he continued to hope to be able to use the presence of the Soviet Army in Manchuria as a bargaining chip for the withdrawal of U.S. troops from North China and also to appease Washington with his apparently cooperative attitude toward Chungking. On November 30 in Changchun, Malinovskii and Chiang Ching-kuo agreed upon January 3, 1946, as the new date for the withdrawal of Soviet troops.[259]

In view of its imminent withdrawal, the Soviet Union sought to derive as many economic advantages as possible for itself from Manchuria. On the one hand, it continued to dismantle industrial equipment as war booty. On the other, Malinovskii's economic adviser, Mikhail Sladkovskii, began to exercise pressure on his KMT negotiating partner in Changchun—in vain—to transfer all Japanese companies and a total of 80 percent of Manchurian heavy industry to Sino-Soviet joint enterprises. He promised that it would be easier to resolve political problems in this way.[260]

The postponement of the withdrawal deadline is not without its piquancy when one considers that the Party leadership in Yenan had instructed its Northeast Bureau on October 6 to negotiate with the Soviet Union immediately. They were also supposed to arrange with the Soviet Union to postpone the withdrawal of the Soviet Army from Manchuria for one to one and a half months to win time for the urgently needed transfer of additional CCP troops to Manchuria or to provide assistance from Zhangjiakou or Suiyuan.[261] The advance of the CCP due to a delay of the withdrawal schedule apparently occurred as a reaction to the fact that the Soviet leadership had informed the CCP in mid-September that it wanted to withdraw the Soviet troops from Manchuria by the end of November, in keeping with the July 11 agreeement.[262] It came too late, however, because the Soviet ambassador in Chungking, as mentioned earlier, had already announced on October 1 that the withdrawal would be completed by December 1 at the latest. Moscow now found itself in the comfortable position of being urged by both sides in the civil war to postpone the withdrawal. Both parties expected to derive advantages from a postponement.

At the end of October, Yenan was of the opinion that the "key to all the problems involving Manchuria" lay in keeping the KMT army from landing during the following two months. But this could only be achieved successfully with the assistance of the Soviet Army. The CCP leaders urged the Soviet side at that time "1. to postpone the deadline for the withdrawal, 2. prohibit the introduction of Chiang's troops into Manchuria during November and December, and 3. allow our side to take over governmental power and the autonomous governments elected by the People immediately." If they could gain time until November, they could mobilize 60,000 regular troops and 20,000 cadres. Communist units then began to take over larger numbers of large and small towns, administrations, and garrisons.[263]

From mid-November 1945 through February–March 1946, it looked as though Stalin was prepared to make peace with the KMT government, hand over Manchuria to it as called for in the treaty, and sacrifice the Chinese comrades on the altar of the new cooperation with Chiang Kai-shek out of consideration for the tense relations with the United States. Thousands of KMT soldiers were flown to Changchun and Harbin. Two divisions of the Nationalist army landed in Mukden with Soviet permission in early December. On November 24, Petrov went so far as to apologize to Wang Shijie for Soviet behavior in Manchuria during the previous months.[264]

The shift in Moscow's Manchuria policy began to have a negative effect on Soviet policy toward the Chinese Communists. The Soviet Army prohibited them from introducing their own currency in early November. In Kirin, it obstructed a Communist takeover of the telegraph office. In the middle of the month, the Soviet commander of Changchun, who had observed the Communist seizure of power in the city at the beginning of the month without intervening, prohibited the—mostly Communist—city organizations from making anti-KMT propaganda. At about the same time, the Soviet commander in Mukden ordered Peng Zhen, head of the Northeast Bureau of the CCP, to ensure that the bureau left the city within a week, and stated that no Communist troops could in future be stationed near the airport or along the railroad lines. If the Communists resisted, they were driven off with tanks. On his return from this talk with the Soviets, Peng Zhen cried out angrily: "The army of one Communist Party using tanks to drive out the army of another Communist Party! Something like this has never happened before. Can this kind of action be acceptable?"[265]

CCP functionaries in other Manchurian cities also received orders similar to those delivered in Mukden. The Soviets threatened to disarm CCP units if they did not withdraw from the vicinity of railroad lines. Soviet commanders instructed CCP officers not to obstruct the advance of the KMT Army. In many cases, Communist units had to withdraw to the mountains and forests of eastern Manchuria. In Harbin, the Soviet Army drove the provincial governor, mayor, and chief of police—all Communists—from their positions. The new Soviet policy led to the removal of Communists from leadership positions practically all over Manchuria. Gao Gang later confirmed to the Soviet ambassador in China that at the time Peng Zhen had viewed the surrender of the cities to the KMT government as a "betrayal by the Soviet Army."[266]

The Chinese Communists suffered a further serious setback with the military defeat in the battle with the KMT troops advancing along the land corridor via Shanhaiguan. The leadership in Yenan did not want to come to terms with the fall of Shanhaiguan. Despite the hopeless situation, it ordered Lin Biao to make a counterattack on November 20. Yenan worked on the basic assumption that it needed to hold out for only another half a month. During this time, 100,000 soldiers could be moved to Manchuria and thus obstruct the incursion of the KMT; two months after that, another 90,000 soldiers could be dispatched. This must have made the Soviet Army's announcement to the Northeast Bureau of the CCP that they would hand over all of the territory along the Changchun Railroad to the KMT seem all the more disappointing. They could not fight where there were Soviet troops, it was ordered. The Communist units had to withdraw fifty kilometers. Moscow reminded the Chinese Communists that their interests had to be "the highest interests of the Communists all over the world" and ordered them to follow the Soviet orders. The CCP troops were later able to arrange, however, for the security clearance to be reduced to twenty kilometers.[267]

Yenan bowed to the inevitable. Internal Central Committee information indicates that the KMT based its claims on the August 14 treaty. The "central problem in the world," it said, was "the struggle between the United States and the Soviet Union" as it was reflected in the struggle between Chiang Kai-shek and the CCP. The Soviet Union had altered its policy under pressure from Washington and Chungking. Moscow was afraid of the United States. The USSR therefore had to withdraw from Manchuria in keeping with the treaty, it argued, and hand over power to the KMT. The CCP had to respect the foreign policy of the Soviet Union, it said, and could only do what did not harm Soviet diplomacy—for example, infiltrate the authorities in preparation for the withdrawal of the Soviet Army. The Party could not count on Soviet aid in the upcoming struggles, it said, and had to rely on its own strength. If the Chinese Communists did not give the Soviets too much trouble in their negotiations with the KMT, local Red Army commanders would still ultimately help the CCP to develop its authority in Manchuria. The CCP Central Committee ordered its Party organizations and military units operating in Manchuria to obey the Soviet instructions and withdraw from the cities. The local Party functionaries and officers obeyed the instructions from Yenan. They did not resist the Soviet commanders' orders, although they often sought to circumvent them. This occasionally led to conflicts.[268]

When discussing the Soviet instruction that CCP troops should not resist the takeover of the area along the Changchun Railroad, Yenan observed resignedly: "If the other [the Soviet] side wants it that way, we can only obey and leave the Changchun Railroad and the cities there to Chiang's troops." It would be impossible for CCP forces to conquer Manchuria on their own, the Central Committee concluded. The Party leadership instructed the Northeast Bureau first to occupy the middle-sized and small cities outside the area of the railroad and only later the railroad itself and the open spaces. Hopefully, the Party would not have to surrender the cities on the west coast of Liaoning, such as Jinzhou and Huludao, and north Liaoning, so that its troops would be able to stop the Nationalist army's advance toward Mukden.[269]

But not even this hope was fulfilled. The KMT units advanced along the coast to the northeast. The Soviet Army informed the Chinese comrades that it could not provide any assistance with respect to the areas around Huludao, Jinzhou, and south of Mukden, because it had already withdrawn from there. The KMT army conquered Huludao on November 22; three days later, Jinzhou, approximately 200 kilometers from Mukden, fell. It received the order to halt there. At the beginning of December, it renewed its advance and reached Mukden on January 26, 1946.[270]

The Soviet Army only distanced itself from the CCP as far as seemed absolutely necessary to preserve appearances, however. It continued to cooperate wherever there was a good chance that it would not become public. Thus, for example, the Soviets—apparently secretly—supported a Conference of People's Delegates of Manchuria (*Dongbei renmin daibiao huiyi*) convened by the CCP for November 22 in Changchun and did not insist on the withdrawal of the CCP's "civilian" propaganda and other organizations. They delivered weapons from Dairen, Port Arthur, and Korea to the Communist troops and made a radio transmitter available to them to maintain contact with the Soviet armed forces. In the end, they agreed to allow the Chinese comrades not to clear the lines of communication north of Liaoning so that they were able to block the advance of the KMT troops.[271]

At the beginning of December, the Soviet Army extended the deadline for its withdrawal for a second time to February 1. This was again done at the request of Chiang Kai-shek, who hoped that this would allow the Soviets to hand Manchuria over to the KMT government in an orderly fashion.[272]

During this period, the Nationalist Government was able to establish a firm position in larger cities such as Changchun, Mukden, Harbin, Qiqihar (Tsitsihar), and Siping, as well as in southern Manchuria, with the assistance of the Soviet Army. It controlled almost all of Liaoning and established a series of regional governments in the areas it dominated. The Soviet Army, however, continued to provide for order and security in the cities it captured. CCP troops no longer resisted the KMT army's advance in this period. The Communists generally dispersed to the areas surrounding the cities, and the Soviet Army handed over important industrial areas, such as Liaoyang, Anshan, and Benxi, to them. The Soviets also let the Chinese comrades know that they should prepare for a "takeover on a grand scale [*daliang jieshou*]" when the Soviet Army finally withdrew.[273]

On December 27, 1945, negotiations, which President Truman hoped would lead first to a cease-fire and then finally to the peaceful unification of China under a proposed National Conference of Delegates, to which Chiang and Mao had already agreed in principle, began in Chungking between the KMT and the CCP under the American mediator, George C. Marshall. It seems that the Chinese Communists had some hope that Marshall's mission would result in the recall of Patrick Hurley, whom they perceived as their archenemy, and that they also speculated that the negotiations might lead to a cease-fire and thus also put an end to the KMT troops' advance into Manchuria for the time being. Yenan received a confidential recommendation from Moscow at the beginning of 1946, however, asking the CCP to keep the Manchurian question out of the negotiations and consult the Soviet leadership on all issues involving Manchuria. This would make it easier for the USSR to seek to participate in the negotiations later. This again demonstrates the Soviet Union's primary interest in maintaining its special rights in Manchuria while attempting to avoid letting the issue come to an open conflict with Washington. Zhou Enlai, the chief Communist negotiator in Chungking, at first took the advice from Moscow to heart. He suggested to Marshall that the question of Manchuria be excluded and recognized the Nationalist Government's right to take over Manchuria.[274]

Yenan was concerned, however, that the exclusion of Manchuria from the peace negotiations might lead the KMT to demand that Jehol and Chahar also be excluded, occupy these two provinces, and cut the Communist troops off from the Soviet Union, Outer Mongolia, and Manchuria. The CCP leadership therefore urged Moscow—in vain—to participate in three-way negotiations with it, Washington, and Chungking. Chungking did in fact ask for Jehol and Chahar to be excluded from the cease-fire and based its request on the August 14, 1945, treaty, which called for all areas occupied by the Soviets to be handed over to the Nationalist Government.[275] The treaty in fact covered only the "three eastern provinces of China [*Zhongguo dong san sheng*]"—that is, Manchuria in the narrow sense.[276] Chungking yielded under American pressure and gave up the attempt to capture the cities of Jehol and Chahar, which had been occupied by the Chinese Communists.[277]

The cease-fire Marshall mediated on January 10 prohibited any military movement north of the Yangtze—with the exception of Manchuria, into which KMT troops could move "for the purpose of reestablishing Chinese sovereignty."[278] Yenan agreed under Soviet pressure to the regulations governing Manchuria. Stalin continued to consider it necessary to fulfill the obligations of the August 14 treaty—at least publicly. He also continued to fear U.S. intervention in Manchuria. Moscow warned the Chinese comrades in January that they could not "under any circumstances fight" in Manchuria, as there would otherwise "be a very great danger that American troops would be drawn into Manchuria."[279]

Mao then explained to the troop commanders in Manchuria that a continuation of armed conflict would lead to defeat; they would have to reach a compromise with the government in Chungking. Mao also ordered the commanders always to remain close to the Soviet Army as it withdrew so that Communist troops could secretly move into the areas the Soviets abandoned. They should do everything possible, he said, to coordinate CCP policy with Soviet foreign policy.[280]

As it would be impossible to fight a long-term civil war in Manchuria alone, given the overall domestic and foreign policy situation, an instruction from Yenan at the time stated that it had to convince the KMT to "recognize certain legal positions" of the CCP as a precondition for the reforms they were to implement jointly and to make peaceful development possible. All opinions in the Party that opposed peaceful cooperation with the KMT in Manchuria should be suppressed. Should the KMT attack, however, the CCP should fight back and annihilate it.[281] The demands the Chinese Communists made at the negotiating table in Chungking reveal what the Central Committee meant by "certain legal positions." Among other things, they demanded that the KMT army dissolve its Manchurian headquarters and replace it with an administrative committee that included representatives of all parties and mass organizations. The KMT should also recognize the Eighth Route Army, the (Communist) self-administering troops, and the "local governments" (most of which were formed under Communist influence); carry out "democratic elections"; and limit the number of troops to be sent to Manchuria to 100,000 to 150,000 men.[282]

The most recently announced deadline for the withdrawal of the Soviet Army, February 1, passed without the Soviet troops having left Manchuria. Moscow blamed unfavorable weather conditions for this and named March 1 as the new deadline. This provoked indignation in Chinese public opinion and led to student demonstrations in several cities. At the end of February, Malinovskii's staff announced that most Soviet units had already left Manchuria and that the rest would withdraw without delay.[283]

When Washington began to criticize Soviet pressure on the Nationalist Government to agree to grant the USSR additional economic influence in Manchuria in February 1946, the Soviet leadership apparently began to suspect that this was another sign that the United States was using its support for the Open Door policy as an excuse to establish itself in Manchuria. Stalin now felt it again necessary to take a hard line in Manchuria and increase the power of the CCP as the USSR's representative in order to block the extension of U.S. influence. Shortly before, he had warned Chiang Kai-shek's son Ching-kuo in talks in Moscow not to tolerate an American military presence in China—whereby he meant in Manchuria in particular.[284]

The Soviet leadership let the Chinese comrades know—probably at the end of February—that Moscow would react strongly if Washington tried to send troops to Manchuria on the pretext of the Open Door policy. The USSR would officially keep itself out of the conflict between the KMT and the CCP, however. Yenan should transfer yet more troops to Manchuria from the main body of its armed forces. At the beginning of March, the Soviet Army informed the Northeast Bureau of the CCP Central Committee that it would withdraw from the cities of Fushun, Mukden, and Kirin and that it would not hand these cities over to the KMT; the Chinese Communists could move about there freely. Following the rapid withdrawal of Soviet armed forces from southern Manchuria, the Communists could take over numerous southern Manchurian cities and industrial areas. At the same time, the CCP began under Soviet influence to refuse the U.S. request to establish a consulate and other American organizations in Harbin.[285]

The CCP leadership did not follow any uniform policy line in Manchuria between the end of February and middle of March. Zhou Enlai was very impressed by Marshall's nonpartisan efforts. As he led the negotiations in Chungking, he felt considerable pressure from the student protests and stated publicly that the sooner the Soviets left Manchuria, the better. He recognized that the Chinese Communists' credibility in China depended in large part on how well they could convince Chinese public opinion that they resisted all foreign intervention effectively—even from the Soviet Union. The delay in the withdrawal of the Soviet Army, Soviet looting of Manchurian industrial equipment, and Moscow's pressure to acquire still further economic concessions in Manchuria moved Zhou at the end of February to make the following bitter statement: "After all these years this is too much!" He conceded to Marshall that the KMT was entitled to occupy the Changchun Railroad and the areas from which the Soviets had withdrawn in order to create bargaining conditions at the same time for a solution of the political question and consultation on the introduction of troops.[286]

Yenan did not agree with Zhou Enlai's position. Ye Jianying had already stated previously that the CCP and the KMT did not need any foreign mediation. Zhu De criticized Zhou's demand for the withdrawal of the Soviet Army as false; Bo Gu defended Soviet policy in Manchuria firmly; and a Central Committee spokesperson praised the USSR's support for the forces of the people. The Communist media defended Soviet actions. The CCP lost popularity in China as a result and played into the hands of anti-Communist forces in the KMT around Security Chief Dai Li, who considered any kind of negotiations with the CCP an abomination.[287]

The CCP leadership did, as has been shown from internal documents, feel it necessary to accept terms for the cease-fire that involved renouncing at least the occupation of the cities along the railroad between Mukden and Harbin and handing them over to KMT troops—even if, as they informed Zhou Enlai earnestly, it was on the condition that the KMT negotiated with the CCP honestly and recognized its position in Manchuria.[288] The leadership stated, however, that the Party should under no circumstances recognize that the KMT had a right to occupy the entire Changchun Railroad or all the areas abandoned by the Soviets. The Communists, it said, should seek to take control of a large part of the Central Manchurian Railroad and southern portions of the Southern Manchurian Railroad. The Soviets had already handed over a large part of the areas outside the two lines as they withdrew; another part would follow. "Of these, we are not handing over any areas other than Fushun and Benxi," CC instruction said.[289]

The Handover of Manchuria to the Chinese Communists

When it became clear that the Soviet Army was serious about withdrawing from northeastern China, the final conflict for Manchuria became inevitable.

The situation became increasingly tense during the month of March. Because Soviet troops still had not left Manchuria completely after the new deadline announced for the withdrawal passed on March 1, Chungking accused the USSR on March 6 of violating the treaty and ordered it to withdraw its troops immediately. On the following day, the Soviets suddenly abandoned Mukden, which was occupied by KMT troops. Three days later, the U.S. State Department also protested the continuing occupation of Manchuria —and of parts of Iran—by the Soviet Army. Moscow reacted on March 11 by reporting that the withdrawal had already begun and on March 22 by announcing that it would be complete by the end of April. Increased Chinese protests do not appear to have been as responsible for the final withdrawal as the growing Western mistrust of Soviet expansionism that appeared to be visible in among other things the Soviet troop presence in Manchuria—and in Iran. At the beginning of March, Churchill called for the creation of an Anglo-American alliance against the Soviet Union in his now famous Iron Curtain speech. Chiang felt more confident after assurances from the White House that the United States now considered its conflict with the Soviet Union to be global and that it would support the KMT in all its future conflicts with the USSR. George C. Marshall's departure for Washington on March 11 seemed to confirm that the negotiations for a political solution in Chungking had reached a stalemate. Washington let it be known that it would be prepared to transport additional contingents of KMT troops to Manchuria. Chiang Kai-shek now seemed determined to seek a decision in Manchuria by military means. The skirmishes there, which had been sporadic up to then, increased and the advancing KMT army encountered occasional resistance from the Communists.[290]

In mid-March, Yenan made the decision to begin armed conflict in Manchuria. Factors other than the imminent withdrawal of the Soviet Army also played a role. Washington, for example, despite its continued political and material support for the KMT, was gradually reducing its military presence in North China, until at the end of April it had only approximately 30,000 men. The fact that there was increasing resistance to Zhou Enlai's negotiations with Marshall and the KMT from the majority of the leaders in Yenan, and the military leaders in the CCP leadership with Zhu De at their head in particular, played a major role. The battle readiness of commanders on the front in Manchuria, such as Lin Biao and Zhang Xueshi, supported their basic position that the issue should be decided on the battlefield in Manchuria and not at the negotiating table in Chungking. It is unclear what position was taken in these discussions by Mao Zedong, who had been criticized in the fall of 1945 by parts of the leadership in Yenan for his negotiations in Chungking.[291]

Recommendations from Moscow apparently provided the final impetus.[292] On March 15, the CCP leadership learned that the Soviet Army's final withdrawal from Manchuria had begun. At the same time, the USSR urged the Chinese comrades to take more offensive action. Moscow criticized them for being too polite to the United States and encouraged them to strike out freely and in a grand style in all the areas the Soviet Army abandoned, including Mukden and Siping, and to hold their positions north of Changchun and Harbin unconditionally.[293]

In an instruction on March 15, the Chinese CP leadership observed: "The Soviet Army has already withdrawn from Mukden and the surrounding area. In Manchuria, the conflict between the two armies of the KMT and CCP will now begin." All positions were ordered "to make military preparations." At the same time, the leadership gave permission to attack the important railroad junction at Siping, on the border between Liaoning and Kirin, because the Soviet Army had left on the same day. It was captured on March 15–16. The Northeast Bureau received the order to establish a connection with the Soviet Army quickly to "disperse our troops along the entire Central Manchurian Railroad (including Harbin), to occupy it forever and not to allow the KMT to station even a single soldier [*bu rang guomindang jinzhu yi bing yi zu*]." On March 17, the Party leadership observed that most of Manchuria (*Dongbei dabu*) was already in Communist hands. It resolved to send troops into the areas abandoned by the Soviet Army north of the Changchun Railroad. In this way, the CCP was able to take control of the cities and railroad lines located north of Mukden.[294]

Zhou Enlai continued to seek a compromise in negotiations with Marshall and the KMT government until March 20, but the rest of the Party leadership increasingly disavowed him. During a five-day stay in Yenan, the leadership apparently criticized him for being excessively flexible and put him on a short leash. On his return on March 25, he effectively represented the hard line resolved in Yenan. He subsequently sought to block political solutions and pointed out that the instructions from Yenan formed the basis for the military situation reports from Manchuria.[295] The willingness for compromise also disappeared in Chungking in the face of the news from Manchuria. Lieutenant General Alvan C. Gillem, the ardent anti-Communist who represented Marshall after his departure for the United States on March 11, made little effort to restrain the KMT government. In the following weeks, the negotiations in Chungking ground to a halt. In mid-April, a spokesperson for the CCP observed publicly that control of Manchuria would be decided by force.[296]

The CCP press had essentially treated the United States benevolently since Hurley's recall and Marshall's arrival. The tone now grew hard again. Upon his return on April 18, Marshall was greeted with the warning that general civil war would break out in China, and world peace would also be threatened, if Washington resumed Hurley's anti-Soviet and anti-Communist policies.[297]

The Chinese Communists were no longer interested in American mediation and felt strong enough to seek a decision on the battlefield. This could be seen for the first time in the battle for Siping, the southernmost city held by the Communists. There, units of the Manchurian formations that had been commanded by Lin Biao since the fall of 1945 abandoned their previous tactic of avoiding larger-scale military conflicts with enemies who had superior forces. When it attempted to break through to Changchun on April 1 at Siping, the KMT army encountered serious resistance from Communist troops, which took until May 19 to overcome. In the most significant battle China had seen since 1937, the Communists suffered a heavy defeat. The Soviet troops, which had helped them to conquer Siping, withdrew during the defense of the city. They allegedly felt dissatisfied with their Chinese comrades' military achievements and stated the opinion that the Chinese Communists should take care of themselves. Malinovskii is supposed to have recommended to Moscow not to grant any further support to the Communists. Peng Zhen had to act as the scapegoat for the defeat at Siping. He was replaced by Lin Biao as first secretary of the Northeast Bureau of the CCP in the summer of 1946. Mao Zedong told Liu Shaoqi that the USSR could not help much as it did not at that time feel able to challenge the United States. As it withdrew to the north, the Soviet Army stopped first at Sungari, south of Harbin. This was the beginning of heavy fighting in various parts of Manchuria. Although both sides may initially have hoped at the time to be able to limit the military conflict to Manchuria, one can see the fighting around Siping as the first battle of the upcoming civil war.[298]

From mid-March to May 3, the date of its final withdrawal, the Soviet Army was busy leaving as many of the Manchurian areas it abandoned as possible to the Communist troops. On the other hand, it hardly made any attempt to leave such areas under the control of the KMT army, although this occasionally happened—as, for example, in Changchun.[299] Despite the fact that the KMT government had been asking Moscow to present a withdrawal plan since February 1, the deadline for withdrawal, it was presented only at the beginning of April—much too late. Even after this, Soviet troops in many cases continued to keep their movements secret from the KMT, while they informed Communist units a few days early in each case, so that the latter could arrive before the Nationalist army.[300] In the second half of April, Communist troops occupied important large cities in the north and the center such as Changchun, Kirin, Harbin, and Qiqihar. When the Soviet withdrawal was finished, the larger part of Manchuria was in Chinese Communist hands. The KMT army at this time controlled primarily a corridor in western Liaoning along the railroad line from Peiping to Changchun and Mukden and the surrounding area.[301]

Stalin, who apparently continued to underestimate the Communists' and overestimate the KMT's military strength, may have considered the following scenario to be a desirable alternative at that time: a Communist-ruled northeastern Manchuria would function as a buffer state while the Manchurian south remained under the domination of the Nationalist Government; then would follow a North China under the Communists and a Central and South China again ruled by the KMT.[302] This scenario is reminiscent of the accusation raised frequently by the People's Republic of China—to this day without proof—that Moscow sought to create a divided China after the Second World War.

The Soviet Army had used the time that remained before its final withdrawal to dismantle the most valuable equipment from the highly developed Manchurian industry, which had an estimated value of US$900 million, and carry it off to the Soviet Union as war booty.[303]

The Nationalist army conquered Changchun and Kirin in an offensive in May 1946. Following a cease-fire that lasted from June to October and was limited to Manchuria, it was able to capture almost all the Manchurian cities by the end of the year. At the beginning of 1947, however, the Communist troops, who called themselves the People's Liberation Army (PLA) from July 1946, went on the offensive and conquered all of Manchuria by the end of 1948.[304]

In conclusion, we see that Soviet policy in Manchuria made a considerable contribution to the Communist victory in China, despite Moscow's initially hesitant and later changeable behavior there. Stalin sought to impose the greatest possible Soviet influence in Manchuria—beginning in August 1945 through the public presence of the Soviet

Army and later by secretly assisting the Chinese Communist Party, who he hoped would act as a kind of representative of the Soviet Union, to seize power. To achieve this, the Soviet Army had to delay its withdrawal for as long as possible so that the CCP could transfer large numbers of troops and civilian cadres to Manchuria to bring the revolution to the cities from the countryside with Soviet support—or at least toleration. The hidden Soviet intervention in Manchuria made it impossible for the KMT to take over power quickly and made available to the CCP a "springboard for the conquest of China proper."[305]

The significance of the Japanese weapons and ammunition the Soviet Army handed over to the Communist troops—and later the economic, technical, and medical assistance provided to the areas the Communists occupied—to the victory of the CCP should not be underestimated.[306] We do not know the total volume of this support, but it was considerable. By Soviet accounts, the Soviet Army secretly handed over more than 3,700 pieces of artillery, 600 tanks, 861 aircraft, approximately 12,000 machine guns, 680 different warehouses, and the warships of the Sungari fleet to CCP units on two of the three fronts in the initial phase alone.[307] The weapons apparently derived primarily from Japanese arsenals, but some produced in the Soviet Union and Czechoslovakia were also provided free of charge.[308] The provision of weapons—in particular of those produced in Communist countries—was kept as secret as possible. Even after the surrender of Fu Zuoyi in January 1949, all Soviet weapons were taken from Lin Biao's troops and replaced with American ones before Communist troops marched into Peiping in order, as Mao Zedong commented sarcastically, to show "how Chiang Kai-shek supplies the troops of the People's Liberation Army with American technology."[309]

Soviet assistance in the reconstruction of the railroad network from the end of 1946 and the exchange of goods with the USSR, the total volume of which increased from 413 to 916 million rubles between 1947 and 1949, were also beneficial to Communist Manchuria. In comparison, the volume of Soviet trade with those parts of Manchuria under KMT control was pathetic; exports from the USSR in 1948, for example, came to only 35 million rubles.[310]

From the Chinese Communists' perspective, however, things looked significantly different. As mentioned earlier, the Soviets dismantled industrial equipment in great quantities and transported it back to the USSR after they occupied Manchuria. The industrial production of Manchuria, the most developed industrial region in all of China, suffered catastrophic damage. An American investigative mission assessed the value of the dismantled equipment, as already mentioned, at US$900 million, but according to the same data it identified only about one-tenth of the damage actually done. Despite attempts by the Chinese Communists to downplay in public the value of the goods stolen,[311] we can assume that the looting left a bitter taste in the mouths of the CCP leadership. Peking might have drawn up an alternative account of this kind for themselves when they assessed Soviet aid to China in the 1950s in particular.[312]

From Truce to Civil War in China: The Four Protagonists' Actions

Following this description of the postwar developments in Manchuria that are of particular significance to the general theme of the present book, let us return now to examine the general developments in China between the end of the Second World War and the creation of the alliance between the USSR and Communist China in 1949. Let us focus on the position taken by Moscow's relations with the CCP within the strategic quadrangle.

On October 11, 1945, Mao Zedong returned to Yenan after six weeks of negotiations with Chiang Kai-shek, in which Stalin had forced him to participate.[313] After another four weeks, the negotiations failed because the two sides could not reach agreement on the continued existence of the Communist regional administrations or the incorporation of Communist units in the KMT army. Ambassador Hurley considered this to mean that his mission had also failed and attributed the responsibility for this to opposition to his policies from parts of the State Department. On November 26, he stepped down from his position.[314]

At this time, the behavior of the CCP leadership showed that it was prepared to yield *nolens volens* to pressure from Moscow despite its disappointment with Moscow's tactical policy reversal in Manchuria, which took place during November.[315] It increasingly felt pressure from the KMT army, which was supported by the Americans, and did not want to risk a conflict with the Soviet Union at the same time. On the contrary: the Communist leaders took Soviet interests into account in a way that bordered on self-denial. It can be demonstrated that they assumed that Chiang Kai-shek wanted to neutralize the Soviet Union in the struggle with the CCP. At the same time, Moscow was trying to keep the KMT neutral in its conflict with the United States. The CCP had to demonstrate to the Soviet leadership that its struggle with Chungking had nothing to do with the USSR.[316]

The CCP leadership also sought to neutralize the U.S. influence in its struggle with the KMT. At the beginning of November, Mao Zedong, who had withdrawn from day-to-day political work at that time, even toyed with the idea of again playing the American card with respect to the long-sought-after united front tactic in order to drive a wedge between Washington and Chungking. Following the close military cooperation between the U.S. and KMT armies in North China in the capture of Shanhaiguan on November 20 in particular, Liu Shaoqi and other leading functionaries saw little point in this, however. To receive any kind of assistance at all, the Party would have to improve its relations with the Soviet Union. As mentioned above, the CCP leadership—probably under pressure from Moscow—thought the Party should "reduce its present tensions with the United States." It ordered local commanders to avoid incidents with U.S. troops and thus "to reduce the possibility for U.S. provocations." The success of the struggle against the KMT depended, it said, on keeping the Americans out of the military conflict.[317]

Put simply, Moscow and Yenan presented the following picture around this time. The Soviet leadership perceived the United States as its main opponent and feared above all escalating American intervention in Manchuria and North China. It therefore sought to appease Washington by sticking to the treaty with the KMT and thus also to keep Chungking from forming a military alliance with Washington. It did not allow the Chinese comrades to interfere in the implementation of the treaty of August 14, 1945. For the Chinese Communists, the primary enemy was the KMT, and the main danger to be avoided was an offensive by Chungking that had the support of the U.S. Army. They could expect aid from the Soviet Union if necessary and thus could not antagonize Moscow. Yenan and Moscow felt linked by the goal of obstructing an alliance between Washington and Chungking.

Discussions of future U.S. policy toward China in Washington were overshadowed, as mentioned earlier, by problems that had arisen in relations with Moscow due to the USSR's policies in Europe (the sovietization of the Central and Eastern European states) and in northeastern Asia (among other things, the demand for a Soviet occupation zone in Japan). The United States had already become suspicious of the USSR's intentions in China when the Soviet Army continued to advance in Manchuria weeks after the Japanese surrender. Alongside the growing American distrust of the USSR's policy toward China, the U.S. military units that landed in China increasingly served to reinforce the KMT and help its troops disarm the Japanese army. In this way, President Truman wished to create the China that would be friendly to America and free from Soviet expansionism that his predecessor, Roosevelt, had already dreamed about during the war. Washington's distrust increased when Moscow used the repeated extension of the deadline for withdrawing its troops from Manchuria to the end of April 1946 to dismantle Manchurian industry extensively and allow CCP troops to occupy northern Manchuria.[318]

In December 1945, the future policy of the United States toward China became clearer. In a statement on China on December 15, President Truman confirmed his support for the Nationalist Government, criticized the Chinese Communists for maintaining a separate army, announced that he would continue to station U.S. troops in China, and stated his intention to contribute to ending the civil war. The nomination of George C. Marshall as "special representative" of the president with the rank of ambassador on December 27, which came in the wake of Ambassador Hurley's resignation, was designed not least to achieve these goals. Marshall based his views on the following premises, however: if Yenan would not make any concessions to Chungking, he would transport KMT troops to the north; if Chiang Kai-shek was unwilling to compromise, the United States still would not drop him.[319]

The resignation of the outspokenly KMT-friendly Hurley and Truman's statement were welcomed in Yenan. The leadership saw both as signs that Washington's military support for Chiang Kai-shek was coming to an end and that the U.S. Army would soon withdraw from China—even though it was assumed that the United States would in principle continue to take the KMT's side. The CCP leaders arranged for clearly friendly tones to be struck both in public statements about the United States and in personal contacts between Communists and Americans. Accordingly, the head of the American observer group in Yenan got the impression at this time that the Chinese Communists would throw themselves in the lap of the United States if General Marshall accepted their points of view.[320] There were only tactical reasons for the friendly attitude toward the United States, however. The leadership continued to state within the CCP that Truman's statement on China was yet another expression of his policy of *fu Jiang, fang Su, ya gong* (support Chiang, keep the Soviet Union under control, suppress the Chinese Communists).[321]

On December 27, the Communist delegation in Chungking resumed negotiations with the KMT, which took place under Marshall's chairmanship. The CCP leaders apparently hoped to be able to achieve legal status by abolishing the one-party government of the KMT, replacing it with a reasonable participation by the CCP in a coalition government, and then winning the upper hand through peaceful competition. Chiang Kai-shek continued to be interested primarily in subordinating the Communist troops to his command. Moscow advised the CCP leadership at the end of November to take a constructive position. Stalin also let the Chinese comrades know that he wanted to collaborate more closely with them again and had obstructed an advance by U.S. troops into Manchuria.[322]

The United States, the USSR, and Great Britain agreed at a conference of foreign ministers in Moscow at the end of December that it would be desirable for both Soviet and American armed forces to withdraw from China as soon as possible. The three states also called for an end to the civil war and for a "unified and democratic China" under a KMT government that was to be extended to include other "democratic elements"—whose character remained undefined. In public, the Nationalist Government welcomed the results of the conference, but KMT foreign minister Wang Shijie and other KMT leaders in fact viewed it with skepticism. Wang feared that the great powers had again reached an agreement at the expense of his government, as had already happened once at Yalta. In Yenan, the results were welcomed with conviction and contributed to the new sense of optimism that was emerging.[323]

The quality of the relations between the USSR and KMT China, and the problems that were involved at any given moment, can be seen—like in a snapshot—from the talks that Chiang Kai-shek's eldest son, Ching-kuo, held with Stalin at the end of 1945 and the beginning of 1946. This meeting took place at the highest rank of any between the Moscow and KMT leaderships after the end of the Sino-Japanese war. A protocol has become available to researchers only very recently.[324] It demonstrates that Chiang Kai-shek and Chiang Ching-kuo did not always tell the truth about the content of the talks. We will discuss the talks extensively in the following pages and examine what they indicate about relations between Moscow and the Chinese Communists in particular.

Chiang Ching-kuo's father sent him to Moscow primarily to assure Stalin of the Nationalist Government's unchanged friendly solidarity, ask him for assistance in controlling the Chinese Communists, and discover what the Soviets would want in exchange for cooperating in Manchuria. In the written instructions he gave his son to take with him, Chiang Kai-shek asked him to express his thanks for the Soviet contribution to the victory over Japan, insist on the withdrawal of the Soviet Army from Manchuria according to the timetable, and call for Sino-Soviet-American economic cooperation, whereby Chiang Ching-kuo should not advocate American or other foreign participation in the development of Manchuria. A Chinese-controlling majority should be guaranteed in Sino-Soviet joint enterprises in Manchuria. Chiang Ching-kuo should also inform Stalin that his father had ruled China according to the doctrines of Sun Yat-sen's political philosophy during the two decades of the anti-Japanese war. The CCP therefore had nothing to fear from him, his instructions said. Chiang Kai-shek would be willing to include the Communists in a coalition government under two conditions: they could neither endanger Chinese unity nor question the KMT leadership of China. Chiang Kai-shek asked Stalin to please support the unity of China under his government and advise the Communists not to challenge his power with military force.[325]

Chiang Ching-kuo, thirty-six years old at that time, acted as his father's trusted—possibly the most trusted—adviser on Soviet affairs. He could pass for an expert on the

Soviet Union because he had lived in the USSR from 1925 to 1937, had learned to speak Russian fluently, and had married a Russian. In the Soviet Union, he took the name Nikolai Vladimirovich Elizarov—a gesture of respect for Lenin's older sister, Anna Il'inichna Elizarova. When he finished his studies at the Sun Yat-sen University in Moscow in 1927, the Soviet authorities did not allow him to return to China or even to send letters. He was permitted to return to China only after Stalin had decided to support Chiang Kai-shek as the national leader after the Sian incident in 1936. When he returned, he behaved loyally toward his father. During the war, he often expressed his father's wish to Soviet diplomats in Chungking for friendship and cooperation with the USSR and, together with his wife, sought to make friends with the Soviet diplomats. The latter avoided his approaches, however, out of fear of the Narodnyi Komissariat Vnutrennikh Del (NKVD).[326] At the end of the war, Chiang Ching-kuo had an opportunity to gain practical experience with Soviet foreign policy. He participated in the Moscow negotiations for the Sino-Soviet treaty of alliance and negotiated, together with Marshal Malinovskii, the withdrawal of the Soviet Army from Manchuria as his father's special envoy in Changchun between October and December.[327]

Chiang Kai-shek would certainly have preferred to meet with Stalin in Moscow himself to discuss the future of Manchuria and the situation in China. Stalin apparently did not dare to invite him, however, because he was concerned about the sensitivities of Mao Zedong, whom he did not want to see in Moscow, either, given the USSR's obligations in the treaty of alliance of August 14, 1945. This was the reasoning behind Stalin's willingness to receive Chiang Ching-kuo as his father's "personal representative." It remains unclear, however, whether he invited him—according to Chiang Ching-kuo, unexpectedly—of his own accord, or whether he was reacting to a request put forward by Chiang Kai-shek. We can assume that the initiative came from Chiang Kai-shek. This is suggested by the fact that Stalin asked his guest at the first meeting what he wanted to say. Chiang Ching-kuo flew to Moscow on December 25.[328]

A report prepared for Stalin by Deputy Minister of Foreign Affairs Solomon Lozovskii on December 29, 1945, provides revealing insights into current Soviet interests.[329] It was presented to Stalin and Minister of Foreign Affairs Molotov in preparation for the first talk with Chiang Ching-kuo the following day. Lozovskii took the position that Ching-kuo was "one hundred percent" loyal to his father. "As a former member of the CPSU(B)," he would maneuver back and forth between Chiang Kai-shek and the Soviet leadership and would claim to be a "selfless friend of the USSR," even going so far as to criticize his father in order to achieve the goals the latter had set for him. He was in fact a "man of the middle [*seredniak*]," the report said, and in this respect not to be compared with the "hard dealer [*delets*]" T.V. Soong. After all, he had not been charged with making any kind of agreement with the Soviet leadership; he was merely supposed to hold talks "and possibly prepare the ground for a meeting with Chiang Kai-shek himself."[330]

In his report, Lozovskii assumed that Chiang Ching-kuo would want to talk about topics like the political situation in Manchuria, the use of the Japanese factories there, the participation of Soviet capital and Soviet technicians in the development of Manchuria, and the placing of all of Manchuria under the civilian and military administration of the Nationalist Government. He had apparently been charged with reassuring the Soviet leadership about the role of the United States in China and explaining that Chiang Kai-shek would not grant any advantages to the United States over the USSR—even though he might possibly have given similar assurances to Washington about Moscow, the report said. One thing, in any case, was clear, it added: Chiang Kai-shek wanted to exploit both the Soviet Union and the United States and in particular the contradictions between them, to strengthen the Nationalist Government.

Lozovskii developed the following scenario: Chiang Ching-kuo would seek to make the Soviets "convince the Communist troops by force or through moral influence to give up the fight for Manchuria and North China." He would request a slower withdrawal of Soviet troops. The Chinese knew, Lozovskii continued, that the Soviet Union has carried off a rather large amount of equipment from Manchuria. They would ask only for the return of the equipment for such Japanese enterprises as they wish to operate jointly with the USSR. The Soviets could expect, the scenario went, that if they refuse to do so the Chinese will threaten to import American equipment, which would involve the infiltration of American capital into Manchuria. The same would be true of the participation of Soviet capital in the economic and technical development of Manchuria. Chiang might possibly ask Stalin to exert a calming influence on the rebels in Sinkiang and ensure that Outer Mongolia refrains from ever intervening in the affairs of Inner Mongolia. Chiang might, Lozovskii said at the end of his scenario, possibly mention the very different interpretations of the agreement on Dairen. Lozovskii was referring to the fact that the Soviet Union, in keeping with the treaty, had seized entire control of Dairen (and also of Port Arthur) and prohibited both Chinese and foreign ships from using Port Arthur.[331]

Lozovskii recommended to Stalin that he insist on the rapid recognition of the Mongolian People's Republic. He had the impression that Chungking wanted to delay recognition because it hoped, among other things, that relations between the Soviet Union and the United States would worsen, and this might put China in a position to get around having to recognize Outer Mongolia. He recommended further that Stalin insist that the Soviet troops assigned to protect the CCR be allowed to continue in service for another two or three years, as railroad traffic might otherwise be disrupted and the trains robbed. They should also make it clear to Chiang Ching-kuo, he suggested, that the Soviet Union would not tolerate it if the United States attempted to project its influence into Manchuria. To obstruct the introduction of foreign capital into Manchuria, Moscow would have to commit itself there economically. They should therefore propose, among other things, to operate former Japanese enterprises jointly. This should also include the large repair dock in Dairen.

Chiang Ching-kuo held talks with Stalin on December 30 and January 3 for approximately one and a half hours each time. Molotov and the Chinese ambassador in Moscow, Fu Bingchang (Fu Ping-chang), also participated, and an official from the Foreign Ministry attended as interpreter. At the beginning of the first talk, Chiang presented a letter from his father,[332] in which the latter expressed his admiration for Stalin and his regret that he had been unable to visit him much sooner because of the war. He felt bonds of friendship for the Soviet Union and was prepared to discuss all the problems in their bilateral relations.

In the following paragraphs, we will summarize the contents of the two talks, insofar as they are of importance to the topic of this book, on the basis of the Soviet protocol.

Chiang initiated the substantive portion of the talk on December 30 with the observation that his father assumed that Sino-Soviet relations would become even closer after the war. They would be strengthened, he said, if "there were complete mutual agreement" between Chiang Kai-shek and Stalin, "his personal friend." Chiang then turned to the question that most concerned his father: relations with the CCP. He used almost half of the first talk in an attempt to convince Stalin to influence the Chinese Communists to subordinate themselves politically and militarily to the Nationalist Government. The KMT and the CCP could coexist, he argued. He assured the Soviets that the KMT did not plan "to liquidate" the CCP and went so far as to claim that there were "with respect to the political line . . . no contradictions" between the two parties. But the CCP should not plan to liquidate the KMT, either.[333]

Stalin referred to the communiqué of the Moscow conference of foreign ministers that the USSR had also signed, according to which China should be united and democratized "under the leadership of the Nationalist Government," "other democratic elements" should be included in the government, and the civil war ended. Chiang insisted that the Communists could participate in the government, but the Communist regions would have to submit "to the center." Chiang Kai-shek and Mao Zedong had agreed in Chungking that the Communists had to decrease the size of their army and could retain only sixteen to twenty divisions. Here, too, they would have to submit to a "unified command," however. Chiang thus repeated the demands that had led to the failure of the Chungking negotiations on October 11. Chiang reminded Stalin that the latter had espoused the "unification of China under the hegemony of the KMT" and the participation "of broad democratic forces" to T.V. Soong.[334]

Stalin, who apparently sensed that Chiang was interested in a Soviet mediation, replied evasively: The Soviet Union had withdrawn its three representatives from Yenan because of differences of opinion with the Chinese Communists, he said. "The Chinese Communists are not subordinate to the Russian Communists. The Comintern no longer exists." It would be difficult for Moscow to mediate, as it would not wish to provide advice that would then be rejected. Furthermore, the Chinese Communists had not requested Soviet advice.[335]

Chiang continued to urge Stalin to put pressure on Yenan: "The authority of Generalissimo Stalin will force the Chinese Communists to follow his advice." Stalin acted as though he had no idea what was happening: He had no idea what plans the Communists had and why the negotiations in Chungking had failed, he said. He did not know the reason for the difficulties between the KMT and the CCP. "Perhaps the problem is that the two leaders do not trust each other. Maybe Zhu De or Mao thinks that Chiang Kai-shek is deceiving them." Stalin said that Soviet headquarters had not allowed the Communists into Manchuria when they had tried to push forward there. The Soviet government, he said, was dissatisfied with the CCP's behavior. The Communists had not asked for any advice from Moscow since August, he said. Chiang asked naively how Stalin would behave if Yenan did ask for advice. Stalin's reply was vague: he would answer in the spirit of what he had already said to Chiang, he said. Stalin continued to repeat that the Communists did not want to ask him for advice, until Chiang stopped insisting—not without pointing out sharply that some people thought the Soviet Union was standing behind the CCP, which was struggling against the Nationalist Government. Stalin said dryly that they thought wrong.[336]

Chiang implored Stalin that they should return to the state of Sino-Soviet relations that existed in 1923–24, when Sun Yat-sen was still alive, and referred to the Soviet military assistance for the KMT with the keywords "Whampoa school."[337] (The Whampoa school was a symbol for the close Soviet-KMT cooperation in the mid-1920s.) Stalin replied that conditions for cooperation were now even better than that: "The Soviet Union is prepared for the closest relations." Chiang declared in his father's name that China would "under all circumstances and in any situation participate in nothing that was directed against the Soviet Union." Stalin made the reciprocal statement for the USSR. He used the occasion to speak disparagingly about the United States and Great Britain. Their secret services were disseminating the disinformation that there would soon be another war between their two countries, on the one hand, and the Soviet Union, on the other—apparently in an attempt to intimidate the USSR and China. Chiang assured him in his father's name that China would first consult with the Soviet Union on international affairs in order to act in concert. Stalin then accused China of having sided against the USSR at international conferences up to then. Chiang countered that China would in future act in concert with the USSR.[338]

Chiang then addressed another key topic: Manchuria. He first thanked the Soviets in his father's name for the assistance that Soviet headquarters had provided there to "reestablish the organs of power."[339] He assured Stalin that this area would never again be used as a corridor for an attack on the Soviet Union. China was prepared to demilitarize the border area. His father suggested that they "follow the policy of the open door, with the Soviet Union maintaining the leading role in the economy." Stalin acted modest: China had to be master in Manchuria, he said; the Soviet Union was not seeking a dominant position (*dominiruiushchee polozhenie*) there. Chiang replied that the Chinese government wanted to put the USSR in such a position. Stalin tacitly accepted the offer by thanking him for it. In response to a suggestion from Chiang, Stalin confirmed that the USSR would purchase goods in Manchuria and provide "as much economic assistance as possible" there.[340]

Chiang now raised another sensitive question. He said that Soviet headquarters in Manchuria considered all enterprises there to be booty enterprises. Stalin objected: according to martial law, only enterprises that had worked for the Japanese Guandong Army were booty enterprises. Chiang said that his father proposed that no enterprises should be declared booty enterprises. Instead, China offered to hand over half of all enterprises to the USSR in view of Soviet losses during the war against Japan and in recognition of Sino-Soviet friendship. Stalin objected, pointing out that Moscow had acted in a similar fashion in Poland and had confiscated only enterprises that had previously been German, even then and had left half of the equipment alone. He would, however, think about Chiang Kai-shek's suggestion and act so that "China is not offended by anything." Chiang suggested that they establish several joint enterprises in Manchuria and not—as in the Japanese period—a single one, in order to avoid the impression that they were copying the Japanese model. Stalin agreed. He asked whether the Chinese government would ask for the deadline for the withdrawal of Soviet troops to be postponed again. Chiang said no: the request that it be postponed to February 1 was the last, he said. He then asked Stalin to mediate in the conflict between the Nationalist Government and the rebels in northern Sinkiang, to which the latter agreed in principle.[341]

Chiang shifted the conversation to the topic of Sino-American relations. His father, he said, was interested in "cooperation," an "alliance" between China, the Soviet Union, and the United States. The American interlocutors in Chungking, he said, had never once said anything negative about the USSR. This was especially true of General Marshall, Chiang added, who had expressed his complete faith in Stalin. Stalin agreed to what had been said. Chiang said that China's position was "closer to the Soviet Union for historical and geographical reasons," however. It expected economic assistance from the United States, but would not surrender its political independence. Stalin agreed. As if to justify the event, Chiang stated that the U.S. marines had flown to North China, where there were no KMT troops, only to disarm the Japanese there. They had then been withdrawn. Stalin said that the Soviet government did not want to have U.S. troops move into Manchuria. "That is a Soviet zone [*Eto—sovetskaia zona*]." No American, English, or other foreign troops would be allowed into Manchuria, Stalin added. Chiang assured him that U.S. troops would not move into Manchuria and would withdraw from China altogether as soon as they had completed their tasks.[342]

Stalin complained that the United States was not moving against the Japanese decisively enough. It had not taken the Japanese army or even the general staff prisoner, he said. It behaved as it had to the German officer corps and general staff after the First World War. When Chiang replied that the Chinese people would never forget the Japanese, Stalin let slip the provocative observation that the Chinese people were good, but their leaders also had to be good. Chiang ignored this. Stalin sought to point out what they had in common regarding the United States, and observed that China and the Soviet Union knew what Japanese or German occupation meant and understood that they had to render Japan unable to fight. "The Americans do not understand this."[343]

Chiang said that he had posed all the questions he had. Molotov asked that China recognize the Mongolian PR. Chiang promised that this would be done at the beginning of February, when the government moved to Nanking.

At the beginning of the second talk, which took place, with the same participants, on January 3, 1946, Stalin surprised Chiang with the news that the Soviet military insisted that the Japanese enterprises that had worked for the Guandong Army had to be treated as booty enterprises. The fact that the Chinese side did not want this to happen insulted the military. They had spilled blood. The booty enterprises, Stalin continued, should be operated on equal terms by the USSR and China. He did not accept Chiang's objections. He did, however, consider that it might be possible to exclude some heavy industrial enterprises, as Chiang proposed, and hand them over to China.[344]

Chiang was acting obsequiously when he asked Stalin to express his opinion about the policy the Nationalist Government had adopted in recent times. His father, he said, wanted Stalin to express his doubts and tell him where he disagreed—a request that Chiang later repeated. Stalin acted modest. He could not express his dissatisfaction with just anyone, because he did not know who was guilty. He did not know the facts, so how could he state an opinion? He only had questions—such as, for example, why it was taking the

Nationalist Government and the United States so long to disarm the Japanese, and why Chiang Kai-shek and Mao Zedong could not reach an agreement. "Mao Zedong is a strange man and a strange Communist. He moves through the villages, avoids the cities, and is not interested in them." Stalin repeated that the Soviet Union's policy toward China was one "of friendship and support for the Nationalist Government of China." He considered China's policy of friendship with the United States to be correct because Chiang Kai-shek expected assistance from the Americans. "The Soviet Union cannot provide much assistance to China." Chiang pointed out that the difficulties in disarming the Japanese were due to the fact that the KMT army was located in the south of China. They had been unable to reach an agreement with the Communists in Chungking, because the two leaders did not trust each other. The Chinese people, he said, were extremely interested in an agreement because civil war was horrible. Stalin agreed: the Soviet Union knew what civil war was, he said.

When asked what he thought of the form that democracy took in China, Stalin answered hypocritically: In the USSR, he said, there were no classes that were enemies to one another, and therefore a one-party system was possible. Were there any other parties other than the KMT and the CCP in China? Stalin and Molotov took great pains to explain extensively to Chiang that the USSR had an effective two-chamber parliament. When asked whether he considered a system like the one that existed in Yugoslavia and Poland to be acceptable, Stalin replied evasively that they both had, as in France, a two-chamber system. Stalin sought to convince Chiang that there was no difference in principle between the Western democracies and the political systems that existed in Central and Eastern Europe (Romania, Bulgaria, Poland, and Hungary). When Chiang asked how he assessed the balance of power between the KMT and the CCP at that time, Stalin replied that it was hard to tell what the people thought because there had not yet been any elections in China. "In all probability, the KMT would have to obtain a majority," but how great it would be he did not know. He considered that it would be possible for the KMT and the CCP to coexist. The two parties could get along with (*uzhivat'sia*) one another just as easily as, for example, the Soviet Union gets along with the American and English capitalists.[345]

When Chiang asked what he thought about the KMT, Stalin made the accusation that pamphlets were being distributed in the name of the KMT in Manchuria that called on people to massacre (*rezat'*) the Russians. The KMT had two faces—the one legal and the other illegal. In Manchuria, the KMT called illegally for the Russians to be thrown out of that territory. Chiang objected that his father had ordered the KMT organizations in Manchuria that carried out anti-Soviet agitation to be dissolved and their members arrested. Stalin claimed that the KMT had not dissociated itself from such organizations publicly. He asked whether the CCP was really stronger than the KMT. Mao Zedong proclaimed that he had a million troops; the Americans based their assumptions on half a million. Chiang replied that these numbers were of course exaggerated. Stalin took the position that the KMT would continue to exist as a national liberal party: "Those who think that the Communists will eat the KMT up are mistaken. The KMT is a broader and more influential party than the CP."[346]

Chiang shifted to the topic of Chinese economic development. Stalin pontificated that if China wanted to emerge from its semicolonial situation, it would have to construct its own industry and could not simply engage in trade. China possessed raw materials and a hardworking people. It had to produce petroleum. It would have to accept foreign credit to construct its industry rapidly, but it should not allow its creditors to determine how such credit was used on their own. It was necessary to import from abroad, but they could not allow foreigners to impose any conditions. The foreign powers naturally opposed the development of an independent Chinese industry.[347]

Chiang wanted to know what Stalin thought about the Open Door policy. The latter responded that the foreign powers had wanted the USSR to open its doors, but the Soviet government had told them to go to hell. But China, as a weak country, had to state its approval of the Open Door policy formally—a policy that they required of semicolonial countries. But China should later close its doors to develop its own industry. Washington had already turned to the Soviet government about such a policy in Manchuria and had been amazed when the Soviets pointed out that they were not the rulers of Manchuria, but China. In response to a corresponding question from Chiang, Stalin replied that in Yalta, the USSR had not called for open doors in China, but had said that it had no objection if China agreed to it. Chiang stated that he believed that no country other than the Soviet Union wanted China's rebirth. Stalin promised that the USSR would help China to develop its industry, and trade with the Chinese, and purchase soybeans, rice, cotton, several raw materials, and some tungsten, among other things. In exchange, it could make workbenches and machines available to China, and provide specialists. Stalin agreed in principle to Chiang's request to allow Chinese personnel for the CCR to be trained in the Soviet Union. He approved a trip to the USSR by a Chinese economic delegation. Chiang asked whether it would be possible to send Soviet specialists to Sinkiang again. Stalin pointed out that they had been recalled to Moscow after Sheng Shicai had begun to arrest them. The USSR could send specialists to Sinkiang if they would be treated well.[348]

At the end of the talk, Stalin announced that he would give Chiang a letter to take to his father. In this letter, which consisted of three short paragraphs, Stalin expressed his hope that their mutual relations would develop in keeping with the Sino-Soviet treaty. He also emphasized the significance of the results of the Three Power Conference in Moscow for the Far East and also for China and the USSR, which were so very important for both countries.[349]

The two talks reveal that Chiang Ching-kuo acted in an insecure, self-conscious, and at times obsequious fashion in dealing with Stalin, like a schoolboy to his teacher. He often asked Stalin for advice and even asked him to comment critically on the KMT's policies. It never occurred to Stalin to ask anything similar of Chiang even once. One gets the impression that Chiang felt hopelessly inferior—small wonder, given the fact that he had experienced Stalin as almighty dictator during the height of the wave of terror that spread fear and horror everywhere as a simple member of the Russian Communist Party (Bolshevik), or RCP(B), living in the USSR only eight years earlier.

Chiang passed over in silence several sensitive topics in their bilateral relations, which Lozovskii, in his report for Stalin, had thought that he would address. These included, for example, the Soviet war booty in Manchuria. Chiang did address the question of booty enterprises; but he did not demand, as Lozovskii had expected, the return of the equipment from enterprises that were to be operated jointly but had already been carried off to the Soviet Union. The question of booty enterprises remained open, as had been the case during the negotiations for the treaty of alliance in the summer of 1945. Contrary to his father's instructions, Chiang also made no claim to a controlling majority in the future joint enterprises. Chiang did not address Moscow's behavior in Dairen (and Port Arthur) in violation of the treaty, which Lozovskii thought he would criticize. On the other hand, Stalin revealed a considerable degree of impudence, given his support for the Chinese Communists in Manchuria in violation of the treaty, when he accused the KMT of anti-Soviet behavior toward the Soviet Army—to which Chiang reacted meekly and defensively.

Given the fact that the Soviet Army was willing to hand over individual Manchurian cities to the KMT only after considerable pressure and only from mid-November, it sounds paradoxical for Chiang to thank Stalin for the assistance that Soviet headquarters gave to "reestablish the organs of power" of the KMT, and otherwise—contrary to Lozovskii's prediction—to say nothing about the obligation to subordinate all of Manchuria to the Nationalist Government. Lozovskii's prediction that Chiang would request a slower withdrawal of the Soviet Army from Manchuria was also incorrect, although Chiang did not, as his father requested, insist that the Soviets withdraw by the deadline, either. Stalin must have been extremely satisfied with Chiang's assurance—which he volunteered freely—that China wanted to help the Soviet Union assume a "dominant position" in Manchuria. The same was true for Chiang's guarantee that his government would not let any U.S. troops move into Manchuria—which, however, he made only after Stalin had requested it in general terms.

Lozovskii's prediction that Chiang would pressure Stalin to convince the CCP to adopt a more pacific attitude toward the Nationalist Government proved accurate. This was the first topic that Chiang addressed, and it took up the most time in the talks—a sign of how anxious the KMT leadership was about it. Stalin's response that he had absolutely no influence over the Chinese Communists and had little contact with them was pure hypocrisy. What he said about Moscow having withdrawn its contact people from Yenan in the meantime, albeit only in November 1945, was true.[350] But the CCP leadership continued to have radio contact with Moscow and also maintained close contacts with the Soviet embassy in Chungking. In August 1945, the CCP leadership had bowed to the Soviet request to negotiate with Chiang rather than begin the final conflict with the KMT. Even afterward—at the very time that Chiang was having his second talk with Stalin—the latter was giving advice to the Chinese comrades.[351] For example, as mentioned above, he recommended to the CCP leadership that it exclude Manchuria from the talks during the cease-fire negotiations with the KMT that were mediated by Marshall—and the leadership followed his recommendation.[352]

It is striking that neither Chiang nor Stalin criticized the United States—with the exception of Stalin's observation that the Americans (for lack of personal experience), unlike China and the USSR, did not understand what it meant for a country to experience enemy occupation and that their secret services disseminate disinformation. Remarkably, Stalin even welcomed China's policy of friendship for America. The decision not to play the American card reveals respect for the United States, with whom neither Stalin nor Chiang Kai-shek wanted problems at that time. We can thus presume that Chiang Ching-kuo's claim after the fact that Stalin had attempted to get him to turn against Washington is not accurate. He reported that Stalin said: "You Chinese must understand that the Americans have been using you as a tool for their own interests. They will sell you down the river whenever it suits their purpose to do so."[353] There is no reference to such a statement in the Soviet protocol. Chiang Ching-kuo apparently invented it following the open outbreak of the civil war to put Washington in line against Moscow. It is also striking that neither Stalin nor Chiang Ching-kuo made any sharp criticisms of the Chinese Communists.

Chiang Kai-shek's later claim that Stalin had said that he hoped to be able to meet with him is clearly also untrue.[354] This was presumably the real reason, Chiang went on, that Stalin had invited his son to Moscow. He had decided to reject Stalin's invitation, he said, in order not to arouse any distrust on the American side. There is no basis for these statements in the Soviet documents, in the protocols of the talks, or in the letter that Stalin gave Chiang Ching-kuo to take to his father.[355] The invitation would have had to have been made at least in the latter.

Stalin's letter to Chiang Kai-shek, which was dated January 4, 1946, sounds matter-of-fact, formal, and restrained—especially in comparison with the emotional tones of the letter Chiang had written to Stalin and that the young Chiang had delivered on December 30. In that letter, the Chinese generalissimo had pledged his admiration for the Soviet dictator and the friendship of China for the Soviet Union. By writing the way he did, Stalin signaled to KMT China that his attitude was friendly but distant and designed to promote cooperation based solely on self-interest. Chiang Kai-shek in turn was seeking the closest possible relations both with the Soviet Union and with the United States, from which he expected support against the Chinese Communists and assistance in the economic development of China. Chiang Ching-kuo's servile behavior toward Stalin did not help to increase Stalin's respect for KMT China. Lozovskii was correct in his assessment that the younger Chiang's character was not aggressive (no "hard dealer") like T.V. Soong's. In this respect, it was not a particularly good idea for Chiang Kai-shek to send his son to Stalin. The weak figure Chiang cut in Moscow probably also explains why the Chinese protocols of the talks still are not accessible today.

At the beginning of January 1946, as mentioned before, differences arose between Yenan and Moscow. The CCP leadership—like the KMT government—did not approve of having the United States mediate on its own between it and the KMT and wanted the USSR and Great Britain to participate. Only in this way, according to an internal CCP instruction, could they use the slogan "use barbarians to control barbarians [*yi yi zhi yi*]" to guarantee China's independence—a new use of a classical Chinese strategy as a model for the Communist united front tactic. "If China were dominated by one power," the instruction continued, "China would not exist." Great Britain should therefore also be included in the mediation mission.[356]

Stalin rejected the CCP leadership's request that the Soviet Union participate in the mediation attempt. He urged Yenan to accept Washington in the role of sole mediator. The Chinese Communists should also "learn from France's experience." (The French

Communists had formed a coalition government together with other parties.) Moscow thus repeated its August recommendation that the CCP should form a coalition government with the KMT. The CCP, the recommendations from Moscow continued, should seek to have the problem of Manchuria excluded from the mediation process. The Soviet leadership insisted on being consulted only on all military issues that involved Manchuria. This advice suggests two things: Moscow continued to be interested primarily in maintaining its privileges in Manchuria and wanted to avoid open conflict with the United States in China on other issues.[357]

Soviet policy toward China in early 1946 reveals contradictory tendencies. On the one hand, Moscow urged the CCP leaders to negotiate under an American chairman to form a coalition government with the KMT. On the other, it sought to impress Yenan with the horrible scenario that the United States was planning to "change China into an anti-Soviet battlefield in the future" and called on the Chinese comrades to foil this plan.[358] Although Soviet fears did not take on quite such dramatic proportions and may have involved Manchuria in particular, they were apparently still an important reason why the USSR was finally prepared to reinforce the CCP's military position in Manchuria from mid-March. The fact that the Soviet Army was beginning its final withdrawal from Manchuria at that time also played a role in this decision. If we take into account the fact that the United States supported the Nationalist Government in large part because it wanted to retain its influence in China, we will have to see the conflict between the CCP and the KMT as a kind of "war between representatives" of the USSR and the United States. Because the United States did not decide to make an open military intervention in support of the KMT, the Soviet Union saw no grounds to engage openly on the Communist side. A tacit agreement apparently developed in Moscow and Washington not to make China into any more of a scene of open rivalry.[359]

When the aforementioned cease-fire agreement was signed on January 10, 1946, a Political Consultative Conference convened for the first time in which half of the seats were assigned to the KMT and half to the CCP, which the latter shared with other groups. On January 31, the two sides finally agreed on a coalition government.[360]

Internal CCP documents reveal that the Party leadership considered the results of the negotiations in Chungking a political and military success. In their view, they constituted a "great victory."[361] A "development toward peace and democracy," it was claimed, had begun throughout the country.[362] This result had been achieved on the basis of the resolutions of the Moscow Conference of Foreign Ministers in December 1945, which in turn were due to the strength of the CCP and of the Soviet Union. The guns would now have to fall silent, and the struggle should be fought by parliamentary means. The Party leadership became convinced that "the Soviet Union will not help us in a civil war, but it will indeed help us to reach a compromise with the Kuomintang."[363] The CCP warned all troop units at the time to respect the cease-fire agreement strictly; anyone who did not, it said, would be punished.[364]

The CCP leaders' willingness to make compromises can in all probability be attributed in large part to advice from Moscow. The Soviet ambassador in Chungking also assessed the results of the negotiations as a "great victory." He advised the Chinese comrades to learn from the experience of France and now to endeavor to win over the masses.[365] This would have meant for the Chinese Communists to abandon armed conflict and take the parliamentary path. The Party leadership at that time did in fact consider adopting this assessment for a while. An internal top leadership document states: "Perhaps the victory of the democratic revolution will allow itself to be won even without any further civil war."[366]

The negotiations had guaranteed the continued existence of the regions occupied by the Communists at least until the new government was established. It seems that the CCP leadership now sought to use the consultative conference as a vehicle to extend its political influence in the countryside by peaceful means and at the same time avoid involvement in an unequal civil war against KMT forces supported by the United States. A different path lay ahead: "Following the end of the armed conflict, the political struggle and the secret struggle of the spies will break out even more intensely. We must learn these forms of conflict with which we are not yet so familiar, or we will fail."[367]

At the negotiations in Chungking in February, Zhou Enlai went so far as to agree to the evacuation of Communist troops from South China and the incorporation of Communist units stationed in North China and Manchuria into the Nationalist army, which Mao Zedong had rejected strongly in the fall of 1945. This line was resolved at a CC session held from late January to early February and put into a corresponding instruction, which said: "Our troops will be reorganized into official government troops or local security troops and self-defense troops. There will no longer be any political commissars [*zhengzhi weiyuan*], Party cells [*dang de zhibu*], or Party committees [*dangwu weiyuanhui*] in the reorganized troops. The Party will cease to lead the army directly."[368] This resolution was later criticized repeatedly within the Party.[369]

It is difficult to say how seriously the resolution was intended. There are indications that it was seen as the price that had to be paid for an advantageous political agreement. Yenan appeared also to hope that Party discipline would preserve the independence of the Communist units despite their formal integration into the Nationalist army. Chiang Kai-shek did not trust the resolution. He said to George C. Marshall that to ask the Communists to integrate their forces into the Nationalist army was just as difficult as "to negotiate with the tiger for its skin."[370]

By the time the negotiations for the reorganization of the army began in Chungking, Yenan had apparently already become skeptical again. The plan Marshall presented could, as a participant at a CC session put it, place the United States in the position of controlling China's armed forces and dominating China's policies through economic influence. When the armies were united, it was said in the leadership, Chiang Kai-shek would treat the Communist troops less well than his own and abuse the CCP more than his alliance partners. The whole thing was an attack on the Party, which had to be determined to do everything possible to avoid its own destruction.[371]

The agreement on the reorganization of the army was indeed signed on February 25 for the love of peace. But Yenan had already resolved to circumvent it. At the beginning of March, most Communist units received the order to reduce their troop strength by one-third during the following three months. Internally, however, it was made clear that this was only being done to influence public opinion and improve the quality of the army according to the slogan "create elite troops and simplify the structure [*jing bing jian zheng*]."[372]

The view emerged in the CCP leadership during February that there was not all that much more for them to expect from the American mediation mission generally. Internally, it took the position that it was the unchangeable goal of U.S. policy in China "to support the Chiang regime, oppose the Chinese Communists, and contain the Soviet Union."[373] They had to resist this policy, they said, by mobilizing the masses in China and undertaking joint efforts with the Soviet Union.

Despite this, the Chinese Communists did not cease to seek contact and curry favor with Washington for tactical reasons. They apparently assumed that the United States would have to appear to remain neutral—at least publicly—in its role as mediator. Mao Zedong promised Marshall that they would continue to work together. Zhou Enlai told him on February 1 that Mao still hoped to visit Washington. The CCP would remain flexible despite its apparent ties to the USSR: "Of course we will lean to one side. But how far depends on you." Zhou sought to draw up a picture for Marshall of a future democratic Communist China with a free-market economy. The Chinese Communists were not as pro-Russian as they were pro-American, he said.[374]

Stalin's decision to withdraw the Soviet Army from Manchuria was apparently due more to the desire to appease the United States and Great Britain, and thus obstruct the formation of a hostile coalition to counteract growing mistrust of Soviet expansionism in the West, than to growing protests against the continuing occupation from the Chinese government and people. On March 5, Churchill had made his famous speech in Fulton, Missouri, with President Truman at his side, about the "iron curtain" that had been let fall across Europe from Stettin to Trieste, in which he called for the creation of an Anglo-American alliance against the USSR. To numerous Western observers, the Soviet troop presence in Manchuria and Iran was an indication of Moscow's expansionistic ambitions.[375]

Moscow, as mentioned before, had ended its hitherto wavering policy in Manchuria in March 1946 and begun to support the Communist side effectively while at the same time withdrawing the Soviet Army. For the CCP leadership, this was a sign that they no longer had to fear that the Soviet Union would be hurt should Yenan continue to pursue armed conflict. Even the second reason that had restrained the Party up to that time from pushing the armed revolution forward since Stalin's telegrams on August 20 and 22, 1945—fear of a third world war—seemed no longer, or only to a limited degree, to apply. Stalin had reacted firmly to Churchill's speech, and Mao Zedong drew from this the conclusion that the Soviet leader no longer assumed that increased confrontation between East and West would lead to a new world war. The Communists' bitter struggles with the KMT army for the southern Manchurian city of Siping from April 1 were the first clear sign of the CCP leadership's decision to seek a decision on the battlefield.[376]

Mao set out the new view of things in an internal paper that he drew up in April 1946 and circulated among the Party leadership. In it, he stated that the democratic forces in the world were stronger than the reactionary and could dismiss the danger of a third world war. There was a possibility that they could force the leading Western states to compromise only if all the democratic forces of the world struggled resolutely and effectively against the reactionary forces of the United States, Great Britain, and France. In their own countries, the people of the capitalist world did not have to make any compromises: "The people in those countries will continue to wage different struggles in accordance with their different conditions." Mao drew the following conclusion: "The principle of the reactionary forces in dealing with the democratic forces is definitely to destroy all they can and to prepare to destroy later whatever they cannot destroy now. Face to face with this situation, the democratic forces of the people should likewise apply the same principle to the reactionary forces."[377]

The cooperation between the withdrawing Soviet Army and the CCP had this result: on May 3, as the last Soviet soldier crossed the border into the USSR, the Communists dominated most of Manchuria, although the KMT army was able to recapture several Manchurian towns in May. When the cease-fire mediated by Marshall, which had only excluded Manchuria, expired on June 30, 1946, the CCP leaders set the civil war in motion again. On the same day, the Party leadership in Yenan informed Zhou Enlai that they would "both struggle and also negotiate [*bian da bian tan*]" from then on.[378] The "both/and" tactic was apparently due to the fact that not everyone in the Party shared the offensive perspective Mao represented, because they considered the KMT troops with American support to be superior and felt that Communist defeats in Manchuria and in provinces such as Jehol, Shandong, and Hubei confirmed their opinion.[379]

Moscow's and Washington's actions no doubt also had something to do with the fact that the civil war became more widespread and fiercer at this time. The parties to the civil war attacked one another without reservation while the Soviet Union and the United States, which had wanted to stop this very thing, withdrew from China.

In June 1946, the United States reduced its military presence in China yet again, to 25,000 men; by year's end, the total had fallen to 12,000 men. The U.S. Army's decreasing strength and American inability to restrain the KMT in Manchuria made the "American connection" appear worthless in Yenan. From the end of June, the CCP leadership ceased to show any consideration for Washington in its propaganda. It began to attack the "armed intervention on the part of [the] American reactionary clique and its so-called 'aid to China'" with all its guns and claimed that American imperialism was much more dangerous than the Japanese because the former had "a more 'civilized' and 'legal' outlook." At the end of July, Communist units south of Peiping attacked an American convoy and killed several soldiers. Marshall viewed this as an intentional disruption of the relations between Yenan and Washington, as well as a "definite departure" from previous Communist behavior and a "deliberate break" in relations between the two sides. When the Communist side attacked him personally for the first time in mid-August, Marshall felt his mission was over: "This is it. It's all over. We're through in China."[380]

Yenan's behavior toward Moscow at this time was more reserved. At the end of June, the CCP leadership sent a CC member, Ruo Ronghuan, to Moscow with instructions not to appear too demanding to Stalin.[381] The leadership announced that the Party would rely "primarily" on its own strength in the fight for Manchuria and "fully" in conflicts south of the Great Wall. The motives involved here are not entirely clear. It is difficult to assess whether Yenan was still trying to show consideration for Moscow's relations with the KMT and the United States, or whether it was continuing the line of not becoming completely dependent on Moscow that Mao Zedong had already articulated in 1936.[382]

It seems that despite the decreasing American military presence, the CCP leadership no longer assumed in the fall of 1946 that the United States would keep itself out of the Chinese civil war. This may have been due in part to increasing American military and financial aid to the Nationalist Government. In September, the CCP declared publicly that there could be no further doubt about the "imperialist nature" of U.S. policy in China; this came after all hope that the United States would assist China in forming a coalition government made up of representatives of all social groups had failed.[383] Mao Zedong sought to provide a theoretical basis for this view by fitting it into an "intermediate zone theory."[384] According to this theory, American imperialism would first attempt to bring "many capitalist, colonial and semi-colonial countries in Europe, Asia and Africa" located in a "vast zone" between the United States and the Soviet Union under its control. They planned to attack the Soviet Union only after this had been done. This theory should be assessed as a further indication that Mao was becoming less willing to adjust the CCP's policy to the Soviet-American system of coordinates and take Moscow's policy toward Washington into consideration. The struggle against the KMT, which had the support of the United States, took center stage. In November, Mao told a circle of intimates that he was under absolutely no illusions about the KMT: "Chiang's line is: first no freedom, second annihilation."[385]

In the late fall of 1946, the question of the civil war was no longer discussed in China's CP under the slogan "To struggle or not to struggle [*da bu da*]?" but under one reading "To win or not to win [*sheng bu sheng*]?" Mao was optimistic. He calculated to the Party that of the approximately 100 brigades available to the KMT army for an offensive, 25 had already been destroyed during the previous three months. If the CCP's forces could neutralize another 25 brigades in the coming three months or so, they would be able to seize the strategic initiative "and a tremendous change would take place in the relative military strength of the Kuomintang and the Communist Party." They could annihilate 70 to 80 of Chiang Kai-shek's brigades within six months to a year, he said, and destroy in one year what Chiang had collected with the help of the Americans in seven or eight. They would thus bring the power of the two parties into balance and then gain superiority.[386]

In November, the Chinese Communists were apparently losing hope that American-Soviet relations might result in compromise in view of the impending Cold War. They had to continue to take the role of the United States into account in China. Mao Zedong anticipated "trouble [*mafan*]" with Washington for a long time to come. If the Americans ceased to support Chiang Kai-shek and withdrew their troops from China, the "problem can be resolved in three to five years." A war might otherwise develop that would last fifteen to twenty years. In this case, the CCP could not refuse the demand the Americans had again presented to negotiate with the KMT.[387]

After the cease-fire, which was limited to Manchuria, expired in October, KMT troops succeeded in capturing almost all the Manchurian cities by the year's end. At the beginning of 1947, Communist troops, who called themselves the People's Liberation Army from July 1946, struck back in Manchuria and parts of North China. At a Politburo session at the beginning of February, the Party leaders again discussed whether a cease-fire would be possible.[388] From May, the PLA carried out massive offensives and drove the KMT units in front of it. By the middle of the year, the KMT lost about one-third of its troops. With a final effort between September and November 1948, the Chinese Communists brought all of Manchuria under their control. This was also the turning point in the Chinese civil war. Lin Biao's Manchurian army was now able to advance to the south. On January 31, 1949, PLA troops entered Peiping. On April 20–21, they crossed the Yangtze. Nanking was captured on April 24, Shanghai on May 16–17, and Tsingtao on June 2.[389]

A question arises concerning how much the Soviet Union helped its Chinese comrades materially in their victorious march. When the CC of the CCP was forced to leave Yenan under pressure from a KMT offensive in March 1947, Moscow offered aircraft to evacuate the Party leadership to the Soviet Union or Manchuria—an offer that was not accepted.[390] We have already mentioned the transfer of weapons to Communist troops during the conquest of Manchuria and the economic and technical assistance in the development of Manchuria; this, however, must be balanced against the dismantling of a large part of the Manchurian industrial equipment. The Chinese Communists do not appear to have received any other economic or military support from Moscow from the end of the war through 1949.

The CCP's Strategic Leaning to the Soviet Side

All indications suggest that the CCP leadership began to orient the future Communist China's foreign policy strategically toward the Soviet Union in the fall of 1947.[391] This tendency, as recent sources confirm, was so clear that the occasional advances toward Washington and London can be dismissed as halfhearted tactical deviations.

Stalin in turn continued to consider himself bound by the Yalta agreement and the Sino-Soviet treaty of August 14, 1945, to maintain official relations only with the Nationalist Government. The two agreements guaranteed him, among other things, significant influence in Manchuria, which was the primary goal of his policy toward China. He felt able to maintain quasi-formal contact with the Chinese Communists only within the Manchurian sphere of influence, which Washington and Chungking had accepted—that is, in places where the CCP had come to power. Elsewhere, he sought to keep the Chinese comrades at a distance—at least in terms of what became public—despite ideological statements of solidarity with the CCP.

The CCP's efforts to cooperate with Moscow strategically can be attributed to several motives. In the fall of 1947, the Chinese Communists found themselves closer than ever before to the ultimate goal of their struggle: victory throughout the country and the establishment of a Communist-ruled China as a result of their military successes. The PLA declared: "Wherever our troops go, the enemy flees pell-mell before us and the people give thunderous cheers. The whole situation between the enemy and ourselves has fundamentally changed as compared with a year ago."[392] They now felt pressure to set the stage for their future foreign policy. The decision to make a strategic alliance with the Soviet Union their first priority should be attributed to three considerations. First of all, an ideological bond continued to exist with the first and most powerful socialist state. Second, with respect to Chinese interests, there was no evidence that the United States or Great Britain would abandon their preferred partner, the KMT. They could not even completely rule out the possibility that they might intervene militarily in favor of the KMT. Other than the Soviet Union, there was no other potential security partner with international influence to be seen far and wide. Third, the USSR was also the only state from which the CCP leadership could expect aid for the construction of a future Communist China.[393]

The Cold War, which was breaking out openly at this time, accentuated these considerations. On March 12, 1947, President Truman sent a message to Congress calling for a war against the Communist "totalitarian regimes" and guaranteed the help of the United States to all "free peoples" in fighting domestic and external enemies (the Truman Doctrine). The announcement by Secretary of State Marshall on June 5 of an economic aid program for Europe, which was directed against the Soviet Union's hegemonic efforts, pointed in the same direction. The Soviet Union reacted at the end of September by establishing the Information Bureau of the Communist and Workers' Parties (Cominform), a kind of loose successor organization to the Comintern. At the founding session in Szklarska Poreba (Schreiberhau) in Silesia, Soviet politburo member Andrei Zhdanov declared that the main political forces in the world were divided into two irreconcilable "main camps," the one "imperialist-antidemocratic," the other "anti-imperialist–democratic" (the two-camps theory).[394]

To provide one example of the public distance Moscow maintained from the Chinese comrades, the CCP was not represented at the founding assembly of the Cominform. Only the French and Italian CPs attended in addition to the Communist parties from countries making up the Soviet Union's Eastern European bloc—that is, Bulgaria, Romania, Hungary, Czechoslovakia, Poland, and Yugoslavia. Mao Zedong supposedly approved the founding of the Cominform and also welcomed Zhdanov's statement about the two camps.[395] We can assume, however, that the CCP leadership would have been glad to be included in the founding of the Cominform. Liu Shaoqi asked Stalin in July 1949, for example, whether the CCP could join the bureau. Stalin declined, using arguments that were not very convincing.[396]

Following the proclamation of the two-camps theory, the idea of a cease-fire in China appeared more illusory than ever to the CCP leaders. Furthermore, the view of the division of the world into two irreconcilable camps, which both Moscow and Washington accepted equally, exerted pressure on the Chinese Communists to side with the Soviet Union and against the United States more clearly than before. The speculation that the Cold War would increasingly bind the United States to Europe, and that the United States would thus have a decreasing amount of energy to intervene directly or indirectly in China, apparently also played a role in this decision. If Mao had already felt secure from the United States in March 1946 because he counted on a compromise between Moscow and Washington, he now felt strong toward it because—the very opposite—the Soviet-American confrontation had intensified, Moscow's attitude toward Washington had become more aggressive, and American efforts appeared to concentrate primarily on Europe. Mao assumed that it was unlikely that the United States would send troops to China in this situation.[397] "Because America is afraid that the Soviet Union will also send troops," Washington would not otherwise fight an undeclared war against the Chinese Communists. This feeling of strength freed the Party leadership to pronounce its strategic goal openly: "Overthrow Chiang Kai-shek, liberate all of China!"[398] The Party should reject all mediation attempts as foreign interventions.[399] This also applied to attempts by Moscow. On the other hand, there is evidence that Mao did not fully rule out armed intervention, or at least a blockade, by the United States until late 1949.[400]

After the strategic decision in favor of the Soviet Union had been made, the CCP leadership began to argue more clearly than before in public in favor of cooperation between the future China and the USSR, as well as to criticize the view common in the circles of other Chinese parties and movements like the Third Force, the Democratic League, and the left wing of the KMT that China should remain neutral in the struggle between the United States and the USSR.[401]

Stalin, who in any case tended to think in terms of power, was clearly impressed by the Chinese comrades' military successes. It was noticeable that the Soviet media reported more frequently, and on the whole positively, about the revolution in China after December 1947. Phrases like the following were characteristic: "All true patriots are already gathering around the Chinese CP and are closing ranks."[402]

At about the same time—that is, three months after the founding of the Cominform—Mao Zedong celebrated this event, in a newspaper article, as a sign that "all the forces of the anti-imperialist camp" had united. He thus associated Communist China with the camp that had been formed "under the leadership of the Soviet Union [*Sulian wei shou*]." The power of the anti-imperialist world camp was greater, he said, than that of the imperialist camp. The historical period had now begun in which the capitalist world would fall and the socialist world would triumph. At the same time, Mao proclaimed that the civil war in China had now reached its turning point. The "war of liberation of the Chinese People," he said, had "turned from the defensive to the offensive." The People's Liberation Army was on its way to victory, the KMT was drifting toward its downfall.[403]

This article, which was distributed widely in China, was at the same time an important message to the Soviet Union. The document was translated into Russian and sent to Stalin by telegraph.[404] It was meant to show him that his skepticism about the Chinese comrades' chances of winning had been mistaken. Mao later protested that in 1947, "as our troops were winning victories, Stalin insisted on concluding a peace treaty with Chiang Kai-shek, because he doubted the strength of the revolution."[405] Mao's article was also supposed to make it clear to Moscow that the CCP was prepared for an alliance with the USSR. Finally, one could read out of it the CCP leaders' desire for Moscow to prepare to break off relations with the KMT government as soon as possible.

New irritations arose between the Soviet Union and Communist China in the wake of the break between Stalin and Tito in June 1948. Moscow began to suspect at that time that the self-confident Mao Zedong would develop into a Chinese Tito and take China along a path that was independent of the USSR. The CCP leadership heard in September that Stalin could not forget that the Chinese Communists had revealed a "nationalism" during the Second World War that resembled the course currently taken by the Yugoslav CP. Western public opinion at the same time assumed the existence of the Soviet fears described above. Washington hoped they were justified. An analysis of policy toward China by the Planning Staff at the State Department in September 1948, supported by the National Security Council, included "Asian Titoism" as a scenario. Even the Yugoslav leadership took the position at that time that Mao was a "second Tito" and that his declarations of friendship for the Soviet Union were no more than a smoke screen to avoid unpleasantness during a difficult transitional period.[406]

Moscow's conflict with Yugoslavia led the CCP leadership to take the Soviet Union's side clearly in July 1948, just one month after the break between the Soviet Union and Yugoslavia.[407] A little later, an internal instruction described the Soviet Union as "the main force and leader of the anti-imperialist, peace-loving, and democratic front in the world," and concluded that the Chinese people could achieve victory in the revolution only "in a solid fraternal union with the Soviet Union."[408]

At the beginning of November, Mao Zedong wrote a spectacular article for the Cominform newspaper on the occasion of the thirty-first anniversary of the Bolshevik October revolution that pursued the same end of leaving no room for doubt about the Chinese Communists' loyalty to the Soviet Union in the wake of the Yugoslav defection.[409] Mao spoke out clearly against a "third road" here and in favor of proceeding with the USSR:

> Have some thirty-one years of history still not proven this? Everyone who does not agree either with imperialism or with the Soviet Union is completely wrong and has failed totally, and the same is true of all those who seek to follow a "middle line [*zhongjian luxian*]," a "third road [*disan tiao daolu*]," namely one between the imperialists' counterrevolutionary line and the revolutionary line of those who fight imperialism and its lackeys in the various countries.[410]

The same basic line can be seen in Liu Shaoqi's article "Internationalism and Nationalism," which was published a few days earlier. Liu puts it in the following terms:

> Today, the world is divided into two hostile camps: . . . American imperialism has become the bulwark of all the reactionary forces in the world, while the Soviet Union is the bulwark of all the progressive forces in the world. . . . If one is not in the imperialist camp and does not support American imperialism, one must be in the anti-imperialist camp and . . . support . . . the Soviet Union. . . . Not to take one or the other side, to practice neutrality [*shixing zhongli*], is impossible.[411]

Liu later explained that he had written his article to "educate the Party members." He sought to provide arguments for them to use with nationalist Chinese students and workers. Students in Peiping and Tientsin had raised questions criticizing the Soviet Union such as: "Why is the Red Army in Port Arthur?" "Why did the Soviet Union carry off machines from Manchuria?"[412] The effect of his article was that such questions were no longer raised several months later. Mao, Liu continued, had written his article for a similar reason. The Democratic League had suggested that the CCP should follow a middle path and had criticized it for leaning toward the Soviet Union.[413]

In early November 1948, a Communist broadcasting station used tones similar to but more martial than Mao's and Liu's: "The Chinese people cannot but regard leaders of war provocateurs, American imperialistic bellicose elements, as their mortal foes and cannot but regard leaders of world anti-imperialist front, great socialist state, Soviet Union, as their comrades in arms, from first to last."[414]

The American embassy in Nanking carefully registered Mao's and Liu's publications at the beginning of November 1948 and took them seriously. Ambassador John Leighton Stuart observed in a report to Washington that the CCP "had left undone little to subordinate Chinese national interests to those of USSR."[415]

Stalin, however, remained suspicious. In December 1948, he asked his future go-between with Mao, Ivan Kovalev, who had just returned from China, whose side the Chinese comrades were really on in the Soviet-Yugoslav conflict.[416]

Moscow Holds Back: Mao Zedong's Unsuccessful Insistence on a Meeting with Stalin

While the events described above were going on, there was another development that provides striking evidence of the CCP leadership's increased interest in close relations with the USSR: Mao Zedong's efforts to get a meeting with Stalin. This desire did not originate in an awareness of the imminent victory in the Chinese civil war or from increased tension in the Cold War between the USSR and the West, however, but from the weakness of the Party.

The Chinese side first presented the request for a summit meeting to the Soviet leadership at the beginning of 1947. In view of the CCP's temporarily unfavorable military position in the struggle against the KMT government—the Party leadership had actually been driven from its headquarters in Yenan in March—Mao wished to meet with Stalin in Moscow to seek his advice on the best strategy and tactics to employ in the civil war. He especially wanted to ask him for weapons, ammunition, and other armaments, however. Stalin agreed to a visit from Mao in Moscow in mid-June, but he insisted that it be kept secret. He offered to have an airplane pick up Mao in Harbin. Two weeks later, he told Mao that it could have a negative effect on upcoming military operations if he left China. He therefore recommended that Mao "temporarily postpone [*vremenno otlozhit'*)" the trip.[417]

Radio contact between Stalin and Mao on this matter ran through Andrei Orlov (cover name: Terebin). He performed the dual function of doctor and GRU agent with the CCP leadership at that time.[418]

The middle of July 1948 was set as the next date for the visit. But at the end of April, Mao announced to Stalin his intention to travel by train under military cover to Harbin, and from there on to Moscow, before the arranged time and by mid-June at the latest. He wanted to consult with the Soviet Party leadership there on "political, military, economic, and other questions" and to "request instructions." He also wanted to pay a visit to "the countries of eastern and southeastern Europe" to gather information about questions involving the popular front, among other things. He planned to bring two Politburo members, Ren Bishi and Chen Yun, as well as technical staff with him. If Stalin did not agree, he would come alone. A few days later, Stalin informed Mao that he could bring anyone he wanted with him. But in mid-May he again recommended that Mao "postpone the trip a little." The attack the KMT troops were beginning on the North Chinese territories through which Mao planned to travel raised the danger, Stalin maintained, that his absence would have a negative effect on the "course of events" and that his safety could be jeopardized. Mao replied the same day that it would be useful to "postpone the trip for a short time . . . in the current situation." He said that he needed to recuperate a bit and could then come by airplane.[419]

It is conceivable that by requesting to postpone the meeting Stalin was, among other things, seeking to avoid an extensive discussion of the request for assistance that Mao had already made to him in May 1948. Kovalev says that Stalin had shown him the relevant telegram from Mao at the time and told him that the Soviet Union was willing in principle to help.[420]

At the beginning of July 1948, Mao Zedong made another attempt. He informed Stalin that his health had improved significantly. "I have resolved to travel to you in the near future," he wrote. He hoped, he said, that two Soviet aircraft could pick him up on July 25 with an entourage of twenty persons. He could, however, also come by ship or by land. In that case, he would travel north at the middle of the month. In mid-July, Stalin asked that the trip be postponed to the end of November, as the leading comrades had traveled to the countryside "in view of preparations for the grain harvest" and would remain there until November. In a strikingly brief telegram on the same day, Mao agreed to postpone the meeting "until the end of October, beginning of November."[421]

Mao recognized the flimsiness of Stalin's excuse. When Orlov had read Stalin's telegram to him, he asked ironically whether they really considered grain harvests so important in the Soviet Union that leading CC members participated in them. He also posed this rhetorical question to another Soviet interlocutor. Orlov noticed that Mao was dissatisfied with Stalin's decision, even though "in Chinese fashion" he remained friendly on the outside: "It is difficult to see his true soul." Orlov got the impression that Mao had assumed that he would travel at the specified time this time and that he had waited for Stalin's telegram impatiently. The worst he had expected was that Stalin would not send him an airplane or a ship. His suitcases, he said, were already packed, leather shoes had been purchased, and a coat sewn. Mao's bodyguard, Wang Dongxing, confirms that preparations had already been made for the trip. According to Orlov, Ren Bishi had not expected a further postponement, either.[422]

Mao told Orlov at the end of August that he had not needed the trip so urgently in 1947. But the situation was different in 1948. He wanted to travel to Moscow as soon as possible to discuss questions involving relations with the small democratic parties and democratic figures, convening a Political Consultative Conference, the unification of revolutionary movements, relations with the democratic parties and groups in East Asia, and strategy in the fight against the United States and Chiang Kai-shek. He also planned to discuss questions involving the reconstruction of industry—the military and mining industries in particular—and communications systems in China, as well as the existing demand for aid. He also wanted to talk about a loan for US$30 million in silver. Finally, he wanted to reach a consensus with the Soviet leadership on what line he should take with respect to the establishment of diplomatic relations with Great Britain and France. In conclusion, according to Orlov, Mao had stressed this point: "We must be in agreement about the fact that our political course is in complete agreement with the USSR."[423]

At the end of September 1948, Mao Zedong insisted that they keep to the last deadline he had given for the trip—that is, "the end of October, beginning of November." He said that he had "many questions [*xuduo wenti*]" that he absolutely had to discuss with the

Soviet Party leadership and the "commander in chief [Russian: *glavnyi choziain*]." Mao gave the impression of being somewhat submissive by twice stressing his need to receive "instructions"[424] from Moscow. There is evidence, however, that Mao wanted primarily to discuss Soviet aid to the future Communist China. In mid-October, he confirmed the deadline he had given. In mid-November, however, he asked to postpone the trip to the end of December for health reasons and because of his heavy involvement in military operations. At the end of December, he finally informed Stalin that he planned to visit the Soviet Union after the upcoming session of the Party leadership—which would consult about strategic tasks for 1949 and preparations for the Second Plenary Session of the Seventh Party Congress, planned for the spring of 1949—and return to China before the Plenary Session.[425]

3

1949

The Pivotal Year on the Road to the Alliance

Rethinking in Moscow: Is the Victory of the Chinese Communist Party Imminent?

Moscow began to reconsider its position in late 1948. By late October, the PLA had conquered all of Manchuria. It now pushed rapidly and powerfully to the south under Lin Biao's command. By mid-December, Communist groups were already attacking the suburbs of Peiping.[1]

Stalin, who up to that point had doubted the ability of the Chinese Communist Party to establish itself militarily throughout China,[2] now had to realize that he had been mistaken. The PLA's victory seemed to have moved up into the immediate future and it was time to adjust gradually to the idea of China as subject to the Communists. Having recognized that there was no evidence that the West would terminate its support for the KMT government, the CCP leadership had already been seeking to establish a strategic alliance with Moscow since the fall of 1947.[3] According to Ivan Kovalev, Stalin's go-between to Mao Zedong, "at the end of 1948, both leaders must have finally understood that it was necessary for them to meet to put their mutual relations in order as the prospect of an imminent victory of the CCP finally became clear."[4]

Nonetheless, Stalin continued to hesitate to receive Mao in Moscow. On January 10, 1949, he wrote:

> With regard to your trip to Moscow, we are of the opinion that for the aforementioned reasons you should unfortunately postpone your departure yet again, as a trip to Moscow under these conditions will be used by our enemies to discredit the Chinese CP and suggest that it was not independent and dependent on Moscow, which naturally would be disadvantageous [*nevygodno*] for both the CCP and also for the USSR.[5]

The Moscow Politburo discussed Mao's travel plans four days later. Stalin stated that the trip would be inexpedient at the stated time. Mao was playing the role of partisan leader. Even if he wished to travel incognito, his trip could not be hidden from the public. News of his departure from China would certainly trickle through. His trip would no doubt be interpreted in the West as a visit to Moscow for the purpose of receiving Soviet CP instructions (*instruktsii*). Mao would be described as an agent of Moscow. This would damage the prestige of the Chinese CP and be exploited by the imperialists and Chiang Kai-shek against the Chinese Communists. It would, however, perhaps be possible to form a revolutionary government under Mao Zedong in the near future. Mao would not have to travel incognito then and would be able to come officially as a head of government visiting a neighboring state. This, by contrast, would increase the government's prestige and authority and give it great international stature. It would nonetheless be possible to send a Politburo member to China in order to avoid any further delay in discussing the "questions that had emerged." The Politburo, as usual, concurred with Stalin's opinion.[6]

The same day, Stalin sent Mao Zedong a telegram that included the following passage: "We nonetheless insist [*vse zhe nastaivaem*] that you postpone your trip to Moscow for the time being as your presence in China is currently very necessary [*ochen' neobkhodimo*]. If you wish, we can immediately send a responsible member of the Politburo to hold talks on the questions that interest you to you, to Harbin or to another place."[7]

The phrase "We nonetheless insist . . ." appears to suggest that Mao Zedong had insisted on his plan to visit in another hitherto inaccessible message to Stalin, despite the Soviet telegram of January 10.

The reasons Stalin had expressed to Mao in the telegram for postponing the trip were invented—like all the other reasons put forward since the beginning of 1947. This can be seen by simply comparing them with the arguments Stalin used in the Politburo meeting. But even these did not represent the full truth. Stalin was not so very concerned about his Chinese comrades' international reputation. He actually feared more than anything else that an official visit from Mao in Moscow, which indeed could hardly be kept secret, would further strain relations with the United States with respect to the situations in Central Europe and China, which by now were already tense. Washington might have accused the Soviet Union of violating the Yalta accords, in which Stalin had agreed to recognize KMT China. This would have subjected Moscow to the risk that the United States and Great Britain for their part would no longer feel bound by the accords and would withdraw their recognition of the Soviet annexation of southern Sakhalin and the Kuril Islands and the inclusion of the Mongolian People's Republic and Manchuria in the Soviet sphere of interest.

By not allowing Mao to visit Moscow, Stalin also wished to avoid exposing the official Soviet position of neutrality in the Chinese civil war as a lie before world public opinion; after all, Stalin did not invite Chiang Kai-shek to Moscow, either. The Soviet Union's international credibility would certainly have suffered if Stalin had received Mao as the opponent of the KMT government in the civil war, a government with which the Soviet Union was bound by the Treaty of Friendship and Alliance and maintained diplomatic relations. This would have been grist for the mills of the Western propaganda machine, which worked with the theory of a Communist world conspiracy directed out of Moscow.

There is evidence that Mao never fully understood when Stalin used the fact that no Communist government had been formed as a reason for canceling his trip. Zhou Enlai, for example, told Ambassador Nikolai Roshchin in a conversation that took place in mid-November 1949 and served to prepare Mao's trip to Moscow that Mao was of the opinion that "delegations of friends of the USSR from all over the world" would be present on the occasion of Stalin's birthday on December 21, so that his trip would have "a completely open character."[8] This appears to suggest that although Mao knew of Stalin's previous reservations with regard to an official visit in Moscow, he was unaware that these had ceased to be valid when the Chinese CP had formed a government.[9]

Although Stalin continued to reject Mao's trip to Moscow for the aforementioned reasons, he considered it necessary, in the light of the new situation that had arisen as a result of the PLA's victorious advances, not simply to keep Mao waiting as before but to suggest sending a high-ranking Soviet leader to China who would be able to confer with the Chinese comrades about the local situation and future relations between the two Communist parties.

Mao Zedong agreed to this. On January 17, he sent Stalin a telegram that included the following passage:

> I have decided not to come to Moscow at this time. We would be extraordinarily delighted [*shifen huanying*] if you would send a comrade from the Politburo to China. Shijiazhuang seems an appropriate place. The area is relatively hidden, he can drive [from there] to where our Central Committee is located, he can discuss many questions with the five comrades in our Central Secretariat (the primary questions involve the Political Consultative Conference, the coalition government and foreign policy). . . . If you approve this procedure, he could come at the end of this or the beginning of next month.[10]

Stalin sent a Politburo member, Anastas Mikoyan. Having traveled via Shijiazhuang, he arrived in Xibaipo on January 30, 1949, where he conferred with the CCP leadership through February 7.[11]

Affront to the CCP: Ambassador Roshchin Follows the KMT Government to Canton

At the very time that Mikoyan was beginning his discussions in Xibaipo, relations between Moscow and the CCP became strained in an unexpected way.

At the beginning of February 1949, the Soviet ambassador, Roshchin, became the only accredited head of a mission in Nanking to follow the KMT government's request that the foreign missions move to Canton, which became the new seat of the KMT government from February 5.[12] The Soviet Foreign Ministry was informed of the planned move to Canton in mid-January.[13] Roshchin left only a small group of Soviet Foreign Ministry employees behind in Nanking as a kind of branch of the embassy and other Soviet agencies, and they held out there until the proclamation of the People's Republic of China on October 1.[14] After Roshchin's departure, all Soviet missions in Nanking were subordinated to P.A. Shibaev, who held the rank of embassy counselor, but was in fact a member of the Ministerstvo Gosudarstvennoi Bezopasnosti (MGB) or the military intelligence (GRU).[15] At the same time, Shibaev functioned as the Soviet Union's unofficial chargé d'affaires in all the territories the Communists occupied in China.

The Western embassies remained where they were even after the Communists occupied Nanking in April. A high-ranking official in the Foreign Ministry of the KMT government stated that he was convinced that the Soviet step was intended to strengthen the Nationalist Government because Moscow feared that Communist China might develop in a way that was independent of the USSR.[16]

One may also assume that the CCP leadership was extremely shocked by the USSR's China policy as exemplified by this move, which was pronouncedly pro-KMT and went beyond what even the Western governments did. It had to feel duped by its comrades in Moscow. Stalin had not informed it in advance of Roshchin's move.[17] It was reported to the American ambassador in Nanking, John Leighton Stuart, at the time that Roshchin's behavior irritated Mao Zedong.[18]

One may also assume that Mao wished things had gone the opposite way and that the Western embassies had withdrawn from Nanking while the Soviet ambassador had remained. This may also be the basis for Mao's statement to Mikoyan in February 1949 to the effect that the imperialists were not welcome in the new China and would have to wait a bit longer, whereas "our genuine friends" would be able to enter somewhat sooner.[19] Similar conclusions can be drawn from the fact that Mao let Stalin know in late May 1949 that it was not in the CCP's interest for the American and other embassies to remain in Nanking. "We would be happy if all the embassies of the capitalist states would at last leave China and clear off."[20] It is possible that Mao also expressed his interest in having the Soviet ambassador return to Nanking. This may be what Stalin is referring to in a message to Mao delivered shortly thereafter, which stated that "we consider the present moment to be inappropriate for a major demonstration of the friendship between the USSR and Democratic China."[21] In June, the CCP leadership took the position internally that no government should maintain relations with the KMT government.[22]

There is no documentary evidence to explain the motivation behind Stalin's decision to order Roshchin to move to Canton. It is nonetheless plausible to assume that it was the same thing that motivated Stalin repeatedly to postpone meeting with Mao in Moscow. As we have already explained above,[23] Stalin was concerned about demonstrating to the United States and the KMT that he wished to abide by the treaties, in order to leave Washington no excuse for raising doubts about Russian territorial gains and spheres of influence recognized at Yalta and avoid any open conflict with Washington in China. In retrospect, Zhou Enlai assessed Stalin's motives at that time in a similar fashion. He described Roshchin's behavior to Lu Xiao, the Chinese ambassador-designate in Moscow, as follows: The Soviet Union at that time feared that the Chinese civil war might disrupt the arrangement established at Yalta, bring the United States into the game, and endanger the interests of the Soviet Union. Stalin still feared a third world war. His starting point was always to keep the United States quiet strategically in order to gain time for reconstruction.[24]

In response to Zhou Enlai's questions about the reasons for the move to Canton, Mikoyan gave him a detached and statesmanlike response that seemed blasé in February 1949 in Xibaipo. He reported his response in a telegram to Stalin:

> I explained that it was governed by law [*zakonomerno*], as only one government exists so far in China to which our ambassador is accredited and that it was natural that he moved along with the Chinese government to Canton while he left part of the embassy behind in Nanking. Essentially [*po sushchestvu*], this not only did no damage to our common cause, but the opposite, it supported it.[25]

Such statements probably account for Mao Zedong's later opinion that Mikoyan's behavior was arrogant during his visit in Xibaipo.[26] A diplomat in the Nanking branch of the Soviet embassy, Andrei Ledovskii, subsequently stated that Mikoyan's answer had failed to convince Zhou Enlai.[27] It was also unclear, he added, how it was supposed to benefit the "common cause."

In mid-May 1949, the same diplomat—certainly as he had been instructed—explained Roshchin's move to Zhou Enlai's representative in Nanking, Huang Hua, in much the same way as Mikoyan had done in February. He stated that Roshchin had left in connection with the KMT government's transfer to Canton and that government's suggestion that all the embassies should follow suit. In doing so, the Soviet government had behaved in keeping with generally recognized international rules. According to these, an ambassador to a state must be in the same place as the government to which he is accredited.[28]

The Chinese Communists certainly would not have been surprised if they had received such an answer from someone representing Washington. The American ambassador, Stuart, on the other hand, used arguments to justify his continued stay in Nanking to the State Department that the CCP leadership would have been more than happy to hear from its Soviet comrades: "We might lose friendly relations with Chinese people by identifying ourselves in [a] move to Canton with political party which has so completely lost public support. . . ."[29] It was a paradoxical situation.

When Liu Shaoqi asked Stalin in July 1949 why Roshchin had gone to Canton, the answer sounded somewhat different, but certainly no more convincing in Liu's ear: the Soviets had supposedly wished to observe how the KMT would behave.[30] It is possible, however, that there is some truth in this statement. Moscow instructed Roshchin to move to Canton if not primarily then at least also so that he could maintain contact with Soviet secret service officers who had infiltrated the KMT government in order to remain informed of its opinions and its relations with the United States.[31]

The aforementioned Soviet diplomat in Nanking has now presented a credible account of the motives underlying Roshchin's move. According to him, he and Mikoyan repeated at the time the "Soviet government's official version," which was a "very formal version [*sugubo formal'naia versiia*]." The Soviet government had in fact taken into account the very complicated situation in China, which had developed against the backdrop of the increasingly tense international situation. Despite all the sympathy Stalin felt for the CCP from an ideological perspective, he was afraid of becoming drawn too deeply into the Chinese civil war. He was especially concerned that Mao's behavior, which he judged to be "extremist," might provoke the United States into undertaking a direct and broad-based military intervention in China—even into organizing an international intervention as the Western powers and Japan had done against the Soviet Union in the 1920s. This could have placed the USSR in a difficult position. The United States and the KMT were reacting more calmly given the Soviet behavior in Manchuria, because they recognized Manchuria as a "sphere of special interests [*sfera osobych interesov*]" of the Soviet Union according to the Yalta accords and the Sino-Soviet treaty of August 14, 1945.[32]

On the other hand, Ledovskii explained, Moscow behaved cautiously and carefully outside the borders of Manchuria. Any assistance the Soviets gave the Chinese Communists here could lead to reactions that would be extremely unpleasant for the USSR: an American intervention, problems involving the United Nations (UN), and even the annulment by Chiang Kai-shek of the August 14, 1945, Sino-Soviet alliance treaty and the agreements on Manchuria concluded along with it. Stalin therefore did not want to give the KMT government any reason to believe that the Soviet Union would break off relations with it and establish relations with Communist China. Stalin considered it necessary to follow the evolution of events in China very carefully and not to rush forward. He decided to have Roshchin follow the KMT government to Canton for these tactical considerations. They assumed in Moscow that the governments of the United States and the other countries would behave in the same way. In Stalin's opinion, the decision of the other ambassadors to remain in Nanking would not have considerable negative effects on relations between their states and the KMT government. But if the Soviet ambassador had refused to go to Canton, this might have further increased the already deep-seated mistrust the KMT government felt for the Soviet Union's policies. The widespread propaganda from the KMT, the United States, and other Western countries that the Soviet Union was "exporting the revolution" to China also irritated Moscow. Moscow was interested in demonstrating that the USSR had not involved itself in the power struggle between the KMT and the CCP, that it recognized the Chiang Kai-shek government as the only legal government in China and continued to maintain official relations with it. The Soviet embassy's move to Canton—with the ambassador at its head—was supposed to demonstrate this.[33]

The Soviet consul general in Peiping at the time confirmed these statements in one important point. When Roshchin moved to Canton, Moscow did not expect the other embassies to remain in Nanking. In general, Moscow considered the KMT government to be *the* Chinese government of the time and the CCP only a guerrilla movement.[34]

It is not possible to document or deny the hypothesis that Stalin's motives for taking American reactions into account also included the fear that more open support for the Chinese Communists might lead to a new world war, as Zhou Enlai claimed to Liu Xiao. Stalin made contradictory statements on this point. When Kovalev reported on the basis of information from the CCP's military secret service and other sources in February–March 1949 that the United States was planning, along with Japan and the KMT, to attack North and Northeast China and even Siberia—including the use of atomic weapons—in order then to march toward the Ural Mountains,[35] he at first responded very decisively in a telegram to Kovalev that was clearly intended to be passed on to Mao: "A war is not advantageous for the imperialists. They are in crisis and are not prepared to wage a war. They fear the atom bomb, we do not fear them. . . . It now looks like America is less prepared to attack than we are to defend." The next sentence does, however, reveal a degree of skepticism: "That is what it looks like if you make the analysis from the perspective of normal, objective people. But there are abnormal people in history. The secretary of defense in the USA, Forrestal, suffered hallucinations. We are prepared to resist."[36]

At the time, James V. Forrestal advocated that the United States acquire the capacity to carry out a global war—even with atomic weapons—to defend against the spread of Communism. He did not, however, believe that the Soviet Union would be openly aggressive.[37]

In July 1949, Stalin rejected the CCP leadership's request through Liu Shaoqi for military support in the water and in the air on the grounds that this would involve conflicts with the U.S. Navy and Air Force and provide a pretext for unleashing a new world war. If the Soviet leaders did this, the people would not understand them and would "drive us out. For disregarding its distress and efforts during the war and the postwar period. For recklessness . . ."[38]

The quotations presented above could suggest that Stalin had essentially ruled out the idea of an unprovoked attack by the United States but that he nonetheless feared the outbreak of a world war should Washington consider it necessary to respond militarily to an aggressive action by a Communist state. Only in this light can Stalin's statement to Liu Shaoqi and Zhou Enlai's observation to Liu Xiao be taken seriously.

Despite Roshchin's move to Canton, the formal maintenance of diplomatic relations with the KMT government by the Soviet Union, and the fact that—with the exception of Manchuria—Soviet missions officially ceased their activities in areas occupied by the PLA, individual missions secretly maintained relations with the Communist authorities, advised them, and even received material support from them.[39]

Moscow and the CCP at the Beginning of 1949: Creating Political Consensus on the KMT

An exchange of telegrams between Stalin and Mao Zedong in mid-January 1949, which has only recently come to light, suggests how closely these leaders adjusted their policies toward the KMT with one another in an attempt to mislead the Nanking government. This process reveals several different concepts. At issue was the question of how Moscow should react to the KMT government's request of January 8, 1949, that the Soviet Union join the United States, England, and France in mediating between the parties in the civil war. The differences of opinion reflected the different interests of the USSR and the CCP, which had repeatedly come to light in their policies toward the KMT since the end of the war.

Shortly after receiving the KMT's request, Stalin sent Mao Zedong the text of the KMT's note along with drafts of the Soviet response and a message from the CCP to the KMT. Mao was asked to express his opinion if he did not agree with the drafts.[40]

Stalin's draft of the Soviet response to Nanking read as follows:

> The Soviet government was and continues to be in favor of ending the war and establishing peace in China. But before it will accept to mediate, it would like to know if the other side—the Chinese CP—agrees to accept mediation by the USSR. For this reason, the USSR would like the other side—the Chinese CP—to be informed of the Chinese government's peace action and for the other side to be asked to approve mediation by the USSR.[41]

Stalin asked the CCP to suggest a different text if it did not approve of this one. If the KMT asked Mao's opinion, he could reply more or less as follows: The Chinese CP had always been for peace in China, because the civil war had not been started by it, but by the Nanking government, which must also bear the consequences of the war. The CCP agreed to negotiations, but the war criminals, who had provoked the civil war, could not participate in them. The CCP wanted to negotiate with the KMT directly and without any foreign intermediaries. In particular, it opposed mediation by any such foreign power that itself participated in the civil war against the PLA with its armed forces and navy, as such a power could not be recognized as neutral and objective in the civil war when it came to ending the war in China.

Stalin explained the calculations involved in making this statement to Mao as follows:

"As you can see from what was said above, our draft of your reply to the KMT's proposal is designed to allow the peace negotiations to fail."[42]

The KMT would not, Stalin continued, enter into such negotiations without the mediation of the foreign powers and the United States in particular. It would not negotiate without the participation of Chiang Kai-shek and other war criminals, either. It would therefore be safe to assume that the KMT would not accept negotiations on the conditions the CCP proposed. The CCP, however, would appear to be willing to negotiate. One would not be able to accuse it of wishing to continue the civil war. The KMT, on the other hand, would be seen as responsible for the failure of the negotiations. In this way, the peace maneuver put forward by the KMT and the United States would be foiled, and the CCP would be able to continue its war of liberation.

Mao advised Stalin to send Nanking the following reply:

> The government of the USSR has always wished and still wishes to see a peaceful, democratic, and unified China. It is, however, the cause of the people of China to decide for themselves in what way it wishes to achieve peace, democracy, and the unity of China. Relying on the principle of not interfering in the internal affairs of other countries, the government of the USSR considers it unacceptable to participate in a mediation between the two sides in the civil war in China.[43]

Mao explained his suggestion as follows: If the USSR declared that it was essentially willing to mediate, this could give the KMT and the Western powers the impression that mediation was needed. It was, however, necessary for the KMT to surrender unconditionally. The USSR would therefore be providing "enormous support" to the CCP if it acted in keeping with these proposals.

The contents of the telegrams cited above and three further ones suggest that both sides agreed that peace negotiations had to be avoided.[44] The question of whether the USSR or the CCP should take the opportunity to declare that they were willing to seek peace was no more than a trick to distract world public opinion from the fact that the CCP leadership was firmly resolved to force the "unconditional surrender of the Nanking government" (Mao) and "continue its victorious war of liberation" (Stalin). Mao wished to proceed more directly than Stalin did and did not consider it necessary to "to conduct such a maneuver of circumvention yet again." He acted more confident of victory, pointed to the favorable military situation, and predicted crossing the Yangtze and capturing Nanking in the summer.

Mao wished, it appears, to use the phrasing of his draft of the Soviet reply to the KMT and the reference it contained to the Chinese people's right of self-determination, to tell Stalin to please leave decisions involving the treatment of the KMT to the leadership of the CCP. Mao, who had repeatedly run up against Stalin's attempts to placate his opponents in the civil war since August 1945, also wanted, one suspects, to deprive Moscow of yet another opportunity to present itself to Washington and Nanking as an angel of peace.

This is precisely what Stalin originally had in mind as he sought to avoid a civil war that might have put him in the position of having to fight at the side of his Chinese comrades. This again might have led to a direct confrontation with the United States if the Americans intervened in the KMT's favor, which he considered a worst-case scenario. Stalin also recommended an indirect approach that linked the announcement of the willingness to engage in peace negotiations to a demand for preconditions that could not be met in order to lay the blame for the preprogrammed failure of the peace negotiations on the KMT. He showed himself in this way to be a more cunning tactician than Mao.

In the end, both sides made concessions. Mao accepted Stalin's tactical advice and telegraphed Stalin on January 14 that he had proposed eight largely impossible conditions to the KMT government.[45] On the same day, Stalin informed Mao Zedong that the Soviet reply to Nanking would be "drawn up according to the spirit of your wishes."[46] Stalin concluded the exchange of opinions on January 15 with the observation that "a consensus had been established between us on the question of the peace proposal from the Nanking people [*nankintsev*] and that the Chinese CP had already initiated the 'peace' campaign. That means, the matter is to be considered closed."[47]

The exchange of telegrams thus resulted in a compromise. Mao followed Stalin's advice not to reject peace negotiations directly, and Stalin yielded to Mao's pressure to reject participation in any proposed mediation officially and unconditionally. The exchange of telegrams between Stalin and Mao cited above does not support the hypothesis that Stalin had intended to force Mao to accept a Soviet mediation plan, as two Russian authors claim.[48] We will instead have to take the side of historians from the People's Republic of China who interpret Stalin's draft replies to mean that he was still concerned that the Americans might intervene, essentially accepted the peace talks, and would even have liked to play the role of mediator, but that he did not want to say this openly.[49]

Furthermore, it is striking that in early 1949, unlike in August 1945, Stalin sought to avoid giving the impression in the exchange of telegrams that he was treating the CCP leadership like children—as he did in the Comintern. This can be concluded from the following statement in one of Stalin's telegrams: "We ask you to consider our advice as just that: advice that does not obligate you to anything and that you can take or leave. You can feel certain that relations between us will not be influenced if you reject our advice and we will remain the same friends toward you that we always were."[50]

This clearly shows that Stalin was prepared to respect Mao, who was on the road to victory, as an almost equal partner. It had become clear to him that he could not treat him like the leaders of his Eastern European satellite states. He would have acted differently with them in a comparable situation.

As long as Stalin and Mao were unable to agree on tactics, Ambassador Roshchin had to play hide-and-seek. He used sickness as an excuse to avoid handing over the note containing the mediation proposal of the Nationalist Government and attending a meeting of the representatives of the four states addressed in the note proposed by Stuart, the American ambassador.[51]

On January 17, Moscow contacted Nanking along the lines Mao had proposed.[52] The U.S. government had already told the KMT government four days earlier that it rejected the role of mediator because it was convinced that the USSR would not participate in the mediation mission.[53] The State Department based its opinion that Moscow would react negatively to the proposal from Nanking on an analysis of the CCP's eight conditions and publications in the Soviet and other Communist media.[54]

Mikoyan, on the other hand, used the American position as grounds for warning the CCP leadership to maintain better secrecy over exchanges of opinion between Stalin and

Mao Zedong. He did this in early February during his visit in Xibaipo and expressed to Mao his suspicion that the Americans had rejected the proposal because Washington had heard about Stalin's conspiratorial agreements with Mao from "the talkative people who surround the Chinese CP." Mikoyan was clearly hinting that there was a leak in the CP leadership. Mao rejected this idea emphatically. He pointed out that the English and the Americans had already stated previously that they were convinced the USSR and the Chinese Communists would oppose any attempt at mediation. Only the narrowest leadership circle of the CCP, Shi Zhe, the interpreter, and the Soviet contact man, Orlov, had been privy to the communications with Moscow. The reliability of these persons, Mao continued, was not subject to question. Mikoyan showed that Mao's reply did not convince him. Mao, for his part, repeated that an indiscretion from the "environment of the CCP" was out of the question.[55]

The First Top-Level Secret Meeting: Mikoyan in Xibaipo (February 1949)

Stalin had sent Anastas Mikoyan, a Politburo member and one of his closest confidants, on the secret mission to Xibaipo in late January 1949 to hold talks with the Chinese CP leadership. This was the first top-level meeting between the leaderships of the two parties.[56] It constituted a kind of alternative to the visit to Moscow that Mao had been planning since 1947 and that Stalin had repeatedly rejected for a variety of excuses.[57] The meeting reflected Moscow's increased interest in closer collaboration with its Chinese comrades. This was primarily due to the PLA's rapid advances, which promised a speedy Communist victory in the civil war. Nikita Khrushchev remembers that Stalin wanted to send direct assistance to the Chinese Communists at that time and to use Mikoyan's trip to determine what they needed.[58]

Until recently, research into the course and contents of the talks in Xibaipo essentially had to rely on information from two contemporary witnesses: Shi Zhe and Ivan Kovalev.[59] Apparently, neither had access to contemporary documents and thus both had to depend on their memories. A publication based on documents from the Mikoyan trip stored in the Russian presidential archives that appeared in 1995 revealed how poor and unreliable their reports are.[60] The following account is based primarily on this publication, although it does not assess anything like all the material available in the presidential archives and does not always clearly describe the character of the cited document. We will supplement it with Shi Zhe's and Kovalev's memoirs and several Chinese publications.

Mikoyan traveled on January 26 under the code name Andreev (Chinese: *An-de-lie-fu*) from Moscow via Mukden to Shijiazhuang (Hebei Province), where he arrived on January 30. From there, Mao Zedong's bodyguard, Wang Dongxing, and the interpreter Shi Zhe met him with a car and drove him to Xibaipo, the current seat of the CCP leadership.[61] Ivan Kovalev and Evgenii Kovalev, an interpreter from the Central Committee apparatus, accompanied him. Mikoyan conferred under strict secrecy for a week with the members of the Secretariat: Mao Zedong, Liu Shaoqi, Zhou Enlai, Zhu De, and Ren Bishi. On February 7, he began the return journey to Moscow. Ivan Kovalev accompanied him to Xibaipo in his capacity as Stalin's newly appointed go-between to Mao Zedong.[62]

If Mikoyan really traveled to China to determine, among other things, the CCP leadership's position on the Soviet-Yugoslav conflict as a Soviet historian has assumed,[63] there is no evidence of this at least in the three main sources mentioned. The Chinese party leadership had in any case already publicly taken a pro-Soviet position on this. Only once did Mikoyan mention that Mao had stated that Stalin's criticism of the nationalism of the Yugoslav foreign minister, Stanoje Simic, was of particular importance to him.[64] It is plausible to assume, however, that Mikoyan was looking in his observations for evidence that might suggest that the future Communist China was pursuing a separate road à la Tito independent of the Soviet Union.

According to Chinese sources, Mikoyan stated on his arrival on January 30 that Stalin was pleased that the CCP had liberated half of China in numerous victorious battles in the previous two to three years and wished her a final victory as soon as possible. Mikoyan repeated arguments that Stalin had already used,[65] which allegedly militated against a trip by Mao to Moscow. Given the current situation in China, Mao should not leave his command post. The trip led through parts of China that had not yet been pacified and thus meant a security risk. He had come to China in Stalin's place to hear his comrades' opinions of the Chinese revolution, which was very important to Stalin. When he returned, he would report to Stalin. Stalin would decide everything. He then reported on the situation in the USSR and the international situation.[66] Mikoyan reported that Mao informed him of his further military plans in this talk.[67]

Mikoyan held talks until February 7 with the Chinese leadership, which sometimes included all the members of the CCP Central Secretariat and sometimes took place with only a few or even one of them.[68] When all members of the Secretariat were present, Mao Zedong usually spoke. Only seldom did Zhou Enlai or Ren Bishi draw attention to themselves by making suggestions or additional remarks. Shi Zhe characterized the atmosphere of the talks between Mao and Mikoyan as "oppressive at times and not harmonious." Only when they parted were the two "at ease and happy." Shi Zhe never sought to explain why the atmosphere was tense—with one exception. Mikoyan opposed Mao's idea of forming student associations alongside the youth league in China and in doing so aroused Mao's displeasure. He excused himself for intervening and noted that he had no right to state his own opinion, as he had only come to listen. If Mikoyan really expressed himself in this way in this situation, he clearly did so out of courtesy, because Russian archival documents reveal that he generally stated his own or Stalin's opinion frequently and in many cases very forcefully.[69]

Kovalev reported that there were "tense situations or differences of opinion" during the talks. He noted only one concrete example, however, that is essentially confirmed in the Soviet archival documents:[70] Mao expressed the desire that Outer Mongolia be annexed to China.[71]

Tension Due to Roshchin's Move to Canton

The atmosphere first felt strained at the very beginning of the talks. On February 1, Zhou Enlai asked Mikoyan why, unlike the heads of all the other missions, the Soviet ambassador had just followed the KMT government from Nanking to Canton.[72] Zhou was touching on a sensitive topic. Roshchin's action had not previously been arranged with the CCP leadership and had very seriously shocked the Chinese Communists, as already mentioned. This only makes Mikoyan's calm reaction seem all the more remarkable. He granted no more than a brief response; his blasé, statesmanlike tone must have insulted the Chinese further, and the contents of his response must have seemed totally unsatisfactory.[73]

On February 3, another unpleasant situation arose when Mikoyan accused the Chinese leadership of not having sufficiently maintained the secrecy of the telegraphic exchange of opinions between Mao Zedong and Stalin, which in turn probably allowed the United States to learn of the tactics agreed upon by Moscow and Xibaipo with regard to the mission to mediate between the KMT and CCP. Mao vehemently denied this, but Mikoyan suggested that he was not very impressed.[74]

Mikoyan dealt extensively with controversial points—actual or supposed—that emerged in his discussions in a seventeen-page report on his trip to China, which he presented to the Presidium of the Central Committee of the CPSU on September 22, 1960.[75] At this time, the open conflict with the People's Republic of China was just beginning. The Soviet Party leadership had a growing interest in its prior history. This was obviously the reason the leadership asked Mikoyan to prepare an account of his negotiations in Xibaipo at this time. In his report of September 22, 1960, Mikoyan specified seven themes in his discussions in Xibaipo, which he asked the Party leadership to take into account.

Dispute over the CCP's Revolutionary Strategy?

In his first point, Mikoyan claimed that differences of opinion had emerged over the question of the CCP's future strategy with respect to the cities. Mao had explained that it would be "another one to two years before we will be in a position to dominate all of China politically and economically." During this time, the civil war would continue. When the question of capturing big cities such as Shanghai would arise, they would at first hold back. The Chinese CP, which was essentially a party shaped by peasants, had too few trained cadres to administer industry effectively in the cities, which could result in industrial stagnation and unemployment. Only when such cadres had been trained could they think of seizing cities like Shanghai and Nanking. According to Mikoyan, Mao had noted that no big city had been in Communist hands during the previous twenty years and that the Party had only been able to capture cities like Harbin thanks to the Soviet Army, apparently in order to explain the Chinese CP's inexperience in dealing with city problems.[76]

According to Mikoyan's notes, other Chinese leaders also drew attention to their lack of experience and a clear policy with respect to the cities.[77] After they had seized power in some cities, for example, Communist authorities had "handed the confiscated companies over to the workers, distributed their goods among themselves, sold them, and torn down the buildings."[78] Mikoyan reported that he had spoken against this Chinese conception in Xibaipo and explained, in keeping with the instructions he had received in Moscow, that big cities should be captured as quickly as possible.[79] The capture of Shanghai would seriously weaken Chiang Kai-shek and provide the Communists with a proletarian base.

Contemporary Soviet archival documents contradict this version of events presented by Mikoyan. Mao was supposed to have told Mikoyan that Communist troops "were preparing to capture these cities [meaning Nanking and Shanghai] by force after a period of rest and preparation."[80] The contemporary documents available to us suggest that disputes of the kind described by Mikoyan in his report of September 22, 1960, obviously did not take place.

Chinese publications confirm that Mao informed Mikoyan that they would not need much more time once they had crossed the Yangtze River before they could also attack Nanking, Shanghai, and other big cities.[81] After that, they would no longer encounter any major resistance.

In Shi Zhe's estimation, Mikoyan deliberately falsified the facts in his report for the Soviet party leadership in 1960. It was true, Shi said, that the PLA had sought to establish itself in rural areas during the initial phase of the civil war because it had been too weak to capture big cities. And yet it had begun to capture big cities as quickly as possible in 1948 and 1949. Mukden, Peiping, and Tientsin, for example, had already been captured before Mikoyan arrived in Xibaipo; moreover, the PLA had been preparing under the cover of the negotiations with the KMT government to cross the Yangtze and capture Nanking and Shanghai. Mikoyan wanted to present himself as a success in Moscow by giving the Soviet party leadership the impression that Mao had changed his policies in response to Mikoyan's influence.[82]

This hypothesis seems plausible. Even the statement by a prominent Soviet contemporary, Khrushchev, then a member of the Politburo, cannot verify Mikoyan's claim that Mao had spoken out in Xibaipo against the rapid capture of big cities. Khrushchev remembers that the Soviet leadership had been incredulous when the PLA made no attempt to enter Shanghai for several weeks after it had advanced to the city. When Stalin inquired by telegram why he did not capture Shanghai, Mao replied, "There are six million inhabitants there. If we capture [the city], we will have to feed them. But what are we to feed them with?" Stalin had protested angrily in his circle of confidants: "What is that to a Marxist? He considers himself a Marxist, but he does not come to the aid of the Shanghai workers and does not want to assume responsibility for their fate."[83] Khrushchev's observation, however, deals with the battle for Shanghai, which took place in April and May, and not with the talks between Mikoyan and Mao in early February.

Mikoyan had probably already given the impression in the oral report he made to Stalin upon his return from Xibaipo that Mao had shown himself to be reluctant to capture big cities quickly. Khrushchev recalls that Stalin had discussed China's problems in his inner circle of confidants after Mikoyan returned and had angrily asked: "What kind of person is this Mao Zedong? He has some strange peasant attitudes if he fears the workers and keeps his army away from the cities."[84] In this context, remember that Stalin had already previously suspected that Mao was primarily interested in the revolution in the rural areas and was circumventing the cities. In early 1946, he angrily told Chiang Kai-shek's son Ching-kuo that Mao was a "strange kind of Communist," who moved through the villages and avoided the cities.[85] It is possible that Mikoyan's purpose in presenting biased reports was also to confirm Stalin's prejudices against Mao as a kind of flattery. The cliché view of Mao as an "agrarian revolutionary" was widespread in Western capitals as well as in Moscow during the 1940s. It is also conceivable that Mikoyan wanted to make a propaganda statement by polishing up the old cliché in his 1960 report at a time when Soviet criticism of Maoism was becoming increasingly intense.

In recommendations that he had Mikoyan deliver to the Chinese leadership in February 1949, Stalin pointed out the need to take greater care of workers, the youth, and women in the cities and to accelerate the training of the necessary cadres. Stalin recommended that they should "not prohibit strikes as the CCP would otherwise lose the workers' confidence." This suggestion, Mikoyan reported to Stalin, had provoked "noticeable surprise and displeasure [*zametnoe nedoumenie*] from Mao Zedong and the members of the Politburo present."[86] Taken together, however, Mikoyan continued, Stalin's recommendations had left a good impression. Mao and the Politburo members had agreed unanimously to follow the advice. In reporting this to Stalin, Mikoyan significantly slipped into the Comintern style and spoke of "carrying out [*budut vypolneny*]" the advice rather than "following [it]" or "taking [it] into account."[87] If the discussion really took this form, it is easy to understand the Chinese comrades' "noticeable surprise and displeasure" in response to Stalin's recommendation to allow strikes in China. They could hardly have failed to notice that the Soviet leadership had consistently drowned strike movements in the USSR in blood, beginning with their suppression of the Kronstadt Revolt. It is not known what motives moved Stalin to make such a suggestion.

The second point in Mikoyan's 1960 report was also based on the accusation that the Chinese Communists neglected the city proletariat. Mikoyan argued that Mao generally attributed too little importance to the proletarian stratum (*proletarskaia prosloika*) in the Party and in the cities, as well as to the working class, in comparison to the peasantry. This attitude had begun when the Party and the army operated in the mountains and far from working-class areas. The times had changed, but relations with the workers had remained the same.[88]

Mao had "pointed out with pleasure" that the CCP ruled without opposition in the countryside. Among the working classes in the cities, on the other hand, the KMT was stronger, although the CCP exerted considerable influence over the student body. When the CP had legally become active at the end of the Second World War, for example, only two-fifths of the workers in Shanghai had sided with the CCP; the others had sympathized with the KMT. Mikoyan found it remarkable that Mao could state that the Chinese peasants were "more conscious than all American and many English workers." He apparently decided not to contradict Mao on this complex of topics, although he was of a different opinion. It is possible that he did not wish to unleash another of the conflicts of principle that had repeatedly soured Moscow's relations with the wing of the party Mao had led since his retreat into the Jinggang Mountains in 1927.

Here, too, it is conceivable that by exaggerating the worst aspects of the observations he had made in Xibaipo, Mikoyan was contributing to the ideological denigration of Maoism in the context of the aforementioned anti-Chinese propaganda that was just beginning.

Differences of Opinion About Forming a Government

The third topic Mikoyan mentioned in his 1960 report is very suggestive for an assessment of Stalin's China policies at this time. Mikoyan argued that Stalin pressured Mao to speed up the process of forming a government during his stay in Xibaipo. Mikoyan's report, which seems credible on this point, stated this as follows:

> As instructed in directives from the Central Committee, I sought to convince Mao Zedong not to postpone the formation of a revolutionary government, to form it quickly on the basis of a coalition, which would be beneficial. Let us say, the formation of a new revolutionary government could be announced immediately after the capture of Nanking or Shanghai. This would also be beneficial from an international perspective (the Communists would then no longer present themselves as partisans, but as a government), and this would facilitate the rest of the struggle against Chiang Kai-shek.[89]

And, Mikoyan could have added, this would have allowed the USSR to stop playing its double game, enter into diplomatic relations with the newly formed Communist government of China, and assist it officially.[90] There is, however, no evidence that Mikoyan ever said anything of the kind. Why not is unclear. On another occasion, Stalin employed an additional argument to convince Mao to form a government quickly. He pointed out that Chiang Kai-shek, with U.S. help, could take the initiative in forming a coalition government and accept representatives from other parties in this government, which would broaden the KMT government's social base.[91] It was clearly his meaning that Mao should anticipate any such attempt by forming his own coalition government.

Mikoyan claimed that Mao did not accept Stalin's recommendations with respect to the speed of forming a government while he was in Xibaipo. Mao took the position that one could not be overly hasty on this matter. It was even more beneficial for him, he said, to live without a government. If there were a government, there would be a coalition and one would have to be accountable to the other parties, which would lead to difficulties. The Communists now presented themselves as a revolutionary committee that was independent of the parties and yet in contact with them. This would help the CCP in cleansing the country of counterrevolutionary elements. Mao insisted that the government should not be formed until June or July and not already in the wake of the planned capture of Nanking in April. But Mikoyan did not deviate from his position, either.[92]

During Liu Shaoqi's visit in Moscow in the summer of 1949, as Mao considered establishing a government on January 1, 1950, Stalin again urged him to hurry—this time with the argument that "the enemy" could use the "lack of a government" as a pretext to intervene.[93] The government was ultimately formed on September 30, 1949.

Mikoyan's report would appear ultimately to suggest that whereas differences of opinion arose during the talks in Xibaipo with respect to when a government should be formed, both Mao and Stalin assumed that the Chinese government that would be formed would have to be a coalition government. This had not always been the case, however. On November 30, 1947, Mao Zedong had informed Stalin by telegram: "When the Chinese revolution is finally victorious, all political parties but the CCP will have to leave the political scene, following the example of the USSR and Yugoslavia, and this will strengthen the Chinese revolution considerably."[94]

In his reply of April 20, 1948, Stalin contradicted him:

> We do not agree with this. We believe that several political opposition parties that represent the middle classes of the Chinese population and oppose the Kuomintang clique will continue to

> exist for a long time and the CCP will be forced to involve them in the fight against the Chinese reaction and the imperialist powers once it has achieved hegemony, by which we mean the leading position. It is possible that some representatives of these parties will have to be accepted into the Chinese People's Democratic government and that it will be necessary to declare the government itself a coalition government in order to broaden the government's base among the population in this way and isolate the imperialists and their Kuomintang agents.[95]

When Mikoyan quoted these sentences from Stalin's April 20, 1948, telegram in his report for the Presidium of the Central Committee of the Soviet Communist Party, he passed over another passage in the telegram. In it, Stalin spoke out even more forcefully on the question of the coalition government: "One must bear in mind that following the victory of China's People's Liberation Army, the Chinese government's politics will—at least for a period of time after the victory, the length of which is difficult to determine now—be nationalist, revolutionary, and democratic but not Communist." This would mean, Stalin continued, that they would have to wait not only to nationalize the entire country, abolish private ownership of landed property, confiscate the property of the entire trade and industry of the petty and big bourgeoisie, and confiscate the property of big landowners, but also the middle and small landowners who worked with hired labor. They would have to wait some time before they could carry out these reforms. Stalin also corrected the Chinese comrades, verbally wagging his finger: "For your information, other parties that belong to the People's Front also exist in addition to the Communist party in Yugoslavia."[96]

Stalin thus recommended to Mao the model of the people's democracy, which had already worked well in the Soviet satellite states in Eastern Europe. Walter Ulbricht described the Communists' de facto seizure of power in the Soviet occupation zone in Germany at that time with amazing precision as follows: "It's quite clear—it's got to look democratic, but we must have everything in our control."[97]

In his report of September 22, 1960, Mikoyan claimed that Mao changed his current policy toward the bourgeois parties in response to Stalin's advice and assumed in their talks in Xibaipo that they would be forming a coalition government.[98] This would mean that he had accepted the idea Stalin put forward in the meantime.

Some Chinese sources might almost give the impression that it was Stalin who opposed and Mao who favored the formation of a coalition government "with the participation of all democratic parties, people's associations, and individuals with no party affiliation."[99] Mao was supposed to have informed Mikoyan that this form of government would not be exactly the same as that in the USSR and Eastern European states, but that it would by its nature and goals lead to socialism under the leadership of the CCP in China.[100] One of these sources even made the claim that Mao's discussions of the situation in China had emphasized the question of forming a coalition government because this question was the one concerning Stalin the most (*zui wei guanqie de*): Stalin had concerns about allowing "democratic figures with close ties to the USA" to participate in the new government.[101]

The exchange of telegrams on this topic between Stalin and Mao in 1947 and 1948 cited above tends to suggest that Mikoyan's version is correct. If this is true, the Chinese version may have been designed to give the impression that the idea of a coalition government—which, after all, did become reality on September 30, 1949, when the new government was formed—had originally been pursued by Mao and had been implemented despite Stalin's opposition, and not the opposite.

Peking's official party historiography felt pressure to justify its position after Ledovskii published the Soviet archival documents on the Mikoyan visit—however incompletely. Although it did publish Mikoyan's report to the Soviet Party leadership of September 22, 1960,[102] it noted that no telegram from Mao Zedong to Stalin dated November 30, 1947, with the contents cited by Mikoyan could be found in the Chinese archives.[103] Furthermore, the opinion attributed to Mao in the alleged telegram did not correspond to the attitudes he had "constantly [*yiguan*]" expressed on this question. Only one single statement by Mao Zedong from 1947 was offered as evidence to support this. The Declaration by the Chinese People's Liberation Army of October 10, 1947, drafted by Mao, stated that all oppressed classes should unite to form a united national front to overthrow Chiang Kai-shek's dictatorial regime "and establish a democratic coalition government [*minzhu lianhe zhengfu*]."[104] This, like all the other declarations that Mao had made before and after, the official party note suggested, showed that Mikoyan's information was untrue.

Shi Zhe also claimed that Mikoyan falsified Mao's opinion by citing Mao's statements in the November 30, 1947, telegram out of context.[105] All the documents in the possession of the Central Committee of the CCP at that time, as well as the relevant passages in Mao's statements, suggested a desire to attempt an "alliance with the democratic parties." Shi Zhe suggested that Mikoyan was seeking to flatter Stalin and place himself in a favorable light with him by giving him the impression that Mao had submitted to the Soviet arguments.

This line of argument from Peking does not seem particularly convincing. Even if it is true that the contested November 30 telegram from Mao Zedong is not in the Chinese archives, this is no proof that it did not exist. A large proportion of the documents was burned when the Party leadership had to surrender Yenan in the spring of 1947 under pressure from approaching KMT troops.[106] Besides, Mao took the only position that was documented by Party historians publicly, namely in his declaration of October 10, 1947. This declaration probably thus served above all in the search for allies in the united front tactic with whom to fight the common primary enemy, the KMT, and says nothing a priori about what Mao was thinking within the Party about the treatment of other parties and organizations after the Communists seized power. Finally, and in conclusion, the passages Mikoyan cited from the telegrams exchanged between Stalin and Mao Zedong are very clear and plausible.

Port Arthur and the Chinese Changchun Railroad

The questions Mao Zedong raised on February 4 about the Sino-Soviet treaty of August 14, 1945, the Port Arthur agreement, the Chinese Changchun Railroad, and the two Mongolias were particularly sensitive. They constituted the fourth group of topics in Mikoyan's 1960 report. Some of the terms used to report the discussion were imprecise. It mentions both the "Treaty of August 14" and the "Treaty on Port Arthur" and the

"Treaty on the Chinese Changchun Railroad" and at times correctly the "Agreement on Port Arthur." Such differences do not affect the agreements' validity in international law, however.

Mao Zedong noted that Chinese public opinion was very interested in the CCP's position on the treaties China had made with foreign states, including the treaty of August 14, 1945. When the CCP demanded the abrogation of the "treasonable treaties," the leaders of the democratic parties asked the leaders of the CCP what they meant by that term. The Communist leadership's answer was that it made distinctions in approaching this problem. It did not demand the abrogation of all the treaties Chiang Kai-shek had concluded, as there were some that were of a "patriotic character," which the CCP wished to maintain. Mao said that he counted the Sino-Soviet treaty of August 14, 1945, among these.[107]

In his cabled report to Stalin about the February 4 meeting, Mikoyan stated that he had asked Mao how he would explain the patriotic character of this treaty in his discussions with representatives of the other parties. He said that Mao had replied laughingly that not he but Chiang Kai-shek had signed the treaty. "I tell them, Mao said, that the Soviet Union came to Port Arthur to protect itself and China from Japanese fascism, because China is so weak that it cannot defend itself without the help of the Soviet Union." When asked why the CCP opposed the American naval base and yet defended the Soviet base in Port Arthur, Mao replied that U.S. imperialism was in China to oppress it, whereas the Soviet Union had placed its forces in Port Arthur to provide a defense against Japanese fascism. When China had become established and was able to defend itself independently against the Japanese threat, "the Soviet Union would itself have no need of the Port Arthur base." Mao added that a representative in the KMT Legislative Yuan had stated that it would be a big thing if the Communists succeeded in getting Port Arthur back from the Russians. This woman, Mao said, understood nothing about politics. Mao concluded his February 4 statements by saying that the Chinese Communists spoke openly of their pro-Soviet leanings. On the occasion of the anniversary of the October revolution, for example, they had emphasized that China would have to take a stand in "the anti-imperialist camp with the Soviet Union at its head. . . . There is no middle road for us."[108]

Mikoyan repeated this passage from the talks in an abbreviated and slightly altered form in his report of September 22, 1960, to the Soviet party leadership. According to him, Mao had stated that a "bourgeois politician" had said to him with reference to Port Arthur that the Soviet Union would no longer have any reason to hold on to their naval base in Port Arthur once a revolutionary government came to power. Mao had distanced himself from this idea. The fact that Communists were in power in both the USSR and China did not make it impossible for the Soviet base to remain. Indeed, it made it possible. The CCP therefore supported the continued presence of this base. American imperialism had come to China to oppress it, but the USSR was in Port Arthur to support China against Japanese militarism. Only when China was strong enough to defend itself against Japanese aggression would the USSR no longer need the base.[109]

Before Mikoyan's departure for Xibaipo, the Soviet leadership had allegedly already decided to annul the agreement on Port Arthur of August 14, 1945, and withdraw Soviet troops from the Liaodong Peninsula in the event of a Communist victory in China.[110]

In 1945, the sole military control rights of the Soviet Union for the defense of Port Arthur were forced on China. The KMT delegation under T.V. Soong had raised no objections to the Port Arthur agreement as such during the treaty negotiations between June and August 1945 and had also stated that it accepted joint use of Port Arthur as a naval base. It had initially demanded, however, that control of Port Arthur and Dairen harbors remain in Chinese hands. Stalin, in turn, had insisted that Port Arthur and Dairen remain under Soviet control. They had finally agreed on a complicated form of joint administration.[111]

In 1949, Stalin felt it would not be possible to justify such an unequal agreement—and one that injured Chinese sovereignty—with a Communist China. This was not so much because a feeling of solidarity with a Communist brother state would forbid this. His treatment of the Eastern European satellite states had demonstrated that any such ideas were alien to him. He treated China differently because Mao had seized power without Soviet assistance. In the light of the independence that emerged from this, as well as China's size and importance, he could not consider reducing the country to the status of satellite, as he had done in Eastern Europe. Stalin's decision may also have been made easier by the expectation that the CCP's seizure of power would change the character of Sino-Soviet relations and the situation in the Far East in ways that would be favorable for Moscow.[112] As a result, there would, in the long run, no longer be any particular need to station troops on the Liaodong Peninsula.

When Stalin learned from Mikoyan that Mao had raised the question of Port Arthur on February 4, he instructed Mikoyan to explain the Soviet position to the Chinese leadership on the following day as follows:

> The Chinese Communist seizure of power has changed the situation completely. The Soviet government has resolved to annul this unequal treaty and withdraw its troops from Port Arthur as soon as a peace treaty has been concluded with Japan and the American troops withdraw as a result from Japan. If, however, the CCP considers it useful for us to withdraw the troops from the area of Port Arthur immediately, the Soviet Union is prepared to comply to this request from the CCP.[113]

Following these instructions, Mikoyan explained to the Chinese leadership on February 5 or 6 that the Soviet side considered this treaty to be "unequal [*neravnym*]." It had been concluded in order to counteract the KMT conspiracy with Japan and the United States against the USSR and the liberation movement in China, he explained. The treaty had had "the well-known benefits" for the liberation movement, but the Communist seizure of power had now completely changed the situation. In this context, the Soviet Union had decided to declare the treaty invalid (*otmenit'*) and withdraw the Soviet troops from Port Arthur as soon as a peace treaty had been concluded with Japan. "If, however, the Chinese Communist Party considers an immediate withdrawal of the troops to be useful, the Soviet Union is prepared to comply to this request." On the other hand, Mikoyan continued, the Soviet side considered the treaty on the Chinese Changchun Railroad "not to be unequal," as this railroad had to a great extent been constructed with Russian funds. It was possible that "the principle of equal rights had not been applied fully" in the treaty. But Moscow was prepared to discuss and resolve this problem "with the Chinese comrades in a brotherly fashion [*po-bratski*]."[114]

Mikoyan reported to Stalin that Mao Zedong and the other Politburo members had found his characterization of the treaty as unequal "so unexpected" that it provoked "open amazement in them." "They began talking about it almost simultaneously and saying that it would not be appropriate at the present time to withdraw the Soviet forces from Liaodong and hand over the Port Arthur base as we would only help the USA in this way." Mao declared that the question of withdrawing from Liaodong should be kept secret and that the treaty could be assessed only when political reaction had been defeated in China, the people had been mobilized to attack foreign capital in order to liquidate it, and the country had been "put in order" with Soviet help. "The Chinese people is grateful to the Soviet Union for this treaty," he reported Mao to have said. When China had regained its strength, the Soviet Army would withdraw from China and they would conclude a Sino-Soviet treaty of mutual assistance along the lines of the Soviet-Polish treaty.[115]

Mao insisted on his position on Port Arthur, Mikoyan wrote in his report to the Presidium of the Central Committee of the CPSU of September 22, 1960. It was clear that he had "certain tactical ideas of his own," but he did not reveal what they were.[116]

Mao also informed Mikoyan that there were minor differences of opinion with respect to possession of the Changchun Railroad that they could resolve on location. Kuomintang troops, for example, had confiscated a portion of the railroad equipment, and the PLA had subsequently given it back to the railroad. The word among the people was that the KMT had taken this equipment in accordance with the Sino-Soviet treaty and that the PLA was giving it back to the railroad—in violation of the treaty, so to speak. Mao suggested that Gao Gang and Ivan Kovalev should deal with this question and report to the leaderships of both parties.[117]

Consensus on the Sinkiang Question

In point five of his 1960 report, Mikoyan claimed that Mao Zedong had not trusted the Soviet Union very far with respect to its plans for Sinkiang. During the talks in Xibaipo, according to his information, Mao raised the question of the existence of an independence movement that was not controlled by Urumchi in the Ili territory, where a communist party existed. As early as 1945, he, Mao, had heard that local rebels in this area had been armed with artillery, tanks, and airplanes of Soviet origin. Mikoyan hurried to dismiss the accusation of Soviet interventionism hidden in these comments. He assured him that the USSR would neither support independence movements nor make territorial claims. It considered Sinkiang to be a part of China for all time.[118]

Conflict over Outer Mongolia

Point six in Mikoyan's 1960 report dealt with an explosive topic that was debated in Xibaipo in a controversial way: Outer Mongolia. The bitterest conflicts during negotiations for the Sino-Soviet treaty of August 14, 1945, had concerned the status of Outer Mongolia and had repeatedly threatened to make the negotiations fail. In the wake of the Yalta agreement, the Nationalist Government had finally had to yield to pressure from Stalin and recognized the status quo that had existed since 1924, which amounted to a de facto Soviet protectorate.[119]

According to Mikoyan, Mao put up a trial balloon in Xibaipo by asking what position the Soviet Union would take on the unification of Outer and Inner Mongolia. Mikoyan, who naturally recognized how explosive the question was, replied, misunderstanding intentionally, that Moscow was opposed, as this would involve the loss of a significant amount of Chinese territory. Mao then felt obligated to show his colors. He said he meant that the two Mongolias should unite and become part of China. Mikoyan replied that this was impossible because the Mongolian People's Republic had already had its independence for a long time and would hardly surrender it voluntarily. If it ever united with Inner Mongolia, the result would probably be a united, independent Mongolia. Ren Bishi countered by referring to the differences in the size of their populations. Three million people lived in Inner and 1 million in Outer Mongolia.[120]

Mikoyan then asked Stalin for instructions and received the following telegram from him:

> The leaders of Outer Mongolia are in favor of the unification of all Mongolian territories with Outer Mongolia as a sign of the independence of a unified Mongolian state. The Soviet government is speaking out against this plan as it would involve the separation of a series of territories from China, even though this plan does not threaten the interests of the Soviet Union. We do not believe that Outer Mongolia will surrender its independence in favor of autonomy within a Chinese state, even if all the Mongolian territories were brought together to form this autonomous unit. Understandably, Outer Mongolia itself will have the last word on this question.[121]

When Mao Zedong became aware of this telegram, he stated that he would take it into account, and added that the Chinese Communists were, of course, not pursuing any kind of chauvinistic Greater China policy and would not raise the question of Mongolian unification.[122] This, however, did not stop him from having Liu Shaoqui raise the problem again in the summer of 1949—this time somewhat more cautiously, as in the case of Port Arthur, packaged in the form of questions raised by "some people in the democratic parties, students, and workers." Liu then let it be known, however, that the Chinese CP leadership was willing to respect the independence of the Mongolian PR.[123]

Stalin's hypocrisy in presenting himself as the protector of the independence of Outer Mongolia is impressive given the fact that this independence was false and was designed to disguise the country's absolute dependence on the Soviet Union. Stalin's chutzpah reached its peak when he also presented himself as a defender of Chinese interests by claiming that he would restrain Outer Mongolia's plans to annex Inner Mongolia. Although Mao did retreat, he must have noticed Stalin's insincerity, which must have left a correspondingly unfavorable impression.

It is amazing how quickly the Chinese CP leadership gave way on the Mongolian question. As late as 1936, Mao had stated that Outer Mongolia "will automatically become a part of the Chinese federation" after the victory of the revolution in China.[124] Mao's retreat is also remarkable when one considers, as already mentioned, that the head of the KMT government's delegation in the negotiations for the Sino-Soviet treaty of August 14, 1945, T.V. Soong, had fought extremely bitterly, if ultimately unsuccessfully, for Outer Mongolia. Chinese negotiating techniques in 1945 and 1949 are comparable in that both times they surrendered Outer Mongolia in order to achieve Soviet concessions in other areas. In 1949, they were interested primarily in Soviet approval of greater assistance for the new China.

Recognition of the Future Communist Government

The last of the topics addressed in Mikoyan's 1960 report was not contested. It involved the question of the recognition of the future revolutionary government by other states. Here Mao mentioned two possible options. Either the foreign states and "first the USSR" would recognize the new government immediately, or China would postpone establishing diplomatic relations by about one year if the foreign states expressed a desire to recognize the new China. Mao obviously favored the second option. It would make it possible, he said, to exert pressure on everything that was foreign in China without having to deal with protests from the foreign governments. Mikoyan apparently did not present a separate Soviet position on this point because he agreed with the procedure Mao preferred. It is known from other sources that Stalin at that time was discouraging the Chinese Communists from establishing diplomatic relations with Western states, whereas he advocated economic relations.[125]

On February 1, Mikoyan and Zhou Enlai again discussed this topic, and Zhou also mentioned these two options. He appeared to prefer the second option and asked whether in this case relations with the USSR should be established as soon as the government was formed or only after a year, along with the other countries.[126]

The Wang Ming Case

What other topics were addressed in Xibaipo? According to Mikoyan, Mao complained about the persecution he had previously endured in the Party, as he reported on the history of the Chinese Communist Party during the first two days of talks. Mao even claimed, clearly exaggerating, that he had been expelled from the Party, had been arrested, and had almost lost his life; and that they (Mao did not say who) had tried to annihilate him. He went into the conflicts with Li Lisan and—particularly extensively—with Wang Ming, who had "played a bad role." Until his political elimination by Mao in October 1941, as has already been described in detail,[127] Wang had been the primary exponent of the Soviet line in the CCP leadership. In an obvious reference among other things to Wang Ming's failed strategy for the defense of Wuhan in 1938, Mao claimed that the Party had been beaten and the army annihilated as a result of his activities.[128] Mao diplomatically avoided criticizing Stalin and Soviet China policy directly in this context and only noted—if clearly enough—that Wang had "supported the Comintern line." Mao did not hesitate to present himself to Mikoyan as an advocate of tolerance in the Party—a claim that is not very convincing given Kang Sheng's brutal purge, which Mao had set in motion during the rectification movement (*zhengfeng yundong*). The "factionists" would have killed leading Party functionaries before. He, Mao, had put a stop to this. He was in favor of showing forbearance within the Party. It was unnecessary, Mao said, to expel comrades from the Central Committee or persecute them for differences of opinion. The Party leadership had also left Wang Ming in the Central Committee. He was thus available to the Central Committee, "although he performed practically no duties."[129]

Mikoyan was careful not to take any position on Wang Ming in Xibaipo, although

Mao sought to entice him to do so by asking "what our position on him [Wang Ming] was and whether we would try to gain support or seek advice from him." Before he left Moscow, the Soviet leadership had agreed that Mikoyan should not attempt to meet Wang. According to Mikoyan, Wang made no attempt to meet him, either; Wang was not present during any of his talks with Mao. Both sides thus attempted to handle the difficult topic of Wang Ming with the greatest delicacy in order not to further burden relations—although Mao had made it clear that he had not forgotten Soviet behavior in this question.[130]

A Masterpiece of Byzantine Flattery

Mao pulled off a masterpiece of Byzantine flattery by repeatedly insisting to Mikoyan on the observation that the CCP had little experience in comparison with the Soviet Communist Party and by asking Moscow for "instructions and leadership." Mao played down his own role as leader and theoretician in the Party, often described himself as "Stalin's pupil," and repeatedly emphasized that he was following a pro-Soviet line. He also portrayed Stalin as the teacher of the Chinese people and of the peoples of the entire world. It should be taken into account, Mao said, that the Chinese Communists were "poor Marxists," that they made "many mistakes," and that they had practically nothing to show for their efforts in comparison with Russia. He himself, Mao said with apparent modesty, had made no contribution to Marxism-Leninism as party leader and could not place himself on the same level with Marx, Engels, Lenin, and Stalin. He had therefore forbidden the provinces to mention his name together with those of Marx, Engels, Lenin, and Stalin, and in doing so had had to overrule the opinions of his closest comrades. This was all misleading information because Mao, as already noted,[131] had been "sinifying Marxism" since 1937–38. He had also ordered, or at least tolerated, the creation of a cult about his person in the course of the rectification movement and allowed the party statutes approved by the Seventh Party Congress to celebrate the thought of Mao Zedong as the "guiding principle underlying all activities [*yiqie gongzuo de zhizhen*]" of the Party.[132]

Mikoyan allegedly saw through Mao's humility and recognized it as a sham. In his 1960 report, he conjectured that it had been "an Oriental maneuver to show signs of modesty" and did not reflect "what Mao Zedong really represents or thinks of himself." In Xibaipo, Mikoyan, who as an Armenian was not unfamiliar with Oriental flattery himself, also pulled out all the stops in contradicting Mao and heaping him with praise. He told Mao his words were a sign of his own and the Chinese Party leadership's modesty. If the Chinese Communists had really been poor Marxists, they would hardly have been able to end twenty years of civil war in victory. More than simply applying Marxism-Leninism mechanically, the CCP had examined the specifically Chinese conditions. Mikoyan went so far as to concede that the Chinese revolution had followed "its own road [*svoi put'*]" and that the generalization of the Chinese experience was of "theoretical value for the revolutionary movement in Asian countries." It is possible that Mao hid his light so far under the bushel precisely in order to provoke such a reaction from Mikoyan.[133]

The CCP's Economic and Social Policies

According to Chinese sources, Mao expressed optimism to Mikoyan about the PLA's strength and predicted that it would soon cross the Yangtze and advance toward the south in keeping with the slogan of the day, "Cross the Yangtze fighting, liberate all of China [*da guo Changjiang qu, jiefang quan Zhongguo*]."[134] After the victory, Mao continued, the Chinese Communists planned to form a coalition government that included "democratic" forces, but would keep power in the hands of the CCP. For the economic reconstruction of the country, they wanted to base their actions on the experience gathered in Russia and the Soviet Union, respectively, following the 1917 October revolution and the Second World War. Economic development could probably proceed rather more quickly in China than in Russia after 1917–18, as they did not have to fear foreign intervention. They would also be able to profit from the Soviet experience in constructing a new state.[135]

In talks about the new China's future economic policies and the treatment of the "national bourgeoisie" in particular, Liu Shaoqi advocated a relatively cautious course, which Mikoyan approved, but according to Liu there were also extremist groups in the Party. Liu anticipated that companies belonging to the "comprador bourgeoisie" would be confiscated as "bureaucratic capital." The question of nationalizing private companies belonging to the national bourgeoisie could be addressed in one to two years. Mikoyan also recommended a cautious policy toward the national bourgeoisie. The Party should not talk about nationalizing their companies at first. It would be better to keep an eye on them and then, when the Communists had consolidated power, raise the question of their future. Liu Shaoqi approved this completely.[136]

It would be important to see whether the small-goods producers tended to lean toward collectivization or capitalism, Liu continued. Lenin had said that the petty bourgeois economy was a source of capitalism. "Our goal is to make a gradual transition to socialism on the way to the reinforcement of the planning principle in the economy." The transition to socialism would take a long time and would involve a gruesome conflict. "We will also be confronted with the question of 'who will beat whom?' [Russian: *kto-kogo*], which Lenin raised in his day." The Communists would only begin the major attack on the capitalist elements in the economy and initiate the nationalization of industrial and commercial companies after ten to fifteen years. They planned to socialize agriculture only when it could be developed to operate on an industrial basis. Various groups in the Party held different views with regard to future policies. Some advocated developing capitalism, granting concessions to capitalist elements, and capitulating to capitalism. They wanted to make China into a "bourgeois capitalist state"; that is, to reestablish the semifeudal and bourgeois order. The others tended to want to build socialism too quickly "in a leftist, adventurous way." Both approaches would damage the alliance between the working class and the peasants. Such approaches would have to be resisted decisively.[137]

In this context, Mikoyan reported, Liu Shaoqi asked the USSR for advice.[138] It is not known whether Stalin or Mikoyan reacted to this.

Mao Zedong also took the position that the Party should implement a cautious, liberal economic policy at first. The CCP had given the peasants land but not the commodities they needed, he argued. If the Communists did not develop industry, they would not be able to provide the peasants with such goods and they would lose their leading position. They should indeed protect workers' interests through labor unions; decrease the prices of grain, oil, and consumer goods through state interventions in trade in favor of workers and the urban population; and protect poor people in the countryside by shifting them into producing and marketing cooperatives. But they should also give private companies an opportunity to develop. It was possible and necessary to allow free competition in China. As long as China remained economically weak and backward, the Communists should use private capital and should not overthrow anything. The Soviet Union had needed twelve years before it could make the transition to the socialist transformation of agriculture, even though prerevolutionary Russia had achieved a higher level of economic development. Given its economic backwardness, the transition period in China would also necessarily be long.[139]

The talks on economic questions took many hours. Mikoyan provided a mixed picture of them in his final telegraphic report to Moscow. His Chinese counterparts, he summarized, were fully competent on general political and Party questions, the peasant question, and general economic problems. They were not very experienced with business and had only a very vague conception of industry, transportation, and banking. They had no data on confiscated assets, nor did they know which foreign companies were most important in China or to which states they belonged. They also had no information on the activities of foreign banks in China. They did not know which companies were part of the "bureaucratic capital" they wanted to nationalize. "All their economic plans seem like general guidelines and there is no attempt to provide concrete details even with respect to what they control in the territories they have liberated. They sit in an out-of-the-way village and are cut off from reality." It became clear in the course of the talks that the Party leaders had no concrete plans about what they wanted to seize (such as large banks or heavy industry) in order to create the economic foundations of the state.[140]

It has not yet been possible to establish whether Mikoyan revealed his critical view of the Chinese leadership's practical economic competence to his counterparts in Xibaipo. If he did, it probably contributed to the aforementioned impression that Mao repeated in the 1958 conversation with Ambassador Pavel Iudin: Mikoyan had indeed come across as very arrogant.

Mikoyan advised Mao Zedong to follow a rather restrictive course in setting Communist China's future policy toward the national minorities: "They should grant the national minorities autonomy and not independence." In his report to Stalin on this talk, Mikoyan noted that Mao had been happy with this advice, "and yet one could see in his face that he had no intention of granting anyone independence, whoever it might be."[141]

The CCP's Foreign Policy

Chinese publications supplement the last point in Mikoyan's 1960 report, which dealt with the recognition of the future Communist government. According to this information, Mao Zedong spoke hesitantly about the new China's future foreign policy. He also did not discuss the question of the new China's diplomatic relations very clearly and, as Mikoyan recounted, put it in the form of two alternatives. It was too dirty in the Chinese house, Mao said: "rubbish, dust, fleas, bugs, and lice." After liberation, the house would first have to be cleaned before guests could be invited in. With a clear reference to the Soviet Union, Mao observed: "Our genuine friends can come in a little sooner and also help us a bit with the cleaning." If the USSR was prepared to offer assistance to the new China, so much the better. But the other guests would have wait a bit longer. The imperialists, with their dishonest intentions, were not welcome. It would not make much sense to liberate Hong Kong and Macao "too hastily." Quite the opposite. These two places could be used "to establish relations with foreign countries and regulate imports and exports." The conquest of Tibet could only proceed very slowly. Resolution of the "Taiwan question" would be even more difficult, as the island was subject to American protection.[142]

According to Shi Zhe, Mao let it be known that he took only the United States seriously as a possible enemy. The imperialists had not yet attempted any serious intervention in the civil war. American troops did not want to become involved directly and had participated only by delivering large amounts of weapons and ammunition to the Nationalist Government. The other imperialist states were "like a clay Buddha crossing a river." They were vulnerable and lacked the strength to confront dangers. "This is also one of the favorable conditions for our final victory in this war of liberation. . . . This is one of the reasons that we cannot ignore this opportunity." After the victory, these states would have to surrender their special rights and settle their debts.[143]

It is striking that Mikoyan called for a restrained and conciliatory policy toward the United States in particular in Xibaipo. In the discussion of what was to be done with foreign capital in China, he recommended on instructions from Stalin that Chinese leaders nationalize Japanese, French, and, where possible, British property. But a cautious policy was called for with respect to American property "to give the Americans the impression that the new political leaders would respect their interests."[144]

At the end, according to Shi Zhe, Mao once again made a statement on the topic of China's foreign friends. This time it was very cryptic. As the Chinese came closer to victory in the war of liberation, they increasingly needed friends: "that means, genuine friends." There are "true and false friends." False friends behaved in a friendly fashion only on the outside. "We avoid this kind." According to Shi Zhe, Mikoyan became somewhat uncomfortable at these words, as if he could not understand them fully, but he said nothing.[145]

Soviet Economic Aid

The question of Soviet aid, including military support for Communist China, played an important role in the talks in Xibaipo. As early as February 1, Zhou Enlai asked Mikoyan in a meeting also attended by Zhu De for antitank guns, of which the PLA only had 150. He also noted that the PLA possessed primarily light tanks, the heaviest of which weighed fifteen tons. They had seized 70 tanks near Xuzhou, but most of these had been heavily damaged. They lacked raw materials. They asked the USSR for TNT for the production of ammunition, skilled workers and equipment for the production of weapons, and advisers for the organization of the army, for military training institutions, and for the construction of an armaments industry.[146]

Mikoyan agreed in principle (*v printsipe*) to provide support for the construction of an armaments industry and to send advisers. He could not say anything about the antiaircraft and antitank guns; he would report to Moscow.[147]

Mao Zedong, Zhou Enlai, Liu Shaoqi, and Zhu De discussed the topic of Soviet aid in general terms with Mikoyan on February 4. In doing so, Mao presented his requests courteously and modestly. On the other hand, he made it clear that his country could, if necessary, get by even without Soviet aid.

The Soviet Union, Mao observed, had provided and continued to provide the CCP with great assistance, for which it was very grateful. When it asked the USSR for aid, it was aware that the Chinese revolution was part of a world revolution and that the interests of a part had to subordinate themselves to the interests of the whole. If the Soviet Union gave nothing at all, the CCP would not be hurt. When Luo Ronghuan, a high-ranking political official of the Communist forces in Manchuria, traveled to Moscow some time before 1947,[148] Mao had instructed him not to present the situation in such a way as to suggest that the Chinese Communists were about to die and had no resources of their own. But it would be desirable to receive adequate support for Manchuria. Lin Biao had often asked the Soviet Union to help in this or that matter since 1947. He had instructed Gao Gang that absolutely everything they received from the USSR would have to be paid for and that, besides, they should get anything that was lacking from the areas occupied by the KMT. The Chinese comrades should deal rationally with Soviet aid. If this aid were not available, they could hardly have achieved their current victories. This did not, however, mean that they did not have to rely on their own strength. But they should realize that military aid in Manchuria, which constituted a quarter of the total aid, had played a very substantial role. "The CCP needs all kinds of assistance from the CPSU(B)."[149]

Liu Shaoqi had already spoken to Mikoyan about basic issues previously. He emphasized that "without the assistance of the Soviet Union and the countries of the Peoples' Democracy" it would be "unthinkable" to create an industrial base. This aid would play a decisive role for China. From the Chinese perspective, aid could be provided in the form of (1) experience with the socialist transformation of the economy; (2) transfer of relevant literature, and assignment of advisers and technicians in the various branches of industry; and (3) transfer of capital. The Chinese Party leadership was of the opinion that the USSR, the Eastern European Communist states, and China should provide each other with economic aid. If China made no mistakes, it would have to be able to achieve socialism quicker with the help of the Soviet Union. China could not reconstruct the Anshan steelworks in Manchuria without Soviet help. The CCP leadership would like to know in advance how much aid it could expect from the USSR in order to take it into account in its economic plans.[150]

Mao Zedong left most of the talks on the concrete organization of civilian aid to Ren Bishi, Liu Shaoqi, Zhou Enlai, and Zhu De. He became involved only at the end of Mikoyan's visit when questions concerning credits and payment methods were discussed.

The Chinese had particularly high concrete expectations for Manchuria, which they wanted to transform into a "workshop of the country's defensive capabilities" (Ren Bishi) that would produce motor vehicles, airplanes, tanks, and armaments. To accomplish this, they considered the following types of Soviet aid: joint Sino-Soviet economic associations, Soviet loans, and the award of concessions to the USSR. They needed Soviet assistance to mine uranium, magnesium, molybdenum, and aluminum, deposits of which were located near Mukden and Jinzhou and in Jehol Province. The Japanese had previously exported a ton of uranium. If the Soviet Union was interested in these mineral resources, the Chinese could consider mining them together or granting the Soviet Union a special concession.[151]

Mao Zedong had already asked Stalin in early January 1949 to deliver the necessary materials for the construction of the planned railroad line north of the Yangtze as soon as possible on credit.[152] In Xibaipo, Ren Bishi proposed constructing a railroad line between the Soviet Union and China across Mongolia.[153] Mao, on the other hand, called for a line through Sinkiang Province. When the question was discussed in Moscow later, Stalin supported the Mongolian alternative because this route was shorter and would be less expensive to construct.

The Chinese side repeatedly requested that Soviet advisers be sent to China. Mao Zedong said the CCP needed two advisers, by which he obviously meant "senior advisers": an economic adviser and a financial adviser. Zhou mentioned "the Party adviser," by which he clearly meant Ivan Kovalev, and observed that the Chinese Party leadership wished to deal with the Central Committee of the CPSU(B) directly on questions of fundamental significance, as such questions would be decided in Moscow. In this context, Mao Zedong suggested that Wang Jiaxiang be sent to Moscow as the future ambassador. Mikoyan raised no objections.[154]

Ren Bishi asked Mikoyan for "no less than 500" specialists in the various branches of the economy to be sent out. He emphasized that highly qualified professionals were needed for the industrial development of Manchuria. They were employing Japanese professionals in the Anshan steelworks.[155]

As Mikoyan reported to Stalin, Mao Zedong requested an interest-bearing credit of US$300 million on February 6 for the delivery of the machines and materials, including silver that was required for the probable production of a hard Chinese currency, petroleum, raw materials, armaments, and 3,000 motor vehicles. Mao said,

> Three hundred million—that is what we need. We do not know if you can give us this sum, or less or more, but even if you do not give it to us, we will not feel that you have insulted us. We are not asking for any assistance free of charge, as this would be exploitation of the Soviet Union by China. We are requesting a reimbursable credit against payment of corresponding interest, which China will be able to pay in future. The latter is important for the Chinese workers, who will know that the credit must be repaid to the Soviet Union.[156]

Mao asked for the credit to be paid out in equal parts over three years, beginning in 1949. He did not mention the rate of interest or the repayment deadline. Weapons shipments, he said, had up to then been made free of charge, but he was aware that the production of these weapons involved the labor of Soviet workers that had to be paid.

About repayment of the credit, Mao said, "It is not clear to us what we are supposed to use to pay this credit."[157] The Central Committee of the CCP, Mao said, had already drawn up a list of the machines and materials it needed, but their total value was unknown. It was also unknown whether the amount of the credit would cover it. If the credit issue was approved, the Central Committee wished to send a delegation to Moscow to sign an agreement. He also expressed the wish that a group of Chinese experts be received in Moscow to familiarize themselves with the banking system.[158]

The US$300 million Mao requested in credit was ten times as much as he had reported that he needed half a year earlier. When he had pressed for a meeting with Stalin in late August 1948, he had included Soviet aid for Chinese industrial development among the topics he wished to discuss with him and mentioned that he needed a credit of US$30 million in silver.[159]

Mao raised the question of Soviet aid and credit in particular once again on the last day: "It pleases us that the USSR grants us support and assistance, but it is not possible to be victorious if one is always hoping for assistance from outside." In order to shorten the period of transition to socialism, China would require economic aid, he said. The CCP thought this aid could come only from the USSR and the "countries of the New Democracy" (i.e., the countries of Eastern Europe under Communist rule). "We need a credit of US$300 million over three years (1949–51), 100 million per year, on which we will pay interest. We wish to use the credit in part to purchase equipment, oil, other goods, and silver, which we need to stabilize the yuan."[160]

An otherwise apparently serious publication from the People's Republic of China states—albeit without indicating its source—that Zhou Enlai asked Mikoyan for a credit of US$400 million and that Mikoyan had considered US$300 million to be possible.[161] According to Kovalev, Mao Zedong was always afraid Stalin would refuse a credit, as Mao was aware of the Soviet Union's current economic difficulties.[162]

With the exception of this reference and the information that Mikoyan had agreed to send Soviet experts to China as requested by Zhou,[163] it is striking that the literature in the People's Republic practically ignores the Chinese requests for aid. Understating in an almost ridiculous fashion, a contemporary Chinese witness, who really should have known better, reported that Ren Bishi came to discuss with Mikoyan the sort of aid they could receive from the Soviet Union "in terms of the specific equipment they would need to restore production" and had stated that they especially needed deliveries of newsprint, medicines, and cotton materials.[164] Party censors clearly wanted up to now to avoid giving the impression that Communist China had played the role of suppliant at that time. The question of aid is also strikingly absent from Mikoyan's 1960 report to the Party leadership. Kovalev, on the other hand, claimed that the talks focused in principle (*v osnovnom*) on Soviet support for the reconstruction and development of the Chinese economy.[165]

The Soviet documents available to us do not show that Mikoyan approved a credit. They suggest that he only advised the Chinese comrades to address their concrete wishes with Moscow. Mao Zedong asked that earlier requests for aid be delivered more quickly. Mikoyan was also informed that the Chinese Party leadership, as already mentioned, planned to send a special delegation to Moscow "a bit later" to discuss further Soviet economic and military aid and credits. This delegation, Mao Zedong said, would conclude a credit agreement and discuss any further problems that had arisen in the course of the rapidly changing situation.[166]

To all appearances, Stalin was not yet inclined at that time to grant the Chinese comrades a credit. When Mao again approached Moscow with a request for a credit in April 1949, he received a dilatory reply. Stalin informed Mao that there would be no problems exchanging goods for goods.[167] This did not even require the approval of the Supreme Soviet. The government, on the other hand, could not itself decide to grant credits, because the Supreme Soviet was responsible for this. Before it could give its approval, a document from the applicant state would first have to be presented. This was, of course, a cheap excuse, which Mao certainly saw through. A credit agreement was signed only in the summer of 1949 during the visit to Moscow of the delegation led by Liu Shaoqi—that is, during the second summit meeting.[168]

In his last talk with Mao Zedong on February 7, according to Mikoyan, Mao expressed his satisfaction with the discussions and "thanked Stalin warmly for his concern for the Chinese revolution." In light of the bitterness the CCP leadership had felt in August 1945, when Stalin urgently advised them to give up the revolution,[169] this praise does not sound altogether convincing from Mao's mouth. Mikoyan reports that Stalin and the Politburo certified that he had fulfilled his mission well on his return.[170]

Results of the Mikoyan Trip

Mikoyan's meeting with the Chinese Party leadership, the first at the top level, was characterized by careful mutual probing. Where they found disagreement, both sides avoided conflict by courteous maneuvering. This was true, for example, of the Mongolian question, when Mao Zedong pretended not to notice the insincerity of the Soviet arguments and made a retreat on the question. And this was also the case when Mikoyan acted as if he had not noticed that Mao's modesty with respect to his opinion of himself was only hypocrisy; he in turn showered Mao with similar insincere compliments. Their social interactions were anything but open. One could question whether this first encounter could have had the effect of significantly diminishing or even dismantling their mutual reservations. If what he told Ambassador Iudin years later is true, Mao found Mikoyan's behavior in Xibaipo unpleasant. Mao said that Mikoyan had behaved "really very arrogantly [*jiazi jiu hen da*]" at that time.[171]

The talks in Xibaipo reveal the areas in which the two Parties could not reach an understanding. On the Mongolian question, Mikoyan made it clear that Moscow was not ready to make any concessions. Stalin insisted to the end that the status quo—that is, the fiction of an independent Mongolian People's Republic that was in fact a Soviet protectorate—had to be maintained. Mao quickly yielded—but only apparently. This can be seen from the fact that he again raised the question of restitution to China with Khrushchev and Nikolai Bulganin in 1954. Mikoyan also signaled that the Soviets were unwilling to yield on the question of the Chinese Changchun Railroad. He stated that this treaty should not be labeled unequal a priori. It would, however, be possible to make improvements, if necessary. Mikoyan clearly advocated the quick formation of a government because this would have allowed Moscow to replace its double game in China with rapid recognition of the new regime. Mao obviously opposed it as he was more afraid of conflicts with uncomfortable coalition partners than interested in accelerating the establishment of official relations with the USSR. Mikoyan's arrogant reaction made Ambassador Roshchin's move to Canton remain an open wound in the relations. The same is true of the Wang Ming problem, which Mikoyan did not even discuss, let alone apologize for.

There were also areas in which they reached agreement. Stalin surprised Mao by his willingness to make significant concessions on the Port Arthur question. Without pressure from the Chinese, he admitted that the relevant treaty of August 14, 1945, was unequal and offered to suspend it. He even declared that he would agree to the immediate withdrawal of Soviet troops. Soviet renunciation of territorial claims in Sinkiang came in the same category. The two sides came to an understanding that the new China should pursue a restrictive policy toward ethnic minorities and that the new government should take the form of a coalition government. Mikoyan and Mao stated that they felt the Chinese Communists should proceed cautiously on economic policy when they seized power and that they should nationalize private companies slowly. Mikoyan reported on the economic competence of the CCP favorably to Stalin, even if he doubted their business abilities in such fields as industry, transport, and banking. Mikoyan and Mao were able to agree that the CCP should pursue a cautious policy toward the United States. Finally, Mikoyan went so far as to admit that Mao had followed his "own road" in the Chinese revolution, which, as he carefully put it, had a theoretical value for other revolutionary movements in Asia.

Mikoyan made promises in a few areas involving future aid from the USSR for the new China, which was of central importance to the Chinese leadership. In others, the Chinese only stated their wishes; Mikoyan apparently did not actually refuse anything. Mikoyan stated that he essentially approved Soviet support for the construction of an armaments industry and sending Soviet specialists to China. He put off the Chinese request for weapons shipments by promising to forward it to Moscow. The same was apparently true with reference to the CCP request for a credit to acquire goods worth US$300 million.

Mikoyan advised the Chinese comrades to present their concrete requests for goods in Moscow. The Chinese Party leadership then announced that they would send a special delegation to Moscow to discuss problems involved in further economic and military aid and the credit directly. This was the background against which Liu Shaoqi arrived in Moscow at the head of a delegation for the second, also secret, high-level meeting at the end of June 1949.

Ivan Kovalev: Stalin's Man with Mao Zedong

As already mentioned, Ivan Kovalev arrived in Xibaipo along with Mikoyan on January 30, 1949. Formally, he was the "representative of the CC of the CPSU(B) to the CC of the CCP."[172] In fact, he acted as Stalin's representative to Mao Zedong. As such, it was his task to inform Stalin "about the CCP leadership and China as a whole and to maintain the connection between Mao and Stalin."[173] His personal connection with Stalin quite rightly gave him the highest rank in the eyes of all Soviet personnel stationed in China. This is reflected, for example, in the comment by the then Soviet consul general in Peiping, Sergei Tikhvinskii: "He was the boss!"[174] Shi Zhe's statement that Kovalev had never played such an important role[175] is refuted by Soviet archival documents that have since been published.

At that time, all important measures involving China policy passed over Stalin's desk, as had been the case since the 1920s.[176] The fact that Stalin for the first time sent an official representative of the CPSU(B) to the Chinese Party headquarters can be taken as a further indication of Moscow's increased interest in the Chinese Communists, which was due to the PLA's apparently unstoppable advance to the south. Kovalev occupied his position during all of 1949. During this period, he sent critical reports on the activities of the Chinese leadership directly to Stalin. He obviously received most of his information from Gao Gang, who also sent his own reports through Kovalev to Stalin.[177] During his time in China, the Chinese considered Kovalev cunning, well organized, authoritarian, vain, and addicted to alcohol.[178]

Kovalev had previously occupied high-ranking civil and military positions in the transportation sector. He was minister of transportation during the Second World War and organized the transportation of Red Army units to the Soviet Far East in 1945 to participate in the war against Japan. In May 1948, Mao Zedong had asked Stalin for a small group of Soviet experts to assist the CCP in the economic development of the big cities and the repair of railroads in parts of Manchuria under Communist rule. In early June, Kovalev traveled to China as leader of this group made up of 300 engineers and specialized workers. To keep his actual mission a secret, he appeared officially as "representative of the Council of Ministers of the USSR to the Changchun Railroad," which was owned jointly by China and the Soviet Union. In addition to restoring the 15,000-kilometer-long railroad network in Manchuria, the Soviet experts helped to develop other sectors of the economy. With his Chinese colleagues, Kovalev worked out a plan for the restoration of the 3,000-kilometer-long railroad network in North China—north of the Yangtze. Mao Zedong praised him for his efforts to Stalin. When Kovalev returned in December 1948, he reported personally to Stalin.[179]

Mao maintained his own telegraphic connection with Stalin until Kovalev returned to China at the end of January 1949. Subsequently, their contact—presumably as Stalin preferred—continued only through Kovalev, who was instructed to direct his radio messages exclusively to Stalin's code name, Filippov.[180] To communicate, Kovalev used the radio equipment belonging to Mao's Soviet doctor, Andrei Orlov, who belonged to the military secret service, the GRU.[181] Shi Zhe claims that Orlov took this equipment, which belonged to the Soviet Army, back to Moscow with him when he left in mid-1949. Up to that point, Orlov had blocked all of Kovalev's attempts to gain total control of the transmitter.[182] This account appears implausible because, so long as Kovalev was acting as Stalin's contact man to Mao Zedong, neither the GRU nor Orlov personally would have dared to put obstacles in his path. It is clear that Kovalev had a direct line to Moscow until at least the beginning of November.[183]

In March, Kovalev moved from Xibaipo to Peiping along with the Chinese leadership and took a house in the Western Hills near Mao's residence. By his own account, he met there with Mao almost every week. During the spring and summer, he participated along with Mao, Liu Shaoqi, and Zhou Enlai in the Politburo Commission on Economic Problems. He accompanied Liu Shaoqi's delegation to Moscow in June and participated in its discussions with Stalin.[184] In the fall, he returned to China at the head of the first group of Soviet experts sent to China. Because he continued to control a direct link to Moscow,[185] he still comported himself as Stalin's personal representative and did not consider it necessary to make arrangements with Ambassador Roshchin. The latter complained about him in Moscow. Stalin sent an investigatory committee to China, and the results of their investigation contributed to Kovalev's subsequent recall. In December, Kovalev accompanied Mao to Moscow at the Chinese leader's request and functioned along with Nikolai Fedorenko as Stalin's go-between to Mao until the beginning of negotiations for the alliance treaty.[186]

On the eve of Mao's departure for Moscow, Kovalev began to write Stalin a critical report of activities by the CCP leadership, which he completed on the train and gave to Stalin on December 24. To win Mao's trust, Stalin handed him the report in late January 1950. At the same time, Stalin gave Mao similarly critical coded telegrams Kovalev had previously sent to him from China and information from Gao Gang meant for the Soviet leader personally. Stalin told Mao that he was distancing himself from Kovalev, whom he accused of high-handedness and political incompetence.[187]

This spelled the end of Kovalev's career in China. He did not participate in the Sino-Soviet negotiations for a treaty of friendship and alliance that began on January 22; by his own account, this was because of a "lingering sore throat."[188] When Zhou Enlai asked Mikoyan and Andrei Vyshinskii in early February 1950 for a meeting with Kovalev to discuss the plan to send Soviet specialists to China, nothing was said to suggest that he was sick. The Chinese request was met with the cold observation that "the Foreign Ministry would discuss the question of specialists."[189] Ivan Arkhipov, who in the meantime had obviously been made Kovalev's successor, had already been appointed "chief adviser" three weeks earlier.[190] There can be no further doubt but that Stalin had sacrificed Kovalev on the altar of the developing Soviet-Chinese friendship.

The CCP's Policy Toward the West and the Role of the Soviet Union: Beginnings of a Strategic Relationship

Stalin's secret diplomacy with the Chinese Communist Party became significantly more intense in 1949 when it became increasingly clear that the Communists would be victorious all over China. This was especially true in Manchuria, where, under Soviet pressure, Yenan took action against the Western consulates—and the American consulate in particular. Stalin's goal was to drive the Western powers out of this part of China, which he considered an exclusively Soviet sphere of influence. Backed up by the Yalta agreement of February 11, 1945, Soviet diplomacy had the courage to act accordingly and was not afraid to associate openly with the Communist provincial authorities.

Outside of Manchuria, however, Moscow held back noticeably from the new rulers of China. This could be seen primarily in terms of diplomatic work with the Communist authorities. Soviet missions in the territories the Communists occupied officially closed their doors. The personnel remained at their posts, however, and unofficially maintained a contact with the new authorities that was far more close-meshed and trustworthy than anything maintained by the Western missions. In Peking, for example, the Soviet consulate general terminated "all official activities" on instructions from the Foreign Ministry in Moscow on February 16, 1950.[191] "At the initiative of Communist leaders and with the approval of the Foreign Ministry," however, the consul general and his deputy met unofficially with independent representatives of the CCP, such as Ye Jianying, Nie Rongzhen, Luo Ruiqing, and Zhu De, and had talks with them that were of "an exclusively friendly character."

Moscow thus behaved in a duplicitous fashion, fearing that the United States and Great Britain would see an official collaboration with the authorities as a violation of the Yalta agreement—which, after all, was based on the Soviet recognition of KMT China. Washington could have taken such a perceived violation as an excuse to withdraw from the treaty and thus terminate the other contractual obligations (Soviet rights in Mongolia and Manchuria and the annexation of southern Sakhalin and the Kuril Islands) and possibly also to begin a major military intervention in the civil war on the side of the KMT, which would have placed Moscow in a difficult position.

But the Soviet side was careful not even to provide advice. Stalin merely let it be known that he was not interested in seeing Communist China establish diplomatic relations quickly with the Western democracies, although he did encourage the CCP leadership to establish economic relations with and accept economic aid from these states.

Mao Zedong was aware of the precarious nature of the situation the Soviet leadership found itself in with the United States and sought to exploit it, albeit in the context of his basic decision in favor of a strategic alliance with the USSR. In early 1949, Mao said within the Party in Xibaipo that the international situation could be compared to two tigers, one white and one red, confronting one another.[192] The CCP was leading the Chinese revolution to victory in the space between them.

Communications between Moscow and the CCP leadership took place for all key questions on the Party channel between Stalin and Mao Zedong.[193] As has been mentioned, Kovalev functioned as middleman at this level from his arrival in China in late January 1949—first in Xibaipo and then from March in Peiping. To all appearances, Kovalev made no or practically no arrangements with Soviet diplomats in China. Coordination between the Party leadership and the diplomatic corps took place in Moscow and, on important matters, between Stalin and the leadership of the Foreign Ministry.[194]

Further lines of communication between Moscow and the CCP passed through other Soviet representatives in China who maintained unofficial links to Communist functionaries, such as Gao Gang in Manchuria, who occasionally sent information to Stalin behind the backs of the Chinese Party leadership. After the Soviet ambassador, Roshchin, moved to Canton in early 1949, this also included contacts that the Soviet diplomats who remained in Nanking maintained with the Communist offices there. They reported to an embassy counselor Shibaev, who was a member of the secret service (MGB or GRU) and at the same time acted as the unofficial chargé d'affaires of the USSR in the parts of China the Communists had occupied.[195] Between early May and mid-September of 1949, the frequent meetings between the first secretary of the embassy, Ledovskii, and Huang Hua, who had been sent by Zhou Enlai to Nanking in April, constituted a special channel of communication.[196]

We are still very far from being able to sketch even a general outline of the secret diplomacy. Documents that have become known in recent years from the Soviet and Chinese archives and reports and information from contemporary witnesses do, however, give us some insight into a few important areas.

There is evidence to suggest that the Chinese Party leadership sought even more intensively than usual to demonstrate its solidarity with its Soviet fraternal party from the fall of 1948. We have already presented evidence of this: the professions of loyalty to the Soviet Union and against a third road that were made from the beginning of November.[197] Further evidence is the harsh treatment of the American consulate in Mukden and—less harsh—of the other Western missions in the city from November 1948. New information based on documents from the CCP archives and materials published by Kovalev end Western speculation about whether and, if at all, to what extent Soviet influence was involved.[198] They demonstrate that the Chinese Communists' rigid position developed under Soviet pressure, or at least out of consideration for Soviet interests. It was no coincidence that these incidents occurred in Mukden, because Stalin saw Manchuria as a Soviet sphere of interest.

Following an initial approach in 1944–45, as described earlier,[199] the Liberated Zone of North China of the CCP again signaled its willingness to establish diplomatic relations with foreign countries in the summer of 1948. Following the capture of Mukden by the PLA on November 1, the Communist administration granted diplomatic protection to the consulates of the United States, Great Britain, and France there. The new Communist mayor made a visit to the American consul general and stated his interest in cooperative relations including cultural exchanges. In doing so, he was acting on instructions from the Central Committee that Zhou Enlai had drafted.[200] These stated that the three consulates should be protected by troops or police and should not be searched under any circumstances. The instructions also stated, however, that the Party should consult (*qingjiao*) the local Soviet diplomatic personnel, but consider their ideas only as advice and to report to the Central Committee on all political questions before acting.[201]

The mayor's conciliatory behavior brought the Soviet Union into the picture. Moscow, as already mentioned, viewed Manchuria as an exclusive sphere of influence and a security glacis for the USSR. The Soviets wanted to drive Western diplomats away and opposed attempts by the Chinese comrades to establish diplomatic relations with Western states. The CCP leadership appeared to yield across the board. Mao Zedong ordered Chen Yun and other leaders of the Northeast Bureau of the CCP to make self-criticism for their conciliatory behavior toward the Western missions. The CCP leadership approved the Northeast Bureau's plan not to recognize the diplomatic status of the consulates of the United States, Great Britain, and France and to force these consulates to leave (*jizou*) Mukden. Mao instructed the Northeast Bureau to inform the Soviet side that the CCP must coordinate all questions involving its foreign policy in Manchuria and China as a whole "in consultation with the Soviet Union and to strive for consensus [*he Sulian xieshang chuli, yi yizhi*]."[202] In mid-November, the city authorities in Mukden interrupted the U.S. consulate general's electricity supply and telephone connections. The consul general, Angus Ward, had previously refused to surrender his radio equipment. The authorities confiscated the equipment, placed the consulate employees under house arrest, interrupted their contacts with the outside world, and thus initiated what came to be known as their Mukden ordeal. In June 1949, the authorities accused the consulate personnel of espionage—baselessly, as recent research by a historian from the People's Republic of China has shown.[203] Ward and four other officials were arrested in October. They were placed in solitary confinement, given only bread and water, and beaten. At the end of November, they were given a suspended sentence of one year in prison and expelled. On December 11, 1949, Ward left China with the consulate personnel.[204]

Chinese archival documents demonstrate that the Communist authorities in Mukden were following Soviet advice when they took the measures against the American consulate general. On November 16, for example, the Soviet consul general in Mukden, Malinin, requested in a telephone conversation with the secretary of the Northeast Bureau of the Central Committee of the CCP, Gao Gang, that all radio equipment held by the three Western consulates be confiscated. He argued that this was an "issue of great importance" for the USSR. When Gao agreed, Malinin was pleased and added that the consulates would decamp from Manchuria for better or for worse when they could no longer perform their duties.[205]

Ivan Kovalev, who led the Soviet advisers in Manchuria between June and December 1948, confirmed that the Ward case was set in motion in response to Soviet intervention. The Chinese comrades had asked him for advice, and, without consulting Moscow, he had urged them "to blockade [the U.S. consulate] and not to let its employees out." They were to limit the mobility of the personnel of the other consulates. If the consulates continued to use their radio equipment, it was to be confiscated. Mao Zedong had scruples, Kovalev stated, about the confiscation of the radio equipment in particular, as this would have required the Chinese authorities to enter the consulate's property. Mao had therefore again asked telegraphically for confirmation from Stalin, but received no answer. According to Kovalev, Stalin subsequently approved Kovalev's behavior. His approval was also based in part on the fact that Kovalev had not advised the Chinese to treat the consulate officials "like enemy personnel in wartime," although, as Kovalev observed in retrospect, this was exactly what he, Kovalev, believed at the time. He had refrained from giving such advice, as he had had the impression that the Chinese did not want to quarrel with the Americans and preferred to leave this quarrel to the Soviets.[206]

The CCP leadership essentially followed the Russian recommendations, as the handling of the Ward case shows. It also differentiated the treatment of the three consulates as Kovalev had recommended. The restrictions it imposed on the British and French consulates in Mukden were not as strict as the ones on the U.S. consulate. Mao informed Zhou Enlai that they should "behave just a little bit more gently" toward England and France than toward the United States and relax the siege of their consulates in Mukden after a while. Zhou Enlai passed this on to the Northeast Bureau in Mukden. The differing treatments reflected instructions from the Central Committee that stated that the measures implemented in Mukden did not mean that the CCP could "not ever form diplomatic relations with imperialist countries, or that no distinctions could be made among imperialist countries."[207] Yenan argued further that it might prove possible to use a differentiated policy to drive a wedge between the United States and Great Britain and France, respectively. Zhou Enlai made it clear to the Northeast Bureau that the policy of "expulsion" might not also have to be applied automatically in the territories outside Manchuria. The CCP in any case yielded to Soviet pressure when it refused to accede to a request from British and French diplomats in Mukden to restore radio contact with their missions in Hong Kong and Nanking.[208]

Developments in the Ward case were accompanied by increasing anti-American propaganda during the following months. In early March 1949, the American ambassador, John Leighton Stuart, informed the secretary of state that "the Chinese Communist Party has revealed an increasingly anti-American sentiment."[209] In mid-March, a Communist radio station assumed a martial tone: "Chinese people cannot but regard leaders of war provocateurs, American imperialist bellicose elements, as their mortal foes and cannot but regard leaders of world anti-imperialist front, great socialist state, Soviet Union, as their comrades in arms, from first to last."[210] On the basis of an analysis of this transmission and the aforementioned speeches by Mao Zedong and Liu Shaoqi during November 1948, Stuart came to the accurate conclusion that the Communist leadership was advancing "its satellite foreign policy still further."[211]

Because the Party leadership had been given the impression that the American consulate in Mukden was involved in espionage activities, it now made the general assumption that Western diplomats and journalists were all trying to sabotage the Chinese revolution. This led to a formal Central Committee resolution in January 1949 that the Party should not be in any hurry to establish diplomatic relations with Western states.[212]

On January 19, 1949, the Central Committee of the CCP issued a directive governing relations of the CCP to foreign states. It was based on the observation that the Party could not recognize the legal status of foreign diplomatic personnel, as they did not maintain diplomatic relations with any foreign state. Foreign diplomats and journalists were to be treated like ordinary foreigners. This made it possible for the Party to reserve the right to undertake diplomatic initiatives without being bound by any humiliating diplomatic traditions from the past. Imperialist privileges in China were essentially to be abolished, but each case was to be treated separately in practice. Fidelity to principle was to be combined with flexibility.[213]

During the next months, Mao Zedong summarized the principles of the new foreign policy in three slogans: (1) "Make a fresh start! [*ling qi luzao,* "build a new oven"]"—that is, do not recognize the status of any diplomat accredited to or any treasonous treaties concluded by the KMT, abolish all imperialist press organs and assume control of foreign trade; (2) "Clean the room before inviting guests! [*dasao ganjing zai qingke*]"—that is, remove all manifestations of imperialist influence in China and only then examine the question of establishing diplomatic relations; and (3) "Lean to one side! [*yi bian dao*],"—that is, ally the new China with the Soviet Union and other Communist-ruled countries without, however, surrendering national independence.[214]

The first two slogans can already be seen in the January 19 directive. The *yi bian dao* slogan was already taking shape there, too. The directive called for the embassies and consulates of the Soviet Union and other "states of the New Democracy [*xin minzhu zhuyi guojia*]," or Communist-ruled countries, were to be treated differently from those of the capitalist states, because the foreign policy of the former was completely different from that of the latter. But, as the Communist states—just like the capitalist ones—had not established "official diplomatic relations [*zhengshi waijiao guanxi*]" with Communist China, one should "also maintain only unofficial diplomatic contacts [*yi zhi zuo fei zhengshi waijiao laiwang*]" with them.[215] One should—it was stated in other documents—treat diplomats and other citizens from the Soviet Union exactly like those from the capitalist countries "in order not to provide the governments and reactionaries of capitalist countries with excuses."[216]

At first glance, a comparison of these statements gives the impression that the CCP leadership had assumed an ambivalent or even contradictory attitude toward relations with the Communist-ruled countries that was shaped by a concern for the Western countries' reaction. In reality, there was only the appearance of "equal treatment."

This could be seen in the first confidential meeting that the de facto foreign minister, Huang Hua, held in early May 1949 with the highest-ranking diplomat among the Soviet embassy personnel that remained in Nanking, Andrei Ledovskii. Zhou Enlai sent Huang to Nanking in late April to establish contact with the foreign embassies and also a link to the American ambassador, John Leighton Stuart. The Soviets considered Huang practically to be the foreign minister of Communist China. Huang first looked up Ledovskii in his home on May 4 and was greeted warmly by him. The two men had last seen each other two years earlier in Peiping, where Ledovskii had been consul general in 1946 and 1947.[217]

After this meeting, the Nanking branch of the Soviet embassy maintained "friendly relations" with the Communist authorities in Nanking until diplomatic relations were established between Moscow and Peking on October 1, 1949. These were kept hidden from Chinese and foreign public opinion and had an "unofficial character." Ledovskii met with Huang Hua twenty times through September 22. In these meetings, he was extensively informed, among other things, about Huang's meeting with Leighton Stuart. Ledovskii reported on these conversations to the Foreign Ministry in Moscow, first in abbreviated form by coded telegram and later more extensively through the diplomatic courier. He received instructions only from Moscow about concrete questions Huang Hua had asked Ledovskii or the Soviet embassy. The Soviet Foreign Ministry sent Ledovskii no concrete rules about which topics he could address or the way he was to do it with Huang Hua, and he did not ask for them. Given his many years of experience in China, Ledovskii said, they trusted him in Moscow.[218]

During their first meeting in Nanking on May 4, 1949, Huang Hua informed Ledovskii confidentially that the Communist authorities would treat the Soviet representatives in Nanking formally by the same rules as all other diplomatic representatives—that is, like foreign private individuals. But they, in fact, wanted to meet with the Soviet diplomats to give them any assistance or support and be able to turn to them to request advice. Huang asked Ledovskii not to be offended by the fact that "outwardly" (Russian: *vneshne*) the Soviet representatives were treated just like the other foreigners. The Nanking authorities would in fact show them "friendly courtesy."[219]

Ledovskii, in turn, informed Huang Hua about the activities of the Soviet embassy in

Nanking, the reasons and motives for the ambassador's move to Canton with most of the employees,[220] Soviet diplomats' contacts with the other embassies, the condition of the Soviet diplomatic corps, and the mood of the foreign representatives.[221]

Huang Hua followed his words up with actions. When he learned from Ledovskii that the Soviet mission was suffering a shortage of money because the courier connection with Moscow had been interrupted, he sent a messenger with "a sack of Chinese money." The messenger was instructed to refuse to accept a receipt and stated that they should let it be known if they needed more money. One of the first questions Huang Hua put to his Soviet counterpart with a request for an urgent answer concerned the rights Soviet Russia had granted to foreigners before relations had been established with their native countries and how these problems were resolved in the Eastern European countries under Communist rule. The new Chinese authorities were facing such questions for the first time, whereas the USSR had great experience in this area.[222]

Huang Hua also asked Ledovskii for advice on several other occasions. In mid-May, he reported to him on a request from the U.S. embassy to permit repairs to a damaged American aircraft at Nanking airport. Huang planned to inform the embassy orally that the new government rejected any responsibility for the damage to the aircraft, which had occurred during the KMT period, and also for any future destruction as a result of KMT air attacks. He planned to approve the repairs only after the embassy had provided information on the further use of the aircraft. In response to Huang's question about how the Soviet government would act in an analogous situation, Ledovskii advised restraint. It was a delicate question and difficult to answer, he said. In the years that followed the October revolution, Soviet diplomacy had behaved with great self-control and fairness in resolving the many complicated problems involving foreign missions that confronted it. It had taken care not to further burden its relations to the Western powers, which were already difficult.[223]

At the same time, Huang Hua asked Ledovskii's advice on handling the *Amethyst* incident. In doing so, Hua revealed considerable uncertainty about the procedures of international law. On April 20, as the PLA set out to cross the Yangtze, a battle had broken out between the British naval frigate *Amethyst* and Communist troops, for which each side made the other responsible. The CCP played up the incident in public opinion and pointed to it as renewed evidence of the Great Britain's old imperialist policies. Within the Party, however, Mao Zedong considered the event a "coincidence" that they should use "to educate the people," but that they did not otherwise have to take very seriously.[224] On May 5, Zhou Enlai ordered representatives of the Nanking Municipal Committee to negotiate in the name of the responsible PLA unit with the captain of the *Amethyst.* If members of the British embassy in Nanking participated in the negotiations, they were not to be treated as diplomats.[225]

Huang Hua reported to Ledovskii about the reaction of the British government, which did not recognize the new regime in China and had instructed its diplomatic and consulate representatives in China to establish de facto contacts with the Communist authorities in order to negotiate about the incident, and which had also sent significant additional military forces to defend Hong Kong. Huang interpreted this behavior as a continuation of the British policy of nonintervention in the internal affairs of a foreign country and, at the same time, as an indication that London would not allow the PLA to capture Hong

Kong. Huang had refused to discuss the *Amethyst* with a diplomat from the British embassy in Nanking, but he had accepted from him a letter from the British ambassador that included the request that the Nanking authorities allow the *Amethyst* to be moved to Hong Kong for repairs and provide assistance in doing this. Huang said that he would next call for negotiations between the commander of the PLA artillery unit involved in the incident and the British captain in order to get the British to admit that they had been the first to open fire and apologize for the incident. If this happened, he would allow the *Amethyst* to be taken out of Chinese waters.[226]

At this point in the discussion, Ledovskii compared the American and British policies toward China with reference to discussions with British and American diplomats. He took the position that increasing differences of opinion had arisen in recent years between Washington and London with respect to policy toward China. In Ledovskii's view these were above all due to the fact that the United States had driven Great Britain from China during the second Sino-Japanese War, and Great Britain wished to take its revenge now as U.S. policy had suffered setbacks in China.[227]

Huang Hua asked Ledovskii to comment on the *Amethyst* incident on the basis of Soviet diplomatic practice in comparable situations in the past. He was particularly interested in the way the negotiations were carried out, the procedures, and the problem of documenting the discussions with the British. Huang wanted to know how the USSR would document an admission of guilt and apology by the British, what value the captain's signature on an agreement would be considered to have legally, and what position Moscow would take should the British refuse to admit guilt and apologize. Ledovskii reported on Soviet negotiating practice, but he apparently refrained from offering concrete advice and recommended a pragmatic response to a possible British refusal.[228]

Ledovskii asked Huang Hua the question that was being hotly debated in the Nanking diplomatic corps: Would the Chinese Communists form a government as soon as they had captured Shanghai or only after they had conquered China as a whole? Huang avoided the question and merely stated that they were currently convening a Political Consultative Conference and two plenary sessions of the Central Committee, which would discuss this question. Communist troops would in any case seize all of China in five to six months. When Ledovskii expressed doubt whether the Communists would be able to manage the "extremely difficult" social and economic problems of a gigantic city like Shanghai, Huang replied that Shanghai was surrounded by PLA troops, who were awaiting the order for a general attack. The situation in the city was very complex. Food and fuel were lacking. Local businessmen concerned about their property were attempting to bribe the KMT general responsible for the defense of the city, Tan Enbo, to surrender the city to the Communists without a fight. The American and British consuls general supported the businessmen in their attempts, as this was in the interest of the "trade and industry circles" in the United States, Great Britain, and other countries that had invested their capital in Shanghai. To Ledovskii's question whether there was an Anglo-American military presence in the city, Huang replied that all foreign warships had been withdrawn from Shanghai harbor and lay at anchor outside. Mao Zedong had previously observed within the Party that the fact that the American and British ships had withdrawn to outside the Wusong estuary on April 26 was "very useful for us [*yu wo you li*]."[229]

When Ledovskii asked about economic conditions in the territories liberated by the PLA, Huang Hua replied that they were particularly difficult in the big cities as a result of the blockade the KMT had imposed on them with the support of the United States. The CCP was therefore interested in establishing trade relations with the Soviet Union through Tientsin, Shanghai, and other captured ports as soon as possible. The Party hoped that Soviet ships of the greatest possible tonnage would come to Shanghai harbor as soon as the PLA had liberated the city. The Communist authorities had already established a contact with the Tientsin office of the trade mission of the USSR and had sent it a list of goods that they wanted to import from the Soviet Union and another list of the goods it proposed to offer in exchange. Discussions on this subject that representatives from Peiping carried out with Soviet trade officials in Tientsin would not have had any positive results because the head of the Soviet trade mission in China had instructed his colleagues in Tientsin not to conclude such contracts hastily and to await further decisions from him.[230]

Ledovskii reacted cautiously. The trade mission could not make decisions on such questions without instructions from Moscow, he said. Caution was also his primary concern when he asked how it would be taken abroad if Soviet ships approached Shanghai harbor. In Huang Hua's view, this would not give rise to any international complications because the Communist authorities had long been doing business with English and other foreign companies through Tientsin, and these connections would be extended after the liberation of Shanghai. Soviet trade organizations could always point to such precedents.[231]

According to his own report, Ledovskii did not seek a comprehensive discussion of Huang Hua's complaints about the lack of progress in the trade talks in Tientsin in his talks with him. There were two reasons for this. As far as Moscow's trade and economic relations with the CCP and the question of providing Soviet aid to the Chinese Communists in their fight against the KMT were concerned, the USSR restricted itself to Manchuria, and it negotiated with provincial Manchurian authorities. The Yalta agreement and the Sino-Soviet treaty of August 14, 1945, made Manchuria a de facto "sphere of influence [*sfera vliianiia*]" of the Soviet Union. The Soviet Union assumed that Chiang Kai-shek and the other signatories of the Yalta agreement would have to observe this. Moscow attributed little weight to the occasional diplomatic demarches regarding this or that Soviet activity in Manchuria. They were convinced that neither Chiang Kai-shek nor the United States would take any more serious steps against the USSR.[232]

But if Soviet organizations outside Manchuria concluded agreements with Communist authorities, Ledovskii continued, this would give the agreement a serious, political character and imply an "open and direct connection [*otkrytaia i priamaia sviaz'*]" between Moscow and the CCP, which was waging an armed struggle to overthrow the legal government of China. This could serve as a reason and excuse for the United States and Great Britain—the signatories of the Yalta agreement—to implement more serious measures against the CCP and the Soviet Union. These actions could lead to a direct and broad military intervention in China, for which the Chiang Kai-shek government was calling desperately and which many KMT supporters were advocating in the United States and other Western states. Outside Manchuria, the Soviet leadership considered itself bound to follow a cautious tactical line in its contacts with the CCP.[233]

For this reason, Ledovskii concluded, Soviet diplomatic, consular, and other official missions officially terminated their activities quickly wherever the Communists seized power. Their personnel remained in place, however, and had instructions "essentially [*po sushchestvu*]" to continue their current activities—but without appearing as Soviet officials. The consulates general in Peiping and Tientsin thus concluded their official activities on February 16, shortly before the PLA moved in. The Soviet Foreign Ministry ordered the signs taken off the doors. The consulate general in Shanghai and the Nanking branch of the embassy acted in the same way. Soviet missions in the Manchurian territories constituted the only exceptions. In Harbin, Dairen, and other cities that had not been occupied by KMT troops, they remained open and maintained official relations with the Communist authorities. In this way, Stalin expressed the fact that he considered Manchuria a Soviet sphere of influence.[234]

Developments in the Ward case provided the CCP leadership and Mao Zedong personally with grounds to obligate local Party and administrative organizations not to take any steps of importance to foreign policy without first seeking the approval of the central leadership.[235] In this way, the leadership already secured control over diplomacy before the People's Republic was founded. One can therefore assume that any behavior by local Communist bodies toward foreigners that led to problems from the end of 1948 and the beginning of 1949 on was most probably not an accident but the expression of a centrally steered policy. The same was true of comparable events during the rest of 1949, such as, for example, the case of a young American consulate official who was arrested, beaten, and treated in an inhumane fashion without any recognizable reason in prison in Shanghai in early July.[236]

In practice, the Communist authorities followed the January 19 instruction. Washington protested repeatedly in vain. In Peiping, Nanking, and also Hong Kong, the Communist offices responsible for foreign relations refused to accept the relevant demarches, left them unanswered, or sent them back. The Chinese Communists' behavior, and Ward's arrest in particular, provoked considerable indignation in American public opinion. Washington discussed and ultimately rejected various repressive responses ranging from a military intervention through an economic blockade in the fall of 1949. In early October, immediately after the People's Republic of China was established, the State Department declared that it would continue to maintain diplomatic relations with the KMT government and did not recognize the Communist regime, as the latter in turn did not respect its international obligations. In mid-November, President Truman advocated blockading the coal transport ships along the China coast and even sinking ships until Secretary of State Dean Acheson dissuaded him. Acheson observed in mid-November that Washington would not consider diplomatic recognition until the Americans held in Mukden were freed.[237]

The Party did not go so far as to break off all connection to these countries or to rule out all possibility of establishing diplomatic relations with them—including the United States. In individual cases—for example, in Peiping and Tientsin—it allowed Western journalists to maintain contacts with Communist diplomats. In Nanking, Huang Hua even held exploratory discussions—unofficially—on behalf of Zhou Enlai with the American ambassador, Stuart.[238]

Stalin had no objection if Communist China wished to establish economic relations with Western states—quite the opposite. Before the end of the war, he had already told American politicians that the United States should play the greatest role in getting China back on its feet after the conflict. It was the only country, he said, with sufficient capital and manpower, whereas the Soviet Union had to take care of its own reconstruction.[239] During the second half of the 1940s, Stalin clearly encouraged Mao Zedong repeatedly to seek a dialogue with the United States and accept credits from the Western states and trade with them.[240] In this way, Stalin apparently hoped primarily to spare the weakened Soviet economy from providing greater assistance to Communist China. If it was Washington that provided the assistance, Stalin apparently also welcomed it as an additional economic burden on the United States. In early 1949, the Soviet ambassador to China told an Indian diplomat that the Americans should go ahead and pour all the money they wanted into China; this would only make them weaker.[241] American diplomats, on the other hand, speculated that Soviet economic aid to Communist China "would impose drain on Soviet economy perhaps serious enough weaken its activities among satellites."[242]

Throughout 1949, the Soviet leadership followed the line we have described and encouraged the Chinese comrades to seek trade and economic aid primarily from the West. When the CCP leadership asked for a credit from Moscow in April, Stalin remained very reserved, as already mentioned, and demanded that they first request assistance from the Supreme Soviet of the USSR "in a document signed by state representatives."[243] At the same time, he advised the Communist leadership not to reject "credits and trade with the capitalist countries,"[244] but rather to intensify trade "with Hong Kong and other foreign capitalists."[245] That summer, he confirmed the correctness of the policy of trading "with the imperialist countries" to Liu Shaoqi, who was visiting Moscow.[246] On the same occasion, he at last for the first time approved a larger amount of Soviet aid, which Mao had repeatedly requested since August 1948.[247] He approved a credit for goods for US$300 million and sent 220 Soviet specialists to China.[248] The fact that Moscow had already sent several hundred Soviet experts to Manchuria the previous year, where they worked primarily in the railroad sector, was heavily dictated by Soviet self-interest, given Russia's or the USSR's traditional vital interest in this area.

Stalin reacted quite differently when the CCP appeared to be attempting to establish political contacts and diplomatic relations with Western states. Here he advised restraint—obviously because he suspected that the Chinese comrades might make agreements with the West behind his back.[249] In April 1949, he advised Mao only to establish such relations when these states agreed "officially" to give up "military, economic, and political support for the KMT government of Chiang Kai-shek,"[250] although he knew full well that this would not happen soon. In July, Stalin again advised the Chinese Party leadership not to urge the Western powers to grant diplomatic recognition.[251]

Kovalev indicated that any steps by the CCP leadership in this direction would have provoked an "intensified reaction [*obostrennuiu reaktsiiu*]" from the Soviet side—which was understandable, Kovalev added, given that the Cold War was at its peak. Such steps probably included contacts like those the CCP leadership established through Huang Hua with the American ambassador, Stuart, in Nanking between March and June 1949—which were presumably designed to obstruct relations between the Nationalist Government and Washington.[252] The distrustful Kovalev had advised against these contacts. The CCP leadership apparently made further secret exploratory probes without Moscow's knowledge of the United States and Great Britain—also for tactical purposes. The Soviet side at about the same time noticed with indignation that the CCP leadership had made no attemps to drive the U.S. Navy out of Tsingtao harbor. Kovalev felt this consideration for the United States was being "really too sensitive [*slishkom uzh delikatno*]." The Chinese comrades' behavior obviously irritated Moscow.[253]

On the whole, however, the CCP leaders behaved the way Stalin advised them to behave. In the spring and summer of 1949, the Party signaled to Washington and London through various channels that it was interested in trade, credits, and technical support.[254] On the other hand, Mao suggested that he was not interested in establishing diplomatic relations "with all countries," as he assumed that "the imperialists" at least would certainly not give China equal rights very quickly, which was the prerequisite.[255] In late April, the CCP leadership informed its military units that they could consider whether diplomatic relations should be established with the United States and England if they broke off relations with the KMT.[256]At the same time, Mao made the establishment of such relations conditional on the corresponding government severing relations with the remnant KMT forces and withdrawing its armed forces from China.[257] In mid-June, he formulated his conditions in a similar way.[258] The American ambassador, Stuart, for his part told Huang Hua that the United States would not consider offering diplomatic recognition to China as long as the Communists had not formed a "national government" and did not fulfill their international obligations (i.e., recognize the diplomatic status of embassy and consulate personnel).[259] There appeared to be no solution to the stalemate in sight.

The question naturally arises whether the CCP leadership behaved this way because Stalin advised it to do so—and if so, to what extent—or if it was acting of its own accord. It remains impossible to answer this unambiguously even now. On the basis of an assessment of Chinese archival documents, Chen Jian took the position that Ward's harsh treatment was designed, on the one hand, to create anti-American feeling in the general Chinese public. But it was also supposed to demonstrate to Stalin that Communist China stood securely on the Soviet Union's side and desired to collaborate strategically with it—because it saw the USSR as the leader of the international proletarian movement, on the one hand, and as "the only great power from which the new China could expect to receive strategic assistance," on the other.[260] This thesis implicitly assumes that the CCP leadership decided to take such harsh action in the Ward case after it had recognized that this was what Stalin or Kovalev wanted. If this is true—and obviously it is—the thesis would very probably also be valid with regard to Stalin's wish that the leadership not establish political or diplomatic relations with the Western powers in the short term. Chen Jian's argument seems plausible.

It is not so far-fetched to suppose, however, that the leadership vacillated on the question of whether and, if so, on what conditions it should seek to establish diplomatic relations with the United States—without offending the Soviet Union. It is possible to demonstrate on at least one occasion that China expressed its interest in doing just this to the American side without stating all the otherwise usual conditions Stalin required. It was in a message that Mao Zedong, Zhou Enlai, and Ye Jianying had Chen Mingshu, chairman of the Kuomintang Revolutionary Committee in Shanghai, deliver to Ambassador Stuart in early July.[261] In it, the establishment of diplomatic relations—this time not subject to any conditions—was named as a precondition for halting the anti-American propaganda, as demanded by Washington. In this case, one can assume with a particularly high degree of probability that this action by the CCP leadership had not been coordinated with Moscow.

In the view of Secretary of State Dean Acheson, the Ward case precluded any possibility of American recognition of Communist China.[262] If this and the arguments in the previous paragraph are correct, Soviet pressure, to which an essentially compliant CCP leadership yielded because it was dependent on close cooperation with Moscow, obstructed the reduction of tension between Washington and Communist China.[263] Chen Jian is convincing to the extent that he concludes that "strategic relations" between Communist China and the Soviet Union began with the Ward case.[264] Other evidence addressed in this section that has nothing to do with the Ward case also suggests that this hypothesis is correct.

Was Stalin Seeking to Divide China into Two Parts?

We can thus confidently assume that the stage had been set for a strategic partnership between Moscow and the CCP by the beginning of 1949. It is nonetheless easily conceivable that Stalin would have preferred to deal in the future with a China in two parts.[265] Truman was even convinced that the Soviet Union would have preferred a divided and weak Chinese neighbor to a united China under a dynamic Communist leadership.[266] The idea of a Communist North Chinese buffer state that was dependent on the USSR between the USSR and an American protectorate in a KMT-led South China would have been attractive to Moscow from the perspective of security. This state, could in turn, have been divided into two security zones: the first would include Manchuria, with an increased Soviet influence along the lines of a traditional sphere of influence; and the second, the area from the Great Wall to the Yangtze line. Tensions between these two Chinas would have put Moscow in a position to exploit them to its own advantage. A unified China under a Mao Zedong, who had repeatedly demonstrated his unwillingness to be subservient to Moscow, must have seemed less attractive in comparison.

It is a completely different question to ask whether Stalin could have attempted to make the CCP implement a model of this kind in fact. This would have meant dissuading it from crossing the Yangtze to move south. We must rule this out a priori. In August 1945, Stalin had already urged the CCP once not to pursue the civil war. Mao accused him of betraying the Chinese revolution. It is difficult to imagine Stalin doing the same thing yet again four years later, at a time when Mao was determined to conquer all of China and his troops' victorious advance toward the south could apparently no longer be halted. Such behavior would have involved risking a total rupture in relations with the new China.

An analysis of political interests thus provides an unambiguous result. It would seem superfluous to consider this hypothesis any further if such important contemporary witnesses as Mao Zedong and Zhou Enlai had not claimed that Stalin had advised the CCP to break off its advance at the Yangtze. This is disputed in the literature, and not least in that of the People's Republic of China.

As early as the spring of 1949, Mao said: "Some friends abroad half believe and half disbelieve in our victory. [They are] urging us to stop here and make the Yangtze River a border with Chiang, to create 'a Northern and a Southern dynasty.'"[267]

In 1957, Mao was somewhat more specific:

> Until 1949, when we already had the crossing of the Yangtze in sight, there were people who wanted to keep us from it and said that we could not under any circumstances cross the Yangtze, as if we did we would cause the Americans to send troops and "a Northern and a Southern Dynasty [*beichao, nanchao*]" might be established in China. We did not listen to them. We crossed the Yangtze, the Americans did not send any troops and no "Northern and Southern Dynasties" were established in China. A division of this kind would certainly have taken place if we had listened to these warnings. Our comrades later met with the man who wanted to restrain us from crossing the Yangtze, and the first thing that he said was: "Victors do not have to accept any rebuke [*shenglizhe shi bu yinggai shou zebei de*]."[268]

Mao did not claim that the advice came from Moscow. He did not state when or in what circumstances "people" gave it. The quote he gave at the end does, however, suggest that he meant Stalin by "the man." According to an eyewitness report, when Stalin first met Mao and his entourage on December 16, 1949, in Moscow, he said, "One does not judge victors. One cannot condemn victors [*bu neng qianze shenglizhe*]. This is a general principle."[269] Stalin had allegedly already said something similar to Liu Shaoqi in July 1949.[270]

In 1955, when Zhou Enlai instructed the ambassador-designate to the Soviet Union, Liu Xiao, in his future duties, he said the following about Soviet policy toward China in early 1949:

> At that time, the political and military situation was very good. We were just preparing to move across the Yangtze and liberate all of China. The Soviet Union had its own views on this. It demanded that we halt the civil war, which in effect would mean that they encouraged "a Northern and a Southern Dynasty [*bei nan chao*]," two Chinas. . . . At that time, our assessment of the international situation and of our ability to liberate all of China was different from that of the Soviet Union.[271]

Some authors from the People's Republic of China state that it was Mikoyan who, on instructions from Stalin, advised the CCP leadership to halt the advance at the Yangtze and leave South China to Chiang Kai-shek during his visit to Xibaipo in early 1949.[272] Other writers in the People's Republic of China and some Western historians assume that it was Stalin or the Soviet leadership, if not necessarily Mikoyan, who gave the advice.[273] Various motives are assumed to explain this alleged Soviet action.[274] Stalin supposedly did not believe the PLA to be sufficiently strong; he was not interested in a strong China under a united leadership; or he feared that the United States would intervene militarily if the PLA's victorious advance continued and that the USSR might be drawn into a war with the United States in this way.

The disputed thesis cannot rely on any statements by contemporary witnesses to the extent that it involves transmission by Mikoyan. Mao and Zhou did not mention Mikoyan. Kovalev, who was asked about this explicitly, did not remember that Mikoyan had given any such advice.[275] Shi Zhe denied that Stalin wanted any such thing and even claimed that Mikoyan expressly encouraged the Chinese comrades in Xibaipo to "free all China completely."[276] It is striking that Shi Zhe and Kovalev, who took part in talks with Mikoyan, both agreed on this point although they often opposed one another sharply on other subjects. In his report on the talks in Xibaipo, Mikoyan himself did not mention that the question played any role at all.[277] It is easier to deduce the opposite from the report. Mikoyan urged Mao to form a government as quickly as possible—"let us say, after the capture of Nanking or Shanghai"—and thus tacitly assumed that he would cross the Yangtze. As far as they have already been examined, the other documents in the Russian presidential archives do not contain any evidence that could be used to support the disputed theory, either.[278] Authors from the People's Republic of China who base their work on documents from the Party archives in Peking and interviews with contemporary witnesses take the position that neither Mao nor any other member of the Chinese Politburo has claimed that Mikoyan ever gave such advice.[279]

To the extent that Mao actually claimed that the Soviet leadership had advised him to halt his advance at the Yangtze, it would be easy to defend the position that he contradicted himself. In March 1956 and July 1958, he outlined the mistakes in Stalin's policy toward China from the 1920s to the Soviet ambassador, Iudin, in a long list of sins.[280] The list includes the accusation that he had sought to prohibit the revolution to the Chinese comrades in 1945. Mao also criticized Mikoyan for appearing arrogant in Xibaipo. But he said nothing to claim that Stalin or Mikoyan advised him not to cross the Yangtze.

A historian at the Research Institute for Party History (*dang shi yanjiushi*) of the Central Committee of the CCP stated tersely that the Chinese who claim that the Soviet Union had advised against the crossing of the Yangtze could refer only to Mao Zedong and to no one else. Despite an energetic search, he had found no other evidence in the Party archives.[281] Mao's personal physician and confidant also only referred to his conversations with Mao when he observed that Stalin had urged the Chinese Communists to halt their advance north of the Yangtze and grant control over the south to the KMT.[282] The same is true for Mao's bodyguard, Wang Dongxing, who notes that Mao told him in late December 1949 in Moscow that Stalin had insisted under the influence of Yalta that the PLA should not cross the Yangtze and should first take it as a border and negotiate with the KMT.[283]

It can, on the other hand, be demonstrated that Stalin advocated crossing the Yangtze to Mao Zedong in April 1949. In a telegram that he sent Mao at that time, Stalin advocated a "unified China under the leadership of the Communists."[284] This message cannot, however, be taken as conclusive evidence that Mao's accusations against Stalin were not true for that time, at least, as it cannot be established whether the telegram was sent before or after the Yangtze was crossed (April 20). However, an exchange of telegrams between Stalin and Mao Zedong between January and April 1949 cannot serve as evidence that Stalin did not advise Mao against crossing the Yangtze. There is only one reference to this question by Mao in the telegrams that have been published in full or in part up to now.[285] Mao informed Stalin in mid-January 1949, a full two weeks before Mikoyan arrived in

Xibaipo, of his plans to cross the Yangtze "already that summer."[286] Stalin, who was asked by Mao to say something if he did not approve Mao's statements, did not indicate any objections in his reply on this point.[287] He did not address this topic in subsequent telegrams—to the extent that they have been published—either. Nonetheless, the Russian historian Sergei Tikhvinskii is not accurate in claiming that the exchange of telegrams he published between Stalin and Mao between January 10 and 15 contradicts the thesis presented in several American books on the history of Sino-Soviet relations that the Soviet Union had advised the CCP to halt at the Yangtze and seek an agreement with the KMT about dividing the country.[288] Tikhvinskii would only have been correct if he had stated that Stalin did not provide such advice in the telegrams he presented.

The thesis that the KMT was pursuing a plan to divide China along the Yangtze line appears to be accurate. Shi Zhe indicated that a KMT representative made such a proposal during negotiations in Peiping in April 1949. The representative, Shi said, also raised the question of "a Southern and a Northern Dynasty [*nan bei chao*]" in order to halt the PLA's advance at the Yangtze and divide China along the river. Mao rejected this proposal, Shi added, and declared that the PLA would in any case cross the Yangtze.[289]

It is possible that Shi Zhe was referring to a conversation that Mao had with Zhang Zhizhong on April 8. Zhang had come to Peiping as head of a KMT delegation sent by Li Zongren and expressed his hope to Mao that the bloody battles would end at the Yangtze. He said that he hoped "both sides, KMT and CCP, would restrain their own troops [*guo gong shuang fang yueshu ziji de budui*]" and that they would draw the border along the river (*hua jiang er zhi*). The American ambassador, John Leighton Stuart, advised his government to support Li Zongren in his attempt to convince the CCP to divide the country along the Yangtze.[290] Mao rejected Zhang's proposal and commented that the revolution had to be played out to the end. At best, the two delegations could agree on a deadline at which the PLA would cross the Yangtze.[291]

It is also true that shortly before the PLA crossed the Yangtze, there was widespread hope in KMT circles that Stalin would stop the Chinese Communists. As he was negotiating with the Communist Party leadership in Peiping in April, Zhang Zhizhong was quoted as saying that "if the CP has resolved to fight on, it is not because of their relations to the Soviet Union. The Soviet Union only advised them to halt the advance of the troops at the Yangtze."[292] In early April 1949, Wu Tiecheng, general secretary of the KMT, confided to an American diplomat in Canton that he and other KMT leaders believed the Soviets would hold the Chinese Communists back from crossing the Yangtze, if they really controlled the CCP. Moscow did not want to risk any conflict with the Western powers, which were about to come together to form the North Atlantic Treaty Organization (NATO). A conflict of this kind could result from possible Chinese Communist action in Southeast Asia.[293]

At first glance, statements by a historian from the People's Republic of China appear to support the thesis that Moscow had sought to restrain the PLA from crossing the Yangtze. He claims that Ambassador Roshchin had reached an agreement with President Li Zongren concerning three conditions for Soviet mediation, despite the Moscow declaration—reached under pressure from Mao Zedong—which stated that the USSR would not participate as a mediator in peace negotiations between the CCP and the KMT.

He had thereby provided support for Li's attempt "to set the boundary at the river [*hua jiang er zhi*]" and thus establish "a Northern and a Southern Dynasty [*bei nan chao*]." And yet the idea that Roshchin pursued a line that differed from Moscow's should be ruled out. Even if it were possible to find sources to provide evidence of Roshchin's alleged behavior, the resulting conclusion with respect to the *bei nan chao* question would go way beyond its target.[294]

The hitherto published contents of a "long telegram" sent by Stalin to Mao in April before the Communists crossed the Yangtze cannot support the disputed theory, either. In it, Stalin commented on the "military activities after crossing the Yangtze." He warned the CCP not to attack imperialism "in order to protect the islands and Asia." This could lead to anything from a blockade to military conflict. The danger that English or American troops would land would be very great after an advance by the main forces of the PLA toward the south. Stalin therefore suggested that they not rush into anything and prepare the PLA advance to the borders of the neighboring states carefully. Nothing in this would suggest, however, that Stalin sought to keep the Chinese comrades from crossing the river in this telegram. On the contrary, the passages that have become known give the impression that Stalin saw the advance across the river as an established fact that had to be taken into account in making further plans.[295]

In conclusion, it clearly would have been a reasonable security calculation for Stalin to have been interested in dividing China. It does not appear plausible, however, that he asked the CCP leadership to participate in the realization of this plan and not to cross the Yangtze. Furthermore, it cannot be proved by the available sources. If Mao nonetheless became convinced that Stalin was seeking to divide China at the Yangtze,[296] it was not on the basis of a corresponding request from the Soviets but rather of an interpretation or misinterpretation of Soviet statements or behavior that continues to remain unknown to us. One could perhaps imagine that the CCP leadership heard of the secret talks between Ambassador Roshchin and the Nationalist Government about the possibility of a cease-fire at the Yangtze and thus think of at least a preliminary division of the country.

The Second Top-Level Secret Meeting: Liu Shaoqi in Moscow (Summer 1949)

On June 26, 1949, a Party delegation led by Liu Shaoqi arrived in Moscow.

Background and Goals

As has already been shown in detail above,[297] Mao Zedong had been pressing for a summit meeting in Moscow since 1947, but the meeting had repeatedly been postponed to a later date. In late January 1949, Stalin had instead sent Mikoyan on a secret mission to China to a first top-level meeting, which took place in Xibaipo, the current seat of the CCP leadership.[298] On this occasion, the Chinese leadership had asked Mikoyan for economic and military assistance on credit and indicated its intention to send a special delegation to Moscow "after some time" to conclude the relevant treaties and also to discuss other current questions.

Following the Second Plenary Session of the Central Committee and the transfer of the Central Committee apparatus to Peiping in March 1949, and as the Party began to discuss the issue of establishing the People's Republic of China, it appeared increasingly urgent to the CCP leadership to hold another top-level discussion with the Soviet Party leadership to reach a consensus on the next steps and conclude concrete agreements on Soviet aid. This was even more the case when the PLA crossed the Yangtze on April 20 and appeared to be on the verge of achieving victory in the whole country. Mao wanted to lead the Chinese delegation himself and asked Kovalev in late April to discuss a timetable for the visit with Stalin. Kovalev claimed that Mao was very much worried about it, because he was concerned about his prestige and was afraid of the meeting with Stalin. Stalin replied: "You should not hurry to travel to Moscow. You cannot leave China and give up the leadership of events [*rukovodstvo delami*] in connection with the complicated character of the situation in the south and in connection with the fact that China essentially has no government, and this involves a certain degree of danger for the cause of the revolution."[299]

Stalin also gave advice in his telegram about establishing the future Chinese government. After Mao had read it, according to Kovalev, he jumped up from his chair, threw his arms in the air, and shouted out three times: "Stalin ten thousand years!" Kovalev suspects that he was so happy because he had read into the telegram that Moscow now recognized him as the leader of the Chinese revolution.

It is open to doubt whether this interpretation is correct. Mao Zedong's reaction could also be interpreted as irony. Given the fact that Stalin had rejected Mao's attempts to arrange a meeting in Moscow so often and at times with such flimsy excuses, it must actually have hurt Mao more that Stalin refused again and had at the same time yet again admonished him condescendingly that he really could not leave China at that time—as if he were denying that Mao was able to know best when he had to be in China and when not. The real reasons for Stalin's refusal to receive Mao in Moscow, which have already been presented in detail,[300] remained valid.

Months later Zhou Enlai confirmed to Ambassador Roshchin that Mao had himself wanted to travel to Moscow during the summer at the head of the delegation that was led by Liu Shaoqi. In doing so, Zhou, obviously in order to save face, passed over the fact that Stalin had refused Mao and at the same time acted as if the reasons Stalin had given for refusing Mao were actually valid and were the result of the CCP leadership's own deliberations. He said that at that time the Party leadership had only just come to Peiping and the government had not yet been established. They had a mountain of work in front of them that could not be done without Mao, "who was staying in Peking for that reason."[301]

Apparently the decision to send a delegation to Moscow under the leadership of Liu Shaoqi was already taken in May. Stalin gave his approval.[302] In sending Liu's delegation, the Chinese Party leadership obviously intended to remind the Soviet comrades of the Chinese revolution's peculiarities and its significance for the world revolution. It was to bind Moscow more clearly than before to the China represented by the CCP, gain assurances of diplomatic recognition for the People's Republic of China, which was about to be established, prepare Mao Zedong's first official visit to Moscow, ask the Soviet Union for "all-around assistance and support [*ge fangmian de zhichi he yuanzhu*]" for the new China, and support the Party's efforts to obtain greater political and moral support internationally.[303]

The CCP leadership was concerned not only because the Soviet ambassador, Roshchin, was the only head of a mission to move from Nanking to Canton at the beginning of the year, but also because Moscow continued to negotiate with the KMT government about extending the contract for a joint airline company even after the capture of Nanking by Communist troops. An author from the People's Republic of China describes the Soviet Union's behavior toward the KMT and CCP at that time as a "'tightrope walking' policy ["*zou gangsi*" *zhengce*]." Liu Shaoqi's trip was designed to clarify these questions.[304]

According to Chinese sources, Zhou Enlai had already asked Mikoyan for a credit of US$400 million for economic and technical aid in early February 1949, and was told by him that he thought US$300 million was possible.[305] Soviet archival documents state that Mao Zedong requested US$300 million, but that Mikoyan did not make any firm commitment.[306] Chinese Party leaders had also already stated concrete requests for economic and military aid and the appointment of Soviet specialists to Mikoyan, who had promised that some of them would be met.[307] Mao Zedong again asked Stalin about a credit in April and received an evasive answer.[308] In early May, Mao yet again asked Stalin to send out experts to assist in economic development.[309] Mao explained his request by saying that the Chinese Communists could complete the military task—the final annihilation of the enemy—on their own. But "we very much need your help [*hen xuyao nimen de bangzhu*]" to complete the second task, the development of the economy, which, he said, in the long run was even more important.

The trip of the delegation led by Liu Shaoqi was kept secret—so secret that Western researchers heard of it only in the late 1980s.[310] The undertaking, as Zhou Enlai put it, "took place under the cover of Gao Gang's Manchurian mission."[311] Zhou's comment referred to a report in *Pravda* on July 31, that a delegation "of Manchurian People's Democratic authorities led by Mr. Gao Gang" had visited Moscow to hold talks on trade questions; a one-year trade agreement had been concluded between Manchuria and the USSR.[312] The delegation was reported to have begun the return trip to China on July 30. The KMT protested this on the grounds that the conclusion of the agreement involved a violation of the Sino-Soviet treaty of August 14, 1945.[313] Foreign Minister Vyshinskii drafted a Soviet response to contest this, but it was decided instead not to answer the Chinese note immediately but to await a Chinese request.[314]

The issue of camouflaging the delegation arose immediately after its arrival when Liu Shaoqi made the request that his delegation be allowed to familiarize itself with the organization and activities of the Soviet state apparatus. Stalin feared a loss of secrecy. He clearly did not see how he could guarantee the secret character of the Chinese mission, which was extremely important to him, and disguise its true purpose, if state institutions were also involved. He therefore proposed "legalizing [*legalizovat'*]" the delegation. "If you agree, we will publish a report in the press to the effect that a trade delegation under the leadership of comrade Gao Gang has arrived in Moscow and then you will have a broad opportunity to see everything. . . ." The Chinese side agreed, as evidenced by the aforementioned publication in *Pravda* on July 31. The fact that the Manchurian People's Government was established and Gao Gang elected as its chairman only four days earlier[315] may have been connected with the need to "legalize" Liu Shaoqi's delegation.[316]

Formally, the Soviet Foreign Ministry had nothing to do with the Chinese delegation's visit. Foreign Minister Vyshinskii, who was not a member of the Politburo, apparently did not participate in any of the working sessions attended by Stalin and made a special appearance only on Stalin's orders, during which he instructed the Chinese delegation at its own request about the structure and activities of the Foreign Ministry. Secrecy was so strict that no one from the Foreign Ministry other than Vyshinskii knew anything about the Chinese visit. Even the Far Eastern Division, which was responsible for China, and the Soviet embassy in Nanking knew nothing. Shi Zhe was the sole interpreter at the talks.[317]

It is not known whether the CCP leadership had insisted before the trip on allowing the visit to take place openly—which was probably important to it.[318] It is also unknown whether Mao Zedong had sought to convince Moscow to break off diplomatic relations with the KMT government and establish relations with Communist China. If this was the case, it can be assumed that Stalin took the position that the Soviet Union could establish official relations with the Chinese Communists at state level only after they established a Communist-led Chinese government—a position that the U.S. ambassador, Stuart, had also taken in a talk with Huang Hua with respect to relations between the United States and the Communist regime in China.[319]

After receiving extensive instructions from Mao Zedong, the delegation led by Liu Shaoqi traveled on June 21 by train from Peiping via Mukden to Dairen, from there in Soviet military aircraft through North Korean airspace to Khabarovsk, and then by stages across Siberia to Moscow, where it remained from June 26 until August 14.[320] It used the railroad on the return trip. It arrived in Mukden on August 25, departed again on August 28, and reached Peiping on August 29 or 30. Among others, the delegation included Gao Gang and Xu Jiefan, who joined in Mukden, along with Wang Jiaxiang, Deng Liqun, Ge Baoquan, Huang Weiwen, Wu Zhenying, and Shi Zhe as interpreter. Ivan Kovalev also joined the group at Mao Zedong's request, as did an attaché of the Soviet embassy. Gao Gang returned to Manchuria on July 30 because of differences with Liu Shaoqi.[321] Deng Liqun flew from Moscow to Kulja (Yining) in mid-August to prepare the conquest of Sinkiang by the PLA in collaboration with the Soviet consulate general.[322] Wang Jiaxiang remained in Moscow after the departure of the delegation as liaison between the leaderships of the two Parties[323] and was nominated as the first ambassador of the People's Republic of China to the USSR on October 20.[324]

Liu was the second-highest-ranking person in the Party at that time. He was followed in rank by Gao Gang, who was a member of the Politburo of the Central Committee and of the Northeast Bureau of the Central Committee—the "uncrowned King of Manchuria," as he was often called then.[325] On his arrival in Moscow, he was not yet, as a Russian author assumes,[326] the head of the Manchurian government, because the government there was established only one month later. Wang Jiaxiang, who had studied at Moscow's Sun Yat-sen University during the 1920s, was a member of the Central Committee.[327] The other members of the delegation were staff.

The following attempt to reconstruct the course and results of Liu's mission to Moscow is based on the aforementioned documents from the Russian presidential archives, information that has recently been published in the People's Republic of China, and statements by two contemporary witnesses, Shi Zhe and Ivan Kovalev, the latter basing his account on his own records. Kovalev noted that a comparison of his own and Shi Zhe's notes of the July 27 session at the time had shown that they did not always agree: "We heard the same thing, but noted down something different."[328] The published versions by the two contemporary witnesses do in fact differ on many points, but Kovalev generally appears more credible.[329]

Given the sources, which still remain unsatisfactory, it is not possible to reconstruct the Liu mission completely. It is not even possible to determine beyond all doubt how many working sessions Stalin attended. In one place, Shi Zhe mentioned "four to five," and in another place six meetings, each lasting four to five hours, only two of which he dated;[330] another Chinese author with access to the CCP archives gave the number as five.[331] Only four meetings can be dated: the ones on June 27, July 11, July 27, and August 5.[332] It can only be shown that Stalin received Liu twice in his office in the Kremlin: on June 27 and August 5.[333] We have minutes from only the first of these sessions and have to reconstruct the other three from information provided by Shi Zhe and Kovalev. There is considerable evidence that only the four meetings that can be dated took place as actual working sessions with Stalin. In this case, the difference with the number mentioned in the Chinese sources may be that the other "sessions" were in reality the receptions and banquets to which Stalin invited the delegation that are mentioned in several publications. Between the sessions, and in particular in the period from July 15 through 26, Liu Shaoqi, among other things, held talks with Foreign Minister Vyshinkii, the heads of the State Bank (Gosbank), the State Planning Committee (Gosplan), the ministries of Finance, Foreign Trade, and Chemistry, and the Moscow Municipal Soviet.[334] He also toured large companies to obtain information on business administration and the role of the Party and labor union organizations there, and visited collective farms, the university, museums, and exhibits.

When Liu's delegation arrived in Moscow on June 26, it was housed in a Soviet Communist Party guesthouse, a luxurious villa from the czarist period.[335] Apparently in order to create a favorable atmosphere for the talks, Mao Zedong published what was up to that time the most decisive public declaration in favor of an alliance with the USSR shortly after the Chinese delegation arrived.[336] Stalin recognized this step shortly afterward by having the declaration printed in *Pravda*.

Shi Zhe remembered that Stalin received the members of the delegation, after they had rested a bit, in his dacha in Kuntsevo for a welcoming banquet lasting almost four hours. The Politburo members Kliment Voroshilov, Molotov, Georgii Malenkov, Bulganin, Beriia, Lazar Kaganovich, and Mikoyan also attended. This had been an opportunity for informal conversation, but some important political questions were also raised.[337]

The First Session: Questions of Soviet Aid

On June 27, one day after their arrival, the delegation was received at about 11 P.M. for a one-hour talk in Stalin's office in the Kremlin.[338] In addition to Stalin, the Politburo members Molotov (vice-chairman of the Council of Ministers), Malenkov (secretary of the Central Committee and vice-chairman of the Council of Ministers), Mikoyan (minister of foreign trade), and Ivan Kovalev attended on the Soviet side. On the Chinese side, Liu Shaoqi, Gao Gang, Wang Jiaxiang, and Shi Zhe (as interpreter) were present.[339] Stalin began by inquiring about Mao Zedong's health. Liu Shaoqi thanked him and gave Stalin a letter from Mao in which he expressed his gratitude for the "enormous help" the Soviet Union had given China and asked him to receive the Chinese delegation.[340]

It is striking that according to Kovalev's record Stalin was practically the only one to speak during the substantive portion of the talk that followed. His statements sound like responses to Chinese requests, as is noted explicitly in three cases. It is unclear, however, whether these Chinese requests had already been presented before the talk—for example, in the form of a written wish list that is still unknown to us—or Liu presented them in the talk. The former is more probable, for Kovalev would have had to take note of the fact if the requests had been expressed during the talk. Liu did not present any requests or questions according to Kovalev's record, however, and only made a few observations that will be described below.

The talk dealt almost exclusively with the future aid the Soviet Union would provide Communist China: a goods credit, Soviet specialists, and aid in the military sphere. The situation in Sinkiang was also addressed.

Stalin opened the substantive talks with the statement that the Central Committee of the CPSU(B) had resolved to grant the Central Committee of the CCP a credit of US$300 million. This was the first time in history, he continued, that a contract of this type had been concluded between two Parties. The credit was subject to interest at the rate of 1 percent per annum and was to be made available to China in equal parts of US$60 million each over five years in the form of equipment, machines, materials, and goods. The credit was to be repaid over ten years beginning after it had been paid out fully. Stalin pointed out that Mao had informed him by telegraph that 1 percent was very little for such a credit; that it should be more. The Soviet Union had granted the "Western democracies"[341] credits at 2 percent interest, Stalin explained. It was now asking China for 1 percent, because, unlike in the Western democracies, where there was no war and the economy was already established, in China there was war and the destruction continued, and so China needed greater aid on more favorable terms. Then Stalin laughed and joked, "Now, if you insist on paying higher interest, that is your affair, we can also take higher interest."[342] The protocol continues:

> On the question of *signing the credit agreement*, comrade Stalin said there were two alternatives: one, the agreement could be signed by authorized representatives of the Central Committee of the CPSU(B) and of the CC of the CCP, and two, it could be signed by authorized representatives of the Soviet government and of the government of Manchuria that already existed and then to formalize [*oformit'*] the agreement later when an all-Chinese democratic coalition government was established in contracts between the governments of the Soviet Union and China.[343]

If the first of the alternatives Stalin had suggested was taken, the agreement had to remain secret in order not to impair diplomatic relations between the USSR and the Chinese Nationalist Government. If the second alternative was taken, publication of the agreement could be considered because a Manchurian Communist government actually existed and the Western powers considered Manchuria to belong to the Soviet sphere of interest in the Yalta agreement and the Sino-Soviet Treaty of August 14, 1945.

Ledovskii claims that they decided to employ the "Manchurian alternative" for this agreement, which the Soviet and Chinese sources sometimes refer to as a credit and sometimes as a trade agreement.[344] This is formally true. Chinese references, however, paint a more complex picture. They suggest that a committee was established to prepare the credit agreement.[345] Its members included Mikoyan, Kovalev, Liu Shaoqi, Wang Jiaxiang, and Gao Gang. On July 25, Liu demanded from Stalin a copy of the "commercial treaty" "that had been promised during a personal meeting with you."[346] The text was made available two days later at the latest and examined by the Party leadership in Peiping.[347] Stalin proposed that Gao Gang sign the contract as representative of the Manchurian government, which Liu Shaoqi rejected.[348] Liu and Malenkov signed the agreement on July 30 in the Kremlin.[349] Stalin later confirmed that the agreement had been concluded at the Party level.[350] The Soviet press reported on July 31 that an "agreement for the exchange of goods" between the USSR and Manchuria had been signed.[351] Internally, the agreement was labeled the "Soviet-Chinese Trade Agreement."[352]

On June 27, 1949, Stalin then raised the question of the Soviet specialists, which Mao had previously requested in his correspondence with Stalin and discussed with Mikoyan at the beginning of the year in Xibaipo. The Soviet side would provide specialists and was ready to send the first group requested by the Chinese soon, Stalin said. "But we must come to an agreement on the conditions of the specialists' salaries. We believe that the payment and possibly also the food, if you provide that to your own specialists, must be at the highest level of salaries provided for your specialists, not lower but also not higher." Because the Soviet specialists received high salaries, Stalin added, they could if necessary receive something additional at the Soviet state's expense.[353]

Stalin asked his Chinese interlocutors to let him know if any individual Soviet specialist behaved incorrectly; after all, it was said that there was a monster in every family and that one person among good people could turn out to be bad. Poor behavior would heap disgrace on the Soviet Union, however; therefore "we take steps involving prevention, education, and if necessary punishment"—the Soviet side would not tolerate Soviet specialists who "looked down on the Chinese specialists and Chinese people and behaved in a condescending manner." Liu Shaoqi replied to this that there were foreign specialists in China who had nothing to do with the imperialists' activities and who drew salaries that were much higher than those paid to the Chinese specialists. Stalin said that the Soviet State had opinions and ways of doing things that were different from those of the capitalist countries "and we wish to hold ourselves to them."[354]

Stalin stated, "We have *selected 15 specialists* [*for Shanghai*] *and can send them at your request at any time.*" The Chinese side should consider that there were many Chinese specialists and qualified workers in big cities—Shanghai in particular—who would be able to provide not less but more assistance than the Soviet specialists. They should therefore be involved in active work.[355]

Stalin then declared that he was prepared to assist in clearing mines from the waters around Shanghai and to send out specialists and minesweepers for the purpose. To maintain secrecy, he proposed to do this via Manchuria. The USSR could, for example, "sell Manchuria some minesweepers and teach Chinese sailors how to clear mines in Dairen, Port Arthur, and Vladivostok, and the Manchurian government could then 'sell' them to the Central Government," he said, laughing.

Stalin then brought the discussion to Sinkiang Province. "The conquest of Sinkiang should not be delayed, as a delay could involve intervention by the English in Sinkiang's affairs. They could mobilize the Muslims, among them the Indian ones, to continue the civil war against the Communists, which would not be desirable, as there are great oil and cotton reserves in Sinkiang that China urgently needs." Stalin pointed out the small proportion of Han Chinese and recommended that the proportion of Han be increased through resettlement from its current level of 5 percent to 30 percent. This was necessary to promote the "general appropriation" and would reinforce the protection of China's borders. "In general, all border regions should be settled with Chinese in order to increase the protection of China."

Stalin then turned to the troops of the KMT general Ma Bufang, who was in charge of the neighboring province of Qinghai at that time and would have been able to block access to Sinkiang from there. In Stalin's opinion, the CCP overestimated the strength of these troops, as Ma Bufang's cavalry could easily be beaten by artillery. "If you wish, we will give you 40 fighter aircraft, which will help you to beat and disperse this cavalry quickly."

China, Stalin continued, had no fleet of its own. But had it not, as it appeared, already seized a few ships from the KMT? China had to have a fleet of its own. "We are prepared to assist you in creating a fleet. We could, for example, salvage sunken trading ships now and help you to repair them. With reference to your request for reinforcements for the defense of Tsingtao, we could send our squadron on an official visit in Tsingtao harbor, once an all-Chinese government has been established." In referring to Tsingtao, Stalin was probably reacting to a request from the CCP for support in the defense of the harbor against a possible attack from a strong U.S. naval contingent that was cruising in the vicinity. If this assumption is correct, Liu could not have been very satisfied with a visit by the fleet.[356]

Liu Shaoqi then thanked Stalin for all the "enormous assistance" that was currently being given to China in all areas of life. The assistance was granted on the most favorable conditions ever seen in history. The CCP had instructed all Party organizations to create working conditions for the Soviet specialists that could not offend any of them. Stalin replied, "*We are also formulating instructions for the Soviet specialists so that they do not offend the Chinese specialists, either.*"[357]

Stalin agreed to satisfy the Chinese request to establish "*a flight connection between Moscow and Peiping*" and said this could be done immediately. He promised the Chinese that the USSR would build an aircraft assembly and repair shop. "*We can give you at your choice Czech or Russian fighter aircraft of the most modern* type"[358] for use in training flight crews, he said. Stalin accepted the request "*to listen to*" a Chinese report on the military, political, and economic situation in China "*in the Politburo*" and to debate a series of important questions as soon as this report was ready, "in three of four days."

Stalin assured them, "*We are prepared to assist you comprehensively in studying the state apparatus, industry, and anything you wish,*" but to do this it was necessary to "*legalize you* [*Vas legalizovat'*]." He then made the proposal mentioned previously that the Chinese guests be described in a press release as a trade delegation from Manchuria under the leadership of Gao Gang—which, as has also already been mentioned, was done in this way.

At the end of his account, Kovalev noted that the Chinese comrades asked "to be given an opportunity to reply after they had consulted with comrade Mao Zedong."

Yi bian dao: *Mao's Definitive Profession of Loyalty to the Soviet Union*

Four days later, on July 1, the twenty-eighth anniversary of the founding of the CCP, Mao Zedong published an article with the title "On the People's Democratic Dictatorship."[359] He presented his most definitive and most extensively substantiated public profession of loyalty to the Soviet Union to that time. Mao argued passionately in favor of the concept of "leaning to one side [*yi bian dao*]": to the side of the Soviet Union and the socialist camp it led.

Publication of the article had presumably not been timed at random. Mao was clearly interested in using the oath of allegiance to Moscow it contained and the call for a tight alliance between the two states to put Stalin in a positive mood for the meeting with Liu Shaoqi. He probably also intended to increase Stalin's willingness to provide support—not least material—for the future Communist China.

Mao had been stating his allegiance to the "anti-imperialist camp" led by the Soviet Union since early 1948, and in November of that year had rejected the idea that China had to follow a "middle road [*zhong jian luxian*]," or "third road [*di san tiao daolu*]," between "imperialism" and the Soviet Union.[360] In April 1949, Mao again expressed his position rejecting the concept of neutrality toward Washington and Moscow in a discussion with Zhang Zhizhong, who at that time was attending peace negotiations in Peiping at the head of a KMT delegation and who later cooperated with the CCP. Zhang regarded the United States and the USSR as equally important to China, advocated the policy of holding back equally from both states, and argued in favor of "requesting assistance also from England and America." Mao rejected this and pointed out that two great blocs had emerged in the world following the Second World War: one capitalist, monopolistic, and aggressive bloc led by the United States, and the other a socialist bloc led by the Soviet Union. He then said, "We can only lean toward [*dao*] the bloc led by the Soviet Union, we cannot lean toward the bloc led by the USA. . . . There is no third road [*di san tiao lu*]. . . . We combat also the illusion of a third road [*di san tiao lu de huanxiang*]. . . . Internationally, we belong to the side led by the Soviet Union and can only receive real assistance from the Soviet side."[361]

Mao mentioned to Zhang that he was writing an essay at that time in which he discussed Zhang's concept of a third road, which others also shared. The publisher of the record of the discussion suggests that he was referring here to the article "On the People's Democratic Dictatorship" of July 1, 1949.[362] Regardless of whether Mao really referred Zhang Zhizhong to an article he was still preparing or inserted a fictitious event when he "revised" the minutes of the talk in order to show continuity and consistency, the July 1 article does in fact contain a sharper analysis of the views Zhang represented about a third road. On June 22, Mao had already made a strong protest directed at Washington.[363] It appeared to indicate Mao's intention to intensify the conflict with Washington and to that extent fits the picture.

In his July 1 article, Mao observed:

> "You are leaning to one side." Exactly. The forty years' experience of Sun Yat-sen and the twenty-eight years' experience of the Communist Party have taught us to lean to one side [*yi bian dao*], and we are firmly convinced that in order to win victory and consolidate it we must lean to one side. In the light of the experience accumulated in these forty years, all Chinese without exception must lean either to the side of imperialism or to the side of socialism. Sitting on the fence will not do, nor is there a third road. We oppose the Chiang Kai-shek reactionaries who lean to the side of imperialism, and we also oppose the illusions about a third road [*di san tiao lu*]. *Not China alone, but the whole world leans without exception toward socialism, if it does not lean toward imperialism. Neutrality is hypocrisy, there is no third road. . . .*
>
> "Victory is possible even without international help." This is a mistaken idea. In the epoch in which imperialism exists it is impossible for a genuine people's revolution to win victory in any country without various forms of help from the international revolutionary forces, and even if victory were won, it could not be consolidated. . . . If the Soviet Union had not existed, . . . the international reactionary forces bearing down upon us would certainly be many times greater than now. In such circumstances, could we have won victory? Obviously not. And even with victory, there would be no consolidation. . . .
>
> "We need help from the British and U.S. governments." This, too, is a naive idea in these times. Would the present rulers of Britain and the United States, who are imperialists, help a people's state? Why do these countries do business with us and, supposing they might be willing to lend us money on terms of mutual benefit in the future, why would they do so? Because their capitalists want to make money . . . —it is not a matter of helping the Chinese people. . . . Only once in his whole life did Sun Yat-sen receive foreign help, and that was Soviet help. . . . Internationally we belong to the side of the anti-imperialist front headed by the Soviet Union, and so we can turn only to this side for genuine and friendly help, not to the side of the imperialist front.[364]

Mao Zedong was apparently pursuing three main goals with his article. First of all, he wanted to influence future alliance partners in China, such as the left wing of the KMT and the Democratic League, which, like Zhang Zhizhong, supported a neutral Chinese position between the Soviet camp and the leading Western powers.[365] Second, he wished again—and particularly strikingly—to convince Stalin that he was securely on the side of the USSR and that he was far from sharing Tito's unreliability. Kovalev confirmed that Mao was making great efforts at that time to dispel Stalin's distrust.[366] And last, as mentioned previously, Mao wanted to place the Soviet leadership under an obligation to provide the aid he sought.

He was able to achieve the last of these goals. Regardless of whether Stalin was particularly excited about the publication of Mao's article, *Pravda* published it a few days later.[367] In doing so, the Soviet leadership demonstrated two things. On the one hand, it finally identified itself in an open public forum politically and ideologically with its Chinese fraternal party after its hesitant behavior of previous months and thus jumped over its own shadow, which arose from its existing diplomatic relations with the Nationalist Government. On the other, by reprinting the article, the Soviet leadership implicitly indicated its willingness to provide the assistance that Mao had requested for Communist China.

In addition to the three main reasons mentioned, two other motives were clearly operating that can be derived from these reasons.[368] First, there is evidence that Mao still—or even increasingly—feared that the United States might intervene in the Chinese civil war on the side of the KMT, or at least blockade Chinese harbors completely, and wished to protect himself by making a recognizable connection with the USSR.[369] Mao expressed the latter fear in a public speech in mid-June.[370] Deng Xiaoping established a causal connection between the blockade and the *yi bian dao* slogan in mid-July, when he cited Mao's argument that they were choosing to lean on their own initiative in order to avoid being forced to lean in a particular way in future (*zhe yang shi zhudong de dao, miande jianglai beidong de dao*).[371] It is striking that Deng—once again making reference to Mao—stressed a second basic principle. In domestic politics, he said, one had to "rely on one's own strength [*zili gengsheng*]"—that is, construct the New Democracy over a long period of time.[372] This can be seen as a reference to the Party's intention not to commit itself totally to the Soviet Union.

To establish a causal connection between the perceived threat and the *yi bian dao* slogan, we must point to a few somewhat mysterious facts. The CCP's military espionage branch supposedly believed—in February and March 1949, at least—on the basis of information they had acquired from Chiang Kai-shek's staff that the United States planned an "Asian version" of a third world war. This called for atomic strikes on targets in Manchuria, the Soviet Far East, and Siberia, followed by a joint landing with Japan and KMT China of an army numbering 3 million in North and Northeast Chinese harbors and an advance toward the Ural mountains.[373] Because no other sources have emerged to confirm these views, which sound so fantastic, it cannot be excluded that the CCP leadership invented this information in order to pass it on to Stalin to increase his willingness to cooperate militarily.[374]

Mao's second additional motive for his decisive emphasis on the *yi bian dao* policy was clearly that he was concerned that if the Chinese Communists did not distance themselves further from the United States, his domestic opponents—including those he had brought into the united front—might become stronger and ultimately be in a position to break the Communists' power with American assistance.[375]

The publication of the *yi bian dao* article also meant that a representative of the top CCP leadership for the very first time mentioned the topic of Soviet aid for the Chinese revolution publicly at a time when the Communist revolution was practically already victorious. Liu Shaoqi mentioned this fact to the Soviet chargé d'affaires in Peking in late December 1949. Mao had previously always emphasized, Liu stated, that the Chinese Communists had to rely solely on their own strength and not on foreign aid. Liu interestingly revealed that he was still sympathetic to the latter position and deleted the quotation from Mao Zedong's July 1 article that suggested that Soviet aid had been essential for the victory of the Chinese revolution from the manuscript of a speech that Deng Yingchao was to deliver in December 1949 to the Asian Women's Conference in Peking. This could suggest that some Party leaders had reservations about Mao's *yi bian dao* policy.[376]

The fact that Liu Shaoqi at least had reservations can also be seen in statements Liu made a few months later to a Soviet interlocutor. But if Liu was speaking for the Party leadership in general, his words document yet again how insulted the Chinese Communists felt that the USSR maintained official relations with the KMT government, and organized its public statements accordingly, right up until the People's Republic of China was established. Liu complained:

> During the past year, we have begun to speak about our friendship with the USSR in public more frequently. Why didn't we emphasize the issue of internationalism as much previously as we do now? Simply because, up to October 1, 1949, i.e., the establishment of the PR China, the Soviet Union also issued no propaganda[377] regarding China [i.e., naturally Communist China in particular]. Only after October 1 did the Soviet Union begin to publish a broad range of materials on China, whereas the American newspapers had already written a great deal about China before October 1. The Chinese CP had also asked itself why the USSR wrote nothing about China. . . . In China, quite the contrary, people know much more about the Soviet Union than they know about China in the Soviet Union. . . . In discussions with Soviet comrades, one gets the impression that only a few people understand China's problems. . . . The victory of the CP came as a surprise for the Soviet people and for Party comrades (not for all of them, of course).[378]

The Americans understood Mao's *yi bian dao* article correctly. Ambassador Stuart noted that no article "could have aligned CCP closer to USSR. Contrarywise, it could hardly have demonstrated greater hostility toward Western democracies."[379]

On July 9, three days after he made this assessment, Stuart received a message from Mao Zedong, Zhou Enlai, and Ye Jianying through Chen Mingshu, in which they sought to play down the one-sided orientation toward the Soviet Union announced in Mao's *yi bian dao* article: The article had been intended for internal Party consumption and should not be seen in a purely political light. Mao's speech on June 20,[380] which reflected a willingness to establish diplomatic and economic relations with all countries, and his speech of July 7, which struck notes that were on the whole friendly to the United States, the three authors of the message continued, were more important. As this last speech offered to establish diplomatic relations without preconditions despite Soviet recommendations, one can assume that it was not coordinated with Moscow.[381]

The message from Mao, Zhou, and Ye was clearly an attempt to alleviate the devastating impression Mao's July 1 article had made in Washington in order not to provoke the United States to provide increased support for the KMT and not to block the desired economic relations with the United States completely. Ambassador Stuart also had a similar interpretation and sent a telegram to Washington stating that the message was an "expression of naive hope that despite political relations, economic relations can and will be established between China and the West."[382] Other Communist leaders also made, in isolated cases, attempts after publication of the *yi bian dao* article to keep relations with the United States from failing completely.[383]

Liu's Written Report to Stalin

In the June 27 session, Stalin stated that he would be ready to receive reports from the Chinese delegation about China's military, political, and economic situation in a Politburo session "in three or four days."[384] According to Shi Zhe, it was Wang Jiaxiang's idea to present a comprehensive report in writing. The delegates believed that in this way they would help their Soviet hosts to better understand China's problems, discuss them more systematically, and avoid repeating arguments. They were able to make use of materials that had been prepared in Peiping and could thus write the report quickly.[385]

Liu consulted with Mao by telegraph.[386] On July 4, he sent the four-part report, signed with his own hand, with a letter "to the CC of the CPSU(B) to Comrade Stalin I.V."[387] Stalin underlined certain passages, which are italicized in the following quotes, and added handwritten comments that he either wrote diagonally across the text or noted in the margins and assigned running numbers.[388] The Chinese literature also refers to the contents of the report, and Shi Zhe apparently had full or partial access to the Chinese original.[389]

The two first sections dealt with the Communist revolution's status and future prospects in China under the headings "The Current Status of the Chinese Revolution" and "The New Political Consultative Conference and the Central Government." The third section bore the title "Questions of Foreign Relations" and the fourth dealt with bilateral issues ("Questions of Soviet-Chinese Relations").

"The Current Status of the Chinese Revolution"

One of the most important goals Liu was pursuing in the first section was apparently to convince Stalin that the victory of the Communist revolution was imminent. The report thus begins with this sentence: "The revolutionary war of the Chinese People is currently already won in principle [*v osnovnom*] and will shortly end in total victory." At the end of May, it continued, the PLA already occupied 30 percent of the total territory—namely, the richest part—with 57 percent of the total population, or "*275 million people.*" In the course of three years, the PLA had destroyed "*5.59 million KMT troops.*" The KMT army included only another 1.5 million troops; but the PLA, on the other hand, had "*3.9 million men.*" The report provided also some details: "*There are 7,500 men in the air force*, and *7,700 men in the navy.*"[390]

The report is based on an assumption that the geographical core of China would be conquered by the winter of 1949–50. The PLA would liberate Taiwan, Sinkiang, and Tibet in the course of 1950—Taiwan and Sinkiang perhaps even earlier. "The question of Tibet must be resolved politically and not by military actions." The CCP leadership wished to occupy Sinkiang as soon as possible, but hostile troops were blocking the way, the means of transportation were lacking, and the PLA had to cross a desert between Gansu and Sinkiang with no provisions or water. "In our opinion, there can no longer be any doubt about the victory of the Chinese revolution now." It will nonetheless still take some time "to achieve total victory."[391]

The subsequent text reveals that the Chinese Communists continued to feel insecure about the possibility of an American intervention in China. They made a distinction between an armed intervention and a blockade. The CCP leadership, the report stated, had always considered the possibility of a "military intervention by the imperialists against the Chinese revolution." Instructions from the CPSU(B), with which they were in complete agreement, had made them even more watchful in this respect. It was altogether possible that the imperialists would send 100,000 to 200,000 soldiers to capture three to four Chinese harbors or carry out various diversionary measures. They had already made some preparations for such an eventuality. True, they had no navy or air force and no coastal defenses. An imperialist attack "can cause us difficulties and do damage, but our armed forces will not be defeated." The Chinese people and its army would rise up against the imperialists and drive off the interventionists' troops.[392]

Following these strong words, the report concluded the topic of intervention by dismissing the problem and providing an optimistic prediction: "As a result, we believe that if we proceed correctly the imperialists might not attempt to carry out a direct military intervention in China. We must be careful not to provide the imperialists with a pretext for a military interference." There is, however, "a high degree of probability that the imperialists will begin a policy of imposing a blockade." They had actually already begun to do so, and this had caused some problems in the fields of shipping and foreign trade in particular.

At the end of the first section, the report emphasized the issue of help "from the world proletariat and people's democratic forces and from the Soviet Union in particular" for the Chinese people, which was the "the decisive condition for the victories of the Chinese revolution." Its authors, however, also demonstrated a considerable self-confidence with respect to Moscow when—not for the first time[393]—they emphasized the value of their independent revolutionary approach as revealed in the "Sinification of Marxism" and suggested that the "Chinese road" could be taken as a model: the CCP's experiences in "organizing an anti-imperialist national united front," the agrarian revolution, "armed struggle in rural areas," "encircling the cities," "illegal and legal struggle in the cities," and its connection with the actions of armed forces and the creation of a Marxist-Leninist party "in a country like China"—all these experiences would "probably be useful in other colonial and semicolonial countries."

"The New Political Consultative Conference and the Central Government"

Liu then informed Stalin of the CCP's plans for the creation of the power structure of the new China and for the establishment of a government in particular. He also described which social strata the leadership wished to include in the new structure and how this concept should be understood in the context of Marxist-Leninist class analysis.

First the report stated that the CCP's next major tasks would be to terminate the war more quickly, cleanse the country of the remnants of Chiang Kai-shek's KMT, develop the economy, and learn how to develop and rule the country. It then said that the Central Committee of the CCP had resolved to establish "*a new Political Consultative Conference (PCC)*" and a coalition government "*in August of the same year.*"[394] Preparations were already in progress. Twenty-three organizations were preparing the PCC, including the political parties, people's organizations, and representatives of the national minorities and overseas Chinese. There were only 43 Communists among the 134 members of the preparatory committee. Forty-eight of them, including 15 "undercover Communists [*zakonspirirovannye kommunisty*]," were "progressive individuals, who would support us without reservation," and 43 centrists; of these, 12 belonged to the right wing. It was certain that the CCP would dominate the preparatory committee and its 21-member presidium. The preparatory committee had resolved that the PCC would be made up of 510 representatives from 45 organizations, and that the Communists would again constitute a majority.[395]

"It is our opinion that the PCC is the appropriate form for the united national front in the Chinese revolution, and the masses are very familiar with it. We therefore intend to retain the PCC as a permanent working body and wherever necessary to establish local PCCs locally." The CCP would pass a general political program to which all the parties, groups, and organizations would subscribe.

The report then provided concrete information on the future government and a list of the planned ministries, commissions, and committees. This particularly interested Stalin, as can be seen from his underlining and marginal comments. "*The central government*[396] will be elected by the PCC. . . . *Comrade Mao Zedong is designated for the chairman of the central government*[397] and Zhou Enlai for the prime minister. Comrades Liu Shaoqi and Ren Bishi will not be members of the government."[398]

The report then discussed the way power would be exercised in the new China. It described the state as a "people's democratic dictatorship, based on an alliance of the workers and peasants under the leadership of the proletariat." This dictatorship would be established against the imperialists, feudal forces, and *bureaucratic capital.*[399] It emphasized that "*the people's democratic dictatorship is not a dictatorship of the bourgeoisie, but neither is it a dictatorship of the proletariat.*"[400] The people's democratic dictatorship in China, the report continued, would have elements in common with that of the workers and peasants that Lenin discussed during the 1905 revolution through 1907. What they have in common is that the proletariat would provide the leadership and the alliance of workers and peasants would constitute the basis of this dictatorship.[401] But the people's democratic dictatorship in China would also be made up of representatives and political groups from the liberal bourgeoisie that desire to fight against imperialism, feudalism, and bureaucratic capital. "*This constitutes the difference.*"[402] This could be explained by the fact that China "*was a semicolonial state and that in the phase of the revolution and after its victory* we will have to concentrate all our efforts on fighting imperialism and its agents for a long time."[403] This could also be explained by a peculiarity of the Chinese national bourgeoisie and was precisely what Comrade Stalin had said in 1926 in his speech to the session of the Chinese Committee of the ECCI of the Comintern when he pointed out that the revolutionary power in China would in future be "mainly anti-imperialistic."[404]

The people's democratic dictatorship in China, the report continued, would take the form of a regime by assemblies of representatives of the people, and yet not the form of a regime of bourgeois parliamentarianism. It would be more like the Soviet regime, but differed from the dictatorship of the proletariat in the form of soviets[405] in that it included representatives of the national bourgeoisie.[406]

The report moved on to the future relations with the bourgeoisie in particular:

> Some say that "after the collapse of the Kuomintang power or when land reform is complete, the contradictions between the proletariat and bourgeoisie will become fundamental contradictions and the struggle between the workers and capitalists will immediately become the main struggle." We take the position that this view of the question is not correct, because if the existing power

> should direct its main fire against the bourgeoisie, . . . this procedure would have the effect of forcing the national bourgeoisie, which can now work with us, into the imperialist camp. *It would be a dangerous and adventurous policy* to carry out such a policy *at this time*.[407] In February of this year, comrade Andreev[408] approved the CCP's policy of bringing the national bourgeoisie over to our side in a conversation with Mao Zedong and the Central Committee of the CPSU(B) later also indicated the necessity of bringing the national bourgeoisie over to our side. We are in complete agreement with these instructions.[409]

The objective contradictions between work and capital would gradually become stronger after the collapse of the KMT. The working class would therefore have to enter into a respective struggle with the bourgeoisie. "At the same time, however, they have to make compromises to reach the necessary and appropriate agreement and to form a bloc with the national bourgeoisie in order to concentrate their forces against the external enemy and overcome China's backwardness."[410] It would be a relatively long time from now before the national capital could be nationalized completely. "We are of the opinion that it will take 10 to 15 years to achieve this."

No unified economic system had yet developed in China because of its backwardness, its underdeveloped transportation system, the previous imperialist spheres of influence, and the feudal fragmentation of the country. "And we are therefore currently forced to grant *local governments* relatively comprehensive rights of self-administration in order to increase local activity and initiative in this way.[411] *We believe it would be wrong and harmful to implement a system of excessive centralism.*"[412]

At this point, the report modestly stated: "We would like comrade Stalin and the CC of the CPSU(B) to tell us whether the views we have presented above are correct."

The report then provided some details on the "democratic groups and parties" in China, whose leaders had for the most part already come to Peiping. There were "approximately ten" such organizations with a membership of altogether not more than 30,000 people, of whom 20,000 were members of the Democratic League. The league exerted a comparatively strong influence on a certain part of the intelligentsia.[413]

> The remains of the Kuomintang and agents of imperialism are attempting to infiltrate these parties and groups in any way in order to acquire legal status. We have therefore severely warned these parties and groups. . . . Some of them share the CCP's views on land reform with great difficulty and some reveal clear nationalist tendencies on the question of relations with the USSR. . . . All democratic parties and groups accept the program developed by the Communist Party in an official context and officially declare themselves to be willing to recognize the CCP as the leading power.[414] With the exception of the agency of imperialism, Chiang Kaishek's KMT, there is no other fully developed party that represents the national bourgeoisie in China. *Up to now, none of the reactionary bourgeois parties that exist in the European countries are present in China.*[415]

At the end of the second section, the report stated that the "mass organizations of the people" that participated in the PCC played an important role in practical political life in China. "In February, we received instructions from the CC of the CPSU(B) to intensify work among workers, employees, youth, and women. We agree with these instructions." The Party was currently moving forward on this activity in the cities. And yet, because there was a long interruption in the past, it was now necessary to train cadres.

There were schools now to train cadres in each organization with hundreds or thousands of trainees. It was therefore not possible to move forward quickly at that time. "We would like some experienced Soviet comrades to come to China to assist with the implementation of this task. . . ."

Stalin obviously agreed in principle with Liu's analysis of the situation in China as it was presented in the first two sections of the report. This can be seen from the fact that he made numerous affirmative comments and no critical comments in the margins, which can be checked in the notes to this section. Stalin's interest in the position Mao would take in the state ("de facto the *president?*") and in the state structure ("And will there *also be provincial governments?*") is striking. Was he toying with the idea of playing such governments off against the central government in Peiping as he had already done with respect to Manchuria and Gao Gang with some success?

Mao had already advocated the idea of a coalition government as it was presented in Liu's report to Mikoyan at the beginning of the year in Xibaipo. The CCP leadership thus again confirmed its acceptance of the ideas already put forward by Stalin in 1948 and for a second time de facto rejected Mao's concept of a one-party state, which he had communicated to Stalin on November 30, 1947. Liu thus sought to give the impression that the concept of a coalition government was Mao's idea and that Stalin had been convinced to support it.[416] Stalin indicated complete agreement with the Chinese idea of not fighting the "national bourgeoisie" for the present but including it in the conflict with the Western states.

"Questions of Foreign Relations"

The third section dealt primarily with the question of how the CCP leadership should behave toward the Western powers. In doing so, it revealed uncertainty in particular areas.

The leadership, the report stated, fully agreed with the last instructions on foreign policy, foreign trade, and the loan from the CC of the CPSU(B). After the imperialists had suffered a defeat in their policy of providing full support to the KMT in its fight against the Chinese revolution, they employed the following methods to continue the fight against the Chinese revolution:

1. They continued to provide aid to the KMT and other potential forces in the fight against the Chinese revolution.
2. They employed all possible methods to break into the revolutionary camp in order to undermine it and damage the Chinese revolution.
3. They sought to employ any means to obstruct relations between the Chinese revolution and the USSR and the international Communist movement.[417]
4. They sought in various ways to express their wish to approach the CCP and at the same time to tempt the CCP to approach the imperialist states.[418]

The CCP leadership had recognized these intentions clearly and would under no circumstances allow them to be realized. It had already warned all the democratic parties and groups about them and mobilized the "vigilance of the people."

We are of the opinion that we must cling to the following principles in establishing foreign policy at the present time and in the future:

1. Fight against the imperialist states in order to achieve the complete independence of the Chinese people.
2. Take a position in international affairs alongside the USSR and the countries of the New Democracy, combat the dangers of a new war, and protect peace and democracy all over the world.
3. Exploit the contradictions between the capitalist countries and within these countries.[419]
4. Develop China's trade relations with foreign countries and the USSR and the people's democratic countries in particular on the basis of the principle of equal rights and mutual benefit.[420]

The report admitted that the Chinese Communists did not have access to the latest data on the imperialist countries' investments in China. It therefore made use of Japanese data from 1936. According to this, Great Britain, the United States, France, Germany, Italy, and Belgium had made investments in the amount of US$1,828 million. Of this, $1 billion came from Great Britain, $220 million from the United States, and $210 million from Germany and Italy. Railroad and political loans amounted to *US$500 million.* The sum of $450 million had been invested in foreign banks and financial institutions, $380 million in the import-export trade, and $360 million in mining and other branches of industry. "Following the *defeat of Japan, Japan's, Germany's, and Italy's investments in China have been confiscated, however.*"[421] The investments of England, France, and Belgium had also suffered some losses, but American investment had increased slightly. "At the present time, the following foreign companies exist in China: banks, insurance companies, export-import companies, ships, moorings, and warehouses. . . ."[422]

The following passage from the report reveals in detail how the CCP was going to treat foreigners in China and how uncertain it was regarding the complicated problems connected with international recognition of the future Communist China:

Although we are implementing a few necessary restrictions, we are currently allowing companies of the imperialists to continue their activities in China as a general rule.[423] Nonetheless, some capitalists have already withdrawn their capital from China or are adopting a passive wait-and-see position. We do not yet have any solution to the problem of when and by what means the issue of investment of the imperialists in China should be decided: should it be confiscated or should we do something else? We are currently very busy with military affairs and other work and have not yet dealt with finding the solution to this problem. Nonetheless, we must prepare to resolve it.[424]

The imperialist countries possess *propaganda organs and cultural establishments* in China. . . . The CC of the CCP has already decided to suspend publication of foreign newspapers and magazines and the activity of foreign news agencies and correspondents. This resolution is already being implemented locally. As some objects of great economic significance are in the hands of the imperialists in Shanghai, the CC of the CCP has approved the advice of our Shanghai comrades *first to refrain* from implementing the aforementioned resolution; we nonetheless do plan to implement it in future. With reference to schools and hospitals belonging to the imperialist countries in China, we have been allowing them up to now to continue to operate on condition that they obey our laws. But we will not allow them to open *new schools or hospitals* in China. In future, when we are in a position to take over these schools and hospitals, we will place them under our management.[425]

We allow foreign religious organizations to continue their activities on condition that they obey our laws; but at the same time we make antireligious propaganda. *We confiscate* the lands belonging to religious missions and *churches and distribute them with the approval of the believers.*[426] We also confiscate and distribute the landed property belonging to other foreign institutes and organizations.[427]

We do not recognize the employees in the diplomatic service of foreign states in China and view them as foreign residents. As a result of our policy on this question, the people understand that China has already risen up and is not afraid of the imperialists. This also makes it possible for us to avoid many inconveniences, as the members of democratic parties and groups no longer dare to maintain contacts or meet with employees in the diplomatic services of the imperialist countries and simple people do not even dare to meet with these foreigners from imperialist states.[428]

The employees in the diplomatic services of the imperialist countries are coming to us everywhere with requests and are seeking to establish relations with us in order achieve de facto recognition from us. No one in the history of China has dared to implement the policy that we are *at this time* implementing with respect to foreigners in China,[429] but as a result of our implementing it large numbers of foreign residents who are feeling various inconveniences are turning to us with the request that they be allowed to leave China. At the same time, we are also feeling certain inconveniences in this connection.[430]

There are approximately 120,000 foreign residents in the territory of liberated China, more than 65,000 of whom live in Shanghai alone. There are 54,000 foreigners in Manchuria. Most of these are Soviet citizens and *after them the Japanese take second place.*[431]

When the new central government is formed, the questions of establishing diplomatic relations with foreign states and our cooperation with the UN and other international organizations as well as international conferences will immediately arise. It is possible *that the imperialist states will for a while either ignore us or present us with demands that would subject us to restrictions in exchange for recognizing us.*[432] *What should our policy* on this be?[433] We naturally will not agree to accept conditions that would impose restrictions on our activities, but should we undertake active measures to gain recognition from these states, so that we receive a *legitimate starting position for playing an international role? Or should we perhaps wait and not hurry*[434] *to gain recognition* from these states and deal instead with cleansing *the home front in order to avoid inconveniences?*[435] *Could the USSR and the countries of the New Democracy quickly recognize the new government of China* after the establishment of the new government of China,[436] even if the imperialist countries chose to ignore us? If the imperialist countries should initiate a policy of recognizing the new government of China, *we intend in this case to establish diplomatic relations with these countries, and then it would be our wish for the USSR to recognize us before the other countries.*[437]

It is our intention to examine every treaty and every agreement that the Kuomintang government concluded with a foreign state. In doing so, we will employ the following principle: We are prepared to recognize *and continue to observe* all treaties *that are in the interests of the Chinese people, peace, and democracy all over the world, such as, for example, the UN charter, the Cairo declaration, and the treaty of friendship and assistance between China and the USSR.* We are prepared to abrogate all treaties that are detrimental to the Chinese people, peace, and democracy all over the world, for example, the treaty on trade and shipping between China and the United States. We plan to recognize some treaties after the necessary corrections have been made.[438]

. . . We will doubtless be in agreement with the USSR in our policy regarding international relations and we have already provided some explanations about this to the democratic parties and groups.[439] Some non-Party members criticized our policy for leaning solely to the USSR, but Comrade Mao Zedong told them that our policy only leaned to the USSR because it would be a mistake not to take a stand alongside the USSR in the fight with the imperialist front and to seek to follow a middle road. Following these explanations, all the democratic parties and groups signed and published the declaration condemning the North Atlantic Pact along with the CCP.[440]

We request instructions from the CPSU(B) and comrade I.V. Stalin on the various foreign policy questions.[441]

It is again striking in the third section that Stalin expressed approval many times by writing "Yes!" in the margin while he did not indicate disapproval a single time. It is worth noting that Liu included the Sino-Soviet treaty of August 14, 1945, among the treaties China would recognize and continue to observe—a point we will go into in detail in the context of the fourth section.[442] It is also striking the way Liu, unlike when he discussed domestic policy questions, several times took the position of a pupil to the teacher by occasionally asking what the Chinese leadership should do and, when he had mentioned all the problems, asking the Soviet fraternal party and Stalin for "instructions" (Russian: *ukazaniia*). In the first two sections, on the other hand, he asked only whether Stalin and the CPSU(B) considered the Chinese views to be correct once and never asked for instructions. This shows Mao Zedong's self-confidence as victor in the civil war and his resulting determination not to allow Moscow to continue interfering in domestic problems.

"Questions of Soviet-Chinese Relations"

The fourth and last section constitutes the most sensitive part of the report.

Liu Shaoqi began by emphasizing in impressive words the fundamental significance of friendly relations between the Soviet Union and China:

> The firm friendship between the two great peoples of the USSR and China is of extraordinarily great significance for our two countries and for the entire world. It is of decisive significance particularly for the independence and construction of China. The Central Committee of the CCP fully understands the importance of this issue. The CCP certainly will not spare any effort to consolidate the friendship between our two peoples.[443]

Liu then addressed the sensitive issues that had already been discussed by Mao and Mikoyan at the beginning of the year in Xibaipo: the Sino-Soviet Treaty of Friendship and Alliance of August 14, 1945, the status of Dairen and Port Arthur, and the independence of Outer Mongolia.[444] Liu had already mentioned the treaty of August 14, 1945, in the previous section of the report as one of the treaties the CCP wanted to "recognize and continue to observe," because it was "in the interests of the Chinese people, peace, and democracy all over the world"—a sentence Stalin had underlined.[445]

In the fourth section, the report now said: "The Soviet-Chinese Treaty of Friendship and Alliance has already brought the Chinese people great benefits in the past. The new government of China will observe this treaty, and this will be an even greater contribution to the peoples of China and the USSR and to the Chinese people in particular. We want to observe this treaty completely."[446]

In assessing the treaty in this way, Liu Shaoqi was expressing the opposite of what the CCP leadership had felt when it first heard that the treaty had been concluded in August 1945. At that time, the Chinese Communists had reacted with dismay; they had felt that Moscow had sold them out to Chiang Kai-shek.[447] Nonetheless, the Chinese Party leadership had also already stated during the talks with Mikoyan in February 1949 in Xibaipo that the treaty of August 14, 1945, was one of the "patriotic" treaties that the CCP wanted to retain.[448] The only even somewhat plausible explanation for the Chinese behavior appears to be that it was an attempt essentially to show goodwill in order to prepare the ground for achieving their main goal of obtaining help from the Soviet Union in the conflict with the KMT and the United States and in the economic construction of China.

It is striking, however—and this conflicts with what had been said up to now in his report—that Liu Shaoqi did not simply wish to leave the treaty as it was. The report stated that when diplomatic relations had been established it would be necessary to "deal with this treaty." Generally speaking, one could proceed on the basis of "one of the following three principles":

1. The new government of China will declare that it fully approves this treaty and that it continues to be valid without any changes.
2. Representatives of the two sides will again conclude a new treaty of friendship and assistance between the USSR and China in the spirit of the original text in order to insert, on the basis of the new situation, a few changes in style and content.
3. Representatives of the governments of the two countries exchange notes to the effect that the present treaty will remain as it is for the present but that they are ready to review it again at the appropriate time.

Which of the three alternatives presented above is good?[449]

Stalin wrote the following handwritten comment in the margin next to this paragraph: "17) Resolve this problem when Mao Zedong comes to Moscow."[450]

After listing the three alternatives for a solution to the question of the treaty, Liu seized another hot potato: "Some people in the democratic parties, students, and workers were discussing the questions of the presence of Soviet troops, the independence of Outer Mongolia, and the removal of mechanical equipment from Manchuria by the Soviet Union." Stalin felt unpleasantly attacked by the last part of this observation in particular. Across the entire paragraph, he wrote in pencil: "We have only transported a portion and by no means all of the Japanese capital from Manchuria."[451] The report continued:[452]

> We gave these people the following explanations: If we had not approved the stationing of Soviet troops in Port Arthur, this would have helped the imperialists as long as we are not in a position to defend our coasts ourselves.
>
> On the question of the MPR.[453] We said that in keeping with the principles of the self-determination of nations the Mongolian people had called for independence and that we must recognize the independence of Mongolia.[454] If the MPR wanted to unite with China, however, we would welcome this. Only the Mongolian people has the right to decide this question.[455]
>
> On the question of the removal of mechanical equipment from Manchuria by the Soviet Union, we said that these machines belonged to the Japanese and that the Soviet Union had removed the equipment *as booty for use in its socialist* construction and on the grounds that this equipment should not fall into the hands of the reactionary Chinese party, which would have used it in its fight against the Chinese people. The Soviet Union's actions were altogether correct.
>
> *Are these explanations correct?*

Stalin could be satisfied with the results of these arguments. He expressed his approval by drawing a line in the margin next to the paragraph on Mongolia and writing "Yes!" next to it. He wrote a "P" ("Right") in the margin next to the last question.[456] Stalin also had nothing to say about the statement that the machines that had been removed were "booty," as can be seen from the fact that he underlined this passage without comment.[457] One can doubt, however, that the Chinese side had expressed its real opinion of the dismantling of Manchurian industry.

It is in fact true that patriotic Chinese had repeatedly asked the Communists why the Soviet Army was in Port Arthur and why it had removed machines from Manchuria.[458] It is also true that representatives of "democratic" parties (i.e., parties with which the Communists felt they could form a coalition) exerted pressure on the Communists to negotiate with Moscow for the return of the booty and thus placed the CCP in an unpleasant position. The CCP did in fact use the arguments presented in Liu's report to explain the dismantling. Thus, for example, a publication appearing in Shanghai in late 1949 argued that the factories were being dismantled so as not to fall into the hands of the KMT.[459] In one case, the "democrats [*minzhu renshi*]" were to be informed that the Soviets had dismantled the installations so that they would not fall into the hands of the "American imperialists" and that they could negotiate for a restoration after a democratic government had been established.[460]

The Mongolian question put the CCP leadership in a similarly unpleasant situation because the cession of Outer Mongolia by the KMT government in early 1946 had led to hefty protests in Chinese public opinion. As has already been described in detail, Mao openly expressed his interest in a united Mongolia that was a part of China to Mikoyan in February 1949 in Xibaipo, which aroused opposition from Mikoyan and Stalin.[461] In the end, Mao had yielded and declared that the Chinese Communists were not pursuing a chauvinistic Greater China policy and would not raise the issue of uniting Mongolia. This did not stop Liu from raising the question a second time in his July 4 report and again emphasizing the CCP's interest in attaching the MPR to China—albeit making use of the formula Stalin had used in February, with the important restriction that the people of the MPR would have to make their own decision. Liu naturally knew that they could not expect a pro-Chinese decision in Ulan Bator given the existing circumstances, so his arguments were designed to promote recognition of the status quo.

As Mao had done to Mikoyan in February 1949 with respect to Port Arthur,[462] Liu raised the sensitive issues mentioned above in his report in a clever package by not stating them as his own questions but placing them in the mouths of third parties (representatives of political parties, students, and workers). It is nonetheless striking that Liu raised them at all. In this way, he gave to understand that the Party leadership in Peiping considered the questions problematic and as yet unresolved. With respect to the dismantling of Manchurian industry, he may also have intended to remind Stalin that the USSR had enriched itself at China's expense—and for that reason as well was obligated to provide aid to the new China.

It is striking that the Liu report did not raise the question of the Chinese Changchun Railroad. It is possible that it was considered too sensitive because the two sides' positions were too far apart on this question, as would be shown in talks between Stalin and Mao Zedong on January 22, 1950. Mikoyan had stated in Xibaipo that the Soviet side did not consider the CCR agreement to be unequal, unlike the agreement on Port Arthur, although the principal of equality might not have been observed fully.[463] For this reason, Stalin had repeatedly asked Kovalev by telegraph what the Chinese comrades' true opinion of the CCR agreement was and whether they really considered it unequal.[464]

In his July 4 report, Liu expressed the Chinese interest in establishing postal, telegraphic, railroad, and air connections between the USSR and China and founding a mixed Sino-Soviet airline. "How should this be done?"[465]

This was followed by another question: "*What is the best way to resolve the problem of establishing relations* and trading *with the countries of the People's Democracy in Eastern Europe?*" In the margin next to this paragraph, Stalin put two vertical lines and this comment: "By means of direct negotiations with them." Across the text and heavily underlined, he wrote, "We will help."[466]

The following paragraph raised an urgent request for advice and help:

> We have spent a long time in village surroundings as we were waging a partisan war and therefore have very little familiarity with foreign affairs. We must now administer such a great state, promote economic construction, and carry out diplomatic activities. We have a lot to learn. In this respect, instructions [*ukazaniia*] and help from the CP of the USSR are of very great significance. We urgently need these instructions and this help. In addition to sending Soviet specialists to China to help us, we would like you to send Soviet teachers to China to teach and that delegations be sent from China to get to know and study the Soviet Union. Furthermore, we would like to send students to the USSR for training.[467]

Liu now turned to the future relations between the parties. Here it is the CCP's confession of its unrestricted willingness to subordinate itself to Moscow that is surprising. "The CPSU(B) is the main staff of the international Communist movement and the CCP merely represents a staff of one direction. The interests of a part must be subordinated to the international interests, and therefore *the CCP subordinates itself to the resolutions of the CPSU(B),*[468] although the Comintern no longer exists and the CCP does not belong to the Information Bureau of the European Communist Parties. If differences of opinion should arise between the CCP and the CPSU(B) on various questions, the CCP will submit and resolutely *carry out the resolutions of the CPSU(B),* after it has presented its point of view.[469] We consider it necessary to establish *the closest possible relations between the two parties,* exchange *appropriate politically responsible representatives,*[470] in order to make decisions on questions that interest our two parties and also achieve a greater mutual understanding between our parties."[471]

Liu now turned to Mao Zedong's visit to Moscow, which had repeatedly been postponed since 1947: "Comrade Mao Zedong would like to visit Moscow, but cannot come to Moscow secretly at this time, and he has no alternative but to wait until the establishment of diplomatic relations between the USSR and China, at which time he will be able to visit Moscow legally. We therefore ask you to advise us when Mao Zedong can come to Moscow and how best to do this."[472]

After that, Liu stated in a separate paragraph: "We therefore request that you provide instructions [*ukazaniia*] on all the questions mentioned above." By this Liu obviously meant the bilateral questions.

Liu then returned briefly to the US$300 million credit that Stalin had approved during the two delegations' meeting on June 27. He declared his "complete agreement" with the conditions on which it was to be provided and expressed his thanks for the help the Soviet Union had shown to the Chinese people.[473]

As in the previous section on foreign policy and unlike in the two sections on domestic policy, it is striking that Liu requested instructions from Moscow in the fourth section. The difference can serve as an indication that the Chinese CP leadership essentially was not ready to allow Stalin to influence domestic politics. On the other hand, the final sentence in the report again expressed the modest attitude assumed by the Chinese side—this time in relation to the entire report: "We would like the CC of the CPSU(B) and comrade Stalin constantly and without any embarrassment to provide instructions and criticize the work and policy of the CCP."[474] Liu Shaoqi signed the report in his capacity as "leader of the delegation of the CC of the Chinese CP."

Traditional Chinese rituals of courtesy probably played the main role in Liu Shaoqi's obsequious conduct. In the Chinese original of the report, Liu obviously used the formula "*qing geiyu zhishi*" for "request to provide instructions."[475] As has just been suggested, it will also be necessary to explain Liu's behavior by the fact that he was playing the role of suppliant in Moscow. But even taking all this into account, it must remain amazing that Liu went so far as to state that the CCP was prepared to submit to the resolutions of the Soviet fraternal party, even when it was of a different opinion. Stalin clearly found this totally unexpected and obviously felt that it was going too far. He therefore reacted logically by expressing his decisive rejection at this point—the only time in the entire report.

It is too much to assume that the Chinese gesture of subordination was intended seriously. The Chinese Party leadership had found the Comintern's attempts to restrict it unpleasant too often in the past to be able to be nostalgic for this status. It is more probable that Liu correctly assumed in making his offer that Stalin would reject it and thereby confirm that the days of compulsory instructions as had existed under the Comintern had passed for good. Liu probably also had a second goal: He was trying to use his gesture to emphasize the CCP's loyalty to Moscow in the context of Russian suspicions—of which the Chinese CP leadership was aware—that Mao was a Chinese Tito. With respect to Stalin's reaction: as much as the Chinese confession of its desire to subordinate itself may have pleased him, he could not but emphasize to the Chinese comrades the independence of all Communist parties that at least theoretically existed since the Comintern was disbanded in 1943. Stalin and Liu had animated debates on this subject on two subsequent occasions.[476]

One can assume that Liu's obsequious behavior both flattered and also irritated Stalin. This found expression in the discussion of Liu's report on July 11 when Stalin rejected the request for instructions, which Liu had apparently repeated orally, with a reference to the two parties' equal rank.[477] Stalin had a similar—if more sophisticated—discussion with Zhou Enlai three years later. When Zhou asked Stalin to give the Chinese CP "instructions [*ukazaniia*]" with respect to relations with Asian Communist parties, Stalin, who had apparently become nervous, was unable to restrain himself and asked, "Instructions or advice [*ukazaniia ili sovety*]?" Zhou replied sophistically that Stalin might perhaps see it as advice but the Chinese saw it as instructions. Although Stalin emphasized that the Soviet advice was not binding, Zhou insisted on his opinion: The Chinese saw them as "instructions and indeed valuable instructions," which they would not obey blindly, however. Stalin now found himself forced to show extreme modesty as well, and said that the Soviet side knew little about China and would therefore "be careful not to give instructions." Zhou remained insistent and again asked Stalin if he would not give any instructions. Stalin submitted and consented to give something that he called "advice."[478]

As suggested above, the problem in the case of both these conversations with Liu and

Zhou was not least an intercultural phenomenon—a misunderstanding between two Chinese of the old school interested in extreme courtesy that bordered on flattery, and a Russified Georgian who was not familiar with Chinese rituals. The situation was made more complicated by the fact that in China an educated Chinese who wanted to be particularly courteous could "ask" someone who was not superior to him "for instructions [*qing shi* or *qing geiyu zhishi*]," when he really only wanted to receive good advice.

If Stalin was expressing a degree of anger and not merely amazement when he asked "Instructions or advice?" this could be because it reminded him of the Comintern period, when the ECCI had issued directives (in Chinese: *zhishi*)[479] that were binding on the Communist parties, which he knew the Chinese Party leadership under Mao had seldom enjoyed and sometimes had not obeyed. As already mentioned, Stalin considered it very important to emphasize that each individual Communist Party was independent since the dissolution of the Comintern in 1943—which was, of course, mendacious in the postwar era with regard to the Eastern European satellite states or the Mongolian People's Republic. If such associations played a role with Stalin, he could also have understood Zhou Enlai's request for instructions as a sarcastic reference to the "good old days of the Comintern."

Liu's Wish List

Two days later, on July 6, Liu Shaoqi sent Stalin a letter containing a series of questions and requests.[480] Stalin underlined certain passages, which are italicized in the quotes that follow. The letter was organized into eight points and dealt with such disparate topics as the threat of a third world war and Chinese requests to examine the operations of the Soviet state, establish a Soviet teaching institution to train Chinese cadres, and for Soviet teachers to be sent to China. Plans for transportation links between the USSR and China, Chinese requests for help in constructing a navy, conquering Sinkiang, resolving economic problems in Manchuria, and future cultural relations also played a role.

Point one referred to Liu's July 4 report and observed that a series of requests had already been made in that report, which included a request for instructions. Stalin had already given instructions about the credit and the specialists that would be sent to China. With this observation, Liu was referring to the information that Stalin had given the Chinese delegation at their meeting on June 27.[481]

Before they left China, the July 6 letter continued, Mao Zedong had asked Liu to request instructions from Stalin "about his assessment and analysis of the international situation, the threat of a new war, and mutual relations between the USSR and the imperialist countries, the United States and England," which the CCP would use to form an assessment of the international situation.[482]

In his second point, Liu requested that the Chinese delegation be allowed to observe certain important areas during its stay in Moscow.[483] In the field of public administration, Liu specifically mentioned the structure of government bodies at the various levels; interrelations between central and local political bodies and between Party, government, and mass organizations; and the organization of the armed forces, people's courts, security organs, and financial institutions. They also needed information on the structure of cultural and educational institutions and the Foreign Ministry. In the fields of economic policy and administration, the Chinese were interested primarily in the coordination of the development of industry, agriculture, and trade; the state budget, communal budgets, and the budgets of enterprises, institutions, schools, and kolkhozy; relations between state enterprises, local enterprises, supplemental firms of individual enterprises, institutions, schools, and cooperative enterprises; the organization and role of banks and cooperatives; the maritime customs and the Foreign Ministry; the operation of the tax system; and the structure of the transportation system. They also asked to observe the structure and operations of cultural organizations, the Party, mass and youth organizations, and trade unions.[484]

Liu requested that his delegation be allowed to visit "responsible comrades" of the following institutions among others: the Council of Ministers of the USSR, the Ministry of the Interior, the Ministry of State Security, the Ministry of Education and other cultural organizations, the Foreign Ministry, the State Planning Committee (Gosplan), the State Bank (Gosbank), the ministries of domestic and foreign trade, the Ministry of Finance, the Central Council of the Labor Unions, and the CC of the Komsomol. He requested that the delegation be allowed to meet with members of the Organizational Department of the CC of the CPSU(B), the heads of the Moscow city and regional committees of the CPSU(B), and of the Moscow Municipal Soviet. Finally, he requested permission to visit enterprises, kolkhozy, and sovkhozy with his delegation and to hold talks with managers, secretaries of Party organizations, and chairpersons of labor union organizations.[485]

In the third point in the letter, the Soviet government was asked "to establish a special school similar to the former University of the Toiling People of China to train cadres for the new China, who are needed for the development and administration of the state and the enterprises." They were thinking of up to 1,000 students at first. The school should include the following departments, among others: industry, trade, banking, law, and education. Training courses should last one, two, and three years. Courses could be taught through interpreters. Nothing should be changed in the existing teaching provided to Chinese students in various Soviet schools.[486]

Liu requested permission to send "various responsible colleagues" to the USSR to gain experience for three to four months. They could visit objects of interest to them there, attend lectures, hold talks, and improve their knowledge in the fields of state and economic administration. On the other hand, the Soviet side was asked to send teachers to China who could help to train cadres in various fields.[487]

The fourth point in the letter dealt primarily with questions of the development of transportation links between the Soviet Union and China. The Chinese side requested talks and resolutions on the following topics: "*the establishment of a postal and telegraphic link between the USSR and China, ocean links* between the USSR and China, *air connections,*" and air routes from the USSR via Harbin and Mukden to Dairen, from Mukden to Peiping, and from Peiping via Ulan Bator to the USSR. Also included in the topics to be discussed were the founding of a Soviet-Chinese airline, Soviet aid for building aircraft construction and repair facilities, and the training of cadres for the air force.[488]

In the following point, Liu tersely requested assistance in training cadres for the navy, the use of minesweepers, and aid in raising sunken ships and constructing coastal defenses.[489]

The sixth point included a request for help with the "liberation of Sinkiang." In a telegram, Mao Zedong expressed his intention to occupy Sinkiang as soon as possible and to ask in Moscow for "*help from Soviet aviation and in transporting troops by air.*" Liu requested materials on the situation in Sinkiang and a decision about the provision of help by air.[490] (Because the Soviets were the de facto rulers in Sinkiang for many years, they were, therefore, very familiar with the situation over there.)

The seventh point dealt exclusively with Manchuria. Liu reported that Manchuria could export from 800,000 to 1 million tons of provisions to the USSR during the winter of 1949–50 in exchange for machinery. He recommended "*introducing a single currency in Dairen and Manchuria*" to encourage economic relations between the two, reduce the burden on the Chinese currency, and "reconstruct industry in Dairen even better." He asked about opening Dairen harbor for exports of, among other things, coal and salt to Hong Kong. Could Dairen be used for the export and import of goods from and to China in the case of a blockade of Shanghai and Tientsin? Could American and English merchant ships be allowed in such a case to load and unload in Dairen? Finally, Liu raised the issue of a hydroelectric power station on the Yalu, for the construction of which China had invested 75 million and Korea 50 million Japanese yen. He reported that the Manchurian authorities were asking for the power plant to provide half of its electric energy to Manchuria and were requesting that the USSR assist them in resolving this problem.[491]

The eighth and final point dealt with cultural relations. These were described as an important means to bring the two states closer. Liu requested talks "with the responsible Soviet comrade" about the telegraph agencies; the film industry; a Soviet-Chinese cultural association; an exchange of delegations of workers, peasants, and scientists; the training of Russian-speaking cadres in China and Chinese-speaking cadres in the USSR; the translation of political, economic, and literary texts; the construction of a library of Russian-language literature in China; opening bookstores and publishing newspapers, magazines, and books for the USSR and the "countries of the New Democracy."[492]

At the end, Liu requested "instructions" on all these questions or that a specific person to be provided to deal with them. The letter ended with the phrase "With a Bolshevist greeting!"[493]

Although Stalin did not write handwritten notes on the letter to express acceptance or rejection or comment on the text—as he had on the July 4 report—one can see what he found particularly important (or sensitive) by the passages he chose to highlight: in particular, questions of military and economic cooperation.

Liu Shaoqi's July 4 report and his additional questions on July 6 were discussed along with a few other topics in a joint session with Stalin and other Soviet leaders that took place on July 11. The discussion of some of these questions was continued in a joint session on July 27.

Discussion of Liu's Report and Wish List: The July 11 Session

The July 11 meeting took place at 10 P.M. in the Politburo meeting room in the Kremlin and took the form of an extended Politburo session. As far as we know, in addition to Stalin, the Politburo members Molotov, Malenkov, Beriia, Mikoyan, Kaganovich, and

Bulganin, along with the Politburo candidate Nikolai Shvernik and Ivan Kovalev, attended on the Soviet side. As military questions would be addressed, there were also several high-ranking military officers present, including the first vice defense minister, Vasilii Sokolovskii.[494]

According to Shi Zhe, the atmosphere was serious but friendly. On the Soviet side, Stalin was almost the only person to speak. He praised the clarity of the Chinese July 4 report. He himself and the other Soviet leaders, he said, had no difficulty understanding it.[495]

Liu acted modest. The report merely discussed a few essential subjects, he said, and was by no means entirely accurate. He said that he hoped Comrade Stalin would provide instructions (*Si-da-lin tongzhi zhishi*). Stalin protested: When had he ever given instructions (*zhishi*)? The two parties were brother parties. There was no hierarchy. How could he give instructions?[496]

According to Shi Zhe, Stalin then discussed specific points in the report. We do not know whether he commented on the first section, which dealt with the current situation in China. If he did, it was certainly simply to state his approval, because he had not made any comments that suggested that he disagreed with it on the report itself. Stalin clearly indicated his agreement with regard to the second section, which dealt with establishing the government. This is not surprising, because he had expressed his agreement in numerous comments in the margin of Liu's report itself. On July 11, Stalin thus explicitly approved the CCP's policy of including the national bourgeoisie in the government. In the Eastern European countries and Germany, he stated, many representatives of the bourgeoisie had collaborated with Hitler and then accompanied the German occupation forces when they retreated; but this was not the case in China with regard to Japan. Taken together, the Sino-American trade and shipping treaty had hurt the interests of the Chinese bourgeoisie. It had therefore taken a position in opposition to the United States and Chiang Kai-shek. It was correct for the Chinese CP to exploit the bourgeoisie's anti-American resentment and pursue "a policy of long-term cooperation [*changqi hezuo de zhengce*]" with it to bring it into the anti-imperialist camp.[497]

The CCP should offer the national bourgeoisie favorable conditions; that is, it should impose protective tariffs. There was indeed an objective contradiction between capital and labor. But to avoid allowing this to destroy the long-term collaboration between the bourgeoisie and the CCP, it was necessary to conclude contracts between the capitalists and the workers that protected the workers' interests. The Party had to induce the capitalists to treat the workers well. Stalin expressed his approval of the term "people's democratic dictatorship" used by Liu in the report to describe the nature of the Communist use of power in China.[498]

According to Shi Zhe, Stalin indicated concerns during another session about at least one point that dealt with the planned coalition government.[499] The inclusion of other parties in the government, he said, could have the result that state secrets could not be kept and could even be betrayed "to the enemy." Stalin therefore suggested that the Chinese comrades should ensure that they achieved a majority in general elections and formed a one-party government. He advised them to hold the elections in 1954 and to link them to the promulgation of a new constitution, which would have to replace the proposed "Common Program." This constitution should not be socialist in nature, but should reflect existing social conditions. Both of these, the elections and a new constitution, were also necessary to counteract accusations "by the enemy" that the Communists had come to power by military means in China and lacked legitimacy. The constitution should include the right of the entire population to elections, of entrepreneurs and rich peasants to hold property, and of foreigners to do business in China.[500]

It is not yet possible to document whether Stalin commented extensively on the third section of Liu Shaoqi's July 4 report, which dealt with foreign policy. We only know of answers to questions on two points raised by Liu in his report. Stalin advised the CCP not to lump the "comprador bourgeoisie [*maiban zichanjieji*]" together with the "national bourgeoisie [*minzu zichanjieji*]," or to confiscate enterprises with capital investments from the imperialist states in China overly hastily. Instead, they should use strict labor legislation to keep these enterprises under control. In keeping with his comments in the margin of the July 4 report, he also recommended that the CCP first do business with the imperialist countries and later discuss with them the question of recognizing the new China. "You should not be concerned about recognition by the imperialist states and even less about how they behave toward you. You have a good measure of behavior—trade with the imperialist countries. The economic crisis has already begun with them. I think this will accelerate recognition. But you now have to trade with them."[501] Stalin had already commented in the margin of the corresponding passage in Liu's report that the Communist government should not hurry to establish relations and that China should not provide commercial privileges to any country that does not recognize it.[502]

Stalin apparently did not provide any explanation here of why he was advising the Chinese Party leadership not to establish diplomatic relations with the Western powers quickly. As already mentioned, he apparently wanted to avoid allowing the new China to be placed in a position of pursuing a trilateral policy through direct political contacts rather than remain on the Soviet Union's short leash in its international relations. Alarmed—or should one say excessively alarmed—by Kovalev's reports, he observed the CCP leadership's actions with suspicion. Any suggestion that the Chinese comrades were not following his recommendation provoked a "heightened reaction" in him.[503] In a report that he delivered to Stalin in December 1949, Kovalev complained that the Chinese Party leadership was "infected by the illusion of quick recognition" by the Western states, although these continued to support Chiang Kai-shek, whereas Stalin had advised the CCP only to establish diplomatic relations with the Western states after they had ceased to support the KMT.[504]

Stalin responded positively when Liu asked in the same context in the July 4 report whether the USSR and the other Communist-ruled states would be able to recognize the new Chinese government as quickly as possible after it was established. It was consistent with the tone of his comments in the margins of the report that Soviet recognition would follow immediately. The CCP leadership would have to discuss what the other countries would do next with them directly, but the Soviet Union would help the Chinese.[505]

If, as can already be seen from his handwritten comments on Liu's July 4 report, Stalin essentially agreed with the foreign policy section of the report as much as with the two first sections, this was not the case with the fourth and final section. This section, as has already been described, contained some hot potatoes, including the Sino-Soviet treaty of August 14, 1945, the questions of Port Arthur and Dairen, and the Mongolian question.

According to Chinese sources, Stalin said during the July 11 session that the Soviet side admitted that the Sino-Soviet treaty of August 14, 1945, was unequal (*chengren zhege tiaoyue shi bu pingdeng de*);[506] he had already had a discussion with Mao Zedong on this topic. The USSR, he stated, had been forced to pursue this policy at that time because of its relations with the KMT. The three alternative proposals Liu had made in his July 4 report were not necessary, however.[507] They would settle this question, Stalin continued, when Mao Zedong came to Moscow. This would happen as soon as the new government had been established, at which time the USSR would recognize it immediately. If this were not possible, a Soviet delegation would travel to China.[508]

It is not clear to which discussion of the 1945 treaty with Mao that Stalin was referring. But it is clear that during his talks in Xibaipo in early 1949, Mikoyan, speaking for Stalin only, described the agreement on Port Arthur, and not the Treaty of Friendship and Alliance itself, as "unequal."[509] In comparison with that, Stalin's July 11 statement represented a concession to the Chinese.

According to Chinese sources, Stalin said at the July 11 session that the Soviet troops were in Port Arthur as "forces for freedom" to frighten off the armed forces of the United States and Chiang Kai-shek, protect the Soviet Union, and at the same time serve the interests of the Chinese revolution. The USSR would station troops there to counter any threat from the United States or Chiang Kai-shek until a peace treaty was signed with Japan and the United States withdrew its troops from Japan. If, however, the Chinese CP wanted the USSR to withdraw its troops, it would do so immediately (*mashang*). Dairen and Manchuria could be unified administratively. Dairen was a free port, Stalin added, and both sides could continue to use it as before until a new government and diplomatic relations had been established.[510]

Kovalev's report described the discussion in essentially similar terms. Stalin addressed the question of the continued stationing of Soviet troops in Port Arthur "in the form of a test" and said:

> Our government is of the opinion that the Soviet Union could consider withdrawing its troops from Port Arthur after the withdrawal of American troops from Japan. But if the CCP considers the immediate withdrawal [*nemedlennyi vyvod*] of the troops from Port Arthur to be necessary, the USSR is prepared to withdraw its troops from Port Arthur immediately [*seichas*] so that the CCP can derive political advantages from it.[511]

According to the information available up to now, Liu did not answer Stalin. In doing so, his behavior differed from that of the Chinese leadership to Mikoyan one-half year earlier in Xibaipo. At that time, Mikoyan had declared on instructions from Stalin that the agreement on Port Arthur was unequal.[512] The Soviet Union would either abolish the agreement and withdraw its troops as soon as a peace treaty had been concluded with Japan and the U.S. troops had withdrawn from Japan, or it would withdraw the troops immediately, if the CCP wished this. At that time the Chinese Party leaders had hurried to assure Mikoyan that they were not interested in an immediate withdrawal, because this would only help the

United States. As we see, Stalin made an offer to the Chinese side on July 11 that was similar to the one made during Mikoyan's visit to Xibaipo. The only difference was that on July 11 Stalin apparently did not characterize the agreement on Port Arthur to Liu as unequal, and he also did not explicitly discuss the possibility of abolishing it. This suggests—but does not necessarily have to mean—that he had changed his position. Because the Soviet naval presence constituted the core of the agreement, the termination of the existing agreement would, so to speak, be an obvious consequence of a withdrawal of the troops.

If Liu really did say nothing, it is unclear why. The explanation Sergei N. Goncharov, John W. Lewis, and Xue Litae offer in their book, which is based essentially on Kovalev's unpublished memoirs, is not very convincing.[513] They argue that Stalin saw the Port Arthur naval base as an "indivisible part of his security belt in the confrontation with the United States and as a point of leverage against Japan." Stalin's offer to withdraw the troops was therefore not meant seriously and constituted a kind of test. If Liu had accepted Stalin's offer, the Chinese Communists would have been revealed in Stalin's eyes as unreliable partners and Stalin "would have dealt with them accordingly." This was the reason Liu had remained silent, Goncharov, Lewis, and Xue suppose, although the Chinese Party leadership was interested in an immediate troop withdrawal.

The following hypothesis seems more plausible. If the Soviet Union did in fact have the strategic interest described, and there is some evidence for this, Stalin's offer was risk-free only insofar as it involved an immediate withdrawal, because Stalin could presume that the Chinese would not want this, as they continued to fear U.S. intervention. The fact that the Chinese Party leadership had immediately rejected a corresponding offer in Xibaipo supports this interpretation. Port Arthur's strategic significance would diminish significantly as soon as the U.S. troops were withdrawn from Japan following a peace treaty with Tokyo. This part of the July 11 Soviet offer should therefore be taken seriously, and this was also the reason that it encountered no objections from the Chinese in Xibaipo. We must nonetheless continue to ask why Liu remained silent rather than reject the immediate Soviet withdrawal as the Chinese leadership had in Xibaipo. This would have been the plausible alternative to remaining silent and not, as Goncharov, Lewis, and Xue assumed, acceptance of Stalin's offer to withdraw the troops immediately. Goncharov, Lewis, and Xue could not refute Kovalev, as they did not know the archival documents on Mikoyan's visit in Xibaipo.

It is hard to understand how Kovalev was able to reach his interpretation. He had been present during Mikoyan's negotiations in Xibaipo and therefore had to know that the Chinese leadership had spoken out spontaneously and decisively at that time against a quick withdrawal of the Soviet troops. It is most probable that Kovalev's extremely deep-seated mistrust of the Chinese leaders' motives played a role and led him to ascribe profound anti-Soviet feelings to them. From this perspective, Mao had to be intensely interested in getting rid of the Soviet troops as quickly as possible. Kovalev had to suspect any statements to the contrary to be no more than lies designed to get into the desired Soviet feed trough.

Any discussion of Liu's lack of response to Stalin's offer of an immediate troop withdrawal should, however, take into account the fact that Liu had already spoken out, if indirectly, against a rapid withdrawal of the Soviet Army in his July 4 report. As long as the Chinese CP, it said, was not in a position to protect China's coast itself, it would only help the imperialists if it did not agree to the stationing of Soviet troops in Port Arthur. This is the answer they had given to "some people in the democratic parties" who had asked about the Soviet troop presence in Port Arthur.[514]

It is not known whether Stalin made any statement on another sensitive topic in their bilateral relations that Liu had addressed in his report on July 11: the status of the Mongolian People's Republic. If this was the case, he may have approved the explanation provided by Liu in his report that China had to recognize the independence of Outer Mongolia and that only the Mongolian people could decide about unification with China. Stalin had already noted his approval by hand on the report. It is hard to assume, however, that he addressed on his own initiative the most sensitive point, the Chinese Changchun Railroad, which Liu had intentionally passed over in his report.

Stalin then took a position on Liu's statement in his report that the CCP was prepared to "carry out the resolutions of the CPSU(B)" as it was the "main staff of the international Communist movement." He said:

> In your report, the Chinese delegation stated that the CCP would subordinate itself to the resolutions of the CPSU. This seems strange to us. The party of one state subordinates itself to the party of another. Nothing like that has ever happened before and it is not permissible. Both parties must bear responsibility for their peoples, provide advice on certain questions, help each other, and stand firm together when difficulties arise—that is right. Today's Politburo session with your participation was a special kind of link between our parties. That is the way it must be.[515]

Shi Zhe reports this statement by Stalin in a similar way: The Communist Party of each country was responsible to its own people. If there were problems, they could discuss them together. If there were difficulties, they should help each other. Their relations should be close, but this did not mean that one party had to obey any other.[516]

After Stalin had responded in this way to Liu's July 4 report, Liu Shaoqi posed some questions that had in part been covered in the first point of his July 6 letter to Stalin. Liu asked about the probability of a third world war, the possibility of avoiding one and pursuing peaceful reconstruction, and the prospects for international development. Stalin replied to the first question in an optimistic way. He considered a new world war to be improbable in the near future, he said. The peoples of the world did not want one and the United States was not prepared for one. But it was hard to predict how long the peace would last—whether fifteen or twenty years. They would have to use this period for economic reconstruction: "If we really become very strong, a war can be avoided." And yet there were those who were mad about war. An American military commander had become convinced that the Soviet Army had landed in Alaska and had jumped off a building to his death. One had to be careful about such crazy people who did not care either about themselves or about their people. If they could get twenty years for peaceful construction and development, the possibility of a new war would be minimized. They must nonetheless be prepared for one. The ideas about war and peace that Stalin provided his Chinese guests remind us in a certain way of Lenin's theory of the "breathing space [*peredyshka*]" the Soviet Union needed before it would reemerge to make a renewed attempt at world revolution.[517]

A further question from Liu concerned the status of the international Communist and workers' movement.[518] It was apparently in this context that Liu inquired whether the CCP could join the Cominform.[519] This had been founded by Stalin in late September 1947 as a kind of successor organization to the Comintern—which had been dissolved in 1943—that would control the Eastern European Communist parties and promulgate the Soviet propaganda line to Communist parties all over the world. After Yugoslavia left the Cominform in June 1948, it was made up of the representatives of the Communist parties of the USSR, Bulgaria, Romania, Hungary, Poland, Czechoslovakia, Italy, and France.[520]

According to Kovalev's record, Stalin replied to Liu's inquiry that although it was possible, it was "absolutely unnecessary" for the CCP to join. For one thing, China was subject to pressure from imperialism, whereas this was not the case in the "countries of the New Democracy" in Eastern Europe. For another, the bourgeoisie in Eastern Europe had collaborated with fascism and had withdrawn with it, so that the proletariat could with full right confiscate their companies, establish its dictatorship, and follow the road to socialism. In China, on the other hand, the bourgeoisie had not surrendered to the Japanese and also had not collaborated with the Americans or Chiang Kai-shek. There was thus no reason to move against it and establish a dictatorship of the proletariat. Instead, a "democratic dictatorship of the workers and peasants" existed in China. These not inconsiderable differences made the entry of the CCP in the Cominform inappropriate at the present moment.[521]

But as the situation in the East Asian countries was similar to that in China, Stalin continued, one could consider founding a "union of the communist parties of East Asia [*Soiuz kompartii Vostochnoi Azii*]." This was "more necessary and at this time more appropriate" than the entry of the CCP in the Cominform, but it could still be too early for this. Because the USSR was a country located in both Europe and Asia, "it will participate in a Union of the Communist Parties of East Asia."[522]

The answer that Stalin gave on July 11 to Liu's question about the possibility of the CCP joining the Cominform sounds artificial, scholastic, and not very convincing. Stalin also contradicted himself by describing the status of the Eastern European countries as a "dictatorship of the proletariat," and then saying that these countries did not "de facto" constitute dictatorships of the proletariat, but "people's democracies," and then finally again claiming that dictatorships of the proletariat did "de facto" exist in them.

In this context, it is worth noting that Mao Zedong took the position to Stalin in May 1949 (in response to a question from Stalin or of his own accord?) that it was still too early to establish an "Eastinform" (Russian: *Vostochnoe informbiuro*). The CCP maintained contact with only five Communist parties in the twelve countries in the East: these were in the Mongolian PR, Siam, Indochina, the Philippines, and Korea. It had no permanent links to the other countries—even to the Japanese or Indonesian CPs—and the CCP leadership knew little about the situation in these countries. It would therefore be better, Mao continued, first to establish such connections, learn about the situation, and only then undertake to found an Eastinform. At that time, it would be appropriate for the CCP to limit itself to establishing reciprocal radio connections that it could use to exchange ideas on concrete current problems.[523]

When Stalin made his argument on July 11 in response to the question about the

CCP's entry in the Cominform, one suspects that he was interested in something other than what he polemically stated to Liu. On the one hand, he wanted to keep Communist China out of European politics. He was determined to operate the Cominform primarily as a Soviet supervisory authority over Communism in Europe. China, with its tendency toward independence and its weight as a quasi great power, could easily prove a disruptive element in these plans. On the other hand, it seemed tempting to Stalin to allow China to employ its revolutionary experience to spread socialism in Asia as, so to speak, Moscow's executive assistant without the Soviet Union having to become involved in the confrontation that could be expected with the Western powers. To achieve this, it could prove useful to establish an "East Asian Cominform" in which Moscow had the right to participate.[524]

Stalin's Chinese guests probably had mixed feelings when they heard Stalin's plan to make Moscow the seat and voice in a future Asian Cominform on the basis of the Soviet Union's Eurasian character while China was to remain excluded from the Cominform. Nothing has become known about Liu Shaoqi's answer to Stalin's proposal. On the other hand, the CCP leadership could interpret Stalin's statements to mean that it had been given a green light to develop its own strategy for implementing the revolution in Asia. The outlines of such a policy, which focused on armed conflict with imperialism in Asia, took shape in a speech that Liu Shaoqi delivered in December 1949 to a labor union conference of the Asian countries in Peking.[525]

It is easily conceivable that Stalin made another comment in the context of what he said about the "CCP and the Cominform," which was also recorded by Kovalev and dated July 11, in which he emphasized the role of Communist China as an example in Asia to lessen the blow to the Chinese Party leadership of being excluded from the Cominform. He said that the Chinese CP had to understand its important position and the fact that it was fulfilling an unprecedented historic mission. To China with its population of 475 million should be added the populations of India, Burma, Indonesia, and the Philippines, whose peoples would listen to the Chinese—as would the Japanese, probably. "The peoples of Asia are looking to you with hope." There was no other Party in the world, Stalin said, that had such far-reaching prospects; the Chinese CP had numerous pupils.[526]

Shi Zhe claimed that Stalin then admitted to Liu Shaoqi that he had made mistakes in his policy toward the Chinese CP.[527] When Liu turned to developments in China after the Second World War, he reported that Stalin asked him: "During your process of struggle, have we disturbed you or have we been a hindrance to you?" Shi Zhe commented that Stalin was referring to his behavior in August 1945, when he had urged Mao Zedong by telegraph to negotiate with the KMT.[528] Liu denied this and granted only that it had been unnecessary for Mao Zedong to travel to Chungking. Zhou Enlai would have been sufficient. But Chungking had resulted in a success. Stalin, Shi Zhe continued, objected that it had been dangerous for Mao to travel to Chungking. He might have been murdered. Stalin then asked whether the CCP had not lost anything in the American peace campaign[529] and if the Soviet Union had harmed it. Liu again denied this. The peace campaign had isolated the United States and Chiang Kai-shek, he said, and had thus ultimately had positive effects for the CCP.[530]

Stalin replied:

> The victors will not be on trial. All victors are always correct. And we should not blame you. You Chinese comrades are too polite to express your complaints. We know that we have made ourselves a hindrance to you, and that you did have some complaints, but you would not tell them to us. You should certainly try to judge if our statements are correct or wrong, because we may give you erroneous advice as the result of lacking understanding of the true situation in your country. Whenever we have made a mistake, you should let us know, so that we may notice the mistake and correct it.[531]

Shi Zhe added that Stalin said this with a guilty conscience, self-critically and as if he were apologizing. This had surprised the Chinese delegation, which had not planned to criticize Stalin. In Shi Zhe's opinion, Stalin meant his self-criticism seriously. Shi even went so far as to claim that Stalin had shown himself to be a great revolutionary leader in this way and had earned greater respect and admiration. The delegation immediately reported the event to Peiping.[532]

The situation in Sinkiang was also discussed during the July 11 session. Stalin had already urged during the first meeting on June 27 that the province should be conquered quickly, as there would otherwise be the threat of a British intervention there. Liu had agreed with this but pointed to logistical and other difficulties. Stalin had then approved sending forty fighter aircraft to fight General Ma Bufang's troops.[533] In his July 4 report, Liu again referred to the logistical problems. Liu's July 6 letter to Stalin renewed the request for Soviet air support. On July 11, Stalin warned for the second time that there could be a foreign intervention: he had learned from secret service sources that the United States and Great Britain were provoking ethnic unrest in Sinkiang to provide a pretext for extending their presence there.[534] Stalin again agreed to assist in conquering the Northwest Province as quickly as possible.

On July 11, Stalin also provided a positive response to further requests for aid that Liu had presented in his July 6 letter. He stated that he was prepared to establish a naval academy in Port Arthur.[535] He also approved assistance for the construction of coastal defenses and a railroad from Ulan Bator to Zhangjiakou (Kalgan). It is not yet known whether or to what extent he, on this occasion, also approved other requests that Liu had expressed—for example, regarding the training of cadres for the air force, economic assistance for Manchuria, opening Dairen for Chinese imports and exports in the case of a blockade of Shanghai and Tientsin, and the development of diverse cultural relations between the two countries.

Stalin also rejected one Chinese request during the July 11 session: the request that the Soviets provide protective cover to support the conquest of Taiwan. In June 1949, Mao Zedong ordered the responsible officers to make the necessary military preparations during the summer and fall to conquer Taiwan that winter. In this way, Mao sought to avoid allowing the KMT to obstruct Chinese coastal trade and bomb mainland cities like Shanghai, Nanking, Fuzhou, and Xiamen with aircraft from Taiwan, as had already happened. He was also convinced that the danger of a military intervention from the United States, which at this time he still feared,[536] would not be put to rest until all of China—including Taiwan—was in Communist hands.[537]

Mao had already repeatedly asked Stalin by telegraph for naval and air support before Liu's departure for Moscow. When Liu Shaoqi again presented the request on July 11, Stalin rejected it. The Soviet economy, he said, had suffered colossal damage during the war. The country had been devastated from the western border to the Volga. Support for an attack on Taiwan would involve conflict with the U.S. Air Force and Navy and provide a pretext for unleashing a new world war. "If we, the leaders, did anything like this, the Russian people would not understand. And more than that. They would drive us out. For disregarding their distress and efforts during the war and afterward. For recklessness."[538]

Stalin's reference to the presumable reaction from the Soviet population was, of course, pure hypocrisy; he did not have to fear any opposition from this quarter. Even Kovalev, otherwise such a loyal follower, spoke of this as demagoguery.[539] But the concerns Stalin expressed about a third world war were genuine. The Soviet government constantly feared becoming embroiled in another world war during the postwar years. Khrushchev reported that Stalin trembled at such a prospect, knew the weakness of the Soviet Union, and "never did anything that might provoke a war with the United States."[540] Stalin had no intention of subjecting the Soviet Union to such danger for the sake of the Chinese Communists.

Liu apparently did not raise any objection. Stalin proposed that they again address the question at the next session planned for July 27, to which some leading military leaders would also be invited.[541]

The discussion of Soviet military support for an invasion of Taiwan is treated as a state secret in the literature in the People's Republic of China: it is not raised. Shi Zhe even went so far as to deny explicitly that this topic was discussed in Moscow; Kovalev had invented it, Shi maintained.[542]

We can explain Shi Zhe's actions theoretically in two ways. His memory could have failed him, which is more than unlikely given the significance of the issue. The second alternative seems more plausible. He intentionally kept the discussion secret because it was taboo in the official Chinese Party historiography. It became taboo because it was considered a loss of face in Peking that Mao Zedong despite repeated requests in Moscow had remained unsuccessful in obtaining Soviet military support—a state of mind that persists to this day. Furthermore, this historiography attempts to portray Mao Zedong as someone who was always very concerned about the sovereignty of the People's Republic of China and rejected the stationing of foreign military forces on Chinese territory, particularly during Mao's conflicts with Khrushchev over military cooperation during the summer of 1958. The request for Soviet support in the conquest of Taiwan would not have fit this picture.

And yet there is a comment in Shi Zhe's memoirs that can be taken as indirect confirmation of the fact that the Taiwan issue was discussed.[543] In describing the discussion at the July 11 session, Shi Zhe noted that the KMT had carried out air raids on Shanghai in May and June and had caused great damage. He wrote that Mao had then asked Stalin by telegraph to provide air support to counter these attacks. The two Soviet aircraft squadrons that had been sent from Port Arthur had ensured that no more KMT aircraft appeared over Shanghai. The connection between the planned conquest of Taiwan and the air raids carried out from there against Shanghai and other coastal cities has already been described above.

It is not possible to confirm the truth of this claim that the Soviet air force intervened over Shanghai. But it is more than improbable that Stalin authorized this, because he feared that military encounters with the KMT could develop into a major confrontation with the United States and even a third world war, as mentioned previously. Why would he categorically refuse Soviet participation in the conquest of Taiwan out of fear of a major confrontation with the United States and then intervene militarily in the case of Shanghai?

The July 11 session ended with Stalin inviting his Chinese guests to a film presentation.[544]

A Fact-Finding and Inspection Tour

At the first session on June 27, Liu Shaoqi had expressed the desire to familiarize himself with the organization and working methods of the Soviet state apparatus. As has been presented in detail elsewhere, this is why his mission became legalized—albeit in a misleading way. In his July 6 letter to Stalin, he indicated the wide range of areas and issues that interested him particularly.[545]

Kovalev organized a program of activities to meet this request that was carried out for the most part between July 15 and 26. The Chinese delegation visited the State Planning Committee (Gosplan); the Finance and Economic Affairs committees; the ministries of finance, foreign trade, and chemistry; the State Bank (Gosbank); factories producing ball bearings, precision instruments, and automobiles; collective farms; the Moscow Municipal Soviet; the university; the Lenin Library; the Historical Museum; and several industrial exhibitions. The Chinese familiarized themselves with questions involving production, management, training, and the role of the Party and labor unions in the factories. Kovalev also sent the delegation numerous Soviet specialists to speak on these issues and answer questions. Thus, for example, Liu asked to speak with Mikoyan about questions involving trade networks in the domestic market, the organization of foreign trade, and the problems involved in cooperation and trade among the USSR and China and Manchuria.[546]

Liu had asked Stalin on July 6 to teach him about "the structure of the Foreign Ministry and diplomatic struggle."[547] He met with Foreign Minister Vyshinskii for a three-hour talk on July 30—without Gao Gang, who by this time was probably already sitting in the train to China. Kovalev, Fedorenko, Wang Jiaxiang, and Shi Zhe also attended the meeting. Liu asked Vyshinskii to inform him about the ministry's structure, working methods, and cadre policy; the status and working methods of Soviet representations abroad; and the basic principles involved in "conflict in the diplomatic arena." He told Vyshinskii to make suggestions about the diplomatic activity that the future central government of the new China should pursue.[548]

Vyshinskii explained his ministry's organizational structure and mode of operation to Liu and indicated how it differed from the U.S. State Department and the British Foreign Office. The principle of one-man management (*edinonachalie*) applied, he said, and the minister's orders were binding on all employees. The minister presented specific questions to the government for examination. The Party organization, having no control function in this respect, did not intervene in the ministry's affairs, dealt solely with political and educational work, and resolved organizational and Party problems. The Party organization was headed by the Moscow committee of the CPSU(B). The ministry leadership cadres were subject to the Central Committee. The ministry had approximately 1,500 employees, 95 percent of whom were Party members. There were no Party organizations in Soviet representations abroad, but there were labor union organizations. He had never heard that Party organizations existed in the embassies of any of the People's Democracies in Moscow. Vyshinskii also answered Liu's general questions about the foreign representations' mode of operation. He reported on the diplomatic cadres' training, which was the responsibility of the Party, and, in response to Liu's question about the extent to which they employed "old specialists," stated that they still did not do without "old non-Party specialists" such as experts in international law, consultants, and other lawyers, although the ministry apparatus had been "completely renewed" under the Soviets. These individuals would "not remain passive observers in the process of constructing a new society" if they were provided with conditions in which to work creatively.[549]

Liu asked "what Soviet suggestions and desires" he could transmit to Mao Zedong on issues involving international relations. Vyshinskii replied that one of the most important international questions involved the peace settlement with Japan. The new Chinese government should ensure that this question was not discussed by a conference of eleven or twelve states, as Washington and London proposed, but by the Council of Foreign Ministers.[550] Vyshinskii warned Liu that the "Chinese democrats" should be particularly careful about attempts by "the American ruling circles" to exploit the situation for their own "imperialist purposes." They had to retain the freedom to maneuver in foreign affairs. Diplomacy was "the natural sister of military strategy." They had to be constantly concerned "to exploit contradictions in our enemies' camp." This was Stalinist diplomacy (*stalinskaia diplomatiia*).[551]

Finally, Liu mentioned that Mao Zedong had a great desire to visit the Soviet Union at an appropriate time and to familiarize himself in person with that country's successes in constructing socialism. He then invited Vyshinskii to visit China. The latter thanked him for the invitation. Liu asked whether a group of Chinese democrats could come to the USSR for diplomatic training, or, if this was acceptable, whether Soviet specialists in the training of these cadres could be sent to China. Vyshinskii replied that in principle this could be done, but, he added carefully, this question "would have to be arranged between the highest authorities." Liu understood and informed him that he would write a letter on this matter "to comrade Stalin in person [*na imia tovarishcha Stalina*]." It became clear during the talk that the Chinese did not possess sufficient reference works on questions involving international relations. Vyshinskii agreed to send them a few copies of the new *Dictionary of Diplomacy (diplomaticheskii slovar')*, which was published in Moscow.[552]

Plans for Military Cooperation: The July 27 Session

Another session that can be dated precisely took place on July 27 in the form of an extended Politburo meeting in the Kremlin. About the participants, we know only that Stalin, Liu Shaoqi, Gao Gang, Wang Jiaxiang, Defense Minister Bulganin (a Politburo member), Minister of the Army Alexandr Vasilevskii, Ivan Kovalev, and Shi Zhe were present. We can assume, however, that other Politburo members also attended.[553] As Stalin had suggested during the previous session on July 11,[554] they discussed the question of Soviet military support for the planned attack on Taiwan during this session for a second and final time. We do not know how the discussion developed, but we know its result. Stalin insisted categorically on his decision to refuse. Liu ultimately agreed with Stalin and said that he had been given authority to state that the Politburo of the CCP withdrew its request with respect to Taiwan and Hong Kong[555] and that the question was thus settled.[556]

Liu's behavior made it clear that he had consulted with Mao Zedong and the Chinese Party leadership in Peiping on the Taiwan issue. In the process, there was a moment of irritation. On July 25, two days before the meeting, Mao Zedong had sent Liu Shaoqi a telegram, which the latter forwarded to Stalin with a request that the Soviet leader "familiarize himself with its contents and provide instructions." Among other things, this telegram included the following:

> The anti-imperialist movement in Europe and in other parts of the world may be moving forward. It is also possible that an economic crisis is breaking out in America and England. If we use help from the USSR to conquer Formosa in this case (i.e., given the fact that we are asking the USSR to help us train pilots and to sell us airplanes and that we also have to ask the USSR to send us specialists from the Soviet air force[557] and navy and pilots to participate in the military operations)—will this not damage mutual relations between America and the USSR?[558]

Mao Zedong had received a report by telegram from Liu Shaoqi on July 18 about the course of the July 11 meeting, during which Stalin had rejected the request to provide military support for an attack on Taiwan. It is hard to imagine that Liu would not have reported on this important topic, especially as it had remained open. But if Mao knew of Stalin's intention to reject the request, it is somewhat amazing that the Chinese leader nonetheless repeated his request in the telegram to Liu on July 25 and asked him "to report to Comrade Stalin on this [i.e., on the contents of the entire telegram], so that he can think about how it could be implemented in practice."[559]

In this case, Mao revealed a remarkable degree of stubbornness and also exhibited considerable chutzpah when he inquired with apparent concern whether Soviet military aid in the conquest of Taiwan would damage relations between Moscow and Washington, knowing full well that Stalin had even said that he feared it could lead to the outbreak of a new world war.

On July 26, Mao apparently instructed Liu not to continue to insist if Stalin stood by his rejection in the July 27 meeting despite the July 25 telegram. In this case, Liu should concentrate his request to Stalin during the July 27 meeting around an additional option that the CCP leadership had already developed internally on July 10. This included adopting considerations at Zhu De's suggestion to send three or four hundred men to the USSR to be trained for the air force. In addition, it should be considered whether China should acquire some one hundred Soviet fighter aircraft. These, along with the already existing air force, could form an "offensive unit [*gongji budui*]" to provide cover when they crossed the sea in connection with the conquest of Taiwan, which was planned for the summer of 1950.[560]

It has also been documented that the Military Commission of the CCP Central Committee resolved on July 26 to form a new air force command and send Liu Yalou, the chief of the general staff of the Fourth Field Army and apparently the designated commander of the air force, to Moscow to negotiate the acquisition of aircraft, the dispatch of Soviet experts and advisers, and support for the construction of a naval academy in Port Arthur, which Stalin had approved on July 11.[561] On the same day, Liu Shaoqi was informed of this and asked to negotiate with Stalin the purchase of Soviet aircraft and the invitation of Soviet trainers to China in connection with the Chinese air force unit (*kongjun zhandou budui*) that was to be formed that year.[562] If Stalin agreed, they would send Liu Yalou to Moscow with a small delegation.[563]

Mao had already presented his concrete wishes to Stalin in the telegram of July 25:

> The difficulties have constantly been growing worse in Shanghai since the blockade. But in order to break the blockade Formosa must be conquered. But it would be impossible to take it without an air force. We would like you to exchange ideas with General Stalin to see whether the USSR can grant us aid in this area, i.e., whether it can train 1,000 pilots and 300 technical workers in the flight repair services for us in Moscow within six months to a year. In addition, whether the USSR can sell us 100 to 200 fighters and 40 to 80 bombers to be employed in the military conquest of Formosa. We also request the USSR to help us in the field of developing a navy. We are assuming that in the second half of next year, i.e., at the time of the attack by our troops on Formosa, we will have occupied the entire territory of the Chinese mainland with the exception of Tibet.[564]

At the July 27 session, on instructions from Mao, Liu gave Stalin a paper requesting assistance from the USSR in modernizing the PLA and the Chinese defense industry. There was no further mention of Taiwan. In keeping with his instructions from Peiping the previous day, Liu presented the request for the Soviet Union to sell aircraft to China to build up a Chinese air force, train pilots and technical personnel in Moscow, and send experts to China, and in doing so clearly based his statements on concrete information included in the telegram he had received from Mao on July 25. He asked Stalin to approve the sending of a delegation with Liu Yalou at its head. It is possible that he also requested the delivery of antiaircraft guns. Stalin approved Liu's requests immediately.[565]

Trouble Due to Gao Gang

According to Kovalev, a serious clash occurred during the July 27 session between Liu Shaoqi and Gao Gang in connection with the debate over the planned attack on Taiwan that should be seen in the context of Gao's pro-Soviet leanings. Gao was in all probability aware of Stalin's true intention not to withdraw the Soviet troops from Manchuria immediately through his confidential contacts with Kovalev. In order to support Stalin's position, he went further than Stalin did himself. At the July 27 session, he proposed reinforcing the Soviet troops in Dairen and stationing Soviet naval units in Tsingtao to protect Manchuria from American attacks and transform it into an even more dependable base for the advance to the south and the final defeat of the KMT army.[566]

In this context, Gao may have said that the KMT and the United States would not dare to attack Manchuria if it was a Soviet republic. Stalin immediately recognized that Gao had made his statements without authority from Peiping and distanced himself from Gao. Liu was furious at Gao's performance and accused him of treason after the session. Stalin then sought to mediate between Liu and Gao, and even put in a good word for Gao to Mao Zedong. Mao complied with Liu's suggestion to recall Gao from Moscow early for his treasonous behavior. Gao traveled back to Manchuria on July 30—but not without having previously offered to inform the Soviet side confidentially of the anti-Soviet behavior of certain Chinese Party leaders, which he subsequently also did.[567]

As this was a full-blown incident, we will discuss it in greater detail below along with other events associated with Gao Gang that burdened relations between Stalin and Mao in 1949.

Gao Gang, Politburo member of the CCP since 1945, held practically all the key positions at Party and state level in the Northeast region in 1949 and was considered at the time the uncrowned king of Manchuria.[568] To all appearances, he felt a particular sympathy for the Soviet Union and for Stalin, which the latter reciprocated. In the second half of 1948, the connection between Stalin and Gao Gang ran via Kovalev, who at that time was in Manchuria as the leader of the group of Soviet experts. Suspicions raised at the time, and even later,[569] that Stalin wished to exploit Gao's affection to transform Manchuria into a Soviet satellite should, however, be dismissed to the realm of fable. Reports from the contemporary witness, Kovalev, have recently helped to shed some light on relations among Stalin, Mao Zedong, and Gao Gang that until then had remained in the shadows of history.

Stalin allegedly praised Gao Gang in a conspicuous way and treated him with particular respect during the stay of the delegation led by Liu Shaoqi in Moscow.[570] According to a report by Kovalev, who was present, the details of the aforementioned incident during the joint session of the Soviet leadership and the Chinese delegation on July 27 were as follows.

Gao Gang proposed that they should "declare Manchuria to be the seventeenth Soviet republic." In this way, Gao continued, Manchuria would be preserved from American attack and could develop into a still more reliable base for the advance to the south and the final defeat of the KMT. To the same end, Gao advocated moving the Soviet fleet to Tsingtao and reinforcing Soviet troops in Dairen. According to Kovalev, those present had applauded, but Liu Shaoqi had been visibly furious.[571]

A Chinese author, who apparently had access to the CCP archives, offered a different version. According to this, Stalin asked his Chinese interlocutors whether the Soviet navy could visit Tsingtao. Gao Gang quickly replied that not only was this possible, but Soviet troops could be stationed on the Shandong Peninsula as well. Liu Shaoqi remained silent at this and thereby expressed the fact that he rejected the idea.[572]

Stalin immediately distanced himself from Gao by comparing him to Chang Tso-lin, the former, and to Communist eyes extremely reactionary, dictator of Manchuria. After the session, Liu accused Gao of treason and received a vicious reply from him. Liu immediately sent Mao a telegram in which he requested that Mao immediately order Gao to return to China for treasonous behavior. Gao offered to inform Kovalev confidentially of the internal affairs of the CCP leadership. Many of Mao's colleagues, he said, were attracted to a right-wing Trotskyite heresy. He also wanted to report on the "insincere and anti-Soviet behavior" of certain Chinese leaders toward the CPSU. Kovalev informed Stalin of his talk with Gao. Stalin granted that his criticism of Gao for his comment on Manchuria had been extraordinarily harsh. But it had been necessary, as the Chinese leadership might otherwise have interpreted the situation incorrectly. Stalin forbade Kovalev to accept the information Gao offered because he, Kovalev, would be returning to China and could not allow himself to be drawn into conflicts within the Chinese leadership. They should not permit Gao Gang to return home immediately and should hold him for another few days in Moscow. For all he knew, Kovalev continued, Stalin had instructed another comrade to receive the information from Gao.[573]

Shi Zhe, who interpreted between Stalin and the Chinese delegation, did not say a word about the discussion of Manchuria in his memoirs.[574] When asked about what Kovalev wrote, he suggested that he could not remember any such event. As he certainly would not have forgotten such a significant incident if it had really happened, there could only be one explanation, he suggested: that what Kovalev said was not true.[575] He provided support for this interpretation elsewhere, adding that Gao Gang had been very discontented about the Soviet dismantling of industrial equipment in Manchuria.[576] He had wanted to talk to Stalin about this, but Liu Shaoqi had stopped him. How could Gao Gang, who had such a critical view of Soviet behavior, Shi Zhe asked rhetorically, ever have proposed to transform Manchuria into a seventeenth Soviet republic? On the other hand, one would have to ask Shi Zhe how Gao Gang's supposed attitude toward the dismantling should be reconciled with Gao's fundamentally pro-Soviet attitude, which is generally assumed by researchers.

Shi Zhe's credibility, as can be determined on other occasions, unlike Kovalev's, is not to be rated particularly highly in doubtful cases. Shi made himself fully unbelievable when he made the demonstrably untrue statement that Kovalev could not report on the alleged July 27 incident because he had been present at the talks between Liu Shaoqi and Stalin only once for half an hour; other problems, he claimed, were at play here.[577]

In any case it is inconceivable that Gao Gang would make a proposal as serious the annexation of Manchuria without first consulting with Mao Zedong, the Party leadership in Peiping or at least the leader of his delegation, Liu Shaoqi. A version found in another Soviet source seems most plausible: at the session, Gao Gang merely expressed the view that the Kuomintang and the United States would not dare to attack Manchuria if it were a Soviet republic.[578] Stalin's proposal to Liu of June 27 that he send a Soviet fleet to visit Tsingtao harbor after the establishment of the People's Republic of China could have led to Liu's statement about Tsingtao.[579] Liu had passed over the suggestion in silence. Gao could also have been aware of Stalin's intention to keep Soviet troops in Manchuria for as long as possible through his close and confidential contacts with Kovalev. In this case, it is conceivable that he proposed reinforcing the troops in Dairen in order to accommodate Stalin's wishes, but in doing so, embarrassed him.

Mao approved Liu's request to recall Gao Gang prematurely. On July 30 Gao left in a depressed mood, and he was not accompanied to the airport by a single member of the Chinese delegation. At the farewell reception Stalin arranged in his honor at his dacha in Kuntsevo, Stalin attempted to reconcile Liu and Gao and made them drink to their friendship, which Liu did at Stalin's request, albeit with visible displeasure. Stalin told Liu directly after Gao's departure that he was sorry Gao was leaving as he, Stalin—like Liu—had treated Gao too harshly, and he asked Liu to pass his opinion on to Mao. Kovalev claimed that Mao took Stalin's request into account and—at first—went easy on Gao. He even allowed him to be given high positions in the all-China government established in early October.[580]

Shi Zhe unconvincingly challenged the idea that Gao Gang had been recalled from Moscow because of his conflict with Liu Shaoqi and at Liu's request. Gao, he claimed, had himself decided to return early even before the Chinese delegation departed for Moscow, as he could not be absent for too long owing to the pressure of developments in Manchuria.[581]

Shortly afterward, new difficulties arose in connection with Gao Gang and Manchuria. In September 1949, a Chinese liberal democratic leader, who later became a member of the government of the People's Republic of China, complained to Mao Zedong on his return from a trip to Manchuria that one could see numerous pictures of Lenin and Stalin in Manchuria, but hardly any pictures of Mao. Gao Gang had ordered portraits of the generalissimo to be painted and hung when Soviet visitors had complained about their absence in late 1948. Mao convened a special Politburo session in which Liu Shaoqi and Zhou Enlai accused Gao Gang of being a traitor to the Chinese people, one who wanted to hand Manchuria over to the Soviet Union. At the end of the session, the "Gao Gang line" was condemned, and it was resolved to remove the pictures of Stalin and Lenin throughout China. Stalin placed himself at Mao's side because he had no desire to burden relations with the Chinese Party leadership in any way over Gao Gang. In a telegram to Kovalev, he stated that Mao Zedong, Liu Shaoqi, and Zhou Enlai were right on the question of the portraits. "This is so elementary, that I am amazed that there could be any differences about it. . . . In general, one must say that it would be better to have as few pictures of Lenin and Stalin hanging in China as possible."[582] Stalin also sent a telegram to Mao Zedong in which he approved the Mao-Liu-Zhou line and condemned Gao Gang's actions. After the Politburo session, Kovalev talked to Mao and worked out a compromise. The Stalin pictures should remain in place on Soviet military bases; in areas with concentrations of Soviet citizens; in CCR facilities, where Soviet railroad specialists worked; in joint Sino-Soviet institutions; in buildings of the Association for Chinese-Soviet Friendship; and in the Chinese Party and Youth Organization committees.[583]

When Mao Zedong took the train to Moscow in December 1949, he noticed to his annoyance during a stopover in Mukden that the September Politburo resolution had not been implemented. Pictures of Stalin in the uniform of a generalissimo hung on all major buildings, but there were no portraits of Mao. Mao gave the order for the wagon loaded with presents for Stalin's seventieth birthday that had been coupled to the train in Mukden on Gao Gang's and Lin Biao's orders to be uncoupled. He had half of the presents sent to each of them and told them that he was carrying presents for Stalin from *all* of China, and Manchuria was still a part of China.

In early 1950, Stalin revealed to Mao Zedong, in order to gain Mao's trust, that Gao Gang was his confidant. Stalin gave Mao an extremely critical report and coded telegrams that Kovalev had sent him in which the CCP leadership's behavior was sharply criticized. Kovalev had drawn his information about the internal affairs of the Chinese

Party leadership primarily from Gao Gang. Stalin also gave Mao reports with similar contents that he had received from Gao Gang directly. If up to then Stalin had sought to use his secret contacts with Gao to spy on the CCP leadership, establish special relations with Manchuria, and even replace Mao with Gao, his decision to hand over the file shows that he had by now given up such ideas and begun to count solely on Mao and his supporters.[584]

Kovalev assumed that Stalin's betrayal had sealed Gao Gang's subsequent fate and that Mao was merely waiting for Stalin to die out of consideration for the Soviet leader and his close relations to Gao before he settled accounts with the latter. Ledovskii also attributed the elimination of Gao primarily to his loyalty to the Soviet Union and assumed that before Stalin's death Mao did not dare to begin "dismantling" Soviet influence in Manchuria, which he achieved by eliminating Gao and the other leading Manchurian party cadres.[585]

The most thorough monograph on the Gao Gang case to date, on the other hand, comes to the conclusion that Gao's connections with the Soviet Union and to Stalin were not the primary reason for his downfall. Instead, his attempts to discredit Liu Shaoqi and Zhou Enlai in the Party and to replace Liu as Mao's designated successor played a more important and probably even the central role. The Party leadership criticized Gao Gang sharply along with Rao Shushi (the Gao-Rao affair) in 1953–54. On August 17, 1954, Gao took poison and died.[586]

Recently surfaced Soviet archival documents suggest that a conflict between Gao Gang and Liu Shaoqi may have indeed played a significant role. They contain information from Gao's opponents that had been given to Ambassador Iudin and suggest that a conflict over agricultural policy between Gao and Liu in 1951 played a role. Liu wanted to support those Communists whose "farms extended into farms owned by kulaks," whereas Gao intervened to support the middle-class farms. Mao took Gao's side, but no resolution from the Party leadership resulted. Liu Shaoqi and Zhou Enlai accused Gao Gang in February 1954 of having sullied the role Liu had played in the regions controlled by the KMT during the civil war. Instead, Liu and Zhou continued, he had praised the work of the army in the "liberated areas" in order to draw attention to his own activities. Gao thus saw the army cadres as the "real cadres." By propagating such views, Gao had placed the army over the Party and stirred up the danger that the army could slip away from the Party's control. Liu and Zhou argued that Gao had criticized Liu purely for career-related reasons. He had thought that he would take Liu's place if he toppled him.[587]

The "Elder Brother–Younger Brother" Debate

During a reception on the newly constructed upper story of his dacha in Kuntsevo that Stalin in all probability gave in honor of Gao Gang's departure on July 30,[588] a tense debate arose over questions involving prestige in the relations between the two parties. After Mao's wife, Jiang Qing, who at that particular time was vacationing in the Soviet Union, toasted Stalin's health, Stalin began to shower the Chinese present with compliments. He praised their youthful looks, acknowledged with the usual pathos that the Chinese CP now had its childhood and youth behind it, and that having been hardened "in the blazing fire," it had matured into a full-grown Party. "The friendship and solidarity between us two brothers is of the greatest importance and is extremely important for the world revolution." When he said that the Chinese Marxists had matured and the Soviet people and the Europeans should learn from them, he was not flattering them, Stalin continued. Western Europeans had remained behind in their arrogance since the deaths of Marx and Engels. "The center of the world revolution has shifted to the east and is now shifting to China and East Asia."[589]

China and the Soviet Union had to assume greater responsibility in the international revolutionary movement, Stalin said, and practice a form of cooperative division of labor. He suggested that China should help the national-democratic revolutionary movements more in the colonies and semicolonies and vassal states, while the Soviet Union should take more care of Europe and the West. The USSR could not play the same role in Asia, just as China could not have the same influence in Europe as the USSR. The Soviet Union might perhaps be better at Marxism and theoretical issues. It could, in turn, learn from the way the Chinese Communists had implemented Marxism in reality and the diligence of the Chinese.[590]

Stalin took things yet another step further: "Today you call us the elder brother, but the younger brother can catch up to the elder brother and overtake him." This would not only be the wish "of every one of us" but would also be in keeping with the laws of evolution, he said. Stalin then called on all those present to drink to the day when the younger brother would surpass the elder one. Liu Shaoqi protested, to the hosts' amazement. The elder brother would always be the elder, he said, and the younger always the younger. "We will always learn from our elder brother!" Despite all efforts to convince him, Liu refused to accept the toast.[591]

The picture we have presented up to now is based primarily on Shi Zhe's account. Kovalev reported Stalin's statements in a similar fashion, albeit more concisely and in one part with a different emphasis. According to him, Stalin made his remarks in the context of a toast to the victory of the Chinese revolution and the health of Mao Zedong and the other Chinese leaders. He said that when he spoke of the Chinese Communists' successes and stated that "the Soviet people and the peoples of Europe have to learn from you," he was in no way merely flattering them and paying them compliments. The arrogance (*zaznaistvo*) of the leaders of the revolutionary movement in Western Europe had blocked the development of the social democratic movement in Europe after the death of Marx and Engels. "The center of the world revolution has shifted from the west to the east and has now shifted to China and East Asia." The Chinese Communists already played an important role, he said, and they naturally should not become arrogant (*zaznavat'sia*). At the same time, the responsibility that had been given to them had increased even further: "You must do your duty with respect to the revolution in the countries of East Asia." The Soviet people might possibly be somewhat superior to the Chinese Communists on general questions of Marxist theory. And yet in the implementation of Marxist principles in practice, "you have such great experience, that it would be good for us to learn from you. And we have already learned much in the past." One people must learn from the other, he said. Kovalev did not mention the argument over the use of the terms "elder brother" and "younger brother," however.[592]

Stalin's comments on the equality of the two parties and a Eurasian division of labor are linked to previously mentioned statements he had made during the July 11 session. His claims that the Chinese CP would later develop into the "elder brother" of the CPSU were, of course, pure hypocrisy. If he had been able, as he wished, he would have forced the Chinese comrades into a straitjacket, just as he had the Eastern European CP leaderships. Even Kovalev, who remained a faithful Stalin supporter to his death, admitted that Stalin had in reality acted differently: he had sought to force everyone into submission—including Yugoslavia and all the parties of the other socialist states.[593] Stalin knew full well, however, that he had nothing to work with in the case of China, where the CCP was about to achieve victory in the civil war contrary to his predictions and without Soviet assistance. And he was aware of Mao Zedong's relentless pursuit of Chinese national independence and proud conviction that he had developed an independent Chinese revolutionary model that could be applied all over Asia.[594] Stalin therefore did not hesitate to flatter the Chinese and at the same time to embarrass them by doing so.

Mao did some straight talking to Khrushchev when he met him in 1957 in Moscow. He said, "What do you mean by brother parties [*xiongdi dang;* literally, "elder brother–younger brother parties"]? That is just empty talk. In reality they are father-and-son parties, cat-and-mouse parties." The Soviet side, Mao commented on the talk later, had admitted this.[595]

Liu Shaoqi should have seen through Stalin and did not take his compliments very seriously. As already demonstrated, the Chinese leadership repeatedly had reason to mistrust him. On the other hand, there is evidence that Liu and the other members of the Chinese delegation were surprised by Stalin's ostentatious modesty and favorably impressed—as, according to Shi Zhe, had already been the case during the July 11 session when Stalin had confessed to having behaved incorrectly toward the Chinese CP.[596] It was in any case an advantage for the Chinese that Stalin spoke this way. They would subsequently be able to confront him with his own words, if it ever became necessary. Liu's decision to continue to play the role of younger brother was presumably also not meant entirely seriously. The already foreseeable final victory in the civil war had considerably increased the Chinese Party leaders' self-confidence. Liu probably acted modest not least because he wanted to alleviate Stalin's suspicion that Mao would follow Tito's path. He may also have been conscious of playing the role of suppliant in this situation.

Final Session and Farewell

The final session took place in Stalin's office on August 5. The participants were Stalin, Bulganin, Malenkov, Mikoyan, Kaganovich, Kovalev, Liu Shaoqi, Wang Jiaxiang, and Shi Zhe.[597] Stalin began by once again addressing the question of the experts who were to be sent to China and discussed their living conditions there in greater detail than before.[598] He then shifted the conversation to questions involved in a development strategy for China. He stressed the priority of economic development. Here they should advance in a solid and disciplined way, rather too gradually than too quickly; they should not seek to skip phases, and should take one step after the other, lest the economy fall into disorder. These suggestions appear to foreshadow the Sino-Soviet conflict over people's communes and the Great Leap Forward of 1958. Stalin then drew attention to the great significance not only of the natural resources, but also of modern technology and well-trained technical cadres and skilled workers. In China they should make use of the experience gathered by specialists, technicians, and skilled workers of the old society as they had in the Soviet Union. They should modernize existing industrial structures involved in, for example, building ships, locomotives, and vehicles and in addition create new branches of industry to build such things as aircraft, automobiles, and heavy machinery. Stalin also recommended that they follow the successful Soviet example and employ German, American, and Japanese technicians.[599]

The Soviet Union was ahead of China in constructing socialism, but its sufferings and sacrifices had been comparatively great: civil war, armed intervention from abroad, boycotts and the war against fascism. If one goes first, such difficulties could hardly be avoided. Given the changing situation, the USSR would take on more obligations from now on, particularly in the international field. It did so joyously and enthusiastically and wished the Chinese Communists a speedy and complete victory so that they would be able to share these obligations with the Soviet Union.[600]

Stalin himself raised the sensitive question of why Ambassador Roshchin had followed the KMT government from Nanking to Canton at the beginning of the year. This had happened, he said, because the Soviets had wanted to see what the KMT would do next. Liu apparently said nothing to this. In February, Mikoyan had given a more honest but also more humilating response to the same question from Zhou Enlai in Xibaipo. He had said that the Soviet action was "governed by law [*zakonomerno*]," as only one government existed in China. The Soviet ambassador was accredited to this government and it was "natural [*estestvenno*]" to move to Canton with it.[601]

Before his departure, Liu Shaoqi was invited to a farewell reception at Stalin's dacha in Kuntsevo. Stalin took the time to show Liu the entire estate, including the gardens. During the meal that followed, he asked when the Chinese planned to establish the government. When Liu mentioned January 1, 1950, as the probable date, Stalin urged them to hurry. The KMT had already been paralyzed. The moment was favorable. They should be aware that the enemy was exploiting "the fact that there was no government" to interfere in China's internal affairs. "You must take measures to counter such a devilish plan." According to Shi Zhe, these observations attracted the attention of the members of the Chinese delegation. They informed the Central Committee and Mao Zedong immediately, and Shi Zhe suggested that he considered it possible that Stalin's intervention influenced the decision to establish the government on October 1, 1949.[602]

Stalin probably had two motives for making his remark. First, he planned to establish legal relations with the Chinese Communists as quickly as possible, which he could do only after a government had been established. And, second, he wanted to use the opportunity to keep alive their distrust of the Western powers and wherever possible to increase it.

According to Shi Zhe, the talks in Moscow took place in a "friendly and sincere atmosphere that overflowed with enthusiasm [*reqing yangyi, youhao zhenzhi de qifen*]."[603] Stalin reportedly always left the initiative to the Chinese side when it came to deciding what should be discussed. Shi Zhe suggested that no misunderstandings or irritations arose and that Stalin's treatment of Liu Shaoqi had been "trusting and respectful." It is possible that Stalin in fact acted this way. One can, however, hardly claim that there were no ill feelings given the clear and irreconcilable disagreement on the question of Soviet support for an attack on Taiwan that Mao considered so important. There is evidence that Liu was extremely courteous—if not obsequious—toward Stalin, as discussed previously; once he made an offer to Stalin "to fulfill every one of your tasks."[604]

Shi Zhe noted slightly critically that during the talks in Moscow Stalin had shown himself not to be particularly familiar with customs and conditions in China or the history of the CCP and the Chinese revolution. Stalin had usually talked only about the Northern Expedition during the 1920s. And yet the suggestions he made for the development of China on the basis of Soviet experience had been of practical significance.[605]

At the beginning of August, Liu informed Stalin that Mao wanted him to return to Peiping.[606] Wang Jiaxiang was to remain in Moscow to support Liu Yalou and Zhang Xuesi, who were expected to arrive in Moscow at this time, in their negotiations for the air force and naval schools. Wang then had to return to Peiping in mid- or late August to attend the meeting of the Political Consultative Conference. Liu himself wished to leave around August 10.[607]

On the day of his return to China, August 14, Liu Shaoqi sent the Soviet leadership a letter in which he stated in emotional words his thanks for the "unusually cordial reception and consideration" his delegation had received in Moscow. The tasks that had been set for the visit had been resolved to their complete satisfaction, "in particular that China immediately receives the credit and the assistance of Soviet experts." On his arrival at the Chinese border station of Manzhouli on August 21, Liu Shaoqi thanked Stalin yet again by telegram.[608]

Military Aid

In the military sphere, the Soviet Union committed itself to support Communist China in the conquest of Sinkiang, deliver Soviet military aircraft, and establish an air force school in Manchuria and naval school in Port Arthur. The Soviet side stated that it was prepared to help develop Chinese coastal defenses and clear the mines in the water around Shanghai. The two parties also reached agreement on assistance for the modernization of the PLA and the Chinese armaments industry, the purchase of Soviet antiaircraft guns and ammunition, and the sending of Soviet military trainers and advisers to China. The Chinese Party leadership had already asked Mikoyan for military aid in February 1949 in Xibaipo.[609] At that time, Mikoyan had approved support for the organization of the armaments industry and the sending of advisers in principle (*v printsipe*).

On June 27, Stalin had urged Liu that the Chinese should conquer Sinkiang quickly and offered to send forty Soviet aircraft. On Mao's instructions, Liu requested Soviet air support on July 6 and stated that they were prepared to occupy Sinkiang quickly. Shortly afterward, Stalin renewed his promise to provide assistance.[610]

On July 25, Mao informed Stalin indirectly that he was prepared to send Deng Liqun, who was a member of Liu's delegation, to Sinkiang with orders to establish a radio link with Peiping. He said that he planned to capture Lanzhou in the fall and Urumchi in the winter. Peng Dehuai, who would assume primary responsibility for these operations, had inquired whether the army could purchase 1,000 to 2,000 motor vehicles in the USSR to transport the troops. Mao, in turn, had asked whether it would not accelerate their advance if they moved forward to Sinkiang on foot that winter to receive the Soviet vehicles via the Ili and Kulja territories. He said that he would forward Peng's answer to Stalin as soon as he received it. A few days later, Mao informed Liu that he had learned from Peng Dehuai that the army could conquer Lanzhou in late August or early September and then prepare for the conquest of Sinkiang.[611]

According to Chinese accounts, the advance to Sinkiang occurred essentially according to the plan Mao had presented to Stalin. Its beginning was moved forward and Peng Dehuai was charged with preparing the campaign. On orders from Mao, Deng Liqun was secretly sent to Sinkiang on August 14 to use intelligence methods with Soviet help to guide the entrance of the PLA. In this way, Sinkiang was conquered in late 1949 and not in 1950, as originally planned. The success of Sino-Soviet cooperation in the conquest of Sinkiang formed the basis of the joint strategic goal of eliminating all Western influence there. Shortly before Liu's trip to Moscow, Mao had instructed Kovalev exhaustively that he considered Sinkiang to be a strategically important province in the case of a "new war with the imperialists" because it constituted a corridor that linked China directly with the Soviet Union.[612]

There is a similar description in another Chinese publication. According to this account, Stalin advocated a rapid advance by the PLA to Sinkiang in his talks with Liu, in order to counter the establishment of an "Islamic republic" with support from the United States, that would have been disadvantageous both for the new China and for the USSR. At the same time, he called for CCP functionaries to be sent into the Ili, Tacheng, and Ashan territories to work with the rebels in an attempt to weaken the KMT troops. Stalin promised Soviet assistance in this. When Liu Shaoqi reported this to Peiping, the CCP leadership resolved to alter its original plan to initiate the Sinkiang campaign in spring 1950 and send troops immediately in order to occupy Urumchi by winter. Deng Liqun flew from Moscow to Kulja in mid-August and worked with the Soviet consulate general there to achieve this goal. The Soviet Union made forty transport aircraft available to accelerate the transfer of the necessary PLA units to Sinkiang.[613]

In January, Mao thanked Stalin for the assistance that a Soviet air force regiment had provided in moving some ten thousand Chinese troops.[614] It is not known whether this was a reference to the forty aircraft the Soviet Union was supposed to deploy in Sinkiang.

Stalin rejected the Chinese request for Soviet naval and air support for the planned conquest of Taiwan in the July 11 and 27 sessions, as described in detail earlier in this chapter. He had previously agreed to deliver Soviet military aircraft, train Chinese pilots in Moscow, and send the relevant Soviet advisers to China. During the July 11 session, he committed himself to assist in establishing a naval school in Port Arthur and constructing coastal defenses. On July 27, he stated his willingness in general terms to assist in modernizing the PLA and the Chinese armaments industry. They also agreed on Soviet support for the creation of an air force school in Manchuria.[615]

In Peiping, Mao Zedong and Zhu De together consulted with Liu Yalou in late July about negotiations with the USSR for the development of an air force.[616] On August 1, Liu Yalou and navy commander Zhang Xuesi traveled to Moscow, where they, along with Wang Jiaxiang, would discuss the concrete details of the planned foundation of air force and naval schools with the Soviet side. Liu Yalou arrived on August 11 and carried out the first negotiations about developing a Chinese air force two days later with Minister of the Army Vasilevskii.[617] When the KMT air force bombed a few cities on the lower course of the Yangtze, the Central Committee instructed the Chinese delegation to insist on moving up delivery of the promised antiaircraft guns. At the same time, it agreed to establish an antiaircraft defense regiment, which was apparently supposed to include Soviet advisers and trainers. The CC also informed Gao Gang that the 360 guns the USSR was to deliver would soon be handed over in Manchuria. Gao and Wu Xiuquan were responsible for accepting them. Stalin had already approved Soviet aid for the construction of coastal defenses in the July 11 session. Mao accepted a Soviet suggestion to send cameramen to China to shoot documentary films at the front during the PLA's planned campaigns beginning in October 1949. On the basis of the agreements reached with Liu's delegation, the Soviet Council of Ministers resolved on September 19 to deliver 334 aircraft along with artillery, including 360 antiaircraft cannon for a value of 166 million rubles (US$26.5 million) to the Chinese Communists. The Council of Ministers passed further resolutions involving deliveries to China on November 4, December 3, and December 19, 1949, and January 7, 1950, that may have included military equipment.[618]

There is no indication in the available publications that Soviet military aid was provided completely or partly free of charge. The Chinese side apparently had to pay for everything at world market prices—in the case of deliveries of equipment and ammunition, by means of the credit that had been granted or in exchange for deliveries of goods; and in the case of the advisers and trainers sent to China, in the form of compensatory payments to the Soviet companies that sent them and by assuming living expenses and paying a salary. By mid-February 1950, both sides agreed that the credit should not be used to pay for armaments delivered to China by January 1 but instead for deliveries of goods in 1950, while they would agree on a method of paying for armaments to be delivered later during 1950.[619]

In agreeing to supply military aid, Stalin was careful to ensure that it could not be recognized publicly outside the borders of Manchuria until the People's Republic of China was established. He believed that the Yalta agreement of February 11, 1945, and Sino-Soviet treaty of August 14, 1945, entitled him to act relatively freely within Manchuria. He therefore also wanted to sell the minesweepers intended for the waters around Shanghai pro forma to the Manchurian government, which would then sell them to the central government. (Stalin's reference to the "central government" was a joke, as after all it did not yet exist.) For the same reason, he wanted to train the crews of the boats in Manchurian harbors or in Vladivostok and not in Shanghai itself. He was also unwilling to send a fleet to Tsingtao before the Communists proclaimed a government.[620]

Stalin rejected the Chinese guests' request to inspect nuclear installations.[621] Instead, he showed them a film of a Soviet atom bomb test. This can only have been a film documenting a preparatory phase to such a test, because the first atom bomb was tested in the Soviet Union on August 29, 1949, after the Chinese delegation had already left. Shi Zhe hypothesized that Stalin had shown the film to demonstrate the Soviet Union's military power to the Chinese comrades and to give them the feeling that they were under the protection of the Soviet Union's atomic umbrella. The presentation greatly impressed the Chinese, who assumed that the Soviet Union had made a breakthrough in the development of new weapons.[622]

Secret Service Cooperation

An agreement was reached during Liu Shaoqi's stay in Moscow to foster closer collaboration between the secret services of the two sides. Liu and Gao Gang had already expressed their interest in such a "joint program against the imperialists and reactionaries" to Kovalev before they arrived in Moscow. They explained this by pointing out that many of the Chinese who worked for the Soviet secret service in China who had either been recommended by the CCP or recruited by the Soviet side itself had withdrawn from the Party's supervision, had been corrupted, and, in some cases, had even worked for the secret services of the United States or the KMT. As an example, Gao explained that the CCP secret service had discovered and confiscated 300 radios in Manchuria, of which 40 were supposed to have been used for the Soviet Union but were in fact being used in the interests of the KMT and the United States. This had been possible because there were not sufficient contacts between the secret services of the USSR and the CCP. They had to put a stop to such uncontrolled actions. Kovalev passed this message on to Stalin on July 7.[623]

Stalin agreed. The situation, he said, called for the efforts of the secret services of both sides to be united. The Soviet side was prepared to begin immediately: "Let us act as a united front!"[624] According to Soviet contemporary witness reports, Stalin revealed his network of agents in China, which dated from the time of the Comintern, to the Chinese Party leadership in the course of the new cooperation—an impressive gesture of trust.[625] This resulted, however, in the execution or imprisonment of hundreds of CCP members loyal to Moscow. This measure sounded relatively harmless the way the Chinese put it. All CCP members who had been secretly active for the Soviet secret services "proceeding from an internationalist ideology" had to reveal this to their Party organization.[626] They were supposedly treated leniently. One can assume that many of them were charged with reporting to Moscow on procedures within the CCP.

The agreement between Stalin and Liu on collaboration between the secret services was confirmed during Mao's visit to Moscow in early 1950.[627] One can assume, however, that the Soviets continued to conduct espionage activities to a certain extent in China.

Economic, Technical, and Cultural Cooperation

The agreements on future cooperation in the areas of the economy, trade, technology, administration, science, and culture were of great significance to the Chinese delegation. We do not have all the details of the talks on these or the agreements achieved.

The most important agreement involved the 1 percent preferred credit for the amount of US$300 million, payable over five years and repayable within ten years from the payment of the entire sum of the credit, which has already been discussed in detail.[628] It was signed on July 30 and was publicly given the form of a one-year trade agreement between the USSR and Manchuria. According to a Soviet press release, it called for Manchuria to deliver soybeans, vegetable oil, corn, and rice, among other things, in exchange for Soviet industrial equipment, motor vehicles, petroleum products, fabrics, medicines, and other goods.[629]

Because the text of the agreement remains unavailable even now and we also lack any other information on its contents, we cannot verify whether this is accurate. One can assume, however, that the agreement merely established which goods would be delivered, but did not establish the amounts. Owing to the lack of an economic infrastructure, raw materials, and expertise, the Chinese Party leadership declared that it was unable at that time to present a comprehensive list of goods in connection with the credit granted. It proposed that the planned Sino-Soviet Economic Commission should be stationed in China. Kovalev should first accompany the most important Soviet experts to China and then, along with the Chinese side, put together the bulk of the list of goods.[630]

In late November, the Chinese leadership requested to review the Soviet plan data, apparently in order to form a statistical basis for their request for goods from the USSR.[631] The Soviet Foreign Ministry suggested that the Politburo make available to the Chinese the materials "on the foreseeable fulfillment of the plan for 1949 and the plan data for 1950." In mid-December, the Chinese side was still having difficulty determining its need for Soviet goods.[632]

It is unclear to what extent the July 30 agreement was implemented during the following months. In early August, for example, Liu Shaoqi sent Mikoyan a request for goods: "The Central Committee of the CCP wishes . . . to purchase automobiles in the USSR on the account of the exchange of goods between Manchuria and the USSR." It ordered thirty automobiles of the Pobeda brand, ten of the Moskvich brand, and twenty buses. Liu asked for prices quoted in U.S. dollars. The automobiles were to be for the use of the Soviet specialists.[633]

In the discussion of how trade should be organized in the future, Stalin recommended that China and the USSR not always use the usual procedure between capitalist countries of making deliveries against payment or credit, as the problem of foreign currency made this very difficult. Instead, they should give precedence to the exchange of goods for goods. The prerequisite for this, however, was that fair prices could be established. In this, China was at a disadvantage, Stalin admitted, as it would deliver primarily agricultural produce and raw materials, the prices of which were low on the world market in comparison with industrial goods. But the Soviet side would keep China's interests in mind. Shi Zhe added to this account that the USSR had not kept this last promise and had used its greater experience in economic affairs to cheat China in trade. This had led to tensions during the 1950s. The two sides were already arguing about world market prices in early 1950 in the negotiations for the trade agreement, he noted.[634]

The source materials available up to now do not allow us to establish with certainty which of the numerous requests Liu presented during his stay in Moscow were transformed into agreements and what the fate of these agreements was—with the exception of the experts who were sent to China.[635] We are aware of certain individual Soviet approvals, promises, and resolutions. On July 11, Stalin stated that he was prepared to support construction of a railroad line from Ulan Bator to Zhangjiakou. This approval appears to have resulted in the November 5 Council of Ministers' resolution, which called for the delivery of rails with the necessary attachments for a value of 33.3 million rubles (US$6.3 million).[636] On June 27, Stalin promised the Chinese delegation that the USSR would assist China in establishing air services, civilian aviation, an aircraft industry, and commercial shipping. He also indicated that he was prepared to satisfy Chinese requests involving the creation of postal, telegraph, railroad, and air connections between the USSR and China and the founding of a joint Sino-Soviet airline.[637] The same was true of the request that the Soviets send specialists and teachers to China to assist in the development of the Chinese administration, economy, and diplomacy and accept Chinese delegations and students in the USSR.

A university in which Soviet teachers would at first instruct up to a thousand students was to play an important role in providing the training "in the development and administration of the state and of companies" that Liu proposed for Chinese cadres.[638] The two sides had initially considered establishing the university in Moscow. Mao's idea was that the Chinese should study with Soviet professors there in order to see how the Soviet and capitalist systems differed. China would assume the costs. Alma-Ata was occasionally also discussed as the location. In the end, they agreed to establish the university in Peiping with Soviet assistance. It is unclear to what extent Stalin looked into Liu's extensive plans for close economic and technical cooperation between the USSR and Manchuria at that time. The same is true for Liu's proposals to institute intensive exchanges in economic and cultural areas.[639]

The first spectacular event that resulted from the agreements between Stalin and Liu on economic and cultural cooperation was the trip to Peiping by a delegation of Soviet intellectuals and scientists that took place at the end of September in connection with the upcoming proclamation of the People's Republic of China. The delegation consisted of forty-three individuals and was led by the prominent writers Alexandr Fadeev, chairman of the Writers' Union, and his deputy, Konstantin Simonov. Mao received the delegation in early October. In an emotional speech, Mao described the Russian October revolution of 1917 and the Chinese revolution of 1949 as the two most important revolutions of the twentieth century. He did not neglect to make reference to the different sizes of the populations in the two countries. When the 500 million liberated Chinese united with the 200 million liberated Soviet people, the balance of power between the two camps in the world would change very significantly, he said.[640]

One could interpret Mao's comment that they had to realize that the Chinese Communists had received no assistance from outside during its eight-year war against the Japanese and had had to rely entirely on their own strength (*wanchuan yikao zili gengsheng*) as a critical reference to the USSR's failure to grant aid before 1945. It is striking that Mao used the concept "by their own strength [*ziji de liliang* and *zili gengsheng*]" twice more in the same context: he described the Chinese people as having also earned its victory in the civil war in this way. China had one advantage today, he continued, namely, a good neighbor, and was therefore not isolated and not without assistance. The words of the well-known intellectual Wu Han appeared to have a similar double meaning when he commented at a press conference a few weeks later in Moscow that the Soviet Union had not armed the Chinese Communists "with cannons and machine guns" but had given them Leninism and "the most powerful weapon—the brilliant works of the great leader of all the toilers in the entire world, comrade Stalin."[641] One senses in Mao and Wu's statements that they were both critical of the Soviet failure to provide aid and proud of what they had nonetheless achieved.[642]

When he completed his speech, Mao asked the members of the Soviet delegation to accompany the Fourth Field Army on its march toward the south to gain an impression of the battles. Apparently only Simonov accepted this offer, and he wrote a novel about his experiences. In the context of a survey of the activities of KMT and American spies in China, Chinese security chief Li Kenong later reported to the Soviet ambassador that he had had over twenty suspicious individuals arrested for crowding around the Soviet delegation at that time.[643]

Liu's Return Trip and the First Group of Soviet Specialists

The importance the leadership of the Chinese Communist Party attached to the sending of Soviet specialists can be seen in the fact that Liu Shaoqi included it along with the award of the US$300 million credit as the only two areas of the planned assistance he mentioned in his letter of thanks to Stalin on the day he returned to Peiping.[644]

Let us recapitulate briefly what had been agreed upon by the two sides on the subject of the specialists on the eve of Liu's departure. At the time of Mikoyan's talks in Xibaipo in February 1949, the Chinese Party leadership had already repeatedly expressed the request that the Soviet Union send specialist workers to China to assist with economic development. During those talks, Liu Shaoqi had mentioned the sending of advisers and technicians in the various branches of the economy in second place when he presented a three-point list of the forms the requested Soviet assistance to China might take. Ren Bishi had specified that they would need "no fewer than 500" specialists to assist in the industrial development of Manchuria alone. Zhou Enlai had also requested that the Soviets send military advisers. There had then been an exchange of letters between Stalin and Mao on the topic of advisers.[645]

During Liu Shaoqi's stay in Moscow, Stalin addressed the topic already in the first session on June 27—immediately after discussing the credit issue. He gave Liu his approval on principle and announced that a first group of advisers would be sent "in the near future." He insisted, however, that the question of salary would first have to be clarified and asked that the advisers be allotted as many provisions as the best Chinese specialists. Because the cost of specialists was high in the USSR, the Soviet side would pay an additional amount if necessary. Furthermore, he would personally ensure that the Soviet experts behaved properly in China. In his written report for Stalin on July 4, Liu briefly addressed the problem of advisers when he requested that the Soviets send specialists and teachers generally. The same is true of Liu's letter to Stalin on July 6. In it, he expressed the request for assistance in training Russian-speaking cadres in China.[646]

As mentioned above, Mao instructed the Chinese delegation in early August to ask the

Soviet Party leadership to ensure that "Kovalev at first bring the most important experts to China"; these experts, together with Chinese specialists, would then determine what goods China required from the USSR.[647] If Stalin agreed, "elder brother Ke [*xiong Ke*]"—the reference is to Kovalev—should first bring specialists in railroads, electrical energy, steel production, mining, the oil industry, and the military. At the same time, Liu asked Stalin in a letter announcing his departure from Moscow "on about August 10" to allow Kovalev to travel with him "if possible with some of the Soviet specialists that had been selected to work in China."[648] The other advisers could travel with Wang Jiaxiang, whose departure was planned at that time for the end of August at the latest. Shortly afterward, the Council of Ministers of the USSR drew up an instruction to govern the activities of the specialists that would be sent to China.[649]

At the farewell meeting of Liu Shaoqi's delegation with the Soviet leadership, they reached a final agreement on sending Soviet experts and technicians. Stalin specified four conditions. First, China must arrange favorable housing, living, and working conditions and generous pay. Second, China was responsible for the maintenance of the experts' family members if they should remain in the Soviet Union and for favorable living conditions if they went to China. Third, the duration of the experts' stay in China should furthermore be determined on the basis of the situation and practical needs. Last, the experts should be subject to Soviet and not Chinese jurisdiction in criminal cases. The Party leadership in Peiping had apparently previously approved a very general "Agreement on the Working Conditions of the Soviet Specialists"[650] and instructed Liu Shaoqi to discuss questions involving the practical implementation of the expert program with Gao Gang and Li Fuchun on his return trip through Manchuria. A dispute later arose over pay for the specialists during negotiations in February 1950.[651]

They agreed that "more than 200 experts" would be sent to China.[652] In the end, 220 boarded the train along with Liu Shaoqi under Kovalev's leadership on August 14 as a first group.[653] They were received enthusiastically by a large crowd at the station on their arrival in Mukden on August 25. Liu concerned himself personally with their well-being. An official reception was held two days later in the guesthouse set aside for them at which Gao Gang, Kovalev, and Liu Shaoqi spoke. On August 28, Liu reported on his Moscow visit to the leadership cadres of the Northeast Bureau of the CCP. In his report, he emphasized that the primary task was to develop China economically and to industrialize it. To do this, they had to learn how to do business, administer finances, and produce steel. The Soviet experts' arrival created good conditions for acquiring such knowledge; but the Chinese also had to make an effort themselves. Liu Shaoqi concluded his report by saying that the vast majority of the experts would remain in Manchuria; the onward journey to Peiping was planned for only thirty to forty of them.[654]

Given the fact that the Soviet Union had already been sending specialists to Manchuria at Mao Zedong's request since June 1948, there were by now more than 600 Soviet development aid workers in China, according to Kovalev. The experts who had previously been sent to Manchuria were pro forma engineers and specialist workers employed in repairing railroad lines and bridges. In fact, this assistance, which was legal because the Soviets were joint owners of the Chinese Changchun Railroad, served as a "very convenient cover" (Ledovskii), allowing Soviet specialists to provide development aid to the newly established Communist regime in Manchuria in a very broad range of areas. During the next year, there appear to have been between 350 and 800 experts in China at the same time. With reference to March 1952, for example, Soviet sources mention 332 advisers and teachers and 471 technical specialists; in September 1953, Zhou Enlai asked Molotov for an additional 172 experts to supplement the 277 already in China.[655]

The experts who traveled with Liu Shaoqi included high-ranking economic and administrative cadres and highly qualified engineers. Given China's low level of development, they could not be employed properly, however, and returned to the USSR shortly afterward and were replaced by less highly qualified colleagues. One case became known in which a dispute arose between Kovalev and Huang Mianfeng, then deputy commander of the railroad engineers in Manchuria, whom Kovalev accused of "anti-Soviet resentment" in a peremptory tone. Huang denied this indignantly. The incident was supposedly resolved peacefully.[656]

Liu arrived in Peiping on August 30 and reported to the Party leadership.[657] On September 3, he delivered a report on his talks in Moscow to high-ranking cadres. With great emotion, he invoked Sino-Soviet friendship and thus initiated the propaganda campaign that continued without interruption throughout the 1950s: "We must learn from the Soviet Union! [*Yao xuexi Sulian*!]" Liu emphasized that the most important task facing the Party now involved economic reconstruction. To do this, they needed the Soviet experts' assistance. Soviet-Chinese friendship was a decisive condition (*yi ju you zhuedingxing de tiaojian*) for economic development. The Chinese comrades had to maintain good relations with the Soviet experts. These were, he said, an aspect of the concrete relations between the two great nations, the Soviet Union and China, the two parties, and the two states. This constituted an important political question, and they could not fail. China had to learn from the Soviet Union in all fields—politically, organizationally, ideologically, technically—as well as in the fields of law, finance, the economy, culture, and education. The Chinese leadership supposedly reacted with particular emotion to Liu's report of Stalin's self-criticism.[658]

The thirty high-ranking Soviet experts who traveled on in the company of Kovalev and Shi Zhe from Mukden to Peiping in late August were received at the station there by Zhou Enlai, Liu Shaoqi, Chen Yun, Yang Shangkun, and other Party leaders. They supplied provisions to the experts free of charge and made available special cooks, who served them Russian dishes. Shortly after their arrival, at Zhou Enlai's suggestion, Mao even received the Soviet conductresses at the seat of the Party and government in Zhongnanhai. In late September, top Chinese leaders visited the experts. In his speech, Kovalev emphasized that the experts "sent by Stalin" were providing "selfless assistance" that was of great importance to the Chinese revolution. Zhou Enlai's address included a remarkable passage that put this statement into perspective. The Soviet assistance, he said, was a concrete expression of proletarian internationalism that obligated the two sides to provide mutual assistance. And when China became strong, it would also mean support and reinforcement to the Soviet Union.[659]

In the course of a reception the Soviet advisers gave for the Chinese leadership on the occasion of the anniversary of the Russian October revolution on November 7, Mao Zedong indicated in a speech that the Chinese side was not prepared to follow all Soviet suggestions blindly. The Soviet advisers, Mao said, had only just arrived and were not yet familiar with China. They should therefore consult with their Chinese comrades in order to understand everything better.[660]

Kovalev acted as leader of the Soviet group of experts until he traveled to Moscow in Mao's company on December 6. He was designated the "general adviser [*zong guwen*]" by the Chinese. He negotiated on the Chinese side with top functionaries in the areas of the economy, finance, and heavy industry. These included, in particular, Chen Yun (Politburo member, one of the four vice-premiers, chairman of the Government Commission of Finance and Economic Affairs, and minister of heavy industry), Peng Zhen (Politburo member and vice-chairman of the Government Commission of Politics and Law), Bo Yibo (Central Committee member, vice-chairman of the Government Commission of Finance and Economic Affairs, and minister of finance), and Li Fuchun (member of the Government Commission of Finance and Economic Affairs). With the exception of Bo Yibo, they had all held high positions in the Party and the government apparatuses of Manchuria before the establishment of the People's Republic of China. Chen Yun was the only one of them who previously had concrete experience in dealing with the Soviets. He had worked with the Comintern in Moscow from 1935 to 1937 and had been the intermediary to the Soviets in Sinkiang in 1937.[661]

Results of the Liu Mission

Liu Shaoqi could essentially be satisfied with the results of his mission. He was primarily concerned to bind the Soviet Union, which was emerging as the only available security partner and donor state, as closely as possible to the new China. Given the Chinese CP's imminent victory in the civil war, Stalin no longer perceived any risk in shifting his saddle to the Communist horse in his China policy. He did wish to wait until the new government was formed to shift his alliance publicly, but treated the CCP de facto as if it were already his new Chinese alliance partner.

During Liu's visit in Moscow, the two sides reached agreements on the political and economic strategy the CCP leadership should pursue. The same was true of the principles that should underlie the new China's foreign policy. Stalin stated that he was prepared to recognize the Chinese CP as a kind of equal partner in world communism. He did oppose accepting CCP membership in the Cominform, but at the same time granted it a leadership role in Asia. Liu achieved in Moscow what Mao Zedong had not even been able to get Mikoyan to promise half a year earlier in Xibaipo: a preferential credit of US$300 million, which corresponded to the amount that Mao had requested from Mikoyan. Stalin also granted Liu extensive assistance in both the civilian and military sectors.

On the debit side, it should be noted that Stalin refused to provide the support by the Soviet air force and navy the Chinese requested for the planned invasion of Taiwan. Liu suggested that the CCP had put the Mongolian problem to rest for the time being. They reached no agreement on the sensitive issue of Soviet privileges in Manchuria. Stalin indicated goodwill with respect to Port Arthur and Dairen, but he did not commit himself to diminishing Soviet privileges either immediately or in the near future. Because Stalin had decided to adopt a hard line on the future of the Chinese Changchun Railroad during Mikoyan's visit in Xibaipo, this constituted the most sensitive issue and was therefore not even made a topic of negotiation. The question of concluding a new treaty of alliance remained open, too. Both sides must have been fully aware that the two last named areas would constitute the most difficult part of the agenda for the planned meeting between Stalin and Mao Zedong in Moscow.

Although these problems continued to exist, the two sides had come considerably closer together. If Mikoyan's trip to Xibaipo had served to establish initial contact at the top level, Liu Shaoqi's talks in Moscow had already set the stage for the alliance both sides by now had come to seek. A Chinese historian correctly observed that relations between the Soviet Union and Communist China entered a new phase with Liu's trip to Moscow.[662]

The CCP's Tactical Exploratory Contacts with the United States and Great Britain and the Role of the Soviet Union

The author assumes that the leadership of the Chinese CP essentially decided during 1947 to seek strategic cooperation with the Soviet Union in the period following the Communist seizure of power. This does not relieve him of the obligation to test all the evidence that could suggest a different interpretation and to disprove the hypotheses that are based on it. This will be done in the following pages.

It is worth noting that Washington also assumed relatively early that there would be a strategic partnership between the USSR and a Communist China. An analysis drawn up in the State Department in September 1948 described China's natural resources as modest, its birthrate as "classic Malthusian," and its prospects for economic development as slight. It came to the conclusion that there were neither economic nor strategic reasons for preserving mainland China from domination by the Soviet Union. After the Cominform condemned Yugoslavia, various parties speculated that China would also follow a road that was independent of the USSR. This scenario of an "Asian Titoism" was incorporated in January 1949 in a recommendation by the National Security Council that exerted a decisive influence on Dean Acheson's China policy. Its goal was to create a "unified, stable, and independent China, friendly to the U.S.," toward which they should use flexibility to keep it from becoming an "adjunct of Soviet power." Because a Communist victory in the civil war became increasingly probable in March, Washington resigned itself to the idea that China would remain allied to the Soviet Union for twenty to twenty-five years. Nonetheless, the United States should employ political and economic means short of military intervention—such as, to a certain extent, trade—in order to exploit rifts in the relations between Peiping and Moscow and also between "Stalinist and other elements" in China. This was also one reason that Acheson accepted the request of Ambassador Stuart in Nanking to establish contacts with the Chinese Communists.[663]

If one takes the basic position mentioned above that the Chinese CP had made a strategic decision in favor of the USSR, the contacts maintained by the Communist leadership with the United States and Great Britain, which were particularly striking in the spring and summer of 1949, can have been established only for tactical purposes. Very diverse individuals made such probes at various levels calling for trade and economic and technical collaboration with the United States and other Western industrialized nations. These individuals included the mayor of Peiping, Ye Jianying; the minister for industry and trade, Yao Yilin; the mayor of Shanghai, Chen Yi; and the Communist representative in Hong Kong, Qiao Guanhua.[664] As will be discussed in detail below, the most spectacular contacts were made with the U.S. ambassador in Nanking, John Leighton Stuart, the American consulate general in Peiping (the Zhou démarche), and the British government.

The debate over the meaning of these contacts led among other things to the emergence of a hypothesis in the West that Washington could have lost a favorable chance to link Communist China to the United States and the Western world (the lost chance school).[665] This hypothesis can be considered to have since been refuted.

As demonstrated above, Moscow and the Chinese Communists had agreed in principle that the CCP should pursue a relatively cautious policy toward the United States and the other Western states. Stalin did not want to be forced into a major conflict with the United States because of his Chinese comrades' careless behavior. He urged the CCP leadership to adopt a hard-line, anti-Western course only in Manchuria, which Roosevelt and Churchill had recognized as belonging to the Soviet sphere of influence at Yalta. With respect to the rest of China, he advised moderate behavior toward the West. He recommended that China establish economic contacts, but opposed establishing diplomatic relations with the Western states quickly because he suspected that Mao Zedong might become a Chinese Tito.

In the following, we will describe and analyze the aforementioned contacts with Stuart, the American consulate general in Peiping, and the government in London in detail. We will be particularly interested in assessing the role played by the Soviet Union. To do this, we will make use of previously unknown Soviet archival documents and contemporary witness reports.

Huang Hua's Contacts with John Leighton Stuart and Andrei Ledovskii

We have access to reports by Ambassador Stuart to the State Department and Huang Hua's memoirs, which essentially agree with those reports, on the talks between Stuart and his personal secretary Fu Jingpo and Zhou Enlai's agent Huang Hua, which took place between May 13 and July 25.[666] Until recently, the great unknown in this story was the role the Soviet Union had played. The first signs of it emerged in 1994 when excerpts from a telegram sent to Stalin by Kovalev in late May 1949 were published. In it, Kovalev informed Stalin about his talk with Mao Zedong in which the latter had reported on the contacts that had taken place to that point between Stuart and Huang Hua. It can be concluded from Stalin's reply that Mao and Stalin had not previously coordinated these contacts—or at least had not previously coordinated them in detail. We do not know if there was any further exchange of opinions between Mao and Stalin on the discussions between Stuart and Huang Hua after late May.[667]

It has recently become known that there were numerous meetings between Huang

Hua and Andrei Ledovskii, first secretary of the Nanking branch of the Soviet embassy during the entire period in which the talks were held between Huang Hua and Stuart or Fu Jingpo. In these, Huang reported extensively on his talks with Stuart. Huang and Ledovskii knew each other from the time when the latter was Soviet consul general in Peiping. Huang arranged their first meeting in Nanking on May 4, 1949, when he unexpectedly visited Ledovskii.[668]

The fact that Stuart had signaled to the Chinese CP leadership through a middleman in Shanghai in late March that Washington intended to discontinue its policy in favor of the Kuomintang in the context of the unresolved Ward case[669] and hoped that the Chinese CP would halt its anti-American propaganda so that the two parties could achieve a rapprochement clearly played a not insignificant role in the Chinese CP's willingness to establish contacts with Stuart. The leadership apparently took this approach seriously. Zhou Enlai reacted with concern to Stuart when soldiers from the PLA broke into the ambassador's residence following the conquest of Nanking in late April. On April 28, Mao Zedong noted in an internal memo that the Chinese Communists could consider establishing diplomatic relations with Washington and London if the United States and Great Britain broke off relations with the KMT.[670]

On March 10, Stuart had asked Acheson for authorization to establish a connection with the Communist leadership with the intention of improving relations with it. Stuart's telegram stated:

> I have the honor to request that I be authorized to discuss with top Communist leaders the points at issue between the United States of America and the Chinese Communist Party in the hope that this would help towards a better understanding. . . . It may be naively visionary to imagine that I or anyone else can influence the Chinese Communist Party to adopt a more broadly tolerant policy. But in view of the fateful questions at stake it is abundantly worth the effort and will at the lowest justify us the more in any sterner measures we may feel called upon to take.[671]

Although Stuart requested a response as quickly as possible, he received only one on April 6. Acheson approved his proposal with reservations. He authorized Stuart to enter into a dialogue with the CCP leadership. The contacts had to be kept secret, however. He was to address the reports on the progress of the talks to Acheson personally. The State Department reserved the right to decide whether they should inform the public about them. This could occur if the talks failed.[672]

On April 24, Communist troops broke into Nanking. The following day, Stuart sent a telegram to Zhou Enlai, who was in Peiping, and asked him to come to Nanking for talks about American-Chinese relations.[673] Stuart apparently chose not to inform Acheson about approaching Zhou Enlai.[674]

Instead of coming to Nanking himself, Zhou Enlai appointed Huang Hua head of the Foreign Affairs Office of the Military Control Committee (*junguanhui waishichu*) of the city of Nanking.[675] In this role, he was to cultivate contacts with the foreign embassies and establish a connection with Stuart. From the Soviet side, Huang was seen in this function as the de facto vice foreign minister—that is, as Zhou Enlai's deputy.[676]

Stuart's personal secretary, Fu Jingpo (Philip Fugh), had been a fellow student of Huang Hua's at Yenching University in Peiping, which was founded by Stuart during the 1930s;

Stuart at that time was rector. Fu Jingpo established contact with Huang Hua unofficially in early May. On May 7, Huang Hua received him. According to Huang, Fu claimed that Stuart had gradually recognized his mistakes about the KMT over the past year. The United States was already no longer providing any further aid to the KMT. Contrary to the KMT's wishes, Stuart had remained in Nanking to establish contacts with the CCP. Fu said Stuart hoped to be able to meet with Huang Hua. They could modify the Sino-American trade agreement, he suggested. The time had come for a change in American China policy. This change could be made with the "old college president [*lao xiaozhang*]," he said; this would be better than with a new person. Huang reported that he replied that he would have to think about the question of a meeting. American assistance to Chiang Kai-shek had, he said, caused enormous losses and endless pain to the Chinese people. The United States at that time had not yet put the ax aside; how could it expect the Chinese people to have friendly feelings for it again? Empty words did not compensate for the reality. The United States must first "do more things that are of use to the Chinese people [*zuo geng duo youyi de yu Zhongguo renmin de shi*]," and then it would gradually find understanding among the Chinese people, he said.[677]

In a report to Secretary of State Acheson on the first meeting between Huang Hua and Fu Jingpo, Stuart began by describing its background. According to the report, Fu called Huang's office and left his name. Huang called back the next day and explained that it would be "inconvenient" for him to visit Fu and proposed that Fu should come to him. He reportedly received Fu "cordially" and asked about their "old college President." Here, as again on later occasions, he avoided using Stuart's official title and declared that he did not recognize Stuart as ambassador nor any other ambassador, because they were accredited to the KMT government. He pointed out twice that it was up to the United States to take the initiative in establishing diplomatic relations with a people's democratic government at the appropriate time. Huang attacked the foreign policy of the United States, its support for the KMT, and its responsibility for the civil war, which were the reasons the Chinese CP considered the United States to be "an enemy." The "New Democracy" that was being presented to the Chinese people differed, he said, from the old democracy of the American and British model. The CCP wanted to eliminate feudalism and American and British imperialism in China. When asked about the Communist soldiers who had intruded into Stuart's bedroom on April 25, Huang passed on a message from Zhou Enlai in which the latter stated that he was distressed by the incident.[678]

According to Stuart, Fu Jingpo proposed that Huang Hua visit his "old college President." Huang Hua agreed, but said he wanted to wait a bit longer. The talk took place in Huang's office in the former Foreign Ministry of the KMT government, lasted one hour, and Huang did most of the talking. Huang reportedly treated Fu in a "most friendly" fashion, accompanied him to the street gate, and told him he would call.[679]

On May 10, or three days after his meeting with Fu Jingpo, Huang Hua visited Ledovskii in his home and reported to him among other things about his talk with Fu. Although Stuart had received no response to his telegram of April 25 inviting Zhou Enlai to Nanking to hold talks, he reported, it was clear to Stuart that Huang's arrival in Nanking had something to do with his telegram.[680]

According to the information Huang Hua provided to Ledovskii, Fu Jingpo informed

Huang on instructions from John Leighton Stuart that he wanted to meet with Huang to discuss the relations of the United States with the Chinese Communists. The United States had changed its China policy. There was evidence to support this. The U.S. government had discontinued its aid to the KMT. He, Stuart, had not gone to Canton along with the KMT and had remained in Nanking, Fu said, to await the arrival of the PLA and the establishment of the new power. This had exerted a certain influence on the other diplomatic representatives, who had followed his example. Washington, Fu continued, was prepared to review the treaty concluded with the KMT government in 1947. Stuart, Fu said, wanted to assure the new Chinese authorities through Huang Hua that he recognized the errors in the earlier U.S. policy toward China. Stuart wished, Fu continued, to dedicate the rest of his life to the goal of doing everything he could to develop the friendship between the American and the Chinese peoples and repay the debt that was owed to the Chinese people for the mistakes of the past.[681]

According to information he himself provided Ledovskii, Huang Hua replied to this that Washington had abandoned Roosevelt's policy and taken a hostile position toward the Chinese people. The United States had helped the KMT government in its war against the democratic forces of China during the past three years. This aid had been the primary cause of the current civil war. Without active support from the United States, Huang reportedly said, Chiang Kai-shek would never have dared to begin this military adventure. The new Chinese authorities, he said, had no reason to believe in the sincerity of Stuart's statements, as there were no facts to indicate any change in American policy. He would have to think about meeting with Stuart and would communicate his answer later.[682]

When Huang Hua asked Ledovskii what Fu Jingpo's visit could mean, Ledovskii replied that the Americans wanted to investigate the CCP leadership's position toward the United States in order to determine its subsequent strategy and tactics in China in view of the KMT's imminent final defeat. Its strategy and tactics would be developed in the context of an increasingly tense conflict within the ruling circles in the United States with respect to the Chinese problem and of contradictions between the United States and the Western European powers and between the United States and Great Britain in particular. According to information from the Soviet embassy, the Americans were particularly interested in the relations between Moscow and the Chinese CP.[683]

A comparison of the three reports on the talk between Huang Hua and Fu Jingpo on May 7 reveals that Huang Hua's article agrees with his own description as reported by Ledovskii on many important points. Huang did, however, suppress from Ledovskii the fact that he knew Stuart and Fu well from his time as a student at Yenching University. He apparently wanted to avoid the appearance of a personal relationship of any kind with Stuart. Accordingly, there is also no mention in Ledovskii's account of the observation in Huang's article that Fu had appealed to Huang to bring about a change in American China policy together with their "old college President." The same is true of Huang's challenge that the United States would first have to do more useful things for the Chinese people before it could find understanding from them. Huang apparently wanted to avoid saying anything to Ledovskii that could suggest that the talks took place in a positive atmosphere and that an understanding with Washington might be possible.

It is striking that Stuart's report to Acheson lacks any mention of Stuart's admission to

Huang of mistakes in American China policy or to his claims that aid to the KMT government had already been halted, that the trade treaty could be revised, or that the time had come for a change in American China policy. As these points are portrayed in the same way both in Huang's article and in his description of the talk to Ledovskii, Fu Jingpo apparently did in fact present them in this way. Why are they missing from Stuart's report to Acheson? Fu Jingpo may have presented them without having been instructed to do so by Stuart, and Fu may therefore have left them out when he made his report to the ambassador. But it is also possible that Stuart had indeed instructed Fu to make these statements although he knew that he was overstepping his mandate, and therefore he decided not to inform Washington. The second hypothesis appears more plausible because Stuart himself made similar statements in his subsequent talk with Huang on May 13—again without informing Acheson, as will be described below. The fact that he was not authorized to do so can be seen from Acheson's dispatch of April 6, in which he authorized Stuart to discuss the topics Stuart had mentioned in his March 10 telegram to Acheson with "top Chinese Communist leaders."[684] To summarize the meaning underlying these topics, which were addressed with such rhetorical flourish, impressive idealism, and scanty precision, Stuart proposed to attempt to convince the Chinese Communists to grant "basic freedoms" in China and pursue "a constructive, friendly foreign policy." There was no mention of admitting that American China policy had been mistaken or of any imminent change in this policy to allow for an opening to Communist China, as Fu Jingpo allegedly claimed to Huang Hua.

In Ledovskii's reproduction of Huang Hua's report of his talk with Fu Jingpo, it is striking that Ledovskii did not give Huang any suggestions about how the CCP should behave toward the United States or the other Western powers. This could be because Ledovskii was being careful not to provide advice without concrete instructions from Moscow.[685]

On May 10, the day of the described talk between Huang Hua and Ledovskii, the Peiping Party leadership gave Huang Hua the green light for a meeting with Stuart to "find out the intentions of the American government." In the meeting, Huang was primarily to listen rather than to express his own opinion. If he nonetheless did express himself, he was to make it clear that Washington had to break off relations with the KMT and discontinue its support if it wanted to establish diplomatic relations with the CCP. The observation Huang had made to Fu Jingpo that the United States would first have to "do more things that are of use to the Chinese people" was false, Huang was told. It gave the impression that the Chinese CP expected aid from Washington and that the United States had already done something useful for China and would merely have to do more. Huang was also to make it clear that these were unofficial talks, given the fact that no diplomatic relations existed. If Stuart behaved in a friendly manner, Huang was also to be friendly, but without showing any excessive cordiality, and should assume a "serious and at the same time obliging attitude [*zhuangzhong er heqi de taidu*]." Huang was to inform Stuart that the CCP welcomed his willingness to cultivate contacts with the Party and to revise the trade treaty.[686]

Huang asked Stuart through Fu Jingbo for a first talk, which took place on May 13 in Stuart's residence and lasted two hours.[687] Before July 25, there were two further talks between Huang and Stuart and several meetings between Huang and Fu Jingpo.

The most essential questions were addressed in the first talk on May 13 and then apparently only repeated—at times with slight variations—in the second meeting on June 6. The third conversation on June 28 dealt primarily with Stuart's planned trip to Peiping. In the meantime, there were meetings between Huang Hua and Ledovskii. Huang Hua and Stuart's information, Huang Hua's report to Ledovskii, and a report by Mao Zedong for Stalin on the contents of the May 13 talk confirm one another on many, but not on all, points.

According to Huang, Stuart said he wanted "to establish new relations with the new China" and hoped that the Chinese government would to a great extent also include democratic figures in its ranks. America, he said, was already no longer supporting Chiang and was prepared to hand over aid goods stored in Shanghai and intended for the KMT, such as grain and cotton, to the PLA. Huang reportedly refused this and declared that if the United States did not wish to interfere in China's internal affairs, it should withdraw its warships from Tsingtao and other places and its ground troops. Stuart reportedly agreed to transmit this to his government. Huang Hua reportedly demanded as a precondition for establishing diplomatic relations that the United States break off its relations with the Nationalist Government, withdraw its armed forces from China, and recognize the independence, sovereignty, and territorial integrity of China. According to Huang, Stuart assured him that Washington wanted to establish diplomatic relations with the new China and already no longer supported Chiang Kai-shek.[688]

Through Kovalev, Mao Zedong reported essentially correctly to Stalin that the talk between Huang and Stuart in early May—in fact, Mao was including two subsequent talks between Huang and Fu[689]—had dealt with three topics: U.S. support for the KMT government, the withdrawal of U.S. armed forces from China, and the formation of a coalition government. Stuart had assured the CCP in the talk, Mao reported, that the Americans "had ceased to support the KMT regime." This could be seen, Stuart reportedly continued, from the fact that the U.S. embassy, with Stuart at its head, had remained in Nanking; the English, French, and other embassies had followed this example. The U.S. troops, Stuart reportedly said, were stationed in Shanghai to protect American property there, which was valued at US$200 million. They would be withdrawn from Shanghai and Tsingtao as soon as the PLA moved in. Stuart reportedly said that he hoped that "all democratic elements as broadly as possible" would be included in the future coalition government. After a coalition government had been established that received the support of the entire people, Washington would recognize and establish diplomatic relations with that government.[690]

Mao added that Stuart's statements conflicted with the behavior of General Douglas MacArthur, who had only recently stationed two regiments in Tsingtao and reinforced the navy in Shanghai. Either Stuart was lying or the military paid no attention to the State Department. Stuart's claim that the United States no longer supported the KMT regime was also a lie, he said. The opposite was true: the support continued. Furthermore, the decision of the American embassy and those of the other powers to remain was not in the interest of the Chinese CP. "We would be happy if all the embassies of the capitalist states would finally leave China and take off home."

It is striking here that Mao reported to Stalin, but did not ask him for advice. In his telegram responding to Mao, Stalin stated his agreement "with comrade Mao Zedong's remark about the American ambassador Stuart" and thanked him in the name of the Central Committee of the CPSU(B) "for the information."[691] He provided no advice, let alone instructions.

As mentioned earlier, we do not know whether there was also such an exchange of opinions between Mao Zedong and Stalin on the further progress of the contacts between Huang Hua and Stuart. Sergei Tikhvinskii, who has access to the Russian presidential archives, stated that Stalin followed the contacts of CCP representatives with the U.S. embassy with great attention and provided recommendations.[692] Tikhvinskii said that Mao Zedong had constantly and precisely reported to Stalin on the latest results of these contacts. Unfortunately, he did not provide evidence to support his claim. The only documentary evidence available up to now—namely, the exchange of telegrams between Mao and Stalin mentioned above—demonstrates that Mao informed Stalin neither constantly nor precisely and that Stalin did not give Mao any recommendations.[693]

There is a further reference to the Soviet response to Huang Hua's contacts with Stuart. It was provided by Kovalev, who reported that apparently on instructions from Mao Zedong, Zhou Enlai had asked him what he thought of such contacts on two occasions in mid-April, the time at which Huang Hua was sent to Nanking. Although Kovalev reportedly led him to believe that he saw no purpose in them, the Chinese-American talks went forward. Kovalev spoke in this context of the "conflicting position" in which the Chinese Communists found themselves. He said they were interested in improving relations with the United States, but did not wish to damage "friendly relations with the USSR" in this way. Kovalev's behavior and comments to Zhou Enlai appear to reflect the ambivalence Stalin felt in following his Chinese comrades' contacts with the United States. On the one hand, he had himself advised them to cultivate such contacts for tactical reasons. But, on the other, he suspected that the contacts could lead to an undesirable closeness between the CCP leadership and Washington, because Moscow did not constantly control these contacts.[694]

The report Stuart drew up for the State Department on his May 13 meeting with Huang Hua placed the emphasis slightly differently than Huang had. According to this report, Stuart explained to Huang that all non-Communists were afraid of the Marxist-Leninist doctrine of world revolution, which implied "subversive interference or armed invasion"—a reference that is missing in Huang's reports. We can only ask whether Stuart actually said this or merely wished to curry favor in Washington by saying that he had. According to Stuart, Huang referred to the "Chinese people's resentment" about American aid to the KMT and other mistakes of American China policy. Stuart described "unity, peace, truly democratic government" in China as his goal. "Huang expressed much interest in recognition of Communist China by USA on terms of equality and mutual benefit." Stuart felt this to be the only proper basis for recognition, but that they should add "accepted international practice with respect to treaties." The U.S. troops in China, and particularly the marines in Shanghai, were there to protect the lives of Americans during civil disturbances and chaotic conditions. Huang, Stuart reported to Washington, had apparently not heard this argument before. "He was obviously impressed." Stuart continued, telling Huang that the Communists still had no government. It was customary to recognize any government when it had clearly the support of the people and was able and willing to perform its international obligations. "Huang expounded upon needs of China for commercial and other relations with foreign countries."[695]

The report that Huang Hua gave Ledovskii two days later about his May 13 meeting with Stuart recorded that Huang had rejected an invitation to lunch in Stuart's residence that Stuart had originally made through Fu Jingpo and had agreed to an informal meeting at the same place. According to this account, the talk, in which Fu also participated, took the following form. Stuart initially stated his intention to discuss the status and future of American-Chinese relations with Huang. He repeated briefly what Fu had already reported to Huang on May 7. He had remained in Nanking to establish contact with the new power and was seeking to consolidate the friendship between the American and Chinese peoples. Huang Hua asked whether Stuart was familiar with the CCP leadership's recently published declaration on the principles of establishing relations with foreign states. When Stuart admitted that he had not yet studied this document sufficiently, Huang explained its contents and dwelt specifically on the question of the conditions for establishing diplomatic relations: (1) observation of the principle of complete equality and mutual respect, (2) respect for the right to the sovereignty and integrity of Chinese territory, (3) the immediate withdrawal of foreign armed forces from China, and (4) the immediate breaking off of diplomatic relations with the KMT government and immediate termination of aid to it. Stuart stated his agreement with the first three points. The United States had already withdrawn its troops from China, he claimed—with the exception of a small naval contingent stationed in Shanghai, which had the sole purpose of protecting the lives and property of American citizens from possible attacks by the KMT. The warships stationed in Tsingtao, he said, had left this harbor permanently. Stuart passed over the problem of breaking off diplomatic relations with the KMT government in silence.[696]

The statement in which Stuart was supposed to have claimed that the American troops in Shanghai served to protect American citizens from attacks from the KMT is missing from Stuart's report to the State Department. One also searches in vain for any evidence that Stuart, as Huang claimed, ever stated that Washington had already ceased to support the KMT. But as there is a reference to this, as mentioned above, in both of the accounts Huang Hua provided and also in the telegram Mao sent to Stalin on May 23, one should assume that Stuart really said it. His decision to leave the statement out of his report for the State Department (as had already occurred in his report on Fu Jingpo's talk with Huang on May 7) is probably due to the fact that he had overstepped his mandate here, clearly in an attempt to appeal to his Communist interlocutors—evidence of Stuart's determination to do business with the CCP at any cost and his willingness to risk going it alone.

The CCP leadership instructed Huang Hua on June 3 to make it clear to Stuart in their next meeting that the Chinese CP had already ceased to recognize the KMT government's right to represent the Chinese people. No government should continue to foster relations, or even negotiate the terms of a peace treaty with Japan, with this government, which was in flight.[697]

We have Huang Hua's and Stuart's versions of their second talk on June 6, which

Stuart and Fu Jingpo attended at Huang's invitation in the latter's office. The two accounts essentially agree where they discuss the same material.

According to Huang Hua, Stuart first turned to his imminent return to Washington. In this context, he mentioned that Roosevelt had agreed in Cairo that Taiwan should be placed under a Chinese mandate when a peace treaty was concluded. Huang countered by saying that Taiwan had always been a part of China and had already returned to China. The Communist leadership, he said, would never allow the KMT to sell Taiwan or permit foreign countries to employ tricks in connection with a peace treaty. Moving on to Sino-American relations, Huang repeated what he had already said previously: that if Washington wished to establish relations with the new China, it would have to break off relations to the KMT government and terminate its aid to the latter. Stuart again referred to the foreign embassies' decision to remain in Nanking and stated obligingly that the American ambassador would not follow the KMT government when it moved on from Canton, as well. But as no new Chinese government had been established yet at that time, they could not recognize it, either. China was currently divided between the two parties, he stated, and the United States could not break off its relations to the old government under international law. Washington was acting cautiously, Stuart said; it could not say which party it supported openly and was remaining passive. It was waiting, he said, until a democratic government was established. When such a government demonstrated that it would observe its international obligations, the question would resolve itself. Not much remained of American aid for the KMT government, he said, and no further aid of the sort would be provided in the future.[698]

Huang Hua countered that the United States was still supporting the reactionary government in its war against the people. Thus, there could be no talk of establishing diplomatic relations. Washington had to break off relations with the KMT government, he said, to demonstrate that it had abandoned its failed policy of interference. In Huang's personal opinion, the Political Consultative Conference would meet when Canton had been taken and announce the establishment of a coalition government. The speaker of the general staff of the PLA, General Li Tao, had stated, he said, that the CCP opposed any American aid.[699]

Huang Hua's version of the way the talk continued goes as follows. Stuart now sought to justify the American China policy, he said. In doing so, he encountered difficulties formulating his arguments and became entangled in contradictions. Like a teacher, he lectured Huang about "true freedom [*zhenli de ziyou*]" and linked the China question to world peace: the United States feared the Communist world revolution would bring about a third world war. Stuart hoped his efforts would contribute to improved Sino-American relations and secure world peace. The assistance of a third party (the reference was apparently to the USSR) could be of use in improving Sino-American relations. At this point, Fu Jingpo added that if these relations were good, they could also serve as a springboard for the improvement of the relationship between Washington and Moscow. Huang Hua objected that no aid was needed from a third party. Stuart emphasized that it would be useful for China to develop its industrialization through free trade with the United States and with American economic aid. He personally hoped that as many democratic figures as possible would be included in the new government. Fu Jingpo asked whether the new government would be interested primarily in industrialization or socialization. Huang Hua refused to reply to these two comments on the grounds that they involved China's internal affairs, and no one should interfere in these from outside.[700]

At the beginning of his report on the approximately two-hour talk on June 6, Stuart gave Acheson his overall impressions. The Chinese Communists were extremely interested in having foreign states, and the United States in particular, abandon the KMT government, which in the Communists' view was doomed to be defeated, and they were extremely sensitive about China's right to make its own foreign policy decisions freely. He said that he had repeated to Huang that there could be no recognition because no Chinese government had yet been established. When such a government was established, the United States would first have to assess whether it really represented the Chinese people and was both willing and able to abide by "hitherto accepted international standards." The United States did not dispute the Chinese people's right to have whatever government they wanted. But if this government obligated itself "to overthrow governments of other countries," this was no longer a purely Chinese affair. Many Americans were afraid of the doctrine of violent world revolution supported by the CCP. Stuart urged in vain for a resolution of the question of the American consulates "in occupied territory," by which he clearly meant primarily the Ward case in Mukden. After the discussion about the need for the economic recovery of China, Stuart asked what he might do to further their mutual relations. Huang Hua replied "almost brusquely" that China wished to work out its own destiny without foreign interference. The only thing the CCP wanted from Washington, he said, was the end of aid to the KMT government and termination of relations with it. Stuart replied evasively that the United States had delayed some shipments and that continuing aid was subject to circumstances over which Washington had no control.[701]

If we seek to summarize the course of the talks between Stuart or Fu Jingpo and Huang Hua to this point, we see an apparently unbridgeable gap. Huang demanded that Washington break off relations to the KMT government and terminate aid to it as an essential precondition for the future establishment of diplomatic relations between Communist China and the United States. Stuart refused to discuss this and referred to the fact that no Communist government yet existed to that date with which relations could be established; moreover, given the CCP's allegiance to world revolution, it could not be guaranteed that any future government would respect existing international law and surrender the intention to seek to overthrow foreign governments. Huang Hua decisively rejected Stuart's cautious attempts to tempt him with economic aid.

Stuart emphasized that the June 6 talk had addressed the issues openly "and at times almost beyond usual Chinese amenities." In any case, Huang and his assistant were very friendly to him, he said, "and expressed their personal attitude of friendliness in various small ways."[702]

The way the two protagonists reported on the talk that took place between Huang Hua and Fu Jingpo on June 8 at Huang's request was unusual and in part confusing.

Stuart reported the next day to Washington that on this occasion Huang had for the first time expressed an interest in economic support from the United States "which will be decisive." He reported that Huang had said that his Party would not pursue an isolationist policy in the reconstruction of China and that it would particularly need American help. The USSR could not give China the necessary aid, he said, because Soviet expectations had not yet been met.[703]

According to his own memory, Huang Hua did not say a single word on the topic of American economic aid on his own initiative in the talk with Fu Jingpo on June 8. Huang Hua reported exclusively on another—if no less exciting—subject that had been discussed on June 8: Stuart's desire to meet with Zhou Enlai in Peiping. According to the report, Fu gave Huang to understand that Stuart wanted to call for improved relations with Communist China in his upcoming visit to Washington and that he could count on the support of Acheson and Marshall. Undersecretary of State James Webb, Fu said, had expressed the hope that Stuart meet with Zhou Enlai in Peiping to learn the views of the CCP leadership. This would strengthen his position in Washington. When asked what topics he wished to discuss in Peiping, Fu again sought to tempt Huang Hua with economic aid, credits, and trade relations. Huang Hua reported that he had refused politely. He pointed out that Washington still had not replied to the CCP's request that the United States break off relations with the KMT government, and so Stuart would receive no answers to his questions in Peiping. Furthermore, it was very difficult to travel to the north because traffic had been so seriously disrupted.

As Huang learned from Peiping after he had made his report, the Party leadership did in fact consider unofficial relations (*fei guanfang guanxi*) better. When it learned that Stuart usually visited Yenching University on June 24 each year to celebrate his birthday there, they had the rector, Lu Zhiwei, send him an invitation, dated June 16, to visit the university. Lu wrote to Stuart that Zhou Enlai had let him know that if Stuart wanted to come to Peiping, he could expect the authorities to approve. Stuart had problems with the way this was formulated and, through Fu, asked Huang on June 27 what it meant, apparently in an attempt to learn whether this was a formal invitation by the CCP. According to a historian from the People's Republic of China with access to the Party archives, this annoyed the Party leadership.[704] Huang reported in his memoirs that the day after he received the relevant instructions from Zhou, he had informed Stuart that he could visit Yenching University.[705]

At about the same time, Stuart learned from a former colleague who was still teaching at the university that Mao Zedong had already made known Stuart's interest in a trip to Peiping, "and I presume the government will give you consent."[706]

During a meeting with Huang Hua on June 11, Ledovskii asked about the most recent status of the talks between Huang and Stuart, aware that "this question interested Moscow." At the center of Huang's answer was the report on Stuart's request to meet with Zhou Enlai. When Stuart answered Huang's question about what topics he wished to discuss with Zhou by saying that this would come out during the conversation, Huang informed the ambassador that it was difficult for him to ask the Party leadership to allow a Peiping trip without knowing any concrete questions. Stuart then named three topics: economic aid for China, Sino-American trade relations, and the decision by the United Nations Relief and Rehabilitation Administration (UNRRA) to keep its office in Shanghai. Huang replied that in his opinion, it would be pointless to discuss the problems mentioned by Stuart because Washington continued to maintain diplomatic relations with the KMT government and provided it with economic and other aid. Stuart had then justified the continuation of the diplomatic relations with the KMT government and economic aid—the latter only to the extent that existing obligations were involved—by saying that there was still no other Chinese government. Huang's answer was that a government would be established as soon as Canton had been taken.[707]

Huang Hua apparently reported the contents of the June 6 talk with Stuart and the June 8 talk with Fu Jingpo in large part correctly to Ledovskii. It is worth noting, however, that he said nothing to him about the interest in American economic aid he had supposedly expressed to Fu on June 8. His remarks, as has already been noted, were not mentioned in Huang's own memoirs, either. Two possible hypotheses suggest themselves to explain this paradox: either Huang did in fact express his interest in support from the United States, but did not mention it either in his memoirs or to Ledovskii; or he never expressed any such interest and Stuart, who was intensely interested in a meeting with Zhou Enlai, gave Washington erroneous information in order to show the prospects for an understanding with the CP leadership in the most favorable light possible.

There is powerful evidence to support the first of these two hypotheses. On May 31, a few days earlier, the Zhou démarche had been presented to the U.S. consulate in Peiping.[708] The underlying structure of the argument in it is the same as the one in Huang Hua's supposed statements to Fu Jingpo on June 8: The Soviet Union was supposedly not in a position to provide the new China with sufficient economic aid, and the CCP leadership therefore hoped to receive assistance from the United States. The Zhou démarche and Huang's remarks could thus have been two coordinated initiatives. Because the official historiography of the People's Republic of China has thus far passed over these events in silence, one can understand that Huang passed over his statements to Fu Jingpo on June 8 in silence in his memoirs as well. The fact that he also did not tell Ledovskii can probably best be explained by the fact that Stalin and Mao had declared the contacts between Huang Hua and Stuart to be a "leadership matter" that the two should discuss through Kovalev. An essentially anti-American message to Stuart that Zhou Enlai wrote during June, but which reached its recipient only the following month, conflicts to a certain extent with Huang Hua's advance on June 8 and the Zhou démarche, however.[709] The discrepancy cannot yet be explained.

Stuart did not report on the proposed Peiping trip to Washington until June 30, or about three weeks after Fu Jingpo first addressed the topic to Huang Hua. According to Stuart's report, Fu Jingpo had asked Huang Hua without his knowledge "in early June" whether it would be possible for Stuart to visit Yenching University in Peiping on his birthday, June 24, as in previous years. On June 18, Huang Hua reportedly asked Stuart through Fu whether he could make a trip to Peiping. On June 28, Huang reportedly assured Stuart in their talk that Mao Zedong and Zhou Enlai had informed him that Stuart would be welcome in Peiping, if he wished to visit the university.[710]

Stuart considered this to be a "veiled invitation from Mao and Zhou to talk with them." He reported that he had avoided making a direct reply and had pointed out obstacles such as the difficulty of a railroad trip for a "feeble old man." Huang objected that one could reach Peiping by rail in less than three days and that the Chinese would arrange a comfortable rail trip. Stuart would have preferred for prestige reasons to travel in his own airplane. Huang Hua also considered it conceivable that a flight might be approved.

Huang's conciliatory and almost urgent reaction led Stuart to suspect, he reported, that "Mao, Chou and Huang are very much hoping that I make this trip, whatever their motives." Stuart told the State Department that he would be willing to make such a trip and at the same time pointed out the possible advantages and disadvantages. Among the advantages, he counted the fact that it "would be [a] step toward better mutual understanding and should strengthen more liberal anti-Soviet element in CCP." He considered as a disadvantage primarily the probability that his trip would be criticized in American public opinion and by the heads of other missions in Nanking, who could see it as a violation of the agreed-upon "united front policy" toward the Chinese Communists. The trip would greatly increase the national prestige of the Chinese Communists and of Mao Zedong personally and constitute a second step—the first step was Stuart's decision to remain in Nanking—toward recognition of the Communist regime.[711]

In his memoirs, Huang Hua reported that he informed Stuart during their June 28 discussion that his hope for a meeting with the "authorities [*dangju*]" in Peiping would probably be fulfilled. Stuart reportedly seemed extremely pleased (*ji wei gaoxing*) by this news. Huang confirmed that Stuart mentioned reasons that could stand in the way of a trip at this time, among other things the fact that he would certainly be criticized if he traveled to Peiping now. He said that he would present the issue to Acheson for a decision.[712]

When Huang Hua described his talks with Fu Jingpo on June 27 and with Stuart on June 28 to Ledovskii and Embassy Council P.A. Shibaev, he presented essentially the same picture that can be found in his memoirs. Unlike his description in the memoirs, Huang reported to them that Lu Zhiwei had indicated in his letter to Stuart inviting him to Peiping that Mao Zedong had "expressed his desire to meet with Stuart." Huang reportedly informed Stuart on June 28 that the "democratic authorities would not raise any objections to his trip" if he really planned to visit Peiping. Stuart reportedly confirmed his desire, but added that he must unfortunately distance himself from the planning of the undertaking "given the existing circumstances." He gave as his reason the fact that the U.S. Congress was criticizing the government's China policy and his own activities severely at that time. Under these circumstances, he, Stuart, considered the trip inexpedient at the present time. He wished instead to send Mao Zedong a written message and asked Huang Hua to travel to Peiping along with Fu Jingpo and hand Mao the message. Huang reportedly refused to do this.[713]

There is no such clear cancellation of the trip to Peiping either in Stuart's report or in Huang's memoirs. The same is true of the request Stuart reportedly expressed that Huang Hua and Fu Jingpo hand a letter drawn up by Stuart to Mao Zedong. The discrepancy between the versions cannot be explained convincingly. It is also inexplicable that Huang reported to Ledovskii on Stuart's preparations for his trip to the United States during his next meeting on July 11, but did not say a single word about the fact that Acheson had prohibited the trip to Peiping on July 1.[714] Huang Hua reported only two weeks later, on the basis of a report that he had received from Fu Jingpo on July 24, that the impetus for Stuart to make a trip to Peiping had come from George C. Marshall "some time ago." When Stuart asked Acheson specifically, the latter had told him that he considered the trip inexpedient, as it could result in difficulties for the State Department from Congress, which was meeting at that very time.[715]

Stuart's behavior is striking in various respects. Regardless of whether it is true that Fu Jingpo mentioned his planned trip to Peiping to Huang Hua on June 8 without Stuart's knowledge, which, given the importance of the subject, appears rather unlikely, it appears more than unusual that Stuart passed over the subject in silence in his report to Acheson on the June 8 meeting. It is, however, easy to understand why he would try to disguise the fact that the topic had been addressed in the June 8 talk with the words "early June" in his June 30 report—as he would otherwise have subjected himself to the unpleasant question of why he had withheld such an important point earlier. He also did not mention at all in his June 30 report that Fu Jingpo had on June 8 supposedly given as the reason for the planned Peiping trip that Stuart wanted to meet with Zhou Enlai there.

The optimism that Stuart expressed in his cable to Washington of June 30 with respect to his trip to Peiping essentially seemed justified. Huang Hua confirmed to Ledovskii that he had met with Fu Jingpo on June 18 and added that the Party leadership had in the meantime decided to give Stuart a positive response if he should again request permission for a trip to Peiping. He had not yet done this, however, since his return on June 16 from a three-day stay in Shanghai. "Huang Hua stated that in his opinion Stuart's trip to Peiping would have a positive significance for the CCP." It would first of all allow the Party leadership to inform itself in greater detail about the American China policy. The trip would also "reduce the tensions between Peiping and Washington somewhat and give the Americans the impression that they can count on good and businesslike relations with the new China." As Huang Hua's report to Ledovskii was made only one day after the meeting between Huang and Fu, it is plausible to assume that Huang had already indicated the interest of the CCP leadership in Stuart's trip to Peiping to Fu on June 18, as Stuart reported to Washington on June 30. Fu also informed Huang on this occasion that the State Department judged Stuart's efforts to establish contacts with the Chinese Communists "extremely positively."[716]

Huang Hua's account to Ledovskii is of great importance because it gives us convincing evidence for the first time that the CCP leadership actually wanted to hold talks with the American ambassador. Western researchers have repeatedly doubted this.

We should not disguise that there is a certain contradiction in the fact that Huang Hua was still expressing interest to Stuart in his Peiping trip with the goal of achieving American-Chinese détente while five days earlier the Communists had abruptly interrupted the contacts with the U.S. consulate general associated with the Zhou démarche. We do not yet have any convincing explanation for this discrepancy—unless it is merely that the CCP leadership considered the secret contacts with the consulate general to be superfluous because they assumed that Stuart would arrive in Peiping.

The American side viewed Stuart's plan to travel to Peiping with suspicion or at best mixed feelings. Consul General O. Edmund Clubb in Peiping advised Stuart to make the trip only on condition that he received a "categorical assurance" that he would meet with Mao Zedong and Zhou Enlai. It would involve "a serious loss to American prestige" for him to travel without such guarantees and not be received by them.[717]

The Policy Planning Staff of the State Department considered Stuart's report on his plans to be even more significant than the Zhou démarche of May 31.[718] And yet they were of the opinion that the Chinese Communists apparently wanted to "try to make as much capital as they could out of such a visit. All the objections raised in the message[719] are valid." If Stuart made the trip, the staff suggested, they could only counter the significant reaction they could expect in American public opinion if he first flew to Mukden, took Consul General Ward and his wife—who had been imprisoned by the Communists—on board, and only then went on to Peiping to talk to Mao Zedong and Zhou Enlai. Not only would the United States "make a lot of face" in Asia in this way, but it could also serve as justification for Stuart's visit in Peiping in the eyes of American public opinion. This idea would probably be unacceptable to the CCP leadership, however, and disrupt the possibility of talks.[720]

The State Department made its decision on July 1. Secretary of State Acheson instructed Stuart "under no circumstances to make visit Peiping," and did so for the reasons Stuart had himself raised in his June 30 report. There had first been a discussion of the report at the highest level—thus presumably with the president. The strong anti-American position Mao Zedong took in his famous *yi bian dao* article of July 1 played no role in Acheson's decision, as it was not yet known in Washington at this time.[721] The trip was apparently not allowed primarily because of criticism from Congress and unfavorable public opinion, and Acheson was apparently himself concerned that Mao and Zhou wanted to entice Stuart to Peiping in order to humiliate him there. Huang learned from Fu Jingpo on July 2 that Acheson had forbidden Stuart to make the trip to Peiping. This supposedly did not surprise the CP leadership.[722]

A cautious assessment of the events just described should be based on the assumption that Stuart wanted to travel to Peiping with the honest intention of relieving tensions in the relations between the United States and Communist China. Even after the State Department had prohibited the trip, he received information that convinced him Zhou Enlai would have been happy to receive him for talks in Peiping and that relations would have been improved as a result.[723] He said this despite the fact that he had been shocked a few days earlier by Mao's *yi bian dao* article, in which the latter spoke out more clearly than ever before in public in favor of a one-sided alliance with the Soviet Union.[724]

Peiping apparently did not expect very much from Stuart's travel plans, but hoped to use his visit to probe U.S. intentions. Within the Party, Zhou Enlai in any case stated that regardless of whether Stuart traveled to Peiping, the CCP was under "no illusions about the American imperialists changing their policy." He wanted it made clear that Stuart was not coming to Peiping at the CCP's invitation but on his own initiative, so that the Americans could not use this "as an example for propaganda."[725] There is evidence that the CP leadership suspected that Stuart wanted to use his contacts with the CCP to drive a wedge between the Soviet and Chinese Communists.[726] Their approval of Stuart's Peiping trip supposedly also arose out of a tactical attempt to demoralize the KMT leadership.[727]

Zhou Enlai's June Message

Another fact points us in the same direction. In June—that is, still before Stuart's Peiping trip was cancelled—Zhou Enlai sent Stuart a message via General Chen Mingshu, chairman of the Revolutionary Committee of the Kuomintang in Shanghai, which, however, Stuart received only on July 9 (henceforth called the June message).

It was in itself discouraging. Zhou criticized American China policy in harsh words and contrasted it to the Soviet Union's positive behavior. America's many good points pertained to small matters, whereas those of Soviet Russia to big matters. His, Stuart's, function made him a representative of imperialism. The United States wanted to find relief for itself in its own crisis and therefore exported goods in large quantities. Its stomach is sick with glut, ours with hunger, he wrote. China's economic prospects were favorable. China was self-sufficient in many branches of the economy, production would be increased, and the opportunities for export were good. "In conclusion, I do not depend upon you, you depend upon me." Nonetheless, he said, if Stuart wanted to come to Peiping in a private capacity, "it is possible that you would be able to meet a person in a responsible position."[728]

The picture Zhou Enlai painted of a United States that was dependent economically on China was not accurate. The opposite was the case: Between 1946 and 1948, more than half of China's imports came from the United States, and approximately a quarter of its exports went to the United States. American exports to China constituted less than 5 percent of total American exports and American imports from China made up only 2 percent of total imports.[729] Clearly Zhou Enlai was poorly informed or he would hardly have behaved so arrogantly and at the same time made himself look so ridiculous.

Putting that aside for a moment, the Communist leadership's basic attitude as revealed in Zhou Enlai's June message practically excluded a political compromise with Washington—even if there had been a meeting in Peiping. This was apparently also the way it was seen in Washington when they became aware of the June message there in mid-August. A comment by Paul Nitze, at that time a high-ranking official in the State Department, was at least placed in a causal relationship with it. Nitze observed, "At one point, we thought we had established contact . . . but the Chinese Communists cut it off. They dropped the boom on all negotiation."[730]

On August 2, Stuart returned to the United States without having softened the hardened fronts between Peiping and Washington. A diplomat in Canton took over his business as chargé d'affaires.[731] After the State Department published a white paper on China, Mao Zedong called out derisively to Stuart in the middle of the month in the title of an article: "Farewell, Leighton Stuart!" In the article, he wrote: "Why can't we live without the United States? . . . Leighton Stuart has departed and the white paper has arrived. Very good. Very good. Both events are worth celebrating."[732]

The CCP Leadership's July Message

In a confidential message to Stuart (the July message) also delivered through Chen Mingshu, Mao Zedong, Zhou Enlai, and Ye Jianying employed a tone that differed from the one they had used in this pamphlet intended for the public. The message is undated, but, to judge from its contents, was based on a talk the senders had with Chen on July 7 or 8. The message was couched in strikingly friendly terms, recognized the friendship Roosevelt, Marshall, and Henry A. Wallace had shown for China, and stressed that the Chinese CP considered the future of Sino-American relations to be important. The senders gave Stuart assistance in interpreting more recent Communist positions. They made an effort to play down the significance of Mao's *yi bian dao* article of July 1, in which he had pledged himself clearly to an alliance with the Soviet Union and argued polemically against the United States.[733] They had intended the article to be read within their own Party, the message suggested. Mao's June 20 speech,[733] which emphasized their willingness to establish diplomatic relations with all countries, and the July 7 message to the Preparatory Committee for the New Political Consultative Conference, which used tones that were primarily friendly to America, were more important. As soon as diplomatic relations were established between Communist China and the United States, the Communist propaganda would change.[735]

We will in fact have to assume that the message's main purpose was to relativize the devastating impression Mao Zedong's *yi bian dao* article left in Washington. The claim that the Communists were willing to establish diplomatic relations unconditionally should not be taken seriously. It reflected neither the recommendations that Stalin had made nor the statements the CCP made elsewhere on this topic.

For unknown reasons, Stuart did not pass the July message—as well as Zhou Enlai's June message, which Chen Mingshu delivered at the same time—on to Washington immediately and merely announced that he would report on them after his return.[736] He did not seem particularly impressed by the July message, possibly because he was influenced by the uncompromising character of the June message, and observed that it merely expressed the Communists' naive hope to be able to establish economic relations with the West despite the poor political relations. The State Department learned about the message from Mao, Zhou, and Ye only five weeks later, on about August 15, when Stuart delivered it there.[737] No definitive reaction from Washington has become known.

The Zhou Démarche

At about the same time that Stuart was establishing contact with Peiping through Huang Hua, there were two other much less transparent probes. They were initiated by the CCP and aimed at Washington and London, and had essentially the same goal.

The first involved a message that Zhou Enlai supposedly passed secretly to David D. Barrett, assistant military attaché at the American consulate in Peiping. It is generally known as the Zhou démarche in the literature. The oral message was transmitted by an unknown intermediary, who had supposedly spoken with Zhou Enlai and his secretary, "Zhang," to the Australian UPI correspondent Michael Keon, who forwarded it to Barrett on May 31, 1949. The message stated that Zhou asked for the message to be forwarded to the "highest American authorities" and also to "the British." This should be done at the top secret level. Zhou's name should not be mentioned, as he would otherwise disavow any knowlege of it.[738]

When Clubb proposed seeking another channel of communications to the middleman through Keon, he informed them "Gestapo-like" on June 23 that the Communist side was not willing to accept such a message in such a way. At this point, Clubb saw no sense in continuing to pursue the contact further. In a secret meeting on August 15, Zhang recommended urgently that Keon leave Peiping as quickly as possible. There would otherwise be a danger that Bo Yibo, who was under pressure from the Soviet Union, would have him convicted and sentenced to a long prison term for espionage. The Zhou faction reportedly had little chance of winning in the extremely bitter factional conflict within the CCP leadership. With Mao's approval, however, Zhou could arrange for Keon to be deported, albeit under humiliating circumstances.[739]

The most significant statements in the Zhou démarche described the CCP leadership as divided. Zhou Enlai, it claimed, stood at the head of a "liberal wing" while Liu Shaoqi led a "radical wing." Mao Zedong was keeping out of the conflict. The radicals supposedly wanted an alliance with the Soviet Union similar to the one that existed between the United States and Great Britain. The liberal wing advocated American and—if possible—British aid for China, because the USSR could not help. The United States had a genuine interest in the Chinese people, which could become the "basis of friendly relations between two countries." If Mao realized his concept of private capital, China might not be a Communist country for a long time. De facto working relations had to be established speedily between the new China and the United States, the Zhou démarche continued. The Soviet Union's foreign policy was "crazy" and risked a war. If the United States helped the new China and established good working relations, China would be able to mediate between the Western Powers and Moscow and "do its best make USSR discard policies leading to war."[740]

Great Britain received a similar message at the same time. As Foreign Minister Ernest Bevin reported to the British cabinet on August 23, 1949, the government had received a message "with some delay" through a third party from "a high Communist authority" in China. The Foreign Office later identified Zhou Enlai as the sender. The message stated that there was a conflict in the Chinese CP between "fanatical supporters of the USSR" and those who wanted to establish normal relations with the West and trade with it. In this case, the new China would not necessarily vote with the Soviet Union in the United Nations. Bevin did not cast doubt on the message's authenticity and came to the conclusion that Great Britain should keep a foot in the door in China and remain there as long as possible. It cannot be determined whether the message reached Bevin through Washington, as Zhou Enlai requested. We know only that the State Department did not want to forward it to London in mid-June and insisted that the "names of intermediary and source are to be revealed" in the meantime.[741]

While the American reaction to the Zhou démarche was not uniform, it was predominately cautious. Consul General Clubb in Peiping considered it possible that there were two wings in the CCP leadership in the context of Titoist tendencies. But he also raised the possibility that the message constituted an action organized by the Party leadership and perhaps even approved by the Soviet Union. The Chinese Communists could be seeking to "continue diet Soviet political bread" and at the same time "eke out diet with American economic cake." In this case, the United States should make them pay economically or politically for any deliveries.[742]

Embassy minister-counselor Lewis Clark in Canton reacted with suspicion to the message. The démarche smacked more of tactics than strategy, he argued. It would be a fatal mistake for Washington to fall for a trick and help the Chinese Communists out of their desperate need until they could get along without the United States.[743]

Stuart had the most positive views on the message, which is not surprising. He was firmly convinced that there was a power struggle between the "wholly pro-Soviet" and anti-American Liu Shaoqi, on the one hand, and Zhou Enlai, on the other, and viewed the message as a "call for help." He proposed an answer that would reiterate America's traditional willingness to provide aid to China, but on the condition that the U.S. consular staff, and the staff, in Mukden in particular, be treated correctly and the anti-American propaganda terminated.[744]

Washington had grave reservations about the Zhou démarche. President Truman decided "to be most careful not to indicate any softening toward the Communists but to insist on judging their intentions by their actions."[745] The State Department drafted a reply in this spirit, arguing that the United States hoped for friendly relations but that these could not be established until the CCP terminated its anti-American propaganda and released Consul General Ward, his staff, and two American pilots who had been taken prisoner. The State Department instructed Clubb to formulate the reply on plain paper without any signature, and to seek to transmit it directly through a completely reliable intermediary in order to reduce the danger of exposing the contacts.[746]

After the Communists broke off contact, Clubb considered the political part of the Zhou démarche to be insincere and an attempt to make the Communist leaders appear more appealing in order, if possible, to make the United States provide trade and direct aid.[747] At the same time, and in the same context, the American chargé d'affaires in Moscow, Foy D. Kohler, stated that he was convinced the Chinese Communists were interested in both Soviet and U.S. economic aid. But if they did try the more effective U.S. aid first, this did not mean that they would give up their orthodox Communist principles. Kohler recommended that the United States give no aid at first and wait until Moscow did so. If the Soviet aid remained unsuccessful, one could hope that "Mao and company" would toy with the idea of taking a road that was independent of the Soviet Union.[748]

To this day, Western researchers continue to discuss the Zhou démarche particularly intensely in the context of the debate over the alleged lost chance in China.

It is not yet possible, however, even to prove the authenticity of the message to determine whether it really came from Zhou Enlai—that is, whether the unknown intermediary had really met with Zhou Enlai as he claimed. There is disagreement about whether we can assume that two factions pursuing different policy lines toward the Soviet Union existed. It is questionable, furthermore, whether such a dramatic advance as an authentic Zhou message could have been made behind Stalin's back.

Although he has previously argued that there were in fact two factions at that time,[749] the author now assumes, after assessing new materials, that there was unity in the CCP leadership on the fundamental questions concerning the future course of foreign policy. With the exception of the Zhou démarche and the parallel démarche to the British government, we can find no evidence that a pro-American Zhou faction and pro-Soviet Liu faction existed.[750] As explained earlier,[751] the Party leadership had pursued a clear course in favor of a future strategic alliance with the USSR since 1947. Furthermore, Zhou Enlai, who from the Zunyi Conference of January 1935 until his death was always distinguished for his loyalty (not to say servility) to Mao Zedong, would presumably hardly have dared to take such a spectacular step without Mao's knowledge.

If this is true, it seems plausible to see the Zhou démarche as a tactical maneuver that Mao Zedong carried out with the participation of the most important Party leaders. Stalin was apparently not informed of this. A Russian historian, Ledovskii, with access to the Russian presidential archives found no reference to it in the exchange of telegrams between Stalin and Mao in this period; his research in the archives of the Russian Foreign Ministry also remained unsuccessful.[752] The same historian, who at that time worked as a diplomat in the Soviet embassy in China, pointed out that he personally, the embassy, and the Soviet Foreign Ministry had all known nothing about the démarche at the time.[753] Tikhvinskii, who was Soviet consul general in Peiping during this period, also indicated that he had known nothing about it at the time.[754] He claims that Mao Zedong kept Stalin "uninterruptedly" and "precisely" up to date on the results of the CCP's contacts with the Americans, but, on the other hand, as already mentioned earlier, was only speaking of the contacts with the "American ambassador"—that is, Stuart in Nanking—in this context.[755]

Although there was thus in all probability no operational coordination between Moscow and the CPP leadership with respect to the Zhou démarche at the time, it was nonetheless fully consistent with what Mikoyan had agreed upon with Mao Zedong on instructions from Stalin during his visit in Xibaipo in early 1949, as mentioned earlier: the conciliation of Washington in order to obviate a military intervention by the United States in China.[756] As has already been demonstrated, Stalin, who was interested in protecting the Soviet Union's scarce resources, had also repeatedly approved one of the message's primary goals: namely, American economic aid.[757] Stalin actually urged the Chinese comrades to look in this direction. It is also striking that the message did not suggest establishing diplomatic relations with Washington but only "de facto working relations," which is also in keeping with the policy Stalin had recommended to the CCP leadership.[758]

Mao Zedong apparently wished to use the Zhou demarche to probe how willing Washington would be to establish trade relations and provide credits and technical assistance. "De facto working relations" would have sufficed to arrange such aid. This is consistent with the fact that Huang Hua, as mentioned earlier, also expressed a strong interest in American aid in his contacts with Ambassador Stuart through his personal secretary only a few days later, on June 8, and observed that the Soviet Union was not in a position to provide such aid.[759]

The CCP leadership was setting out bait when it sought to suggest to the White House that a "liberal Zhou faction" existed that, if it was successful, could distance the Party from Moscow and pursue a pro-American policy. In this way, Mao was responding to desires that Western governments and public opinion had at the time. The hope was that such factions existed in the Chinese CP leadership and that Mao would probably develop into a "Chinese Tito"—a hope that Chiang Kai-shek repeatedly warned against.[760]

Tikhvinskii takes a similar position today: American diplomats in China at the time were carefully looking for signs that differences of opinion existed between the Soviet and Chinese Communists, he argues. They would therefore have eagerly "bitten at the version of the 'Zhou-En-lai démarche' that had probably been thrown out to them by the secret services [*spetssluzhbami*] of the CCP."[761] If Tikhvinskii was suggesting by this that the secret services had operated alone, this is not a serious hypothesis. The extremely centralized and disciplined foreign policy of the CCP makes it impossible to believe that such a maneuver could have been carried out without Mao's and Zhou's knowledge.[762]

Given the uncertainty that can repeatedly be seen within the Party leadership about whether the United States might not still intervene in favor of the KMT, or at least impose a blockade on Chinese harbors,[763] it is probable that there was an interest in investigating whether the Americans really had any such intention and, if they did, in attempting to placate them. The CP leadership was naturally also interested in how Washington would react to the new China generally.

It seemed reasonable to leave Zhou Enlai as the supposed sender. Ever since his time as the representative of the CCP in Chungking during the war, Zhou had—sometimes deservedly—had a reputation in Western circles of being an example of a liberal, cultivated type of Chinese Communist, one who would be willing to negotiate. It also fit here that Mao at this time had charged Zhou with sounding out the Soviet side on the "most delicate issues."[764]

At this time, it is possible only to speculate about why the Communists broke off contact with Barrett so abruptly. The fact that the American consulate general took three weeks before it promised an answer might have led the CCP leadership to conclude that the other side was not going to rise to the bait. And the fact that the consulate proposed an alternative channel of communications might have played a role, as this might have endangered the secrecy. The CCP leadership must have considered strict secrecy to be necessary to avoid the risk of compromising the contact. This can be seen from the Zhou message itself, which requested that the fact that the sender was Zhou Enlai be kept secret or he would otherwise distance himself from it. The Party leadership must have feared that if the Zhou démarche were published, the contradictions between its contents and the strong official anti-American propaganda in public opinion would have made it appear untrustworthy. An internal observation Zhou Enlai made in late June in connection with the planned visit of Ambassador Stuart to Peiping fits in this context. It must be made clear, he said, that Stuart was coming to Peiping of his own initiative and not at the invitation of the CCP leadership because the American side could otherwise use it as "an example for propaganda."[765]

If Moscow was not privy to the action, this would have been another reason for the Chinese fear that Clubb's efforts to find new channels of communication could have exposed the contacts.

The fact remained that the Zhou démarche of May 31 sent out signals that were similar to those sent by Huang Hua's message to Fu Jingpo on June 8, and that the contents of both stood in sharp contrast to Zhou Enlai's June message to Stuart. The first two involved an offer to cooperate with Washington, whereas the latter sought to distance the CCP from the United States. The available sources do not allow us to resolve this contradiction yet.

July 1 as Turning Point

July 1, 1949, constituted an important marker in relations within the Moscow-Peiping-Washington triangle in two senses. This was the day on which Mao Zedong announced more explicitly than ever before, through his *yi bian dao* article, the course the Party had already been following since 1947 toward forming a strategic alliance with the Soviet Union. It is a strange coincidence that this was the same day that Secretary of State Acheson—unaware of Mao's article—instructed Ambassador Stuart not to travel to Peiping on any account to establish contacts with the CCP leadership. It was now even more obvious in Washington that the stage was set for lasting conflict in the relationship between Communist China and the United States—even though the CCP leadership sent Ambassador Stuart a confidential message in early July in which it again extended friendly feelers, as has already been described in detail.[766]

From this time—that is, from July—the Chinese Communists' actual behavior caused relations to become even worse, as has already been mentioned. In early July, the Shanghai police arrested a young American vice-consul in public, beat him, threw him in prison and treated him very badly there. He never recovered from it. The Nanking authorities blocked Stuart's departure until early August. After that, they were no longer willing to maintain even unofficial relations with the American diplomats. In late October, Consul General Ward was arrested along with some of his staff. In prison, the Americans were beaten and made to suffer hunger and cold. Acheson instructed Consul Clubb in Peiping to hold talks about this with Zhou Enlai or some other high-ranking Party functionary. The Communist side rejected every attempt by Clubb to make contact. The same was true of similar attempts in Hong Kong. Ambassador Stuart was indeed able to raise the question with Huang Hua, but he hit hard rock. Truman considered the idea of exercising pressure by blockading the coal transports along the Chinese coast and in the process sinking ships, if necessary. Acheson and his staff dissuaded him from doing this. On November 21, a case was brought against Ward and the other imprisoned officers, who had no legal counsel. They were condemned to prison sentences of several months and were finally allowed to leave China, along with the remaining consular staff, on December 21.[767]

As has already been discussed in detail earlier, Soviet recommendations played a significant role in the hard anti-American position the Chinese Communists took, and Soviet pressure was even more clear in the Ward case.[768]

If one compares the Chinese CP's contacts with the United States with those it had with the Soviet Union, one can see a qualitative difference. Whereas Mao Zedong coordinated his activities with Stalin at brief intervals—at times daily—by telegraph, communication with Washington, as has been described, was limited to infrequent unofficial talks by Party functionaries with American diplomats, supplemented with messages carried by third parties.

Washington became accustomed to the imminent close and one-sided connection between Communist China and the Soviet Union as a fait accompli—albeit only as a temporary one—during the second half of 1949. As time passed, the U.S. government adopted the opinion presented by the leading China diplomats that the United States should wait patiently until the Chinese had had enough of the Soviets: "Let CCP learn by experience that USSR has little aid to offer, that Soviet friendship is always one-sided, that China will lose much more than it will gain by such association, that it will receive no assistance from US as long as it is satellite of USSR."[769]

In late December 1949, Truman, Acheson, and the National Security Council adopted a paper outlining the principles underlying their Asia policy and established on the basis of experience in dealing with the Chinese Communists that the United States would "continue to recognize the Nationalist Government of China" and "avoid recognizing the Chinese Communist regime" as long as this lay in America's best interests. Most states announced that they would follow this policy. A few, such as India, Pakistan, and Great Britain, constituted the exception; London recognized Peking on January 6, 1950.[770]

Following the violent occupation by the Chinese authorities of U.S. military barracks in Peiping in January 1950, the United States closed all its diplomatic establishments on the territory of the People's Republic of China by April.[771] Washington's efforts to establish dialogue with Communist China had come to a definite end. Relations between Peking and Washington sank to a new low as a result of Chinese participation in the Korean War. The Cold War that became established here between "Red China" and "American imperialism" led to a deep sense of alienation not only at government level but in the consciousness of public opinion in the two states as well.

Conclusion: No Lost Chance in China

Our analysis of relations between the Chinese Communists and the United States in 1949 has shown that when the CCP leadership probed Washington in May, June, and July 1949 in Nanking and Peiping, these were merely tactical gestures designed to appease the United States, hold open the door for economic relations, and explore America's further intentions. The Party leadership was not pursuing a third road policy in doing so and certainly had no pro-American leanings. And yet even these probes, with their modest goals, were destined to fail as a result of the Communist side's wild anti-American propaganda and mistreatment of American diplomats in violation of international law.

The two states' mutual perceptions were marked by a conflict of values that was irreconcilable at that time. On the one side, the Americans failed to understand the emancipatory and anticolonial goals of the "revolutionary nationalism" advocated by the Chinese Communists, and the Communists failed to understand "bourgeois" norms of international law, such as diplomatic immunity (the Ward case), and deeply offended the self-esteem of the American great power by disrespecting them. In this context, Chen Jian speaks of "ethnocentrism" on both sides.[772] This assessment is not fully accurate because it does not take into account the chiliastic aspects of Chinese Communism, which have nothing to do with Sino-centrism, and also underestimates the messianic aspects of the American conception of the "rule of law," which are shared by the other Western democracies. The impossibility of reconciling these two positions can be seen particularly clearly in the confidential message that Mao Zedong, Zhou Enlai, and Ye Jianying sent to Stuart in early July.[773] In it, they offered to terminate the anti-American propaganda as soon as diplomatic relations had been established. Washington demanded that they reverse the order and first terminate the propaganda (and respect international law in the treatment of American diplomats) and only then establish diplomatic relations.

There can be no talk of a lost chance in China. American diplomats were being completely realistic when they, by March 1949 at the latest, estimated to the contrary that the Chinese Communists had decided to pursue a strategic cooperation with the

Soviet Union. Furthermore, they correctly assumed that this cooperation would fail in the medium term.

In retrospect, two aspects of this appear remarkable: the casualness with which the Truman administration accepted the developments that had occurred as inevitable in the short term, and the accuracy of the prognoses by high-ranking authorities such as the State Department, the National Security Council, and the Joint Chiefs of Staff that had been developed beginning in March 1949 and that the president accepted, according to which it would be about a quarter of a century before Washington would again be able to pursue a successful and active China policy.[774]

Establishment of Diplomatic Relations: Problems with the Soviet Ambassador

On October 1, 1949, Mao Zedong, as newly elected chairman of the Central People's Government (CPG), proclaimed the establishment of the People's Republic of China in Tiananmen Square in Peking.

This step had originally been planned for a later date, possibly January 1, 1950. In mid-August, Stalin advised Liu Shaoqi not to lose any time in taking it so that "the enemy"—whereby he apparently meant primarily the United States—could not carry out its "devilish plan" to exploit anarchic conditions in China to justify an intervention. It is possible that Stalin's argument contributed to the decision to anticipate establishing the government.[775]

When Zhou Enlai, as foreign minister, called on the Soviet Union—as he did all other governments—to establish diplomatic relations on October 1, Andrei Gromyko, as vice foreign minister, informed Zhou Enlai, as the representative of the government of the People's Republic of China, on instructions from the government of the USSR on the following day of the USSR's desire to establish diplomatic relations and exchange ambassadors with the People's Republic.[776] As Zhou Enlai gratefully acknowledged in his telegram in response to Gromyko's, the Soviet Union was thus the first state to recognize the People's Republic of China.[777]

Stalin had taken care to ensure that the Soviet satellite states would not precede him in recognizing the People's Republic of China. For example, Gromyko advised Stalin to give a positive response to a corresponding inquiry by the Czechoslovakian chargé d'affaires in Moscow—but only after Gromyko's note about Soviet recognition of the People's Republic of China had been published.[778] Stalin apparently approved Gromyko's proposal because Bulgaria, Romania, Czechoslovakia, Poland, and Hungary in fact recognized Peking only on October 3.[779]

On October 2, Gromyko published a statement in which the Soviet government described its relations with KMT China as terminated. The "government of Mr. Yan Xishan," it stated, no longer exercised any power in the country and had become a "provincial government of Canton."[780] At the same time, Gromyko proposed to Stalin that he ask the British government to protect Soviet property in Canton.[781]

Stalin did not, however, consider it necessary to send a message to congratulate Mao Zedong on the proclamation of the People's Republic of China, although he sent personal congratulations to the president of the German Democratic Republic on the establishment of the first German socialist state on October 7, 1949. Mao made it very clear to Ambassador Roshchin that he had definitely noticed this affront.[782] Roshchin reported to Moscow that Mao had "taken this very much to heart [*kraine perezhivaet*]." The Soviet press honored the proclamation of the GDR and not the People's Republic of China as the most remarkable event in postwar history. As he had received no instructions from Stalin to do so, Kovalev did not attend the banquet held in Peking on the occasion of the founding of the state.[783] This constituted a further slight to Mao.

Peking must also have been displeased to note that Zhou Enlai's October 3 telegram thanking Gromyko was not published in the Soviet press. Stalin had decided to do it this way, although Gromyko had recommended that he have the complete telegram published on the first page of the central newspapers on October 4.[784] Instead of this, a brief TASS report was to appear about the event first.[785] The significance of this was relativized by the addition of a report on the nomination of the two ambassadors. The TASS report finally appeared in this form on page 2 of the October 5 edition of *Pravda.*

Stalin's insulting behavior apparently reflected his continuing discomfort, or at least ambivalence, about seeing Communist China ruled by a leader as independent minded and self-confident as Mao Zedong. Stalin also expressed his lack of respect in yet another way. It was customary at that time for *Pravda* to print news from the "fraternal states" in a full column on the far right-hand side of the first page. A month passed after the establishment of the People's Republic of China (which, after the USSR, was the most powerful socialist state) before it was given the honor of appearing in this column for the first time—and even then in last place. Dwarf states such as Albania were mentioned frequently during the same period. Peking presumably also noted this.[786]

With respect to international recognition of the new China, the government of the United States let it be known at this time that it would not hurry to establish diplomatic relations and would wait to see how Sino-Soviet relations developed. Great Britain and India, on the other hand, expressed a certain interest in granting recognition.[787]

Sergei Tikhvinskii, up to that time consul general of the USSR in Peking, was temporarily nominated to be chargé d'affaires. He was present on October 5 at the foundation of the Chinese-Soviet Friendship Society, over which Liu Shaoqi presided and at which Song Qingling, Sun Yat-sen's widow, gave a speech. The foundation was staged as a major event, attended by 1,500 people. Liu Shaoqi first assumed the presidency of the society. Song Qingling later took his place.[788]

Stalin's October 5 decision to send Nikolai V. Roshchin as the first ambassador to Peking cast a further shadow over the recently established diplomatic relations between Moscow and Peking.[789] When Tikhvinskii sought *agrément* from Foreign Minister Zhou Enlai for Roshchin at the seat of the Party and government, Zhongnanhai, Zhou's "face reflected the disappointment he felt about Stalin's choice for a moment."[790] Zhou asked Tikhvinskii to wait for a while and went to Mao Zedong's office, which was not far away. After a quarter of an hour, he returned and informed Tikhvinskii that the Chinese government would accept his credentials.[791]

The Chinese government was apparently dissatisfied with Roshchin's nomination for several reasons. Clearly without consulting the CCP leadership, he had been the only one of the accredited ambassadors in Nanking to follow the KMT government's recommendation to move to Canton, where it officially sat from February 5, 1949.[792] He represented the USSR there until the last day—that is, until September 30, 1949. Mao Zedong would have preferred the Soviet side to honor the revolutionary change in China by sending out a different ambassador. Roshchin, who was originally named Razumovskii, had long worked for the GRU, the Soviet military secret service—a further reason for Chinese mistrust. He had been military attaché at the Soviet embassy in Chungking with the rank of major general from 1943 to 1945 and knew Zhou Enlai from that time. He was nominated as the Soviet ambassador to China in 1948. Finally, Roshchin did not hold any high office in the Soviet Party hierarchy, as from the Chinese perspective should have been necessary for a position as prestigious as ambassador.[793]

Although he had Zhou Enlai greet Roshchin on his arrival in Peking, Mao Zedong expressed his displeasure with this choice by not ordering any of the other honors that were available according to protocol. Mao essentially ignored Roshchin as long as he was ambassador in Peking. During his stay in Moscow, Mao attempted in vain to have him recalled. Foreign Minister Vyshinskii reacted by saying that "the Soviet Politburo and not Mao will decide who the right person is to be ambassador in Peking."[794]

On October 16, Roshchin, dressed in diplomatic uniform, presented his credentials to Mao Zedong in the presence of the general secretary of the government, Lin Boqu, the chief of the general staff, Nie Rongzhen, and Foreign Minister Zhou Enlai in a stateroom in the government quarter of Zhongnanhai. The procedure followed a ceremony on which Tikhvinskii had advised the Chinese Foreign Ministry, which had no experience with these matters. In the following discussion, in which Zhou Enlai and Consul General Tikhvinskii participated, Mao recognized in a short speech the fact that the Soviet government had been the first to establish diplomatic relations after the establishment of the People's Republic of China and that it had quickly sent an ambassador to Peking. He emphasized his belief in the development and strengthening of Sino-Soviet friendship and that the collaboration between them would become ever closer—and would also involve their common fight for a lasting peace. He raised a toast to Stalin's health and stated that he was convinced that "the two governments would be able to resolve all problems that might in future arise between China and the USSR in the spirit of true friendship."[795] He also made reference to his last meeting with Roshchin in 1945 in Chungking—an ambivalent reference, because at that time Mao had traveled there to negotiate with Chiang Kai-shek only under pressure from Stalin.[796] Roshchin congratulated Mao for having taken Canton and raised a toast to him in his position as chairman of the Revolutionary War Committee. The banquet that was supposed to follow had to be canceled because of a technical problem.[797]

Mao informed Stalin on October 20 of his decision to send Wang Jiaxiang as ambassador to Moscow. Wang would at the same time be responsible for China's relations with the "states of the New Democracy" in Eastern Europe and represent the CCP in making contacts with the CPSU, he said.[798] Mao clearly wished to signal to Stalin in this way that he was determined to discontinue the Soviet Union's previous dual-track approach, which separated diplomatic and—the more important—Party contacts. To all appearances, he wished at the same time to make clear Communist China's desire to be treated from that point on as a state enjoying equal rights, one that was no longer willing to submit to "recommendations" from Moscow that passed through Party channels in the style of the Comintern or the Cominform. Mao had already expressed his intention to send Wang to Moscow as future ambassador to Mikoyan at the beginning of February in Xibaipo. Mikoyan had not raised any objections.[799]

Stalin ignored this signal from Peking—or there were other reasons why Mao's intentions were not implemented. It is in any case striking that after his arrival in Moscow, as far as is known, Wang Jiaxiang acted as a diplomat and not as a Party representative. He forwarded messages from Zhou Enlai to the Soviet Foreign Ministry, and the ministry made arrangements with Stalin internally through Gromyko.[800] In addition, there continued to be contacts at Party level. Thus, for example, Mao communicated with Stalin either through Kovalev or directly.[801]

Wang Jiaxiang was well equipped to perform his planned dual function. Unlike Roshchin, he held high positions in the Party and the state. He was a Central Committee member in the CCP and a deputy foreign minister. Wang arrived in Moscow in late October and presented his credentials as ambassador on November 3. He had attended Sun Yat-sen University in Moscow from 1926 to 1930 and spoke Russian. Along with Wang Ming, Bo Gu, and Luo Fu, he was one of the most prominent of the "twenty-eight Bolsheviks" who had returned to China in 1930 under the leadership of the Comintern representative Pavel Mif and replaced the existing CCP leadership under Li Lisan with comrades loyal to Moscow.[802]

When Mao informed Mikoyan in Xibaipo of his intention to send Wang Jiaxiang to Moscow as ambassador, he added in the usual Communist jargon that Wang Jiaxiang had previously "permitted some mistakes in connection with the Wang Ming group." This was a clear reference to Wang's loyalty to Moscow during the conflicts between the Comintern and the CCP dating from the early 1930s. But Wang had returned to China in July 1937, Mao continued, in response to instructions from the Comintern, which had approved the political line the CCP was pursuing and helped the Party to overcome its mistakes.[803]

In honor of the thirty-second anniversary of the Russian October revolution on November 7, the Soviet embassy organized a large reception in its building in the Dongjiaomingxian, which Mao Zedong, Zhou Enlai, and Zhu De attended with their wives.[804] The official cooperation had a promising beginning and was carried out with goodwill. Thus, for example, Liu Shaoqi, as president of the Friendship Society, offered to make available to Roshchin any informational material the Soviet embassy desired and to have the best professors and professionals present lectures on developments in China.[805] In response to a request by Zhou Enlai, the Soviet Foreign Ministry in turn stated that it would instruct any Chinese who were supposed to hold various foreign positions but had no relevant knowledge in Moscow on the history of diplomacy and diplomatic practice.[806]

Irritations in Anticipation of Mao's Moscow Trip

New irritations arose in connection with the Conference of Trade Unions from the Countries of Asia and Oceania that took place in Peking from November 16 through December 1, 1949. As had already happened once before in the context of the dispute over whose pictures were hanging in Manchuria,[807] Stalin decided in this case to accommodate the Chinese side in order not to burden relations with Mao Zedong unnecessarily just before meeting with him in Moscow. The conference was convened by the Soviet-led World Federation of Trade Unions, which had been founded in Paris in October 1945 and was the successor to the Red Trade Union International (Profintern), which had been a section of the Comintern since 1920. It was attended by 117 delegates from thirteen Asian countries—including the Soviet Union.[808]

The Chinese Communists from the beginning had opposed holding the planned Asian trade union conference, as well as a Conference of Women of the Asian Countries, that was convened by the also Soviet-led Democratic World Association of Women in Peiping. The CCP leadership had informed the two organizers of its intention to reject the conferences in February 1949. Nonetheless, the organizers decided to hold the two conferences at the end of the year in Peiping. Liu Shaoqi later explained the CCP's general intention to reject the conferences as follows: Peiping at that time was not yet fully in Communist hands. The trade union and women's movements were still in their initial stages in China. The Chinese Communists did, if necessary, support illegal conferences. Public conferences would only lead to increased suppression of the trade union and women's movements in Asia by the imperialists. Furthermore, they would have to keep all resolutions reached by such conferences secret because the CCP would necessarily call for armed conflict against the imperialists, "the actual rulers of the Asian countries."[809]

Kovalev claimed in a report to Stalin in December 1949 that Liu Shaoqi and Li Lisan had spoken out against participation—clearly meaning participation in a public conference—by trade union representatives from such countries as Japan and India, which were ruled by the "Anglo-American bloc" or were allied to it, so as not to provoke any "irritation [*razdrazhenie*]" in the United States or Great Britain.[810] He interpreted this as part of a conciliatory policy designed to improve the chances that London and Washington would recognize China. Kovalev's claim cannot yet be verified from other sources.

Liu Shaoqi adopted an extremely militant position at the Peking trade union conference. In his opening speech, he described the conference's primary task to be the fight against imperialism, with the goal of achieving national independence in the countries of Asia and Australasia. In many colonies and semicolonies, this conflict had to take the "main form" of "armed conflict" and was to be carried out by a "regular revolutionary army," he said. He repeatedly and almost insistently recommended that the colonies and semicolonies in which conditions similar to the ones in China predominated follow, as the "main road," the same road to this goal that China had taken: "This road is Mao Zedong's road." References to armed conflict and the "Chinese experience" stood at the center of his report. It did not mention Stalin at all and the Soviet Union only once—as a force that would take the colonies' and semicolonies' side in their struggle for liberation, along with "the countries of the New Democracy and the forces for peace and democracy all over the world."[811]

Liu Shaoqi's speech was received negatively and condemned as "ultra-leftist" by the members of the delegation from the Soviet Federation of Trade Unions present, who were led by the organization's secretary, Leonid Solov'ev. Solov'ev convinced the other foreign delegates to adopt this opinion. They expressed their opposition to Liu's proposals, and to the idea of armed conflict in particular. The Soviet line was to support armed conflict only in those Third World countries that were seeking to liberate themselves from colonial rule. On the other hand, Moscow sought to establish good relations with the governments of states in which this had already happened, such as, for example, India, Burma, and Indonesia, in order to bring them over to the Soviet side in the context of international politics.[812]

Liu did not accept the pro-Soviet criticism. In his opinion, the critics may have been familiar with European conditions, but they had only a "weak conception" of the specific situation in the Asian countries. Until the power of imperialism there had been broken, there could be no talk of fighting for the eight-hour workday, social security, or "other democratic freedoms." The critics, Liu reported, were afraid to commit themselves to armed conflict, as this might damage their activities in France and other countries. Instead, they pointlessly suggested that Asian workers should protest about their lack of rights and difficult material conditions to the UN.[813]

At Solov'ev's suggestion, the members of the executive committee proposed that Liu should not seek to publish his speech. Liu was not willing to do this at first, but ultimately gave in, against his convictions, to the pressure exerted on him.[814]

Stalin had already received the text of Liu's speech through Kovalev two weeks earlier. Kovalev was furious that Mao Zedong had up to then ignored the recommendations that Stalin had made to Liu Shaoqi in Moscow about establishing an association of Communist parties of Asia. The text of Liu's speech, he told Stalin in his accompanying comments, constituted clear evidence of the Chinese Communist's pursuit of hegemony in the region. Stalin also expressed his dissatisfaction with the speech, which he saw as an expression of a "deviation to the left," but kept, in Kovelev's opinion, his criticisms to himself in order not to burden relations with Mao. On the contrary, he criticized Solov'ev's behavior on November 20 in a telegram to the trade unionist that Kovalev handed to him, in which the Soviet leader described him as "two-faced [*dvurushnik*]." Solov'ev had made a "gross political error" when he took a stand against the publication of Liu's speech, which he, Stalin, considered "correct and timely," he said. Stalin instructed Solov'ev to approve publication of the speech immediately and to show the full text of his telegram to Mao Zedong. Solov'ev did as he was ordered and underwent self-criticism to Mao Zedong. Mao then asked Stalin not to condemn Solov'ev; he had done no damage, Mao argued. In his unpublished memoirs, Kovalev described this incident as evidence that Stalin was willing to sacrifice a loyal ally—as he had Gao Gang that summer in Moscow—to ensure Mao's loyalty to him.[815]

By approving publication of his speech, Liu Shaoqi later wrote, Solov'ev established "normal practical working relations" between the functionaries of the World Federation of Trade Unions and the Chinese comrades. The Chinese media and the Cominform journal published the speech, but it was not published immediately in the Soviet media. Although Liu Shaoqi had already made the text of the speech and a Russian translation available to the Soviet side, Moscow demanded a second—that is, a "revised [*proverennyi*]" and "certified [*doverennyi*]"—translation from Liu in late December. The speech was first published on January 4, 1950, in *Pravda,* after a delay of almost two months. As a "nice gesture,"[816] its publication coincided strikingly with the breakthrough in Mao Zedong's negotiations at the same time in Moscow. *Pravda* published the other important Chinese papers as soon as they were presented.[817]

Another dispute arose at the end of the conference. Liu Shaoqi took a definitive stand against the text of the resolution formulated by the members of the executive committee. In his opinion, the resolution was "toothless" because it did not call on the working classes of the Asian countries to wage armed resistance against imperialism. Liu was then asked to submit a draft resolution of his own. At the center of his resolution, Liu called on labor unions in those Asian countries that already had national armies of liberation to fight to have them strengthened. Armies of liberation should be established in countries that did not already have them. The members of the executive committee and the general secretary, Louis Saillant, in particular reacted by observing that they might as well pack their suitcases and move from Paris to Moscow immediately if they accepted a resolution like that. It would no longer be possible to carry out any further activity on the territory of imperialist countries. Furthermore, it was not acceptable to arouse the peoples of Indochina, Burma, and other countries to wage armed warfare while they themselves continued to live peacefully in Paris, London, or other cities of the imperialist states.[818]

In the end, Liu said, the Chinese comrades found themselves forced to approve the executive committee's draft—not because they agreed with it completely, but because it did not contain "any fundamental mistakes." They nonetheless considered the resolution to be bad and one that "was of no use to the working classes in the countries of Asia."[819]

According to Liu Shaoqi, the Chinese delegates also approved another suggestion by the functionaries of the World Federation of Trade Unions at the end of the conference that they had rejected at the beginning: the establishment of a "liaison office of the countries of Asia." They agreed to approve this office, however, only when they were given the leading role in it—as the people most familiar with the situation in the Asian countries. They agreed that this office would have to work "in strict secrecy" in order to protect the functionaries of the Asian trade union movements from any possible repression from the imperialists.[820]

Differences also arose at the Asian women's conference. Unlike the French members of the executive committee, the Chinese side was of the opinion that women's work in Asia could be done only in strict secrecy, like the trade union activities, because the imperialists would otherwise annihilate it. It opposed the executive committee's call in the final resolution to fight legally for the prohibition of child labor, protection of women's work, social security, and training free of charge, among other things. This would be a "fantastic and unrealistic" demand, it said, as the abolition of child labor would condemn millions of Asian children to hunger and squalor. The resolution should instead contain a call for armed struggle against imperialism in the Asian countries, and for women to support the armies of liberation and participate in illegal forms of struggle. In this context, the Chinese criticized the executive committee of the Democratic World Association of Women for repeating the mistakes of the World Federation of Trade Unions. Both cases had revealed a "lack of knowledge of the specific conditions of the countries of Asia," they argued, where the most important thing was "armed struggle under the leadership of the Communist parties."[821]

Liu Shaoqi had originally crossed out a quotation from Mao Zedong's article "On the People's Democratic Dictatorship" of July 1, 1949, in which Mao referred to the necessity of Soviet aid for the victory of the revolution in China, from the text of the speech delivered by Chinese delegate Deng Yingchao, Zhou Enlai's wife. By doing so, Liu explained later, he had wished to keep the women of Asia from placing all their hopes on aid from the USSR, rather than relying on their own strength. For this reason, as well as in order not to provide any grounds for claims that the Soviet Union and the People's Republic of China were interfering in the internal affairs of other Asian countries, Liu continued, he had also opposed any mention of the USSR or the victory of the Chinese revolution in the resolution. But when some members of the executive committee put this so that it seemed that the Chinese women comrades did not wish "to acknowledge the role played by the Soviet Union in world history in ensuring the victory of the people's liberation movement in the countries of Asia," Liu had approved inserting the Mao quotation into Mrs. Deng's speech.[822]

The Chinese delegation to the conference at first spoke out decisively against the establishment of a "liaison office of the women of the countries of Asia," as it was not at that time possible for women to become active legally in the countries oppressed by imperialism, and the All-Chinese Women's Association was not yet in a position to establish a leadership center for illegal women's work in the Asian countries.[823]

Liu Shaoqi's experiences with his European comrades at the two conferences in Peking made a negative impression on him. In a description of the course of the conference that he gave the Soviet chargé d'affaires in Peking shortly afterward, he said that people in Paris, as in Europe generally—he was presumably referring not least to Moscow—were not very familiar with the situation in the Asian countries.[824] Shortly afterward, he said something similar, if even more decisive, and with a direct reference to Soviet ignorance about China.[825] It is possible that observations Liu had made during his stay in Moscow during the summer of 1949 also played a role in his criticism.

4

Stalin and Mao Zedong in Moscow

The Breakthrough to the Alliance

The Treaty of Friendship, Alliance, and Mutual Assistance that was signed in Moscow on February 14, 1950, established the alliance between the Soviet Union and the People's Republic of China. This historic event took place at the end of Mao Zedong's visit to Moscow between December 16, 1949, and February 17, 1950, during which he met Stalin for the first and only time. This was also Mao's first trip abroad.

Mao's Trip to Moscow: Motives, Planning, and Arrival

Mao Zedong had been planning a meeting with Stalin in Moscow since the spring of 1947, but had repeatedly postponed it—as a rule at Stalin's request.[1] According to Chinese sources, Stalin informed Liu Shaoqi in July 1949 that Mao should come to Moscow immediately after he had established the new Chinese government and that they would also resolve the issue of the treaty of August 14, 1945, at the same time.[2] In a slightly different vein, Zhou Enlai stated in a private discussion in November that Stalin had expressed to Liu that he would be happy "to meet with Mao Zedong in Moscow at any time."[3]

On November 5, Stalin informed Mao through his middleman in China, Ivan Kovalev, that he would be glad to see him at any time.[4] In the same talk, Mao said that he would like to travel to Moscow in December to wish Stalin, who turned seventy on December 21, "a happy birthday in person." As delegations of friends of the USSR from all over the world would be present there at the same time, his trip could "have a completely open character." He said that he also wanted to rest and recuperate in the Soviet Union.[5] On November 8, Mao sent Stalin a telegram through Kovalev in which he again described the kind of trip he wanted.[6] In it, he asked Stalin to determine when he should leave Peking.[7] He apparently also expressed his wish to begin the trip in early December. At the same time, he asked Stalin to consider carefully and then decide *(zhuoding)* whether Zhou Enlai should travel with him or if it should only be decided after Mao arrived in Moscow whether Zhou should come later, and, if so, when.[8]

After having reached agreement on principle through Party channels, Mao proposed to the Soviet side at diplomatic level to begin his trip in December. On November 10, Zhou Enlai had an extensive talk with Ambassador Nikolai Roshchin on instructions from Mao in which he presented Mao's conception of the Moscow trip and requested that Roshchin forward them to Stalin. Zhou began the talk by observing that Mao had long held the dream of visiting the Soviet Union. Roshchin asked "what the character of Mao Zedong's visit will be" and who would accompany him. Zhou replied that in addition to the fact that Mao wished to establish "personal friendly contact with Stalin," he "apparently will discuss the problem of the Chinese treaty." Zhou did not specify any other topics of discussion. In this way, he signaled to Stalin that the treaty question took first place on Mao's agenda. No one would accompany Mao to Moscow from the government, Zhou continued. But if "a new Chinese-Soviet treaty were negotiated as a result of Mao Zedong's trip," he, Zhou Enlai, would be able to come to Moscow by airplane to sign the treaty at short notice. Zhou also inquired whether Roshchin would accompany Mao to Moscow. The latter replied that he would do as Moscow instructed him.[9]

Mao thanked Stalin on November 12 for approving his trip and announced that he would leave on one of the first ten days of December.[10] At the same time, he requested that Kovalev be allowed to accompany him. As Kovalev had already assigned jobs to the Soviet specialists, his absence would not disrupt their work, he said. The Soviet Foreign Ministry suggested that Stalin allow Ambassador Roshchin to come as well.[11]

It is not yet possible to find conclusive evidence of how Stalin reacted to Mao's query whether Zhou Enlai should travel with him or come later, when a new treaty was signed. It would appear that Stalin stated that Mao could come without Zhou.[12] On November 12—that is, on the same day that Mao thanked Stalin for approving his travel plans—Ambassador Wang Jiaxiang reported as if he were apologizing to the Soviet Foreign Ministry that they feared in Peking that "at the present time, as the entire structure of state power is being organized, the simultaneous departure of two of the most important [*osnovnykh*] leaders might have an effect on the work."[13] Andrei Gromyko, then deputy foreign minister, suggested to Stalin on the same day that Wang be informed orally that the Soviet Union approved "whatever decision the government of the People's Republic of China took" on this question, which had clearly raised doubts on the Chinese side. It is not known whether Stalin approved Gromyko's suggestion.[14]

From the Chinese perspective, the routine exchanges on the question of whether Zhou Enlai should accompany Mao or come later involved more than a question of protocol. By proposing to come to Moscow to sign a new treaty, Zhou wanted to signal for a second time the Chinese leadership's intense interest in concluding a new treaty. By not coming immediately and being absent from Moscow at the beginning, he was clearly seeking to provide Stalin with a constant reminder of the Chinese leadership's desire for a new agreement. To this extent, the reasons Wang Jiaxiang gave for Zhou's failure to come were merely a pretext, as Mao had not expressed any reservations in his alternative proposal to Stalin about the idea of traveling to Moscow together with Zhou. And the reasons Wang Jiaxiang gave for Zhou's not coming to Moscow with Mao—that is, the need to organize "the entire structure of state power"—would still have had to be valid if Zhou came a few weeks later. The Chinese side apparently wanted to save face in this way after Stalin had made it clear that he would prefer Mao to come without Zhou.

One week after Zhou Enlai's talk with Ambassador Roshchin, on November 17, the head of the Chinese secret service, Li Kenong, told Roshchin that one of the primary goals pursued by the American espionage service in China was to undermine the friendship between the USSR and the People's Republic of China by, among other things, "attacking the Sino-Soviet treaty of 1945, which it characterized as an unequal and imperialistic treaty, and as a result of which China had lost the Guandong Peninsula[15] and all the industrial equipment in Manchuria."[16] It is easily conceivable that Li Kenong—certainly not without Mao Zedong's knowledge—was seeking in this indirect way to give Stalin a further signal that Moscow should consider abolishing this "unequal treaty."

During the November 25 session, the Politburo agreed that Liu Shaoqi should lead the Central Committee and Zhu De the Military Commission of the Central Committee during Mao's trip to Moscow.[17]

There are contradictory reports about the planning for Mao's trip. Kovalev indicated that Mao wanted to remain abroad for a total of three months.[18] One month was to be set aside for talks with Stalin, the conclusion of the treaty of friendship in Moscow, and visits to Leningrad and Stalingrad. In the second month, Mao wished to visit Poland, Czechoslovakia, and Romania, and he planned to vacation at the Black Sea spa of Sochi during the third month. Mao, on the other hand, informed Ambassador Roshchin in Moscow that he had originally wanted to remain in the USSR for three months.[19] Shi Zhe merely stated that Mao was supposed to vacation in the USSR after the birthday festivities and that Zhou Enlai would carry out the negotiations for the treaty.[20]

In retrospect, Mao himself described "the conclusion of a treaty of friendship, cooperation, and mutual assistance" to replace the old treaty of August 14, 1945, which was concluded with the KMT government, as "one of our primary goals."[21] Wu Xiuquan, who arrived in Moscow in January 1950 along with Zhou Enlai, went even further and emphasized that their primary goal had been to sign the treaty.[22] Other authors from the People's Republic of China also describe Mao's wish to conclude the aforementioned treaty so that it reflected the new character of relations between the two states as the most important goal.[23] They say that he also planned to ask the Soviet leadership for aid and to congratulate Stalin on his seventieth birthday. The arguments Mao used to Stalin in preparing his trip in November, as described above, and Mao's behavior toward Stalin in Moscow, as described later in this chapter, confirm this impression completely: The most important goal of the visit in Moscow was to conclude a new treaty.

For understandable reasons, we can only speculate about the thoughts, hopes, and feelings that went through the minds of the leaders of the two largest states under Communist rule as they looked forward to their first meeting. There can hardly be any doubt, however, that mixed feelings predominated. They both considered themselves to be allied to one another by Marxist-Leninist ideology with its eschatological belief in the victory of the world revolution. They were, however, both also conscious of the irritations that had arisen in their bilateral relations in the past because both sides had again and again given priority to their own national interests—as they also continued to do.[24] Kovalev observed correctly that both had personal reasons to distrust each other.[25] Stalin, at that time, stood four years before his death and was in any case pathologically suspicious: "He considered everybody a spy or a traitor."[26] Something similar was true of Mao, who, being fourteen years younger, was then age fifty-six. Both were extremely power-hungry, cunning, and unscrupulous. During the Great Purge of 1937 and 1938 alone, Stalin had had almost three-quarters of a million people killed, including the greater part of the CPSU leadership elite. Mao had refrained from mass liquidations in the CCP, but nonetheless had driven many undesirable comrades to their deaths during the rectification movement *(zhengfeng yundong)* between 1942 and 1944. When thinking about the treaty they were about to conclude, one has to picture both as having similar underlying attitudes comparable with the calculations Stalin described himself and Hitler as having made in signing the Ribbentrop-Molotov pact on August 23, 1939: "Of course it's all a game to see who can fool whom. I know what Hitler's up to. He thinks he's outsmarted me, but actually it's I who have tricked him!"[27]

For a long time, Stalin had refused to recognize Mao as leader of the Chinese CP and favored Wang Ming instead. In his short- and medium-term planning, he had given priority to his partnership with the KMT, the strongest political force in China, and had given it aid—including military aid—during the Second World War, while at the same time ignoring the Chinese CP's requests for deliveries of weapons. Mao, for his part, had ignored Stalin's repeated requests to tie Japan down in China by cooperating more closely with the KMT and to take a more offensive military stance against the Japanese occupation in order to restrain the Japanese from attacking the USSR. He had ignored Stalin's insulting advice to renounce the revolution temporarily in August 1945. And for years, Stalin refused to receive Mao in Moscow because the USSR maintained diplomatic relations with KMT China—a relationship that endured until the end of September 1949.

Stalin viewed Mao's victory in the Chinese civil war—like all other political victories Communist parties achieved without Soviet aid—with a mixture of distrust and respect. He suspected that Mao would develop into a "Chinese Tito." Mao had the following to say about this later: "When the war had ended in victory, he again suspected that it was a Titoist victory. . . ."[28] He felt respect not only for the CCP's victory, which he had not anticipated would come so quickly, but also for the physical weight of the newly established Communist state with its approximately 500 million inhabitants—more than twice the population of the USSR. Mao, in turn, may have been impressed by the great prestige that Stalin enjoyed in the world as leader of the first socialist state in world communism and by the Soviet Union as a great power.

Mao's most important interlocutor after Stalin was Vyacheslav Molotov, one of Stalin's closest confidants.[29] At that time, as previously during the negotiations for the Sino-Soviet treaty of August 14, 1945, Molotov, followed by Anastas Mikoyan and Andrei Vyshinskii, acted as the highest authority after Stalin on questions involving China.[30] Years later, Mao recalled that Molotov had not behaved any better than Stalin toward the Chinese side.[31] Molotov took the same line as Stalin, Mao said. He had practically had to negotiate more with Molotov than with Stalin. Molotov behaved like a merchant, "which means he bargained with us."

Kovalev claimed that Mao had been afraid of his meeting with Stalin. Because Mao was aware of the difficult economic conditions in the Soviet Union, he feared that Stalin would reject his request for a credit. This was why he had had Kovalev summoned one week before his departure for Moscow: to portray to him the difficult conditions in the Chinese economy. The version of events Kovalev presented appears exaggerated because Stalin had already approved a credit for US$300 million to Liu Shaoqi that summer.[32]

Mao Zedong's travel plans apparently encountered opposition from non-Communist partners in the coalition government such as the Democratic League and the left wing of the KMT. It was argued that "foreign barbarians" had traditionally always waited on the emperor and never the opposite. Concerns were also raised that the trip might complicate relations with the United States, Great Britain, and France and decrease their willingness to provide economic aid.[33]

On December 6, Mao traveled by train from Peking to Moscow. He was accompanied only by a small entourage, all close confidants, but notably not a single person who was at that time one of the twelve members of the Politburo. Stalin, on the other hand, brought three Politburo members along to the talks, the minutes of which are accessible. The only person accompanying Mao who could conceivably be considered somewhat prominent was Chen Boda, who had been Mao's chief secretary for political and ideological affairs since 1937 and had in the meantime been awarded the title of professor. Chen acted as one of Mao's most important ghostwriters, had been a member of the Central Committee since 1945, and was also deputy head of the Propaganda Department of the Central Committee and deputy director of the Institute for Marxism-Leninism. He was the only person accompanying Mao that the Soviet press mentioned by name when the delegation arrived in Moscow. Ye Siluo acted as Mao's personal secretary for secret affairs and Wang Dongxing was responsible for security. Shi Zhe acted as interpreter. The entourage also included the Soviet ambassador in Peking, Nikolai Roshchin, a few lower-ranking Chinese, and—at Mao's request—Kovalev and the interpreter Nikolai Fedorenko, who was then working in the Soviet embassy in Peking.[34]

Kovalev suspected that Mao had only brought a small entourage of close confidants rather than one made of the leading Party and state functionaries because he wanted to speak to Stalin without witnesses. He saw the fact that Zhou Enlai came to him shortly before his departure with the request that Fedorenko be allowed to travel with Mao to interpret at his talks with Stalin and other Soviet leaders as confirmation of this assumption. Mao, he reported, was concerned about his authority and feared that this might be damaged in the eyes of his Party comrades. Kovalev also had the impression that Mao was concerned that he might not be able to fulfill the high expectations the Chinese Party leadership placed on his talks in Moscow with repect to politics and the economy. There may be some truth in these assumptions. And yet they are not accurate with respect to the role that was allegedly to be played by Fedorenko, because Mao also took Shi Zhe along to interpret during the two talks with Stalin, for which we have minutes.[35]

The trip was kept secret until the delegation arrived in Moscow. The strictest security measures were taken along the Chinese section of the route between Peking and the border station at Manzhouli. Mao's train was escorted in front and behind by two trains, each occupied by fifty soldiers. Guards were posted on both sides of the track along the entire route. Each time the train stopped, only two individuals from the Party, the government, or the military were allowed to board the train. No one was allowed to get off the train or ask for anything. The train was carrying selected agricultural and artisanal products collected specifically by Mao's wife, Jiang Qing, and Wang Dongxing as presents for Stalin. These included white cabbage, radishes, leeks, and pears from Shandong; white turnips from Weifang; winter bamboo shoots; the famous Dragon Well tea from Hangzhou; high-proof Maotai liquor from Guizhou; cigarettes from Shanghai; silk; Jingdezhen porcelain from Jiangxi; and embroidery and cushions from Hunan. Gao Gang accompanied Mao from Mukden to the border.[36] Deputy Foreign Minister Anatolii

Lavrent'ev was waiting for the Chinese on the Soviet side of the border. They were greeted by a high-ranking Soviet dignitary at each station. Foreign Minister Vyshinskii inquired about Mao's health and wishes by telephone in Novosibirsk. In Iaroslavl', Wang Jiaxiang, the Chinese ambassador in Moscow, boarded the train. Mao had caught a cold while the train was still in China and did not feel well during the trip.[37]

Cool Reception in Moscow

Mao Zedong arrived in Moscow around noon on December 16. He was met at the Iaroslavl' Station, then the Northern Station, by Politburo members Molotov and Nikolai Bulganin, who by protocol figured as vice chairmen of the Council of Ministers of the USSR. The Soviet media described Mao as "Chairman of the Central People's Government" and—like the Chinese media—mentioned only "Professor Chen Boda" from his entourage by name. In addition to Molotov and Bulganin, minister for foreign trade Mikhail Men'shikov, Deputy Foreign Minister Gromyko, the Moscow city commandant, Lieutenant General K.R. Sinilov, senior officials of the Soviet Foreign Ministry and the heads of diplomatic missions from all the Eastern European satellite states, North Korea, and the Mongolian People's Republic all attended to greet him. Stalin did not come, as was customary Soviet practice at the time. Foreign Minister Vyshinskii was also absent. Molotov and Bulganin categorically refused Mao's invitation to join him at an opulently prepared table in his train car to taste some Chinese specialties with the comment that there was no mention of this in the protocol. Pointing to Mao's ill health and the cold weather, they explained that they had abbreviated the welcoming ceremony on the railroad platform. They asked Mao to review the military honor guard but then not to give a speech, should he have prepared one. It would be enough for him to give the text of any such speech to the press.[38]

Even a head of state less conscious of prestige than Mao would have found such a welcome hurtful. Kovalev's observation that Mao was "unmistakably embittered [*iavno ogorchen*]" by the reception's coldness appears credible. A photograph published in *Pravda* shows Mao Zedong and his Soviet reception committee with stony expressions on their faces as they reviewed an honor guard. Molotov, Lazar Kaganovich, and Gromyko can be recognized in the front row. The reception for Mao was thus much less friendly than the one for the KMT foreign minister, T.V. Soong, when he had arrived in the Soviet capital to negotiate the Sino-Soviet Treaty of Friendship and Alliance on June 30, 1945. The atmosphere of the reception on that occasion was described as "cordial."[39]

Molotov and Bulganin used arguments based on protocol to refuse Mao's invitation to accompany him to his lodgings in his car.[40] Mao was lodged in the Blizhniaia ("the Near One"),[41] the closest of Stalin's dachas, where Stalin himself usually lived when he was not staying in the Kremlin.[42] It was located in Volynskoe, a little outside the city limits of that time, near today's Kutuzovskii Prospekt, and could be reached in approximately fifteen to twenty minutes by car from the Kremlin on the Staryi Arbat, which at that time was still passable. Mao, Ye Zilong, and Shi Zhe lived on the ground floor; Chen Boda and Wang Dongxing on the upper one.[43] Kovalev did not accept Mao's offer to move with his wife into the same dacha.[44]

In the manuscript of his speech that was printed in the Soviet press the following day, Mao Zedong acknowledged the Soviet Union as the first socialist state in the world and spoke of the "deep and strong friendship" between the Chinese and the Soviet peoples. He described both China and the USSR as great powers. He did not mention the need to conclude a new treaty to replace the one concluded with the KMT on August 14, 1945. There can hardly be any doubt, however, that his reference to the Soviet government as the first to have abolished the unequal treaties czarist Russia concluded with China was intended as a very unsubtle hint.[45] The Chinese media, incidentally, gave the impression that Mao had in fact delivered his speech on the station platform.[46] It must remain an open question whether this was done out of ignorance of the way things actually occurred or with the intention of saving face.

Mao could for the most part be satisfied with the reports in the Soviet media. *Pravda* celebrated Mao's arrival as the event of the day and the relevant article occupied about half of the front page. It is striking that it referred to Mao as "Mr. [*gospodin*] Mao Zedong." The same was true for Zhou Enlai, when he arrived in Moscow on January 20, 1950.[47]

On the occasion of his participation in the central celebrations for Stalin's seventieth birthday, on the other hand, Mao was given the titles chairman of the CC of the CCP and "comrade [*tovarishch*] Mao Zedong."[48] Shi Zhe was surprised when Stalin also addressed Mao as "Mr." at the December 16 talk, although it was the Soviet custom for two Marxists always to address one another as "comrade" and only non-Marxist friends as "Mr."[49] The Soviet side, he reported, always referred to Mao as "Mr." on place cards and in prepared statements, too. He reported that he, Shi Zhe, had at first not dared to ask Stalin about this. Only at the January 22, 1950, talk did he dare to raise the question with him, at which time Stalin had observed, smiling, that it was true, Mao was indeed a comrade, and they should start addressing him as "comrade" immediately. In the Chinese report of Mao's arrival in Moscow, he was given the title "chairman [*zhuxi*]," while they exalted Stalin strikingly with the title "generalissimo [*yuanshuai*]"—a rank awarded to Stalin in 1945 for his services in the Second World War and which he was subsequently eager to employ.[50]

One Chinese author suspected that the Soviet behavior was due to the fact that Stalin did not yet recognize Mao as a Marxist in December 1949.[51] Soviet interactions with representatives of other states under Communist rule—to the extent that they presented themselves as state representatives—reveals, however, that this assumption is not accurate. In those cases, such representatives were also addressed as "Mr.," according to Soviet protocol.[52] Everything suggests that Moscow was careful to distinguish between foreign Communists acting in a Party capacity and those in a state capacity in official protocol.

Stalin Takes a Firm Stand: The December 16 Talk

Stalin received Mao Zedong for a two-hour talk in the Kremlin at 6 P.M. on the day of his arrival. Politburo members Molotov, Georgy Malenkov, and Bulganin, Foreign Minister Vyshinskii, Mao's interpreter, Shi Zhe (Russian name: Karskii), and Stalin's interpreter, Nikolai Fedorenko, were also present.[53] The latter was then a member, and was acting temporarily as the head, of the First Far Eastern Department of the Foreign Ministry, which was responsible for China.[54]

Until recently, researchers had to depend on published reports by each of the interpreters to understand the course of the talk. In his reports, Fedorenko usually described the external conditions and atmosphere of the talks in long-winded, flowery, and at times even childish language, while remaining silent or at best vague about political substance.[55] His behavior may be attributable to loyalty to Stalin or the Soviet regime and/or to his inability to remember due to advancing age.[56] Fedorenko's colleague Shi Zhe did occasionally go into political content, but generally lost himself in the anecdotal.[57] The reason for this may also be found in his obvious loyalty to Mao and the Party and/or in his fading memory and/or in the influence of official Party censorship.[58] Furthermore, as can easily be seen, neither interpreter had any or much access to archival documents in drawing up their memoirs. This is certainly an additional reason for some of the demonstrably inaccurate information they provide. Let one example suffice: Shi Zhe claimed that the entire Politburo (at that time consisted of twelve individuals) and Foreign Minister Vyshinskii had all been present at the December 16 talk,[59] while Fedorenko gave the impression that only Shi Zhe and he himself had participated in addition to Stalin and Mao.[60] Neither account is accurate.

Our sources on this question have recently been significantly improved as two key documents have become accessible: the Soviet minutes Fedorenko kept (henceforth: Fedorenko minutes)[61] and a report on the talk that Mao wired to Peking following the interview (henceforth: Mao report).[62] Both documents confirm that the descriptions Fedorenko and Shi Zhe had published previously were to a great extent incomplete and partially false.[63] Fedorenko's recollection of his work interpreting between Stalin and Mao reveals the terror Stalin spread around him at that time.[64] On one occasion, for example, he had had difficulty translating a classical citation Mao had used and had to ask about it repeatedly. Stalin then intervened suspiciously and asked how long this conspiracy *(konspiratsiia)* was going to continue. Fedorenko commented that Stalin's question had struck him like a bolt of lightning. He had great difficulty overcoming Stalin's mistrust and had then been able to feel the cut of the guillotine move away from his neck "almost physically."

Mao reported to Peking by telegram about the atmosphere of the talk, saying that the "friendly feelings" had been "genuine [*qingyi kenqie*]"—a characterization that for reasons that will be discussed below cannot be taken at face value.[65] Shi Zhe's report that there was a degree of irritation at the very beginning, during the usual exchange of polite conversation, when Stalin stopped Mao from protesting about Stalin's drawn-out disrespect in the past by commenting, "One does not judge victors!" is particularly not very credible because Shi Zhe observed that the atmosphere at that time was "completely enthusiastic and moving [*shifen relie, dongren*]."[66] There is no mention in Fedorenko's report of the remark Stalin is supposed to have made.

When the talk turned to substantive matters, Mao Zedong said that China would need a period of three to five years of peace to reconstruct its economy. When he asked how long the international peace would last, Stalin replied that there was no immediate threat for China at that time. Japan had been too severely weakened to wage war. The United

States feared war more than anything else. Europe was afraid of war. In fact, there was no one who wanted to fight with China—or was Kim Il Sung going to march into China? The peace, he said, would depend on "our efforts." If China and the USSR continued to maintain friendly relations, the peace could last twenty to twenty-five years and perhaps even longer.[67] In his report on the talk to Liu Shaoqi, Mao observed that the Chinese side shared Stalin's assessment that it was very improbable that a war would break out *(zhang shi hen nan da qilai de)*.[68]

Without further transition, Mao then addressed the delicate problems associated with the Sino-Soviet Treaty of Friendship and Alliance of August 14, 1945. By his own admission, as mentioned above, the conclusion of a new treaty was "one of the primary goals" of his trip to Moscow.[69]

Liu Shaoqi had raised the topic briefly during his visit to Moscow. He had mentioned three alternatives in his report to the Soviet Party leadership on July 4, 1949: they could retain the treaty, conclude a new one, or retain the old one for the time being and at the same time state their intention to conclude a new one later.[70] Stalin had admitted to Liu in Moscow that the treaty was unequal and had said that he had already corresponded with Mao Zedong about it.[71] The USSR, Stalin stated, had been forced to act as it had in 1945. Stalin had proposed to Liu that they wait to resolve the question until Mao came to Moscow.

On December 16, Mao Zedong introduced the discussion of the treaty by observing tersely that the Chinese Party leadership had discussed the question after Liu Shaoqi's return from Moscow.[72] It is striking that Mao did not give the impression of making any demands on this point that he considered so central, but rather remained passive and waited. Stalin, however, grasped the problem by the hair and took on the issue directly: They should address this question and decide.[73] They had to be clear about whether they wished to declare that the existing treaty continued to be in force, to announce future changes, or implement these changes immediately. He thus repeated—if in a different order—the three alternatives Liu Shaoqi had named five months earlier.

Stalin clearly advocated the first alternative. The treaty *(dogovor)* in force at that time, he argued, had been concluded on the basis of the results of the Yalta agreement *[Ial'tinskoe soglashenie]*, which had regulated key points in the agreement, such as those governing the Kuril Islands, southern Sakhalin, Port Arthur, and so on. "This means that the treaty in question had been concluded with the approval of America and England, so to speak," he said. "Keeping in mind this circumstance, we have resolved in our inner circle not to change any of the points of this treaty for the time being, because a change in even one single point [*izmenenie khotia by odnogo punkta*] could give America and England legal grounds to raise the question of changing also the points in the treaty that involve the Kuril Islands, South Sakhalin, etc." For this reason, Stalin continued, the Soviet leadership considered it possible "formally to retain" the existing treaty "but in fact to change it, i.e., to retain the Soviet Union's right to station its troops in Port Arthur formally, but to withdraw the units that were there at the suggestion of the Chinese government." They could do the same thing with respect to the Chinese Changchun Railroad (CCR) by "taking into account the wishes of the Chinese side, [and] retain the respective points in the agreement formally while changing it in fact." But if the Chinese comrades were not satisfied with this combination, they could make their own proposals.[74]

The Mao report repeated these statements by Stalin in abbreviated but essentially identical form: On account of the Yalta agreement, it would not be appropriate *(bu yi)* at the present time to touch the legal status of the existing Sino-Soviet treaty. If they did conclude a new treaty, it said, the United States would have grounds to take the Kuril Islands *(meiguo jiu you liyou yao na Qiandaoqundao)*. They could not at that time alter the point by which Port Arthur was leased to the USSR for thirty years formally, but the Soviets would in fact withdraw their troops and allow China to station its troops there.[75]

The mention of Port Arthur and the Chinese Changchun Railroad shows that when Stalin spoke of the "treaty," he meant not only the treaty itself but everything involved with it, including the supplementary agreements on Port Arthur, Dairen, and the CCR.

According to the portion of the Fedorenko minutes we have just reproduced, Stalin confuses the Yalta Agreement Regarding Entry of the Soviet Union into the War against Japan of February 11, 1945, at one point ("the question of changing also the points in the treaty that involve the Kuril Islands, southern Sakhalin, etc.") with the Sino-Soviet treaty of August 14, 1945. The treaty—or more precisely the three agreements signed along with the treaty—regulated, among other things, property relations in Port Arthur, Dairen, and concerning the Chinese Changchun Railroad, but not such questions as those regarding the Kuril Islands and southern Sakhalin. As Stalin himself had observed correctly, these problems—along with the aforementioned Manchurian questions—were the subject of the Yalta agreement that Stalin, Roosevelt, and Churchill had signed.[76]

The aforementioned confusion in the Fedorenko minutes can only have been made by Fedorenko himself, because he was not sufficiently familiar with the complicated subject matter and confused the *soglashenie* of Yalta with the Sino-Soviet *dogovor*. This is also suggested by the fact that Mao Zedong understood Stalin correctly, as can be seen from his reply below.

It is clear what Stalin meant. He did not want to put the territorial gains in the Kuril Islands and southern Sakhalin that he had made in Yalta at Japanese expense at risk by changing the terms of the Sino-Soviet treaty that related to Manchuria. As the treaty had been reached by implementing relevant stipulations in the Yalta agreement, Washington and London could have seen such changes as a violation of the Yalta agreement. The United States and Great Britain would then have had the right to declare that they also considered the agreement obsolete and demand a change in the stipulations involving southern Sakhalin and the Kuril Islands.

On the other hand, Stalin foresaw no such problems if the complex of Sino-Soviet agreements remained formally unchanged for the time being and they merely modified the way they were implemented. They could reach agreements with the Chinese Communists privately on the best way to do this. Stalin, in any case, ran no risk in promising to withdraw the troops in fact—although he did not want to do this soon. Mao's reaction to the proposal to withdraw the troops immediately that Stalin had made through Mikoyan in February 1949 in Xibaipo had made him aware that the Chinese side was not interested in this. He was essentially in debt to the Chinese Party leadership, however. The Soviet Party leadership had, to all appearances, already resolved before Mikoyan's stay in Xibaipo to abrogate the agreement on Port Arthur and withdraw the troops from Liaodong in the event of a Communist victory in China as soon as a peace treaty was concluded with Japan, and this had been communicated to the CCP leadership. Stalin had made a similar declaration to Liu Shaoqi in July. But when it came to actually doing so, he began to have concerns about taking such a step in public and formally revising the system of agreements because of Yalta.

It is striking that Stalin avoided describing the treaty of August 14, 1945, as "unequal" to Mao Zedong, as he had to Liu Shaoqi five months earlier.[77] He had already described the agreement on Port Arthur as unequal to Mao through Mikoyan at the beginning of the year.[78] On the same occasion, he had let Mao know that the Soviet government had resolved to declare the agreement invalid and withdraw its troops from Port Arthur after a peace treaty was concluded with Japan and U.S. troops had withdrawn from Japan—or even immediately, if the Chinese side considered this appropriate. The Chinese Party leadership had spoken out against an immediate withdrawal at that time. During the summer, Stalin had again offered to withdraw the Soviet Army to Liu—"after the withdrawal of American troops from Japan" or even immediately, if the Chinese side considered this necessary.[79] Liu had apparently made no response to this. But it is striking that Stalin did not describe the agreement on Port Arthur as unequal—as Mikoyan had in Xibaipo—and also did not mention the possibility of abolishing it. As has already been mentioned, however, this does not necessarily indicate that he had changed his position.

One might have expected Mao to contradict Stalin at this point on December 16 and insist on concluding a new set of treaties, as was his intention. But this did not happen, according to both sources available to us. His response to Stalin's statement about Soviet special privileges in Manchuria was similar to the one the Chinese leadership had given Mikoyan in Xibaipo. According to the Fedorenko minutes, he stated that the existing situation with regard to the railroad and Port Arthur satisfied Chinese interests because the Chinese troops were not yet in a position to resist imperialist aggression effectively. Furthermore, the CCR was a training site for Chinese railroad and industrial cadres. The Mao report plainly states that Mao had objected that it would not be in the Chinese interest for the Soviet troops to withdraw too soon. Despite all their other differences, both sources allow us to conclude that Mao spoke out against an immediate troop withdrawal.[80]

It nonetheless remains possible to doubt that all Mao's statements reflected his real opinion. It is easiest to believe he was expressing his true convictions with respect to the Soviet troop presence in Manchuria. The CP leadership had feared an American intervention in the civil war on the side of the KMT since spring 1949, although this fear had apparently diminished somewhat by the end of the year.[81] The situation was different with respect to the CCR. Here there was apparently a consensus in the CCP leadership that the Soviet leadership was obligated to return the railroad to China immediately.[82] Mao apparently did not wish to raise this demand openly because he was concerned about the strategic partnership China desired with Moscow and its urgent need for Soviet economic aid. In any case, he did subsequently demand that Stalin reduce the period of validity and change the existing legal status of Soviet rights to the railroad in the talk about principles on January 22, 1950.[83]

It is also possible that Mao did not contradict Stalin's refusal to conclude a new treaty immediately on December 16 because he was not prepared for Stalin to reject the conclusion of a new treaty so rigidly, given the more flexible position Stalin had taken through Mikoyan in the spring and to Liu Shaoqi in the summer of 1949, and had therefore not prepared a good defensive strategy.

Stalin then found himself compelled to explain his approval of a troop withdrawal. According to Fedorenko's minutes, he stressed that the proposed withdrawal did not imply that the Soviet Union did not want to help China, if that became necessary. At the same time, he said, the Soviet leadership, as Communists, did not feel comfortable stationing troops on foreign soil, in particular in a friendly country. If Soviet troops were stationed in China, he asked, why should not, for example, the British station their troops in Hong Kong and the Americans theirs in Tokyo? They had a lot to gain internationally from a conjoint troop withdrawal from Port Arthur, he argued. And the CCP would gain prestige in the eyes of its national bourgeoisie, which it had to take into account, because it would give the impression that the Communists had achieved what Chiang Kai-shek had been unable to achieve. The USSR had the right by treaty to station its troops in Port Arthur, but it could surrender this right and withdraw its troops at the request of the Chinese government. If the Chinese government did not desire a withdrawal, the Soviet troops could remain two, five, or ten years—whatever China wanted. "They should not get the impression that we want to run away from China [*khotim bezhat' iz Kitaia*]. We can also leave our troops there for twenty years."[84]

The Mao report states plainly that Stalin said that it should not look as though the Soviet Union were leaving its Chinese comrades in the lurch if it withdrew its troops. The two sides could publish a joint statement in connection with the withdrawal from which China could draw political capital.[85]

He knew how Mao had reacted to a similar offer in Xibaipo: the Chinese leadership had not been interested for reasons that have already been explained in detail (see chap. 3, pp. 144–45). He could therefore make this generous gesture without any risk that it might have an undesired effect. Stalin's suggestion that it was particularly painful for him to station troops in a "friendly country" was hypocrisy, given the massive Soviet troop presence in the "friendly" satellite states of Eastern Europe at that time. His behavior here is incomprehensible, as Mao must also have been aware of the true situation in Eastern Europe.

The following exchange is exciting because Mao, who had apparently recovered from his surprise, expressed, according to his own report, albeit indirectly, his objections to Stalin's intention to retain the treaty of August 14, 1945. As he had already done in previous dealings with the Soviets, he did so by placing criticisms in the mouths of third parties. He said that the Yalta agreement did indeed have to remain in force, and yet that there was a general perception in Chinese public opinion that the treaty then in force—Mao was clearly referring to the entire set of agreements—had lost its significance *(shiqu le yiyi)* because it had been concluded with the KMT government, which had since been overthrown.[86]

One searches in vain for any similar argument in the Fedorenko minutes. Instead, Fedorenko's account said that Mao admitted that they had not taken the American and British positions on the Yalta agreement into account in China, adding: "We must act in the way that is best for the common cause."[87] They would have to reconsider the issue further, Mao is supposed to have said: "It is, however, already becoming clear now that it is not appropriate to alter the treaty immediately [*seichas*], just as it is not appropriate to hurry to withdraw the troops from Port Arthur." Mao apparently used the formula "the common cause" to signal some distance from Stalin's position without contradicting him directly. He expressed himself in a similar way when he wanted to indicate to Stalin that he did not agree with him at one point in the talk on principles on January 22, 1950: they would have to take into account the interests of both sides—both China's and the Soviet Union's.[88]

Stalin's next comment gave the impression of being an accommodating reply to the request for a new treaty that Mao was expressing indirectly: they would have to improve *(xiugai)* the treaty that was in force after approximately two years, Stalin reportedly said, and they would have to "undertake fairly major corrections [*xu zuo xiangdang da de xiugai*]." This comment is included only in the Mao report, however.[89]

The two sources are thus not in agreement on this. They do, however, at least both suggest that Stalin and Mao took different positions on the question of a new treaty. If Mao, as Fedorenko indicated in the minutes, in fact stated that he agreed not to alter the treaty for a while, this would imply an impressive amount of self-denial. If it was true, Mao might have suppressed this passage in his report to Liu because he feared the Chinese Party leadership would accuse him of servility. On the other hand, it is not possible to explain why Fedorenko failed to mention Mao's objection that the treaty had lost its significance.

According to the Fedorenko minutes, Mao did, however, insist on the question once again by asking Stalin whether Zhou Enlai should come to Moscow "in connection with the decision on the treaty question." Stalin answered that Mao had to decide this question himself, as Zhou might be needed in regard to other matters. According to the Mao report, Stalin said that Zhou—literally "the foreign minister [*waizhang*]"—did not have to come to sign. Mao put a preliminary end to their discussion of the treaty at that point and said that he would think about it *(kaolü yixia)*.[90]

If these two sources reproduce this part of the discussion at all accurately, one can only assume that Stalin misunderstood Mao Zedong intentionally. Mao had again raised the question of concluding a treaty, but Stalin apparently failed to hear what he said because he mentioned only a "statement," which Zhou would not have to sign.

In his telegram to Liu Shaoqi on December 18, Mao called for ("the prime minister") Zhou Enlai to come to Moscow when there were agreements to sign together on such things as trade, credits, and air transport. He asked the Politburo to make a suggestion *[ti yijian]* about how the treaty question should be resolved. On December 19, he again called urgently for an answer from Peking. The answer came after the Politburo meeting on December 21. Liu Shaoqi, Zhu De, and Zhou Enlai wired to Mao that everyone at the meeting had taken the position that Zhou did not have to travel to Moscow if the Soviet side was not yet ready to sign all three agreements on credit, air transport, and trade, and merely wished to issue a statement about the troops stationed in Port Arthur and on general political questions. "The Politburo asks you to reconsider this and provide instructions."[91]

Mao's December 18 report to Liu Shaoqi provides further evidence that Mao saw the treaty question as his primary interest. With the exception of the first three introductory sentences that dealt with peace and war, this was the only subject he considered important enough to report on to Peking.

This makes it even more amazing that Fedorenko did not acknowledge Mao's indirect request for a new treaty with a single word in his minutes. At the same time, the precedents suggest that the very indirectness with which the request was presented makes Mao's report to Liu Shaoqi appear credible.

Mao's December 18 telegram to Liu Shaoqi demonstrates the unreliability of Shi Zhe's memoirs on many points with respect to the treaty discussion. Shi gave the impression that there had been no mention of the treaty problem in the December 16 talk. Mao, he said, had wanted to turn the discussion to the treaty by using the allegory that they had to achieve something that both looked good and also tasted good *(ji hao kan, you hao chi)* when Stalin asked what he expected from his visit. Although Shi had attempted to interpret this phrase in a manner that was good in terms of both form and content, Stalin failed to understand the reference.[92]

Shi Zhe essentially confirmed what the Fedorenko minutes said about the issue of whether Zhou Enlai should come to Moscow. When Mao observed that he wanted Zhou to come, Stalin replied that he did not know why Zhou should come if they had not yet been able to decide what they actually wanted to achieve.[93] The fact that Mao Zedong had already asked Stalin before he departed for Moscow whether Zhou Enlai should accompany him or come later, and Stalin had at that time left the decision to him but hinted that he would prefer for Mao to come alone, suggests that this portion of the description of the December 16 talk is accurate.[94] The Chinese side had at that time already established the connection between Zhou Enlai's trip and the possibility of signing a new treaty. According to Shi Zhe, Stalin and Mao had yet another disagreement in this context. Mao wanted to have the two prime ministers, Stalin and Zhou Enlai, sign any future treaty, whereas Stalin wanted to sign it with Mao.[95] If this was true, it is difficult to say whether Mao's intention was the result of protocol considerations alone or that he wanted to avoid responsibility should any problems arise from the treaty in the future.[96] If Stalin really wanted to sign any future treaty together with Mao, it is plausible that he was angry that Mao insisted that Zhou Enlai should sign it.

When they had finished discussing the treaty issue, the December 16 talk turned to other topics, including credit and trade, air traffic connections, military aid, and the treatment of foreign property in China. We depend completely on the Fedorenko minutes for the details of this part of the talk. The Mao report merely mentions the possible conclusion of credit, trade, and air traffic agreements generally. Neither Fedorenko nor Shi Zhe discuss this in their memoirs.

Mao proposed concluding an agreement at the government level at that time on the Soviet credit in the amount of US$300 million[97] that Stalin and Liu Shaoqi had already agreed upon at Party level during the summer of 1949, and wished to do so immediately because this would make a good impression in China.[98] Stalin agreed.

Mao informed Stalin that it was necessary to regulate questions involving trade and Soviet trade with Sinkiang in particular.[99] Nonetheless, China could not present an "exact outline of trade operations" for this region. Stalin did not mention at this time that a draft of a decree of the Council of Ministers of the USSR on Soviet trade with Sinkiang, drawn up by Molotov, Mikoyan, and Vyshinskii, had already been presented to him.[100]

We should probably describe the situation in Sinkiang at this point. Mao's particular interest in trade with Sinkiang was due to the fact that this region's economic relations with the USSR had been interrupted since 1942. At that time, the ruler of Sinkiang, Sheng Shicai, who had been pro-Soviet since 1933, changed sides and allied himself with the KMT. Under pressure from him, the Soviet Union closed its trade missions in 1943, withdrew its advisers and technicians, and interrupted all contact. This resembled an economic blockade, as nearly all industrial goods had been imported from the Soviet Union before that time or had been produced in firms under joint Sino-Soviet management. Production of furs, wool, cotton, and silk fell off by 50 percent. There were few transportation links with the rest of China. This resulted in anti-Chinese unrest. The Soviet government rejected the KMT government's attempts to restore economic relations between Moscow and Urumchi. The situation in Sinkiang became even more critical after the PLA marched in, as that army, which, along with the incorporated KMT troops in Sinkiang, came to nearly half a million men bearing arms, now also had to be fed and clothed. The province itself could cover only 30 percent and the Central Government could cover 40 percent of this demand. (Another 30 percent should, according to the wishes of the Chinese side, be covered by credited trade with the Soviet Union.) The area's size and poorly developed transportation network made the situation more difficult. The cost of transporting the grain that had to be taken from the south to the north was ten times as great as the cost of the grain itself. There was galloping inflation, and in a short time prices increased a hundredfold and more.[101]

In his reply, Stalin did not address the situation in Sinkiang specifically. He merely said that the USSR had to know what equipment China required because there were no reserves in this area and they would first have to order industrial goods ahead of time. Mao said it was difficult to order equipment "as we do not yet have a clear picture of industry." Stalin insisted, however, because in the Soviet Union requests for equipment had to be submitted at least one year in advance. He then approved Mao's request for assistance in establishing air transport routes and proposed that such routes could be created via Sinkiang and the Mongolian People's Republic. He also approved Mao's requests for support in the construction of a naval force, approval he had already expressed to Liu Shaoqi, and said that the Soviet Union could train Chinese for this in Port Arthur and make ships available: "You give the people, and we give the ships." After they had been trained, these people could return to China on these ships.[102]

When Mao requested direct military support for the conquest of Taiwan, Stalin refused. The Chinese comrades had already confronted him with this issue during Liu Shaoqi's stay in Moscow during the summer of 1949. Liu had predicted in his written report for Stalin at the beginning of July that the Chinese Communists would conquer Taiwan in the course of 1950.[103] As some of the KMT troops stationed there might desert to the Communists, they might be able to count on an even earlier date. Mao had already let Stalin know at the end of July that the Communists could not break the blockade of Shanghai without conquering Taiwan, which, however, they could not do without an air force.[104] He had asked Stalin if the USSR could help by training 1,000 pilots and 300 ground staff in Moscow and delivering 100 to 200 fighter aircraft and 40 to 80 bombers to be deployed in the conquest of Taiwan. Similarly, Mao had requested aid from Stalin to create a navy. In this, he had ultimately gone a step further. He had said he might also have to ask the Soviet Union to allow Soviet aviation and naval specialists and pilots to participate in the conquest of Taiwan. At the end, Mao had asked with hypocritical concern whether it would not damage Soviet-American relations if the USSR participated in the conquest of Taiwan.

To Liu Shaoqi, Stalin had refused to allow the Soviet air force or navy to provide direct support to the PLA for an attack on Taiwan at that time by pointing out the inevitability of conflicts with the U.S. Army, which Washington could use as a pretext for unleashing a new world war.[105] Mao had then withdrawn his request. On the other hand, Stalin approved the Chinese request that the Soviet Union provide aircraft and antiaircraft guns, send experts, and assist in creating a naval school.[106]

Mao remained obstinate. On December 16, he once again presented Stalin with a request for direct military participation by the USSR in an attack on Taiwan. He made the transition to this request by observing that the KMT had established both a naval and an air force base on Formosa. The fact that Communist China had no navy or air force complicated the military conquest of the island, he said. "In this context, some of our generals advocate the idea that we should take up aid from the Soviet Union, which could send us their pilots as volunteers [*letchiki-volontery*] or secret military units to accelerate the conquest of Formosa."[107]

Stalin replied that aid could not be excluded, but that they would have to think about what form it should take. He repeated what he had already used as his central argument to Liu Shaoqi: "The main thing here is not to give the Americans any pretext for an intervention." The USSR could make staff officers and instructors available at any time. "We will think about the rest." This was a polite refusal. Mao did not insist any further. When Stalin asked him about it, he replied that China controlled an assault landing regiment that had deserted from the KMT. Stalin then suggested, "A company of landing forces could be selected, trained in propaganda, sent to Formosa, and used to organize an uprising on the island."[108]

The Chinese leadership kept Stalin up to date on its plans with respect to Taiwan. In mid-January, Mao Zedong informed Stalin that the plan to break through the blockade around Shanghai and conquer Taiwan would be finished in three to five days. General Liu Yalou would then bring it to Moscow. Mao asked Stalin to order Generals Kotov, Prutkov, and A.A. Avseevich, as well as Admiral Kuz'min to accompany General Liu. They worked as military advisers in Peking; Kotov held the position of military attaché.[109]

Mao then mentioned that the PLA had reached China's borders with Burma and Indochina. Great Britain and the United States were nervously asking themselves whether the Communist troops would halt their advance there or cross the borders. Stalin advised Mao to spread the rumor that the PLA would cross the borders into Burma and Indochina "in order to frighten the imperialists a little [*popugat' imperialistov*]."[110]

Mao pointed out that several countries, Great Britain in particular, were very actively seeking recognition of the People's Republic of China. The Chinese leadership did not want to hurry, however. It would first have to restore order in the country and consolidate its position, he said. Only then would it speak with the foreign imperialists. Stalin agreed. It was not necessary to create conflict with the British and the Americans, he continued. If it became necessary to exert pressure on Great Britain, they could find a way out by creating conflict between Guangdong Province and Hong Kong. Mao could then play the role of mediator to resolve the conflict. "The main thing is not to rush anything and to avoid conflicts." [111]

Stalin then asked Mao about the economic situation in China. Mao stated that banks in Shanghai were still working for bourgeois Chinese and for foreign firms, which they had no plans to touch at first. The Japanese firms had been nationalized. The government controlled the customs services. Stalin advised Mao to keep an eye on the customs services because the state could derive considerable revenues from them. With respect to foreign influence in the country, Mao said, the British dominated investment in the fields of the economy and trade, and the Americans in the fields of culture and education. The Chinese CP leadership had already been completely successful politically and militarily. But they had not yet liberated China from foreign influence in the fields of culture and the economy. In response to a question, Mao told Stalin that the government did have controllers or agents in the foreign firms. They controlled, among other things, the Kailan mines and the electrical power plants and waterworks in Shanghai. Stalin advised the Chinese to install government controllers who would be able to work legally and also to tax foreigners at a higher rate than the Chinese.[112]

Stalin was interested in knowing who would handle the mining of tungsten, molybdenum, and oil (answer: the government) and advised Mao to have the output of minerals and oil in particular developed. He did not do this disinterestedly, as his later insistence on the delivery of strategic raw materials, including tungsten, shows.[113] The same is true of later plans to extract oil jointly in Sinkiang.[114] Stalin then sought to demonstrate his familiarity with the situation in China by advising Mao to construct an oil pipeline from Lanzhou (in Gansu Province) to Chengdu (in Sichuan Province), and then to transport the oil from Chengdu by water. Mao acknowledged that the Party leaders had not yet decided which areas they should develop first in China, the ones on the coast or those located inland, as they were unsure of the prospects for peace. Stalin replied plainly: oil, coal, and metal are always necessary, regardless of whether there will be war or not. He asked if rubber trees could be planted in South China (Mao replied that this had not yet been done successfully) and whether China had a meteorological service. When Mao replied that it did not, Stalin advised him to create one.[115]

At the end of the talk, Stalin asked for a list of Mao's works that could be translated into Russian. Mao replied that he was currently revising his works that had been printed by various different local publishers and contained large numbers of errors and distortions. He hoped to complete his revisions by about the spring of 1950. Nonetheless, he would be happy to take advantage of the assistance of Soviet comrades. First, he would like to work on the texts with Russian translators; and second, he would welcome assistance in editing the original Chinese works. Stalin approved the assistance, but asked whether Mao really considered it necessary to edit his works. Mao said he did and asked Stalin to select a qualified comrade: "Let us say, one of the members of the CC of the CPSU(B)." Stalin said it could be done if necessary.[116]

It is worth noting Nikita Khrushchev's interpretation of Mao's behavior. He said that Mao had been "somehow putting himself down [*samounichizhenie*]" by asking Stalin for assistance with the revision of his works. He had thus wanted to give Stalin the illusion that he was willing to see issues involving the theory and practice of Marxism with Stalin's eyes and was not claiming to have his own conception of the construction of socialism in China. This, Khrushchev continued, contradicted everything that occurred in the course of China's subsequent history.[117]

In late March 1950, Mao returned to his request. He informed Stalin that they were planning to begin work on an edition of his works in Peking in the near future.[118] Stalin had promised, he said, to send "a comrade with a strong theoretical background" to China to assist with this. He was asking Stalin to inform him when this comrade would arrive. Stalin did not send Mao a Central Committee member or even the leading Soviet ideologue at the time, Mark Mitin, but another well-known expert on ideology, Pavel Iudin, who had played an important role in the dispute with Yugoslavia in 1948. Iudin spent many months in China between July 1950 and October 1951 and helped to edit the first and second volumes of the *Selected Works of Mao Zedong*, the first three volumes of which were published in Chinese in Peking between 1951 and 1953. A Russian translation was published in Moscow in 1952–53. Mao returned the favor by having a Chinese edition of Stalin's works published. Ambassador Wang Jiaxiang requested Soviet approval for this project in late February 1950.[119]

In retrospect, Mao himself interpreted his request for a Soviet reader at that time differently. He had not had so little self-confidence that he had had to ask for help from Moscow in revising his works, he told Iudin in 1958 (by that time, Iudin had become Soviet ambassador in Peking). No. He had wanted in this way to have someone come to China who could determine "whether China was correct or half correct and half false Marxism." From the context, we can assume that Mao was playing on Stalin's mistrust that Mao might be a Chinese Tito.[120] It is difficult to say what Mao's true motive was. It may have been a combination of all three motives mentioned.

Shi Zhe claimed that the December 16 talk ended in a depressed atmosphere. He said Mao was disappointed by the results of the discussion of the treaty. To him, the replacement of the old treaty with a new one was a matter of principle—the negation of the old China's diplomatic legacy and a decisive part of the new China's diplomacy. Stalin's rejection, Shi said, created a psychological barrier between them. Kovalev observed that they had not had the trusting talk that Mao had desired. These assessments appear credible in the light of all the information that has become available since. The fact that Mao claimed in his report to Liu Shaoqi that "the friendly feelings" had been "genuine" during the talk[121] does not contradict this. In all probability, he did not want to admit to the comrades in Peking that he was not making good progress with Stalin.[122]

To summarize, we should note about the December 16 talk that there was only one fundamental disagreement. This involved the treaty question. Stalin avoided describing the treaty of August 14, 1945, as unequal, as he had to Liu Shaoqi in Moscow in July. He was not prepared to confirm that the agreement on Port Arthur was unequal, either, as had been done during Mikoyan's visit with the Chinese leadership in Xibaipo in February 1949, or to promise to abolish it once a peace treaty had been concluded with Japan, as had also been stated both in February and July. He instead stated categorically that the Kremlin leadership had resolved "not to change any of the points of this treaty for the time being." By this, he meant the entire set of agreements, for he observed explicitly that he wanted to retain the agreements on Port Arthur and the CCR de jure. At the same time, he offered to withdraw Soviet troops de facto at the request of the Chinese government, as he had already done through Mikoyan and to Liu, and also to allow real changes with respect to the CCR.

As the Chinese Party leadership had already done during Mikoyan's visit to Xibaipo, Mao rejected the offer to withdraw the troops "too early." He probably meant this seriously, for there is evidence that he still feared a military intervention by the United States in favor of the KMT at that time. Mao was apparently—and probably correctly—convinced that Washington would not intervene in Manchuria at least as long as the Soviet Army was present there. It remains inexplicable why he—according to Fedorenko—is supposed to have stated that the existing situation with respect to the CCR reflected Chinese interests when we know from other sources—or at least the talk between Stalin and Mao on January 22, 1950—that this was not his real opinion. Mao made it clear on December 16 that he was seeking to abolish the entire bundle of agreements and conclude a new treaty. This was an issue of principle for him. In this way, he wished to document the end of the colonial era in China and at the same time set a precedent for the abolition of the other "unequal treaties" that China had concluded with the Western powers.

There is no evidence of why Stalin—contrary to Mao's expectations, certainly—refused at the beginning of the December 16 talk even to discuss either abolishing the bundle of agreements immediately or at least altering it. The simplest, and at the same time most plausible, hypothesis appears to be that he was telling the truth when he said that he was afraid that Washington and London might see this as a violation of the Yalta agreement and use it as a pretext to terminate other Yalta agreements, such as the surrender of southern Sakhalin and the Kuril Islands to the USSR. In any case, Stalin did agree in response to Mao's cautiously formulated objections to undertake some "fairly major corrections" to the bundle of treaties after about two years.

The fundamental disagreement, as it appeared on December 16, was that Stalin wanted to allow the whole set of agreements to remain in force unchanged at first, but offered to withdraw Soviet troops in fact, whereas Mao wanted to abolish the agreements immediately and replace them with new ones, but was not interested in an immediate withdrawal of the Soviet Army—as Stalin knew.

Six years later, Mao Zedong described Stalin's behavior on December 16 to the Soviet ambassador in Peking as follows: "In my first talk with Stalin, . . . I proposed concluding a treaty at state level, but Stalin avoided giving an answer."[123] This is not true. Stalin did not avoid anything at all. He stated clearly and explicitly that he opposed concluding a new treaty and was even opposed to any formal changes in the old one. He merely suggested that some actual changes could be made with respect to certain aspects of the bundle of agreements.

"Stewing in His Own Juices at the Dacha": Act I

Mikoyan and Vyshinskii visited Mao Zedong on December 18 and had dinner with him.[124] Two days later, Molotov and Mikoyan—also together—visited him and discussed organizational questions involved with Stalin's birthday celebrations.[125] None of the substantive issues that so urgently interested Mao were apparently discussed on either occasion. Kovalev, who also mentioned Bulganin as a visitor, described the meetings as "brief official talks."[126]

Fedorenko visited Mao every day.[127] Kovalev also remained in constant contact with him. He observed that the latter was "in a bad mood and worried [*rasstroen i bespokoitsia*]." When he informed Stalin of this, the latter replied that many foreign guests were coming —he obviously meant because of his birthday celebrations—and it would not be appropriate to treat Mao any differently from them.[128]

The Chinese delegation gave presents to Stalin on his seventieth birthday, including embroidered portraits of Stalin, Chinese editions of his publications, and ivory carvings. These were exhibited on Stalin's orders along with other presents in the Pushkin Museum in the first room, which was decorated with the Chinese five-star banner and calligraphy by Mao: "Lucky as the Eastern Sea, a long life like the Southern Mountains [*fu ru donghai, shou bi nanshan*]."[129]

The telegram Mao sent on December 19 to congratulate Stalin on his seventieth birthday was brief and formal. It consisted of a single sentence: "Mr. Chairman! On the occasion of your Excellency's seventieth birthday, I have the honor of sending you my warm greetings and desire that the Soviet Union—the bulwark of peace in the entire world—may grow ever stronger and develop further with every day under your Excellency's leadership."[130] The telegram thus resembled the congratulatory telegrams from Western statesmen such as British prime minister Clement Attlee, Italian prime minister Alcide De Gasperi, and French president Vincent Auriol that also consisted of a single sentence. It was altogether different from the telegrams that arrived from the capitals of the Communist states of Eastern Europe, which were up to six times as long and drenched in Byzantine flattery. Mao may possibly have consciously sought to differentiate himself from these and thus to emphasize China's special, independent status. It is also possible that he was seeking to express his dissatisfaction with Stalin's behavior in this way.

The reasons for Mao's dissatisfaction became clear in a talk to which he invited Kovalev on December 20 at the dacha. Mao asked Kovalev to pass a record of the talk, at which Fedorenko interpreted, on to Stalin. He requested another "regular meeting [*ocherednaia vstrecha*]" with Stalin on December 23 or 24 and put two alternative agendas for the upcoming talks up for debate. On the one hand, they could discuss the Soviet-Chinese treaty, credit, trade, air traffic agreements, and other issues, including the recognition of China by Burma. In this case, Zhou Enlai would have to be brought to Moscow to negotiate and sign the treaties. While Zhou came to Moscow, Mao himself would take a trip to Stalingrad and Leningrad. The second alternative consisted of discussing the same topics but without concluding formal agreements. In this case, Zhou would not need to come at that time and could come later when the various agreements would have to be signed. It is very obvious that Mao once again placed the treaty question in first place on the agenda.[131]

On the same occasion, Mao expressed to Kovalev his desire to pay visits to Politburo members Molotov, Bulganin, Mikoyan, and Nikolai Shvernik. He also requested that Kovalev and Fedorenko remain constantly at his disposal, as there were several questions he wished to discuss with them. In his report to Stalin, Kovalev added that he had not visited Mao every day and that this had led Mao to protest that he, Kovalev, did not want to make himself constantly available to provide Mao with necessary advice. Mao had requested, he said, that Kovalev and Fedorenko remain at his side during Stalin's seventieth birthday festivities.

Mao's talk with Kovalev on December 20 reveals how much the fact that Stalin had not considered it necessary to suggest a schedule for the continuation of their substantive talks after December 16 bothered Mao. The fact that Mao asked to be invited for talks with the aforementioned Soviet Party leaders also suggests that he felt neglected. He had, as we have previously described, already met with the Politburo members Molotov, Mikoyan, and Bulganin and with Foreign Minister Vyshinskii; but these, as we know from Kovalev, had merely been short talks held according to the rules of protocol. There was the additional fact that these Soviet leaders visited Mao at his dacha, while Mao considered it very important to receive invitations to visit them. Kovalev described Mao's situation as follows: after the December 16 talk with Stalin, Mao "stewed in his own juices [*tomilsia*] for a few days in the dacha."[132]

Khrushchev, who had been a Politburo member since 1939 and had been transferred from the Ukraine back to Moscow in December 1949, where, as newly appointed Central Committee secretary and at the same time Moscow city and regional Party secretary, he had close contact with Stalin and was "therefore kept up to date on issues involving China,"[133] put it in a similar fashion. After the first talk, he reported, "Stalin did not meet with Mao at all for several days. And he did not meet [with him] himself; none of us went to Mao, either." The latter was not in fact entirely true, but Khrushchev's observation nonetheless suggests that Molotov's, Mikoyan's, and Bulganin's visits had been made on Stalin's orders. The description Mao gave of his own situation in a telegram he sent to the Politburo in Peking on December 19 sounds euphemistic in this context: "We are doing well here. We have rested a bit in recent days."[134]

Although Kovalev gave Stalin a report on his talk with Mao on December 20, Mao's situation did not change significantly, Kovalev said in retrospect: he remained as isolated as before. Mao took his revenge, he said, by refusing to receive Ambassador Roshchin, who wanted to speak with him about the Japanese CP.[135]

Stalin's Seventieth Birthday and Mao's Role

Mao could not complain of any lack of respect at the main celebration of Stalin's seventieth birthday that took place on December 21 in the Bolshoi Theater, however. Stalin distinguished him from all the other guests by placing him at his right hand on the stage; the Politburo member Kaganovich stood at Mao's right side.[136]

Mao was also given the honor of being the first foreign guest to be asked to speak. He was restrained in his brief congratulatory address, which was translated by Fedorenko. According to the official Soviet report, he was interrupted five times by "tumultuous applause [*burnye aplodismenty*]" and the entire audience rose up from its seats at the end. He only once employed an epithet of praise in connection with the jubilee, and even that seemed comparatively simple. Most of the other thirteen representatives of foreign parties and states outdid one another with Byzantine flourishes—"the wise and beloved Comrade Stalin" (Hungary); "our beloved teacher and leader" and "our dearest friend" (Bulgaria); "the great genius" (Czechoslovakia); "the brilliant teacher and leader" (Romania); "the wise leader and teacher" (Mongolian People's Republic); "the brilliant helmsman" and "beloved leader of the peoples" (German Democratic Republic). Most of the Soviet speakers took pains to express their devotion in still more moving words, celebrating Stalin as the "sun, that stands high above the earth" (Byelorussian SSR), as "our beloved father" (the writer Aleksandr Tvardovskii), as the "father of the fatherland" (Ukraine), or as the "great genius of humanity" (Karelo-Finnish ASSR). Stalin frequently addressed Mao in the course of the festivities. And yet, as much as he and Kaganovich tried, they were unable, Shi Zhe reported, to win a smile from Mao. He remained silent, serious, and withdrawn the whole time.[137]

Fedorenko claimed that Mao was involved in an unpleasant incident in connection with his speech. When he had finished and handed the microphone to Fedorenko so that he could translate, Stalin told Mao that he could at least tell people that his translator was going to speak now and not another speaker. Mao became confused and started searching for Fedorenko's name in his papers, and Stalin said that he should not be so hectic and told him the name. Mao then announced the name "in a trembling voice" and the translation began. It seems difficult to believe, however, that Stalin would have made such a gross faux pas that could only humiliate Mao publicly. It must certainly be an invention that can be attributed to Fedorenko's fatal love of the theatrical.[138]

Following the ceremony, they ate together and watched an artistic program during which Mao sat with Stalin in the czar's loge. After the performance, the audience turned and raised an ovation to Stalin and Mao: "Stalin! Mao Zedong!"[139] On the following evening, the Soviet government presented a banquet with a cultural program in honor of Stalin's birthday in the Kremlin, at which Mao was seated next to Stalin at the main table.[140] This implied yet another honor, as did the fact that Shvernik—who, as chairman of the Supreme Soviet, was heading the banquet—after toasting Stalin, the Bolshevik Party, the Soviet government, and so on, began a series of toasts to the leaders and peoples of the other socialist states as follows: "To the Chinese people, to the representatives of the People's Republic of China who are present in this hall, to the leader of the Chinese people, comrade Mao Zedong!"[141]

According to Wang Dongxing, Mao did not enjoy the banquet, which lasted from 8 P.M. until 1 A.M. He was unaccustomed to festivities that lasted so long, Wang said, and had become tired. There had been nothing good to eat, Mao said sullenly afterward, and they had not been shown anything beautiful. They would not copy anything like that at home.[142]

Stalin was certainly not acting altruistically by honoring Mao Zedong publicly. It could only heighten his own prestige in World Communism for him to demonstrate a personal closeness to Mao, who had just achieved a triumphant victory in the most populous country on earth and could be considered the most "outstandingly successful non-Russian Communist of the day,"[143] before the numerous representatives of Communist parties, states, and movements present.[144]

Stalin's seventieth birthday was celebrated at great expense in China. The day before the Chinese-Soviet Friendship Society gave a reception, which was attended by Liu Shaoqi,

Zhu De, and Zhou Enlai, the highest-ranking leaders after Mao. On Stalin's birthday, some 20,000 people, sent by 210 "people's organizations," gathered in front of the Soviet embassy in Peking to congratulate him. Millions of Chinese enrolled on lists of persons offering their congratulations.[145]

Stalin Makes Mao Wait: The December 24 Talk

On the day of the banquet, December 21, Stalin informed Mao Zedong that he approved the proposal Mao had made two days earlier to hold the second talk on December 23 or December 24. Mao immediately cabled the news to Party headquarters in Peking and added: "We can determine the course after this talk. I will inform you by telegraph."[146] On December 23, Stalin informed the CCP leadership through Wang Jiaxiang that the Politburo and Stalin would meet with Mao for a talk on a date to be determined in advance, at the earliest the following day.[147]

According to the Chinese, the talk took place on December 24 and lasted approximately four hours. In addition to Mao, Chen Boda, Wang Jiaxiang, and Shi Zhe attended on the Chinese side, the latter as interpreter. As already mentioned, Mao had informed Stalin through Kovalev four days earlier that he was interested primarily in a talk about the future of the treaty of August 14, 1945, and about credit, trade, and air transport agreements. To his disappointment, Stalin did not address the treaty problem at all on December 24. His main topic was the condition of the brother parties in Vietnam, Japan, and India. Mao reported that his Party wanted to recognize the Democratic Republic of Vietnam and support Ho Chi Minh's resistance movement. Stalin promised to follow the CCP on this path. Stalin again did not propose any concrete date for their next talk.[148]

Years later, Mao made the following statement to the Soviet ambassador in Peking about the contents of this talk, which contained an indirect reference to the request for a new treaty he had expressed to Stalin at the December 16 talk:

> I returned to this question during the second talk, after I had showed Stalin a telegram from the CC of the CCP expressing this same request. I proposed summoning Zhou Enlai to Moscow to sign the treaty. Stalin used this proposal as a pretext for a rejection and said that it would be "inappropriate to proceed in this way as the bourgeois press would cry out that the entire Chinese government was located in Moscow."[149]

Kovalev's Denunciatory Report to Stalin

As has already been described in detail, Stalin had stated unmistakably during the first talk with Mao on December 16 that he did not want a new treaty but would be willing to consider making changes in the old treaty, either immediately or later. The fact that he affronted Mao during the talk on December 24 by showing absolutely no willingness to discuss the treaty or the agreements the Chinese wanted could be due, among other things, to the fact that his distrust of the Chinese Party leadership, which had never quite disappeared, had been revived when he apparently read a written report drawn up by Kovalev with the title "On Some Problems of the Policy and Practice of the CC of the CCP" immediately prior to the talk with Mao.[150] Kovalev had begun the report approximately two weeks before his departure for Moscow and completed it in the train. After Kovalev, who feared the explosive character of the report might result in unpleasantness, had attempted in vain to convince Fedorenko to co-sign it, he delivered it to Stalin on his own responsibility on December 24.[151]

In his report, Kovalev, who was overwhelmed by an incurable distrust of the Chinese Party leadership, sought to suggest to Stalin that the leading comrades in Peking were unreliable because they exhibited bourgeois, pro-Western, and anti-Soviet tendencies. The report presented the quintessence of the information that Kovalev had apparently received primarily through the indiscretions of Gao Gang. Kovalev had regularly forwarded this information to Stalin during the previous months; according to Khrushchev, he out-and-out "showered" Stalin with it. Stalin leaked some of the information within Soviet Party leadership circles, which explains how Khrushchev, a Politburo member, was aware of it.[152]

In the report, Kovalev accused the CCP leadership of disregarding Stalin's suggestions and seeking recognition as quickly as possible by the United States and Great Britain, although the latter still actively supported Chiang Kai-shek. In doing so, the Party leadership was succumbing to pressure from "bourgeois democrats and other capitalist elements" in China, who, by achieving rapid recognition by Washington and London, wanted to put the Chinese bourgeoisie in a position to halt the further democratization of China and keep the Sino-Soviet friendship from becoming stronger. There were allegedly people in the CCP and among the Central Committee members who had been pro-American and anti-Soviet in the past. The CC leadership was now supporting these people. The proportion of democratic figures in the central government was very large, the report claimed, and the government had in fact become a federation of all the various parties.[153]

Kovalev himself stated that he also directed some "rather sharp critical remarks at Chinese leaders" in his report.[154] He apparently attacked Zhou Enlai and Liu Shaoqi primarily.[155] Kovalev had previously reported to Moscow that Zhou Enlai had revealed a "negative attitude" by opposing the sending of Soviet specialists to Tientsin and Shanghai, because powerful American and British interests were concentrated there. Liu Shaoqi and Li Lisan, Kovalev reported, had opposed the participation by representatives from Japan, India, and other states in the bloc controlled by the British and Americans in the Asian conference of the World Federation of Trade Unions in Peking in November–December 1949 so as not to irritate the United States and England. Liu was, Kovalev maintained, making baseless criticisms of Gao Gang. Kovalev apparently excluded Mao Zedong from his criticisms in his report and in the information that he had previously sent to Moscow about activities in the Chinese Party leadership.[156]

Years later, Mao spoke of the report as "Kovalev's notorious document about anti-Soviet tendencies in the CCP leadership," in which Stalin's distrust of the Chinese Communists had been expressed.[157]

"Stewing in His Own Juices at the Dacha": Act II

It is not very probable that a further five-hour meeting took place between Mao and Stalin on the evening of December 26. Mao's bodyguard, Wang Dongxing, was the only person to make such a claim. Wang did not provide information about the contents of the alleged talk. On the following day, Wang continued, Mao indicated to him that the Soviet side still did not understand the Chinese revolution although Liu Shaoqi had made great efforts during his stay in Moscow the previous summer to explain it fully to Stalin. In fact, Stalin still had doubts about the abilities of the Chinese Communists, Mao said, and still had hopes for the KMT. If there were problems in China they could ask the "elder Soviet brother [*Sulian lao dage*]" for help, but they should under no circumstances surrender their sovereignty.[158]

In his dacha, Mao Zedong dealt with questions of Chinese domestic and foreign policy—for example, with Burma's offer to establish diplomatic relations; the willingness of Poland, Czechoslovakia, and the German Democratic Republic (GDR) to establish trade relations; and with plans for the conquest of the island of Hainan. He remained in active contact by telegraph with the Peking Party headquarters on these and other issues. He passed the rest of his time taking walks on the grounds of the dacha, which were deeply covered with snow, and watching movies his host made available to him, which included films about outstanding figures in Russian history, such as Peter the Great, Mikhail Kutuzov, and Alexandr Nevskii, and about Napoléon. The films also included a Charlie Chaplin movie and one about the battle of Stalingrad. Mao submitted to a routine examination by five Soviet specialists, who observed that his organs were functioning normally but that he suffered a degree of nervous exhaustion. They prescribed vitamins and sleeping pills and advised him to stop smoking, drinking alcohol, and eating fatty foods and to get as much exercise as possible to avoid obesity.[159]

According to information Mao himself provided years later to Pavel Iudin, the Soviet ambassador, the situation became even tenser after the second talk—that is, apparently after December 24: "After that, Stalin refused to meet with me at all. For my part, I tried to reach him at home twice by telephone, but they told me that Stalin was not at home and advised me to meet with Mikoyan. All this hurt me greatly." He said he decided to react by making no further attempts and continuing to wait at the dacha.[160]

Chinese accounts present a different and rather implausible version. It was clearly designed to gloss over the humiliating situation Mao Zedong found himself in at that time and to save face. According to these accounts, Stalin called Mao several times in the days following the December 24 talk and inquired about his health and plans. In a call on December 25, Stalin was supposed to have assured Mao that he would have been happy to visit him himself, but that the security staff and doctors would not let him leave the house.[161]

The report by Ambassador Wang Jiaxiang's wife, who was staying with her husband in Moscow at that time, sounds more convincing. According to this report, the Sino-Soviet talks had reached a dead end. Stalin was nowhere to be seen and "had deposited [Mao] there [*lengluo zai zheli*]." Mao was concerned. Stalin was someone, Wang's wife said, one had to ask and beg a thousand times, but who was unwilling to leave the Kremlin.[162]

Mao continued to stew in his own juices at the dacha. His uneasiness must have increased the longer Stalin refused to receive him, and the longer he felt that Stalin was putting off discussing the new treaty and other substantive questions. The request Mao had made on December 20 to be received by individual Soviet leaders continued to be ignored, too. On January 2, 1950, he again complained that he had still had no opportunity "to go out and talk with any of them alone, face to face."[163]

Toward the end of the month, Mao gave his displeasure free rein. He protested to Kovalev, who had visited him with Fedorenko as interpreter, that they had had him come to Moscow and now nothing was happening. What had he actually come there for—just to eat, shit *(lashi)*, and sleep every day? Or was it just to congratulate Stalin? He had come to achieve something *(banshi)*! There were further important questions about their mutual relations that had to be discussed, he said. It was probably about this talk that Kovalev recalled "meeting a pained, peevish Mao who was shouting '*Bu hao, bu hao*! [Bad, bad!]' and pointing his fingers toward Moscow. When asked what he meant Mao answered that he was angry at the Kremlin, not Kovalev."[164] Shi Zhe reported, however, that Mao was happy after the Soviet visitors left and told him pleasantly that he had finally given Kovalev a lesson to make him report about "the situation *[qingkuang]*" to Stalin.[165]

Years later, Mao confirmed to Iudin that he had made a scene in front of Kovalev:[166] "In Moscow, Kovalev took care of me, Fedorenko was the interpreter. I got furious [*fa piqi*], I banged on the table. I said that I have three tasks here, first to eat, second to sleep, and third to shit [*lashi*]."

Mao had given Iudin a less drastic description of the incident two years earlier.[167] After he had decided not to do anything further and to remain at the dacha in response to Stalin's refusal to hold talks, Mao said, "there was an unpleasant talk with Kovalev and Fedorenko, who suggested that I should take trips in the countryside. I rejected this proposal sharply and replied that I preferred to rest up at the dacha."

In addition to the reports of Mao's dissatisfaction Stalin received through Kovalev and Fedorenko, Mao's dacha—in which, according to Mikoyan, Mao was practically a prisoner—was bugged.[168] In this way, Stalin learned that Mao was telling his Chinese confidants privately that he was not sure he would ever be able to return to China. He therefore wanted either to meet with Stalin or to leave soon.

Khrushchev also heard about this threat at the time. He reported in his memoirs, "Thus Mao began to show his dissatisfaction that he was sitting behind lock and key in a residence that had been assigned to him, that they did not show him anything, that no one met with him, and he declared that if things continued that way, he would leave."[169]

It is easily conceivable that Mao assumed he was being bugged and threatened to leave in order to put pressure on Stalin. Stalin reacted[170] by sending Molotov to see Mao at the dacha on one of the last days of December, saying to Molotov that he should after all take a look to see "what kind of guy he is [*chto za tip*]." Mao asked Molotov when he could meet with Stalin. Molotov judged Mao to be a clever peasant leader, a kind of Chinese Pugachev, who "naturally" was far from being a Marxist. Mao had confessed to him that he had never read Marx's *Das Kapital*. This episode is typical of the blasé attitude with which Stalin and his closest advisers, who considered themselves the only legitimate heirs and interpreters of the Marxist classics, looked down on Mao and most of the Chinese comrades. Following the talk, Molotov recommended to Stalin that it would be appropriate to receive Mao.[171]

According to a Chinese publication, Mao introduced Molotov to the history of the Chinese revolution and the problems of conflict within the Party during the talk and thereby bridged the intellectual gap that separated the two sides.[172] Shi Zhe confirmed that these subjects had been discussed, but added that Molotov had not shown much interest in Mao's accounts and had remained passive.[173]

Mao's behavior toward Ambassador Roshchin when the ambassador visited him on January 1 on behalf of Foreign Minister Vyshinskii was more conciliatory than it had been toward Kovalev and Fedorenko. And yet here, too, Mao insisted on meeting with Stalin and other Soviet leaders.

Roshchin characterized the initial formal part of the talk as "friendly and warm." He reported that Mao also discussed the state of his health. He was feeling better after two weeks of rest. In the past four days, he had slept the normal eight hours each day without any medication. And yet he could not go out walking for more than a quarter hour each day without becoming dizzy, he said. He would therefore recuperate with one more week of complete rest. He would then like to meet with Stalin "for substantive talks [*dlia delovykh razgovorov*]." He also wished to visit Comrades Molotov, Kliment Voroshilov, Lavrentii Beriia, Malenkov (all Politburo members), Shvernik (Politburo candidate), Minister of the Army Aleksandr Vasilevskii, and Foreign Minister Vyshinskii to hold "simple talks with them," not ones dealing with "special topics" or "substantive issues." He had in mind holding a not very extensive talk with each of them, one on each day, and preferably after 5 or 6 P.M.[174]

He had originally wanted to spend three months in the Soviet Union, Mao said. But "circumstances involving his work in China" now forced him to shorten his stay to at most two months. He wanted to leave Moscow at the end of January, he said, and be back in Peking on February 6. He planned to use any time he had left before he returned home after completing the substantive talks with Stalin to lay a wreath at the Lenin mausoleum, visit some collective farms, and go to the theater. He would abstain from visiting factories, appearing before large audiences, and giving public speeches because such activities would be too stressful given his health and might lead to a return of his sleep disorders and attacks of dizziness. He had previously wanted to take trips to various places in the Soviet Union, but he would give these plans up because of the condition of his health and the long return journey ahead of him.[175]

During the rest of the talk, Mao reported on the Burmese and Indian offers to establish diplomatic relations with the People's Republic of China, on indications that England and other Commonwealth countries were preparing to recognize Peking, and on the military situation in China. Roshchin came to speak of the Soviet government's desire to turn over for judgment several hundred Japanese war criminals who had committed crimes in China. In view of the fact that the justice system in China was still underdeveloped and the necessity of judging KMT criminals came first, Mao asked that they postpone handing over the criminals until the beginning of the last six months of 1950.[176]

Following the discussion with Mao and a dinner together, Roshchin learned in a private talk with Shi Zhe, who had interpreted for them, that Mao Zedong had really felt better in the past three days, was sleeping normally and without medications, made jokes, was happy, and talked with everyone. And yet, as before, he could not stay out for very long and continued to be plagued by attacks of dizziness. The doctors were to visit him again the following day.[177]

The record of the talk reveals that Mao did not exhibit to Roshchin any of the annoyance he had expressed—intentionally or not—shortly before in his talk with Kovalev. The reference to the state of his health as the reason for canceling the trips within the USSR was clearly a pretext, for as early as the following day, on which a solution to the treaty problem appeared for the first time, he arranged excursions to Leningrad and Gorkii with Molotov and Mikoyan.[178]

Mao viewed the talks that he desired to hold with Stalin as something that had to be carried out at the highest Party level: between Stalin and Mao himself. This is suggested by the desire he expressed to Roshchin to speak about the "substantive issues" only with Stalin and not with the other Soviet leaders. This was apparently also the reason why Mao did not discuss with Roshchin the treaty, the other agreements under consideration, or the "TASS interview"[179] that had been arranged for that same day. Instead, he merely used him as a messenger to carry his request to meet with Stalin one week later. Mao's deep-rooted dislike of Roshchin (about which we have already reported) probably also played a role.[180] The different quality of Mao's behavior to Kovalev revealed that he saw him here—as he previously did in China—as Stalin's representative to the CCP leadership and correctly assumed that he had direct access to Stalin.

One can assume that Mao's dramatic scene in front of Kovalev and his threat to leave did not fail to have the desired effect on Stalin. Nonetheless, other events (which are described below) would apparently play a decisive role in changing Stalin's mind.

Mao's "TASS Interview"

About two weeks had passed since Mao's arrival in Moscow and nothing had been said in public about the progress or results of the negotiations. Mao had not been mentioned again in the Soviet media since Stalin's seventieth birthday, or for some ten days. This led to some speculation abroad—among other things, to a report spread by a British press agency—that Stalin was holding Mao under house arrest.[181]

The Chinese delegation was "very upset [*hen zhaoji*]."[182] Stalin was probably no less upset. He, too, was interested in issues of personal image and the prestige of international communism. Khrushchev, who was a member of the Politburo at the time, remembered the mood surrounding Mao's stay in Moscow: "Stalin was anxious to create the impression that we were on the best of terms with Mao and firmly on the side of the Chinese people."[183] On January 1, the two sides finally agreed that they had to refute the rumors publicly. It was supposedly Wang Jiaxiang who succeeded in convincing Molotov in a talk to do this in an "interview" with Mao Zedong by the Soviet press agency TASS. This interview, which consisted of only three questions and three short answers, was published very prominently on the first page of *Pravda* on January 2 and took the wind out of the speculations.[184]

Mao began the interview with a brief comment on the favorable developments in

China. He then answered a question about the length of his stay in Moscow: he had come for a few weeks, and the duration of his stay depended in part on how long it would take to resolve the problems that interested the People's Republic of China. His answer to the last question about what these problems were constituted the core of his message. Mao said: "These problems include ones in particular such as the existing Treaty of Friendship and Alliance between China and the USSR, the problem of a Soviet credit for the People's Republic of China, the problem of trade, and a trade agreement between our countries and others." Mao added that he also wished to visit certain Soviet regions and cities to familiarize himself better with the economic and cultural construction of the Soviet state.[185]

Kovalev brought the text of the interview to Mao Zedong on January 1 on instructions from Stalin in the form of a draft that Stalin had signed. After he had read it, Mao said that the "form and content" of the interview were "very good," and that he had no comments or additions to make. He added with reference to his travel plans, which had been mentioned in the interview, that he wanted to visit only Leningrad, as he had already said, and he would otherwise remain in Moscow until January 22 or January 25. During this time, he wanted to have "a few meetings with Comrade Stalin to discuss the questions mentioned in the interview and others." Mao asked Kovalev to tell Stalin that he wished, as he had already requested earlier, to visit Molotov, Malenkov, Beriia, Mikoyan, and Bulganin and also to meet with Vyshinskii and Vasilevskii, and that he was asking for advice about how he could meet with them either individually or together. He was also interested in a talk with Palmiro Togliatti, he said, general secretary of the Italian CP, who attended Mao's birthday celebrations.[186]

Mao Zedong seemed very satisfied with the new situation that arose as a result of the interview. He drank to Stalin's health at dinner with Kovalev and at the same time expressed his gratitude for Stalin's assistance in the creation of the Chinese CP and the execution of the Chinese revolution. He showered Stalin with flattery—naturally fully aware that Kovalev would report this immediately to Stalin. The "twelve conditions," Mao said, that Stalin had formulated with reference to the German Communist Party and his postscript to the *Short History of the CPSU(B)* had provided the program and guidelines for the CCP. He behaved in an underhanded way in the next sentence, however: the two documents had helped in destroying left- and right-wing Trotskyite deviations "that had been led by Wang Ming for a while," and in making the CCP powerful and united. Hardly anyone knew better than Stalin that Wang Ming had always felt obligated to implement Stalin's line in the Chinese CP. Mao ended his toast with an emotional phrase, which at the same time included a tribute to the Soviet leader: "The shots of the October revolution brought the teachings of Marx, Lenin, and Stalin to China. If there had been no CPSU(B), there could have been no victory of the revolution in China. Long live the teachings of Lenin and Stalin!" After the dinner, Shi Zhe, who had acted as interpreter, informed Kovalev "confidentially" that Mao would be happy if one of the Soviet Politburo members would visit him at the dacha.[187]

Mao could in fact be satisfied with the new situation that arose as a result of the interview. He had been given the opportunity to express to world public opinion the fact that he was dissatisfied with the existing treaty and that the treaty question was his highest priority. Stalin's approval of the interview's published wording indicated his willingness finally to put the treaty question on the agenda. This rendered his statement to Mao on December 16, 1949—that the Soviet leadership wished "not to change any of the points of this treaty for the time being"—obsolete. The publication of the TASS interview thus marked the decisive breakthrough in the Moscow talks. Ambassador Wang Jiaxiang's wife, who experienced the event as a contemporary witness, confirmed this assessment: "The stalemate had been broken, the relations became more active, and a new situation arose."[188]

Years later, Mao revealed to Ambassador Iudin in Peking that the fact that they had kept the reference to talks in Moscow about a Soviet-Chinese treaty in the interview had been a "considerable step forward." Mao added that Stalin possibly had been moved to change his position by the fact that the Indians and British recognized China in January 1950.[189]

Mao was not only correct about his first observation, as has been substantiated above. His supposition in the second sentence also appears to be accurate. If true, Mao himself appears to have contributed intentionally to this result. On January 1—thus, in the course of the same day on the evening of which Kovalev brought him Stalin's draft of the TASS interview—Mao had, as mentioned above, received Ambassador Roshchin and pointed out to him—strikingly at the very beginning of the talk—the imminent recognition of China by Delhi and London.[190] On the following day, Mao made available to Stalin the wording of a telegram directed to him in which Zhou Enlai had reproduced Foreign Minister Jawaharlal Nehru's telegram in which India recognized the People's Republic of China.[191] The options that were developing here for Peking also to improve its relations with the West in this way could in fact have provided Stalin a significant reason—in addition to the one already mentioned—to decide finally to make concessions to Mao on the treaty question.[192] Stalin, who had been made sensitive to this by Kovalev's reports, was always looking suspiciously for signs of any undesirable rapprochement between Communist China and the United States and Great Britain. He certainly did not wish to run the risk of stimulating such a process by alienating the Chinese Communists and took care to play a role in determining the quality of the relations between Communist China and the Western states.

It is, however, conceivable that Stalin had yet another reason to change his mind, in addition to the attempt to oppose foreign rumors of the failure of Sino-Soviet negotiations and hinder any excessive rapprochement between the new China and the Western powers. It is possible that Stalin had learned through intelligence sources that American East Asian policy in Washington had been altered in a National Security Council document of December 30, 1949, classified as "strictly confidential." The text of the document, which was made public by President Truman in a press conference on January 5, 1950, revealed that the U.S. administration was not prepared to defend Taiwan and wished to keep out of the Chinese civil war in the future. Stalin could have interpreted this to mean that the United States had come to terms with the Communist victory in China. He may have found support for this view of events from Mao, who, at the end of 1949, no longer seemed to assume that the United States would intervene in China.[193] Stalin could also have concluded that Washington had come to terms with Moscow's alliance with

Communist China, and so he no longer had to expect negative consequences for the Yalta agreements if Moscow and Peking replaced the treaty of August 14, 1945, with a new one.[194]

The Breakthrough

The publication of Mao Zedong's TASS interview had already indicated a shift in Stalin's thinking. On the evening of the same day that it was published—that is, on January 2—Stalin did things properly. He sent Molotov and Mikoyan to Mao to address the treaty issue in this way for the first time since December 24. The two men opened the talk, which began at 8 P.M., by asking Mao what his position was on the Sino-Soviet treaty and other problems. Mao presented three alternatives:

1. They could conclude a new treaty of friendship and assistance.
2. They could publish a brief communiqué through the press agencies in the two states in which the two leaders agreed on important questions involving the old treaty and other problems without going into further detail.
3. They could sign a declaration that would define the most important points in their bilateral relations without concluding a new treaty.[195]

Mao left no room for doubt that he favored the first alternative. It involved "extremely great advantages [*ji da liyi*]," he said. Sino-Soviet relations would be strengthened. Workers, peasants, intellectuals, and the left wing of the national bourgeoisie in China would be enthusiastic, which would allow them to isolate the right wing. They would be able to move more effectively against the imperialist states on the international scene and revise the treaties that had previously been concluded between China and the imperialist powers. The second alternative, however, would put the problem off for many years, and it would only have to be dealt with again later. In this case, as well as if the third alternative were adopted, Zhou Enlai would not need to come to Moscow.[196]

Molotov immediately approved the first alternative and called for Zhou Enlai to be summoned to Moscow. When, to be sure, Mao asked whether the old treaty would be replaced with a new one, Molotov answered yes. Mao then drew up a timetable. If Zhou came to Moscow on January 19 and the negotiations for the new treaty took ten days before signing, he, Mao, and the prime minister could return to Peking in early February. Mao once again requested to meet with the Soviet leaders individually.[197]

Molotov's immediate approval of the first alternative reveals that Stalin had already made this decision previously. When the TASS interview was published, it apparently became clear to him that world public opinion would see Mao's second and third alternatives as a failure of the negotiations and thus also involve a significant loss of face for him, Stalin. The Soviet leader thus made an about-face with respect to the position he had taken on December 16. At that time, as already described in detail, he had absolutely refused to conclude a new treaty.

We owe our knowledge of the course of the talk between Mao Zedong, Molotov, and Mikoyan on January 2 primarily to a telegram Mao sent that same evening to the Central

Committee in Peking. It begins with the following sentences: "There has been an important development in the work here in the past two days. Comrade Stalin has agreed that Comrade Zhou Enlai should come here and that a new friendship and alliance treaty and credit, trade, air transport, and other agreements should be signed."[198]

Mao Zedong had thus achieved a breakthrough. The two humiliating weeks of stewing in his own juices at the dacha were over. Zhou Enlai would perform the remaining substantive negotiations after he arrived in Moscow. Mao's relief can be seen in the fact that at the end of the talk with Molotov and Mikoyan he took the opportunity to discuss an extensive travel and sightseeing program that he wished to undertake in the meantime. He planned, among other things, a visit to the Lenin mausoleum, excursions to Leningrad and Gorkii, and visits to a kolkhoz and—at Molotov's and Mikoyan's suggestion—to the Moscow subway and an arms factory.[199]

Mao had already expressed the desire to carry out a sightseeing program in and around Moscow to Roshchin the day before. At the same time, however, he had stated with reference to his allegedly faltering health that he would have to forgo the trips he had originally wished to undertake in the USSR.[200] The fact that he arranged the excursions to Leningrad and Gorkii—actually, only the trip to Leningrad took place—now reveals that the alleged reason for his earlier change of plans had been a pretext. It is probable that the real reason was that up to that time it was unclear whether a treaty would be concluded. By refusing to make the trips up to then, Mao clearly wished to signal to Stalin that he was unwilling to demonstrate that their relations were normal—as he would be doing if he undertook such trips—so long as Stalin was unwilling to conclude a new treaty.

On January 11, one day after Zhou Enlai boarded the train to Moscow, Mao Zedong began his sightseeing program. He visited the Lenin mausoleum and was then received, along with Chen Boda, Ambassador Wang Jiaxiang, and Shi Zhe as interpreter, by the Soviet head of state and chairman of the Presidium of the Supreme Soviet, Nikolai Shvernik.[201] On the following day, Mao visited the Moscow subway, an arms factory in a Moscow suburb, and a kolkhoz.[202]

Mao Zedong originally planned the trip to Leningrad for the time from January 13 through January 15. He asked to visit the Winter Palace, the Smolnyi Institute, and a factory, and then, if possible, to make an excursion to Kronstadt.[203]

His departure was delayed by one day. On January 14, Mao Zedong boarded the night train to Leningrad, along with Chen Boda, Wang Jiaxiang, Shi Zhe, a Chinese major general as aide-de-camp, another assistant, and the Soviet doctor Mel'nikov. On his arrival, he was greeted by the chairman of the City Soviet and the Regional Party secretary. He then took a car trip across the heavily iced-over Gulf of Finland toward Kronstadt and visited the Winter Palace, the legendary cruiser *Avrora*, and the Kirov machine factory. When he was told during a visit to the Hermitage that it would not be possible to show him a Chinese room as originally planned because it was being renovated, he commented derogatorily to Shi Zhe that the Russians were probably embarrassed to show him the room, because the czar had stolen so many things from China. He was then shown the defensive lines from the time of the siege of the city during the Second World War. In the evening, he attended a banquet and saw a ballet at the Kirov Theater. He spent the night in the Smolnyi Palace and arrived back in Moscow on January 16 at about 5 P.M.[204]

Deputy Foreign Minister Gromyko, Ambassador Roshchin, and the responsible department head in the Foreign Ministry, Kurdiukov, greeted Mao at the station, which was decked with the national flags of the People's Republic of China. The Soviet newspapers published prominent reports of Mao's trip to Leningrad—a sign of Stalin's need to demonstrate to the world that relations with his prominent guest from China were normal, perhaps even cordial, after the almost complete absence of news during the second half of December.[205]

Stalin's Advances to Mao Zedong

After Stalin made the decision on January 2 to yield on the treaty question, he made further gestures to improve the mood of his guest from Peking. On January 4, *Pravda* published Liu Shaoqi's opening speech at the Peking Conference of the Trade Unions from Countries of Asia and Oceania, publication of which had been delayed in the Soviet Union for more than one and a half months on flimsy pretexts.[206]

In the speech, Liu Shaoqi had described the strategy of armed struggle by a regular army as it had been developed and practiced successfully by the CCP as a model for the "liberation of other peoples in the colonial and semicolonial countries" with similar conditions. This was the "main road," he claimed; "Mao Zedong's road." When it was presented at the trade union conference, Stalin had apparently rejected Liu Shaoqi's thesis, which he viewed as "left-wing radical." Unlike the leadership of the pro-Soviet World Federation of Trade Unions, which had criticized Liu sharply at the conference and had decided against publishing the speech, Stalin did not express his rejection of the thesis openly in order maintain good relations with Mao, who was about to travel to Moscow.[207]

Stalin's decision to publish Liu Shaoqi's speech in *Pravda* on January 4, 1950, could be seen as a signal that he recognized the "Chinese road" in principle as the revolutionary model for most developing countries. As if not to leave any room for doubt that this and nothing else was meant, *Pravda* printed an article on the role of the dictatorship of the proletariat in the People's Democracies the next day that implicitly approved Liu Shaoqi's thesis.[208] It stated specifically: "The Chinese people's struggle is an example for all colonial and dependent peoples who are struggling for their independence." The Mao quotations that followed should also be seen as a salute by Stalin to his high-ranking Chinese guest in Moscow.

The first in a series of eighteen articles from the pen of the prominent writer Konstantin Simonov describing the victorious Communist revolution in China in heroic tones appeared under the title "Struggling China" in the same issue of *Pravda.* The last article in the series ended with an emotional tribute to Mao: "And as it expresses the people's most heartfelt dreams, the battle-hardened Communist Party of China with its leader Mao Zedong at its head leads the great Chinese people with a firm hand on the path of peace and creative activity."[209]

The fact that the People's Republic of China was honored with a report in the column of the first page of *Pravda* reserved for news from the "brother countries" on each of the five days beginning on January 5—on four of these as the lead story—can also be seen as a friendly gesture to Mao Zedong.[210] This had occurred a total of only five times in the three months since the People's Republic of China was established.[211] Given the strict press censorship common at that time, the new preferential treatment of the People's Republic of China could only have been the result of instructions from the highest level.

Stalin went yet another step in his attempt to win Mao's favor. In late January, he gave Mao the denunciatory report on the CCP leadership's politics he had received from Kovalev on December 24. At the same time, he handed over to Mao coded telegrams, including similar critical materials that Kovalev had sent to Stalin from China and that had apparently been based on information from Gao Gang, and files of information that Gao Gang had made available to Stalin personally. Note that Gao did not in fact actually accuse Mao himself of anything. Kovalev first learned that Stalin had done this in a confidential report from a Chinese source in February 1950.[212]

According to Khrushchev, who was a member of the Politburo at the time, Stalin gave Mao the aforementioned documents with the words: "Here, you might be interested in these." Khrushchev explained that Stalin was seeking "to win Mao's trust and friendship." Stalin, who trusted no one—not even himself—had apparently assumed that Mao would ultimately have caught Gao Gang anyway, Khrushchev added. Stalin justified his actions at the time—apparently to his closest colleagues—by calling them a "friendly gesture." Khrushchev condemned Stalin's act as a betrayal of Gao Gang comparable to Peter the Great's betrayal of Kochubei. The latter had informed Peter of treasonous plans harbored by Matsepa, the leader of the Cossacks. To win Matsepa's favor, Peter informed him of Kochubei's denunciation. As a result, Matsepa had Kochubei executed. In the 1970s, Mikoyan claimed that he had not been able to find any explanation or justification for Stalin's actions at this time.[213]

When he had handed the materials to Mao, Stalin could hardly avoid distancing himself from Kovalev. Stalin claimed that he had only glanced at Kovalev's report.[214] Kovalev, he said, had drawn up the report on his own initiative and not in response to hints from the Soviet leadership. He was a railroad engineer, Stalin said, who did not understand politics, had no political experience, and was a complete outsider in the field of politics. When he became involved in politics, he was like a mouse who had become stuck in the bellows. When Mao asked whether Stalin could send Kovalev to China as an expert, Stalin refused and said he would send more competent people. Kovalev himself indicated that he was admitted to the hospital in January for a throat illness aggravated by neglect and was not able to participate in the talks that were held after Zhou Enlai's arrival.[215]

There is no clear evidence of how Mao interpreted Stalin's unusual decision to give him Kovalev's report and the other documents that also incriminated Gao Gang. Stalin's claim that Kovalev was politically incompetent, if it was really made, could not have been very convincing. How could Stalin have sent him as his contact man with Mao and thus entrust him with such a highly responsible political mission? Nonetheless, it can be assumed that Mao still accepted Stalin's gesture in principle as what it was supposed to be: an attempt to establish greater trust. Bo Yibo, who at that time was a member of the Central Committee and the minister of finance, noted that Stalin had "improved mutual understanding" by handing over the report, although there was still some residual mistrust that the two sides were never able to eradicate completely.[216] Years later, however, Mao assessed the event completely differently. He told the Soviet ambassador in Peking that Stalin had apparently sought to express his distrust of the CCP by handing over the document.[217] We should not attribute much meaning to this assessment, which was made *post festum.*

Bo Yibo's observation that the Chinese and the Soviets had not yet been able to eradicate their mistrust completely was also confirmed in a conversation between Liu Shaoqi and a Soviet interlocutor in late January 1950. Liu's assessment was probably also affected by experiences he had had in Moscow during the summer of 1949 and in Peking during the Asian trade union conference in November 1949.[218]

Liu observed that misunderstandings existed between the CCP and the CPSU(B) that had roots reaching far into the past. Progressive individuals in China understood the situation in the Soviet Union, but progressive people that he and his comrades met in Europe and the Soviet Union had only an unclear and naive conception of the situation in China, he said. The reactionaries understood China better, he added, which was naturally an advantage for them. In discussions with Soviet comrades, it was noticeable that few had any understanding of China's problems. The Soviet people and most Party comrades had not anticipated the victory of the CCP and it was therefore incomprehensible to them.[219]

The foreign comrades, Liu continued, did not know or had only superficial knowledge of the fact that the CCP's struggle had been harder and more dangerous than that of the Soviet CP. They were aware primarily of the 1927 revolution, but there had been many other dangerous moments and threatening defeats since. The comrades—among them the Soviet comrades—knew almost nothing of the violent struggle within the CCP's Central Committee. Differences arose with the Soviet comrades, because the latter believed Li Lisan had made serious mistakes, he said. But this could not be explained so easily. High-ranking Soviet and Chinese comrades should study this question together for years. It often happened, he said, that the Soviet comrades considered someone to be good while the Chinese considered him bad. And yet the Chinese had been able to see that he was bad for many years, and they were certain of their judgment.[220]

It is striking that Liu mentioned only Li Lisan as a controversial figure in this complaint about differences of opinion between China and the Soviet Union. It would have been more to the point for him to mention Wang Ming in this context.[221] The fact that he did not do so shows how sensitive and current the conflict over Wang Ming's role was still considered to be at that time.

The United States: Invisible Partner at the Negotiating Table

The striking concern Stalin showed for Mao Zedong from January 2, 1950, was accompanied by Soviet efforts to drive a wedge between Peking and Washington. It seems that ever since he had perceived the changes in American East Asian policy, Stalin had begun to fear that Mao Zedong might wish to achieve a rapprochement with the United States and other Western countries. He saw this, among other things, as a threat to his goal of keeping Manchuria within the Soviet sphere of influence for the foreseeable future. In addition to the new American East Asian policy, there were two further reasons for Stalin's mistrust: new emphases in British China policy and the exaggerated interpretation of a talk that Kovalev had with Mao at that time.

On January 5, 1950, President Truman, who found himself facing strong Republican opposition in a Congress that was friendly to the KMT, emphasized in a press conference that the United States would not provide military aid to the KMT on Taiwan and would do everything it could to avoid becoming embroiled in the Chinese civil war. Truman based his position on a top secret National Security Council document dated December 30. On January 6, the British government declared that it was breaking off relations with KMT China, recognized the People's Republic of China, and was prepared to establish diplomatic relations with Peking. Both events, which he apparently suspected were connected, alarmed Stalin. Truman's declaration, he felt, might lead the PLA to undertake the conquest of Taiwan immediately, because it no longer had to fear American intervention. But if the Communists conquered Taiwan and the United States did not intervene, one of the important obstructions that still stood in the way of establishing diplomatic relations between Peking and Washington would disappear. The normalization of relations between Peking and London would also further the process of Sino-American rapprochement.[222]

This shift in American China policy, about which Stalin might already have heard through secret service channels at the end of December, may, as has already been discussed in detail earlier in this chapter, be one important reason for his decision on January 2 to allow Mao to negotiate a new treaty, which he had hitherto refused to permit. The same was true of the imminent recognition of the new China by London and Delhi, about which he had just heard from Mao and Kovalev on January 1.[223]

Stalin must have feft that his fears that Peking might wish to seek a rapprochement with the West had been confirmed when he heard Kovalev's report on a talk he had had with Mao at the beginning of January. Mao was reported to have said that China would allow the establishment of diplomatic missions in China only after a treaty of friendship, military, economic, and political alliance and agreements on credit, trade, and other matters had been concluded with the Soviet Union. When the treaties had been concluded with Moscow, Mao continued, "we can immediately proceed to revise and abolish the unequal treaties that the Chiang Kai-shek government concluded with the imperialist countries." In his report, Kovalev interpreted Mao's statements as follows: Mao was pressuring the Soviet Union "to surrender all its interests in China." It was striking, Kovalev continued, that Mao had not followed Stalin's advice to make it a condition that the United States renounce its support for the KMT government.[224]

One can assume that Kovalev, for whom even the appearance that the CCP leadership was undertaking an independent policy that was not based on Soviet guidelines implied a betrayal of Moscow, was reading too much into Mao's statements. Mao gave Stalin no other reason to suspect betrayal. Peking did accept J.C. Hutchison, who had been nominated chargé d'affaires ad interim by the British, as London's representative to negotiate possible diplomatic relations. But Stalin could hardly object to this, as he had always said in his advice to the Chinese comrades that they should make the establishment of such relations dependent on the breaking off of relations with KMT China—which, in the case of London, had after all been done. The media Peking controlled nevertheless continued to spread pro-Soviet and anti-Western propaganda as before.[225]

Stalin nonetheless adopted Kovalev's suspicious interpretation as his own. This is suggested by Kovalev's comment that the assessment of Mao Zedong's views he had sent to Stalin had been "our common opinion." This can also be seen from the fact that Stalin took steps designed to obstruct the rapprochement that was supposedly being planned between the People's Republic of China and the United States. One day after Truman's press conference, *Pravda* began to publish articles that were intended to give the impression that there was secret military cooperation between Washington and the KMT government on Taiwan.[226] At the same time, Stalin changed the position he had hitherto taken with the Chinese comrades on the Taiwan question. In the summer of 1949, he had still voiced serious reservations to Liu Shaoqi and rejected his request to provide Soviet military support for the planned attack on Taiwan.[227] There is evidence that from January 1950 he encouraged the Peking leadership to conquer Taiwan and supported it—if secretly—in making the necessary military preparations, among other things by allowing it to use half of the approved US$300 million credit for the purchase of Soviet armaments.[228] Stalin then supposedly also sought to convince Mao Zedong to attack Hong Kong.[229] All of these steps were shaped primarily by an attempt to stir up the People's Republic of China and the KMT government on Taiwan against one another and drive a wedge between Peking and Washington.

Events in another context reflected the same trends. On the night of January 6–7—or one day after Truman's statement to the press—Foreign Minister Vyshinskii, one of the most despicable and most unscrupulous leadership figures in Soviet history, went to see Mao on instructions from Stalin. As attorney general during the second half of the 1930s, Vyshinskii had staged the political show trials and placed the burden of proving their own innocence on the accused. Vyshinskii came to his main point quickly with Mao: they needed to take coordinated steps designed to lead to the expulsion of KMT China's representation from the United Nations. He proposed to Mao that the People's Republic of China should make a statement to the UN Security Council that described the presence of the KMT representative in the council as illegitimate and demanded his expulsion.[230] The USSR would support this statement and, for its own part, demand the expulsion of "the representative of the KMT group," then Jiang Tingfu. If the latter nonetheless remained in the Security Council, the USSR would boycott council meetings until he was expelled. Mao approved this action "one hundred percent."[231]

The United Nations received the text of a telegram based on a draft drawn up by Mao and bearing Zhou Enlai's signature on January 8. It stated that the People's Republic of China was of the opinion that the presence on the Security Council of the representative of "the remnant reactionary gang" of the KMT was illegal and that this representative should be expelled from the Security Council.[232]

At the January 10 Security Council session, the Soviet delegate, Adam Malik, did in fact support the Peking government's position and himself made a request that the UN expel the KMT government representative—a proposal that had not been included on the established agenda. He also threatened that the Soviet delegation would no longer take part in sessions until Jiang Tingfu had been removed. Jiang, who was chairman, did not permit the question to come to a vote and requested that it be discussed at a special session. His request was carried with eight votes for it (including the United States, Great Britain, and France), one dissent (USSR), and one abstention (India). Malik then left the session. At the request of Yugoslavia and India, it was resolved that the Soviet demand would be discussed at the following session. At this session, which took place on January 12, Malik made a move to leave the hall, but then took his place again when the Cuban delegate took the chair. Malik was given an opportunity to present his position on the Chinese seat on the Security Council in detail.[233]

Vyshinskii reported these events to Mao in a talk on January 13. Vyshinskii said that they had been able to make Jiang Tingfu give up the chairmanship of the Security Council for the duration of the debate on China. To achieve this end, the Soviet representative had had "to leave the session." Vyshinskii asked whether it would not be useful "in order to complete our struggle" for the Chinese government to nominate a representative of its own for the Security Council and present him to the United Nations to force it "to treat the question at a concrete level." Mao agreed. He saw the current deputy foreign minister, Zhang Hanfu, who had studied in the United States, as the most suitable candidate. And yet, until a majority was established on the Security Council on this issue, the candidate from China would not be "legally authorized" and would be "forced to sit in Peking." Vyshinskii said that Peking's nomination of a representative would nonetheless have "great political significance." Some council members were also upset that the presence of Jiang Tingfu, "this Kuomintang cadaver," constituted grounds for the USSR to discontinue its work, "which would in practical terms lead to the collapse of the U.N."[234]

When Mao asked whether the United States and Great Britain were interested in the existence or the collapse of the United Nations, Vyshinskii replied that the former was correct. They needed the United Nations as a cover for their imperialist plans. The collapse of the United Nations would reveal all their intentions and plans and deprive them of the means that enabled them "to hide the aggressive plans and deceive international public opinion." Mao again stated that he was willing to nominate his own representative but wished to coordinate all aspects of this step with Zhou Enlai, who was expected to arrive in Moscow on January 19 or 20. He asked if the Soviet side would be willing to accept this delay. Vyshinskii felt that they could postpone the issue if Mao considered it necessary to coordinate with Zhou Enlai. The amazement we seem to hear in this answer is probably due to the fact that Stalin would certainly never have attempted to coordinate with Vyshinskii in a comparable situation and would have immediately made his decision alone.

Mao reported on his talk with Vyshinskii to Liu Shaoqi by telegraph and asked the Central Committee to nominate a Chinese delegation head for the Security Council and tell him who it was. The final decision would be made after Zhou Enlai's arrival in Moscow.[235]

Mao did not withhold his decision until Zhou Enlai's arrival, however. He coordinated his plans on January 18 by telephone with Zhou, who had by then reached Sverdlovsk. He then telegraphed to Liu Shaoqi that Zhang Hanfu was in any case suitable only to become deputy UN delegate, as he had lost prestige. Mao proposed the Politburo member Zhang Wentian (Party name: Luo Fu) as delegate. Zhou, Gao Gang, and Li Fuchun shared his opinion, Mao stated. At the same time, he sent Liu the draft of a telegram for the United Nations. If the Central Committee approved his proposals, Mao added, the telegram informing the UN of Zhang Wentian's nomination should be sent the following day and published. In the telegram it was also asked when the KMT representative would be expelled and when the Peking delegate could begin his work. The Peking leadership then acted as Mao had proposed.[236]

The Soviet request made on January 10 to expel the KMT representative from the Security Council was rejected on January 13 with six votes against the proposal (including the United States and France), three votes for (the USSR, Yugoslavia, and India, the latter of which had already established relations with Peking), and two abstentions (Great Britain and Norway, which had both recognized Peking). Malik repeated his January 10 statement that the Soviet delegation would no longer participate in Security Council sessions until the KMT representative was expelled. He then left the hall with his delegation. From January 16, the representatives of the USSR and the People's Democracies also ceased to participate in sessions of the other subordinate UN organizations. On the same day, the Soviet media began to support the Chinese demand for the expulsion of the KMT representative from the United Nations in reports that appeared daily.[237]

Stalin could not, however, have seriously presumed that it would be possible to force KMT China out of the Security Council in this way. Given the structure of the majority on the council, he could not assume that a vote would be favorable. This, and the fact that Stalin had provoked Mao Zedong to go on the offensive in the Security Council one day after Truman's speech, leads us to suspect that this was a further tactical maneuver designed to stir up conflict between Peking and Washington and obstruct the rapprochement he feared. Stalin was probably not even interested in the expulsion of Taiwan and preferred to continue the conflict in the Security Council. By taking a position of solidarity in the United Nations, he could demonstrate to Mao that the Soviet Union was the most powerful friend the new China had while the United States continued to appear to be the strongest enemy. Western observers, including the British UN delegate, Sir Alexander Cadogan, already attributed this and similar motives to Stalin at that time.[238]

A speech by Secretary of State Dean Acheson to the National Press Club in Washington on January 12 fed Stalin's mistrust further. In it, he repeated what Truman had already said on January 5 about the disengagement of the United States with respect to Taiwan. He emphasized this concept by drawing out a new defense perimeter for the United States in the Pacific that included Japan, the Ryukyu Islands, and the Philippines, but bracketed out South Korea and Taiwan. Here in Asia we see already the first outlines of a global containment policy toward perceived Soviet expansionism that would be given concrete form as the Truman doctrine about one year later.[239]

Acheson also declared that the Soviet Union was in the process of annexing portions of North China, namely, Outer and Inner Mongolia, Sinkiang, and Manchuria. He said:

> The Soviet Union is detaching the northern provinces of China and is attaching them to the Soviet Union. This process is complete in Outer Mongolia. It is also nearly complete in Manchuria, and I am sure that in inner Mongolia and in Sinkiang there are happy reports coming from Soviet agents to Moscow. . . . The fact that the Soviet Union is taking the four northern provinces of China is the single most significant, most important fact, in the relations of any foreign power with Asia.[240]

Acheson no doubt intended to drive a wedge between Peking and Moscow with this part of his speech.[241]

The American administration had long been making this argument a standard part of its tactical attack, as can be seen, for example, in a memorandum drawn up in August 1949 by the undersecretary of state in the Department of State to the National Security Council that assigned the following propaganda line to the United States Information Service, which included the Voice of America: ". . . to foster possible rifts between Chinese Communists and the USSR to emphasize the imperialistic aims of the USSR in China as evidenced in Manchuria, Sinkiang, and Mongolia. . . ."[242]

There is, however, no evidence that Mao had any intention of seeking a rapprochement with Washington at this time. If he had, Mao would certainly not have adopted the measures he did one day after Acheson's speech, such as the confiscation of American—also French and Dutch—diplomatic property in Peking, including the Marine Legation barracks, and the property of the U.S. Economic Cooperation Administration in Shanghai, which could only have significantly damaged Sino-American relations. The confiscations in Peking had already been decided on January 6, but were only carried out on January 14. They met with a decisive reaction: the United States closed all its consulates in China and recalled all diplomats from China—a step that Mao described to Liu Shaoqi as "extremely favorable for us [*dui women ji youli*]."[243] One further consequence was that many Western states that had planned to recognize Peking put these plans on ice for the time being.[244]

Mao Zedong informed the Soviet side of these steps in advance. On January 13, he told Vyshinskii that before establishing diplomatic relations with Washington it would be necessary "to gain time to establish order in the country." Mao said that he was determined to carry out two measures: to confiscate the former barracks of the foreign armed forces that the foreigners had at one time received by means of an unequal treaty for their own use and the provisions and property of the Economic Cooperation Administration in Shanghai, which the Americans had used to help Chiang Kai-shek, as was well known. The two measures would allow China at the same time to confiscate the buildings of the American consul general in Peking and in other cities, drive the consular representatives from China, and confiscate significant stores of provisions in Shanghai. This action would provoke "great enthusiasm" in the Chinese people and help to isolate the Chinese bourgeoisie's pro-American right wing.[245]

Vyshinskii's reaction was noticeably reserved. Even if it could not be excluded that the Chinese would be able to achieve these goals, he said, they should nonetheless study the best way to implement such measures from all sides. In thinking about the U.S. decision to recognize the People's Republic of China, they had to keep the political goals the United States was pursuing in mind and connect this question with the question of fighting communism, not only in China but in other Asian countries as well. Vyshinskii did not express his views any more concretely. It is unclear what he meant.[246]

Mao also had to realize that another action he was taking—Peking's recognition of Ho Chi Minh's government and the Democratic Republic of Vietnam on January 18—would further worsen relations not only with the United States but also with France.[247] Mao approved this step, which came in response to a request from the Vietnamese Communists on January 15, on January 17.[248] In supporting Communist Vietnam, he was, among other things, seeking to relieve the military pressure on his forces along the Vietnamese-Chinese border, where the KMT had strong troop concentrations, and establishing a fraternal and allied regime on the southern periphery in order to increase China's security.[249] According to CIA estimates, Ho at this time had the support of 80 percent of the population and already controlled two-thirds of the country.[250] In supporting the Vietnamese Communists, Mao was not merely pursuing Chinese interests. He had been closely allied with Ho Chi Minh since the 1920s and felt obligated to show him revolutionary solidarity—as he did to all the other revolutionary movements in Asia whose leader he felt himself called to become.

Mao also demonstrated his interest in cooperating with Moscow in developing an anti-American policy in Asia by adopting the Soviet line and publicly criticizing the Japanese Communist leader Nosaka Sanzo for his pro-American activities, which he carried onto despite Mao's long-standing close relationship with him.[251] Peking also continued its polemics against the United States and Taiwan without any hesitation even after mid-January with grossly aggressive caricatures in the media.[252] One caricature must have pleased Stalin particularly. It depicted a man labeled "American Imperialism" who had collapsed with a sigh as he examined a map of Eurasia on which the USSR and the People's Republic of China were both accentuated in black ink—an indication of the formation of a Sino-Soviet bloc against the United States.[253]

Despite this, Stalin found the portion of Acheson's speech on January 12 dealing with the Soviet Union's alleged annexation plans in China in particular so alarming that he considered it necessary for the USSR, the People's Republic of China, and the Mongolian PR to provide a coordinated response. His behavior suggests that Acheson had hit a particularly sensitive spot. Stalin was apparently pursuing two goals in formulating a coordinated reaction. It was designed, first, to refute the American claims—which would sound much more convincing coming from a Chinese than a Soviet speaker. And second, Peking was to be forced to announce its solidarity with Moscow on this question publicly, which would decrease China's options for maneuvering with the United States.

On January 16, Mao returned to Moscow from his one-day visit to Leningrad. On the following day, Molotov and Vyshinskii appeared at his dacha. Molotov raised the question of Acheson's January 12 speech,[254] cited the passage in it that alleged Soviet plans to annex portions of Manchuria and Sinkiang, and asked Mao to take a position. Mao replied in the way the Soviet side hoped he would: Acheson was merely seeking slanderously to sow dissension, primarily in order to distract attention from the U.S. claim to Taiwan. Mao sought to gain Molotov's and Vyshinskii's trust by assuring them that the Chinese leadership had ignored the feelers Washington had recently extended toward Peking. Molotov and Vyshinskii then proposed that Peking and Moscow should issue coordinated official statements to refute Acheson's slander.[255] Mao stated his agreement, and statements to this effect appeared on January 21 in the Moscow and Peking press.[256]

The atmosphere between Stalin and Mao became strained, although the two sides technically issued their statements simultaneously. They had in all probability coordinated the general contents of their statements, as can be seen, among other things, from the fact that both statements made the former American consul general in Mukden speak as the primary witness through his statement of January 14, in which he denied that there was any indication of Soviet control in Manchuria, with the exception of the joint administration of the Changchun Railroad.

It is striking, however, that the Soviet statement included a clear denial of Acheson's comment that Moscow wanted to control all of North China: there is "not a single word of truth [*net ni slova pravdy*]" in this, it said. The Chinese statement lacks any such clear denial. It merely states that Acheson's claim could only lead to the following results: it could "arouse the scorn of the Chinese and Soviet peoples [*jiqi zhongsu liangge renmin fenkai*]" and "consolidate the friendship and cooperation between China and the Soviet Union [*jiaqiang zhongsu liangge de youhao gongzuo*]." One could assume that Mao Zedong felt that Stalin's long refusal to conclude a new treaty confirmed his suspicions that Stalin wanted to maintain the USSR's privileges in Manchuria and influence in Sinkiang. The leadership of the CCP—like the KMT leadership—had in any case never completely gotten over the loss of Outer Mongolia. In this way, Mao could have perceived Acheson's attack as welcome support in the upcoming treaty negotiations with Stalin.

One may assume that the Chinese failure to provide a clear denial of Acheson's aforementioned claims—which particularly affected Stalin, and the significance of which Molotov emphasized to Mao on January 17--annoyed Stalin. In a brief talk in the Kremlin during the last days of January to which Stalin invited Mao Zedong and Zhou Enlai—Zhou had arrived in Moscow on January 20—and in which Molotov and Shi Zhe also participated, however, he decided not to argue about the contents and merely criticized the Chinese side on the formal grounds that it had not satisfied protocol. Together with Molotov, he accused Mao of failing to fulfill the January 17 agreement whereby each of the foreign ministries of both sides was supposed to publish an official statement. Mao had disguised his statement in the form of a fictitious interview that Hu Qiaomu, head of the Information Department of the Chinese government, supposedly gave to a Xinhua correspondent,[257] whereas the Moscow statement had been published in Foreign Minister Vyshinskii's name.[258] According to Stalin, however, Hu Qiaomu's statement could not be considered official according to international usage. This sort of thing should not be allowed to happen again. One could not act any way one pleased. "I believe we should maintain our agreements, coordinate precisely, and match our steps [*budiao yizhi*]; only in this way can we gain in power."[259]

Stalin was correct at least in saying that Mao, Molotov, and Vyshinskii had finally agreed on January 17 that both sides should have the statements published by their foreign ministries. Mao had previously proposed having the new Xinhua agency publish the Chinese text. He had accepted Vyshinskii's objection, however, that Acheson's speech would have to be answered in an official statement, as it represented a statement from the State Department at this protocol level.[260]

It is not clear why Mao disregarded the January 17 agreement. The other circumstances governing the Chinese publication do not suggest that the Chinese intended to assign the incident a low protocol level—the opposite is more the case. The Chinese statement was placed on page 1 of *Renmin ribao*, whereas the Soviet text appeared only on page 2 of *Pravda*. The same impression is given by the fact that *Renmin ribao* published Vyshinskii's statement in unabridged form and on page 1, whereas *Pravda* reprinted Hu Qiaomu's interview only with abridgments and on page 4.[261]

It is possible that Mao had not taken the agreement particularly seriously because he was not yet sufficiently familiar with diplomatic usage. And Molotov's and Vyshinskii's dogmatic insistence may perhaps also have annoyed Mao and made him very consciously ignore the agreement for this reason. One may assume that Stalin was on the whole less upset about the form of the Chinese statement than by the lack of a clear denial. But he did not want to address this explosive topic and released his anger by criticizing the procedural issue, which was less delicate. And yet he was probably in fact really angry that Mao had ultimately not taken Vyshinskii's objections seriously and casually ignored the agreement he had made.

According to Shi Zhe, Stalin's accusations annoyed Mao, and he did not say a single word. Only Zhou made a few reserved statements. After the talk, Stalin invited his Chinese guests and Molotov to visit him at his dacha and took Mao and Shi along in his car. Mao instructed Shi not to mention to Stalin that he had promised to make a visit to Mao's dacha. Stalin apparently understood what was being discussed from the context. As a result, "the atmosphere again was as heavy as lead [*qifen chenzhong de you xiang guanshang le qian*]." At Stalin's dacha, Politburo members Malenkov, Bulganin, Beriia, and Kaganovich, Foreign Minister Vyshinskii, and Ambassador Roshchin joined them. Thus, exactly half of the Politburo was gathered together. The depressed mood continued during dinner. Mao remained silent the whole time.[262]

After the meal, Shi Zhe continued, Stalin turned on some music, and at his suggestion everyone present formed pairs and danced together—with the exception of Stalin himself and Mao Zedong, who, despite repeated insistence, could not be convinced to dance. This did nothing to improve the atmosphere, either. At first glance, it seems difficult to believe this portion of Shi Zhe's report. In his memoirs, however, Khrushchev did mention that in 1953 Stalin had set up a record player at a New Year's Eve party at his dacha, to which Stalin had invited his closest colleagues, and told everyone to dance to Russian and Georgian music. Stalin, Mikoyan, Voroshilov, Kaganovich, and Khrushchev had all danced, albeit each individually.[263]

On the other side, there is evidence that Mao sought to counter Stalin's attempts to bring the People's Republic of China more into line with respect to the United States with his own attempts to drive a wedge between Moscow and Western states. We can take as evidence of this the fact that Mao asked Liu Shaoqi on January 17 to forward Ho Chi Minh's January 15 message, in which Ho stated his desire to establish diplomatic relations with all the governments of the world,[264] through the Foreign Ministry in Peking to the Soviet Union and its satellite states.[265] Peking's actions in solidarity with Communist Vietnam, which had already taken the form of immediate diplomatic recognition on January 18, put pressure on Stalin also to show solidarity by recognizing the Democratic Republic of Vietnam. On the other hand, recognition of the Ho Chi Minh government would place a significant strain on the Soviet Union's relations with France, which was fighting a war against Communist Vietnam. Moscow was interested in good relations with Paris, because the French government was at that time opposing American plans to rearm West Germany and bring it into NATO.[266]

Stalin ultimately felt himself constrained for ideological reasons to follow the Chinese example. Mao Zedong had already told Stalin on December 24 that China planned to recognize Communist Vietnam and support Ho Chi Minh's resistance movement, and Stalin had promised him that the USSR would follow the CCP on this road.[267] At the beginning of January 1950, Stalin assured Mao: "I am convinced that comrade Ho Chi Minh is a solid and intelligent [*tverdyi i tolkovyi*] Communist, does his business well, and deserves every assistance."[268] The USSR finally recognized the Democratic Republic of Vietnam on January 30, 1950.[269] Stalin, who did not believe that the Vietnamese Communists would be victorious, often regretted later that he had recognized them: "We were too hasty," he complained. "It's too early for recognition."[270] We do not know whether Mao only put pressure on Stalin by providing the Chinese example or also sought to influence him during their talks in January 1950.[271]

When Ho Chi Minh visited Moscow secretly in February 1950, Stalin was not very accommodating to him. Apparently out of consideration for Moscow's relations with Paris, Stalin avoided Ho's request to conclude a treaty of alliance with Communist Vietnam similar to the one Stalin had concluded with China.[272] He refused Ho's request to transform his stay into an official visit.[273] At Ho's request, Stalin first signed an issue of a Soviet periodical, but then had the MGB steal it back from him secretly.[274] When Ho asked him for quinine to treat the malaria that was raging in Vietnam, Stalin granted him the ridiculously small amount of one-half ton, although the USSR was already producing this medication in massive quantities at that time.[275] When Ho asked him for military aid, Stalin referred him to Peking. Mao Zedong and Zhou Enlai stated that they were willing to provide aid and sent the first Chinese military advisers to Vietnam in April.[276] These events also reflect the desire Stalin had already expressed to Liu Shaoqi in the summer of 1949 to leave the leadership role in the revolutionary movements in Asia to the Chinese comrades. At the same time, however, he claimed the right to determine the shape of the People's Republic of China's relations with the United States and the West.

Stalin's decision to recognize the Communist regime in Vietnam came as a surprise to Washington. Along with Soviet attempts to expel the KMT government from the United Nations, it was taken as a sign that the Soviet Union was attempting to seize the initiative aggressively in Asia and "to accelerate the revolutionary process in Southeast Asia."[277] We can see the first outlines of the future domino theory. Secretary of State Acheson interpreted Soviet recognition of Vietnam by insisting that the United States should no longer have any illusions that Ho Chi Minh's goals were "nationalistic": Ho was the "mortal enemy of native independence in Indochina."[278] After France established Vietnam, Laos, and Cambodia as autonomous states within the French Union in late January, Washington recognized the Democratic Republic of Vietnam under Bao Dai and the kingdoms of Laos and Cambodia diplomatically.[279]

Acheson was pursuing several goals in his calculation: he was seeking to encourage the non-Communist national aspirations of colonial peoples in Southeast Asia to establish non-Communist governments on the borders of China, to support France, and to provide "a demonstration of displeasure with Communist tactics, which are obviously aimed at the eventual domination of Asia, working under the guise of indigenous nationalism."[280] The recognition of Communist Vietnam by Peking and Moscow led to a recommendation by the State Department in early February to undertake strong measures to obstruct China's advance to the south and to a congressional resolution to increase aid for Korea and Taiwan—which involved a certain deviation from the strategic line Truman had announced on January 5, 1950.[281] The American administration's position toward both Moscow and Peking became firmer, and its suspicion that the two Communist great powers were pursuing expansionistic policies in Asia increased.[282] One could go yet another step further in judging the reaction: the recognition of Communist Vietnam reinforced already-existing attempts to form Washington's Asian containment policy and thus exacerbated the Cold War in Asia.

The events analyzed in this section thus demonstrate, as a standard work on the topic of the present book shrewdly put it, that "the United States had become the third invisible partner in the Stalin-Mao dialogue."[283]

Chinese Requests for Aid

In late 1949 and early 1950, the Chinese Party leadership asked Stalin for aid in several concrete areas.

They made an urgent request for a hydroelectric power plant on the Sungari near Kirin in Manchuria, where there was a threat of a possible dam break. If the dam broke, Mao Zedong told Stalin on January 2, this would threaten industry in Manchuria, destitute several million inhabitants of the Sungari Valley, and flood the cities of Harbin and Kirin. Mao asked Stalin "to instruct the relevant Soviet bodies to send Soviet experts in dam construction and hydroelectric power plants so that they could assess the situation on the spot and take the necessary measures."[284]

Stalin reacted promptly. Four days later, he promised in a letter to Mao Zedong that a special committee under Molotov's leadership would make a positive decision.[285] On the same day, Foreign Minister Vyshinskii informed Mao that the Soviet government had resolved to send four experts to China for a month within five days to assess the status of the system of hydroelectric installations and to take the necessary measures to relieve the dangerous situation.[286] Mao thanked him and stated that "the provision of aid by the Soviet Union on this occasion is of enormous significance to the economy of China." On January 12, Vyshinskii informed Mao that a further seven specialists would travel to Kirin in the near future.[287]

The Chinese also made a request in the military field. In a telegram to Stalin in late December, Liu Shaoqi referred to an existing agreement that the Soviet Union would deliver 93,000 tons of high-quality and 38,000 tons of less-high-quality gasoline, 10 percent of the necessary lubricating oil, and a corresponding amount of other materials in connection with plans to train PLA pilots. As the flight school had already begun to train pilots, they asked that the aforementioned materials be delivered as soon as possible. Liu sent a copy of the telegram to Mao, who forwarded it to Stalin. Stalin and Liu Shaoqi had reached an agreement on the training of Chinese pilots by Soviet instructors at the end of July.[288]

Stalin also fulfilled this request, if not completely. On January 6, he stated in a letter to Mao that the request for the delivery of "goods of various types," as they had been described in the telegrams sent to Stalin via Mao from Peking, was being examined by the special committee led by Molotov and that he had no doubt they would reach a positive decision. On the same day, Vyshinskii informed Mao that they planned to send Liu a reply by telegram,[289] stating that an assessment of need according to the norms of the Soviet Army had revealed that only 13,400 tons of high-octane gasoline, 5,270 tons of low-octane gasoline, and 1,315 tons of lubricating oil were needed for the stated purpose. These would be delivered during the first six months of 1950, beginning in January. Payment methods could be established in the talks about the exchange of goods during 1950. Mao indicated his agreement with the contents of the reply telegram, thanked the Soviet government, and joked about the Soviet reduction of the delivery that "our people always like to get more," that they had to watch them closely. He was, he added, grateful to the Soviet government in this case for reassessing their needs and agreed completely. It was really necessary, he said, to be strict about the use of fuel; this was in China's own interests, as it would obligate the Chinese to be more careful about using products provided through foreign aid. On the next day, he sent a cable to "Zhou Enlai and the Central Committee" that was brief and not entirely accurate. It would be possible, he stated, to satisfy "our request for the purchase of aircraft fuel [*manzu women goumai feiji de yaoqiu*]."[290]

The Chinese also made a request involving Sinkiang. It was included in a telegraphed report that Peng Dehuai had sent to Mao Zedong on December 29 after his return to Peking from Sinkiang. Mao forwarded the request to Stalin on January 2. Peng began his report by recapitulating how the economy of the northwest province had declined after the ruler of Sinkiang, Sheng Shicai, had broken off relations with Moscow in 1942. The situation had become even more critical following the arrival of the PLA. Inflation had achieved dramatic proportions. It no longer seemed certain that the Communist troops could be fed. Peng feared a deterioration of relations between the PLA and the overwhelmingly non-Chinese population if aid were not provided soon. But extremely difficult logistical conditions made it, he said, practically impossible to help Sinkiang from China.[291]

For these reasons, Peng Dehuai continued, "I consider it unavoidable that significant aid be provided by the Soviet Union in Sinkiang to overcome the difficulties in the present and build the future." The troops had to be supplied with such things as uniforms, technical materials, and means of transportation. "Normal trade" had to be established immediately between Sinkiang and the USSR, he wrote, to exchange local products for mass-market commodities of Soviet origin. The Soviet Union had to introduce economic cooperation in Sinkiang in order to allow it to mobilize its own resources gradually. In the past, when Zhang Zhizhong had been in Sinkiang, there had been talks with the Soviet Union "about the creation of a Chinese-Soviet oil company and a Chinese-Soviet company for the mining of rare and nonferrous metals." In Peng's opinion, it would be "extremely desirable at the present moment to reach an agreement with the Soviet Union about this and to establish these two companies." At the end of his discussion, Peng stated that the northwest region's great size and inadequate communications system greatly influenced "the organization of the work." It was therefore necessary to hold talks with the Soviet Union about extending civil air connections from Alma-Ata, Urumchi, and Hami to Lanzhou, Sian, and Peking, as called for in the agreement.[292]

It is not known to what agreement Peng Dehuai was referring. On December 16, 1949, however, Stalin had already agreed with Mao to provide help in creating air routes through Sinkiang (and the Mongolian PR).[293] The most striking part of Peng's telegram was his call for the creation of two joint companies. This means that the initiative for these companies, which were then in fact established on March 27, 1950,[294] came from the Chinese side. Researchers have assumed up to now that Stalin forced them on the Chinese comrades with gentle pressure. It is striking that Stalin did not mention the requested aid package for Sinkiang at all in his letter to Mao Zedong on January 6, in which he reacted to the telegrams and promised aid in two concrete cases.[295] Mao must have subsequently raised the issue of aid for Sinkiang again at least with respect to the delivery of grain, as Stalin himself raised the topic in the talk on principles on January 22, 1950.[296]

Preparations for the Negotiations

It only became possible to begin negotiations for the treaty and supplementary agreements when Zhou Enlai arrived in Moscow on January 20. Up to that point, the two sides prepared separately. They occasionally also probed one another—as, for example, in the talks Mao Zedong held with Vyshinskii and Kovalev. The negotiations began on January 22 and ended with the signature of the treaty and several supplementary agreements on February 14 and the conclusion of further agreements in March and April.

The preparations for the negotiations began immediately after Mao Zedong's talk with Molotov and Mikoyan on January 2, which had resulted in the breakthrough.

Mao Zedong took the first steps that same evening. They involved preparations for Zhou's trip and set the tone for informing the state leadership bodies in Peking of the goals of the negotiations. This can be seen from a telegram in which Mao—apparently after a telephone conversation with Stalin[297]—gave Peking Party headquarters the relevant instructions.[298]

The Government Administrative Council had to approve Zhou's trip, it said. They had to inform this council and the Central People's Government Council that the trip was necessary to carry out negotiations and sign the following agreements:

- A new Treaty of Friendship and Alliance. There could be partial changes *(bufen de biangeng)* in this treaty with respect to Port Arthur and Dairen.[299] They had yet to discuss the concrete contents of the treaty. The new treaty would continue to be shaped by the spirit of the clauses involving defense against possible aggression from Japan and its allies and recognition of the independence of Outer Mongolia.
- A credit agreement. The Chinese side had proposed US$300 million, it said, spread out over several years *(ji nian)* and no longer, as it would be more advantageous now and for numerous years *(shu nian)* to take out smaller rather than greater credits.
- A civil aviation agreement. This would be useful in helping China create its own aviation industry.
- A trade agreement. An agreement establishing the extent of Sino-Soviet barter trade would be useful both in developing Chinese production and also in negotiating trade agreements with other countries.

The Party leadership should make it clear to the two councils named above that this action by the People's Republic of China was designed to put China in a still more favorable bargaining position. It would force the capitalist states to play by the rules China established. It would favor the unconditional recognition of China by various countries as well as the abolition of the old agreements and the conclusion of the new treaties. The capitalist countries would be restrained from reckless undertakings.

Mao Zedong's instructions reveal that he attributed a significance to the treaty they were seeking with Moscow that was far greater than just the stabilization of China's future relations with the Soviet Union. The treaty and agreements were to be concluded on the basis of equal rights and were to provide a precedent for abolishing all the unequal treaties with other states and replacing them with treaties based on equal rights.[300] Immediately after the arrival of Mao's telegram, the Central Committee in Peking gave its approval to the agreement reached between Mao and Molotov, resolved to send Zhou to Moscow, and instructed Mao of its resolution by telegraph.[301]

On January 5, Mao warned the comrades in Peking that they should prepare for the upcoming negotiations well. They would have "to express our position clearly [*jiang women de guandian shuo qingchu*]" during the negotiations. The Central Committee would have to continue to prepare after Zhou's departure and report its findings constantly to Moscow. This was also true of the preparation of materials in the field of trade if the time that remained before Zhou's departure did not suffice. The Central Committee should ask the Government Administrative Council and Central People's Government Council to treat discussions of the new treaty in confidence. The members of the Zhou delegation were to be obligated before departure to observe the strictest discipline and follow instructions.[302]

Zhou and his staff continued to prepare for the negotiations on the train to Moscow between January 10 and January 20.[303] Mao Zedong occasionally became involved in this by telephone. Nothing beyond Mao Zedong's aforementioned statements has yet come out about the contents of the Chinese preparations, however.

Given the clear agreement to conclude a new treaty that Mao and Molotov had reached on January 2, it is striking how much Foreign Minister Vyshinskii emphasized, in a talk in Mao's dacha four days later, the alleged difficulties such action would involve. Kovalev, Fedorenko, Ambassador Wang Jiaxiang, and Shi Zhe also took part in this forty-five-minute talk.

The main reason for Vyshinskii's visit was, as already mentioned,[304] Stalin's attempt to convince the People's Republic of China to make a statement to the United Nations Security Council that the USSR would support, in which it asked the council to expel the representative of the KMT government. After discussing this issue, Mao Zedong turned to the approximately 1,000 Japanese being held prisoner in the Soviet Union who were accused of war crimes that they had allegedly committed in China.[305] Mao Zedong finally raised the topic of the upcoming treaty negotiations with Vyshinskii. He was increasingly coming to the conclusion that they would have to conclude a "new Treaty of Friendship and Alliance." This was made necessary by the fact that "completely new conditions" had arisen between the two states since the victory of the popular revolution in China. Mao then introduced a new argument. A revision (in Russian, *peresmotr*) of the existing treaty was all the more necessary as "two important components" of the treaty had changed radically: Japan no longer existed militarily, and the KMT had been beaten. As he had told Stalin on December 16, Mao also mentioned to Vyshinskii that "a certain portion of the population" had expressed its dissatisfaction with the existing treaty, and that it was therefore in the interests of both states to conclude a new treaty.[306]

Vyshinskii replied that the issue of the new treaty seemed complicated to him because any attempt to sign a new treaty or revise the old one and introduce modifications could be used as a pretext by the Americans and British for a revision in those portions of the treaty *(dogovor)* in which changes could damage the interests of the Soviet Union and China: "This is not desirable and cannot be allowed to happen." Mao replied that this factor "no doubt" would have to be taken into account "in establishing a formula to regulate this question."[307]

Two things are strange in the dialogue reproduced in this paragraph. For one thing, Vyshinskii used an argument that Stalin had employed with Mao Zedong on December 16 to explain why he refused to change "even one single point" in the existing treaty.[308] He had not even considered concluding a new treaty in this talk, in which Vyshinskii had participated. The position Stalin had taken at that time could be considered obsolete by January 6, however, as Stalin had informed Mao—through Molotov and Mikoyan—of his willingness to conclude a new treaty four days earlier. There is only one plausible explanation for Vyshinskii's strange behavior: Vyshinskii, who was not a member of the Politburo, did not yet know that Stalin had changed his position because the new policy had to that point been discussed only in strict confidentiality and in the most restricted circle, that is, among Stalin, Molotov, Mikoyan, and—possibly—the other members of the Politburo dealing with the China problem. On the other hand, it is nonetheless amazing that Mao decided not to contradict Vyshinskii and refer to his talk with Molotov and Mikoyan of January 2. It is possible that he did this because he had noticed that Vyshinskii was not a member of the most intimate circle.[309]

Mao Zedong again raised the topic of the treaty in a discussion with Kovalev that took place that same day. He did this in connection with some statements he made about the recognition of the People's Republic of China by "England, Burma, India, and other countries assigned to America and England." These countries were in a hurry to recognize China, Mao said, primarily because they wished to preserve their economic and political interests in China and wanted to become active against the new regime underground. Under the leadership of the United States, they were attempting to split the coalition government and create a "third power" out of the national bourgeoisie and other elements to help in overthrowing the democratic—read: Communist—regime. They already received support in this from the "right wing of the pro-American national Chinese bourgeoisie." This wing, which the Chinese Communists were seeking to isolate from the masses, thus constituted the greatest danger to China since the collapse of the KMT regime.[310]

On the basis of what has just been observed, Mao Zedong continued, the Chinese leadership would protract the approval process for the exchange of diplomatic missions with the countries that had recognized China: "We would like to allow diplomatic missions in our country . . . only after the conclusion of treaties of friendship; of military, economic, and political alliance; of credit; and, of trade, among other things, with the

Soviet Union." With the exception of the right wing of the pro-American bourgeoisie, which in this case was isolated, such treaties had the support of the entire Chinese people. Mao concluded the topic with this fundamental statement: "On the basis of the treaties with the Soviet Union, we could . . . proceed immediately with the revision and abolition of the unequal treaties that the Chiang Kai-shek government concluded with the imperialist countries." This is precisely the same argument that Mao had used in his telegram to the Peking Party leadership on January 3 to justify the treaty he was seeking; in that same communication, he gave instructions to present the same arguments to the Central People's Government Council and Government Administrative Council.[311] At the end of the talk, Mao Zedong requested that Kovalev present the problems they had addressed to Stalin on the first possible occasion and to discuss them with him.

On January 9, there was a further talk between Mao Zedong and Kovalev in which Mao discussed the treaty again, and in a similar fashion. He explained that he planned to terminate the trade agreement with the United States that was enslaving China soon. Whenever treaties would be discussed, he said, they would abide by the resolutions that the Soviet side had co-signed at the conferences at Tehran, Yalta, and Potsdam. Mao called for a treaty of friendship and alliance and agreements for a credit and about such things as trade to be concluded as soon as possible with the Soviet Union. He emphasized that China would use these treaties as the starting point for renegotiating the treaties Chiang Kai-shek's government had concluded with the capitalist states.[312]

We have considerably more Soviet than Chinese sources for the internal preparations of both sides for the upcoming treaty negotiations. The Russian Foreign Ministry archives contain copies of several drafts of the treaty of alliance and the agreements on Port Arthur, Dairen, and the CCR.[313] Some of them bear handwritten corrections.

The preparations were in the hands of a commission made up of the Politburo members Molotov and Mikoyan and the foreign minister, Vyshinskii.[314] Vyshinskii—who was neither a member of, nor a candidate for, the Politburo, and also was not a member of the Secretariat but merely a simple Central Committee member—apparently had the first draft drawn up in the Foreign Ministry after January 2.

The development leading from the first to the final draft proceeded as follows:

- The first draft available to us dates from January 5 (henceforth: "first draft").[315] It is apparently a text that was entrusted to an official in the Legal Department of the Foreign Ministry. The draft was intended to include all the significant regulations that were generally important in a treaty of friendship, alliance, and mutual assistance. The official had apparently been instructed to keep the text to the unavoidable minimum and leave out all references to any specific potential aggressor.
- On January 9, Vyshinskii received a consultants' draft that was signed by, among others, Ambassador Roshchin; Kurdiukov, the head of the First Far Eastern Department of the Foreign Ministry, which was responsible for China; and Fedorenko (henceforth: "consultants' draft").[316]
- Two of the authors of the consultants' draft presented Vyshinskii with a draft "corrected on the basis of your instructions" on January 10 (henceforth: "first Vyshinskii draft").[317]
- An undated draft followed including further handwritten corrections (henceforth: "second Vyshinskii draft").[318]

- Vyshinskii presented a very slightly corrected version of this draft to Molotov and Mikoyan on January 12 as the "draft prepared by the Ministry of Foreign Affairs [MID]"[319] (henceforth: "MID draft").[320]
- Vyshinskii sent the text, modified by corrections made by Molotov and Mikoyan, to Molotov on January 16, along with the draft of a protocol and alternatively a declaration about the CCR (henceforth: "Molotov-Mikoyan draft").[321]
- Molotov, Mikoyan, and Vyshinskii presented the final result, which included a few further corrections by Molotov, to Stalin for review *(na rassmotrenie)* on January 22, along with one protocol each about Port Arthur, Dairen, and the CCR, as well as, drafts of other documents, as "the drafts prepared by the commission," and had already given them the form of a draft resolution for the Central Committee of the CPSU(B) (henceforth: "commission draft").[322]

The procedure we have described demonstrates, as we might have expected, that Molotov bore primary responsibility after Stalin for the treaty negotiations. It can be assumed that he was the one who ultimately put the Molotov-Mikoyan draft in the form of the commission draft. In the Politburo, which at the time had twelve members, Molotov and Mikoyan, along with Stalin, were primarily responsible for the Soviet-Chinese negotiations, followed by Beriia, Malenkov, Kaganovich, and Bulganin. Foreign Minister Vyshinskii also played a prominent role.

Let us now turn to a detailed discussion of the production of the Soviet draft. As an analysis of the contents and handwritten marginalia shows,[323] the draft was modeled on the Soviet-Chinese Treaty of Friendship and Alliance of August 14, 1945,[324] and the first-generation bilateral treaties of alliance that the USSR concluded with Eastern European states between 1943 and 1949.[325] With respect to the latter, the marginalia refer to the treaties concluded with Romania (February 4, 1948), Hungary (February 18, 1948), and Bulgaria (March 18, 1948) in particular.[326] In one case, a provision in the Soviet-Mongolian Treaty of Friendship and Alliance of February 27, 1946, served as the model.[327]

No reports on the discussions of the individual drafts have yet been made available to us. The partially handwritten suggestions for corrections and the actual corrections of individual passages or sentences allow us, however, to reconstruct the production of the Soviet draft and draw conclusions about the thought processes that went into it. We begin our analysis of the drafts with a few thoughts on the title of the treaty and the designation of the partner states in the text of the treaty.

In almost all drafts, the title of the treaty is given as the "Treaty of Friendship, *Cooperation*, and Mutual Assistance." It thus had the same title as the treaties concluded with the Eastern European states between 1943 and 1949.[328] The title read only "Treaty of Friendship, *Alliance*, and Mutual Assistance" in the final commission draft of January 22, which was edited by Molotov.[329] This title was ultimately adopted in the final version the two parties signed on February 14, 1950. The Sino-Soviet treaty of August 14, 1945, bore the title "Treaty of Friendship and Alliance." The title of the February 14, 1950, treaty thus constitutes a combination of the designations given to the treaties concluded by the USSR with the Eastern European states, on the one hand, and the Sino-Soviet treaty of August 14, 1945, on the other.

Stalin and Mao clearly did not want to use a designation that sounded in any way less binding than the one used in the Sino-Soviet treaty of 1945. The term used there, "alliance" (in Russian: *soiuz*; in Chinese: *tongmeng*), suggests a tighter bond than "cooperation."[330] It is possible that the Chinese side had expressed a strong interest in retaining this term before January 22. Mao Zedong had used the designation "treaty of friendship, alliance, and mutual assistance" in a talk with Stalin on December 16, 1949,[331] but had then spoken of a "treaty of friendship and alliance [*youhao tongmeng tiaoyue*]" on the model of the Sino-Soviet treaty of 1945 in telegrams to the Party leadership in Peking referring to his talk with Molotov and Mikoyan on January 2, 1950.[332] In international law, it is not the title of a treaty that matters, however, but its substantive contents.[333]

We should note another marginal phenomenon that, though perhaps equally irrelevant legally, may possibly have had important consequences politically and on the atmosphere of the talks. The parties to the treaty are called the "Contracting Parties" in the Sino-Soviet treaty of February 14, 1950. In the treaties of alliance and friendship that the USSR concluded with socialist states that the author examined, the parties to the treaties are generally termed more respectfully the "High Contracting Parties." This is also true of treaties with Asian partner states such as the Mongolian PR (January 15, 1966) and Vietnam (November 3, 1978).[334] The treaty with the Mongolian PR of February 27, 1946, the agreement with North Korea of March 17, 1949, and the treaty with the GDR of September 20, 1955, constitute the exceptions.[335] They also employ the less-respectful term "Contracting Parties." In the June 12, 1964, treaty with the GDR, however, the signatories use the term "High Contracting Parties."[336] No rationale can be seen for the choice between these different terms.

But for two reasons it does not appear to be entirely arbitrary that the Sino-Soviet treaty of February 14, 1950, speaks of "Contracting Parties." First, it should not be forgotten that the more-respectful traditional term "High Contracting Parties" had been used in the earlier Soviet treaty concluded with the KMT government on August 14, 1945. Second, it is striking that all the Soviet drafts of the 1950 treaty used the same term, "High Contracting Parties." The traditional formula appears to have been abandoned in response to actions by the Chinese, who may have wanted to present themselves as revolutionary.[337]

We will essentially pass over a detailed discussion of the introductory general clauses (the preamble), the legal validity of which is disputed.[338] These clauses were rewritten repeatedly in the process of drawing up the treaty, but as a rule the changes were only stylistic. In the clauses, both parties stated their commitment to friendly and good neighborly relations, expressed their conviction that these relations would do the most to further the economic development of both states, and proclaimed their determination to collaborate in the interests of peace and general security.

Only two things stand out among the contents of the general clauses. First, Vyshinskii inserted a reference to the 1945 treaty in the consultants' draft of January 9[339] at the very beginning of the process. It read: "Developing and amending the Treaty of Friendship and Alliance . . . of August 14, 1945 . . ." (the first Vyshinskii draft).[340] He crossed out this reference again in the next version (the "second Vyshinskii draft").[341] It also did not reappear in the following versions. Vyshinskii apparently became aware in the meantime that such a statement would be like waving a red flag in Mao's face. Mao had repeatedly stated forcefully that he wanted to conclude a completely new treaty that should have absolutely nothing to do formally with the old one.

Second, the formula "in accordance with the principles and purposes of the United Nations" was inserted for the first time by hand in the second Vyshinskii draft in connection with the declaration of the two parties' desire to cooperate "in the interests of general peace and security." The formula, which also appeared in the treaties the USSR concluded with Romania, Hungary, and Bulgaria in 1948, was included in all following versions up to the final text of the treaty of February 14, 1950.

The following subjects were treated in the articles that followed the general clauses in the drafts and also in the final treaty:

1. Joint measures to prevent aggression (henceforth: "prevention of aggression")
2. Participation in international action to ensure peace and security (henceforth: "international action for peace and security")
3. Mutual guarantees of assistance should either side be attacked or become involved in warfare (henceforth: "clause of assistance")
4. Renunciation of participation in any alliance, coalition or other activity that was directed against the treaty partner (henceforth: "prohibition of alliances")
5. Renunciation of separate negotiations with Japan
6. Consultation on important international issues (henceforth: "clause of consultation")
7. Development of economic and cultural relations, mutual economic assistance (henceforth: "economic and cultural cooperation and mutual assistance")
8. The future of Port Arthur, Dairen, and the Chinese Changchun Railroad (henceforth: "Port Arthur, Dairen, and the CCR")
9. Provision for termination

In the following sections, we will discuss the settlement of the aforementioned subjects in the various drafts.

Prevention of Aggression

In the drafts, these provisions were based on Article 1, Paragraph 1 of the Soviet treaties of alliance with Romania, Hungary, and Bulgaria in 1948 and Article 3, Paragraph 1 of the Sino-Soviet treaty of August 14, 1945. In the treaties concluded with the three Eastern European states, the treaty partners committed "to take all joint action in their power to obviate any threat of renewed aggression by Germany or any other Power which might be associated with Germany." The provision in the 1945 Sino-Soviet treaty obligated both sides "to take jointly all measures in their power to render impossible a repetition of aggression and violation of the peace by Japan." The Molotov-Mikoyan draft referred in handwritten marginalia both to the three treaties concluded with Eastern European states and "Article 3, beginning" of the 1945 Sino-Soviet treaty.

This provision is found in Article 1, Paragraph 1 in all drafts. The formula employed in the treaties with Romania, Hungary, and Bulgaria essentially served as model. The first draft did not specify any potential aggressor, however, and the clause therefore only spoke of "any threat of aggression." The later drafts through and including the Molotov-Mikoyan draft retained the wording of the provision as it was formulated in the three treaties concluded with the Eastern European states. The only deviation consisted in their replacing Germany with Japan as potential aggressor. The commission draft also specified Japan as the potential aggressor, but used the formula "all measures dependent on them." The difference between the two formulas appears to have to do with style and not content.

International Action for Peace and Security

Here, too, a corresponding provision in the treaties of alliance concluded by the USSR with Romania, Hungary, and Bulgaria in 1948—specifically Article 1, Paragraph 2—provided the model. In it, both sides declared their intention "to participate, in the sincerest fashion, in all international action for insuring the peace and security of peoples and will fully contribute to the realization of these lofty aims."

This provision was included in Article 1, Paragraph 2 of all drafts. As a variation of this, a connection was established with the contents of Paragraph 1 in the first draft with a sentence that begins as follows: "To achieve this goal. . . ." Instead of "declare their intention . . . to participate," the first draft stated more definitely "will participate."[342] The consultants' draft and subsequent drafts reverted to the original formulation used in the treaties concluded with the three Eastern European states. Vyshinskii introduced two small corrections by hand in his second draft that were adopted in all subsequent drafts but did not affect the contents.

Clause of Assistance

This provision is of central importance in all treaties of alliance. With the exception of the first draft, it was always based on the clause included in the treaties of alliance concluded between the USSR and Romania and Bulgaria in 1948. This clause reads: "Should either of the High Contracting Parties be involved in hostilities with Germany which might seek to renew its policy of aggression or with any other State which might be associated with Germany in a policy of aggression either directly or in any other way, the other High Contracting Party shall immediately extend to the Contracting Party involved in hostilities military and other assistance with all the means at its disposal."[343] The formula employed in the treaty with Hungary deviated slightly from this.[344]

The clause of assistance appeared in Article 2 in all drafts. In the first draft, it read: "The governments of the Union of SSR and the Chinese Peoples Republic undertake, in the event of a military attack on one of the Contracting Parties, to render each other every possible assistance, including military assistance." This formulation is literally equivalent to the corresponding provision in Article 2 of the Soviet-Mongolian Treaty of Friendship and Mutual Assistance of February 27, 1946.[345] In both cases, the name of the aggressor is not mentioned, and there is therefore also no mention of the aggressor's allies. The obligation to provide assistance is thus—unlike in the Soviet treaties with

Bulgaria, Hungary, and Romania in 1948—stated more comprehensively in terms of the potential aggressor. Unlike in these other treaties, however, the obligation to provide the assistance "immediately [*nemedlenno*]" is missing.

The consultants' draft and all subsequent drafts adhered strictly to the formulation in the treaties with Romania, Hungary, and Bulgaria, however, while specifying Japan rather than Germany as the aggressor and that the assistance was to be provided "immediately." These drafts were thus equivalent to the assistance clause as stipulated in the Sino-Soviet treaty of 1945, except for the fact that it did not specify any obligation to provide assistance even against states allied with Japan and that the act of being drawn into military conflict had to be the consequence of Japanese aggression.

It is striking that one sentence that was included in Article 2 of the clauses of assistance in the treaties concluded with Romania, Hungary, and Bulgaria in 1948 was missing in all drafts: "The present Treaty will be implemented in conformity with the principles of the United Nations Charter." And yet a reference of this type to the UN charter in the assistance clause included nothing to which the partner states were not in any case already entitled or obligated. It was merely designed to demonstrate that the respective pact involving the provision of assistance was consistent with the law in the UN charter. Moscow's decision not to include the phrase in the treaty with the People's Republic of China probably reflected the fact that the People's Republic of China was not represented in the United Nations at that time, as the KMT government on Taiwan represented China. If this assumption is correct, the Soviets did not apply this policy consistently. As we have already mentioned,[346] later Soviet drafts included a passage in their general clauses whereby the contracting parties would act consistently with the principles and goals of the United Nations in their attempt to collaborate closely in the interests of peace and security.[347]

Prohibition of Alliances

This clause is part of the classic repertoire used in treaties of alliance and friendship. It constitutes Article 3 in the Soviet drafts and follows the model of the treaties concluded with Bulgaria, Hungary, and Romania (Article 3) in 1948. The clause states there that both sides commit themselves "not to enter into any alliance or take part in any coalition or in any action or measures" that target the other side.

The first draft once again constituted an exception. In it the phrase "action or measures" were missing.[348] In doing so, it corresponded literally with the formulation of the clause prohibiting alliances in Article 4 of the Sino-Soviet treaty of August 14, 1945. This was no coincidence. A marginal note in the Molotov-Mikoyan draft documented that Article 4 of the 1945 treaty had been taken into account in the considerations.[349]

In all subsequent drafts, the clause prohibiting alliances appeared in the form in which it was included in the treaties of alliance concluded with the three Eastern European states.

Renunciation of Separate Negotiations with Japan

This clause was included in all the Soviet drafts. It constituted Article 4 and was based on Article 2 of the Sino-Soviet treaty of August 14, 1945. The clause stated that both sides committed themselves "not to enter into separate negotiations with Japan and not to conclude, without mutual consent, any armistice or peace treaty either with the present Japanese Government or with any other government or authority set up in Japan which do not renounce all aggressive intentions."

The first draft adopted this formulation literally and merely discarded the term "armistice," which had become obsolete with the end of the war. The consultants' draft only mentioned the obligation "not to enter into negotiations with Japan or conclude, without mutual consent, any peace treaty either with the present Japanese government or with any other government or other authority that could be set up in Japan." This involved a greater restriction on policies the parties to the treaty could pursue with Japan, as according to the new formulation separate negotiations or the conclusion of a separate peace treaty were not permitted under any circumstances, even if Japan should "clearly renounce all aggressive intentions."

In his first draft, Vyshinskii provided an abbreviated version of Article 4 that discarded further padding and reduced the provision to its basic substance. It obligated the parties to the treaty clearly and simply "not to enter into negotiations with Japan or conclude any peace treaty without mutual consent." The second Vyshinskii, MID, Molotov-Mikoyan, and commission drafts then also adopted the article in this form.

Clause of Consultation

This provision also was a classic element in treaties of alliance or friendship. It was included in Article 5 in all drafts. The first drafts followed the clauses included in Article 4 of the Soviet treaties with Romania, Hungary, and Bulgaria in 1948: "The High Contracting Parties will consult together on all important international questions involving the interests of both countries." The Molotov-Mikoyan draft for the first time included the following handwritten passage to supplement the previous text: "with a view to strengthening peace and general security." The commission draft adopted this wording.

Economic and Cultural Cooperation and Mutual Assistance

The clause of consultation was followed in all drafts by Article 6, in which the contracting parties agreed to engage in cultural and economic cooperation, to respect each other's national sovereignty and territorial integrity, and not to intervene in each other's domestic affairs.

There are models for this in the Soviet treaties with Romania, Hungary, and Bulgaria in 1948. In Article 5, which employed almost the same wording, the contracting parties declared that they "act in a spirit of friendship and cooperation for the further development and strengthening of the economic and cultural ties [*sviazi*] between the two States, in accordance with the principles of mutual respect for their independence and sovereignty and of noninterference in the domestic affairs of the other State."[350]

A comparable provision was also included in the Sino-Soviet treaty of August 14, 1945, however. The parties agreed in Article 5 that "having regard to the interests of the security and economic development of each of them," they would "work together in close and friendly collaboration after the coming of peace and to act according to the principles of mutual respect for their sovereignty and territorial integrity and of noninterference in the internal affairs of the other High Contracting Party." Article 6 stated that the contracting parties "agree to render each other every possible [*vsiu mozhnuiu*] economic assistance in the postwar period with a view to facilitating and accelerating reconstruction in both countries and to contributing to the cause of world prosperity."

This subject was treated in Article 6 of all drafts. The first draft stated that the contracting parties, "convinced that the interests of security and prosperity [*protsvetaniia*] of the Soviet and Chinese peoples require developing and strengthening a solid and lasting friendship, will secure friendly cooperation [*sotrudnichestvo*] between the two countries in accordance with the principles of mutual respect of their independence and sovereignty, and of nonintervention in the domestic affairs of the other state." It is striking that whereas the general subject was cooperation, as in the Sino-Soviet treaty of 1945, it did not mention cultural assistance in particular—unlike the 1948 Soviet treaties with the three Eastern European countries—and "every possible economic assistance," as in the treaty of 1945. There was also no reference to the signatories' respect for each other's territorial integrity (Article 5 of the Sino-Soviet treaty of 1945).

The consultants' draft supplied some of the missing pieces and formulated the provision, as follows: The contracting parties "declare that they shall develop and strengthen the economic and cultural ties [*sviazi*] between the two states in accordance with the principles of mutual respect for each other's independence and sovereignty and nonintervention in their domestic affairs." And yet, here, too, there was no additional obligation to provide mutual economic assistance or any reference to their respect for each other's territorial integrity.

After "will establish. . . ," the first Vyshinskii draft added "and will render each other all possible [*vsiu mozhnuiu*] economic assistance." The passage was inserted on Vyshinskii's initiative. As can be determined from a handwritten marginal comment on the consultants' draft, this was a conscious reference to Article 6 of the 1945 Sino-Soviet treaty.[351] It is difficult to understand why the Soviet side had not planned to use this formula from the very beginning; after all, they had already agreed to provide Soviet assistance to China during Mikoyan's visit to China in January and February 1949 and confirmed the decision during Liu Shaoqi's stay in Moscow during the summer of 1949.

A reference to respect for territorial integrity was now also introduced into the second Vyshinskii draft. This passage, which was corrected by hand, read "observing the principles of mutual respect for state sovereignty, territorial integrity, and nonintervention in the domestic affairs of the other Contracting Party." The MID draft, the Molotov-Mikoyan draft, and the commission draft all adopted the provision in this form.

Port Arthur, Dairen, and the CCR

The most delicate subject in the treaty involved the future resolution of the three great Manchurian issues: Port Arthur, Dairen, and the CCR. An analysis of the various drafts of the treaty of alliance and related agreements reveals that the Soviet side was very uncertain about these issues.

With the necessary brevity, let us look back to August 14, 1945, on which day the KMT government felt compelled to sign the Treaty of Friendship and Alliance with the Soviet Union, which Stalin had agreed to conclude in the Yalta agreement on February 11, 1945, with Roosevelt and Churchill. Three agreements were signed on the same day as the August 14, 1945, treaty. According to one of these, both sides were to use Port Arthur as their exclusive naval base and the USSR was to be responsible for its defense. In a further agreement and a protocol, Dairen was declared an international free port, China was obligated to lease half of the port facilities and equipment to the USSR free of charge, and it was established that the harbormaster was to be a Soviet and his deputy a Chinese citizen. A third agreement established that the CCR belonged in equal parts to China and the Soviet Union. According to the agreement, the president of the board of directors, which was to be made up half of Chinese and half of Soviet citizens, was to be a Chinese, and the vice president was to be a Soviet citizen. The opposite was true of the manager of the railroad: he was to be a Soviet citizen while his assistant was to be a Chinese. The three agreements were to remain in effect for thirty years.[352]

The Chinese Communists, with whom Stalin had not cooperated during the negotiations, had been depressed about their results, because they—not unlike the KMT government—considered them to be an attack on China's sovereignty in the traditions of czarist colonialism. Stalin had made offers to Mao Zedong to withdraw the Soviet troops from Port Arthur through Mikoyan in February 1949 and to Liu Shaoqi in the summer—after the conclusion of a peace treaty with Japan, but also immediately if the Chinese side desired this. Both sides, Stalin stated, could use the free port of Dairen until the new government was established. In his December 16, 1949, talk with Mao, Stalin expressed the desire to retain the agreement de jure in order to respect the Yalta agreement. At the same time, he had repeated his offer to withdraw the Soviet troops de facto "at the request of the Chinese government"—this time without mentioning a peace treaty with Japan. With respect to the CCR agreement, which Mikoyan had described as not being unequal in early 1949 in Xibaipo, Stalin said on December 16 that they could act in a similar fashion and allow it to remain in force formally while "changing the respective points in fact." On December 16, Mao had replied—much as he had to Mikoyan in February 1949—that the current situation fully reflected Chinese interests given China's weak defensive capabilities; that it would not be useful to withdraw the Soviet troops too soon. Dairen was not mentioned in this talk.[353]

If this was the status of the Sino-Soviet talks at the beginning of January 1950, the settlement the Soviets proposed in the first draft of the treaty is perfectly consistent. Article 7 stated simply that both sides recognized that the agreements between the Soviet Union and the Chinese Republic concluded on August 14, 1945, on the CCR, Port Arthur, and Dairen harbor remained in effect.[354] According to this version, the Soviet troops were, so to speak, tacitly supposed to be withdrawn before the end of the thirty-year period.

Article 7 of the consultants' draft discussed Port Arthur and Dairen in a similar but somewhat more strongly worded provision: the parties to the treaty "declare that in accordance with the present Treaty of Friendship, Cooperation, and Mutual Assistance . . . the Agreement on the harbor of Dairen of August 14, 1945 and the Agreement on Port

Arthur of August 14, 1945, with all of their appendices, remain fully in effect [*sokhraniaiutsia v polnoi sile*] as an inseparable part of the present treaty."[355]

The consultants' draft did not mention the CCR. This leads us to conclude that they already intended at that time to treat this subject separately. No draft of this kind was to be found in the Russian Foreign Ministry archives. There was also no reference to any such draft in the letter addressed to Vyshinskii on January 9 that accompanied the consultants' draft.[356]

At this moment, the Soviet leadership changed its opinion about the treatment of Port Arthur and Dairen. On Vyshinskii's instructions, the existing Article 7 was crossed out.[357] According to the wording of the version that followed in the first Vyshinskii draft of January 10, the contracting parties agreed to examine the issue of the existing agreements on Dairen harbor and Port Arthur benevolently following the conclusion of a peace treaty with Japan. A second paragraph was planned that would call for the Soviet forces that were currently in place in Port Arthur and Dairen to be reduced or withdrawn after a period of two, three, or more years after the treaty came into effect.[358] The CCR once again was not mentioned. And no draft of a CCR agreement was stored at this place in the archive or mentioned in the January 10 letter to Vyshinskii that accompanied the draft of the treaty discussed in this paragraph.[359]

The second Vyshinskii draft introduced a further revision. Above all, it no longer linked the planned assessment of the agreement chronologically with the conclusion of a peace treaty with Japan. In it, Article 7 read: "The High Contracting Parties have agreed that the Soviet military units currently in place in Port Arthur and Dairen will be withdrawn after a period of 2-x-3-x [handwritten correction: 2–3] years from when this treaty takes effect and that at the end of the aforementioned period, the agreements on Port Arthur and Dairen harbor concluded between the USSR and China on August 14, 1945, will be examined [handwritten supplement: and also the protocol[360] of the agreement on Dairen harbor shall be examined]." This draft did not address the issue of the CCR, either.

The MID draft of January 12 adopted the settlement of the issues of Port Arthur and Dairen included in the second Vyshinskii draft with the handwritten corrections. The CCR continued not to be mentioned.[361]

Article 7 of the Molotov-Mikoyan draft of January 16 called for the Soviet troops "in the course of a period of 2–3 years from when this treaty takes effect" to be "withdrawn completely [*polnost'iu*] to the territory of the USSR," and for the troop withdrawal to begin in 1950.[362] On the issue of when the agreements were to be examined, this draft returned to the first Vyshinskii draft, as it was agreed "that the Agreements concluded between the USSR and China on August 14, 1945, on Port Arthur and the harbor of Dairen and on Port Arthur and the Protocol to the Agreement on the harbor of Dairen shall be subjected to renewed examination following the conclusion of a peace treaty with Japan."[363]

A protocol on the confirmation of the validity of the agreement on the Chinese Changchun Railroad, described as the "first variant," and a "declaration" on the same subject (the "second variant") are included in the appendix. Both versions established that "the validity of the Agreement on the Chinese Changchun Railroad, concluded on August 14, 1945, in Moscow, is confirmed in its entirety for the period indicated in this Agreement." The phrase "in its entirety" was crossed out by hand in the declaration.[364]

The commission draft adopted the contents of the resolution of the issues involving Port Arthur and Dairen as arranged in the Molotov-Mikoyan draft. The members of the commission did, however, remove it from the treaty (the earlier Article 7 now included only the final clause) and changed it into a document titled "Protocol concerning the Soviet-Chinese Agreements on Port Arthur and the harbor of Dairen."[365] A second protocol dealt with the Soviet-Chinese agreement on the Chinese Changchun Railroad.[366] It repeated the observation that had been made in the aforementioned protocol—or in the declaration—of the Molotov-Mikoyan draft, but with two modifications. First, the new version did not use the phrase "in its entirety." And second, it added a half sentence stating that "in the interests of guaranteeing the successful operation of the railroad" both treaty partners could add "a few supplementary provisions."[367] The removal of the provisions from the text of the treaty and treatment of the question in separate protocols had no influence on their legal validity.[368]

An analysis of these drafts of the treaty leads us to the following conclusions about the process by which the Soviet leadership formed its opinion on the complex of issues involving Port Arthur, Dairen, and the CCR between January 5 and January 20, 1950. The Soviets initially expected only to confirm the validity of the 1945 agreements on all three subjects in the future treaty. Any troop withdrawal, to the extent that it was planned at all, was clearly supposed to take place de facto and without any modification of the treaty as Stalin had already proposed. By January 10, this opinion involving Port Arthur and Dairen took the explicit form that the Soviet troops would be withdrawn within two to three or more years after the treaty became effective and that the 1945 agreements would be examined following the conclusion of a peace treaty with Japan. The Soviet leaders subsequently adopted the position that the agreements should be examined at the end of this period and not following the conclusion of a peace treaty with Japan. By January 16, they set the deadline for withdrawing the troops definitively at two to three years and for beginning the withdrawal at 1950. At the same time, they returned to their old position that the examination of the agreements should be linked chronologically to the conclusion of a peace treaty with Japan. At this time, they were still insisting that the validity of the 1945 agreement on the CCR should be confirmed. It was not until the final draft was submitted on January 22 that they yielded somewhat on this settlement by supplementing the existing agreement slightly to give both parties the right to request "supplementary provisions" designed to ensure the business success of the railroad.

Final Clause

The first draft called for the treaty to be ratified and the documents ratifying the agreement to be exchanged in Peking as soon as possible. All subsequent drafts adopted the contents of this provision. The treaty was to remain in effect for thirty years. If neither of the parties gave notice that it wished to terminate the treaty one year before it ceased to be in effect, it was to remain in effect for an unlimited period and be subject to termination on one year's notice.

The first draft followed the model of the Sino-Soviet treaty of 1945 in setting the period of duration and the procedures for giving notice. The treaties of alliance concluded with Bulgaria, Romania, and Hungary in 1948, however, were to remain in effect for twenty years and to be extended for five years each time if notice had not been given to terminate them one year before the end of such periods.

The consultants' draft retained the provisions of the first draft dealing with notice, but reduced the term of the treaty to twenty-five years. All subsequent drafts again called for the treaty to remain in effect for the original period of thirty years.

The Soviet side, in the meantime, had to deal with the question of whether and, if so, to what extent the treaty to be concluded with the People's Republic of China should absorb the old one concluded with KMT China on August 14, 1945. Vyshinskii consulted with an expert on international law, Vsevolod Durdenevskii, on this question. He brought an earlier report on the subject up to date and sent it to Vyshinskii on January 14. Durdenevskii came to the conclusion that even if a new treaty dealt with a subject matter that was very similar to one that had been concluded earlier or agreed with it, the old treaty remained in effect unless it was explicitly declared obsolete or notice was given. To avoid any misunderstanding, Durdenevskii recommended at the end of his report that both sides agree "that those portions of the earlier treaty that did not contradict the new treaty should remain in effect." This could be done in the text of the new treaty, in the signing protocol, or in a separate exchange of notes.[369]

Vyshinskii adopted Durdenevskii's recommendation. The materials included in the January 16 Molotov-Mikoyan draft included a draft of a "clause in the signing protocol." It read: "By signing the present treaty, the High Contracting Parties declare that the conclusion of the present treaty does not in any way violate the obligations that the USSR and China have accepted on the basis of treaties previously concluded between them."[370]

It is unclear what Vyshinskii expected to achieve with this clause, which did not reappear later. One can assume, however, that he was not pursuing any concrete goal with it, because Durdenevskii's letter to him stated that "the material and the possible conclusion to be drawn from it" had proved "*more favorable* to our position" than he had assumed on the basis of his recollections.[371]

Molotov, Mikoyan, and Vyshinskii submitted the final drafts of the treaty and the two agreements on Port Arthur, Dairen, and the CCR, along with nine other drafts, to Stalin on January 22 as the commission draft. They sent copies to Beriia, Malenkov, Kaganovich, and Bulganin, the Politburo members who, with Stalin, Molotov, and Mikoyan, were responsible for preparing the negotiations with China—and later for the negotiations themselves. The letter to Stalin began with this sentence: "We are submitting the following drafts on the China questions prepared by the commission for your review [*na Vashe rassmotrenie*]." The list of the twelve titles of the drafts included was accompanied by another equivalent list drawn up in the form of a draft of a "resolution of the Central Committee of the CPSU(B) to be signed by the secretary of the Central Committee."[372]

The term "secretary" was used in this context for Stalin at that time—an only apparently modest form of his actual title of secretary-general. We do not know whether Stalin actually signed the resolution, as the document cited was only a copy of the version sent to him. The original of the document, which has not yet been made available, should be in the presidential archives.

The two lists included the titles of the drafts of the following documents:[373]

1. Treaty of Friendship, Alliance, and Mutual Assistance
2. Protocol concerning the agreements on Port Arthur and Dairen
3. Protocol concerning the agreement on the CCR
4. Agreement on a credit from the Soviet Union to the People's Republic of China to pay for equipment and other goods delivered to China
5. Resolution of the Council of Ministers of the USSR on establishing a Soviet-Chinese joint-stock air transport company
6. Resolution of the Council of Ministers of the USSR on establishing a Soviet-Chinese joint-stock company (Sovsin'metall) for mining nonferrous and rare metals in Sinkiang
7. Resolution of the Council of Ministers of the USSR on establishing a Soviet-Chinese joint-stock oil company (Sovsin'neft) in Sinkiang
8. Agreement on the exchange of goods and payments
9. Resolution of the Council of Ministers of the USSR on trade between the Soviet Union and Sinkiang
10. Agreement on the terms for the payment of Soviet specialists sent to work for institutions, organizations, and companies in China
11. Resolution of the Council of Ministers of the USSR on property administered by Soviet organizations and mixed Soviet-Chinese companies in the territory of Manchuria, in Sinkiang, and the on Liaodong Peninsula
12. Resolution of the Council of Ministers on sending a group of Soviet professors and teachers to China.

The letter to Stalin stated that the first three documents were to be published. They planned also to sign a second protocol dealing with the CCR that would not be published and would clarify "some practical questions (among other things the deadline for the reestablishment of joint administration of the railroad)." The commission would also present a draft of this. They would have to examine separately whether the text or contents "of any of the other documents listed above" should be published.[374] As is clear from the aforementioned formula, the Soviet leadership wanted, in the second protocol, to obligate the Chinese side to restore the parity in the management of the railroad, which had been modified de facto in China's favor some time before.

Only the first three of the twelve draft documents mentioned in the list were to be found in the Russian Foreign Ministry archives—the ones for the treaty of alliance and the protocols on Port Arthur, Dairen, and the CCR that were described above as part of the commission draft.[375]

We do not know whether Stalin examined the commission draft on January 22 before he met with Mao and Zhou the same day. If he did, it is striking that he did not always present arguments on the key issue of Port Arthur that were consistent with this draft in this meeting.[376]

Zhou Enlai's Negotiating Team

Prime Minister Zhou Enlai, who had been entrusted by the Peking Party leadership with leading the negotiations, also participated in the talk on principles that began the negotiations on January 22. Zhou had arrived in Moscow with the Chinese negotiating delegation two days earlier.

When Mao Zedong made the breakthrough in his talk with Molotov and Mikoyan on January 2 and had received Soviet approval of his desire to include Zhou Enlai in the upcoming negotiations, he gave the Party leadership the first instructions about Zhou's delegation that same evening. Preparations for the trip were to be completed within five days. He hoped that Zhou Enlai would leave Peking by train on January 9 along with the minister of trade, Ye Jizhuang, the other colleagues, and all necessary documents and materials. In Zhou's absence, his deputy, Dong Biwu, should act as prime minister. They should only inform the public about Zhou's trip and Dong's nomination after Zhou had arrived in Moscow. Mao asked to be informed if Peking needed one or two more days to prepare the trip and if Li Fuchun "and other comrades" would arrive in Moscow with Zhou.[377]

Mao then agreed with the Soviet leadership that Zhou would arrive with about seventeen people and that his train would leave Peking on January 9. He recommended to the Party leadership in Peking that they also send a high-ranking delegation from Sinkiang to Moscow to arrive either on the same day as Zhou, or two or three days later.[378]

Zhou's government delegation left Peking on January 10, one day later than planned. It included the following individuals:[379] Li Fuchun, vice-secretary of the Northeast Bureau of the CCP, vice-chairman of the People's Government of Manchuria, and member of the Financial and Economic Commission of the Government Administrative Council; Ye Jizhuang, minister of trade and member of the Financial and Economic Commission of the Government Administrative Council; Wu Xiuquan, head of the Section for the Soviet Union and Eastern Europe of the Foreign Ministry; Lai Yali, acting head of the Ministers' Office of the Foreign Ministry; Cai Shufan, head of the Planning Staff of the Ministry of Industry; Shen Hong, head of the Central Planning Staff for Finances and the Economy; Su Nongguan, secretary of the Ministry of Foreign Trade responsible for confidential information; Lü Dong, vice-minister of industry of Manchuria; Zhang Huadong, vice-minister of trade of Manchuria; Ouyang Qin, secretary of the Municipal Committee of Dairen and Port Arthur; Cheng Mingbi, head of the Department of Electricity of Manchuria; Chang Yanqing, head of the Ministers' Office of the Foreign Ministry of Manchuria; Wang Xun, vice-director of the Manchurian Anshan Steelworks; Nie Chunrong, deputy head of the Department of Machinery of Manchuria; and Luo Wei, head of the Planning Staff of the Coal Office of Manchuria.

It is striking here, first, that Zhou's delegation—just as in the case of Mao's entourage—did not include a single Politburo member with the exception of Zhou himself. Second, the large proportion of functionaries from Manchuria leaps to the eye. This reflected the extraordinary importance of this region for relations between Moscow and Peking, which it had because of the Soviet Union's special rights there. At the same time, it emphasized the key role ascribed to Manchuria in the development of a modern Chinese industry with Soviet assistance. Finally, it is remarkable that Mao did not order the number one man in Manchuria, Gao Gang, to come to Moscow—presumably a sign of Mao's growing distrust of Gao, who, as Mao would learn at the end of January, had been delivering secret reports on the situation in the Chinese Party leadership to Stalin through Kovalev since, at the latest, the fall of 1949.[380]

Deng Liqun and Saifudin, who had traveled from Sinkiang and reached Moscow on January 30, after the delegation had arrived from Peking, joined Zhou's delegation. Saifudin met with Mao the same day for a first talk about questions involving Sinkiang.[381]

Deng was secretary-general of the CCP in Sinkiang, but most probably appeared during the negotiations in Moscow in his state capacity as a member of the Government Council of Sinkiang, head of the Department of Foreign Affairs of the People's Government of Sinkiang, and head of the Education and Cultural Department of the same government.[382] The Uighur Saifudin (Uighur: Säjpidin Äzizi; Chinese: Sai-fu-ding; Russian: Azizov) was vice president of the People's Government of Sinkiang and deputy commander of the Sinkiang Military Region. He had apparently joined the Russian CP during his studies in the USSR and had held the position of education minister in the government of the anti-Chinese and pro-Soviet Republic of Eastern Turkestan established in 1944–45. In the fall of 1949, he had taken the side of the Chinese CP and was invited to Peking to attend the Chinese People's Political Consultative Conference in September. In strict secrecy, he was transported on Soviet instructions with a Kazakh and a Han Chinese in a Soviet aircraft from Sinkiang via Alma-Ata and Irkutsk to Peking. Another delegation under the leadership of Ähmät (Akhmedian) was originally supposed to represent Sinkiang at the Political Consultative Conference. The Soviet aircraft carrying it crashed under mysterious circumstances in the vicinity of Lake Baikal in late August, however.[383] When the People's Republic of China was established, Saifudin was elected to the People's Central Administrative Council and the Nationalities Commission of the Government Administrative Council. He was accepted into the CCP only after his return from Moscow at the end of February 1950. He could read Chinese at that time, but he could not write it.[384]

Mao had asked the Central Committee in Peking to send the president of the People's Government of Sinkiang, Burhan (Bao Erhan), to Moscow in addition to the head of the Party, Deng Liqun.[385] It is not known why the Party leadership did not fulfill this request and decided to send Burhan's deputy, Saifudin.

At the Soviet border, Zhou Enlai was welcomed by the administrative head of the Chita region and the general secretary of the Soviet Foreign Ministry; the latter then accompanied Zhou back to Moscow. Zhou called Mao Zedong by telephone from Novosibirsk and Sverdlovsk to discuss the upcoming negotiations. He prepared the speech that he planned to give after his arrival in Moscow and gave it to Wu Xiuquan to be translated. Ambassador Wang Jiaxiang traveled to Iaroslavl' to meet the delegation.[386]

On its arrival in Moscow on January 20, the Chinese delegation was welcomed at the Iaroslavl' station by the Politburo member Mikoyan, who appeared as vice-chairman of the Council of Ministers of the USSR. Mikoyan's position in the de facto hierarchy was below Molotov, who had welcomed Mao. Foreign Minister Vyshinskii, who appeared in uniform, Foreign Trade Minister Men'shikov, Ambassador Roshchin, a vice-chairman of the Moscow Soviet, the Moscow city commander, Lieutenant General Sinilov, and the head of the China Department of the Foreign Ministry, Kurdiukov, also came. Zhou Enlai was described by the Soviet protocol as "premier of the Government Administrative Council and foreign minister." Wang Jiaxiang and Chen Boda were also present, as were the ambassadors of all the Eastern European satellite states, North Korea, and the Mongolian PR.[387]

Pravda devoted a prominent article with a photograph to Zhou Enlai's arrival. It filled about a quarter of the front page. The photograph depicted Zhou delivering his speech into a microphone surrounded by, among others, Vyshinskii, Mikoyan, and Wang Jiaxiang. *Renmin ribao* also published a report on its front page immediately afterward. An article that appeared later stated that the Soviet people everywhere were enthusiastic about Zhou's arrival. A new page was being opened in the history of Sino-Soviet friendship.[388]

In his speech, Zhou emphasized that he had come on instructions from Mao Zedong to participate in negotiations for "good neighborly relations between the two great states of China the Soviet Union." It is striking that he limited the period of friendly Soviet behavior toward China to "a number of years." On the other hand, he saluted the Soviet dictator by attributing this behavior to the "consistent Leninist-Stalinist" policy of providing assistance to all oppressed peoples and confirmed that "Generalissimo Stalin" had followed a correct foreign policy. The fact that Zhou praised China and the USSR three times as "two great states" reveals self-confidence and the need to insist on equality and equal rights in the upcoming relations. At the end of his short address, Zhou stated that no power would be able to hinder the two states from "striding forward hand in hand." He saluted Mao a second time when he quoted Mao, who had celebrated the "friendship and cooperation between China and the Soviet Union."[389]

Mao, Zhou, Li Fuchun, and Wang Jiaxiang met late that afternoon in Mao's dacha to establish guidelines for the upcoming negotiations. Because Zhou was lodged at a very great distance from Mao's dacha, Zhou suggested that he should move in with Mao, which he then did the following day.[390]

On January 21, the Soviet Union celebrated the twenty-sixth anniversary of Lenin's death. The photograph of the main memorial service published on the front page of *Pravda* showed the Soviet leadership on the stage of the Bolshoi Theater. The only foreigners that can be recognized were Mao Zedong and Zhou Enlai. Mao stood in the second row between Bulganin and Mikoyan, and Zhou in the same row between Mikoyan and Aleksei Kosygin. Another ten members of Mao's and Zhou's delegations were present in the hall. There was a half-hour talk among Stalin, Mao, and Zhou at a reception, "during which they agreed on the spirit of the treaty they were to conclude."[391] That evening, Mao watched a film about Lenin with the Soviet Politburo members.[392]

Setting the Stage for the Negotiations: The January 22 Talk

A two-hour talk on principles at the highest level on the evening of January 22 constituted the transition from the preparations for the negotiations to the negotiations themselves. Participating on the Soviet side were Stalin, Molotov, Malenkov, Mikoyan,

Vyshinskii, and Fedorenko; on the Chinese side, Mao Zedong, Zhou Enlai, Li Fuchun, Wang Jiaxiang, Chen Boda, and Shi Zhe.[393] The Soviet press merely reported that Stalin had received Zhou Enlai on January 22 and that Ambassador Wang Jiaxiang and Foreign Minister Vyshinskii had also been present.[394]

Stalin began by distinguishing two issues that should be discussed: the existing agreements [*sushchestvuiushchie soglasheniia*] and current events, involving, among other things, Manchuria and Sinkiang. He proposed beginning with the first issue. "We are of the opinion that we need to change these agreements [*soglasheniia*], although we previously thought that they could be left as they were." It was necessary to change the agreements, including the treaty, because the treaty was based on the struggle against Japan. Now that the war was over and Japan had been beaten, the treaty constituted an anachronism. He asked the Chinese side to take a position on the "treaty of friendship and alliance."[395]

In this way, Stalin confirmed that he had changed his opinion, which he had already signaled on January 2 through Molotov.[396] On December 16, 1949, Stalin had still insisted to Mao that they should not change "any of the points of this treaty for the time being" out of concern for the Yalta agreement. By "treaty" he had meant the entire set of agreements and thus also the supplementary agreements on Port Arthur, Dairen, and the CCR. The reason he gave on January 22 for changing his opinion is not convincing. The treaty of August 14, 1945, did in fact also include an obligation for both sides to fight Japan together until they had achieved "final victory"—but at that time the end of the war was already in sight and was achieved with the capitulation of Japan six days after the treaty was concluded.[397] The treaty thus dealt primarily with the time following the war and was to remain in effect for a period of thirty years. The real reason for Stalin's change of opinion has already been described in detail.[398]

Mao Zedong declared on January 22 that the Chinese side did not possess a fully completed draft, only some outlines. Stalin replied that they could begin by exchanging opinions and then prepare a corresponding draft. Mao then stated that he would find it useful to consolidate the existing friendly relations by means of treaties and agreements. This would elicit a positive response in China and also in the international community. A "treaty of alliance and friendship" should establish everything that would guarantee the flourishing of both states and prevent the repetition of Japanese aggression. It was entirely conceivable, he said, that the imperialist countries would attempt to keep them from flourishing.[399]

Stalin agreed. The Japanese still possessed cadres and would certainly raise their heads again—particularly if the Americans continued to pursue their current policy.[400]

Mao continued by saying that two points should distinguish the future from the existing treaty. The KMT had only talked about friendship. Now, however, "all the conditions for *real* friendship and cooperation" were present. Furthermore, whereas previously there was talk of cooperation in the war against Japan, they now had to think in terms of preventing aggression from Japan. The new treaty had to include issues of political, economic, cultural, and military cooperation. "But the most important issue will be economic cooperation."[401]

There followed a brief discussion about the way the treaty of August 14, 1945, should be changed. Stalin and Mao agreed that the provision included in Article 3 of the current treaty could be dropped. According to that provision, this article would remain in effect until both sides asked to have the responsibility of preventing further Japanese aggression given to the United Nations.[402]

In the first two paragraphs of Article 3, both sides had committed themselves to prevent any renewed aggression by Japan together after the end of the war and provide mutual assistance in the event of a Japanese attack. The provision that Stalin and Mao agreed to cancel constituted the third and last paragraph. This clause had already been canceled in the internal Soviet commission draft.[403] The primary reason for canceling this provision apparently was that the People's Republic of China was not a member of the United Nations.

When Stalin asked what provisions should be included in the new treaty, Mao Zedong mentioned "consultation on international problems." This, Mao said, would strengthen the position of the CCP, as there were objections within the Chinese national bourgeoisie to the policy of rapprochement between China and the Soviet Union on questions of international relations. Stalin agreed.[404] The internal Soviet commission draft already included such a clause of consultation. It constituted Article 3 in that document.[405]

Stalin now addressed the most delicate topic. He asked what suggestions Mao wished to make on the issue of the Chinese Changchun Railroad (CCR).[406]

Despite Stalin's question, the following conversation did not initially focus on the CCR but on Port Arthur and Dairen. It constituted a process of careful mutual probing, which reveals that both sides were conscious of the delicate nature of the subject. In the process, Mao generally tended to be vague, probing, defensive, and at times contradictory. Stalin tended to insist on clarity. Mao only took a firm stand in the case of the CCR. The statements of both sides do not always allow us to draw clear conclusions about their underlying motives, so at best we can only speculate about them.

We should remember here that Stalin had already made an offer to the CCP leadership—during Mikoyan's talks in Xibaipo at the beginning of February 1949, about a year earlier—to abolish the agreement on Port Arthur, which he described as unequal, following the conclusion of a peace treaty with Japan or, if the Chinese side wished, to withdraw the Soviet troops immediately.[407] Stalin thus at least appeared to have realized at that time that he could not afford to commit a Communist China to an agreement he had forced on the KMT government. Years later, Mao described to the Soviet ambassador how he had assessed Stalin's position at that time: "You had Port Arthur and Dairen under control. Why did you take control of them? Because it was the China of the KMT then; you later left of your own account, as it was then the China led by the CP."[408] In Xibaipo, Mao had spoken out against a rapid troop withdrawal and against a quick revision of the treaty.[409] In July of that year, Stalin had described the treaty of August 14, 1945, to Liu as unequal and stated that he was prepared to withdraw the troops after conclusion of a peace treaty with Japan or, if the Chinese Party leadership wished, immediately.[410] Liu's reply revealed that the CCP was interested in a Soviet troop withdrawal only after China had created its own coastal defenses.[411]

When Stalin insisted to Mao on December 16, 1949, that they should not change any of the points of this treaty for the time being, he meant the entire set of agreements and thus also the supplementary agreements, as we can see from the fact that he explicitly wished to retain the agreements on Port Arthur and the CCR. At that time, however, he offered to withdraw the troops de facto, in response to a Chinese proposal, and modify individual points in the CCR agreement, taking into account Chinese wishes. On December 16, Mao had opposed withdrawing the troops too soon. He apparently also stated that the existing situation with respect to the CCR reflected Chinese interests—which, according to everything that we know, was not his real opinion. The disagreement on December 16 had thus consisted in the fact that Stalin wanted to retain the set of agreements de jure but was willing to withdraw the troops de facto and apparently also to allow certain modifications with respect to the CCR. Mao, on the other hand, advocated an immediate abolition of the set of agreements and opposed the immediate withdrawal of Soviet troops—as Stalin knew. We have already described his motives.[412]

On January 22, 1950, Mao responded to Stalin's inquiry about his views on the CCR issue by stating that they should probably operate on the principle that they could retain both the agreements on the CCR and the one on Port Arthur de jure, and yet undertake de facto modifications.[413] Mao's response was surprising because Stalin had gone so far as to state at the beginning of the talk "that we need to change these agreements," whereby it was clear that he meant formal modifications–apart from the fact that Stalin, in February and July 1949, had suggested this as an alternative.

Stalin was clearly surprised and asked if this meant that Mao was willing to retain "the existing agreement" de jure—he was apparently thinking only of the agreement on Port Arthur at this moment—but wanted, as he (or the minute taker) put it somewhat stiltedly, "to employ appropriate de facto changes as the means." Mao's reaction to this appears to indicate that the position he had taken, which Stalin had reproduced correctly, was not intended seriously and was merely designed to shake the bush to see how serious Stalin had been in his earlier statements about formal changes to the agreements. Mao thus did not agree with Stalin and answered formally, stiffly, and somewhat cryptically: "We must proceed taking into account the interests of both sides—both of China and also of the Soviet Union."[414]

Stalin apparently understood this—presumably correctly—as an attempt by Mao to express his dissatisfaction with the status quo and his strong interest in also having formal changes made to the agreement. He hurried to agree with Mao ("Right," he answered) and made an observation that Mikoyan had already made in early February 1949 to the Chinese leadership but that Stalin had not repeated to Liu Shaoqi in July or on December 16 to Mao.[415] Stalin said, "We are of the opinion that the agreement on Port Arthur is unequal [*neravnopravnyi*]." Mao immediately used this observation to his own advantage and called for the revision of the agreements that followed logically from it. With apparent concern, he inquired, "But a change of this agreement goes against the resolutions of the Yalta Conference?!"[416]

Mao was repeating the argument Stalin had used on December 16, 1949, to explain why the Soviet side had decided not to change "any of the points of this treaty for the time being."[417] He provoked a vehement and at the same time unexpected reaction from Stalin, who replied: "Yes, it goes against it—well, and to hell with it [*nu, i chort s nim*]! Once we have taken the position that we change the treaties, this means that we have to go through with it. It is true that this will involve some inconvenience for us and we have to struggle against the Americans. But we have already come to terms with that."[418]

Once again, Mao suggested he was concerned: "The only thing that disturbs us about this is that it could involve undesired consequences for the USSR." Stalin felt compelled by this to make an apologetic statement. As was well known, he said, the Soviet leadership had concluded the existing treaty during the war with Japan. It had not known that Chiang Kai-shek could be taken out of the picture. It had assumed that stationing Soviet troops in Port Arthur would be "in the interests of the Soviet Union and of the cause of democracy in China." Mao did not seem to be deeply moved when he replied, "The issue is clear."[419]

The dialogue we have reproduced in the previous six paragraphs is as exciting as it is ambiguous—the latter with regard to a psychological interpretation of Mao's method of arguing. Were Mao's inquiries about the Yalta agreement simply rhetorical questions, as we have assumed above? In the knowledge of the fact that Stalin was now willing to make formal changes to the set of treaties, was Mao seeking to upset Stalin and provoke him to repudiate or, at the very least, justify the contradictory position he had held on December 16, 1949, and abandoned since January 2, 1950, by twice presenting this position to him?[420] If this was the case, was he seeking to let Stalin know implicitly how much he disapproved of the way he had previously clung to the Yalta agreement—an agreement that the Chinese Communists had always rejected because they saw that Moscow had created privileges for itself in Manchuria as a continuation of czarist imperialism?[421] Or should we agree with two Russian authors who claim that Mao had at first simply not been able to believe that Stalin had really changed his mind, had smelled a rat, and therefore focused on the Yalta agreement to make the matter clear?[422] Each of these interpretations sounds plausible in its own way. They can also be used to interpret Mao's behavior on January 9 when he stated to Kovalev that the People's Republic of China would respect the treaties that the Soviet Union had signed in Tehran, Yalta, and Potsdam in all issues involving the revision of the treaties.[423] It is not very conceivable, however, that Mao acted this way because he was plagued by scruples that the Soviet Union might be hurt by violating the Yalta agreement.

Stalin now offered two alternative solutions: "Declare that the agreement on Port Arthur remains in effect until a peace treaty is signed with Japan, after which the Russian *[sic]* troops would be withdrawn from Port Arthur. Or . . . declare that the existing agreement will be retained, but withdraw the troops from Port Arthur in fact." By this, Stalin clearly meant "withdraw immediately."[424] The Soviet side accepted both alternatives, Stalin said. By presenting the second alternative, which was identical with the alternative of December 16, 1949, Stalin made a last attempt to achieve his preferred solution of retaining the agreement for an unspecified period in the meantime, and thus took a position contrary to his initial statement that they had to change the treaty and all agreements.[425]

It is striking that neither of the alternatives Stalin mentioned was the same as any of the solutions called for in the internal Soviet treaty drafts—not even the one in the draft that Molotov, Mikoyan, and Vyshinskii presented to Stalin as the commission draft on the same day the talk between Stalin and Mao reproduced here took place, January 22. It called for the Soviet troops to be withdrawn from Port Arthur and Dairen by 1952 at the latest and for the agreement to be examined (*"rassmotrenie,"* not meaning "to expire") after a peace treaty was signed with Japan. The question thus arises whether Stalin even examined the commission draft before beginning the talk with Mao—or whether he spontaneously abandoned it in the course of the talk. In any case, Stalin's first alternative was more favorable to the Chinese in terms of abolishing the treaty and less favorable in terms of withdrawing the troops than the solution called for in the commission draft.[426]

Mao replied that they would have to consider these issues. Given his interests as described above, it is not surprising that he chose the former alternative: "We agree with the opinion of comrade Stalin and believe that the agreement on Port Arthur must remain in effect until a peace treaty is signed with Japan; it will expire after it is signed and the Soviet troops will withdraw." Mao did express the desire, however, that Port Arthur should become a base of military cooperation and that China wished to train its navy there.[427]

Stalin raised no objections and turned now to the Dairen question briefly. He declared that the USSR no longer wished to claim any rights there. When Mao asked whether Dairen should remain a free port, he replied that China would have to decide this question herself. Roosevelt had previously insisted that Dairen remain a free port. Mao asked whether it would be in the American and British interest to retain it as a free port. Stalin replied: "Naturally. It is a house with open gates."[428]

Stalin's lack of interest in Dairen may have come as a surprise to Mao. The Soviet leader did not, however, say how he thought the question would proceed—either in terms of the troop withdrawal or of the expiration of the agreement. And yet he gave the clear impression that he wished to repeal the agreement on Dairen before it expired. The internal Soviet commission draft called for the same thing as for Port Arthur: withdrawal by 1952 at the latest and examination of the agreement after the conclusion of a peace treaty with Japan.[429] Stalin's disinclination to allow Dairen to continue as a free port was clear, however.

Mao now again stated the Chinese interest in Port Arthur as a base of military cooperation with the USSR and added that Dairen should be developed into a base for economic cooperation. There were many companies in Dairen that China would not be able to run without Soviet assistance. Stalin did not discuss Dairen further. With regard to Port Arthur, he repeated: "This means that the agreement on Port Arthur will remain in effect until a peace treaty is signed with Japan. When the peace treaty has been signed, the existing treaty expires and the Russians *[sic]* withdraw their troops." He asked if he had summarized these thoughts correctly. Mao confirmed Stalin's statement: "This is the principle, and it is the very thing we would like to express in the new treaty."[430]

An important impediment was thus cleared out of the way. Mao had essentially achieved what he wanted. He had achieved his goal of terminating prematurely two agreements he considered unequal and discriminatory that had been concluded for a period of thirty years. It is difficult to judge whether he considered the fact that the Soviet troops would at first continue to be stationed in Port Arthur—albeit for a limited period of time—as a blemish that he had to accept.

Stalin had in fact already resolved to abolish the Port Arthur agreement and withdraw Soviet troops from the Liaodong Peninsula in the event of a Chinese Communist victory before Mikoyan's visit to Xibaipo in early 1949. But when the time came, he hesitated to commit himself to this publicly out of fear that this could lead Washington and London to abolish the entire Yalta agreement. This resulted in the disagreement with Mao on

December 16. On January 22, 1950, Stalin overcame his fears, albeit still hesitantly. We have already discussed in detail the probable reasons which led him to change his mind. To summarize briefly, they were Stalin's fear that his negotiations with Mao might fail and concern that there could be a rapprochement between the People's Republic of China and the Western powers. In all probability, the change of position also had something to do with the fact that the United States, as appeared to become clear after Secretary of State Dean Acheson's speech to the National Press Club on January 12, 1950, at the very latest, had resolved to disengage from the Asian mainland and come to terms with an alliance between the USSR and the People's Republic of China. In view of the last-mentioned fact, Stalin came to the conclusion that he no longer had much to lose in East Asia if he ceased to tailor his activities to the "spirit of Yalta."[431]

When the January 22 discussion of Port Arthur and Dairen had been concluded, Stalin proposed to discuss the question of the CCR. This was the most delicate question of all. Stalin revealed that he was aware of this in his introductory statement: "Tell us, as good Communists should, what doubts do you have?"[432]

Whereas Stalin was prepared to make formal modifications to the CCR agreement, he clearly wished to retain the situation as it existed in fact and at best to allow minor modifications. This had already become clear in February 1949 when Mikoyan, on instructions from Stalin, had admitted in Xibaipo that the Port Arthur agreement was "unequal" but observed immediately afterward that the Soviet side considered the CCR treaty "not to be unequal," as the railroad had been constructed in large part with Russian funds. Expressing himself cautiously, Mikoyan had added at that time that it might be possible that "the principles of equal rights had not been applied fully" in this treaty. Moscow was willing, however, to discuss and resolve this problem "with the Chinese comrades in a brotherly fashion [*po-bratski*]."[433]

The issue of the CCR was apparently never raised during the talks between Stalin and Liu Shaoqi during the summer of 1949. In the December 16 meeting with Mao, Stalin had said that they could formally retain the relevant points of the agreement while taking into account the wishes of the Chinese and modify them in practice. Mao had surprisingly declared that the way things currently stood with the CCR reflected the interests of the Chinese side, as China lacked sufficient forces to combat imperialist aggression successfully. Furthermore, the CCR constituted a school in which to train Chinese railroad and industrial cadres. This must have sounded very reassuring to Stalin's ears.[434]

In the meantime, Stalin had become wary: A few days before the talk on principles on January 22, 1950, he had heard that the Chinese had begun to create new—and, for the Soviet Union, embarrassing—conditions with regard to the railroad. In a report that reached Stalin after January 10 and clearly before January 20, Kovalev pointed to signs that in his opinion demonstrated that the Chinese side wished to alter in its own favor the organization of the CCR as it stood at the time and had been established in the Sino-Soviet agreement of August 14, 1945. Ever since Chinese technicians under Kovalev's leadership had restored the CCR, the railroad had "in fact been managed by the Chinese," who also controlled the finances, while the Soviet personnel found themselves "in the positions of adviser-consultants" on the greatest part of the railroad network.[435]

If Kovalev's information was correct, the Chinese behavior he described did indeed constitute a violation of the 1945 CCR agreement. This agreement called for the manager to be a Soviet citizen and the assistant manager a Chinese citizen.[436]

According to Kovalev's report, Mao Zedong, Liu Shaoqi, Zhou Enlai, and Gao Gang often described the CCR agreement as "just and good, totally satisfying the interests of the Chinese people." Before he left for Moscow (on January 10), Zhou, however, had had the director general of the Manchurian railroads, Yu Guangsheng, ask the Soviet manager of the railroad "on what basis the Chinese Changchun Railroad would work in the future." In response to the Soviet counterquestion calling for Chinese views, Yu had let it be known that "that it would be desirable not to single out the Chinese Changchun Railroad from the railway system of Manchuria." This was presumably also the reason, Kovalev continued, that China had not yet nominated its main representatives for the joint management and had not yet introduced an accounting procedure apart from the rest of the Manchurian railroad network for the CCR.[437]

Kovalev ended his report on the CCR by noting that Zhou Enlai's behavior could be explained only as the result of pressure on the Chinese government from the right wing of the national bourgeoisie, which held pro-American views, considered the 1945 CCR agreement to be unequal, and desired to give China sole possession of the railroad and exclude the Soviet side from the territory of Manchuria. Kovalev therefore took the position "that it is not in the interest of the Soviet Union to permit a change in the conditions of the 1945 agreement on the Chinese Changchun Railroad" and that the arrangement established in it should be retained. To this end, he recommended (1) that the conditions in the 1945 agreement should be confirmed; (2) that they should demand that the Chinese side nominate its representatives for the Executive Board, the Audit Committee, and Management Department; (3) that a joint commission should be established to draft the constitution of the board of the railroad and identify the railroad's property; (4) that separate financial accounting should be established on the operation of the railroad within the limits of the agreement, as the USSR's economic interests would otherwise suffer damage every day; and (5) that the Dairen harbor and shipyard should be subordinated to the Chinese Changchun Railroad Society, which was already in operation, and they should not create a separate society for them.[438]

As the first position Mao took on the CCR on January 22 demonstrates, Kovalev's fears were justified. Mao said:

> The underlying idea should be to establish in the new agreement that joint use and joint management will be continued in the future as well. With respect to management, however, the Chinese side must [*dolzhna*] play the leading role. Furthermore, it is necessary [*neobkhodimo*] to examine the question of shortening the duration of the agreement and establish how much each side was to invest.[439]

Molotov intervened for the first time here. He contradicted Mao and lectured him: In the joint management of an undertaking by two states, he said, it is customary for the two sides to participate with equal rights and for them to alternate control of the leading positions. In the existing agreement, the management of the railroad had been in Soviet hands. In the future, however, Molotov conceded, the Soviet side considered it necessary to arrange for the management to alternate, "let us say, every two or three years."[440]

Zhou Enlai replied: "Our comrades are of the opinion that it would be right to remove the existing management of the CCR and the office of manager and put a committee in its place to manage the railroad and in doing so to establish that the positions of chairman of the commission and manager are held by Chinese. We should, however, reconsider this question in the context of Comrade Molotov's suggestion."[441] Stalin stated categorically: "If we are to speak of joint management, it is necessary for the leading positions to alternate. It would be more logical this way. With respect to the duration of the agreement, we have no objection to shortening it."[442] Zhou Enlai interrupted the discussion of this sensitive point by moving on to the second and no less sensitive demand that Mao had initially made: "Should they not alter the ratio of the investments by the two sides and, instead of the current parity, increase the investments by the Chinese side to 51 percent?" Molotov objected: "This would conflict with the existing principle of parity between the two sides." Stalin defended Molotov by referring to Soviet treaties with "the Czechs and Bulgars" that had called for "parity and equality." He added: "And if there is already going to be a joint management, there should also be equal shares."[443]

Mao concluded the discussion of the very difficult complex of CCR issues with a statement that he had already used with reference to Port Arthur to express that no agreement had been reached. He said that they would have to study the question further in order to preserve the interests of both sides.[444] He thereby also signaled that he did not consider Stalin's and Molotov's statements the final word, even though he chose not to argue the question further at that time.

Unlike the discussion on the issue of Port Arthur, on which Stalin had made concessions across the board, the discussion of the CCR on January 22 ended in disagreement on two of the three key issues. Stalin was seeking fully equal rights with respect to management and financing, which would still have meant a loss for the Soviet side in comparison with the status quo with respect to management and the preservation of the status quo with respect to financing. Mao, on the other hand, was seeking precedence for the Chinese in both cases as an indication of the new China's sovereignty. The two sides could achieve general agreement only with respect to the reduction the Chinese side sought in the period of validity.

The discussion now turned to the proposed credit agreement. Stalin began by stating that the agreements that had been reached—whereby he meant the credit agreement Liu Shaoqi's delegation had concluded at Party level in Moscow on July 30[445]—would have to be converted into state agreements.[446] Mao asked whether the credit could be used to finance the delivery of military equipment.[447] Stalin replied that the Chinese side could decide this for itself. It could receive such deliveries on credit or as part of a trade agreement.

Mao raised the concern that only a small amount of credit would remain for industry if military deliveries were made on credit. He suggested that some military goods should be delivered on credit and others compensated with goods. He then wanted to know whether they could not move the date for the delivery of industrial and military equipment forward from five to three or four years. Stalin repeated—as he had already said in his talk with Mao on December 16, 1949—that this depended not least on how quickly the Chinese side presented its lists of orders for industrial equipment. It would be helpful for this to be done as quickly as possible, he said. Stalin advocated dating the credit agreement January 1, 1950, as the deliveries should in fact begin about this time. If the credit were granted for July 1949[448]—when the agreement had been reached at Party level—international public opinion would be unable to comprehend how the Soviet Union could have concluded an agreement with the Chinese, who at that time did not have their own government. This observation again reveals how important it still was for Stalin at that time to maintain the separation between state and Party relations, and how important it was for him to handle the contacts that had taken place secretly at Party level before the founding of the People's Republic of China.[449]

Mao then praised the terms on which the credit was granted. At a rate of only 1 percent, they were "extremely favorable" to China. Stalin commented that the USSR granted its credits to the "countries of the People's Democracy" at the rate of 2 percent interest. He repeated the joke he had already made to Liu Shaoqi in June 1949 in Moscow: "We could also increase this interest rate for you if you wish, comrade Stalin said joking.[450] We naturally assumed that the Chinese economy had been severely damaged." Stalin then praised the Chinese government's plans to employ the army in economic construction. They had also done this in the USSR and had had positive results.[451] Mao played the role of good student here: "We make use of the Soviet comrades' experience."[452]

Stalin now turned to the Chinese request that the Soviet Union deliver "a specific quantity of grain"—Mao corrected: "Wheat and textiles"—to Sinkiang. To do this, he observed, it would be necessary to present specific orders that included numbers. Mao agreed.[453]

Mao asked what they should do to conclude a trade agreement. Stalin replied that up to then there had been an agreement of this kind only with Manchuria, whereby he meant the trade agreement that was concluded on July 30 in Moscow with "Manchurian people's democratic authorities led by Gao Gang."[454] He asked what China's intentions were: Should the USSR conclude separate treaties "with Sinkiang, Manchuria, and other provinces, or a single treaty with the center [the government of the PRC]"?[455]

Mao's reply was somewhat contradictory. China was seeking a single agreement with the central government, but Sinkiang could nonetheless conclude a separate agreement of its own. When Stalin then asked, "Only Sinkiang, and what about Manchuria?" Mao observed decisively that no separate agreement could be made with Manchuria because the agreement with the central government would essentially be met with deliveries from Manchuria. Stalin acted as if he had not heard the categorical character of this observation and said that the Soviet side would like the central government to confirm any "agreements with Sinkiang or Manchuria" and for it to assume responsibility for them. For his part, Mao stood by the special treatment of Sinkiang exclusively and specified further that any agreement with Sinkiang should be concluded "in the name of the central government." He passed over Manchuria in silence. Stalin agreed, and commented that agreements should be concluded in the name of the central government because a provincial government could overlook a lot, but the central government was always more significant [*vidnee*]. This section of the talks could, among other things, mean that Stalin would have liked to legalize the special relations he had already established with Manchuria through Gao Gang in a treaty.[456] If this really was Stalin's intention, he failed as a result of resistance from Mao, who had for a long time considered this particular special relationship to be a thorn in his side.[457]

Stalin, who apparently wanted to reach a conclusion, asked if there were any further questions. Mao replied that the most important problem at that time involved economic cooperation. He specified the "reconstruction and development of the Manchurian economy" as his primary concern in this context. Stalin suggested that the two sides commission Mikoyan, Vyshinskii, Zhou Enlai, and Li Fuchun with making preparations in this area. Were there any further questions? Mao confirmed that the air force regiment the USSR had sent had helped China greatly. It had transported approximately 10,000 people. He thanked Stalin for the assistance and asked him to continue to keep the regiment in China so that it could help in transporting provisions for Liu Bocheng's troops, who were preparing the offensive against Tibet. Stalin observed: "It is good that you are preparing the offensive. The Tibetans must be taken in hand [*Tibettsev nado vziat' v ruki*]." They would discuss the air force regiment and give the Chinese side an answer.[458]

In summary, the talk on principles of January 22, 1950, revealed that there was agreement that the Treaty of Friendship and Alliance of August 14, 1945, had to be replaced with a new treaty. The two sides were also able to achieve basic agreement on the sensitive issue of Port Arthur, calling for the current agreement to be annulled and Soviet troops withdrawn following conclusion of a peace treaty with Japan. Stalin thus returned to the position he had already presented to the Chinese government one year earlier, in February 1949. The two parties did not discuss the Mongolian question.[459] Liu Shaoqi had already let it be known during his stay in Moscow in July that the People's Republic of China no longer wished to lay any claim to the Mongolian PR.[460] The Chinese and the Soviets could not reach agreement on two of the three points on the agenda that involved the CCR. Stalin did agree to shorten the duration of the CCR agreement of 1945, but he did not agree to Chinese proposals designed to transfer management of the railroad to China and increase the share of Chinese capital to 51 percent. Stalin sought full parity. It was left to later negotiations to resolve the controversies involved in the CCR question, and also to find concrete solutions in the other areas in which they had found agreement only in principle and to transform these into treaty texts that could be signed.

Survey of the Negotiating Process

In the talk on principles of January 22, Stalin and Mao Zedong agreed to give the two foreign ministers, Zhou Enlai and Andrei Vyshinskii, responsibility in the negotiations that were about to begin for the treaty of alliance.[461] Mikoyan, Vyshinskii, Zhou, and Li Fuchun were to work out the agreement on economic cooperation. In the Soviet Politburo, Molotov, Malenkov, Beriia, Mikoyan, Kaganovich, and Bulganin were responsible along with Stalin for the negotiations—just as they had been for their preparation. They received copies of all records of the talks and other relevant documents.

They were unable to fulfill the original plan of completing the treaty of alliance in January.[462] If we disregard the two agreements on postal, telegraph, and telephone traffic concluded on February 7, the first set of agreements, which included the most important political issues, was signed on February 14. The first set comprised the treaty of alliance, with a secret supplementary agreement; an agreement on the Chinese Changchun Railroad, Port Arthur, and Dairen, with a secret supplementary protocol; a credit agreement, with a secret supplementary protocol; and an exchange of notes on the annulment of the set of treaties of August 14, 1945, and the recognition of the independence of the Mongolian People's Republic. Three days later, Mao and Zhou Enlai returned to China.

Several members of the Chinese delegation remained in Moscow to negotiate further under the direction of Ambassador Wang Jiaxiang and consulted with Zhou Enlai on important questions by telegraph. Their counterparts were, among others, Mikoyan and Deputy Minister of Foreign Affairs Gromyko. They worked on agreements that could not be signed with the others on February 14 because the Chinese side had not yet approved the existing texts. As Zhou Enlai informed Vyshinskii on February 13, these texts still required "further clarification and coordination."[463] The agreements involved the establishment of mixed joint-stock companies to mine raw materials in Sinkiang, civil aviation, trade, and a few other economic agreements associated with the trade agreement, which were all signed in March and April. A delegation from the People's Bank of China that arrived in Moscow after Mao and Zhou departed participated in the negotiations on the settlement of the Soviet credit.[464]

We do not know why Mao and Zhou departed although some of the agreements had not yet been fully settled. One could conceive of urgent reasons that required their presence in Peking. This is contradicted by the fact that instead of traveling on to Peking directly after he crossed the border at Manzhouli, Mao spent about a week in various different places in Manchuria. It is also possible that he felt professional economists could easily negotiate the final versions of the remaining agreements, which involved economics more than politics.

While Mao and Zhou were in Moscow, Mikoyan, Vyshinskii, and Zhou generally carried out the negotiations at the highest level, as far as can be determined. Gromyko, Roshchin, Li Fuchun, Wu Xiuquan, Wang Jiaxiang, Saifudin, and General Liu Yalou in particular participated at the working level. At this level, according to Mao Zedong, "they bargained in the truest sense of the word."[465] Stalin, Mao continued, approached matters in an unpleasant way and clearly expressed his disbelief in and mistrust of the CCP. This characteristic, which Mao attributed to the Moscow negotiations as a whole, is certainly valid for the portion of the talks that dealt with Soviet privileges in Manchuria and also for the portion that was economically significant. The available protocols of the talks give the general impression that the Chinese side presented itself as modest—although occasionally stubborn—in the actual negotiations, whereas the Soviet side generally presented a clearly more self-confident attitude and acted in every respect like a great power with considerable diplomatic experience.[466] This difference reflected the real situation to the extent that the Chinese were in fact generally demanding Soviet assistance and that the new China did lack diplomatic experience.

Mao had already convinced Stalin on January 22 that the old set of treaties should be replaced with a new one. This can be seen in two ways in the results of the negotiations. On the one hand, new treaties, agreements, and protocols were signed and notes exchanged. On the other, as stated in the joint communiqué of February 14, 1950, the corresponding treaty including agreements concluded between China and the Soviet Union on August 14, 1945, was annulled in an exchange of notes between the two foreign ministers on February 14, 1950.[467]

The text of the exchange of notes remained secret. The fact that there had been such an exchange of notes, however, was mentioned in a joint communiqué on August 14, 1950.[468] Zhou Enlai proposed the exchange of notes on January 31.[469] The draft text he presented at the same time is in the Russian presidential archives and has since become accessible.[470] The draft, which was addressed to Vyshinskii and which we can assume became effective in this form, stated that "in view of the signing of the aforementioned treaty and agreement the [reference is to the treaty of alliance and the agreements on the CCR, Port Arthur, and Dairen], the parties to the treaty agree to annul the treaty of friendship and alliance and the agreement on the Chinese Changchun Railroad, Port Arthur, and Dairen and all related agreements." If the Soviet side confirmed it, the draft stated at the end, the note would become a part of the Treaty of Friendship, Alliance, and Mutual Assistance.

Until recently, we still had absolutely no conception of the way the negotiations proceeded. Improved access to the Russian archives has since opened a considerable crack in this area; we are now familiar with the many draft texts exchanged in the course of this process and with the written reports Vyshinskii prepared for Stalin on the negotiations with Zhou Enlai. The new material enables us to verify or impeach information in the memoirs of Chinese contemporary witnesses, correct mistaken interpretations in the secondary literature, and reconstruct key aspects of the negotiating process.

In broad strokes, this process proceeded as follows according to the new information until the treaty of alliance and the other agreements signed on February 14, 1950, were concluded. The Soviet side apparently presented the first draft of the agreements in each case. There were two sessions on January 23: one involving Mikoyan, Vyshinskii, Roshchin, Zhou Enlai, Wang Jiaxiang, and Li Fuchun,[471] and another later in the evening at which Zhou Enlai—assisted by Shi Zhe alone—negotiated with Molotov, Malenkov, Beriia, Bulganin, Voroshilov, Kaganovich, Khrushchev, and Mikoyan.[472] At one of the two sessions, the Soviet side handed over a draft of the treaty of alliance (henceforth: "first Soviet treaty draft").[473] It has not yet been made accessible.[474] We can, however, confidently assume that the text of this draft was the same as the internal commission draft submitted to Stalin on January 22 insofar as it concerned the actual treaty of alliance.[475] The commission draft already included both of the modifications to the August 14, 1945, treaty that Stalin and Mao had agreed to on January 22.

After the Chinese side had examined the first Soviet treaty draft, they reworked at Zhou Enlai's dictation their own draft, which they had already prepared.[476] In addition to Mao, Wang Jiaxiang, Chen Boda, and Li Fuchun were involved.[477] The resulting draft was handed to the Soviet Foreign Ministry late in the evening of January 24.[478] Vyshinskii forwarded a Russian translation of the draft to Molotov and Mikoyan on January 26 and to Stalin on January 28. On January 29, Vyshinskii sent a revised version of the Chinese draft treaty to Mao and Zhou through Kovalev.[479]

We know that the Chinese delegation completed the draft of an agreement on Port Arthur, Dairen, and the CCR on January 25.[480] It would appear that Zhou Enlai did not at first submit this draft, because he wanted to wait for the Soviet draft. Vyshinskii made it available to the Chinese side on January 29, along with other drafts of the protocol governing the transportation rights for the Soviet troops stationed in Port Arthur, the credit agreement, and a protocol to the credit agreement obligating China to deliver strategic raw materials to the Soviet Union.[481]

The draft of a civil aviation agreement was submitted to the Chinese side on January 26. At about this time, but by February 10 at the latest, Vyshinskii also delivered drafts of other—but apparently not all—documents contained in the list presented to Stalin on January 22.[482] This was true of the trade agreement and an agreement on the salaries of Soviet specialists. The draft of an agreement on the establishment of mixed joint-stock companies to exploit raw materials in Sinkiang became available on February 12 at the latest.[483]

At a session on January 31, Zhou Enlai suggested corrections to the Soviet drafts of the treaty of alliance, the agreement on the CCR, Port Arthur, and Dairen, the credit agreement, and two attached protocols that had been presented.[484] Also on January 31, Zhou submitted to Vyshinskii a draft of an exchange of notes on annulling the treaty and the agreements of August 14, 1945, and on the independence of the Mongolian PR.

On February 1, Vyshinskii handed Zhou Enlai a corrected draft of the treaty of alliance that took into account the corrections Zhou demanded, along with corrected drafts of the two agreements and apparently of the two protocols as well, at a session also attended by Mikoyan. A furious conflict arose over the protocol for the agreement on the CCR, Port Arthur, and Dairen (transportation rights for the Soviet troops).[485]

At the February 2 session, Mikoyan, Vyshinskii, and Zhou Enlai agreed on the text of the treaty of alliance, and the agreement on the CCR, Port Arthur, and Dairen. They could not yet reach any agreement on the protocol—to the credit agreement (delivery of strategic raw materials) or the protocol—to the agreement on the CCR, Port Arthur, and Dairen. Zhou did address the question of joint exploitation of metals and oil in Sinkiang, but it would appear that neither side had as yet presented any drafts for agreements on this subject. On the following day, the two sides compared their texts of the treaty and the agreement on the CCR, Port Arthur, and Dairen.[486]

At a dinner with Molotov and Roshchin at Mao's dacha on February 5, Mao again raised questions involving the CCR, Port Arthur, and Dairen.[487] On February 11, Mikoyan, Vyshinskii, and Zhou Enlai agreed on the wording of a "supplementary agreement" prohibiting foreign concessions in specific border areas between the two states. Zhou approved the existing text of the civil aviation agreement and raised the hope that the trade agreement and all the documents could be signed on February 14 or February 18. They were still unable to find an agreement on the payment of the Soviet specialists.[488]

On February 13, thus one day before the signing, Zhou and Wang Jiaxiang separately informed Vyshinskii that the trade agreement, the agreement on the establishment of a civil aviation company, and the agreement on the exploitation of raw materials in Sinkiang were not yet ready for signature because the relevant text required "several clarifications and adjustments." Vyshinskii suggested that they limit themselves to initialing these documents. The Chinese side agreed.[489]

If one disregards the two technical agreements on postal, telegraph, and telephone traffic of February 7,[490] the negotiations had produced the following agreements (the titles have been shortened or paraphrased), arranged in three categories on the basis of the degree of publicity.

Agreements Published at the Time

- Treaty of Friendship, Alliance, and Mutual Assistance of February 14, 1950[491]
- Agreement concerning the Chinese Changchun Railroad, Port Arthur, and Dairen of February 14, 1950[492]
- Agreement concerning the grant of a credit of February 14, 1950[493]

Agreements Not Published at the Time, but the Conclusion of Which Was Announced Publicly

- Exchange of notes of February 14, 1950, on annulling the treaty of August 14, 1945, with agreements, and on the independence of the Mongolian People's Republic[494]
- Agreement on the establishment of a mixed joint-stock company to exploit oil and natural gas reserves in Sinkiang of March 27, 1950[495]
- Agreement on the establishment of a mixed joint-stock company to mine nonferrous metals in Sinkiang of March 27, 1950[496]
- Agreement on the establishment of a mixed joint-stock company for civil aviation of March 27, 1950[497]
- Trade agreement of April 19, 1950[498]
- Protocol on the exchange of goods of April 19, 1950[499]
- Protocol of April 19, 1950, on deliveries of Soviet equipment and materials to China to be made from 1950 to 1952, on the basis of the credit agreement of February 14, 1950[500]

Secret Agreements

- Supplementary agreement (to the Treaty of Friendship, Alliance, and Mutual Assistance) on the prohibition of foreign actions in Sino-Soviet border regions of February 14, 1950
- Protocol (to the agreement on the CCR, Port Arthur, and Dairen) on the supply rights of the Soviet troops stationed in Port Arthur of February 14, 1950
- Protocol (to the credit agreement) on the delivery of strategic raw materials to the USSR of February 14, 1950
- Protocol (to the agreement on the founding of two mixed joint-stock companies to exploit oil and natural gas resources and mine nonferrous metals) of March 27, 1950
- Agreement on the working conditions of the Soviet specialists in China of March 27, 1950

- Protocol of April 19, 1950, on the joint conditions for the exchange of goods between the foreign trade organizations of the People's Republic of China and the Soviet Union in 1950[501]

It is not possible to determine with absolute certainty why particular agreements were to be published, others merely reported, and still others kept secret. There were differing opinions on the question. According to an internal document, the Soviet side originally wished to publish only three agreements: the treaty of alliance, the agreement on Port Arthur and Dairen, and the agreement on the CCR.[502] On February 13, Zhou Enlai expressed the desire to have all the signed documents published. If this did not happen, the Chinese delegation would have to explain the decision specially to the members of the Government Administrative Council in Peking. "In our opinion," Zhou said, "the publication of all signed agreements [in Russian,[503] *soglasheniia*] would be in the interests of China, the Soviet Union, and also of all democratic countries[504] as this would increase their authority in the entire world even further." Zhou is apparently referring here to the fact that many of the unequal treaties between foreign states and China had been concluded as secret treaties in the past. To distance itself from this colonial practice, Soviet Russia had withdrawn from all secret czarist treaties concluded at China's expense in July 1918.[505] Vyshinskii replied evasively that the two sides had not yet reached an agreement that "all agreements and other documents" should be published. They would have to discuss this point further.[506]

Zhou's demand for total publicity is not completely comprehensible because the Chinese negotiating delegation actually must have itself been in a dilemma over this point. On the one hand, it certainly did reflect their basic position that they should break with the notorious tradition of secret treaties and in this way signal to world public opinion the dawn of a new Chinese foreign policy based on equal relations, and at the same time create a precedent. On the other hand, Mao must have been interested in keeping certain agreements secret. These would include the agreements in which he found himself forced for opportunistic reasons to accept terms that were based de jure or de facto on inequality at China's expense, as, for example, in the protocol on the supply rights of the Soviet troops stationed in Port Arthur and in the supplementary agreement to the treaty of alliance. If these were published, the CCP leadership would find itself in an awkward position in Chinese public opinion.

Zhou's statement that they would have to explain any secret agreements they might conclude "specially to the members of the Government Council" should not be taken literally but as an indication of the difficulties involved in the composition of the Government Administrative Council. If Mao wanted to be somewhat certain that nothing would be leaked to the public, he could inform only the formal and secret CP members and a few particularly trustworthy individuals among the other members (those belonging to "democratic parties" and those with no party affiliation). The two sides agreed on which agreements were to be kept secret after the discussion between Vyshinskii and Zhou on February 13, which we just described, and during the following night. No information is available about the course the discussions took, but we know their result. They agreed to publish the credit agreement in addition to the agreements the Soviet side had originally planned to publish (the treaty of alliance and the agreements on Port Arthur, Dairen, and the CCR). At 5 A.M. on February 14, apparently immediately following the night session, Mao gave Liu Shaoqi and Hu Qiaomu strict instructions by telegraph to delete references to the secret agreements that had been included in a draft of a lead article prepared by Xinhua on the conclusion of the set of agreements. These included the supplementary agreement to the treaty of alliance, the protocol on the transportation rights of the Soviet troops stationed in Port Arthur, and the protocol on the delivery of strategic raw materials. Mao made it clear at the end of the telegram how important he considered it to be that his instructions be carried out fully. He asked Liu and Hu each to proofread the article to ensure that his instructions had been followed. He advised them that they could not allow themselves to make the slightest mistake [*wushi haowu yihan*]. "I ask you urgently to pay attention; it is extremely important, extremely important [*wu qing zhuyi, zhi yao, zhi yao*]!" The "whitewashed" article appeared on February 15 in *Renmin ribao*. They did not make any decision at that time about how the agreements that were signed only after February 14 should be handled publicly.[507]

In the following section, we will describe the course of the negotiations for the agreements in detail. We will first show how the two sides reached the agreements signed on February 14.

Negotiations for the Agreements Signed on February 14, 1950

The Treaty of Alliance

When the first Soviet treaty draft, which apparently had the same text as the internal Soviet commission draft of January 22, 1950,[508] was delivered to the Chinese side on January 23, Zhou Enlai responded on the evening of the following day with his own first Chinese treaty draft.[509] A comparison of the Chinese draft with the first Soviet treaty draft leads to the following result.

Both sides were in agreement about the title of the treaty. It constituted an amalgamation of the titles of the treaties the USSR had previously concluded with the Eastern European satellite states ("friendship, cooperation, and mutual assistance") with the treaty concluded on August 14, 1945, with KMT China ("friendship and alliance"). Mao had already used the wording "friendship, alliance, and mutual assistance" with Stalin on December 16, 1949, to describe the treaty they had to conclude.[510] All internal Soviet drafts bore the title usually employed in the treaties with Eastern European states—with the exception of the last one, the commission draft, which bore the title "Treaty of Friendship, Alliance, and Mutual Assistance."[511] Mao and Stalin were thus in agreement about the title of the treaty from the beginning of the negotiations. The title was supposed to express a particularly close relationship between the two states by emphasizing the obligation to provide assistance in the case of war and aid in peacetime.

Shi Zhe is therefore not correct when he claims that the wording "and mutual assistance" was included in the title of the treaty only in response to Zhou Enlai's insistence.[512] One Chinese author even celebrates this alleged event as a "victory."[513] Shi Zhe, who should really have known better, was apparently attempting here—as on other occasions—to demonstrate that the Chinese side repeatedly asserted itself forcefully against Stalin. The first Chinese draft does, however, establish that the term "the two Contracting Parties," which was regularly used in the final version of the treaty, really was the response to a Chinese suggestion. Both the Sino-Soviet treaty of 1945 and the Soviet treaties with the Eastern European states read "the High Contracting Parties." Did the Chinese side—the question has already been raised earlier in this chapter—wish to appear revolutionary by giving up this "bourgeois flourish"?

A comparison of the general introductory clauses, the legal validity of which is contested, reveals that they are given a more general wording and refer more decisively to the bilateral relationship in the Soviet draft, while the Chinese draft addresses the situation in the Far East more directly and already calls for combating potential danger from the Japanese in the first paragraph—which does not occur in the Soviet draft and did not absolutely have to be included because the contracting parties' obligations with respect to Japan were stipulated concretely later (in Articles 1 and 2 of the Soviet draft and in Article 1 of the Chinese draft in particular). The style of the Soviet draft sounds more emotional, as it speaks of "good-neighborliness and friendship" and "close cooperation"—words that do not arise in the Chinese draft, which adopts more businesslike language. It is surprising that the Chinese—unlike the Soviet—draft contains no discussion of the promotion of economic construction, although Mao had pointed out that economic cooperation will be "the most important issue" at the January 22 session.[514] Both drafts instead emphasize the goal of peace and security in the world (the Chinese draft states "in the Far East and in the entire world") and appeal in this context to the principles and purposes of the United Nations.

In the following section, we will examine how the two drafts treated the individual terms of the treaty, which have already been classified in the context of the internal Soviet drafts above.[515]

Prevention of Aggression

This provision constitutes the first section in Article 1 in both drafts. The drafts obligate both sides equally to take measures to prevent aggression from Japan or any state allied with it. As in the Sino-Soviet treaty of 1945, the Chinese draft rigidly specifies in this context "all the means at its disposal." The Soviet draft—somewhat more cautiously—states "with all measures dependent on them."

International Action for Peace and Security

The contents of this clause are identical, and its wording almost identical, in both drafts. In both cases, it constitutes the final section of Article 1.

Clause of Assistance

The two drafts used almost the same wording in the clause of assistance, which constitutes the core of any treaty of alliance. The obligation to provide "military and other assistance" "immediately" "with all the means at its disposal" is identical, as is the definition of the aggressor. Unlike the Soviet treaties with Eastern European states in 1948 (which, of course, referred to Germany, not Japan), the drafts also specified potential Japanese allies in addition to Japan alone as aggressors, which was directed primarily against the United States. The Soviets and the Chinese defined the precondition for a *casus foederis* differently. The Soviet draft speaks of one side being "drawn into military hostilities" with Japan or its allies and thereby retained wording that had been used in the Soviet treaties with Romania and Bulgaria and also *cum grano salis* in the Sino-Soviet treaty of 1945. The Chinese draft, on the other hand, describes the precondition for a *casus foederis* as when one of the two sides was "in a state of war" with Japan or its potential ally. The regulations governing assistance constitute Article 2 in the Soviet draft and the second clause of Article 1 in the Chinese, so that its contents are associated with Clause 1 ("prevention of aggression"). It is difficult to assess whether the Chinese version was seeking to give the clause of assistance greater importance in this way or if it was merely an editorial question.

A comparison of the two clauses of assistance demonstrates that Wu Xiuquan, who was present in Moscow during the negotiations, was not telling the truth when he claimed that the Soviet draft had provided that in the case of an attack on one side, the other side had the opportunity *(deyi)* to provide assistance. This, according to Wu, had been too vague for Zhou Enlai. He then allegedly insisted that they use the wording "immediately with all the means at its disposal [*ji jin qi quan li*]." Wu misled serious authors like Goncharov, Lewis, and Xue, to whom the two drafts were not available with this information. It led them to develop an imaginative hypothesis about the motives for which the Soviet leadership had supposedly opposed including the clause "with all the means at its disposal" in the treaty: that is, the Soviets did not want to be embroiled in a war that would involve atomic weapons that was caused by Mao's risky actions, which they feared.[516]

Because they had no access to the two drafts, Goncharov, Lewis, and Xue suggested a further untenable hypothesis. According to this hypothesis, it was Stalin, after he had been unable to determine the form in which assistance would be provided, who had insisted that a *casus foederis* should exist only if one of the two sides should "find itself in a state of war" rather than simply "drawn into military hostilities." The three authors suspected that Stalin wished to create a higher barrier to participation by the USSR in Asian military conflicts in this way. As described above, the exact opposite was true: Zhou Enlai proposed the wording that used the term "in a state of war." Goncharov, Lewis, and Xue also mistakenly base their further hypotheses with respect to the Korean War on this inaccurate version of the facts.[517]

Prohibition of Alliances

This provision constitutes Article 3 in both drafts. The two versions are nearly identical. The Soviet draft goes somewhat further than the Chinese one in that it prohibits "actions or measures" in addition to "alliances" and "coalitions" directed against the other side.

Renunciation of Separate Negotiations with Japan

The Soviet draft regulates this subject in Article 4 and the Chinese draft in Article 2. Both drafts prohibit the conclusion of a peace treaty without mutual consent. The Soviet draft also prohibits all negotiations with Japan.

Clause of Consultation

The two sides agreed to include a provision of this kind on Mao Zedong's suggestion at the joint meeting on January 22. It constitutes Article 5 in the Soviet draft and Article 4 in the Chinese draft. The provision regulates consultation "on all important international questions" in identical terms. One distinction stands out. The Soviet draft uses the wording employed in the treaties with the Eastern European states and reads "questions involving the interests of either country," while the Chinese draft speaks of questions that "affect the *common* interests of China and the USSR."[518] Was the Chinese side seeking to use this limiting adjective to keep open a back door in order to avoid unwanted cooperation from the USSR in certain—for example, Asian—affairs? The Soviet draft also included the addition "with a view to strengthening peace and universal security."

Economic and Cultural Cooperation and Mutual Assistance

The Soviet draft regulates this subject in Article 6 and the Chinese draft in Articles 5 and 6. Both drafts called equally for the development of economic and cultural ties and for mutual economic assistance. The Chinese draft also mentions "economic cooperation." Both drafts point out that cooperation must respect the principles of mutual respect for state sovereignty and territorial integrity (these two are combined in the Chinese draft as "territorial sovereignty") and nonintervention in the internal affairs of the other party. It should be noted that the Chinese draft also mentions the principles of "equality" and "mutual interests." This shows Mao's attempt to demonstrate publicly that a new era had dawned in China's relations with foreign countries against the backdrop of China's bitter experiences with the "unequal treaties" of the past.

Final Clause

Both drafts call for the treaty to remain in force for thirty years and to be revocable after a one-year period of notice before the date of the treaty's expiration, and for the treaty to remain in force for an undetermined period of time should no notice be served. Should the treaty remain in force, the Soviet draft called for the treaty to remain in force for at least one year and the Chinese draft called for at least five years, whereby notice could be served one year before the treaty was to expire.

The Soviet draft calls for the treaty to become effective immediately after ratification and the Chinese draft immediately after signing. The Soviet draft assumes that the treaty will be ratified "as quickly as possible"; the Chinese draft does not indicate any time frame. Both drafts call for the two sides to exchange the instruments of ratification in Peking.

The Soviet side revised the first Chinese draft between January 24 and January 28.[519] This involved Vyshinskii, Mikoyan, Molotov, and Stalin. The resulting second Soviet draft,[520] with which Stalin himself became involved on January 28,[521] was delivered to Mao Zedong and Zhou Enlai through Shi Zhe on the night of January 28–29.[522]

The second Soviet draft was based on the organization proposed in the Chinese draft; only Articles 5 and 6 of the Chinese draft were contracted into a single article (Article 5).

With the exception of the Soviet wording "the High Contracting Parties" and a few modifications in the general clauses designed to make them stronger or more precise, the only significant corrections the Soviet side introduced involved the prevention of aggression (Article 1, Section 1, Clause 1), the clause of assistance (Article 1, Section 1, Clause 2), the renunciation of separate negotiations with Japan (Article 2), and economic and cultural cooperation and mutual assistance (Article 5).

In the area of prevention of aggression (Article 1, Section 1, Clause 1), Stalin adopted the more rigid Chinese wording "all necessary measures within their power" and thereby obligated himself more to China than he had to Romania, Hungary, and Bulgaria. Stalin replaced the wording "to remove any threat of a renewed unleashing of an aggression by Japan or any other state that might join with Japan directly or in any other form" "to prevent a repetition of aggression and breach of the peace by Japan or any other state which might directly or indirectly join with Japan in acts of aggression."

In the clause of assistance (Article 1, Section 1, Clause 2), Stalin manipulated definitions to resolve the discrepancy that had arisen between the two drafts with respect to the *casus foederis*. According to the original Soviet reading, the alliance should be triggered when one of the two parties was "drawn into military hostilities" with Japan or its potential allies. The Chinese text required one side to be in a "state of war" with them for this to happen. Stalin reworded the phrase so that a *casus foederis* would arise when one side was "attacked" and "thus found itself in a state of war." His original version said nothing about the cause of the military actions. This meant that the other side would also have to provide assistance if one side should end up at war as the result of its own aggressive acts against Japan or its potential ally. Was Stalin seeking to use his new wording to protect the USSR from becoming drawn into an Asian war that might arise as a result of Chinese aggression? It would have been worth debating Goncharov's, Lewis's and Xue's aforementioned hypotheses if they had supported them with the concept of aggression introduced by Stalin.[523]

Stalin ignored what was contained in both drafts on the renunciation of separate negotiations with Japan (Article 2). He turned the obligation to refrain from concluding a peace treaty that had not yet been negotiated with Japan (and, according to the Soviet version, also all negotiations with Japan) into an obligation to positive action: both sides were obligated, "by common agreement, to strive for the conclusion at the earliest possible date, in conjunction with the other powers which were their allies during the war, of a Peace treaty with Japan."

With respect to the prohibition of alliances (Article 3), the second Soviet draft repeats the original Soviet provision that prohibited "actions or measures" in addition to participation in coalitions.

In the clause of consultation (Article 4), the Soviet draft adopts the Chinese wording of "common interests" but insists that both sides had to have "a view to strengthening peace and universal security."

In the area of economic and cultural cooperation and mutual assistance (Article 5), the Soviet side corrected its first draft and thereby accommodated the Chinese side. The changes involved the passage in which the two sides obligated themselves to observe a list of defensive principles in international law. From the Chinese draft, Stalin adopted the wording "in the spirit of friendship and cooperation" and the principles of "equality" and "mutual interests," which the Chinese side understandably considered very important. In the Russian translation, Stalin replaced the term *ravenstvo* (literally "equality"), which corresponds essentially with the Chinese term *pingdeng*, with the legally more precise term *ravnopravie* (literally "equal rights"). The Russian version of the text of the treaty signed on February 14 accordingly employs the term *ravnopravie*. The Chinese version, on the other hand, retains *pingdeng*, although the Chinese term *pingquan* exists, which probably corresponds more closely to *ravnopravie*. The Chinese side probably preferred the term "equality" precisely because it seemed to it more general and more comprehensive and stressed the concepts of equal rank, value, and birth. Stalin replaced the term "territorial sovereignty," which the Chinese delegation had apparently employed because it lacked familiarity with the terminology of international law, with "state sovereignty and territorial integrity," the usual wording in international law, which had already been included in the first Soviet draft.[524]

The second Soviet draft did not adopt the demand in the Chinese draft (Article 6) that the mutual assistance and economic cooperation serve "the facilitation and acceleration of economic construction" in the two countries and "the flourishing of the entire world." Stalin also limited the general Chinese demand for rendering each other "all possible assistance" by referring to "all possible economic assistance"—also a return to the first Soviet draft.

Zhou Enlai declared his agreement in principle with the "second Soviet draft" on January 31 and suggested only two insignificant further stylistic changes and one minor modification of content. He wanted, first of all, to replace the concept "friendly relations" with "friendship" in the second paragraph of the preamble. Vyshinskii sought to convince him that the Soviet term was preferable—which was not without a degree of absurdity, as the term "friendship" had already been employed once in the preamble of the first Soviet draft and had subsequently been eliminated in the Chinese draft. Second, Zhou suggested making "during the war" in Article 2 more precise by specifying "during the Second World War." Finally, he argued that it should be stipulated in Article 6 that the treaty would become effective "immediately after its ratification." Zhou was thereby returning for no apparent reason to the wording that had been contained in the first Soviet draft and rejected in the Chinese draft. Vyshinskii immediately approved the latter two suggestions.[525]

As a glance at the version of the treaty signed on February 14 reveals,[526] the Soviet side also adopted Zhou's first suggestion along with, as mentioned above, the less baroque Chinese wording "the two Contracting Parties." This made the treaty of alliance practically ready for signature on January 31. The Soviet side apparently sent Zhou Enlai the last version of the draft treaty on February 1, because Zhou stated that the treaty would be approved the following day.[527]

The Secret Additional Agreement on the Prohibition of Foreign Actions in Sino-Soviet Border Areas

This agreement was required by the Chinese side and was perhaps the one that most profoundly and most persistently aroused the rancor of the Chinese Party leaders, because it—as in the case of the transportation rights of the troops stationed in Port Arthur—revealed a relapse into colonial modes of thinking and was seen as humiliating. Mao is supposed to have counted the agreement as one of the "bitter fruits" that Stalin had forced him to swallow.[528]

In a talk with the Chinese delegation that apparently took place on February 7 at the latest,[529] Stalin, according to Shi Zhe, expressed the wish that the Chinese government would prohibit "citizens of third states" from entering and settling in Manchuria and Sinkiang. When Stalin unexpectedly expressed this idea, which the Chinese side perceived as intervention in its internal affairs, it was at first met with bewildered silence. Zhou, apparently somewhat irritated, then asked whether he also meant to include the 200,000 Koreans who lived in Manchuria but did not yet hold Chinese passports and the Mongols who lived there but came from abroad. Stalin raised his hands and said that they should only keep Americans, Japanese, Englishmen, and citizens of other imperialist countries from becoming active in Manchuria.[530]

We do not know whether Mao had ever mentioned to Stalin before this talk that the Chinese side wished also to allow nonsocialist countries to engage in economic actions in China, including Manchuria and Sinkiang. In a telegram sent to Mao at the beginning of January, Liu Shaoqui had proposed, with reference to "the comrades' opinion" in Peking, the idea of not only asking the Soviet Union and the other "new democratic countries" to found joint Sino-Soviet companies to exploit oil and nonferrous metals in Sinkiang as well as in other parts of China, but also making an effort to organize joint companies together with "groups and industries from the capitalist countries."[531] It is not known whether Mao and Zhou supported this idea. If this was the case, and if they had presented this idea to the Soviet side, they were blocked by Stalin's veto—at least with respect to Sinkiang and Manchuria.

On February 11, Mikoyan, Vyshinskii, and Zhou agreed on the text of an agreement. It was based on a Soviet draft created following discussions between the two sides and presented to Zhou the previous day. It read:

<u>ADDITIONAL</u> AGREEMENT

> between the Government of the Union of Soviet Socialist Republics and the Central People's Government of the Chinese People's Republic <u>to the Chinese-Soviet Treaty of Friendship, Alliance, and Mutual Assistance</u>
>
> In conjunction with the signature today of the Treaty of Friendship, Alliance, and Mutual Assistance by the Presidium of the Supreme Soviet of the Union of SSR and the Central People's Government of the Chinese People's Republic and in the interests of the defense of both countries, the Government of the USSR and the Central People's Government of the Chinese People's Republic have agreed on the <u>present additional agreement</u>.[532]
>
> No <u>right</u> to concessions will be granted to foreigners and no actions will be allowed by industrial, financial, trade, or other companies or by institutions, companies, or organizations in which capital from third countries or citizens of these states participate either directly or indirectly either on the territory of the Far East region[533] or of the central Asian republics[534] of the USSR or on the territory of Manchuria or Sinkiang of the Chinese People's Republic.
>
> Issued in Moscow on "..." February 1950 ...[535]

On February 11, Zhou proposed only two modifications, indicated by underlining (italics) in this draft, which were in principle accepted by the Soviet side because they did not change "the meaning and substance" of the Soviet draft, as the Soviet side stated in this context. On the one hand, Zhou considered it important for the agreement to be identified as an additional agreement to the treaty—apparently because he wanted to see the principles of equal rights and mutual interests established in the treaty also extended formally to the agreement. The sense of the second modification is not quite so clear ("right to concessions" instead of simply "concessions" as before). It is possible that Zhou wanted to leave open a small backdoor by not recognizing any general or a priori right to concessions for foreigners, but nonetheless did want to offer them the possibility of obtaining a concession in individual cases and on the basis of an application. We can assume that the agreement was signed in the above-cited form on February 14.[536]

Stalin used this agreement to guarantee the Soviet Union a monopoly position in the traditional Russian areas of interest in Sinkiang and Manchuria. This involved first of all the economic field, in which foreign competition was to be ruled out. Stalin apparently also wanted to prevent the operations of foreign news agencies that would be able to spy on Soviet actions, not least any military actions in Manchuria, and the situation along the border with the USSR.

If we follow Shi Zhe, Stalin had only proposed to exclude foreigners from the two Chinese border areas. The regulation extended only to the Soviet Far Eastern region and Central Asia when the Chinese insisted on reciprocity—this cannot yet be demonstrated in the documents—which was a farce, given the fact that Stalin had already cleared the foreigners out and closed these two areas before the war. The Chinese delegation cannot have been unaware of this. It nonetheless insisted on Soviet reciprocity to save face. They were more successful here than they were when they attempted unsuccessfully to include the transportation of Chinese troops through Siberia to Sinkiang in the secret agreement on the supply rights of the Soviet troops stationed in Port Arthur.[537]

For the reasons given above, the additional agreement under discussion here had de facto the character of one of the ill-reputed "unequal treaties." If it had become public, Mao and the Chinese leadership would have become burdened in Chinese public opinion with the onus of having contributed to a further limitation to China's sovereignty. This would furthermore have provided grist for the U.S. propaganda mill. Secretary of State Acheson had after all stated on January 12 that the USSR was seeking to annex portions of North China—including Sinkiang and Manchuria.[538] These were presumably the reasons for which Mao and Stalin agreed to keep the agreement secret.

In his telegram to Liu Shaoqi and Hu Qiaomu on February 14 ordering that references to secret agreements be purged from the Xinhua article on the conclusion of the set of treaties, Mao explicitly demanded that the reference to the additional agreement be deleted. This was disguised in an apologetic observation that the additional agreement to the treaty constituted a "necessary measure" for "guaranteeing the defense interests of both states [*baozhang liangguo guofang liyi*]," which would be "beneficial for both China and the Soviet Union."[539]

According to Russian accounts, the Chinese side delayed implementation of the additional agreement by taking the first steps only after three years and also only with respect to Manchuria. In February 1953, negotiations with Japan began for the repatriation of the approximately 30,000 Japanese living in Manchuria. The first contingent was repatriated the following month. This action did not involve the 1,800 Japanese war criminals, who were to be judged in China, or "a portion of the progressive personalities and specialists," who remained in China to work or study.[540]

Mao appears to have considered the additional agreement particularly humiliating. It is possible that the Soviet side was not sufficiently sensitive to recognize this. Thus, Khrushchev did not suggest to the Chinese side that it be canceled after Stalin's death—as he did in the case of the agreement on mixed companies in Sinkiang. When Mikoyan visited Peking in April 1956, Mao complained that he had always seen this agreement as strange and an infringement of China's sovereignty. In response, Molotov signed a resolution at the end of the month stating that the agreement did not represent the spirit of Sino-Soviet relations. On May 10, 1956, the Soviet government proposed to the Chinese side that it be annulled. Zhou Enlai agreed on May 29.[541]

Mao allegedly observed contentedly at that point that he had returned all the bitter fruit that Stalin had forced him to swallow—by which he supposedly meant the mixed companies and the additional agreement.[542] According to more recent Russian information, the Chinese still did not grant a single foreign concession in China during Mao's lifetime after the additional agreement was annulled, and there was not a single Chinese company that had foreign participation during this period.[543]

The Agreement on the Chinese Changchun Railroad, Port Arthur, and Dairen

As described above, the questions involving Port Arthur, Dairen, and the Chinese Changchun Railroad (CCR) were originally supposed to be resolved in the treaty itself according to internal Soviet drafts. The commission draft presented to Stalin on January 22 separated these subjects out of the treaty for the first time and changed them into two independent protocols: one dealing with Port Arthur and Dairen, and the other with the CCR. The draft protocol on Port Arthur and Dairen called for "renewed examination [*novoe rassmotrenie*]" of the agreement following the conclusion of a peace treaty with Japan and the withdrawal of troops within a period of two to three years, and stipulated that the withdrawal was to begin in 1950. The draft protocol on the CCR called for the CCR agreement of August 14, 1945, to be allowed to remain in force and for both sides to be allowed to make only minor additions. We see from the letter that was presented to Stalin along with this draft and other documents that "several practical steps" were to be specified, including "the restoration of joint administration of the railroad,"[544] in a second CCR protocol that was not to be published.

The talk on the principles of January 22 between the two delegations rendered the aforementioned Soviet drafts obsolete. Stalin and Mao agreed that the agreement on Port Arthur should expire and that the troops should withdraw at the same time following the signature of the peace treaty with Japan. There was a dispute over the CCR. Stalin and Molotov insisted on the restoration or retention of full administrative and budgetary parity. Mao and Zhou, on the other hand, called for a shorter period of validity and insisted that the management should in the future be headed by China and that the proportion of Chinese capital should be increased to 51 percent.[545]

On January 25, the Chinese side completed its draft of an agreement on Port Arthur, Dairen, and the CCR,[546] but apparently kept it to itself as it wished to give precedence to the Soviet side. The expected Soviet draft[547] was apparently delivered to the Chinese delegation on January 29[548] (henceforth: "January 29 draft"), along with a draft of a protocol for this agreement to regulate the transportation and supply rights of the Soviet troops stationed in Port Arthur. A comparison of this draft with the last internal Soviet draft texts on the subjects of Port Arthur, Dairen, and the CCR, as contained in the commission draft of January 22, 1950, leads to the following results.

The most striking difference is found with respect to the CCR. It contained a compromise that was on the whole in the Chinese favor. Stalin began the retreat, as he had already shown in the talk on the principles of January 22, with respect to the period of validity of the CCR agreement. According to the draft of the "Protocol Concerning the Soviet-Chinese Agreement on the Chinese Changchun Railroad" of January 22 (commission draft), as mentioned above, the validity of the CCR agreement of August 14, 1945, was to be confirmed and both sides could add only "a few supplementary provisions . . . in the interests of guaranteeing the successful operation of the railroad."[549] This meant that Soviet privileges based on equal ownership and management rights were to be guaranteed until the year 1975.

Clearly impressed by the severity with which Mao Zedong and Zhou Enlai demanded the shortening of the period of validity, an increase in the proportion of Chinese capital to 51 percent, and Chinese leadership in the railroad's management during the talk on the principles of January 22, Stalin stated in the January 29 draft that he would be willing to transfer all Soviet rights to the CCR to China without compensation by 1952 at the latest—thus, twenty-three years earlier than had been called for in the 1945 agreement. The concession that Mao accepted was that the status quo ante, which the Chinese side hated but which China had already altered de facto in its own favor,[550] would remain in force through 1952. It should be assumed that the compromise was achieved in the negotiations that took place between January 23 and January 29. No documents are available to us to support this conclusion. However that may be, Stalin demonstrated an amazing willingness to compromise on the CCR, which may not have been easy for him. His generosity put Mao in a position to present the results to the Chinese public as a significant success in the negotiations. In any case, on February 1, Mikoyan threatened (as will be described later in this chapter) to extend the period for the return of the CCR by two and a half years, in connection with the unilateral right to transport military supplies through Manchuria demanded by the Soviet side. This moved Mao to make a further concession and renounce his demand for reciprocity. This was done in a secret agreement, as the two sides feared a negative reaction in China.

Stalin did make concessions to China, but he did not go anywhere near as far as the Bolshevik government had on July 25, 1919. At that time, Deputy Minister of Foreign Affairs Lev Karakhan had declared that the Soviet government was returning the East China Railroad along with all associated concessions without compensation—a noble gesture, which, however, was immediately withdrawn.[551]

A comparison of the January 29 draft with the internal Soviet draft of the "Protocol Concerning the Soviet-Chinese Agreement on Port Arthur and Dairen" of January 22 (commission draft) and with the agreement that Stalin and Mao reached on January 22 leads to the conclusion that the Chinese side was also able to improve its position with respect to Port Arthur. The January 22 draft called for the August 14, 1945, agreement on Port Arthur to be "subjected to renewed examination" following conclusion of a peace treaty with Japan and for Soviet troops to withdraw by 1952 at the latest. In the January 22 talk on principles, Stalin and Mao had decided that the agreement would expire when a peace treaty was concluded with Japan and that the Soviet troops would withdraw at the same time. The January 29 draft called for the troops to withdraw "immediately after the conclusion of a Peace Treaty with Japan, but not later than the end of 1952."[552]

It was an admirable success for the Chinese side to negotiate an agreement for a fixed withdrawal date "not later than the end of 1952." This went beyond the agreement reached between Stalin and Mao on January 22, although the January 22 internal Soviet draft had already called for it—which Mao could not know. There was no further discussion of examining the 1945 agreement following conclusion of a peace treaty with Japan, as had been specified in the January 22 draft. The 1945 agreement was annulled in the aforementioned exchange of notes on February 14, 1950, and replaced by the new agreement called for in the January 29 draft. Stalin's willingness to seek compromise with Mao may have been increased by the fact that Mao had stated his interest on January 22 in developing Port Arthur as a base of military cooperation.[553]

There were further improvements in China's favor over the 1945 agreement. Unlike the old agreement, which in Article 3 had called for a Joint Military Commission to be made up of three Soviet and two Chinese representatives, the January 29 draft established parity.[554] The regulation in the old Article 4 whereby China "entrusted . . . the defenses of the aforesaid naval base" to the Soviet government and left it a free hand to construct any installations it required for this end was cancelled without compensation. In its place, the Soviet side proposed joint military use of Port Arthur in the event of an attack by Japan or its ally in the January 29 draft. The Chinese side could accept this without hesitation because it was worded as a possibility and was also subject to the condition that the Chinese government had to suggest the action. It is striking that this agreement was not limited in time.

Zhou Enlai understandably raised no reservations in principle to the Soviet January 29 draft on January 31. Almost all the corrections he suggested were stylistic in character. Only two turned out to be significant in terms of content. First of all, Zhou Enlai considered it to be important to mention the subject of the Chinese Changchun Railroad in the first place in the title of the agreement. He thereby again made clear that this issue, which had hitherto been the most controversial, was of prime importance to the Chinese leadership. The second correction he sought involved Article 2, Section 1 of the Soviet draft, which in the new arrangement proposed by Zhou read: "the high contracting parties have agreed that the Soviet troops shall be withdrawn immediately after conclusion of a peace treaty with Japan, but not later than the end of 1952, from the *jointly used* naval base Port Arthur and shall *hand over* the military installations in this *area to the government of the Chinese People's Republic*." Zhou made a corresponding addition in Section 2: "Pending the withdrawal of the Soviet troops *and the transfer of the aforesaid installations*. . . ." Zhou considered it important to make certain that the Soviets did not dismantle their military installations and carry them back to the USSR. He may have also been remembering bitterly that the Soviet Army had carried off practically all important industrial equipment to the USSR before its withdrawal from Manchuria in 1946.[555]

At the January 31 session, the Soviet delegation stated that it agreed to practically all the corrections Zhou Enlai wanted.[556] Stalin, who may have actually been considering dismantling the installations, could not accept Zhou's addition in Article 2, Section 1 as it seemed to imply tacitly a transfer of military installations without compensation. He added by hand at this point: ". . . whereby the cost will be repaid to the Soviet Union."[557]

The corrections Zhou had demanded were taken into account in the version signed on February 14.[558] In return, the Chinese side accepted Stalin's demand for repayment. It was included in the phrase "the Government of the People's Republic of China shall compensate the Soviet Union for the expenditure incurred by the Soviet Union since 1945 in respect of the reconstruction and erection of the said installations." Mao and Zhou certainly did not agree to this enthusiastically but must have believed that they had to make this concession in the light of the Soviets' obliging behavior overall and in view of the overall set of treaties. Nonetheless, not everybody in China approved of the delegation's behavior. Thus, for example, Wu Xiuquan, who participated in the negotiations in Moscow, still considered it necessary in 1990 to defend the regulation involving Chinese compensation payments against "certain people [*you ren*]" who argued that the argument that the armaments that the Soviet Army left behind when it withdrew in 1955 were not scrap materials but modern weapons.[559]

A terminological correction proposed by Stalin was also made. It was legitimated on logical grounds: Stalin replaced the concept "Chinese people" with "China" in the phrase "State independence and territorial integrity . . . of the Chinese people" in the preamble.[560]

The Chinese achieved a major success by establishing that the Soviet troops in Port Arthur and Dairen would in the future be subject to Chinese jurisdiction. At least, there is no indication that any secret agreement was reached, or even merely sought by the Soviet side, to the effect that Soviet courts would be responsible for those troops. Stalin had forced an agreement to this effect on KMT China and a secret protocol had been signed in August 1945.[561] As no comparable agreement was reached in the case of the Soviet specialists sent to China, either,[562] the set of treaties signed on February 14, 1950, with its follow-up agreements and protocols, also signaled the end of Soviet special jurisdiction that China found so humiliating.[563]

The Soviet side was not prepared to allow all its rights with respect to Dairen, unlike Port Arthur, to expire by the end of 1952 at the latest. This can be seen in the January 29 draft, which called for the question of Dairen to be examined following the conclusion of a peace treaty with Japan. This conflicts with the statement Stalin had made shortly before to Mao: that the Soviet Union had no intention to claim any rights in Dairen.[564]

However, assets administered or leased by the Soviets were, according to the same draft, all ready to be handed over to the Chinese government. In the talk on principles of January 22, 1950, Stalin had let it be seen that he opposed the continuation of Dairen's status as a free port because this only served the interests of the United States and Great Britain.[565] He had his way, as can be seen from the January 29 draft and the agreement signed on February 14.

The last internal Soviet draft for an agreement on Port Arthur and Dairen (commission draft) of January 22 did not mention the assets the Soviet Union controlled. This draft did, however—like the regulation for Port Arthur—propose the withdrawal of Soviet troops from Dairen by 1952 at the latest. It is surprising in this context that the 1945 agreement included no plans at all to station Soviet troops in Dairen. The Soviet January 29 draft did not mention the Soviet troops and clearly sought to leave the troops stationed there de facto for the time being and—possibly—to allow the Soviets to continue using the harbor. It is unclear what else was meant by "the question of the harbor of Dairen/Dal'nii/," which was supposed to be revised following conclusion of a peace treaty with Tokyo. De jure, none of the rights the USSR had obtained to Port Arthur and Dairen in 1945 could have been meant because this agreement, as mentioned before, was to be annulled by a Sino-Soviet exchange of notes. The draft therefore was also not referring to the Dairen agreement that was to be revised following the conclusion of a peace treaty with Japan—as was still the case in the internal Soviet commission draft of January 22. Instead, it meant very vaguely the "question of the harbor of Dairen/Dal'nii/."

Zhou Enlai raised no objections to the section on Dairen in the Soviet January 29 draft at the January 31 session, with the exception of a requested correction to Article 3, Section 2 that made no significant change in the content (instead of "China," it should formally be called "the government of the People's Republic of China").[566] The same is true of the February 2 session.[567] Here Zhou merely indicated that he was not interested in having China take over all the property previously controlled by the Soviet side. He suggested that when the treaty came into effect, the proposed Joint Military Commission should examine which companies the two sides should operate jointly. When Mikoyan asked which companies he meant concretely, Zhou said that the repair docks should "certainly be operated jointly." They agreed to discuss the question further, and Gromyko was made responsible for this on the Soviet side.

The answer to why Zhou accepted that the Soviet presence in Dairen—unlike in the case of Port Arthur—was not limited precisely in time may perhaps be revealed in an observation that Mao Zedong had made shortly before to Stalin. In the talk on the principles of January 22, Mao had said that it was his opinion that Dairen could serve as "a base for Sino-Soviet economic cooperation."[568] However that might be, the section on Dairen was signed—with Zhou's stylistic correction—on February 14 as the Soviet side had conceived it in the January 29 draft.[569]

If it was already unclear why the agreement set no fixed date for the expiration of Soviet rights in Dairen, although the agreements remained in force for the other two subjects "not later than the end of 1952," it is equally unclear and indeed highly amazing that Dairen was returned sooner than the CCR and Port Arthur: at the end of 1950, both sides signed a protocol at the Chinese suggestion that the Soviet Union should return Dairen to China by the beginning of 1951, which apparently also happened.[570]

The CCR was transferred to China on December 31, 1952, as agreed.[571] The Port Arthur problem, on the other hand, developed differently. During his stay in Moscow in August and September 1952, Zhou Enlai mentioned to Stalin that he should extend the deployment of Soviet troops there in connection with the Korean War.[572] Stalin replied that the Chinese side would have to make any such demand: "We are guests there, and guests do not make such demands." They agreed to handle the issue in an exchange of notes. In his démarche to Vyshinskii on September 15, 1952, Zhou asked the USSR to extend the deadline for the withdrawal of Soviet troops provided in Article 2 of the agreement in view of Japan's unwillingness to conclude a peace treaty with China and the USSR "until such time as peace treaties are concluded between the Chinese People's Republic and Japan and between the Soviet Union and Japan" in order to secure the peace and on the basis of the treaty of alliance.[573] Vyshinskii agreed. During his visit to China in the fall of 1954, Khrushchev proposed to the Chinese that they should finally take Port Arthur back. On October 12, both sides declared in a joint communiqué that in view of the end of the Korean War, the restoration of peace in Indochina, and the improvement of the defensive capabilities of the People's Republic of China, they had agreed that the Soviet troops would be withdrawn from Port Arthur by May 31, 1955, and that the installations would be handed over to China without compensation.[574] The Soviets, however, appear to have nonetheless received reimbursement for at least some of the installations.[575]

The Secret Protocol on the Right to Supply the Soviet Troops Stationed in Port Arthur

Finally, Stalin had managed to come to the decision to annul the agreements on Port Arthur, Dairen, and the CCR, to return the railroad to China without compensation by the end of 1952, and to withdraw the troops from Port Arthur by the end of 1952 at the latest. In addition to the provision governing the return of the railroad, the agreement on the CCR included only the general regulation that "the present status of the joint Soviet-Chinese administration" of the railroad would be maintained, whereby, however, control of the leadership positions would alternate between the two sides.

They could naturally have explained the phrase "present status of the . . . administration" a priori in a general sense and understood by it all existing provisions established in the 1945 agreement. The Soviet side wanted nonetheless to make perfectly clear that one point was included in this continuation: that is, Article 11 of the August 14, 1945, agreement on the CCR, which guaranteed that goods would be transported duty-free from the Soviet border to Dairen and Port Arthur and back.[576] To ensure that this rule would be maintained at least for the supply of Soviet troops in Port Arthur, Vyshinskii handed the Chinese delegation a draft of a protocol to the agreement on Port Arthur, Dairen, and the CCR on January 29.[577]

This document will be reproduced below in the form in which it was filed in the Russian Foreign Ministry archives:[578]

Draft

PROTOCOL

of " " 1950

The Central People's Government of the Chinese People's Republic and the Government of the Union of the Socialist Soviet Republics have agreed as follows:

1. Materials and goods that are sent from the USSR on the Chinese Changchun Railroad or also by water through the harbor of Dairen to the Soviet military units stationed in the area of the Port Arthur naval base and also materials and goods that are sent by the Soviet military command from Port Arthur to the USSR, will not be burdened by the Chinese People's Republic with customs duties or other taxes or dues.

2. The materials and goods mentioned in Point 1 are not subject to any customs inspection in the territory of the Chinese People's Republic. Materials and goods that are sent with the Chinese Changchun Railroad will be transported in sealed wagons on Chinese territory.

3. Soviet military units and military goods will be transported unhindered on the Chinese Changchun Railroad at the tariffs valid for Chinese military transports from Mandzhuriia Station to Pogranichnaia Station and back.

Distributed
to comrades Stalin,
Molotov,
Malenkov,
Beriia,
Mikoyan,
Kaganovich,
Bulganin.

A comparison of the draft protocol with Article 11 of the 1945 agreement shows that the draft sought to improve the position of the Soviet side on one point: unlike the 1945 agreement, the 1950 agreement should not grant Chinese customs the right to inspect the transport of Soviet goods.[579]

At the January 31 session, Zhou Enlai did not raise any objections to the aforementioned improvement. But he did make it clear that the Chinese side did not agree with the unilateral character of the regulation the Soviets sought, which reminded the Chinese side of the old-style special foreign privileges. Zhou made a suggestion with which he sought to at least establish reciprocity in one of the areas to be regulated: Article 3 of the Soviet draft should form Article 1, and a new Article 2 should be introduced that would read as follows: "Chinese military units and military goods will be transported unhindered on the Siberian and Turkestan-Siberian Railroads and on the stretch from Alma-Ata–Sinkiang at the tariffs valid for Soviet military transports from Mandzhuriia Station via Chita–Novosibirsk–Alma-Ata to Kulja (Sinkiang Province) Station and back." The Soviet side rejected this suggestion.[580]

Mikoyan and Vyshinskii presented a new draft of the Soviet protocol the following day. It did not take the Chinese demand into account, but specified the conditions for the Soviet troop transports in a limited form: They were seen as an opportunity and a right should the USSR be threatened in the Far East. Mikoyan and Vyshinskii again rejected Zhou's demand for reciprocity. It was not possible to conceive of a situation, they maintained, in which China would be able to make use of a right to transport troops through Soviet territory. First, they could not see how Sinkiang was threatened militarily or might be threatened in the future or by whom. Second, it was well known that the segment of the line from Manzhouli via Siberia and Alma-Ata to Urumchi was longer and less comfortable. It was in any case amazing that China would raise the issue of reciprocity with respect to transports when, as they had understood Zhou, it was practically obvious that the USSR would have transport rights across Chinese territory when the CCR was returned to China. This was a clear reference by the Soviet negotiators to the fact that Zhou had approved the Soviet draft of the agreement on the CCR, Port Arthur, and Dairen the day before, which in Article 2, Section 4 called in principle for joint military use of Port Arthur in the case of one side being involved in hostilities and to the fact that such rights were contained in the clause of assistance of the treaty of alliance.[581]

This argument, according to Vyshinskii's report to Stalin, "clearly embarrassed" Zhou. He now felt forced to reveal the real reason for his insistence. The Chinese side, he said, had not originally wanted to mention the issue of transport rights at all, as these could be derived from the treaty of alliance and they would merely have to reach an agreement at the appropriate time. But, given the Soviet side's concrete suggestion, he had felt compelled to demand the same rights formally because "acceptance of such an article would . . . not be understood in China if China did not at the same time receive the right to transport troops across Soviet territory."[582]

Mikoyan remained adamant. If the Chinese side continued to reject the Soviet suggestion, he threatened, he would withdraw it, but he would then demand that the existing agreement on the CCR remain in force for a shortened period of ten years. This would have meant a period that was two and a half years longer than was called for in Article 1, Section 1 of the present Soviet draft agreement on the Chinese Changchun Railroad, Port Arthur, and Dairen. Zhou gave in. He admitted that the proposed new wording of the clause, which stated that the transports would be carried out only in the case of impending war, constituted an improvement over the original wording. He had no further objections to raise at that time, but he would have to discuss the question with Mao.[583]

Mao was not satisfied. At the February 2 session, Zhou again asked Mikoyan whether it would not be possible to establish in writing that Chinese troops could occasionally be transported on the Siberian railroad from Manchuria to Sinkiang when this should prove necessary. Mikoyan replied that he had already said the last time that a right of this kind was not rejected because such transports would be out of the question "in the case of a real necessity" following an agreement between the alliance partners. He again expressed his "amazement" at the Chinese position on this question: "The Soviet Union as an alliance partner is transferring, without any compensation, tremendous riches—the CCR, Dairen, Port Arthur, all its rights to these areas—and yet the Chinese side does not wish to approve the transportation of Soviet troops in only one direction. What kind of alliance partners are we if the Chinese side does not even want to make such a concession?" Zhou apparently noted this criticism silently and let the matter rest after consulting with Mao.[584]

Publication of the agreed protocol without reciprocal regulations would, as Zhou had already admitted, have put Mao and the CP leadership, who wanted to parade the abolition of Russian/Soviet privileges in Manchuria as a very special success, in a difficult position in Chinese public opinion. It was clearly for this reason that the protocol was ultimately kept secret.

The Credit Agreement

To recapitulate: Mao Zedong first asked the Soviet side for a credit in February 1949. He asked Mikoyan for a three-year credit in the amount of US$300 million to be set aside for deliveries of silver, raw materials, machines, and other goods. At the end of June, Stalin approved a credit for the requested amount at an advantageous rate of interest of 1 percent while meeting with Liu Shaoqi. It was to be payable in equal amounts over five years and repayable over ten years following payment of the entire amount of the credit. The relevant contract was concluded after some back-and-forth at Party level on July 30 and signed by Liu Shaoqi and Malenkov. It was disguised in public by being described as a trade agreement between the USSR and Manchuria for the period of one year.[585]

On December 16, Mao proposed to Stalin that they conclude a government agreement about the US$300 million credit at that time, as this would make a favorable impression in China. Stalin agreed. He again raised the topic on January 22 when he suggested to Mao that they date the credit agreement January 1, 1950, because the deliveries were to begin at that time. Mao suggested that some military deliveries should be made on credit, while some should be compensated for by goods.[586]

The Soviet editorial commission had already presented the draft of a credit agreement to Stalin on January 22; its text is not yet accessible.[587] We can, however, assume that it was identical with the draft the Soviet side sent to Mao and Zhou on January 29.[588]

Zhou Enlai, who took a position on this draft on January 31 and at the same time handed over a text of his own, only called for a few corrections. Thus, for example, he suggested replacing the wording employed at the beginning of Article 1, Section 2—which now read: "In view of the extreme devastation of China as a result of the prolonged military operation . . ."—with new terminology: "In view of China's financial difficulties that were caused by the prolonged military operations . . ." Article 5 was also to call for ratification. In Article 3, Section 1, Zhou wanted to establish that China could also repay the credit with finished products, not only with raw materials (and gold and U.S. dollars). With the exception of the last suggestion, for which the Soviet side requested time to think it over first, all of Zhou's suggestions were initially approved on January 31.[589]

Further negotiations were held later on the credit agreement; no source material regarding these discussions is available to us. The final versions signed on February 14[590] shows that Zhou was able to have his own way only on the demand for ratification. The Soviet side refused to allow China also to use finished products to repay the credit as Zhou had proposed (Article 3). The two sides had been able to agree only that deliveries of tea should be acceptable in addition to the payment methods called for in the Soviet draft.

The Secret Protocol on the Delivery of Strategic Raw Materials to the Soviet Union

Khrushchev remembered that Stalin had once asked his circle of confidants—probably during the first six months of 1949—where China's gold and diamond mines were located.[591] Beriia reportedly answered that China possessed enormous riches that Mao wished to hide from the Soviet Union. If the USSR granted Mao a credit, he said, Mao should do something in return. A connection was thus already created here between the credit and Chinese deliveries of valuable raw materials. As the Soviet leadership consulted on the credit contract during Mao Zedong's stay in Moscow, the idea was formed to create a package designed to use the credit also to acquire strategically important metals that the USSR lacked.

On January 29, the Soviet side presented the Chinese delegation with the draft of a protocol to the credit agreement.[592] The following document is reproduced in the form in which it was filed in the Russian Foreign Ministry archives:[593]

Draft

PROTOCOL

of " " 1950

In the context of the signing on . . . . 1950 of the agreement between the Central People's Government of the Chinese People's Republic and the Government of the Union of Socialist Soviet Republics on the grant of a credit by the Soviet Union to the Chinese People's Republic, both governments have agreed as follows:

Given the fact that the Soviet Union finds itself in a tight situation because it lacks strategic raw materials (tungsten, antimony, lead, and tin), the Central People's Government of the Chinese People's Republic has made a concession to the Soviet Government and has stated its agreement to sell the entire surplus of tungsten, antimony, lead, and tin exclusively to the Soviet Union in the course of the first 10 to 12 years in the context of the annual agreement on the exchange of goods.

Distributed
to comrades Stalin,
Molotov,
Malenkov,
Beriia,
Mikoyan,
Kaganovich,
Bulganin.

Zhou Enlai proposed a correction to the Soviet draft on January 31 in which he sought to ensure that the metals would be sold at world market prices. He also sought to extend the period of delivery. According to his draft, Section 2 would read as follows: ". . . make available the entire surpluses of tungsten, antimony, lead, and tin that China will not use to the Soviet Union for purchase at prices established on the basis of world market prices during the period of 14 years, from 1950 to 1963 . . ." Zhou enquired whether the Soviet

Union wished to purchase all the surpluses, or whether the People's Republic of China would be able to sell some portion of them to the "democratic countries." Czechoslovakia planned to purchase up to 1,600 tons of tungsten ore. Mikoyan replied that the Soviet Union "was prepared" to purchase all the surpluses.[594]

It became clear very soon that the Chinese side would have difficulties delivering the metals. It asked that lead be taken out of the protocol. Mikoyan agreed. Zhou raised no objections to the Soviet desire to have access to 8,000 tons of tungsten in the first two years. He asked Moscow, however, to reduce the requested delivery of 14,000 tons of tin in the first four years by 3,000 tons, because Yunnan Province, where the tin mines were located, had only recently been liberated, and the Japanese had also done considerable damage in the mining regions. If the USSR provided aid, however, the requested amounts could be delivered. Mikoyan approved Soviet assistance, but requested concrete data on what would be required. Zhou also pointed out that the Japanese and the KMT had destroyed the mining regions in Hunan Province, where antimony was mined. He therefore requested that the amounts of the deliveries be reduced and stressed that Soviet aid could lead to the delivery of the amounts originally requested here, too. In this case, Mikoyan insisted on the original amounts, approved Soviet aid, and pointed out—waving his forefinger, so to speak—that these numbers would stimulate the Chinese side to produce more and remind them of their responsibility to meet the agreement. Zhou requested a postponement until he had access to precise information from China about the production of antimony. If this information should turn out to be negative, the amounts of the deliveries would have to be reduced.[595]

The final text of the protocol is not yet accessible.

There is an indication that the Soviet leadership made a close connection between the delivery of the strategic raw materials and the granting of the credit. The draft of a Xinhua article on the conclusion of the set of agreements states: "China will for its part repay the Soviet credit with strategic raw materials that are lacking in the Soviet Union, and this point is also advantageous to the Soviet Union." Mao ordered this sentence deleted.[596]

The Exchange of Notes on Annulling the Treaty and the Agreements of August 14, 1945, and on the Independence of the Mongolian People's Republic

We have already discussed the aspects of this exchange of notes between the two foreign ministries involving the annulment of the 1945 agreements.[597] The other subject of the exchange of notes concerned the status of Outer Mongolia.

Mao Zedong had expressed his interest in reintegrating the territory of the Mongolian People's Republic to Mikoyan in February 1949. Stalin had used a hypocritical argument to reject this demand immediately. Liu Shaoqi again raised the topic ambivalently to Stalin in his July 4, 1949, report during his stay in Moscow: China would recognize the independence of the MPR if the Mongolian people demanded it in a referendum consistent with the principles of the right of self-determination of nations. Should the MPR wish to unite with China, on the other hand, China would welcome this. Only the Mongolian people could decide this question, however.[598]

This last point allowed Stalin to assume that the Chinese leadership had put the Mongolian question aside for the moment, especially as Mao Zedong did not raise it again in the talk on December 16, 1949. Vyshinskii nonetheless instructed Durdenevskii, the leading Soviet expert on international law, to search the relevant documents from the years 1945 and 1946 in the archives of the Foreign Ministry just in case in preparation for negotiations with the Chinese delegation. On January 16, Durdenevskii presented, among other things, the Mongolian government protocol on the implementation of the plebiscite of October 20, 1945,[599] according to which 100 percent of participants had spoken out in favor of the independence of the Mongolian People's Republic; no contrary votes were registered.[600] The protocol reveals, however, that the national referendum was not secret; voters had to provide their signatures or—in the case of illiteracy—a thumbprint. The documents presented also included China's recognition of the independence of the People's Republic of Mongolia, which took place on January 5, 1946.[601]

The vote of 100 percent in favor of "independence" was in all probability correct, as it would have been pretty much suicidal to vote against it given the lack of secrecy in the voting and the Stalinist terror that prevailed at that time. It can be seen from a report by a Soviet spy that some Mongols were of the opinion that the plebiscite had been carried out on Moscow's initiative "to prepare for the following conquest of Mongolia and its incorporation into the USSR."[602]

A few days after the People's Republic of China was established, the government in Ulan Bator offered to establish diplomatic relations with the new Chinese government. On October 16, 1949, Zhou Enlai informed the Mongolian government that his government had resolved to establish diplomatic relations with the Mongolian PR.[603]

Because Mao had not put the Mongolian question on the table during the talk on principles with Stalin on January 22, the Soviet leader was even more surprised when Zhou addressed the topic directly in Mao's absence in the context of another issue. Zhou raised the possibility that the Chinese would make a statement on the independence of the Mongolian PR. Stalin reacted nervously: Had the Mongolian question not been resolved for a long time? Why a statement? He added hypocritically: How could they talk about this question when, after all, the Mongolian comrades were not present? Zhou then explained that the People's Republic of China had to make its position on this question clear because the new China had declared that none of the old treaties and agreements would be recognized, and it had been the KMT government that had recognized Mongolian independence. "Democratic figures [*minzhu renshi*]" in China were also asking the Chinese CP to do this, he said.[604]

Shi Zhe did not reproduce Zhou's last statement correctly; the Chinese declaration he was referring to did not state that all the treaties the KMT had concluded would be annulled, but rather that they would all be examined and then annulled, revised, or renewed.[605] However, Zhou was correct to suggest that given such a statement, they should not leave the government in Ulan Bator in the dark about the results of such an examination.

Stalin seemed relieved by Zhou's explanation and suggested that the Chinese government could make such a declaration first and that the Soviet government would then support it.[606] This is what happened. On January 31, Zhou Enlai handed the Soviet side a draft of an exchange of notes between himself and Minister of Foreign Affairs Vyshinskii that contained the following sentence in Point 2: "Both Contracting Parties recognize that the independent status of the Mongolian People's Republic has already been fully guaranteed as a result of the referendum carried out in Outer Mongolia in 1945, which confirmed its efforts to achieve independence, and as a result of the establishment of diplomatic relations with it by the People's Republic of China."[607] Confirmation of the content of the note by Vyshinskii, it stated, would make the exchange of notes a portion of the Sino-Soviet Treaty of Friendship, Alliance, and Mutual Assistance.

We can assume that the exchange of notes went forward as drafted by Zhou Enlai. Although it was not published, it was mentioned in the joint communiqué on February 14, 1950. The communiqué indicated that the two governments had determined in an exchange of notes between Zhou Enlai and Vyshinskii that "the independent status of the Mongolian People's Republic as a result of the referendum of 1945 and the diplomatic relations that had been established with the People's Republic of China were fully guaranteed."[608]

The two sides thus confirmed the status of the Mongolian People's Republic as an independent state pro forma and a Soviet protectorate in fact. This did not mean, however, that the CCP leadership had come to terms with its loss. A mere four years later, in September–October 1954, Zhou Enlai asked Khrushchev and Bulganin, on the occasion of their visit to Peking, whether Outer Mongolia could not become a part of China.[609] Khrushchev acted skeptical and once again suggested that the government in Ulan Bator was responsible for this question.

Intercultural Irritations

When Zhou Enlai presented Mao's request to undertake "excursions to factories, collective farms, museums, etc. during the next few days" to Vyshinskii on February 2, the Soviet side arranged for Mao and Zhou to visit an aircraft factory, the Stalin automobile works, a collective farm, and the Lenin mausoleum.[610] In addition, Mao visited the Chinese students at Moscow University and the Chinese embassy.[611]

The Soviet side also made suggestions to Mao and Zhou for a cultural program, which Wang Jiaxiang tested. Among other things, Vyshinskii suggested to Mao and Zhou that they should see the ballet *The Red Poppy* [*Krasnyi mak*],[612] which was very popular in Moscow at that time. The piece, first performed in 1927, was the first Soviet ballet on a modern subject. The action took place in revolutionary China, and Vyshinskii apparently believed that it would be of particular interest to the Chinese guests. The ballet told the story of a Soviet marine who met a prostitute in Shanghai, taught her revolutionary ideas, and arranged for Marxism-Leninism to spread in China. When Wang learned about the content of the ballet, he advised Mao not to attend. Mao shared Wang's opinion that this ballet exhibited great power chauvinism that was offensive to the Chinese people.[613]

To avoid offending the Soviet side, they sent Chen Boda with a few others to the performance that had been offered. It was to be a special performance attended by the famous composer of the ballet, the elderly Reinhold Glière. According to Fedorenko, who accompanied Chen, this nearly resulted in a scandal. Chen was offended by the way in which Russian dancers were made up to represent the Chinese—including a pimp—in bright yellow paint. They looked like monsters and bogeymen to him. Furthermore, Chen complained, the title of the ballet was itself offensive to the Chinese because poppies were the raw material for opium, which had been the ruin of the Chinese people for centuries. It was only with great difficulty that Fedorenko was able to keep the Chinese from leaving early. Following the incident, *The Red Poppy* was temporarily removed from the schedule and was given a new title, *The Red Flower,* in 1957.[614]

In his memoirs, Wu Xiuquan described the Chinese actions in the *Red Poppy* affair as an expression of their increased efforts "to preserve national dignity [*weihu minzu zunyan*]."[615] The incident reveals the Soviet hosts' lack of awareness of possible Chinese sensitivities and at the same time the Chinese guests' lack of familiarity with Western art forms. Chen Boda apparently also found it difficult to understand Western art on other occasions.[616] At a performance of *Swan Lake,* he was offended that all the female dancers performed "naked." He dismissed the television performance of a famous Russian bass with a loud laugh and asked why he kept shouting like a bull. Fedorenko, on the other hand, made fun of Chen's reactions in his memoirs. We should not overestimate the significance of these incidents as a "clash of cultures" (loosely following Samuel P. Huntington), but we can nonetheless assume that they had a negative effect on the atmosphere of the negotiations.

Signing the Treaty on February 14, 1950, and Mao Zedong's Return Trip

The treaty, the agreements, and the notes discussed in the last section but one were signed on February 14, 1950. They were originally supposed to have been signed in January.[617]

Difficulties arose in the negotiations during the final days before signing that resulted in the deadline being put off. On February 2, Mao informed the Soviet side that he planned to travel back to China on about February 8 or February 9.[618] But the exact date still had not been established three days before the actual signing. Zhou Enlai informed Mikoyan and Vyshinskii on February 11 that they could sign on February 14 or February 18, because no newspapers would appear in China for three days starting on February 16 due to New Year's festivities.[619] If they could make the trade agreement ready for signature in time, it should be signed on February 18.

On the evening before the signing, Mao and Zhou attended a performance of the ballet *Swan Lake* at the Bolshoi Theater with Li Fuchun, Wang Jiaxiang, Chen Boda, Saifudin, Molotov, Mikoyan, Vyshinskii, and Roshchin. Mao said he was very impressed by the technique of dancing on point.[620]

The signing took place in the Kremlin in formal surroundings. Minister of Foreign Affairs Vyshinskii and Zhou Enlai signed. Stalin and Mao took up positions in the middle behind them. According to Shi Zhe's account, Stalin, who was smaller than Mao, took care to stand two paces further forward in order to appear larger in the photograph—and he was successful, judging by the photograph published in the Soviet press. Zhou took pains to sign in the traditional Chinese style. The necessary writing implements—paper, brushes, inks, and inkstone—were brought from the Chinese embassy specially and set out on the table. Nine of the twelve Politburo members attended the signing ceremony:

Stalin, Molotov, Malenkov, Beriia, Voroshilov, Kaganovich, Mikoyan, Khrushchev, and Bulganin. On the Chinese side, Chen Boda, Wang Jiaxiang, and Saifudin, among others, attended.[621]

During the ceremony, the two foreign ministers made brief speeches in which they characterized the event as "a new, remarkable page . . . in the history of Soviet-Chinese relations" (Vyshinskii) and as "a sign of fraternal friendship and eternal cooperation between China and the Soviet Union" (Zhou Enlai). Unlike Vyshinskii, Zhou mentioned that the USSR was handing over to China without compensation the property taken from the Japanese in Manchuria and the houses on the former Soviet military property in Peking as an "expression of the great friendship of the Soviet government and of Generalissimo Stalin." This action would arouse a "feeling of the greatest enthusiasm" in the Chinese people. Unlike Vyshinskii, Zhou used the occasion of his speech to make an anti-American statement: the scandalous attempts of the "imperialist bloc" led by the United States to destroy the friendship between the two states had "failed definitively." In his speech, Zhou mentioned Stalin's name four times, always in conjunction with the title "generalissimo." Vyshinskii limited himself to mentioning Mao's name once ("the leader of the Chinese people, Mao Zedong").[622]

After the signing, Stalin invited everyone to a brief reception, which was finished at about noon. At 9 P.M., Ambassador Wang Jiaxiang held a reception in Mao Zedong's and Zhou Enlai's honor to celebrate the conclusion of the treaty, in which Stalin participated along with the Politburo members already named and other high-ranking Soviet personalities.[623]

Mao apparently considered Stalin's attendance to be extremely important. Three days earlier, Zhou Enlai had asked Mikoyan and Vyshinskii whether the Chinese side could count on "the presence of comrade Stalin at this reception."[624] If he could not attend, "it would be desirable" for Stalin to visit the embassy before the reception. Zhou asked urgently that this request be presented to Stalin, which the Soviet side agreed to do. Wang Jiaxiang's wife confirms that the Chinese were extremely interested in having Stalin participate, as this would demonstrate China's important international position and do great honor to China.[625] Wang gave Molotov the formal invitation in the name of "Chairman Mao and 500 million Chinese" and reminded him urgently that Stalin's presence at the reception was "the wish of the Soviet and of the Chinese people" and "the hope of the working classes of the entire world." According to Shi Zhe, Mao expressed to Stalin the hope that he would come to the Chinese reception, even if perhaps only for a short time.[626] Stalin replied that he had never before attended a reception of this kind outside the Kremlin, but that he would make an exception and even remain to the end. Fedorenko confirms that a conversation of this type took place, but gives it a slightly different tone.[627] According to him, when Stalin asked in annoyance why Mao did not want to hold the reception in the Kremlin, the Chinese leader replied that the Kremlin was the place for Soviet state receptions and not entirely suitable for China, which was "a sovereign state." When Stalin pointed out that he never attended receptions in restaurants or foreign embassies, Mao asked him urgently to come ("Our reception, and without you . . . No, no, that is unthinkable") and ultimately received Stalin's approval.

The reception, which was attended by some 500 guests, took place in the distinguished hotel Metropol', built in the czarist period and located not far from the Kremlin. Wang Jiaxiang formally acted as host and did the honors with his wife dressed in the traditional *qipao*. Stalin appeared punctually at the head of the participating Politburo members and was greeted—at least according to Wang Jiaxiang's wife[628]—with an embrace from Mao. Zhou Enlai toasted Soviet-Chinese friendship—which would survive for generations, he said—thanked the Soviet Union for its selfless assistance, and stated that China would learn from its elder brother. After a few other toasts, Stalin also rose to make a toast. In it, he agreed with Zhou Enlai but then, surprisingly, attacked Yugoslavia. It had placed itself outside the family, wanted to take an impassable separate path, and would return to the family sooner or later. Fedorenko translated Zhou's speech; Shi Zhe translated Stalin's. Mao limited himself to drinking to Stalin's health and Sino-Soviet friendship. The reception ended at midnight. Mikoyan remained for a while longer to discuss issues involving economic cooperation and trade with Zhou and Li Fuchun.[629]

There was a reception in Peking on February 15 to celebrate the signature of the treaty to which Liu Shaoqi had made the invitations. Approximately 700 people attended, including members of the Chinese government, employees of the Soviet embassy, and ambassadors of other states under Communist rule.[630]

The Soviet and Chinese media celebrated the signing of the set of treaties as an event of the greatest significance. *Pravda* and *Izvestiia* printed the communiqué, the treaty, the two agreements, and the foreign ministers' speeches word for word the following day on the first two pages. The reception in China was downright bombastic. The entire front page of *Renmin ribao* was filled with photographs of Mao, Stalin, Zhou, and Vyshinskii, Red Square and Tiananmen Square, and the following slogans: "We welcome the new era of friendship and cooperation between China and Soviet Union!" "Long live the friendship and cooperation between the two great states of China and the Soviet Union!" "Long live the leader of the toiling people of the world, Generalissimo Stalin, and the leader of the Chinese people, Chairman Mao Zedong!" With the last slogan, Peking clearly acknowledged the USSR's leadership role in international communism. The communiqué, the treaty, the two agreements, and Vyshinskii's and Zhou's speeches were reproduced word for word on pages 2 and 3. Both sides celebrated the event in grandiose commentaries.[631]

On February 17, Mao and Zhou began the journey back to China. The evening before, Stalin arranged for them and the other members of the Chinese delegation a farewell banquet in the Kremlin that was again attended by almost the entire Soviet leadership elite. Ho Chi Minh, who was also present, asked Stalin jokingly why they did not use the opportunity also to conclude a treaty with his government. Stalin avoided the question.[632]

On the day of their departure, which was the Chinese New Year, Molotov drove specially to Mao's dacha to bring him Stalin's best wishes and accompany him to the automobile that would bring him to the station. In the evening, Mao and Zhou ate at the Chinese embassy together with the embassy employees and Chinese studying in Moscow and then drove to the station. Along with Molotov, the Politburo members Mikoyan and Bulganin, Minister of Foreign Affairs Vyshinskii, Minister of Foreign Trade Men'shikov, Ambassador Roshchin, other high-ranking Soviet individuals, and the ambassadors of the states under Communist rule were present on the station platform. Mao's departure was honored with the usual ceremony. He made a brief, emotional speech,[633] in which, as was to be expected, he sought to give the impression that his visit had gone over in unspoiled harmony. It would be difficult to describe in words the "perfect understanding and deep friendship" that had resulted. He had met with warm hospitality in Moscow and Leningrad. The union of the two peoples, he said, was "eternal [*yongjiu*], indestructible, and inseparable." It would influence all humanity in the future and would lead to the victory of justice and peace in the entire world. Mao again referred to the relationship of teacher and pupil: Soviet experiences would provide the model for the construction of the new China. Mao concluded with a salute to the Soviet dictator that would be hard to surpass: "Long live Comrade Stalin, teacher of the world revolution, and best friend of the Chinese people!"[634]

Among others, Chen Boda, Shi Zhe, Wang Dongxing, and Ye Zilong, who had accompanied Mao to Moscow, traveled with Mao and Zhou. They were joined by the secretary of the City Committees of Dairen and Port Arthur, Ouyang Qing, and Ho Chi Minh. Ye Jizhuang, Wu Xiuquan, and others, who had traveled to Moscow with Zhou, remained behind to continue to negotiate the agreements that had not yet been signed.[635]

A Soviet deputy minister of foreign affairs accompanied Mao and his entourage as far as the Chinese border. Mao resolved to look around at every station at which the train stopped. In Sverdlovsk, he visited an engineering works and a combined power and heating plant and met with university professors. In Novosibirsk, the group visited an arms factory and an aircraft plant; in Krasnoiarsk, a plant producing harvest equipment; and in Chita, a kindergarten. After visiting a tea factory in Irkutsk, Mao complained to his Chinese companions about the lack of understanding for tea among the Russians, who crumbled it and made no distinction among the various types. In several places, he attended cultural performances. Mao refused only an invitation to tour the city in Ulan-Ude, capital of the Buryat Mongol Autonomous Soviet Republic, to the disappointment of the local elite. Shi Zhe indicates that this was because Verkhne-Udinsk, as the place was called until 1934, had in Mao's opinion belonged to China before it fell to czarist Russia.[636]

When they crossed the border at Manzhouli on February 26, Mao and Zhou sent telegrams in which they thanked Stalin and Vyshinskii for the "cordial and friendly care" and wished the USSR and China "eternal friendship and cooperation." Mao did not travel to Peking directly, but made short stops in Harbin, Changchun, Shanhaiguan, Tangshan, Tientsin, and other places to carry out inspections. In Mukden, where the group remained for several days, he made a brief visit to the Soviet experts. He had previously ordered that his itinerary be kept secret. On their arrival in Peking on March 4, leading Party and government representatives—under the leadership of Liu Shaoqi and Zhu De—met Mao and Zhou. The Soviet chargé d'affaires and the ambassadors of Czechoslovakia, North Korea, Hungary, and Poland also came. Mao had previously instructed the responsible agencies in Peking that only about 100 individuals could appear to greet him. His arrival was to be kept secret until the last minute.[637]

Negotiations for the Agreements Signed in March and April 1950

Wang Jiaxiang, Li Fuchun, Minister of Foreign Trade Ye Jizhuang; Wu Xiuquan, the head of the department of the Foreign Ministry responsible for the Soviet Union and

Eastern Europe; Liu Yalou; and Saifudin remained in Moscow after Mao Zedong's and Zhou Enlai's departure to negotiate further the agreements that could not be signed on February 14, because the Chinese side had not yet agreed to the existing text. The negotiations involved two agreements on the founding of mixed joint-stock companies to mine raw materials in Sinkiang, an agreement on the founding of a mixed joint-stock civil aviation company, a secret agreement on the working conditions of the Soviet specialists, and the trade agreement.[638]

Because the issues essentially involved economic cooperation, the negotiations were held in the Soviet Foreign Ministry. Ambassador Wang Jiaxiang was in charge of negotiations for the Chinese. His partners were, among others, Mikoyan and Deputy Minister of Foreign Affairs Gromyko. Wu Xiuquan reports that differences of opinion often emerged that made all participants feel very uncomfortable. One of the most controversial questions, according to his account, was the exchange rate between rubles and yuan. The Soviet side had sought to take advantage of the Chinese delegation here. Because the Chinese were the suppliants, Wu continued, they had no choice in the end but to give in and accept the establishment of a rate that was unfavorable to China. It was again a question of money, namely dollars, rubles, and yuan, when a dispute arose over the payment of the Soviet experts.[639]

The two sides were unable to achieve the goal of signing the agreements by March 5 at the latest.[640] The three joint-stock companies were founded on March 27, and the agreement of the Soviet experts was signed on the same day; but the trade agreement and an accompanying agreement and three protocols were not concluded until April 19. We will discuss the negotiations in detail in the following section.

The Secret Agreement of March 27, 1950, on the Working Conditions of the Soviet Specialists in China

The Chinese had already presented their request that the USSR send specialists to China to act as advisers in a wide range of areas to Mikoyan in Xibaipo in February 1949. When Stalin gave his approval in principle at the end of June 1949 to Liu Shaoqi, he insisted among other things that the question of salaries would have to be clarified first. The specialists would have to receive as many provisions as the best Chinese specialists. The Soviet side would if necessary pay something in addition. At the final meeting with Liu in mid-August, Stalin set among other things the condition that the Chinese government would have to arrange favorable living quarters and generous wages for specialists. It would also be responsible for maintaining family members. Should they become involved in criminal acts, the experts would be subject to Soviet and not Chinese jurisdiction—a demand that the Chinese could only find offensive because it was reminiscent of the special privileges for foreigners in the old China. At this time, the Peking leadership had already approved an agreement on the working conditions of the Soviet specialists that was negotiated between Stalin and Liu at Party level and was clearly very general.[641]

On January 22, the Soviet editorial commission presented a draft of an agreement on the salary conditions for specialists who were to be sent to work in Chinese institutions, organizations, and factories to Stalin.[642] The text is not yet accessible. On February 19 at the latest, the Soviet side presented a corresponding draft to the Chinese delegation, which is not available, either. [643]

On February 11, Zhou Enlai presented several questions about this draft to Mikoyan and Vyshinskii for clarification, and also referred in passing to payment problems. The Chinese side, he said, could not understand the proposed regulation in Article 4 of the draft, whereby China would have to pay 2,000 to 4,000 rubles each month in compensation for each specialist. Were these amounts to be paid to maintain the specialist's family, or were they to be paid to the company in which the specialist had worked previously? Mikoyan made clear that the amounts were meant as compensation for the individual companies and would, of course, have to be paid through the Soviet state. Zhou then explained that China would have to raise large amounts of money if the payments were proposed in U.S. dollars. It would be easier for the Chinese government to pay in goods. Mikoyan replied that the draft called for several different solutions for a payment agreement: deliveries of goods, gold, dollars, and pounds. To a corresponding question from Zhou, Mikoyan explained that the agreement referred to all types of specialists (military, teaching staff, etc.). Zhou asked finally whether they could fix the period of time for which the specialists were to be sent out at two to three years. Mikoyan and Vyshinskii pointed out that one year was called for at first, but that it was possible to make extensions.[644]

On the following day, a dispute arose between Zhou and Roshchin over the amount of compensation to be paid for each specialist. Zhou declared that China simply could not afford to pay the equivalent of US$500 to $1,000 per month for each technical adviser, as the Soviet side demanded. This was more, he said, than Chairman Mao and his ministers received. By using the exchange rate of 1:4, this came to the amount of 2,000 to 4,000 rubles. Zhou demanded that the Soviet side should state its agreement to an amount in yuan equivalent to four times his own monthly income. On February 16, one day before his departure, Molotov and Vyshinskii told him that they could not give in on the principle of salaries, but that the Soviet government was prepared to accept monthly compensation in the amount of 1,500 to 3,000 rubles. Zhou thanked them for this concession.[645]

In a further talk that took place in early March, Wang Jiaxiang asked Gromyko whether the numerous Soviet military personnel who were in China should be considered specialists in the sense of the agreement and how they should be paid. Gromyko agreed to an investigation. Wang also proposed a reciprocal regulation of the conditions for recalling specialists: both sides would be entitled to make a recall, and the side recalling the specialist would be responsible for covering the costs. The Soviet draft provided only for recalls by Moscow, in which case the Soviet side would have to cover the costs. Gromyko refused Wang's request and referred to the fact that the regulation in the Soviet draft was the same as the ones that had been provided in corresponding agreements with other countries. For that reason, he said, "it is difficult for us to alter this principle." Wang also requested that a communiqué on the conclusion of the agreement be published, as the Chinese press had already reported on the arrival of Soviet specialists. Gromyko replied that this had not been done in corresponding agreements the USSR had concluded with other states, but he agreed to look into the matter.[646]

The Chinese side was also concerned about how it would be able to create a ruble reserve so that it could make payments to the specialists from it. Wang Jiaxiang proposed to create one from the revenues from the sale of soybeans to the USSR. Gromyko postponed discussion of the topic to the negotiations on the trade agreement.[647]

Both sides bargained hard over payment of the lowest salaries for military advisers. Wang Jiaxiang considered the payment of 1,500 rubles, which the Soviet side demanded for a sergeant (the rank held by most of these advisers), to be set too high and pointed out that they received salaries of only 500 to 600 rubles in the Soviet Union. Gromyko remained immovable.[648]

On March 22, Wang Jiaxiang told Gromyko that the Chinese side agreed to the conditions the Soviet side proposed for the payment of the specialists, "although they are harsh in the section involving compensation of the sergeants." With respect to travel costs (Article 3, Section 2), Wang suggested a solution that would allow Peking to pay the costs of the portion of the trip that took place in China in yuan. The Soviet draft had clearly called for payment of the entire trip in rubles. Wang also wanted to see that Article 8 made clear that the agreement was also effective for the specialists who had been sent to China before the agreement was signed. Wang gave Gromyko a draft to this effect.[649]

On the following day, Gromyko agreed to reduce the sergeants' monthly salary to 1,000 rubles and accepted Wang's other suggestions. The agreement was signed on March 27.[650] The Soviet side later increased the agreed salaries by 15 percent without consultation.[651]

An examination of the text and available documents on the negotiations reveals that the Chinese side was able to get Moscow to make considerable concessions if one compares the results of the negotiations to the original Soviet requests. This includes something that although intangible was very significant to the Chinese: The agreement does not stipulate that the experts would be subject to any special Soviet jurisdiction. Stalin thus withdrew the demand he had made to Liu Shaoqi during the summer of 1949 that the Soviet specialists would have to answer before Soviet and not Chinese courts in case of any infractions.[652]

In this respect, Bruce A. Elleman's thesis that "the USSR's special rights in China were renewed during 1950, when Stalin and Mao signed their own secret agreement granting extraterritoriality to Soviet citizens" is not correct.[653] Elleman bases his statement exclusively on a passage in Sergei N. Goncharov, John W. Lewis, and Xue Litai.[654] It stated that the most humiliating demand that Stalin made with respect to the Soviet specialists in China was the one in which they would be subject to Soviet and not Chinese jurisdiction. Goncharov, Lewis, and Xue in turn base this statement exclusively on Shi Zhe.[655] He in turn does in fact mention "some harsh conditions [*yixie keke de tiaojian*], among them extraterritoriality, that Stalin had made, but does not say anything about when or under what circumstances Stalin made his demands. He was probably referring—without having mentioned this explicitly—to the farewell talk between Stalin and Liu Shaoqi that apparently took place on August 13, 1949, and about which Liu reports in his memoirs.[656] Shi Zhe, however, says nothing about whether the Chinese side met Stalin's demands. Despite this, and without citing any other sources, Goncharov, Lewis, and Xue, who had not seen the text of the agreement, declare that "the Chinese had no choice but to bow to Stalin's demands." The Soviet archival materials that the author has since seen contain no indication that Stalin presented his demand for extraterritoriality again during the negotiations between January and March 1950. If he did in fact do so, the text of the agreement shows that he was unable to implement it, as this subject would necessarily have had to be regulated in the agreement.

If Stalin's demands for extraterritoriality had in fact been met, this would have meant a relapse into a colonial tradition that was humiliating to China. In this context, Shi Zhe says indignantly that Stalin's demand followed a tradition common to the "Western imperialist states" and revealed great power chauvinism.[657]

Shi Zhe should have added that Stalin was also attempting hereby to continue a Soviet tradition. In a treaty with China on May 31, 1924, the USSR had agreed to "relinquish extraterritorial rights and consular jurisdiction."[658] The Chinese government used it as an argument to convince the Western states to surrender their special rights. Washington and London in fact renounced these rights in January 1943. Recent findings have shown that the Soviet Union retained certain special rights de facto even after the treaty of May 31, 1924, and had them confirmed de jure in a little-regarded treaty in 1929 and a secret protocol to a trade treaty in 1939.[659] In the latter case, these rights involved Soviet trade representatives in China. After tough negotiations, Stalin and T.V. Soong agreed on a secret protocol on August 14, 1945, by which Soviet military courts and authorities should have exclusive jurisdiction over all their troops in Port Arthur. It is difficult to say whether Shi Zhe was unaware of these facts or whether he was simply seeking to protect the Soviet Union.[660]

The Council of Ministers of the USSR issued a "directive for the Soviet specialists active in factories and institutions in the Chinese People's Republic" on March 21, 1950.[661] Among other things, it obligated the advisers "to preserve the dignity of the Soviet Union without fail" but also to be conscious of the fact that "the toilers of China feel love and respect for the Soviet Union and its representatives," which bound every Soviet specialist "to treat the Chinese specialists and Chinese population in a friendly manner and without arrogance."

The Chinese were able to reduce the amount of the compensation to be paid by one-fourth during the negotiations in February and March 1950—and in one military category by even more—and to arrange to pay travel expenses incurred on Chinese soil in yuan. The Soviet side also accepted the Chinese suggestion to extend the validity of the agreement to cover the advisers sent out before the agreement was signed. The Chinese delegation was unable to push through its demand for China to have the right to recall the Soviet specialists included in the agreement. Stalin did not yield to the Chinese request to report on the agreement publicly, either. It remained secret.

Soviet specialists, ultimately 1,500 in all, remained active in China until 1960. In the course of the increasingly tense relations between China and the Soviet Union, the Soviet embassy in Peking informed the Chinese government on July 18, 1960, that it would recall all Soviet advisers.[662] As grounds, it stated that the Chinese side used the advisers for anti-Soviet propaganda purposes, ignored their suggestions, and spied on them. The Soviet embassy did not refer to the agreement on the working conditions of the Soviet specialists of March 27, 1950, here. The half-year notice period required by Article 9 of the agreement was not respected.

The Agreements of March 27, 1950, on the Establishment of Two Mixed Joint-Stock Companies to Mine Raw Materials in Sinkiang (with Secret Protocol)

Agreements establishing two mixed Sino-Soviet joint-stock companies were signed on March 27. One company was to drill for oil and natural gas and the other to mine nonferrous metal in Sinkiang, apparently primarily gold, copper, lead, and aluminum.[663] The texts of the agreements are not yet accessible, but they were reported briefly in a coordinated statement in the Soviet and Chinese press at the time.[664]

It stated that the companies had been established on the basis of parity and served the purpose of "promoting Chinese national industry" and "strengthening economic cooperation between the USSR and the People's Republic of China." They would prospect for and extract oil, natural gas, and nonferrous metals; in the case of oil and natural gas, they would also process the raw product. Half of the production and the profit would be for the benefit of each of the two countries. The position of head administrator would alternate between the two sides every three years, beginning with the Chinese; the Soviets would initially nominate the deputy chief. The same was true for the director general and his deputy; in this case, a Soviet would hold the top position first. The two sides would then alternate these positions every three years. Half of the employees would be Chinese and half Soviet citizens, and here, too, the occupation of positions would alternate between the two sides. The agreements were to remain in force for thirty years and were signed by Minister of Foreign Affairs Vyshinskii and Ambassador Wang Jiaxiang.[665]

The Chinese side later occasionally accused Stalin of forcing the agreements on the Chinese to derive unilateral advantages from them. Thus Mao, as mentioned before, is supposed to have said that he had finally returned all the bitter fruits Stalin had forced him to swallow when the companies were dissolved and the secret supplementary agreement to the treaty of alliance was nullified under Khrushchev.[666] Shi Zhe claimed that "Soviet great power chauvinism" could be seen in both agreements.[667]

Khrushchev several times said something similar, too. The treaty, he stated, that the USSR had concluded under Stalin with China on the joint exploitation of the mineral wealth of Sinkiang was a mistake and "even an insult to the Chinese." The French, English, and Americans had occupied China before, "and now the Soviets are also creeping in." This was an unthinkable thing (*nemyslimoe delo*). Stalin had done exactly the same, he added, in Poland, the GDR, Bulgaria, Czechoslovakia, and Romania. Khrushchev said elsewhere that "such mixed companies had been organized by us in all the European countries of the People's Democracy"; the Romanians, for example, had been very upset about it. Peking had viewed the establishment of these companies as an attack on China's independence and an incursion on its territory that bordered on the Soviet Union. They should never have done this with China or with the other brother states, Khrushchev said. The Chinese had accepted the Soviet suggestion without any objections although they naturally were unhappy.[668]

Wu Xiuquan, who translated the texts of the agreements at the time, said that the Chinese side also derived benefit from the cooperation, however. China then did not have the technology to exploit its mineral resources. The Soviet side may indeed have derived great profits from the agreements, but China also benefited.[669]

Khrushchev and Mao's statements give the impression that Moscow had made the suggestion to establish the companies. This is not true. As we have already shown, previously unknown Russian archival documents show that the initiative for establishing the mixed companies came not from Stalin but from the Chinese leadership. In a telegram to Mao, which the latter forwarded to Stalin in early January 1950, Peng Dehuai advocated the creation of Sino-Soviet companies to drill for oil and mine rare and nonferrous metals.[670]

Mao had received a telegram from Liu Shaoqi the same day on which he also advised Mao to ask the Soviet leadership to establish joint companies to extract nonferrous metals and oil in Sinkiang.[671] Referring to the "opinion of the comrades in Peking," Liu did, however, add that they should not form such enterprises in Sinkiang alone or only with the USSR and the "countries of the New Democracy" as the comrades in Sinkiang demanded. It could also be done "with groups and industrialists from the capitalist states." It is not known whether Mao shared Liu's view or, if so, whether he made the telegram or its contents available to Stalin. If he did so, it can be assumed that Stalin rejected it firmly. As we have already shown,[672] he made the Chinese side agree to a secret agreement prohibiting foreign actions in Sinkiang and Manchuria.

It is striking that Stalin agreed to two other requests for aid in his response to Peng Dehuai's telegram and other telegrams Mao sent him the same day, but did not mention Sinkiang or the two mixed companies.[673] One month later, Zhou Enlai raised the issue again.[674] He informed Mikoyan and Vyshinskii that the Chinese side had access to drafts of agreements on the exploitation of oil and metals that Zhang Zhizhong had previously discussed with the Soviet Union. The Chinese side had planned to begin new talks in Urumchi on the basis of these drafts. But because there was a delegation from Sinkiang in Moscow, they could begin preliminary talks there and conclude them in Urumchi. Each of the two sides would have to make half of the capital available for the joint companies. The administration would have to be organized like that of the CCR, that is, the positions would have to alternate between the two sides. The companies would have to be active for a longer period of time than the planned civil aviation company. Vyshinskii's report on this round of negotiations does not include any indication of the Soviet side's reaction.

If the Chinese felt that they were treated incorrectly with respect to these ventures, it could not be because the Soviet side had forced them to create the two companies. The real reason was apparently that the Chinese side had to accept contractual obligations that were disadvantageous to them despite the obligation to divide production and profits stipulated in the agreement. There are indications that allow us to conclude what was involved. First of all, Khrushchev indicates in his memoirs that the entire production of both companies had gone to the Soviet Union.[675] This sounds like exportation free of charge, which, given the clear requirement in the agreement that both sides have equal rights in the companies, at first glance seems somewhat implausible.

The Soviet archival documents on the negotiations made accessible to us reveal what was really involved, and we will reproduce their contents in detail below. They show that the USSR was entitled by contract to purchase "a part" of the Chinese share of the production of oil and nonferrous metals with its dividends. This provision had apparently been forced on the Chinese side with some pressure, as the Chinese naturally had to be interested in selling their share on the world market for hard currency, which they lacked. If we take Khrushchev's claim at all seriously, the USSR could in practice arrange, one can imagine, to purchase a very significant proportion of the Chinese share of the production.

Khrushchev confirms this interpretation indirectly. Speaking of Sino-Soviet mixed companies, he observed:

> I consider it for example insane and a betrayal that all the goods and raw materials that could be sold on the world market [*valiutnye*] that North Korea and China produced and extracted went to the Soviet Union. Each country must naturally have its hard currency [*valiuty*] in order to go on the markets of the capitalist market. . . . China also had such needs, . . . But Stalin was deaf to this; he did not understand and did not want to understand, especially after they had defeated Hitler.[676]

The fact that a secret protocol to the agreement on the Sino-Soviet mixed companies in Sinkiang prohibited Peking from selling the raw materials it produced to third countries without the approval of the USSR also served to restrict China.[677]

In the following pages, we will summarize the most important findings from the Soviet archival documents on the negotiations on the establishment of the companies that are accessible to us. They allow us to see that the Chinese side was able to have its way in part or in full, at least with regard to certain demands.

Molotov, Mikoyan, and Vyshinskii presented the first drafts of the two agreements, which are not yet accessible, to Stalin on January 22.[678] They called for the establishment of two common joint-stock companies with the names Sovsin'metall and Sovsin'neft'. The texts were presented to the Chinese at an unknown time, but while Mao and Zhou were still in Moscow. At Zhou's request, the Soviet side made corrections that would adapt them to the text of the agreement on the establishment of a mixed civil aviation company. After Mao and Zhou left, Wang Jiaxiang; Wu Xiuquan; the Deputy Prime Minister of Sinkiang, Saifudin; and the head of the CP in Sinkiang, Deng Liqun, took part in the negotiations on the Chinese side. Mikoyan, Gromyko, and officials from the Foreign Ministry negotiated on the Soviet side. On February 27, Gromyko urged Saifudin to take a position on this draft as soon as possible so that the agreement could be signed on March 2. Saifudin agreed to reply in the next few days. He was expecting Mao and Zhou to take a position that same day or the next. Gromyko reminded him that Zhou had also expressed the desire that the remaining agreements be signed between March 2 and March 5. Saifudin agreed and said the Chinese side would do its best to sign the agreements as quickly as possible.[679]

On March 2, Wu Xiuquan objected to Gromyko that a map called for in the draft agreement was missing. He admitted that the Chinese side had delayed in taking a position on the draft and said that it would be ready by March 4. Gromyko again urged him to hurry. On March 4, Saifudin received the map Wu had requested. Again he delayed presenting the Chinese position to Gromyko, because Mao would arrive in Peking only on the night of March 4–5, which was later than expected, and they had previously expected him to have given his instructions from Mukden. Four days later, Gromyko became even more insistent and pointed out that "such a long delay in the response" interfered "with our plan for the year 1950."[680]

On March 10, Wang Jiaxiang was finally able to make a statement on the issue. After requesting a few editorial changes, he asked that the specific value—US$350,000—of construction materials the Chinese side was supposed to contribute to the joint-stock company be omitted from Article 5 of the draft, which dealt with the mining of nonferrous metals, because this value might in actuality be greater or smaller. Gromyko refused, but promised to look into the question when Wang insisted. Gromyko refused a Chinese request to delete the passages in Article 14 of both agreements stating that the goods purchased by the Soviet side with its dividends would be duty- and tax-free. But he promised to look into another of Wang's requests: that the Soviet share of the production of oil in Sinkiang could be sold only to the Chinese state trade organizations.[681]

Mikoyan and Wang Jiaxiang agreed on March 15 that the 50 percent share of the production of nonferrous metals that the Soviet side purchased could be exported to the Soviet Union duty- and tax-free, while the corresponding share of the production of oil could either be sold to Chinese state organizations or exported customs- and tax-free to the Soviet Union. On the following day, Wang Jiaxiang sought to arrange that a Chinese export duty could be imposed on the portion of the Chinese share of the production of nonferrous metals and oil that the Soviet side purchased. Gromyko refused. When Wang insisted and referred to instructions from Zhou that dated back to the time when he was still in Moscow, Gromyko pointed out that this was not consistent with the agreement that had been reached with Mikoyan on March 15, but ultimately agreed upon on March 18. Wang proposed adopting an additional provision in Article 8 of both drafts whereby the companies could build or lay railroad and road connections or electrical lines only with the "approval of the Chinese authorities in each individual case." Gromyko considered this provision unnecessary, but also agreed to it on March 18.[682]

The Soviet side accepted the request presented by Wang Jiaxiang to publish a joint communiqué about the conclusion of the two agreements and to make a draft of this document available to the Chinese side before the agreements were signed. This took place on March 23. Wang could not guarantee to Gromyko on March 18 that the agreements could be signed two or three days later, as he wished; he would have to "transmit all the documents to Peking." Four days later, Wang informed Gromyko that the texts would probably be ready for signature on March 25, so that they could actually be signed that day or on March 27. Vyshinskii and Wang Jiaxiang then finally signed them on March 27. The communiqué stated that the negotiations had taken place "in a friendly atmosphere and in the spirit of complete mutual understanding"—which was not quite true. The fact that the Chinese side did not consider publishing the full texts can be seen as evidence not only that the implementation of the agreement favored the Soviet side but that the agreement itself contained provisions—already discussed above—that the Chinese public might consider disadvantageous to China.[683]

The Chinese CP leadership was well advised not to publish the agreements. This was even more true of the secret protocol that was apparently also signed. The report on the conclusion of the agreements described above that was published on March 29 in *Renmin ribao* already provoked protests in patriotic circles in China. On the following day, Liu Shaoqi drafted an internal Party directive stating that "significant unrest" had arisen among Peking students because they feared that the agreements would damage Chinese sovereignty. Numerous members of the Communist youth association asked questions, demanded explanations, and went so far as to attack "the USSR as aggressors and the People's Government as traitors." They wanted to leave the association and submitted petitions to the Central People's Government. Liu took this as an excuse to demand that such enterprises be formed not only with the Soviet Union and with the "countries of the New Democracy," but under certain conditions with capitalist countries as well. They could also lease such enterprises, Liu continued, to capitalist enterprises, just as the USSR had done after 1921 in the context of Lenin's "new economic policy." The Central Committee was preparing a statement on this.[684]

As mentioned above,[685] Liu had already suggested also involving companies from the capitalist states in the exploitation of nonferrous metals and oil in China to Mao Zedong two months earlier. Stalin had then forced the Chinese side to accept a secret agreement prohibiting foreign actions in Sinkiang and Manchuria on February 14. Because Liu was aware of the agreement, his demand could only apply to territories outside the two aforementioned regions. It would appear, however, that his March 30 draft did not become a directive, or that the directive, if it was in fact issued, was not implemented. According to Russian accounts, not one non-Soviet foreign enterprise was active in China until 1978.[686]

We do not know any details of the practical operations of the two companies after they were established. We do know, however, that they were dissolved along with two other mixed companies on January 1, 1955. Khrushchev apparently provided the impetus for this when he visited China in the fall of 1954 at the head of a Soviet government delegation. The joint communiqué on October 12, 1954,[687] that announced the dissolution of the companies stated that their creation had extended the productive force of the Chinese companies and had played a positive role in the reconstruction of the Chinese economy thanks to progressive Soviet experience in economic construction. The two sides agreed to transfer the Soviet shares in the companies to China and pay off their value in the form of goods over many years.

Khrushchev commented on this incident as follows in his memoirs: "After Stalin's death, we liquidated all the companies of this kind. The creation of mixed companies offended national pride [*samoliubie*] and constituted an attack on the material resources of another country."[688]

The Agreement of March 27, 1950, on the Establishment of a Mixed Joint-Stock Company to Provide Civil Aviation

There were apparently fewer complications in the negotiations for the agreement on the establishment of a mixed joint-stock company to provide civil aviation. During his visit to Moscow at the beginning of July 1949, Liu Shaoqi had already expressed the desire to set up a Sino-Soviet airline and establish routes from the USSR via Harbin and Mukden to Dairen, from Mukden to Peking, and from Peking via Ulan Bator to the Soviet Union.[689] In December, Mao expressed the desire to Stalin and Kovalev to establish air routes via

Sinkiang and the Mongolian People's Republic and to conclude an aviation agreement.[690] The Soviet general Avseevich also traveled to China "to prepare concrete measures."[691]

The text of the agreement is not yet accessible. According to the press report published at the time, the company was founded on the basis of parity, with an equal participation in the profits, and was designed to promote "the development of Chinese national aviation" and strengthen economic cooperation between the USSR and the People's Republic of China. Three routes were planned: between Peking and Chita, Peking and Irkutsk, and Peking and Alma-Ata. As in the companies created for the exploitation of raw materials in Sinkiang,[692] the two sides would alternate control of leadership positions, with the head of the administration in the hands of a Chinese and the position of deputy in the hands of a Soviet citizen for the first two years. The position of director general would be occupied in reverse order. The positions would then be alternated between the two sides in each successive two-year period. With regard to the employees of the company, the agreement specified only that they had to be Chinese or Soviet citizens; there was no indication that they had to be in equal numbers or alternate positions, as in the case of the companies established for the mining of raw materials in Sinkiang. The difference can clearly be explained by the fact that the Soviet staff necessarily had to be in the majority because they possessed the most technical training and, in most cases, could not be replaced by Chinese. Unlike the agreements on the Sinkiang companies, which were in force for thirty years, the validity of this agreement was limited to ten years. Minister of Foreign Affairs Vyshinskii and Ambassador Wang Jiaxiang signed the agreement.[693]

Molotov, Mikoyan, and Vyshinskii presented Stalin with a first draft of the agreement on January 22 (like the final agreement, this draft is not yet accessible).[694] Negotiations began four days later. Gromyko and Minister of the Air Force Liu Yalou, who had negotiated with Minister of the Army Vasilevskii in August 1949 in Moscow, headed the delegations during the working sessions.[695] On January 26, the Soviet side presented an initial text, which they referred to as a "working document" that had not yet been approved by the Soviet government and Vyshinskii. The document was based on General Avseevich's report. The Soviet side then changed it in response to suggestions Zhou Enlai made to Mikoyan and Vyshinskii. These individuals agreed on three routes, one of which led across the Mongolian People's Republic.[696]

On February 5, an exciting and at the same time sensitive dialogue developed between Gromyko and Liu Yalou. It turned out that the Soviet side did not at first consider it necessary to ask the Mongolian government to approve the flights across its territory before concluding the agreement. Liu, on the other hand, considered it "appropriate" to ask the Mongolians before they signed the agreement. It is difficult to say whether Liu was really concerned about respecting Mongolian sovereignty or whether he merely wished to embarrass Gromyko by revealing the fact that the Mongolian People's Republic was not a sovereign state but merely a Soviet protectorate. The motive for the second alternative, if it is correct, is clear: resentment of the fact that Stalin was not willing to discuss the idea of changing the status of Outer Mongolia with the Chinese comrades.[697]

The Chinese invested US$3.7 million. The company planned fourteen airports for use in China. Disputes arose over the degree to which the two governments should be responsible for the maintenance of the airports used and over whether only the company's dividends or also its circulating capital should be subject to taxation. Liu Yalou pointed out that two private airlines existed in China at that time, and the Americans held 25 percent of the capital of one of them. If the Sino-Soviet joint-stock company they planned to create was taxed significantly less than the other two companies, the "Democrats included in the government, our national bourgeoisie, and the foreigners would make a lot of noise about our great concessions." The working draft would practically relieve the company of taxation, as only dividends would be taxed. It would therefore be better, Liu said, to raise taxes on the circulating capital, as was done under the KMT, and not the dividends. Gromyko objected that it was common internationally to impose a somewhat higher tax nominally but that the tax did not come to more than 15 percent to 20 percent in fact, so that the 20 percent the Soviet side suggested reflected the usual maximum. We do not know how they finally resolved this question.[698]

One can sense the highly developed Chinese sensitivity to issues involving sovereignty when Liu Yalou demanded that the company be permitted to purchase real estate, factories, and the like only with the approval of the Chinese government. Liu added that this was also necessary "in order not to establish a precedent for other foreign companies." Gromyko considered it unproblematic to include an arrangement of this kind.[699]

There are several striking inconsistencies in the discussion about when the agreement would be ready for signature that reflect the Chinese side's failure to agree on a single position. On February 11, Zhou Enlai informed Mikoyan and Vyshinskii that he considered the existing text of the aviation agreement approved.[700] Two days later, Zhou and Wang Jiaxiang surprisingly included the document among the agreements that the Chinese side did not consider to be sufficiently revised to be ready for signature as anticipated on February 14.[701] On that day, Gromyko and Liu Yalou observed that they at last had reached an agreement on the final version of the document.[702] Liu was all the more surprised when Gromyko informed him that Zhou had told Vyshinskii that he considered it appropriate for the agreement not to be signed then but later, in Peking.

Soon afterward, one of the reasons why the Chinese side did not want to sign yet became apparent. Here, too, questions of sovereignty and national prestige were involved. Liu requested that the names of Chinese cities in the agreement for which an international spelling derived from non-Chinese languages existed be consistently spelled in the Chinese style. Thus, the Chinese called for the agreement to use Shenyang instead of Mukden, Zhangjiakou instead of Kalgan, Ili instead of Kulja, and Dihua instead of Urumchi. They also wanted to include the old Chinese name Kulun in parentheses behind that of the Mongolian capital, Ulan Bator, in the Chinese text of the agreement—perhaps a further indication that they had not yet accepted the loss of Outer Mongolia in Peking and that they might perhaps be signaling this subtly in this way to the Soviet Union. Gromyko agreed to make the requested changes, "as these are the official Chinese names." Liu also informed him that Mao and Zhou had instructed him to sign the agreement "in the next few days" in Moscow and to arrange for "the agreement or a communiqué on this agreement" to be published one to two days after Mao's arrival in Peking. Should they publish the agreement in its entirety? Gromyko promised to provide an answer later. Liu finally stated that he would see to it that the Chinese text of the agreement was ready for signature in two days.[703]

This shows that the two sides began only on February 18 to discuss whether the full text of the agreement or only a communiqué about its signature should be published. In the case of the agreement on the establishment of the mixed joint-stock companies for the mining of raw materials in Sinkiang, the Chinese side had voted against publication of the full text and argued for publication of a communiqué.[704] In the case of the aviation agreement, it considered both forms of publication possible—an indication that there was nothing in the agreement that the Chinese Party leadership had to be ashamed of in the light of public opinion.

Despite their agreement, the Chinese side again did not provide any text that was ready for signature on February 21 and yet still pushed for the agreement to be signed quickly.[705] Gromyko pointed out this contradiction. Again on March 18, Liu Yalou could neither present a final draft nor agree, as the Soviet side wished, to sign on March 21.[706] In Moscow, Gromyko then suggested that they first publish a communiqué on the agreements to establish the two joint-stock companies to produce raw materials in Sinkiang and three to four days later a communiqué on the founding of a civil aviation company, but he did not explain the time differential. And that is what they did. Although the three agreements were signed together on March 27, the statement—there was no joint communiqué—on the signing of the agreements to establish the two mixed joint-stock companies to produce raw materials in Sinkiang was published on March 29, while the one on the signing of the agreement on the founding of a mixed civil aviation joint-stock company was published only on April 2.[707]

The Trade Agreement, Protocol on the Exchange of Goods, and Other Economic Agreements of April 19, 1950

The trade agreement and the protocol on the exchange of goods of April 19, 1950, are also among the agreements that were not published at the time. The conclusion of the agreement and protocol was announced in only four sentences. The coordinated statement focused on the types of goods that were to be exchanged, bluntly reporting that the Soviet Union would deliver equipment and China raw materials.[708]

The trade agreement had a precursor in the Agreement on the Exchange of Goods between the USSR and Manchuria negotiated during Liu Shaoqi's stay in Moscow during the summer of 1949. It was concluded on July 30, 1949, at the Party level and signed by Liu and Malenkov. The Soviet press reported on the agreement in a brief statement at the time[709] in order, as Stalin put it, to "legalize" the Liu delegation's visit at the Party level. The document was in fact a credit agreement. The Soviet Union obligated itself to credit the deliveries of goods to China for the amount of US$300 million. At the time, the agreement was termed internally the Soviet-Chinese Trade Agreement.[710]

Zhou Enlai, Mikoyan, and Vyshinskii initially negotiated the April 19 trade contract and, after Mao and Zhou's departure, the Chinese side continued under the minister of trade, Ye Jizhuang. Among others, Li Fuchun, Minister of Foreign Trade Men'shikov, and the newly named chief adviser in China, Ivan Arkhipov, who had taken Kovalev's place in the economic field, were also involved in this second phase. As early as January 22, Molotov, Mikoyan, and Vyshinskii presented Stalin with the draft of an agreement on the exchange of goods and payments, which was forwarded to the Chinese in the following days—apparently with the title modified to "trade agreement." On February 2, Mikoyan and Vyshinskii approved Zhou's suggestion that they conclude the discussion of the trade agreement and the basic appendixes in the following days and negotiate the remaining questions, above all delivery amounts and prices, with Minister of Trade Ye Jizhuang. Would it be possible to conclude negotiations for the contract and the basic appendixes in the following four or five days? Mikoyan replied that this did not depend on the Soviet side alone, but also on the Chinese side.[711]

Mikoyan addressed two problem areas with Zhou Enlai on February 11. First of all, the Soviet side did not see itself in a position to deliver all the goods Peking had ordered by the desired delivery dates—apparently not least because the Chinese side, as Mao had already admitted once previously, was not in a position to present complete lists of orders in time. The second area involved differences of opinion on establishing prices. The Soviets were just checking the Chinese orders, Mikoyan explained. It was impossible to deliver equipment such as rolling mills and steam turbines during 1950, as the Chinese side wished, because it took one to one and a half years to plan and produce them. The Soviets were checking the delivery dates at that time and would be able to inform the Chinese in the following few days which equipment could be delivered by which time. The Chinese side was apparently not fully informed about the prices, Mikoyan said. Some of the prices they indicated did not reflect world market prices. This had been the result of an examination by the Foreign Ministry of the USSR, which had access to complete data on world market prices. They could make them available to the Chinese side.[712]

On December 16, 1949, Stalin had already insisted that the Soviet side had to know what equipment China needed immediately because it had no reserves, and orders for industrial goods had to be submitted at least one year in advance.[713] Mao had replied that this was difficult because the Chinese did not yet have any clear view of the state of their industries. When Mao asked whether the date for the delivery of industrial equipment and weapons could not be moved forward from five to three or four years on January 22, 1950, Stalin avoided the question and repeated that China "had to hurry and present lists of orders of industrial equipment."[714] The sooner they presented these lists, he said, the better.

The orders the Chinese submitted on the basis of the credit agreement of August 30, 1949, led to the resolutions of the Soviet Council of Ministers of September 4, 1949, November 3, 1949, December 19, 1949, and January 7, 1950, on the delivery of goods—including armaments—to China. The total value of the deliveries called for in the resolutions came to US$31.3 million, of which $8.5 million (for rails and railroad equipment) were to be financed using the credit.[715] We have detailed information on only two of these resolutions so far. The Soviet Union resolved to deliver aircraft and artillery pieces worth US$26.5 million on September 19 and rails worth US$6.3 million on November 5.[716]

Mao apparently applied pressure after Stalin's warning on January 22: the now-completed Chinese lists of orders arrived in Moscow after a few days so that the Soviet Ministry of Foreign Trade could present a detailed list of the goods that were to be delivered to China during 1950 to the Party leadership on February 12—along with a list of the goods that were to be delivered to the USSR by China during the same period. The list of the goods that were to be set aside for export to China comprised primarily industrial equipment and products, including metallurgical equipment, power plants, rolling trains, factories to construct machines, equipment for coal mines, ferrous alloys, rolled steel, pipes, nonferrous metal products, railroad equipment, motor vehicles, agricultural machinery, fuel, lubricating oil, and cellulose. The Soviets did not deliver all the goods in the amounts the Chinese side ordered. The shortfalls included, among other things, seamless pipes (amount ordered: 6,630; deliveries approved: 2,000) and rolled steel pipes (amount ordered: 8,456; deliveries approved: 3,000). The USSR could not satisfy all the Chinese requests from its own production; for example, it delivered 7,400 tons of lubricating oil on the account of its imports from Romania.[717] The list of goods offered by China included primarily raw materials and agricultural products such as tungsten ore, antimony, tin, coal, salt, tea, soybeans, corn, sorghum wheat *(gaoliang)*, millet, meat products, vegetable cooking oil, wool, feathers, raw leather, brushes, and raw silk.[718]

It can be assumed that it was primarily because of disagreements over the world trade prices of the Chinese goods that the trade agreement could not be signed along with the treaty of alliance and other important agreements on February 14. This is suggested by a talk between Zhou and Mikoyan and Vyshinskii on February 11. Zhou said that if the review of the orders was going to take a long time, they could sign the trade agreement alone and "assign the task of reviewing the orders and prices to the foreign trade ministers and other professionals."[719] His Soviet interlocutors were reserved in their response and said that this question could be decided by the Chinese side.

The negotiations were so deadlocked two days later that Zhou did not return to his February 11 suggestion and instead observed that the trade agreement could not be signed, as it required "some clarifications and fine-tuning."[720] One can assume with a high degree of probability that Mikoyan had in the meantime made the results of the Soviet review of the value of the Chinese goods on the basis of world prices available to the Chinese side as promised on February 11. The Ministry of Foreign Trade of the USSR had assessed the value of the Soviet goods internally (including those already delivered in 1949 and early 1950, which were worth US$31.3 million) at a value of US$131.7 million and the value of the Chinese goods at "about US$150.8 million" after reviewing the two lists of goods on February 12. Peking had instead assumed that the Chinese goods were worth US$194.9 million. According to Soviet calculations, the value of these goods thus amounted to about US$44 million less than the Chinese side claimed. The difference came to about 23 percent—an amount the Chinese side was certainly not willing to accept initially.[721]

We do not know in detail how the negotiations proceeded after Mao and Zhou left, because no records of the talks on this subject have yet become accessible. Soviet archival documents do, however, show that the Soviet side insisted on its previous assessment of the value of the Chinese deliveries offered (about US$150 million). If we take into account the US$60 million portion of the US$300 million Soviet credit that was due for the year 1950, China had by these calculations a claim to deliveries of Soviet goods worth a total of US$210 million, and the Soviet deliveries and approved deliveries represented a value of only US$131 million. The Soviets therefore put together an additional list of goods with a total value of US$71 million to be delivered to China—though it remains incomprehensible how they came to this figure because the difference came to US$79 million and the explicit intention was to "cover the value of the Chinese goods and also to guarantee the deliveries on the account of the first year of the credit."[722] The first part of the list presented goods that the Chinese had ordered, but for which the requested amounts had not been approved. They were now to be delivered in the requested amounts. In the second part, additional amounts were offered of goods that had already been approved in the requested amounts and goods that Peking had not yet ordered.[723]

The archival documents accessible to us up to now provide no indication of how the problem of the divergent assessments of the value of the Chinese goods was finally resolved. The memoirs of Wu Xiuquan, who participated in the negotiations, reveal, however, that the Chinese side conceded because it was in the weaker position. Wu presents the conflict over the value of the goods on the basis of world market prices as if it had been a question of the exchange rate between rubles and renminbi, whereas the two sides should actually have been comparing the value of the goods themselves. "The Soviets wanted to take advantage of us here, however," he wrote. "They calculated the value of their ruble very high and assessed the price of our renminbi rather low." The Chinese side had not given way at first. Disputes had arisen and the talks had become deadlocked, Wu continued. But because the Chinese were asking a lot from Moscow, it would not have been to their advantage actually to fight it out over this. They had asked Peking and received permission to make concessions so that they could ultimately reach a compromise. In any case, Wu concluded, the exchange rate was established on the basis of unequal conditions.[724]

The author finds it impossible to assess which side was in the right in the dispute over the value of the goods on the basis of world market prices. A comparison of the Soviet draft of the trade agreement presented to Stalin on February 13 with the final text signed on April 19 reveals that it was also a question of the relationship between the ruble and the yuan.[725] Article 3 of the draft stated that the prices of the goods to be exchanged "were calculated on the basis of world market prices in American dollars." In Article 4 of the final text, "American dollars" was replaced with "rubles." One can assume that this change was made in response to Chinese insistence—as in the case of the agreement on the working conditions of the Soviet specialists, in which the Soviet side had initially called for payment in dollars and the Chinese side in yuan, and they had ultimately agreed on rubles.[726] In the trade agreement, the two sides also changed the denomination of the calculation of any uncompensated payment balances from dollars (Soviet draft, Article 6, Section 2) to rubles (final text, Article 7, Section 2), based on the exchange rate between dollars and rubles of 1:4 current at that time. With the exception of the issue of currencies, the Soviet draft is essentially identical to the final text.

It has not been possible to determine what ruble-yuan exchange rate the two sides adopted during the economic negotiations. We have already mentioned that the Chinese were dissatisfied with this exchange rate.[727] One serious study assumes that the actual exchange rate current from the signature of the trade agreement in April 1950 through the end of 1957 was "approximately 1 to 1, or to be exact 1.03 rubles = 1 yuan."[728] The authors of the study conclude that the ruble was thus overvalued and that the Chinese only remained silent because they urgently needed Soviet aid for the construction of their own industry.

In early February, Zhou Enlai suggested to Mao that China make purchases of Soviet goods for its navy and air force over two years. The US$20 million that had not yet been used of the US$60 million that China had available as a credit in 1950 could be used for the needs of the navy and the air force. The leadership had already agreed at that time that most of the Soviet credits should be used in Manchuria, because its economic development was very important to the entire country.[729]

On February 3, 1951, the trade agreement was extended to the end of 1951. From 1952, trade protocols on the exchange of goods were signed each year on the basis of the trade agreement of April 19, 1950, and thereby extended its validity.[730]

A protocol on the exchange of goods was signed along with the trade agreement of April 19, 1950, and three lists of goods were appended to them.[731] List 1 contained the goods that were to be delivered by the USSR and List 2 those by China during 1950. List 3 contained industrial equipment that the Soviet Union would deliver in 1952 and 1953. The two sides also signed an extensive secret protocol on joint conditions for the exchange of goods between the foreign trade organizations, in which they established, among other things, delivery terms, quality controls, and procedures in the case of complaints.[732] They apparently also signed a protocol on transit questions, which remained secret, too.[733] In it, the USSR granted China free transit of goods not only through its own territory but also through that of Poland, Czechoslovakia, Hungary, Romania, Bulgaria, Albania, and the GDR. In return, the Chinese side guaranteed free transit of Soviet goods not only through China but also through North Korea. These were clearly agreements that affected third parties and reveal the limited sovereignty of the Eastern European satellite states with respect to Moscow and of North Korea with respect to China, and also provide an indication of the division of labor in the Eurasian region between Moscow and Peking, which had already been discussed during the summer of 1949.

The credit agreements of August 30, 1949, and February 14, 1950, the trade agreement of April 19, 1950, and the other economic agreements resulted in a rapid increase in the exchange of goods. In 1948, the USSR occupied ninth place in Chinese exports; in 1949, it was third; and at the end of 1950, it was first. The Soviet share of Chinese import trade rose from about 5 percent in 1949 (fifth place) to about 20 percent in the first nine months of 1950 (second place).[734]

With the signature of the trade agreement and other economic agreements discussed above on April 19, 1950, the entire set of treaties was finally complete. Li Fuchun, Wu Xiuquan, and other Chinese involved in the economic negotiations returned home in April. Minister of Trade Ye Jizhuang stayed behind in Moscow with a few others before also returning to China in May.[735]

Preparations for an Agreement on Secret Service Cooperation

In January or February, Stalin addressed the possibility of cooperation between the secret services with Mao Zedong and Zhou Enlai. He thereby confirmed and also developed something that Stalin and Liu Shaoqi had already agreed upon in principle during the summer of 1949.[736]

According to information he provided years later,[737] Mao first agreed with Stalin "that the USSR would carry out no espionage actions in the new people's democracies and thus also not in China." Nonetheless, Mao protested on the same occasion, Roshchin, the Soviet ambassador in Peking until 1952, had engaged an official in the Chinese Foreign Ministry to act as an informant without informing Zhou Enlai. One could assume, Mao added, that Molotov knew of this.

Stalin and Mao then again agreed to have their secret services cooperate and also began to examine the details of the plan. Stalin pointed out that the Soviets had extremely limited leeway to carry out espionage actions abroad. China's position, he suggested, was much better in this respect. "They watch and obstruct you relatively little," Stalin said, apparently referring to the overseas Chinese. This work was very important to the USSR, he said. Moscow would provide the technical equipment and funding, and both sides could make use of the information.[738]

Mao and Zhou agreed to this. They arranged to leave it to the experts to negotiate the concrete questions and set the results down in a draft agreement. Nothing is known about when and under what circumstances the agreement was signed.

5

Conclusions and Prospects

The emergence of the alliance sealed by the treaty of February 14, 1950, between the Soviet Union and Communist China is one of the most striking events in the international politics of the twentieth century. The alliance constituted one of the basic constants of Western security policy during the 1950s. The same is true of the conflict that shaped Sino-Soviet relations during the two following decades and of the rapprochement that set in at the beginning of the 1980s. Western capitals are today carefully observing the extent to which a "strategic partnership" has developed between the Russian Federation and the People's Republic of China since the spring of 1996. Regardless of the form the changing relations between Moscow and Peking took at any specific time, they could and can be assured of special attention because of the two actors' importance in the international arena.

The emergence of the Sino-Soviet alliance in 1950 could appear at first glance to have been an inevitable process. The USSR and People's Republic of China were, after all, both states under Communist rule, whose leaderships advocated one and the same Marxist-Leninist ideology. And yet, even if one can now assume that Chinese Communists did not consider the American option to be a possible alternative to the alliance with the USSR, one still cannot see the creation of the Moscow-Peking axis as a matter of course. Despite their basic ideological consensus, differences emerged repeatedly before 1950 that reflected differences between the real political interests of the two sides as they perceived them. It would only be a slight exaggeration to say that the 1950 alliance emerged despite the conflicts that had previously been aired.

The path the Chinese CP followed from being simply a section of the Comintern during the early 1920s to become an almost equal alliance partner of the Soviet Union in 1950 was arduous. When Mao Zedong complained to the Soviet ambassador about Stalin's policy toward China in the mid-1950s, he specified three serious legacies Stalin had imposed on the relationship with the Chinese Communists: he had supported Wang Ming, obstructed the revolution in 1945, and treated Mao poorly during his visit to Moscow. Mao could have been more specific about the second point and accused Stalin correctly of having considered the Kuomintang, the CCP's main opponent, to be the Soviet Union's preferred partner in China from the very beginning.

This took place against the backdrop of a gradual change in Soviet foreign policy, which developed away from the original conception of world revolution to one based on Russian nationalism during the first half of the 1920s. At first, security concerns played a role in this process; later, expansionistic goals were the driving force. With respect to China, the new course could already be seen in mid-1919 when Moscow withdrew the offer it had already made to the Peking government to restore the Chinese Eastern Railroad without compensation. Stalin, in whom the various threads of China policy came together after 1922, worked on the assumption that the KMT, the strongest progressive force in the country, would be able to overcome the reactionary warlords, unify China, preserve the country from any non-Soviet foreign influence, and—unintentionally—prepare it for a Communist seizure of power, which would be achieved by the Chinese CP under Soviet leadership.

The first tensions emerged between Moscow and the Chinese Communists when the Soviet leadership required the CCP to form a bloc with the KMT in 1922 in order to infiltrate the KMT Party gradually. Moscow delivered weapons to the KMT while the Chinese comrades received nothing. The fiasco of the bloc policy and the failure of the Soviet strategy of armed uprisings in the cities, both of which became apparent in 1927, were the main reasons for Mao Zedong's and Zhu De's retreat into the mountains of Jiangxi and Hunan to develop a new strategy that rejected cooperation with the KMT and called for carrying the revolution from the countryside to the cities. Moscow observed the consolidation of the soviet regions Mao and Zhu formed that took place without Soviet instructions with suspicion. The same was apparently true of Mao's gradual rise to the position of Party leader, which began in 1935. The long arm of Moscow did not reach far enough at that time to influence this development. Stalin, who favored Wang Ming, began to come to terms with Mao's leadership position only in the second half of 1938. At about this time, Mao began to speak of the Sinification of Marxism and made clear thereby that he wanted to follow a path independent of Moscow in the field of ideology as well. In the fall of 1941, Mao succeeded in isolating Wang Ming and his other opponents in the Party who remained loyal to Moscow. As a result of the process of the emancipation of the CCP that can be seen in these events, Moscow was subsequently never able to keep the Chinese fraternal party fully in line.

Stalin continued to treat the KMT as the preferred party even during the Sino-Japanese War. He was primarily interested in pinning Japan down in China and restraining it in this way from attacking the Soviet Union. To combine the resistance forces against Japan in China, he again pressured the KMT and the CCP to work together. Here, too, he depended primarily on the KMT as the much stronger force and supported it heavily with weapons. He rejected requests for military aid from the Chinese comrades or attached them to unacceptable conditions. On the other hand, he exerted pressure on the KMT not to move against the Communists. By flirting intensely with the United States between the summer of 1944 and March 1945, the CCP leadership sought to gain access to American weapons and drive a wedge between Washington and Chungking. This was a tactical maneuver and not a strategic shift. Nonetheless, this action, which had clearly not been arranged in advance with Moscow, marked the beginning of the CCP's independent foreign policy.

When the United States and Great Britain forced Stalin to become involved in the war against the Japanese to relieve the Western powers militarily in East Asia, Stalin saw a great opportunity to extend Soviet power in this area in order to restore the borders of the czarist empire of 1904 in the Far East and thus also czarist privileges in Manchuria. Behind the backs of the Chinese Nationalist Government and the CCP, he had Roosevelt and Churchill guarantee on February 11, 1945, at Yalta, that in exchange for entering the war, the Soviet Union would see the return of southern Sakhalin and the Kuril Islands and the recognition of its de facto protectorate over Outer Mongolia, Soviet "preeminent interests" in the Chinese Changchun Railroad (CCR) and Dairen, and the lease of Port Arthur. The agreement meant that Washington and London respected the membership of Outer Mongolia and Manchuria in a Soviet sphere of influence. Stalin, in turn, had to ensure that he would recognize the Nationalist Government as *the* Chinese government and grant Chinese sovereignty in Manchuria.

Washington gently pressured Chiang Kai-shek to conclude a treaty of alliance and several supplementary agreements with Moscow on August 14, 1945, thus establishing the rights that had been conceded to the Soviet Union at Yalta. It was particularly difficult for the KMT government to surrender Outer Mongolia de jure, and it accepted this provision only after stubborn resistance. The secret protocol that called for the troops stationed in Port Arthur to be subject to a separate jurisdiction—a new version of the extraterritoriality of the imperialist period, which Washington and London had already renounced two years earlier—was particularly humiliating. The treaty and the agreements were also a bitter pill for the Chinese Communists to swallow both because the Soviets recognized the Nationalist Government in them and obligated themselves to support this agreement alone, and also because it formalized the loss of Outer Mongolia and the establishment of special Soviet privileges in Manchuria.

In keeping with the agreement, the Soviet Army marched into Manchuria in August 1945 and continued to push forward after the Japanese capitulation until it occupied the entire territory. Stalin thus seized the opportunity to create a basis for establishing his claims in Manchuria. He later took pains to employ the withdrawal of the Soviet Army as a bargaining chip for the retreat of American troops from North China. Mao Zedong, in turn, saw the entrance of Soviet troops as the signal for the beginning of a Communist revolution in China, which he would lead to victory with the support of the Soviet Army. He was therefore particularly shocked when Stalin urgently advised him to desist and recommended that he seek a negotiated peace with the KMT. The Chinese Communists viewed this as a betrayal of their revolution, but bowed to the pressure from Moscow because they considered themselves too weak without Soviet support.

We can probably assume that Stalin was already seeking to transform Manchuria into a kind of buffer state between the USSR and China proper in which Moscow would hold considerable power when the Soviet troops marched in. He apparently felt at the time that he would be able to establish these conditions more easily with the Communists than with the KMT. On the other hand, he was obligated by the August 14 treaty to transfer Manchuria to the Nationalist Government. To resolve this dilemma, Stalin had to play a complicated double game. In September 1945, he arranged a strategy with Yenan by which the Soviet Army would hand over the administration of several cities to the KMT government publicly while helping the Communists take military control of the countryside as quietly as possible. The plan worked. At the beginning of May 1946, when the Soviet Army finally withdrew, most of Manchuria was in the hands of the Chinese Communists, and the Nationalist army essentially controlled only a corridor along the railroad line from Peiping to Changchun and Mukden and surroundings.

The KMT army was able to conquer almost all of the Manchurian cities between the end of the cease-fire and the end of 1946. But from the spring of 1947, the Communist People's Liberation Army (PLA), as it had begun to call itself in the meantime, undertook a massive counteroffensive and controlled all Manchuria by November 1948. It now pushed south, marched into Peiping at the end of January 1949, crossed the Yangtze in April, conquered Nanking at the end of the month, and occupied Shanghai in mid-May.

The apparently unstoppable victory march of China's CP placed Stalin in a fresh dilemma. On the one hand, Mao had been pressuring him since 1947 to set the stage for a strategic cooperation and receive him in Moscow to discuss weapons, ammunition, and civilian aid. As the United States and other Western powers were clearly going to continue to support the KMT, the only great power with which the Chinese Communists could seek to form an alliance was the Soviet Union, with which it already felt ideological bonds anyway. The USSR, in turn, was obligated to assist Communist China out of revolutionary solidarity.

Nevertheless, Stalin remained bound by the Yalta accords and Sino-Soviet treaty of alliance to maintain diplomatic relations with the Nationalist Government and provide all aid for China to it. He must have had the greatest interest in maintaining these agreements because they provided the legal basis for reincorporating southern Sakhalin and the Kuril Islands and for the inclusion of the Mongolian PR and Manchuria in the Soviet sphere of influence. He was also afraid that he might become too deeply embroiled in the Chinese civil war if his support for the Chinese Communists became too open. This might have ultimately led to armed conflict with the United States, with which relations were already strained due to the escalating Cold War. The victorious advance of the Chinese Communists might, Stalin feared, provoke the United States to intervene militarily in favor of the KMT.

Soviet policy toward China in the second half of the 1940s was shaped by the fact that Stalin wished to avoid a situation that could lead to military intervention by the United States on the side of the KMT. Here he was not as concerned about the good of the Chinese Communists as he was about the interests of the Soviet Union. An American intervention would have confronted the USSR with the choice between supporting the Chinese Communists and risking direct military confrontation with the United States or not doing so and being revealed to world public opinion—and not least to the eyes of world Communism—as a power that would leave its comrades in the lurch when faced with the threat of destruction. Stalin did not want under any circumstances to be forced to choose between these two options. His strategy toward Mao Zedong, which continued until 1949, was designed to avoid allowing the USSR to become embroiled in the Chinese civil war and to hold Mao to a flexible policy to avoid giving Washington any excuse for military intervention.

As in the case of Manchuria in 1945–46, Stalin first pursued a policy of maneuvering. On the one hand, he put off the visit by the "partisan leader" Mao Zedong (as Stalin called Mao at a Politburo meeting in early January 1949) on a variety of pretexts despite increasingly urgent demands, and instructed the Soviet ambassador in Nanking to follow the KMT government, which was fleeing the PLA, to Canton in early January 1946. He put up with the annoyance that these two actions provoked in the CCP leadership. On the other hand, he increasingly recognized about this time the need to discuss the future of mutual relations with the CCP leadership in light of the Communists' predictable victory in the civil war. Moscow and the CCP leadership were already secretly coordinating their policies toward the KMT in early 1949. Soviet representatives officially closed their office doors in the areas occupied by the PLA, but they secretly established close contacts with the new Communist authorities. They cooperated openly with the CCP in Manchuria and exerted pressure on them to force the Western powers (the United States in particular) out of this area.

Stalin's intention to win China over as an ally after the Communists' final victory in view of increasing tensions with the West clearly also played a role in his growing willingness to accept an alliance with the Chinese Communists. This calculus played an increasing role when the establishment of NATO in April 1949 revived the old fear of encirclement in Moscow and an allied China was increasingly perceived as an important factor in the strategic relief of East Asia.

In early 1949, Stalin sent his close confidant Anastas Mikoyan to Xibaipo to hold the first secret meeting with the CCP leadership at the highest level. In the summer of 1949, he received Liu Shaoqi, the second-highest-ranking individual in the Party, for secret talks in Moscow. This set the stage for strategic cooperation between the USSR and Communist China. The fact that the leadership of the CCP extended feelers to the United States and Great Britain at the same time does not invalidate the basic decision they had made in favor of an alliance with the USSR. A careful review of these initiatives reveals that they were designed as a tactical maneuver to disrupt cooperation between Washington and the KMT government and restrain the United States from intervening militarily on the KMT's side. It is thus not possible to support the revisionist thesis of a lost chance in China that has emerged in the United States. In addition to the factors mentioned above, a chiliastic belief in world revolution on the Chinese side and a firm belief in democratic values such as the rule of law on the American side also stood in the way of any rapprochement.

During Liu Shaoqi's visit in Moscow, Stalin approved generous military and economic aid. The parties agreed to make these agreements public only after a new government had been formed and during Mao Zedong's planned visit in Moscow. The Chinese side did withdraw its claims for a return of Outer Mongolia to China during the meetings in Xibaipo and Moscow. They were not yet able to reach an agreement on Soviet privileges in Manchuria, however. This was done only during Mao Zedong's stay in Moscow from December 1949 to February 1950, when Stalin gave way to pressure from the Chinese leader and discarded his earlier reservations, which were due to the Yalta accords. He concluded a treaty of alliance with the People's Republic of China that replaced the 1945 treaty and declared his willingness in a supplementary agreement to surrender all of the Soviet Union's rights to the CCR and in Port Arthur by 1952 at the latest, which would otherwise only have expired in 1975. The administration of Dairen was transferred to China immediately. Stalin apparently changed his opinion primarily because he had become convinced by the withdrawal of the American defense perimeter in East Asia at the New Year that Washington had come to terms with the Communist seizure of power in China and also the creation of an alliance between the USSR and the People's Republic of China. The United States and Great Britain would therefore, Stalin believed, no longer make any trouble about the Yalta accords.

It is striking how similar the issues involved in the talks and negotiations between the Soviet Union and the KMT government in 1945 and the Communist government in 1949–50 were when one examines them closely.[1] Both Chinese parties were intent on protecting China's sovereignty and preserving or, in the case of Outer Mongolia, restoring its territorial integrity. Both viewed Soviet policy in Manchuria as imperialistic. The heads of both negotiations, Foreign Minister Wang Shijie of the KMT government and Premier Zhou Enlai of the Communist government, compared the Soviet position in Dairen after the entrance of the Soviet Army into Manchuria with the British position in Hong Kong.[2] Both were concerned primarily with the status of Outer Mongolia and Soviet special privileges in Manchuria. We cannot yet explain why Mao immediately gave up the attempt to win back Outer Mongolia after the first attempt in Xibaipo, which Stalin rebuffed decisively, although T.V. Soong fought for it bitterly to the end in 1945. One can assume that Mao sensed that Stalin would not give way and that he did not want to risk failing to get his other requests when he could not choose to form an alliance with any other partner.

One might previously have assumed that the discussions Mikoyan held in Xibaipo and Liu Shaoqi and Mao Zedong held in Moscow dealt primarily—or certainly not least—with problems involving their shared ideology and the world revolution. Today, as archival documents on the talks become accessible, we know that these questions played little or, in the case of Mao's stay in Moscow, practically no role. Neither side was particularly interested in settling ideological disputes or arranging a strategy for world revolution. Instead, they were both concerned with classical national interests.

There are no sources to document Stalin's assessment of the results of the negotiations or his impression of Mao Zedong. One can assume, however, that as a realist Stalin assessed the results as they present themselves to an outsider: as a compromise. With respect to his opinion of Mao, we can believe Khrushchev, who remembers that Stalin "was never once enthusiastic about Mao and made comments that were not particularly complimentary about him" after his meetings with the Chinese leader.[3]

Mao Zedong could be satisfied with the overall results of the negotiations. Despite Stalin's opposition, he had been able to conclude a new treaty of alliance and convince the Soviet Union to surrender its rights to the CCR and in Port Arthur by 1952 instead of 1975. Given this success, it is less important that he had to make some concessions. His request for reciprocity (which would in any case have been apparent only in a secret protocol on the supplies for the Soviet troops stationed in Port Arthur) was unsuccessful. He also had to accept a secret agreement in which China obligated itself to allow only Soviet and no other foreign firms to operate in Manchuria and Sinkiang, where the proposed reciprocity with respect to the Soviet Far East and Soviet Central Asia was a farce. He was also forced to accept a ruble-yuan exchange rate that was unfavorable to China.

Several concrete statements about Mao's opinion have come down to us. When he spoke about the negotiations in Moscow at Party conferences in China, he pointed to his successes with pride without hiding the fact that he had also made compromises. He said that Stalin had refused to place his signature under the treaty in Moscow and had signed it only two months later.[4] There had been differences of opinion: "We demanded the Chinese Changchun Railroad, he did not give it to us. But we were nonetheless still able to snatch the meat from the jaws of the tiger [*laohu kou li de rou hai shi neng na chulai de*]."[5] And they had quarreled over the treaty, the railroad, the mixed companies, and border questions, he said.[6]

On the other hand, Mao had to feel resentment when he looked back on his stay in Moscow—in particular because he felt that Stalin had humiliated him by blocking negotiations for more than two weeks, leaving Mao to stew in his own juices in his dacha during this time until the Chinese leader threatened to leave. Zhou Enlai was referring to this when he later observed that Mao had traveled to the Soviet Union as a victorious statesman, and thus his hosts should have received him particularly cordially.[7] But the atmosphere had not even been as cordial as during Liu Shaoqi's visit. Mao had been coolly tossed aside, Zhou continued. With the exception of one message of congratulations, there had been nothing to talk about. Khrushchev confirms this more or less: Stalin, he reported, had shown Mao "insincere courtesy."[8] One could sense "a kind of arrogance." This had irritated Mao, Khrushchev wrote, but he had not shown this outwardly—except on one occasion.[9] Stalin played the exquisite host to Mao, as he enjoyed doing on such occasions, at only one meal in Mao's honor.

As mentioned earlier in this chapter, Mao himself complained to the Soviet ambassador after Stalin's death that his treatment in Moscow constituted one of the three most serious strains on the relationship between Moscow and the Chinese CP during the Stalin era; the other two were the Wang Ming case and the Soviet betrayal of the revolution in 1945.[10] Stalin, Malenkov, and Beriia had attacked him, he claimed.[11] He had felt the Soviet suspicion that he might be a "Chinese Tito" was even stronger then, he said, than before the People's Republic of China was established.[12] On other occasions, Mao said there had been considerable pressure in 1949 and 1950.[13] Only during the Korean War in the winter of 1950–51 had Stalin put his doubts about the reliability of his Chinese comrade aside.

Newly accessible archival documents show that this was not yet the case during the period of preparations for the Korean War. In late January 1950, when Mao was still in Moscow, Stalin gave Kim Il Sung the green light to attack South Korea.[14]

Stalin did not say a word to Mao about this important geopolitical decision. Years later, Mao complained about Stalin's lack of confidence to the Soviet ambassador.[15] The Soviet leadership, he said, had not consulted sufficiently with the Chinese side about China's involvement in the Korean War. They had agreed on everything involving the Korean question during his stay in Moscow. "There was no discussion of conquering South Korea, but merely of significantly strengthening North Korea." But Kim Il Sung had subsequently been in Moscow and had reached some kind of agreement with Stalin. No one had considered it necessary, Mao said, to consult with the Chinese leadership first on this matter.

Stalin probably did not consult with Mao because he feared that he would raise objections and therefore chose to confront him with a fait accompli. Stalin had previously rejected Mao's request for Soviet assistance in the planned conquest of Taiwan. He may have assumed that Mao would have refused Chinese assistance for North Korea in a military reunification of Korea as long as China was not also reunited.

One can thus assume that Mao looked back on his stay in Moscow with mixed and primarily bitter emotions. The humiliations he suffered there must have increased the reservations he already had toward Stalin and led him to seek to emancipate China even more from the Soviet Union—even though he had to postpone doing so for a while because of his dependence on assistance from the "elder brother." It thus proved impossible to defuse the old mutual resentment between Stalin and Mao in their first and only meeting. On Mao's side, it had essentially the opposite effect.

The seeds of the later conflict between the Soviet Union and the People's Republic of China thus developed in the period preceding the creation of the Sino-Soviet alliance. Mao Zedong's experiences during his two-month stay in Moscow played a particular role in this. Stalin's fears that Mao might be a Chinese Tito were unfounded to the extent that they were based on the doubt that the CCP might take Yugoslavia's side in the conflict between Moscow and Belgrade. He was right, however, to fear that a China ruled by the Communists would not subordinate itself completely to the Soviet Union and would under certain circumstances follow its own path. Khrushchev could sense this clearly when he visited China in 1954.

Despite this, Sino-Soviet relations during the first half of the 1950s were characterized by an attempt on both sides to work together well. When China entered the Korean War, as it did in mid-October 1950 only under Soviet pressure, Stalin's suspicions of a Titoist Mao largely disappeared. The comradeship in arms that emerged in this war created a degree of confidence.

The Chinese social, economic, and legal systems were to a great extent reorganized on the Soviet model during the 1950s. The USSR provided aid in numerous areas, among others in economic planning, the construction or modernization of major industrial companies, health care, and the reform of the justice and education systems. The Soviet Union calculated that Moscow sent out more than 10,000 experts as advisers to China between 1950 and 1959. More than 20,000 Chinese received training in the Soviet Union.[16] During the same period, the Soviet Union granted credits to China at reduced interest rates on the order of nearly US$2 billion and support in the construction of nearly 300 industrial companies and other enterprises, one-third of which were in the military sector. In exchange, China delivered ores and agricultural products to the Soviet Union, and made considerable payments in hard currency. Moscow's later claim that the USSR had helped the People's Republic of China selflessly was not true—at least in that the Chinese had to pay for each Soviet expert and for all technical aid from the USSR.

The propaganda machines on both sides tirelessly polished a spotless picture of the friendship of the age. It became usual in China to refer to the Soviet Union respectfully as "elder brother [*lao dage*]." A composition titled "Moscow-Peking" was commissioned by the Soviet leadership that was performed like an anthem on festive occasions to celebrate the eternal bonds of brotherhood between the two countries. "Six hundred million people have entered the camp of peace," a Chinese poet cheered, and another praised the symbolic value of the bridge the two countries planned to build together with the words: "This bridge of our eternal friendship . . . crosses through 800 million hearts."[17]

This picture did not completely mirror reality—at least at the leadership level—in the two states. After Stalin's death, Khrushchev did attempt to remove some elements of the inequality in their relations by, among other things, participating in the liquidation of the mixed joint-stock companies, which had originally been supposed to last thirty years, during his first visit to China in the fall of 1954. His personal relations with Mao Zedong were nonetheless clouded from the beginning. His report of this visit is typical. He found the atmosphere "typically oriental," "disgustingly sweet," and "nauseating." He said his Chinese hosts had been "incredibly courteous," but that he had seen through their hypocrisy. Khrushchev remembers that he had known from this first meeting with Mao that he would never accept any Communist party that was superior to his own within the world Communist movement. And more than that, Mao had sought hegemony in the world Communist movement, he said. When he returned from China, he told his comrades in Moscow: "Conflict with China is inevitable."[18]

The first concrete grounds for serious tensions emerged in 1956 in the context of de-Stalinization in the Soviet Union. Khrushchev had not discussed his secret speech to the Twentieth Party Congress of the CPSU in February 1956—in which he dealt radically with Stalin's crimes and "personality cult" and which struck like a lightning bolt out of a bright blue sky in the Soviet Union—with any foreign Communist leader in advance and thus also not with Mao, who, after the death of the Soviet dictator, considered himself to be the leading ideological personality in the "socialist camp." Although Mao had, as has already been described in detail, strong reservations about the Soviet dictator because of his policy toward China, he nonetheless saw the danger that any attempt to deal radically with Stalin, the leader of world socialism for decades, would damage the prestige of international communism worldwide. He also felt that Khrushchev's actions attacked his personal authority, albeit indirectly, for a Mao cult had developed in China that was comparable to the cult that surrounded Stalin. Events in Moscow did in fact lend support to Mao's opponents in the leadership. The new Party statutes of September 1956 eliminated the reference to the Thoughts of Mao Zedong as "the greatest achievement of the Party."

The CCP leadership hesitantly adopted moderate criticism of Stalin, but became annoyed that Khrushchev and the other Soviet leaders refused to bear any share of the responsibility for Stalin's "errors."[19] Shortly after the Twentieth Party Congress of the CPSU, Mao informed the Soviet comrades that Stalin's basic path had been correct and that his merits had been greater than his errors, and that the form of the critique was also blameworthy.[20] Within the Party, he pronounced that Stalin had been 70 percent good and 30 percent bad. Other resolutions of the Twentieth Party Meeting that had not been discussed with Peking in advance also irritated China, such as, for example, the promulgation of the doctrine of "peaceful coexistence" as the general guideline in the foreign policy of the Soviet Union and the acceptance of the possibility of a "peaceful, parliamentary road to socialism" in the capitalist states.

The described events were seen as a serious irritation but not a rupture in Moscow and Peking. However, the stage was apparently set for a breakdown in relations in 1958 and 1959. The events described below played a significant role in this.

Without consulting Moscow, the Chinese Communist Party leadership introduced the policy of the Three Red Banners in 1958, which included the notorious campaigns for the Great Leap Forward and the People's Communes. This involved a utopian social experiment for a rapid increase in economic growth that would organize work in quasi-military style, extensively collectivize property, abolish material incentives, and create a "new man." At the same time, the experiment would make it possible finally to take the much-desired Chinese road. The accompanying Chinese commentaries suggested that they had found a way in Peking to catapult China, so to speak, with one stroke into communism as the desired final stage of mankind and thus overtake the Soviet Union. Such words were an incredible provocation in the ears of the Soviet leaders. The Moscow leadership felt that its position was confirmed by the failure of the Chinese experiment, which began to become clear at the end of 1958, and by the "three bitter years [*san ku nian*; 1959–61]" filled with hunger and misery that projected their shadow.

Serious conflicts emerged in the area of military cooperation. In 1958, the Peking leadership rejected the Soviet suggestion that the two countries establish a long-wave radio station together in Fuzhou and build a joint submarine fleet on the Chinese coast. The USSR sought to improve the Soviet navy's operating conditions in the Pacific in this way. Mao reacted very abruptly to Ambassador Iudin and rejected the Soviet suggestions on the grounds that they violated China's sovereignty. The Soviets, he said, could call him a nationalist for this, "and if you do that, I can say: You extend your nationalism to the coast of China."[21] When Khrushchev, very concerned about the dispute, rushed to Peking in July 1958 to calm the waters, Mao stuck by his decision to refuse. Khrushchev gave way and the original Soviet suggestions were filed away.

Other Soviet actions during the 1950s also offended the Chinese.[22] The USSR, for example, did not consider it necessary to apply to Peking for immigration visas for the Soviet specialists and simply sent them to China with service travel documents. They built a Soviet radio station in China without Chinese approval. And the Soviet side sometimes even published "joint statements" without asking the Chinese in advance.

Another incident contributed to the outbreak of the conflict in 1958: the Quemoy crisis. It arose when Peking gave the impression that it would like to take the Taiwanese islands of Quemoy and Matsu by force. The Chinese suspected that the USSR was not prepared to support such a plan militarily.[23] Moscow was instead, to China's displeasure, still trying to improve relations with Washington to preserve the peace. Soviet-American détente aroused the suspicion in Mao Zedong that the two superpowers wished to dominate the world together. Mao, for his part, took the position that the USSR had better rockets and atom bombs than the United States and could therefore afford to intensify the confrontation—even at the risk of a new world war that would be fought with atomic weapons, which China, given its huge population, did not have to fear anyway.

It may in large part have been its experiences in the Quemoy Crisis that led Moscow to take a step with far-reaching consequences. In June 1959, the Soviet leadership refused to fulfill the part of a military aid agreement of October 15, 1957, that required them to deliver the prototype of an atom bomb, along with the necessary materials to construct it, and to assist in the production of plutonium.[24] The Kremlin had apparently become convinced that Peking was interested in military adventures in which they did not want under any circumstances to become embroiled, and that it certainly did not want to deliver nuclear technology to China. The USSR had in fact become one of the founders of the campaign to conclude a treaty dealing with the nonproliferation of atomic weapons. The Chinese, in any case, concluded from the Soviet actions that they could not rely on Moscow.

We can assume that the Soviet refusal to assist in the creation of Chinese nuclear weapons played a significant role in the Peking leadership's decision to seek a confrontation. Its view may have been reinforced further by Soviet actions with regard to the Chinese-Indian conflict of August 1959, in which Moscow took a neutral position pro forma, but de facto clearly supported New Delhi.

By now the situation was becoming tense. When the Soviet Party leadership invited Mao Zedong to talks in Moscow in early 1960, the latter refused to go. Peking published three articles in April that the Soviet side perceived as China's first direct public attack on the Soviet Union. In early June 1960, the Chinese leadership used a meeting of the Communist World Federation of Trade Unions in Peking as an opportunity to attack Khrushchev's thesis that war could be avoided. Khrushchev, in turn, accused the Chinese leadership in a confidential meeting with Peng Zhen at the end of the month at the Congress of the Romanian CP in Bucharest of having no conception of the reality of a modern war and assessing the balance of power in the world incorrectly. Peng Zhen struck back; he pointed to China's experiences in the war against Japan and in Korea and observed that the Peking leadership had no confidence in Khrushchev's analysis of the international situation.[25]

In the course of these talks, Khrushchev attacked Mao Zedong personally. He showered him with gross language, called him such things as "old galosh" and "worn-out shoe"[26] and derided him as a "Buddha who picks his theories out of his nose."[27] Mao heard about this, and the atmosphere was poisoned. Khrushchev later continued repeatedly to attack Mao Zedong personally in unofficial talks with representatives of the CCP and other Communist parties. Deng Xiaoping once told Soviet interlocutors that the Soviet side could have criticized him, Liu Shaoqi, or Zhou Enlai, and a reconciliation might still have been possible. "But you attacked Mao Zedong's authority itself and that cannot be remedied." It is therefore not far-fetched to suspect—especially as it reflects general everyday experience—that such personal insults, which the Chinese side later also made, played at least just as great a role in intensifying the conflict as did objective factors—and possibly a greater one.[28]

In July 1960, Khrushchev recalled the Soviet specialists from China; they were to leave within two months. The relevant note to the Chinese government stated, among other things, that the specialists had increasingly been used for propaganda purposes and spied on.[29] Iurii Andropov, the head of the responsible Central Committee department, did not approve, but he did not dare to suggest any alternatives.[30] Khrushchev probably made his decision primarily because he felt deeply insulted when he was told that Chinese propagandists had described him to the specialists as the "bald-headed idiot"—an insult that had apparently come from Mao himself. Moscow disregarded Peking's request that the recall, if it was unavoidable, take place in a period of one and a half to two years.[31] The recall of the specialists, which caused considerable damage in China, should be seen as a step that played a considerable role in the shift of Sino-Soviet relations from cooperation to conflict.

Both sides were responsible for this development, although the Chinese side later pushed it forward more aggressively and unrelentingly than Moscow did. The conflict ultimately arose as a result of the process of emancipating a regional power with great power ambitions from the tutelage of a superpower. This process should be seen less in terms of ideological than nationalistic interests. As a result of the actual dissolution of the alliance with the Soviet Union, China achieved an independence in its foreign relations that it had never had in the previous 120 years. The price, however, was extensive international isolation, which was only slightly alleviated by the sympathy of several developing countries as the conflict with the United States and its allies continued. On top of that, China now had to survive without assistance from Moscow.

As China confronted both the USSR and the United States during the following decade, the Sino-Soviet dispute presented itself to the nonsocialist world primarily as an ideological dispute over which of the two sides could best claim to be implementing the true will of the classics of Marxism-Leninism. Both sides were convinced of their infallibility. The Soviet leaders based their claim on the fact that the USSR was the "home of the October revolution," that it was the first and most powerful socialist state, and that the Communist Party of the Soviet Union was the "Party of Lenin." The Chinese leadership derived its claim from the fact that the CCP had achieved victory over the Kuomintang against Stalin's recommendations and that it had to follow its own path in the struggle against imperialism on the basis of its national experience.

The public polemics, which for the most part employed ideological arguments in 1963 and 1964, tended to mask the conflict's realpolitik and emancipatory core. The leaderships of both parties each accused the other of betraying the pure teachings, the call for solidarity, and the principle of the unity of the socialist world system in numerous letters. To the extent that ideological differences really existed, they involved in large part the Chinese thesis that the basic contradiction in the present lay not between socialism and capitalism, but between the national liberation movement and imperialism—whereby China implicitly claimed leadership in the worldwide liberation movement.[32]

The Chinese usually took the initiative and were the most aggressive, while the Soviets generally limited themselves to reacting to justify their position. Over time, Peking lost all inhibitions and launched a general attack on Moscow that spared practically no significant aspect of the Soviet interpretation of Marxism-Leninism or of Soviet domestic, economic, or foreign policy. The assault finally climaxed with the thesis that the CPSU had betrayed socialism and that the Soviet Union was no longer a socialist state—an accusation that was retracted only after Mikhail Gorbachev came to power. Mao claimed, for example, that the USSR and the United States were great powers that wished to work together to dominate the entire world (the so-called intermediate zone theory of 1964). The Soviet leadership—formally correctly—countered by accusing the intermediate zone theory of being a non-Marxist concept, because it had abandoned the perspective of class in the analysis of international politics. Many Soviet foreign policy experts in the Central Committee apparatus, among them Andropov, considered it pointless to continue the polemics but did not dare to contradict Khrushchev, who was under the influence of the dogmatist Mikhail Suslov.

Until 1963, neither of the two sides seemed to want to let the situation come to a total break. But a public exchange of opinions during that summer—in particular an open letter from the CC of the CPSU on July 14 that mocked the poverty of life in China—poisoned the atmosphere further and set the stage for an open break. The Chinese leadership rejected the Soviet suggestion that they end the polemics and derided it as a sign of weakness. At that time, Deng Xiaoping refused to greet the leader of the Soviet delegation, Suslov, with a handshake at a round of negotiations in Moscow on the grounds that they should not waste time.[33]

Peking's hope that the Soviet leadership would relent and recognize the correctness of the Chinese line after Khrushchev's fall in October 1964 did not materialize. Zhou Enlai traveled ostentatiously to Moscow at the head of a delegation to take part in the festivities on the anniversary of the October revolution, which Moscow at first took as a sign that the Chinese were willing to seek a reconciliation. Zhou, who was constantly interested in finding a resolution, might even have been willing to move closer to the Soviet Union if he had had his own way.[34] His delegation, however, included Kang Sheng, a close confidant of Mao Zedong who, along with Mao and Lin Biao, was one of the anti-Soviet hard-liners in the Chinese leadership and who had been sent along by Mao as a watchdog. Zhou accordingly demanded no more and no less than that the Soviet leadership retract all of the anti-Chinese positions it had held up to then. This was not only unacceptable to the hard-line majority opinion in the Soviet leadership, which was represented by Suslov; it would have been unacceptable even if a more conciliatory line had predominated. A scandal then also broke out at the festive banquet that further worsened the atmosphere. The Soviet minister of defense, Rodion Malinovskii, who was drunk, provoked Marshal He Long by observing smugly: "We got rid of our idiot [*durak*]; now it is your turn to get rid of yours!" Zhou rejected Brezhnev's attempt to save the situation by saying that what Malinovskii had said under the influence of alcohol should not be taken seriously. Although Leonid Brezhnev later apologized formally, the Chinese delegation left early. Kang Sheng was probably responsible for seeing that Malinovskii's statement was reported to Mao with nothing held back.

In March 1966, the CCP broke off contacts with the Soviet fraternal party. Relations between Peking and Moscow were limited in this way to the bare minimum: to diplomatic business and a marginal exchange of goods.

By constantly repeating its claim that the Soviet Union had betrayed socialism, China confronted Moscow with the strongest challenge it had faced to that time in the international Communist movement. Unlike others, the provocation had a global significance, not merely a regional one involving Asia. The Chinese were unable to seize ideological leadership in world communism in this way as they had hoped, however. Peking's confrontation with Moscow provoked only a few fraternal parties to take the CCP's side openly and oppose the CPSU. Where divisions arose, the Maoist subparties generally became mere splinter groups without major significance.

At the end of the 1960s, only one ruling Communist Party, the Albanian, had clearly chosen the Chinese side. Regional factors affected the choices of the six nonruling, Peking-oriented parties, which were all located in Asia: the Cambodian, Thai, Burmese, Malaysian, Indonesian, and New Zealand CPs. The Sino-Soviet conflict provided some more parties with a welcome opportunity to adopt or strengthen autonomous positions, particularly the Yugoslav, Korean, Japanese, Vietnamese, and Laotian CPs. A regional trend can be seen here, too. The Vietnamese and Laotian parties' decisions to ally themselves with Moscow in 1978 should be attributed to security policy and not ideological considerations.

The excesses of the Chinese Cultural Revolution burdened relations with the Soviet Union even further. The Soviet embassy in Peking was besieged in 1967. Soviet diplomats had to leave the country under humiliating conditions as Red Guards spat on and physically abused them. Under the powerful influence of official propaganda, a wave of anti-Chinese resentment raged in the broad Soviet public and revived the irrational fear of a "yellow peril [*zheltaia opasnost'*]" impending from Asia that had never completely disappeared since the Tartar invasions of the thirteenth and fourteenth centuries. It was easy to mobilize the resulting historical resentment politically during the Cultural Revolution. Even such liberal writers as Andrei Voznesenskii and Evgenii Evtushenko yielded to the temptation to warn of a barbarity from China that threatened to destroy Russian and even European culture in poems that had racist undertones.

Soviet intervention in Czechoslovakia in August 1968 created a new problem. In striking contrast to the Chinese support for the USSR during the bloody suppression of the Hungarian revolution in 1956, Premier Zhou Enlai compared the invasion of Czechoslovakia to Hitler's aggression toward that country in a speech two days later, and described the Kremlin leadership's foreign policy as "social-imperialist" and "social-fascist."[35] It subsequently became customary in China to dismiss the foreign policy of the "renegade Soviet clique" with reference to a Lenin quote as "socialism in words, imperialism in deeds" and to make a distinction between "genuine socialism" and revisionist Soviet "pseudo-socialism." Moscow's implementation of the Brezhnev doctrine of the limited sovereignty of socialist states in Prague had to give Peking the impression that the Soviet leadership would have few scruples about employing military force toward China as well if the Sino-Soviet conflict became even more intense.

The fact that the Soviet Union constantly reinforced its military presence on the Chinese border after 1965 contributed to Peking's increasing sense of vulnerability. There had initially been between seventeen and twenty divisions, but this number was growing to forty by the early 1970s and to around fifty-three in 1985—including the units deployed to the Mongolian People's Republic in connection with the conclusion of the new Soviet-Mongolian treaty of alliance on January 15, 1966.[36] In 1969, a series of bloody border clashes (the Ussuri Conflict), pushed both countries close to the brink of war. A massive Soviet counterattack employing the latest Katiusha rockets that killed some 3,000 Chinese by Soviet accounts followed the initial conflict on Damanskii Island *(Zhenbao dao)*, which was clearly provoked by the Chinese and in which thirty to forty people were killed. Further bloody conflicts followed in other border areas. There is evidence that in the late summer of 1969 the leadership in Moscow was considering whether it would be appropriate to make a limited preventive strike against China's atomic centers using nuclear weapons. At that time, both leaderships, as well as the citizens of the two states, were more hostile to one another than ever before. They had reached the low point in their relations.

Gloomy patriotic films stirred up emotions on both sides. The fact that a selection of anti-Chinese newspaper editorials and articles fills eleven volumes published since 1969 suggests the extent of the increasing flood of anti-Chinese publications in the Soviet Union. Peking's propaganda attacking the Soviet Union did not take second place. The atmosphere began to settle only in the fall of 1969 when Premier Zhou Enlai and Aleksei

Kosygin met in Peking to agree on the continuation of the border talks that had been broken off in 1964.[37] Kosygin's admission at Zhou Enlai's insistence that there were "disputed areas [*spornye raiony*]" on their common border only led to new conflicts, however, for Brezhnev subsequently disputed the existence of such areas.

The border issue developed into a new subject of dispute. The two sides held numerous talks without reaching any agreement. China did not, however, demand, as Soviet propaganda claimed, the return of a territory of approximately 1.5 million square kilometers that it—in its own view—had lost to Russia in the "unequal treaties [*bu pingdeng tiaoyue*] signed during the nineteenth century, but merely recognition by Moscow that the czars had acquired this territory illegally. Peking did, however, claim about 35,000 square kilometers of disputed areas (about the size of Belgium) that Russia had—in the Chinese view—acquired in violation of the unequal treaties. Moscow rejected this demand: the current Soviet border, it argued, was "sacred" and "untouchable." No new border treaty was on the table; the actual borderline could, however, be clarified in particular sections. Moscow's rigid position was no doubt influenced by the effect such a precedent might have on other parts of the Soviet border (especially with Japan and Eastern Europe). China's stubbornness can be explained as a position based in principle and on the consciousness of prestige, coupled with a plan to reveal the Kremlin leaders before world public opinion as the "new czars" who were continuing the tradition of Russian expansionism in Asia.[38]

The tensions with the USSR, which had become an increasingly serious security threat, as the Ussuri conflict had shown, and the intense isolation into which the Cultural Revolution had forced China, led Peking to seek to improve relations with the United States during the course of 1970 in order to end the unfavorable double confrontation with Washington and Moscow. This decision was made easier by the Nixon administration's declared intent to withdraw American troops from South Vietnam and give its Asian allies increased responsibility for their own defense (the Guam doctrine, or Nixon doctrine). Even before President Nixon's first trip to China in February 1972, the change in Chinese policy led to a decisive breakthrough in international politics: In October 1971, the People's Republic of China was accepted into the United Nations.

The Sino-American rapprochement isolated the USSR in the Moscow-Peking-Washington triangle.[39] The Kremlin leadership had therefore been seeking since the early 1970s to conclude a treaty with the United States to prevent nuclear warfare directed against China and, increasingly, to improve Soviet relations with China—without, however, making truly attractive proposals to Peking. The Chinese reacted negatively to repeated and more or less empty Soviet initiatives to improve state relations such as offers to conclude a nonaggression treaty and one banning the use of force, or else they offered package deals that were unacceptable to the USSR. Moscow used Mao Zedong's death in September 1976 as an opportunity to propose reestablishing Party relations with the CCP implicitly. Not only did this advance remain fruitless, but Peking intensified its anti-Soviet position in the following two years by adding a further extremely explosive element. In 1974, Deng Xiaoping compared the Soviet and American "hegemonistic" foreign policies using the "three worlds theory"—as Mao had de facto done ten years earlier in the intermediate zone theory. In 1977–78, the Chinese leadership went one step further. Speaking to non-Communist parties, it threatened to form an anti-Soviet "united front with Japan, Western Europe, China, and the USA" and let it be understood that it was seeking an informal alliance of all the states in the West and the Third World against the Soviet Union.

This dramatic increase in tension was followed shortly afterward by a swing in the opposite direction. For the first time in almost twenty years, Peking let it be known that it wished to relax its relations with Moscow by proposing negotiations with the Soviet government in early April 1979. The policy change in Peking was a result of the policy of reforms enacted by Deng Xiaoping's government in December 1978. It was designed to put an end finally to Mao Zedong's egalitarian socialism based on mobilization and develop China as quickly as possible into an economically and technically modern state. The prerequisite for this was a peaceful international environment that would rule out military entanglements with a superior opponent, make superfluous any further development of China's own military potential, and perhaps even make possible a redistribution of resources from the military to the civilian sector. The establishment of diplomatic relations with the United States, which took place at about the same time (on January 1, 1979), was supposed to serve the same end. It was now no longer necessary for the Chinese leadership to show restraint in its attempt to achieve détente with Moscow, which had previously seemed necessary if Peking was to achieve full normalization of relations with Washington. On the contrary: Improved relations with the USSR promised greater leeway in China's contacts with the United States. In this way, a "sinocentric romantic triangle" emerged.[40]

There is one further reason for China's interest in renewing negotiations with Moscow. The assurances the Chinese government had given Tokyo about preparing a Sino-Japanese treaty of peace and friendship obligated China to serve notice in April 1979 that it wanted to terminate the treaty of alliance of February 14, 1950, which was due to expire in April 1980. To avoid or at least reduce the negative consequences on its relations to the USSR that denouncing the treaty would have, Peking made the aforementioned proposal to begin negotiations toward the normalization of relations in its letter serving notice on April 3, 1979.

Moscow accepted within two weeks, and the two sides held a first round of negotiations in the fall of 1979. This began the process of détente that was interrupted by the Soviet intervention in Afghanistan at the end of year but continued in the fall of 1982 with the establishment of consultations that led to a gradual normalization of relations, including the renewal of economic, scientific, technical, and cultural cooperation, in the course of the following years. This process accelerated when Gorbachev came to power in March 1985. His new foreign policy was particularly helpful in that it promoted military détente. It led to the dismantling of the SS-20 rockets stationed in Asia, the reduction of the number of troops stationed in the Soviet Far East and in Soviet Central Asia, the withdrawal of the Soviet Army from Afghanistan, and the denial of support for the Vietnamese occupation of Cambodia—thus practically removal of the "three obstacles" that Peking had made the precondition for full normalization from the early 1980s. The border negotiations also progressed well after Gorbachev made concessions on a key point.

Gorbachev's trip to Peking in May 1989 attended to the first Sino-Soviet summit meeting in thirty years constituted the high point in the normalization process. Party relations, which had been frozen since 1966, were also revived on this occasion. China's relations with the United States remained friendly during the Gorbachev era. Because Soviet-American relations also improved at the same time, one could speak of a strategic triangle with an equally positive balance on all sides during the period from 1986 to 1989 that resembled the situation following the Second World War through the outbreak of the Cold War.

The USSR's critical and concerned reaction to the Peking massacre of June 4, 1989, again led to a disturbance in Sino-Soviet relations, however. The demonstrating students had given Gorbachev an enthusiastic reception in Peking and embarrassed the Chinese leaders, who did want to reform the economy but nonetheless were determined as political conservatives to cling to the one-party dictatorship. They were even less happy to learn that the Soviet Party leadership had resolved to remove the CPSU's monopoly of power in February 1990. The change in the political systems in Eastern Europe had an even more serious effect on Sino-Soviet relations. The Chinese Party leadership found it shocking and blamed Gorbachev and his "capitulationism" in particular. Sharp and fundamental criticism was, however, only expressed behind closed doors in Peking; in public, the Chinese criticized only specific points and restrained themselves. The same was also true of their assessment of the coup attempt in Moscow in August 1991. The incident was described publicly as an internal affair of the USSR, in which they did not wish to intervene. In reality, the Chinese leadership sympathized with the people behind the coup and stated the opinion internally that they should have simply had Gorbachev and Boris Yeltsin arrested, as had been done in China with the Gang of Four after Mao's death.

After the collapse of the Soviet Union in December 1991, Peking quickly drew conclusions from the new realities. China established diplomatic relations with the successor states and signaled that it was interested in continuing the policy it had initiated with the USSR of close economic and technical cooperation with those states, particularly with the Russian Federation. The Chinese leadership showed thereby that it feared that the fate of the world would now be determined primarily by the United States, which had survived as the only superpower. In Moscow, these views were reciprocated only after the phase of pro-Western foreign policy represented by Minister of Foreign Affairs Andrei Kozyrev had ended. The fallen superpower, the Soviet Union, in its attempt to compensate for its defeat by constructing a multipolar world, and the regional power, China, in its dream of becoming a great power or a superpower, suddenly found themselves in the same boat, allied by a commonly shared anti-Americanism.

Cooperation intensified significantly in the bilateral relationship. Trade became more extensive, and there was more technical cooperation involving in particular Russian participation in Chinese energy plans, including atomic energy. In renewed border negotiations, the two countries agreed on a common border, still one of the most explosive areas of contention during the 1960s, in all but a few short sections, and confidence-building measures were intensified in the border areas. The most spectacular point would seem to be that Russia has become China's most important arms supplier in the meantime. In this way, China has access to modern weapons such as tanks, fighter aircraft, submarines, destroyers, and antiaircraft rockets that the West had refused to deliver since the Peking massacre of June 4, 1989, while Russia earns hard currency and can maintain employment in the armaments industry.

In the meantime, concerns about an unlimited extension of cooperation in the area of armaments appear to be growing on both sides. Concern that China is becoming too dependent on Russia in military affairs is gaining ground in Peking, while in Russia the fear is growing that it is supplying the most modern weapons to a state that will use these same weapons to threaten Russia's security even more effectively in future. The Ussuri conflict has not been forgotten and the horror of a "yellow peril" has constantly been in the bones of all Russians—in their subconscious and ready to be awakened at any time—since the Mongolian invasions of the thirteenth and fourteenth centuries. Rumors of Peking's plans to conquer the unpopulated Russian Far East gradually by allowing masses of Chinese to filter in are eagerly believed.

Such fears alone suffice to limit speculations that would compare current Sino-Russian relations with those in the period of the "great friendship" between the USSR and the People's Republic of China during the 1950s. China has stated credibly that it has no plans to form a military alliance with any state for the time being—and certainly not with Russia, to which Peking feels very superior because of the success of the Chinese economy and the failure of the Russian economy. Furthermore, the fact that both sides accepted the same Marxist-Leninist ideology played a major role in the alliance of the 1950s, a factor that disappeared without a trace in the 1990s. People should, however, expect close pragmatic cooperation for the foreseeable future based on the principle of mutual benefit in the bilateral field and on the calculation that solidarity between two states that both see themselves as underdogs to the United Stataes can only be beneficial to both.

A comparative review of czarist and Soviet Russia's policies toward China provides a remarkable result.[41] Relations under the czars were constructive for about 150 years after their beginnings in the seventeenth century and tense for about 100 years. The relationship developed in the opposite direction during the Soviet period. The relations between the USSR and official China, which have lasted for about three-quarters of a century, were characterized by tensions ranging from minor irritations to military conflict for about 50 years. It appears paradoxical, finally, that the danger that a major war might break out seemed to loom only once during the Soviet period: in 1969, when Communist comrades ruled in China.

Appendix

Document 1
Soviet Draft on Questions Concerning the Chinese Changchun Railroad, Port Arthur, and Dairen of January 5, 1950[1]

The High Contracting Parties recognize that the Agreements between the Soviet Union and the Chinese Republic, which were concluded on August 14, 1945, on the CCR, on Port Arthur, and on the harbor of Dairen, are in effect.

Document 2
Soviet Draft on Questions Concerning Dairen and Port Arthur of January 9, 1950[2]

Both High Contracting Parties declare that in accordance with the present Treaty of Friendship, Cooperation, and Mutual Assistance between the Union of Soviet Socialist Republics and the People's Republic of China, the Agreement on the harbor of Dairen of August 14, 1945, and the Agreement on Port Arthur of August 14, 1945, with all of their appendices, remain fully in effect as an inseparable part of the present treaty.

Document 3
Soviet Draft on Questions Concerning Dairen and Port Arthur of January 10, 1950[3]

The High Contracting Parties agreed that the question of the existing Agreements on the harbor of Dairen and on Port Arthur shall be examined benevolently following the conclusion of a peace treaty with Japan.

The High Contracting Parties also agreed that the Soviet military units currently present in Port Arthur and Dairen shall be reduced or withdrawn 2–3-x years after this treaty becomes effective.

Document 4
Soviet Draft on Questions Concerning Port Arthur and Dairen of January 12, 1950[4]

The High Contracting Parties agreed that the Soviet military units currently present in Port Arthur and Dairen shall be withdrawn in the course of a period of 2–3-x years from when this Treaty takes effect and that at the end of the aforementioned period the agreements concluded between the USSR and China on August 14, 1945, on Port Arthur and the harbor of Dairen and also the Protocol to the Agreement on the harbor of Dairen shall be examined.

Document 5
Soviet Draft on Questions Concerning Port Arthur and Dairen of January 16, 1950[5]

The High Contracting Parties agreed that the Soviet military units currently present in Port Arthur and Dairen shall be withdrawn completely to the territory of the USSR in the course of a period of 2–3-x years from when this Treaty takes effect whereby the troop withdrawal will begin in the year 1950.

The High Contracting Parties also agreed that the Agreements concluded between the USSR and China on August 14, 1945, on Port Arthur and the harbor of Dairen and the Protocol to the Agreement on the harbor of Dairen shall be subjected to renewed examination following the conclusion of a peace treaty with Japan.

Document 6: First Variant
Soviet Draft of a Protocol Confirming the Validity of the Agreement on the Chinese Changchun Railroad of January 16, 1950[6]

1st variant
Draft

PROTOCOL

confirming the validity of the Agreement
on the Chinese Changchun Railroad

The undersigned, acting in the name of the Government of the Soviet Union and in the name of the Central People's Government of the People's Republic of China, by signing the Treaty of Friendship, Alliance, and Mutual Assistance

have agreed on the following:

1. The validity of the Agreement on the Chinese Changchun Railroad, concluded on August 14, 1945, in Moscow, is confirmed in its entirety for the period indicated in this Agreement.

2. The present Protocol, which is done in two copies each in the Russian and Chinese languages, becomes effective on the day of its signing as an inseparable part of the aforementioned Treaty.

Moscow, ... 1950

Document 6: Second Variant
Soviet Draft of a Statement Confirming the Validity of the Agreement on the Chinese Changchun Railroad of January 16, 1950[7]

2nd variant
Draft

DECLARATION

For the purposes of determining the precise legal status of the Chinese Changchun Railway, by signing the Treaty of Friendship, Alliance, and Mutual Assistance between the Union of Soviet Socialist Republics and the People's Republic of China the governments concluding the treaty state hereby that the validity of the Agreement on the Chinese Changchun Railway, concluded on August 14, 1945, in Moscow, is confirmed in its entirety for the period indicated in this Agreement.

In confirmation of this, the undersigned plenipotentiaries of the High Contracting Parties signed and affixed their seals to this Declaration in two copies in the Russian and Chinese languages.

Done in Moscow 1950

Document 7
Soviet Draft of a Treaty of Friendship, Alliance, and Mutual Assistance of January 22, 1950[8]

Draft
Attachment no. 1

TREATY

OF FRIENDSHIP, ALLIANCE, AND MUTUAL ASSISTANCE
between the Union of Soviet Socialist Republics
and the People's Republic of China

The Presidium of the Union of Soviet Socialist Republics and the Council of the Central People's Government of the People's Republic of China,

being convinced that the strengthening of good neighborliness and friendship between the Soviet Union and the People's Republic of China is in accordance with the vital interests of the Soviet and Chinese peoples,

being convinced that the strengthening of friendly relations in optimal fashion will promote the economic construction of both states,

being determined to promote with all our power close cooperation between the peoples of both states in the interests of general peace and security, acting in accordance with the purposes and principles of the United Nations Organization, have decided to conclude this Treaty and have appointed as their plenipotentiaries:

the Presidium of the Supreme Soviet of the Union of Soviet Socialist Republics

. . . .

the Central People's Government Council of the People's Republic of China

. . . . ,

who, having exchanged their full powers, found in good and due form and order, have agreed as follows:

ARTICLE 1

The High Contracting Parties undertake to carry out jointly all measures dependent on them to remove any threat of a repetition of aggression by Japan or any other state that might join Japan directly or in any other form.

The High Contracting Parties confirm that they are determined to participate, in the sincerest fashion, in all international action designed for ensuring the peace and security of peoples and will fully contribute to the realization of these lofty goals.

ARTICLE 2

Should either of the High Contracting Parties be drawn into military hostilities with Japan, should it seek to resume its aggressive policies, or with any other state that directly or in any other form should ally itself with Japan in the policies of aggression, the other Contracting Party shall immediately extend military and other assistance with all the means at its disposal to the other Contracting Party that has been drawn into military hostilities.

ARTICLE 3

Each of the High Contracting Parties undertakes not to enter into any alliance or participate in any coalition or in any actions or measures directed against the other High Contracting Party.

ARTICLE 4

The High Contracting Parties undertake not to enter into negotiations with Japan or to conclude any peace treaty with Japan without mutual consent.

ARTICLE 5

The High Contracting Parties shall consult together on all important international questions involving the interests of either country, with a view to strengthening peace and general security.

ARTICLE 6

The High Contracting Parties declare that they shall develop and strengthen the economic and cultural ties between the two states, with a view to rendering each other all possible economic assistance and observing the principles of mutual respect for state sovereignty, territorial integrity, and nonintervention in the domestic affairs of the other Contracting Party.

ARTICLE 7

The Treaty shall come into force immediately after its ratification and will remain in force for thirty years. If neither of the High Contracting Parties gives notice one year before the expiration of the said period that it wishes to denounce the Treaty, it shall remain in force for an undetermined period and each of the High Contracting Parties shall be able to terminate its effectiveness by informing the other Contracting Party within one year.

The present Treaty shall be ratified as quickly as possible. The exchange of the instruments of ratification shall take place at Peking in the near future.

In faith whereof the plenipotentiaries have signed the present Treaty and attached their seals thereto.

Done at Moscow on " " January 1950 in two copies, each in the Russian and Chinese languages, both texts being equally authentic.

By authorization
of the Presidium of the Supreme Soviet
of the Union of the SSR

By authorization
of the Central People's Government
Council of the Chinese People's Republic

Document 8
Soviet Draft of a Protocol Concerning the Agreement on Port Arthur and Dairen of January 22, 1950[9]

Draft
Attachment no. 2

PROTOCOL

concerning the Soviet-Chinese Agreements
on Port Arthur and the harbor of Dairen

The undersigned, operating on the conviction that the Chinese People's Republic has the ability to strengthen its defensive forces very quickly, acting in the name of the government of the Union of SSR and in the name of the Central People's Government of the Chinese People's Republic, have agreed as follows:

1. In accordance with the changed conditions in China and with the readiness expressed by the Soviet Union, the Agreements concluded between the USSR and China on August 14, 1945, on Port Arthur and the harbor of Dairen shall be subjected to renewed examination following the conclusion of a peace treaty with Japan.

2. The Soviet military units currently located in Port Arthur and Dairen will be totally withdrawn to the territory of the USSR in the course of a period of 2 to 3 years after the present treaty becomes effective, whereby the troop withdrawal shall begin in 1950.

3. The present Protocol, done in two copies in the Russian and Chinese languages, becomes effective on the day it is signed.

Moscow, ____________________ 1950

Document 9
Soviet Draft of a Protocol Concerning the Agreement on the Chinese Changchun Railroad of January 22, 1950[10]

Draft
Attachment no. 3

PROTOCOL

concerning the Soviet-Chinese Agreement
on the Chinese Changchun Railroad

The undersigned, acting in the name of the government of the Union of the SSR and in the name of the Central People's Government of the Chinese People's Republic, have agreed as follows:

1. The validity of the Soviet-Chinese Agreement on the Chinese Changchun Railroad, done in Moscow on August 14, 1945, is confirmed for the period indicated in this Agreement, while a few supplementary provisions can be added, at the recommendation of each of the parties, in the interests of guaranteeing the successful operation of the Railroad.

2. The present protocol, done in two copies in the Russian and Chinese languages, becomes effective on the day it is signed.

Moscow, ____________________ 1950

Document 10
Chinese Draft of a Treaty of Friendship, Alliance, and Mutual Assistance of January 24, 1950[11]

Translation from the Chinese
DRAFT

TREATY

of Friendship, Alliance, and Mutual Assistance Between China and the USSR

The Central People's Government of the People's Republic of China and the Presidium of the Supreme Soviet of the Union of Soviet Socialist Republics

in the wish to prevent jointly, by means of strengthening friendly relations and cooperation between China and the USSR, the rebirth of Japanese imperialism and the renewed unleashing of an aggressive war by other states that might join with Japan,

to ensure long-lasting peace and general security in the Far East and in the entire world in accordance with the purposes and principles of the Organization of the United Nations,

are profoundly convinced that this reflects the fundamental interests of the people of China and the USSR.

As a result of this, the governments of China and the USSR decided to conclude the present treaty, and nominated as their plenipotentiaries:

the Central People's Government of the People's Republic of China—Zhou Enlai, the Premier of the State Administrative Council and Minister of Foreign Affairs

the Presidium of the Supreme Soviet of the Union of the Soviet Socialist Republics—

Both plenipotentiary representatives, having exchanged their full powers, found in good and due form, have agreed as follows:

ARTICLE 1

The two Contracting Parties guarantee to carry out jointly all necessary measures within their power to remove any threat of a renewed unleashing of an aggression by Japan or any other state that might join with Japan directly or in any other form; should one of the Contracting Parties find itself in a state of war with the aforementioned states, the other Contracting Party must immediately extend military and other assistance with all the means at its disposal. The Parties declare their readiness to participate in a spirit of sincere cooperation in all international action designed to safeguard peace and security throughout the world and to devote all their energies to the speediest implementation of these lofty tasks.

ARTICLE 2

Neither of the Contracting Parties shall, without mutual agreement, either individually or jointly with other states conclude a peace treaty with Japan.

ARTICLE 3

Neither of the Contracting Parties shall enter into any alliance directed against the other Party or participate in any coalitions directed against the other Party.

ARTICLE 4

The two Contracting Parties will consult with each other in regard to all important international problems affecting the common interests of China and the Soviet Union.

ARTICLE 5

The two Contracting Parties wish, in a spirit of friendship and cooperation and in accordance with the principles of equality, mutual interests, and also mutual respect for the territorial sovereignty and nonintervention in the internal affairs of the other Party, to develop and consolidate the economic and cultural ties between China and the Soviet Union.

ARTICLE 6

The Contracting Parties agree, in the interests of facilitating and accelerating economic construction in China and the USSR and contributing their part to the flourishing of the entire world, to render each other all possible assistance and to effect the necessary economic cooperation.

ARTICLE 7

The present Treaty shall come into force immediately following its being signed; the exchange of the instruments of ratification shall take place in Peking.

The present Treaty shall remain in force for thirty years. If neither of the Contracting Parties gives notice one year before the expiration of this period that it wishes to denounce the Treaty, it shall remain in force for a further five years and shall thereafter be continued in force in accordance with this provision.

For accuracy:
(signature illegible)

Translated by /Krutikov/
/Fedorenko/
/Kurdiukov/

Document 11
Soviet Draft of an Agreement on Port Arthur, Dairen, and the Chinese Changchun Railroad of January 29, 1950[12]

Draft

AGREEMENT

between the People's Republic of China and the Union of Soviet Socialist Republics on Port Arthur, Dal'nii [Dairen], and the Chinese Changchun Railroad

The Central People's Government of the People's Republic of China and the Presidium of the Supreme Soviet of the Union of Soviet Socialist Republics note that since 1945 fundamental changes have occurred in the situation in the Far East, namely: Imperialist Japan has suffered defeat; Chiang Kai-shek's Kuomintang clique has been overthrown; China has become a People's Democratic Republic; a new People's government has been established in China which is friendly toward the Soviet Union, has unified China in an integrated state, and has proved its ability to uphold the State independence, territorial integrity, and national honor and dignity of the Chinese people.

The Central People's Government of the People's Republic of China and the Presidium of the Supreme Soviet of the USSR consider that this new situation permits a new approach to the question of the Chinese Changchun Railroad, Port Arthur, and the harbor of Dal'nii.

In conformity with these new circumstances, the USSR and China have decided to conclude the present Agreement on the CCR, on Port Arthur, and on the harbor of Dal'nii.

ARTICLE 1

The High Contracting Parties have agreed that the Soviet government shall transfer to the government of the PRC, without charge, all its rights to joint administration of the CCR with all the property belonging to the Railroad. The transfer shall be effected immediately after the conclusion of a Peace Treaty with Japan, but not later than the end of 1952.

Until the transfer is effected, the present status of the joint Chinese-Soviet administration of the Chinese Changchun Railroad shall remain unchanged; however, after this agreement has come into force, the procedure whereby posts are occupied by representatives of the Chinese and Soviet Parties shall be altered and a system of rotation of posts (Manager of the Railroad, President of the Board of Directors, etc.) at fixed intervals shall be established.

The specific means of effecting the transfer shall be agreed and determined by the Governments of the two Contracting Parties.

ARTICLE 2

The High Contracting Parties have agreed that Soviet troops shall be withdrawn from the naval base of Port Arthur immediately after the conclusion of a Peace Treaty with Japan, but not later than the end of 1952.

Pending the withdrawal of Soviet troops the Governments of China and the USSR appoint an equal number of military representatives to form a Joint Chinese-Soviet Military Commission over which the two Parties shall preside and which shall administer military affairs in the area of Port Arthur; specific measures for this purpose shall be determined by the Joint Chinese-Soviet Military Commission within three months after the present Agreement becomes effective and shall be put into force after approval of these measures by the Governments of both Parties.

The civil administration in the aforesaid area shall be under the direct control of the Government of the People's Republic of China. Until the Soviet troops are withdrawn, the zone in which Soviet troops are quartered in the Port Arthur area shall remain unchanged in accordance with the existent boundaries.

Should either of the Contracting Parties become the object of aggression by Japan or any state that might join with Japan, and as a result be involved in hostilities, China and the USSR may, on the proposal of the Government of the People's Republic of China and with the consent of the Soviet Government, jointly use the naval base Port Arthur for conduct of joint military operations against the aggressor.

ARTICLE 3

The High Contracting Parties have agreed that the question of the harbor of Dairen/ Dal'nii/ shall be examined after the conclusion of a Peace Treaty with Japan.

As regards the administration in Dal'nii, it shall be wholly in the hands of the government of China.

All property at present in Dal'nii which is temporarily administered by or leased to the Soviet Party shall be taken over by the Government of China. For the purpose of effecting the transfer of said property, the Governments of China and the USSR shall appoint three representatives from each Party to form a Joint Commission which, within three months after the entry into force of this Agreement, will determine specific measures for effecting the transfer of the property and shall, after the Joint Commission's proposals have been approved by the Governments of both Parties, complete their application in the course of 1950.

ARTICLE 4

The present Agreement shall come into force on the day of its ratification. The exchange of instruments of ratification shall take place in Peking.

Done in Moscow on . . . . 1950, in two copies, each in the Chinese and Russian languages, both texts being equally authentic.

By authorization of
the Central People's Government
of the People's Republic of China

By authorization of
the Presidium of the Supreme Soviet of
the Union of the SSR

Distributed
to comrades Stalin,
Molotov,
Malenkov,
Beriia,
Mikoyan,
Kaganovich,
Bulganin.

Document 12
Treaty of Friendship, Alliance, and Mutual Assistance, Signed on February 14, 1950[13]

Treaty of Friendship, Alliance and Mutual Assistance between the Union of Soviet Socialist Republics and the People's Republic of China

The Presidium of the Supreme Soviet of the Union of Soviet Socialist Republics and the Central People's Government of the People's Republic of China,

Being determined, by strengthening friendship and co-operation between the Union of Soviet Socialist Republics and the People's Republic of China, jointly to prevent the revival of Japanese imperialism and the repetition of aggression on the part of Japan or of any other State that might in any way join with Japan in acts of aggression,

Being anxious to promote a lasting peace and general security in the Far East and throughout the world in accordance with the purposes and principles of the United Nations,

Being firmly convinced that the strengthening of good-neighborly and friendly relations between the Union of Soviet Socialist Republics and the People's Republic of China is in accordance with the fundamental interests of the peoples of the Soviet Union and China,

Have decided for this purpose to conclude the present Treaty and have appointed as their plenipotentiaries:

The Presidium of the Supreme Soviet of the Union of Soviet Socialist Republics: Andrei Yanuarevich Vyshinsky, Minister of Foreign Affairs of the USSR;

The Central People's Government of the People's Republic of China: Chou En-lai, Chairman of the State Administrative Council and Minister of Foreign Affairs of China.

The two plenipotentiary representatives, having exchanged their full powers, found in good and due form, have agreed as follows:

ARTICLE 1

The two Contracting Parties undertake to carry out jointly all necessary measures within their power to prevent a repetition of aggression and breach of the peace by Japan or any other State which might directly or indirectly join with Japan in acts of aggression. Should either of the Contracting Parties be attacked by Japan or States allied with Japan and thus find itself in a state of war, the other Contracting Party shall immediately extend military and other assistance with all the means at its disposal.

The Contracting Parties likewise declare that they are prepared to participate, in a spirit of sincere co-operation, in all international action designed to safeguard peace and security throughout the world, and will devote all their energies to the speediest realization of these aims.

ARTICLE 2

The two Contracting Parties undertake, by common agreement, to strive for the conclusion at the earliest possible date, in conjunction with the other Powers which were their Allies during the Second World War, of a Peace Treaty with Japan.

ARTICLE 3

Neither of the Contracting Parties shall enter into any alliance directed against the other Party, or participate in any coalition or in any action or measures directed against the other Party.

ARTICLE 4

The two Contracting Parties shall consult together on all important international questions involving the common interests of the Soviet Union and China, with a view to strengthening peace and universal security.

ARTICLE 5

The two Contracting Parties undertake, in a spirit of friendship and co-operation and in accordance with the principles of equal rights, mutual interests, mutual respect for State sovereignty and territorial integrity, and non-intervention in the domestic affairs of the other Party, to develop and strengthen the economic and cultural ties between the Soviet Union and China, to render each other all possible economic assistance and to effect the necessary economic co-operation.

ARTICLE 6

This Treaty shall come into force immediately upon ratification; the exchange of the instruments of ratification shall take place at Peking.

The Treaty shall remain in force for thirty years. If neither of the Contracting Parties gives notice one year before the expiration of the said period that it wishes to denounce the Treaty, it shall remain in force for a further five years and shall thereafter be continued in force in accordance with this provision.

DONE at Moscow, on 14 February 1950, in two copies, each in the Russian and Chinese languages, both texts being equally authentic.

By authorization
of the Presidium of the Supreme
Soviet of the Union
of Soviet Socialist Republics:

(*Signed*) A.Y. Vyshinsky
[L.S.]

By authorization
of the Central People's Government
of the People's Republic of China:

(*Signed*) Chou En-lai
[L.S.]

Document 13
Agreement on the Chinese Changchun Railroad, Port Arthur, and Dairen, Signed on February 14, 1950[14]

Agreement between the Union of Soviet Socialist Republics and the People's Republic of China concerning the Chinese Changchun Railway, Port Arthur and Dairen

The Presidium of the Supreme Soviet of the Union of Soviet Socialist Republics and the Central People's Government of the People's Republic of China note that since 1945 the following radical changes have taken place in the situation in the Far East: Imperialist Japan has suffered defeat; the reactionary Kuomintang Government has been overthrown; China has become a People's Democratic Republic; and there has been established in China a new People's Government which has united all China, applied a policy of friendship and co-operation with the Soviet Union, and proved its ability to uphold the State independence and territorial integrity of China and the national honor and dignity of the Chinese people.

The Presidium of the Supreme Soviet of the Union of Soviet Socialist Republics and the Central People's Government of the People's Republic of China consider that because of this new situation a new approach to the question of the Chinese Changchun Railway, Port Arthur, and Dairen is possible.

In accordance with these new circumstances the Presidium of the Supreme Soviet of the Union of Soviet Socialist Republics and the Central People's Government of the People's Republic of China have decided to conclude this Agreement concerning the Chinese Changchun Railway, Port Arthur and Dairen.

ARTICLE 1

The two Contracting Parties have agreed that the Soviet Government shall transfer to the Government of the People's Republic of China, without charge, all its rights to joint administration of the Chinese Changchun Railway together with all property belonging to the railway. The transfer shall be effected immediately after the conclusion of a Peace Treaty with Japan, but in any case not later than the end of 1952.

Until the transfer is effected, the present status of the joint Soviet-Chinese administration of the Chinese Changchun Railway shall remain unchanged; however, after this Agreement has come into force, the procedure whereby posts are occupied by representatives of the Soviet and Chinese Parties shall be altered and a system of rotation of posts (Manager of the Railway, President of the Board of Directors, etc.) at fixed intervals shall be established.

The specific means of effecting the transfer shall be agreed and determined by the Governments of the two Contracting Parties.

ARTICLE 2

The two Contracting Parties have agreed that Soviet troops shall be withdrawn from the jointly used naval base of Port Arthur and that the installations in this area shall be transferred to the Government of the People's Republic of China immediately after the conclusion of a Peace Treaty with Japan, but in any case not later than the end of 1952; the Government of the People's Republic of China shall compensate the Soviet Union for the expenditure incurred by the Soviet Union since 1945 in respect of the reconstruction and erection of the said installations.

Pending the withdrawal of Soviet troops and the transfer of the aforesaid installations, the Governments of the Soviet Union and China shall appoint an equal number of military representatives to constitute a Joint Sino-Soviet Military Commission, over which the two Parties shall preside in rotation and which shall administer military affairs in the Port Arthur area; specific measures for this purpose shall be determined by the Joint Sino-Soviet Military Commission within three months after the entry into force of this Agreement and shall be applied after they have been approved by the Governments of both Parties.

The civil administration in the aforesaid area shall be under the direct control of the Government of the People's Republic of China. Until the Soviet troops are withdrawn, the zone in which Soviet troops are quartered in the Port Arthur area shall remain unchanged in accordance with the existing boundaries.

Should either of the Contracting Parties become the object of aggression by Japan or any State which might join with Japan, and as a result be involved in hostilities, China and the Soviet Union may, on the proposal of the Government of the People's Republic of China and with the consent of the Soviet Government, jointly use the naval base of Port Arthur for the conduct of joint military operations against the aggressor.

ARTICLE 3

The two Contracting Parties have agreed that the question of the port of Dairen shall be examined after the conclusion of a Peace Treaty with Japan.

The administration of Dairen shall be wholly in the hands of the Government of the People's Republic of China.

All property at present in Dairen which is temporarily administered by or leased to the Soviet Party shall be taken over by the Government of the People's Republic China. For the purpose of effecting the transfer of the said property, the Governments of the Soviet Union and China shall appoint three representatives from each Party to constitute a Joint Commission, which shall within three months after the entry into force of this Agreement determine specific measures for effecting the transfer of the property and shall, after the Joint Commission's proposals have been approved by the Governments of both Parties, complete their application in the course of 1950.

ARTICLE 4

This Agreement shall come into force on the date of its ratification. The exchange of the instruments of ratification shall take place at Peking.

DONE at Moscow, on 14 February 1950, in two copies, each in the Russian and Chinese languages, both texts being equally authentic.

By authorization of
the Presidium of the Supreme Soviet
of the Union of Soviet Socialist Republics:
(*Signed*) A.Y. Vyshinsky

By authorization of the
Central People's Government
of the People's Republic of China:
(*Signed*) Chou En-lai

Document 14
Agreement on the Granting of a Credit to the People's Republic of China, Signed on February 14, 1950[15]

Agreement between the Government of the Union of Soviet Socialist Republics and the Central People's Government of the People's Republic of China concerning the grant of a credit to the People's Republic of China

Whereas the Government of the Union of Soviet Socialist Republics has acceded to the request of the Central People's Government of the People's Republic of China for the grant of a credit to China to pay for the equipment and other materials which the Soviet Union has agreed to supply to China, the two Governments have agreed as follows:

ARTICLE 1

The Government of the USSR shall grant to the Central People's Government of the People's Republic of China a credit in dollars in the amount of US$300 million, at the rate of US$35 per ounce of fine gold.

In view of the extreme devastation[16] of China caused by the prolonged military operations in its territory, the Soviet Government has agreed to provide the credit at the favorable rate of interest of 1 per cent per annum.

ARTICLE 2

The credit referred to in article 1 shall be made available over a period of five years from 1 January 1950, in equal instalments of one-fifth of the total credit a year, in order to pay for deliveries of equipment and materials from the USSR, including equipment for power stations, metallurgical and engineering plants and coal and ore mines, railway and other transport equipment, rails and other materials for the reconstruction and development of China's national economy.

The nomenclature, quantities, prices and delivery dates of equipment and materials shall be fixed by a special Agreement between the Parties, the prices being determined on the basis of world market prices.

Any part of the credit remaining unused in any one year may be used in the succeeding years.

ARTICLE 3

The Central People's Government of the People's Republic of China shall repay the credit referred to in article 1 and the interest thereon in deliveries of raw materials and tea, and in gold and United States dollars. The prices, quantities, and delivery dates of raw materials and tea shall be fixed by a special agreement, the prices being determined on the basis of world market prices.

The repayment of the credit shall be effected over a period of ten years in equal annual installments of one-tenth of the credit received, payable not later than 31 December of each year. The first regular installment shall be paid not later than 31 December 1954, and the last not later than 31 December 1963.

The interest on the credit which shall accrue from the date on which each installment of the credit is used shall be payable half-yearly.

ARTICLE 4

For the purpose of keeping accounts of the credit provided for by this Agreement, the State Bank of the USSR and the National Bank of the People's Republic of China shall open special accounts and shall jointly determine the procedure for the settlement and keeping of accounts under this Agreement.

ARTICLE 5

This Agreement shall enter into force from the date of signature and shall be subject to ratification. The exchange of the instruments of ratification shall take place at Peking.

DONE at Moscow, on 14 February 1950, in two copies, each in the Russian and Chinese languages, both texts being equally authentic.

By authorization
of the Government of the Union
of Soviet Socialist Republics:
(*Signed*) A.Y. Vyshinsky

By authorization
of the Central People's Government
of the People's Republic of China
(*Signed*) Chou En-Lai

Document 15
Chinese Draft of a Secret Agreement on the Working Conditions of Soviet Specialists in China of March 22, 1950[17]

Draft

AGREEMENT

between the Central People's Government of the People's Republic of China and the Government of the Union of Soviet Socialist Republics on the working conditions of the Soviet specialists in China

In the context of the readiness of the Government of the USSR to grant the request of the Central People's Government of the People's Republic of China[18] for Soviet specialists to be sent out to China and for the assistance to China in the reconstruction and development of the economy, the CPG of the PRC and the Government of the Union of Soviet Socialist Republics[19] have agreed as follows:

ARTICLE 1

The Government of the USSR shall send out Soviet specialists for the use of the CPG of the PRC at their request to work in companies and also in institutions and organizations of the People's Republic of China.

ARTICLE 2

The duration of the work of the Soviet specialists sent out to China shall be fixed at one year. An extension of the period of their stay in China shall be determined on the basis of mutual agreement of the parties.

The responsible authorities of both sides shall conjointly determine any recall of the Soviet specialists from China, whereby the premature recall of any Soviet specialist who had not been active for more than six months for reasons that do not depend on the CPG of the PRC shall be carried out at the expense of the Soviet side.

In the case of the premature recall of a Soviet specialist by the Soviet side, the latter is obligated to replace him with another specialist. The costs involved in such a recall and replacement shall be borne by the Soviet side.

ARTICLE 3

The CPG of the PRC will pay the Soviet specialists wages in Chinese currency equivalent to the salaries that are paid or will be paid to Chinese specialists with corresponding positions and qualifications, including provision of food and industrial goods and bonuses in addition to wages and the like with which Chinese specialists are supplied.

The CPG of the PRC shall compensate the Soviet side for the following expenditures for sending out the Chinese specialists to China: travel expenses of the specialists and their families, daily allowances, lodging expenses and wages for the period of the return trip on the territory of the USSR, expenses of baggage transportation to the extent of 80 kilograms per person, and travel expenses in the amount of the salary paid in the USSR during the month prior to departure.

The CPG of the PRC shall pay travel expenses, daily allowances, lodging expenses, and wages for the period of the return trip on Chinese territory directly in Chinese currency for the specialist and his family, and vacation compensation in the amount of one month's salary for each year spent working in China, and to Soviet specialists active in Chinese teaching institutions a two months' salary for each working year.

ARTICLE 4

The CPG of the PRC shall pay the Soviet Government between 1,500 and 3,000 rubles for each specialist sent out, depending on his qualifications, in compensation for the loss that the corresponding Soviet institution or factory suffers for sending its specialist abroad.

ARTICLE 5

The CPG of the PRC shall provide furnished apartments with heating and lighting free of charge or pay an amount sufficient for renting a furnished apartment with heating and lighting to the Soviet specialists for the period of their work in China.

ARTICLE 6

The CPG of the PRC engages in the case of the illness of a Soviet specialist not to terminate payment of salary for the entire period of the illness but no longer than three months.

In the event that the illness of a Soviet specialist lasts more than three months, the Government of the USSR is obligated to replace him with another specialist with the same qualifications.

ARTICLE 7

All conflicts arising from this Agreement will be resolved by diplomatic means by the CPG of the CPR and the Government of the USSR.

ARTICLE 8

The working and salary conditions of the Soviet specialists provided for in the present Agreement will also apply to those Soviet specialists who had been sent out before the conclusion of the present Agreement to work in institutions, organizations, and companies in the People's Republic of China.

ARTICLE 9

This Agreement becomes effective on the day it is signed and will remain in effect until the end of a six-month period that begins on the day on which one of the sides expresses the wish to terminate its effectiveness.

Done in Moscow ". . ." . . . . 1950 in two copies, each in the Chinese and Russian languages, both texts being equally authentic.

By the authorization
of the Central People's Government
of the People's Republic of China

By the authorization
of the Government of the
Union of Soviet Socialist Republics

Notes

Note to Preface

1. John W. Garver, *Chinese-Soviet Relations, 1937–1945*; Sergei N. Goncharov, John W. Lewis, and Xue Litai, *Uncertain Partners*; Odd Arne Westad, *Cold War and Revolution*; Chen Jian, *China's Road to the Korean War*; Brian Joseph Murray, *Western versus Chinese Realism*; Michael H. Hunt, *The Genesis of Chinese Communist Foreign Policy*; Eva-Maria Stolberg, *Stalin und die chinesischen Kommunisten, 1945–1953*; Michael M. Sheng, *Battling Western Imperialism*; Odd Arne Westad, ed., *Brothers in Arms*; Richard C. Thornton, *Odd Man Out.*

Notes to Chapter 1

1. The literature on the period from 1921 to 1935 is too voluminous for us to cite it completely in the respective passages. The author will only provide a few select references to support his statements in the following section. Since the 1980s, both the Russian and the Chinese sides have presented publications containing new material and archival documents on this period, such as, for example, Go Khen'iui et al., eds., *VKP(b), Komintern i natsional'no-revoliutsionnoe dvizhenie v Kitae*, vol. 1, *1920–1925*; M.L. Titarenko et al., eds., *VKP(b), Komintern i natsional'no-revoliutsionnoe dvizhenie v Kitae*, vol. 2, *1926–1927*, parts 1–2; *Zhou Enlai nianpu, 1898–1949; Mao Zedong nianpu, 1893–1949*, vols. 1–2; *Liu Shaoqi nianpu, 1898–1969*, vol. 1; Yang Yunruo and Yang Kuisong, *Gongchan guoji he Zhongguo geming*; Yang Kuisong, *Zhongjian didai de geming*; Huang Xiurong, *Gongchan guoji yu Zhongguo geming guanxi shi*, 2 vols. (with an extensive bibliography of the secondary literature published between 1979 and 1987 in vol. 2, pp. 339–72); *Gongchan guoji he Zhongguo geming guanxi jishi 1919–1943; Gongchan guoji yu Zhongguo geming guanxi lunwen ji*; Zhou Wenqi, and Chu Liangru, eds., *Gongchan guoji, Sulian he Zhongguo gongchandang guanxi biannianshi, 1919–1991*; *Gongchan guoji yu Zhongguo geming ziliao xuanji (1925–1927).*

2. Stalin, August 1, 1927, in I.V. Stalin, *Sochineniia*, vol. 10, p. 51.

3. See Article 3 of the statutes and number 16 of the Comintern principles, both adopted at the Second World Congress (July–August 1920), in Günther Nollau, *International Communism and World Revolution*, pp. 336, 340, 343.

4. This was called the Russian Communist Party (Bolsheviks) [in Russian, RKP(b); in English, RCP(B)] from 1918 until 1925, then the All-Russian Communist Party (Bolsheviks) [in Russian, VKP(b); in English, CPSU(B)] until 1952, and then Communist Party of the Soviet Union (in Russian, KPSS; in English, CPSU).

5. Chang Kuo-t'ao, *The Rise of the Chinese Communist Party*, vol. 2, 1928–1938, p. 133. On this paragraph, see Nollau, *International Communism*, pp. 39–107, esp. pp. 95–96, 106.

6. For further details, see Dieter Heinzig, *Sowjetische Militärberater bei der Kuomintang, 1923–1927*, pp. 24–27.

7. Borodin's note, January 24, 1925, on a talk between Chang Tso-lin and Sun Yat-sen on December 5, 1924, cited in Go Khen'iui et al., eds., *VKP(b), Komintern i natsional'no-revoliutsionnoe dvizhenie v Kitae*, vol. 1, p. 515.

8. On Mao Zedong as Chinese nationalist, see Stuart R. Schram, *The Political Thought of Mao Tse-tung*, pp. 161–68.

9. At that time, Minister of Foreign Affairs Georgii Chicherin expressed his satisfaction in a letter to Stalin that the Soviet leader had taken China policy into his own hands. The author is grateful for this information to his Russian colleague Konstantin Shevelev, who had access to Chicherin's note in Rossiiskii Tsentr Khraneniia i Izucheniia Dokumentov Noveishei Istorii.

10. See Chang Kuo-t'ao, *Rise*, vol. 2, p. 114.

11. On this and the following sentence, see the unambiguous statements of the then Comintern represen-

tative in China, Mikhail Borodin, reproduced in Louis Fischer, *Russia's Road from Peace to War*, pp. 118–19. See also Chalidze, *Nikita Khrushchev*, p. 232.

12. Cited in *Die chinesische Frage auf dem 8. Plenum der Exekutive der Kommunistischen Internationale*, p. 124.

13. According to Chang Kuo-t'ao, *Rise*, vol. 2, p. 41.

14. On this paragraph, see Harold R. Isaacs, *The Tragedy of the Chinese Revolution*, passim; Allen S. Whiting, *Soviet Policies in China, 1917–1924*, passim; Robert C. North, *Moscow and the Chinese Communists*, passim; Nollau, *International Communism*, pp. 98–105; Conrad Brandt, *Stalin's Failure in China*, passim; Jürgen Domes, *Vertagte Revolution*, pp. 65–206, passim; C. Martin Wilbur, *The Nationalist Revolution in China, 1923–1928*, passim; C. Martin Wilbur and Julie Lien-ying How, *Missionaries of Revolution*, passim; Tony Saich, ed., *The Origins of the First United Front in China*, passim; V.I. Glunin, *Kommunisticheskaia Partiia Kitaia nakanune i vo vremia natsional'noi revoliutsii (1921–1927 gg.)*, passim; Leong Sow-Theng, *Sino-Soviet Diplomatic Relations, 1917–1926*, passim; Heinzig, *Sowjetische Militärberater*, passim; Bruce A. Elleman, *Diplomacy and Deception*, passim; Kuo Heng-yü, *Die Komintern und die Chinesische Revolution*, passim; Jerome Ch'en, "The Chinese Communist Movement to 1927," pp. 505–26, passim; Lloyd E. Eastman, "Nationalist China during the Nanking Decade, 1927–1937," pp. 116–17; Yang and Yang, *Gongchan guoji*, passim; Huang Xiurong, *Gongchan guoji*, vol. 1, passim; Yang Kuisong, *Zhongjian didai de geming*, passim.

15. On this and the two following sentences, see North, *Moscow and the Chinese Communists*, pp. 117–21; Nollau, *International Communism*, pp. 103–5; Isaacs, *Tragedy*, pp. 272–92; and Chang Kuo-t'ao, *Rise*, vol. 2, pp. 52–53.

16. On this paragraph, see the source references in n. 14 and esp. Isaacs, *Tragedy*, pp. 272–92; North, *Moscow and the Chinese Communists*, pp. 113–21; Nollau, *International Communism*, pp. 98–105; Domes, *Vertagte Revolution*, pp. 219–37, 278–79; Heinzig, *Sowjetische Militärberater*, pp. 285–87; Jerome Ch'en, "Chinese Communist Movement to 1927," pp. 525–26; Ch'en, "Communist Movement, 1927–1937," pp. 183–88; Chang Kuo-t'ao, *Rise*, vol. 2, p. 83; Wilbur, *Nationalist Revolution*, pp. 147–69.

17. On this paragraph, see Domes, *Vertagte Revolution*, pp. 279–80; Ch'en, "Communist Movement, 1927–1937," pp. 183–98; Chang Kuo-t'ao, *Rise*, vol. 2, p. 80.

18. On this, see Zhou Enlai's report to the Central Party School, March 3–4, 1944, cited in Zhou and Chu, eds., *Gongchan guoji*, pp. 415–16.

19. On this and the two following sentences, see Chang Kuo-t'ao, *Rise*, vol. 2, pp. 128–29.

20. On this paragraph, see Richard C. Thornton, "The Emergence of a New Comintern Strategy for China," pp. 90 ff.; E.H. Carr, *Twilight of the Comintern*, pp. 319–28.

21. On this and the following paragraph, see North, *Moscow and the Chinese Communists*, esp. pp. 122–67; Charles B. McLane, *Soviet Policy and the Chinese Communists, 1931–1946*, passim, esp. pp. 5–50; Hsiao Tso-liang, *Power Relations Within the Chinese Communist Movement, 1930–1934*, passim, esp. pp. 13, 23, 125, 128, 320, 343; John E. Rue, *Mao Tse-tung in Opposition*, passim; Chang Kuo-t'ao, *Rise*, vol. 2, pp. 136–37; Wang Chien-min, *Zhongguo gongchandang shi gao*, vol. 2, passim, esp. p. 100; Richard C. Thornton, *The Comintern and the Chinese Communists, 1928–1931*, passim; Dieter Heinzig, "The Otto Braun Memoirs and Mao's Rise to Power," pp. 274–88; A.M. Grigor'ev, "Pavel Mif (1901–1938)," pp. 92–93; Kuo Heng-yü, *Maos Weg zur Macht und die Komintern*, passim; Carr, *Twilight of the Comintern*, pp. 328–65, passim; Ch'en, "Communist Movement, 1927–1937," pp. 168–229, passim; Yang and Yang, *Gongchan guoji*, passim; Yang Kuisong, *Zhongjian didai de geming*, passim; Huang Xiurong, *Gongchan guoji*, vol. 2, passim.

22. Chang Kuo-t'ao, *Rise*, vol. 2, p. 448.

23. Huang Xiurong, *Gongchan guoji*, vol. 2, pp. 164–65.

24. Many of the details of the Zunyi Conference and changes in the top leadership that followed continue to be debated to this day in the secondary literature. On this paragraph, see *Zhongguo gongchandang chuangye sanshi nian (1919–1949)*, pp. 313–14; Huang Xiurong, *Gongchan guoji*, vol. 2, pp. 145–52; Otto Braun, *A Comintern Agent in China*, pp. 95–108, 151; Jerome Ch'en, "Resolution of the Tsunyi Conference," pp. 1–38; Dieter Heinzig, "Otto Braun and the Tsunyi Conference," pp. 131–35; Hu Chi-hsi, "Hua Fu, the Fifth Encirclement Campaign and the Tsunyi Conference," pp. 31–46; Heinzig, "Otto Braun Memoirs," pp. 274–88, esp. pp. 287–88; Benjamin Yang, "The Zunyi Conference, or One Step in Mao's Rise to Power," pp. 235–71; Thomas Kampen, "The Zunyi Conference and Further Steps in Mao's Rise to Power," pp. 118–34; Thomas Kampen, *Mao Zedong, Zhou Enlai and the Evolution of the Chinese Communist Leadership*, pp. 69–77; Ch'en, "Communist Movement in China, 1927–1937," pp. 212–13; John W. Garver, "The Origins of the Second United Front," p. 37; Chang Kuo-t'ao, introduction to *Collected Works of Liu Shao-ch'i Before*

1944, p. vi; Chang Kuo-t'ao, *Rise*, vol. 2, p. 380; Tony Saich, *The Rise to Power of the Chinese Communist Party*, pp. 523–25, 640–54; Freddy Litten, *Otto Brauns frühes Wirken in China (1932–1935)*, passim, esp. pp. 59–81.

25. Braun, *Comintern Agent in China*, p. 151.

26. Michael Sheng states that radio traffic was interrupted in August; Huang Xiurong, in October 1934. Otto Braun assumes that it did not happen before the end of September. See Michael M. Sheng, "Mao, Stalin, and the Formation of the Anti-Japanese United Front," pp. 151–52; Huang Xiurong, *Gongchan guoji*, vol. 2, p. 148; and Braun, *Comintern Agent in China*, p. 79.

27. *Zhongguo gongchandang chuangye sanshi nian*, p. 314. There is a similar assessment in Huang Xiurong, *Gongchan guoji*, vol. 2, p. 148.

28. See the debate between John W. Garver and Michael M. Sheng, *CQ* 113 (March 1988), pp. 37–38, and *CQ* 129 (March 1992), pp. 150–51, 178–79.

29. See Garver, "Origins," p. 38.

30. Mao Zedong in an interview with Edgar Snow on July 23, 1936, cited in Schram, *Political Thought*, p. 419. The interview was not included in the official Peking edition of Mao's selected works; see *Mao Zedong xuanji*, vol. 2, p. 1.

31. On this paragraph, see Kampen, "Zunyi Conference," p. 131; John W. Garver, *Chinese-Soviet Relations, 1937–1945*, pp. 11–13, 64–67; Yang and Yang, *Gongchan guoji*, pp. 440–41, 454; Chang Kuo-t'ao, *Rise*, vol. 2, pp. 565–73; Warren Kuo, *Analytical History of the Chinese Communist Party*, vol. 3, pp. 336–41; *Zhongguo gongchandang chuangye sanshi nian*, p. 375; Frederick C. Teiwes, *The Formation of the Maoist Leadership, from the Return of Wang Ming to the Seventh Party Congress*, passim.

32. On this paragraph, see Chang Kuo-t'ao, *Rise*, vol. 2, pp. 569–80; Garver, *Chinese-Soviet Relations*, pp. 67–70.

33. *Pravda*, July 6, 1938, no. 184 (7509), p. 4.

34. Zhou Enlai indicates that Stalin and Dimitrov trusted or maintained good relations with Wang Ming until at least 1940; see Zhou and Chu, *Gongchan guoji*, p. 414.

35. On this paragraph, see Garver, *Chinese-Soviet Relations*, pp. 74–79, with source references; Yang and Yang, *Gongchan guoji*, pp. 455–58, 460, 468–69; Van Min, *Polveka KPK i predatel'stvo Mao Tszeduna*, passim.

36. Yang Kuisong, *Zhongjian didai de geming*, pp. 361–62; P.P. Vladimirov, *Osobyi raion Kitaia, 1942–1945*, pp. 197–98.

37. For further details, see this chapter below.

38. P.P. Vladimirov, *Osobyi raion Kitaia*, p. 216.

39. On this and the following sentence, see Mao Zedong to Soviet Ambassador Iudin, July 22, 1958, in *Dang de wenxian* 1 (1994), p. 17.

40. Dimitrov to Mao Zedong, December 22, 1943, cited in Zhou and Chu, *Gongchan guoji*, p. 414. Cf. P.P. Vladimirov, *Osobyi raion Kitaia*, p. 248.

41. On this paragraph, see Boyd Compton, *Mao's China*, passim; Dai Qing, *Wang Shiwei and White Lilies*, passim; Gregor Benton and Allan Hunter, eds., *Wild Lily, Prairie Fire*, pp. 7–13, 69–84; P.P. Vladimirov, *Osobyi raion Kitaia*, passim, esp. pp. 122, 195, 288–89; Yang Kuisong, *Zhongjian didai de geming*, p. 362; Garver, *Chinese-Soviet Relations*, pp. 242–47, with source references; Donald W. Klein and Anne B. Clark, *Biographic Dictionary of Chinese Communism*, vol. 1, p. 233; Van Min, *Polveka KPK*, pp. 58–63; John Byron and Robert Pack, *The Claws of the Dragon*, pp. 172–75; A.M. Ledovskii, "Sekretnaia missiia A.I. Mikoiana v Kitai (ianvar'–fevral' 1949 g.)," pt. 1, p. 104. On the Sinification of Marxism, see this chapter below.

42. Zhou and Chu, *Gongchan guoji*, p. 414.

43. For further details, see this chapter below.

44. On this paragraph, see Kampen, "Zunyi Conference," pp. 118–34, especially p. 132; Van Min, *Polveka KPK*, pp. 38–46; "Guanyu ruogan lishi wenti di jueyi," in *Mao Zedong xuanji*, vol. 3, pp. 975–1023; *Istoria Vsesoiuznoi Kommunisticheskoi Partii (bol'shevikov)*.

45. Mao Zedong to Edgar Snow, July 23, 1936, cited in Schram, *Political Thought*, p. 374. On this and the rest of this section, see Raymond F. Wylie, *The Emergence of Maoism*, passim.

46. On this and the following sentence, see Garver, *Chinese-Soviet Relations*, pp. 79–80; Stuart Schram, *Mao Zedong: A Preliminary Reassessment*, p. 37.

47. Cited in *Mao Zedong ji*, vol. 6, p. 261. The term "Sinification of Marxism" was deleted in the official Peking edition of Mao's collected works and replaced by the phrase "to ensure that Marxism was instituted in China" (*shi makesi zhuyi zai Zhongguo jutihua*); see *Mao Zedong xuanji*, vol. 2, p. 497. Cf. Schram, *Political Thought*, pp. 112, 171–75.

48. See Yang and Yang, *Gongchan guoji*, p. 539.

49. Mao Zedong, "On New Democracy," in *Selected Works of Mao Tse-tung*, vol. 2, pp. 339–82, esp. pp. 350–51, 361.

50. Liu Shao-chi, *On the Party*, pp. 8, 29.

51. Ibid., p. 31.

52. See Donald S. Zagoria, *The Sino-Soviet Conflict, 1956–1961*, pp. 14–15.

53. Liu Shaoqi to an American journalist; see Anna Louise Strong, "The Thought of Mao Tse-tung," in *Amerasia* 6 (June 1947), p. 161, cited in Zagoria, *Sino-Soviet Conflict*, pp. 14–15.

54. See also Yang Kuisong, "The Soviet Factor and the CCP's Policy toward the United States in the 1940s," p. 29.

55. Cited in Sheng, "Formation of the Anti-Japanese United Front," p. 159, with source reference.

56. Mao Zedong to Edgar Snow, July 23, 1936, cited in Schram, *Political Thought*, p. 419. The literal citation is provided earlier in this chapter above (section "Mao Zedong's Rise to Party Leader").

57. Mao Zedong, Zhu De, and Wang Jiaxiang in a telegram to Peng Dehuai, May 18, 1941, cited in Yang and Yang, *Gongchan guoji*, p. 534.

58. See, for example, Zhou and Chu, eds., *Gongchan guoji*, pp. 268, 409.

59. The spokesmen in this debate were John W. Garver and Michael M. Sheng. Cf. Garver, "Origins," pp. 29–59; John W. Garver, "Comment," pp. 171–79; Sheng, "Formation of the Anti-Japanese United Front," and Michael M. Sheng, pp. 149–70; "Response," pp. 180–83. We essentially agree with Sheng, who bases his opinion on more recent and stronger sources.

60. Wang Ming and Kang Sheng in a letter to the Politburo of the CC of the CCP, October 27, 1933, cited inYang Kuisong, *Zhongjian didai de geming*, p. 275.

61. Wang Ming's report during a session of the Chinese delegation to the Seventh Comintern Congress, August 25–27, 1935, in Moscow, cited in K.V. Kukushkin, "Komintern i edinyi national'nyi antiiaponskii front v Kitae (1935–1943)," p. 353; similarly, Mao Zedong, December 14, 1935, cited in Yang and Yang, *Gongchan guoji*, p. 353. On this sentence, see Yang Kuisong, *Zhongjian didai de geming*, pp. 275–79, 291; "Wang Ming," in *Jiu guo bao*, November 1935, cited in Huang Xiurong, *Gongchan guoji*, vol. 2, p. 214; Yang and Yang, *Gongchan guoji*, p. 338.

62. Resolution of the CCP Politburo session in Wayaobao (Shaanxi Province), December 25, 1935, cited in *Zhongguo gongchandang chuangye sanshi nian*, p. 330.

63. On this paragraph, see Yang Kuisong, *Zhongjian didai de geming*, pp. 287, 291; Yang and Yang, *Gongchan guoji*, pp. 340, 353–54; Huang Xiurong, *Gongchan guoji*, vol. 2, pp. 207–9; Chang Kuo-t'ao, *Rise*, vol. 2, pp. 416–17; and Sheng, "Formation of the Anti-Japanese United Front," pp. 153–55, with source references.

64. Politburo resolution, December 25, 1934, cited in Zhou and Chu, *Gongchan guoji*, p. 279; see also *Zhongguo gongchandang chuangye sanshi nian*, p. 331.

65. On this paragraph, see Yang Kuisong, *Zhongjian didai de geming*, pp. 284–85; Yang and Yang, *Gongchan guoji*, pp. 347, 373, 375; Huang Xiurong, *Gongchan guoji*, vol. 2, pp. 209–10, 214; Yang Yunruo, *Gongchan guoji*, p. 120; Zhou and Chu, *Gongchan guoji*, pp. 275, 278–79; *Zhongguo gongchandang chuangye sanshi nian*, pp. 330–31; Sheng, "Formation of the Anti-Japanese United Front," pp. 154–55.

66. On this paragraph, see Chang Kuot'ao, *Rise*, vol. 2, pp. 444–47.

67. On this paragraph, see Sheng, "Formation of the Anti-Japanese United Front," pp. 155–56.

68. Garver, "Origins," p. 49.

69. Adam B. Ulam, *Expansion and Coexistence*, p. 323.

70. The text of the Hirota Plan is in Furuya Keiji, *Chiang Kai-shek*, p. 473. Cf. also ibid., p. 474.

71. Resolution of the CCP Politburo, August 10, 1936, cited in *Zhongguo gongchandang chuangye sanshi nian*, p. 342.

72. Mao Zedong to John Service, August 23, 1944, cited in Joseph W. Esherick, ed., *Lost Chance in China*, p. 300.

73. On this paragraph, see Sheng, "Formation of the Anti-Japanese United Front," pp. 156–57; and Huang Xiurong, *Gongchan guoji*, vol. 2, pp. 227, 233. Telegram from the ECCI, August 15, 1935, excerpts reproduced in Huang Xiurong, *Gongchan guoji*, vol. 2, pp. 230–32, and Yang Kuisong, *Zhongjian didai de geming*, p. 288.

74. Yang Yunruo, *Gongchan guoji*, p. 125; Furuya, *Chiang Kai-shek*, p. 508; Garver, "Origins," p. 56; Sheng, "Formation of the Anti-Japanese United Front," pp. 157–58, 162; Stuart Schram, *Mao Tse-tung*, p. 204.

75. From 1927, the CCP's armed forces were referred to collectively as the Red Army (*hong jun*) until

they took the name People's Liberation Army (*renmin jiefang jun*) in July 1946. The Soviet armed forces were called Red Workers and Peasants' Army (*Raboche-Krest'anskaia Krasnaia Armiia*, abbreviated RKKA) from 1918, generally shortened to Red Army, and were renamed Soviet Army (*Sovetskaia Armiia*, abbreviated *SA*) in February 1946. To avoid confusion with the Chinese Communist Red Army (*hong jun*), the Soviet armed forces will be called Soviet Army in this book. See *Bol'shaia Sovetskaia Entsiklopediia*, 3rd. ed., vol. 13, p. 327; ibid., vol. 24, p. 22; Raymond L. Garthoff, "The Soviet Intervention in Manchuria, 1945–46," p. 77.

76. Sheng, "Formation of the Anti-Japanese United Front," p. 163.

77. On this and the following sentence, see Chang Kuo-t'ao, *Rise*, vol. 2, pp. 482–83.

78. The secondary literature on various aspects of the Sian conflict is extensive, and we cannot present it all here. Let us refer solely to Wu Tien-wei, *The Sian Incident*, and to some more recent publications from the People's Republic of China, some of which include references to sources such as Yang Kuisong, *Zhongjian didai de geming*, pp. 286, 294–97; Yang and Yang, *Gongchan guoji*, pp. 385–95; Huang Xiurong, *Gongchan guoji*, vol. 2, pp. 240–55; Yang Yunruo, *Gongchan guoji*, pp. 126–27. The contents of the telegrams from the CCP to the Comintern, December 12 and 13, 1936, are reproduced in Zhou and Chu, *Gongchan guoji*, pp. 306–7. Cf. also Sheng, "Formation of the Anti-Japanese United Front," 129, pp. 163–65; Kuo Heng-yü, *Maos Weg zur Macht*, p. 76.

79. Wang Ming's December 1937 report, cited in Chang Kuo-t'ao, *Rise*, vol. 2, p. 483.

80. *Pravda*, December 14, 1936. The article was published unsigned, but reproduced the opinion of the Soviet leadership. The same tone can be found in other articles in the Soviet press; cf. *Pravda*, December 15–18, 1936.

81. On this paragraph, see Ch'en, "Communist Movement, 1927–1937," p. 226; Chang Kuo-t'ao, *Rise*, vol. 2, pp. 483–84, 498. On the contents of the telegrams, see Zhou and Chu, *Gongchan guoji*, p. 308; *Zhonggong dangshi yanjiu* 3 (1988), p. 78, cited in Sheng, "Formation of the Anti-Japanese United Front," pp. 165–66. The ECCI sent one telegram and Dimitrov signed the other. The latter was coded incorrectly, so the CCP leadership asked for it to be sent again on December 18. Cf. Zhou and Chu, *Gongchan guoji*, p. 308. Two Chinese authors see the Comintern's assessment of Chang Hsueh-liang as "attack and slander *[gongji he wumie]*"; see Zhou and Chu, *Gongchan guoji*, p. 308.

82. On this paragraph, see Zhou and Chu, *Gongchan guoji*, pp. 308–309; *Zhou Enlai zhuan, 1898–1949*, p. 331; *Zhou Enlai nianpu, 1949–1976)*, vol. 1, p. 335; Sheng, "Formation of the Anti-Japanese United Front," pp. 166–67, with source references; Eastman, "Nationalist China," pp. 162–63; Ch'en, "Communist Movement, 1927–1937," p. 226; Kuo Heng-yü, *Maos Weg zur Macht*, p. 77, with source references; Ulam, *Expansion*, pp. 249–50.

83. Chang Kuo-t'ao, *Rise*, vol. 2, pp. 496–98.

84. On this paragraph, see Sheng, "Formation of the Anti-Japanese United Front," pp. 167–69; Kuo Hengyü, *Maos Weg zur Macht*, pp. 82–91; telegram from the ECCI, January 20, 1937, cited in Yang and Yang, *Gongchan guoji*, p. 396, and Huang Xiurong, *Gongchan guoji*, vol. 2, p. 256.

85. Zhou Enlai to Edgar Snow in the summer of 1936; see Edgar Snow, *The Other Side of the River*, p. 76. On this paragraph, see Sheng, "Formation of the Anti-Japanese United Front," p. 169; Kuo Heng-yü, *Maos Weg zur Macht*, pp. 91–97; Kuo, *Analytical History*, vol. 3, pp. 286–89; Conrad Brandt, Benjamin Schwartz, and John K. Fairbank, *A Documentary History of Chinese Communism*, pp. 239–57; Schram, *Mao Tse-tung*, pp. 200–205.

86. P.P. Vladimirov, *Osobyi raion Kitaia*, pp. 7, 9, 28, 35, 59, 212.

87. On this paragraph, see ibid., pp. 5–7, 61, 401, 655, and passim; Iurii Vlasov, "Povest' ob ottse," pt. 1, p. 125; oral communication from Russian sinologist Konstantin Shevelev, who had access to archive materials to the author in 1995. On Vlasov's actions in Yenan, see Zhou and Chu, *Gongchan guoji*, pp. 395–96.

88. P.P. Vladimirov, *Osobyi raion Kitaia, 1942–1945*. The book was subsequently published in several foreign-language editions. The English-language edition appeared under the name P.P. Vladimirov, with the title *The Vladimirov Diaries: Yenan, China, 1942–1945* (Garden City, NY: Doubleday, 1975).

89. Cf., for example, Byron and Pack, *Claws of the Dragon*, pp. 151–53, 169, 174, 188; Wesley M. Bagby, *The Eagle-Dragon Alliance*, passim. John W. Garver, who could not yet have known the comments of the younger Vlasov, considered the "diary" essentially genuine in 1988, but had reservations about "broader interpretative issues." See Garver, *Chinese-Soviet Relations*, pp. 265–66.

90. On this paragraph, see Iurii Vlasov, "Pravda ob ottse"; Iurii Vlasov, "Povest' ob otts," pt. 1, pp. 124–34 passim, and pt. 2, pp. 171–81 passim. The English version of "Povest' ob ottse" is unreliable. Cf. Yuri Vlasov, "The Story of My Father," pt. 1, pp. 189–207, and pt. 2, pp. 193–208.

91. Iurii Vlasov, "Povest' ob ottse," pt. 2, p. 176.

92. Bernard Gwertzman's interview of the contemporary witness John Service, "Literary Detective on Trail of Soviet Diary," *International Herald Tribune*, January 13, 1976. Service, who at that time was an admirer of the Chinese Communists, mentions such facts. He goes so far as to describe the "diary" as "largely an anti-Chinese fabrication," which is most probably an exaggeration. See also the balanced review of the "diary" in Peter M. Kuhfus, "Die 'Wladimirow-Tagebücher' in neuem Licht," pp. 69–77.

93. On this and the rest of the paragraph, see Iurii Vlasov, "Povest' ob ottse," pt. 1, p. 128, and pt. 2, pp. 173–75.

94. James Reardon-Anderson, *Yenan and the Great Powers*, p. 182 n. 5; Russian sinologist Konstantin Shevelev, who has dealt intensively with the issue of the authenticity of this text in correspondence to the author. The Russian sinologist Lev Deliusin was much more critical. He called the "diary" a "classic example" of the "incorrect use of facts and documents in anti-Chinese propaganda during the Sino-Soviet conflict"; see L. Deliusin, "SSSR–KNR: ostavliaia proshloe v proshlom," p. 5.

95. Oral communication from Konstantin Shevelev to the author in 1995.

96. See P.P. Vladimirov, *Osobyi raion Kitaia*, p. 311.

97. Ibid., p. 99.

98. Ibid., p. 634.

99. Ibid., passim, esp. pp. 627, 629.

100. Yang Kuisong, *Zhongjian didai de geming*, p. 286.

101. Braun, *Comintern Agent in China*, p. 159.

102. P.P. Vladimirov, *Osobyi raion Kitaia*, p. 63.

103. Ibid.

104. Braun, *Comintern Agent in China*, p. 264.

105. On this and the following sentence, see Kovalev's telegram to Stalin, July 14, 1948, AP RF, f. 45, op. 1, d. 331, ll. 59–61.

106. On this paragraph, see telegram from Orlov to Stalin, July 14, 1948, AP RF, f. 39, op. 1, d. 31, l. 40, cited in Ledovskii, "Sekretnaia missiia A.I. Mikoiana v Kitai," pt. 1, p. 99; P.P. Vladimirov, *Osobyi raion Kitaia*, p. 655; Iurii Vlasov, "Povest' ob ottse," pt. 1, pp. 126–28, 131–34, pt. 2, pp. 171–72; Shi Zhe, *Zai lishi juren shenbian*, pp. 315, 390; Shi Zhe quoted in Li Haiwen, "Mao Zedong Liu Shaoqi fang Su ji Zhong-Su liang dang guanxi zhong de yixie wenti," p. 95.

107. See E.P. Bazhanov, *Kitai i vneshnii mir*, p. 35.

108. Stalin to Chuikov, in V.I. Chuikov, *Missiia v Kitae*, pp. 57–58. On this paragraph, see Garver, *Chinese-Soviet Relations*, pp. 15–122; John W. Garver, "Chiang Kai-shek's Quest for Soviet Entry into the Sino-Japanese War," pp. 295–316, esp. pp. 315–16; A.M. Ledovskii, "Zapiski diplomata," p. 109.

109. Chinese text in *Treaties Between the Republic of China and Foreign States (1927–1957)*, pp. 486–87; Soviet text in *Sovetsko-kitaiskie otnosheniia, 1917–1957*, pp. 161–62.

110. The data on troop strengths vary. See *History of the Sino-Japanese War (1937–1945)*, p. 172; Edgar O'Ballance, *The Red Army of China*, pp. 125–26; F.F. Liu, *A Military History of Modern China*, p. 135. A recent Russian publication gives the number of Communist troops as "altogether about 60,000"; see Ledovskii, "Zapiski diplomata," p. 109.

111. Ledovskii, "Zapiski diplomata," p. 109.

112. Chuikov, *Missiia v Kitae*, pp. 56–57.

113. P.P. Vladimirov, *Osobyi raion Kitaia*, p. 27.

114. Bagby, *Eagle-Dragon Alliance*, p. 69.

115. Memorandum of November 23, 1943, cited in W. Averell Harriman and Elie Abel, *Special Envoy to Churchill and Stalin, 1941–1946*, p. 261.

116. Stalin to Harriman, June 1, 1944, cited in ibid., pp. 472–73, and Reardon-Anderson, *Yenan and the Great Powers*, p. 71. Similarly, Stalin to Hopkins in May 1945; see Bagby, *Eagle-Dragon Alliance*, p. 213. See also Hurley's report in early July 1945, in *The China White Paper*, p. 99. Sheng took the position—not very convincingly—that Stalin trusted Mao more than Chiang; see Sheng, "Response," p. 182.

117. Hurley's report on his talk with Stalin and Molotov on April 15, 1945, cited in *China White Paper*, p. 95.

118. Bagby, *Eagle-Dragon Alliance*, p. 213.

119. Reardon-Anderson, *Yenan and the Great Powers*, p. 71; see also Barbara W. Tuchman, *Sand Against the Wind*, p. 462.

120. Gordon H. Chang, *Friends and Enemies*, p. 10.

121. On this and the two following sentences, see Molotov's talk with Hurley in late 1944, in *China White Paper*, p. 72; see also Henry Wei, *China and Soviet Russia*, p. 170.

122. Iurii Vlasov, "Pravda ob ottse." Vlasov bases his statement on a talk with former CC Secretary Boris Ponomarev and archival documents.

123. Directive of the CC of the CCP, November 28, 1945, in *Zhonggong zhongyang wenxian xuanji*, vol. 15, p. 455.

124. Henry Wei supposes this. See Wei, *China and Soviet Russia*, p. 170.

125. Harriman and Abel, *Special Envoy*, p. 262.

126. Chang, *Friends and Enemies*, p. 11.

127. See Li Haiwen, "Mao Zedong yu Si-da-lin jiaowang," p. 12. For further details, see this chapter below.

128. On this paragraph, see Howard L. Boorman and Richard C. Howard, eds., *Biographical Dictionary of Republican China*, vol. 1, pp. 81–82; Chang Kuo-t'ao, *Rise*, vol. 2, pp. 374–473.

129. CC resolution, August 6, 1935, cited in Yang and Yang, *Gongchan guoji*, p. 374. On this paragraph, see Sheng, "Formation of the Anti-Japanese United Front," p. 159.

130. Mao Zedong at a Politburo session in Eji (Gansu Province), September 12, 1935, and CC resolution in Eji, cited in Yang and Yang, *Gongchan guoji*, p. 374, and Sheng, "Formation of the Anti-Japanese United Front," p. 159. On the Eji session, see also *Zhongguo gongchandang chuangye sanshi nian*, p. 325.

131. Chang Kuo-t'ao, *Rise*, vol. 2, p. 448.

132. On this and the following sentence, see Zhou and Chu, *Gongchan guoji*, p. 276.

133. Mao Zedong's suggestion on November 23, 1935, cited in Yang and Yang, *Gongchan guoji*, p. 375; Zhou and Chu, *Gongchan guoji*, p. 276. Otto Braun essentially confirms this description. He adds that Luo Fu and Mao Zedong would both have liked to have the First Front Army march toward the Mongolian People's Republic—Luo in an inconspicuous way, Mao in a spectacular fashion. See Braun, *Comintern Agent in China*, pp. 151–52.

134. On this paragraph, see Yang and Yang, *Gongchan guoji*, pp. 347, 375; Zhou and Chu, *Gongchan guoji*, p. 275; Huang Xiurong, *Gongchan guoji*, vol. 2, pp. 209, 214; Yang Kuisong, *Zhongjian didai de geming*, pp. 284–86.

135. Telegram from the CC of the CCP to the four front armies, November 18, 1935, cited in Zhou and Chu, *Gongchan guoji*, p. 275.

136. CC resolution on issues of military strategy, December 23, 1935, cited in Yang and Yang, *Gongchan guoji*, p. 375.

137. Zhou and Chu, *Gongchan guoji*, p. 278.

138. On this paragraph, see Yang and Yang, *Gongchan guoji*, p. 375; Zhou and Chu, *Gongchan guoji*, pp. 278–79; Huang Xiurong, *Gongchan guoji*, vol. 2, p. 209; *Zhongguo gongchandang chuangye sanshi nian*, pp. 330–31; Sheng, "Formation of the Anti-Japanese United Front," p. 159 and p. 159, n. 48; Braun, *Comintern Agent in China*, pp. 151–52. It is unclear whether Moscow approved the Eastern Expedition as such; see also Garver, "Origins," p. 48.

139. On this paragraph, see Braun, *Comintern Agent in China*, pp. 157–60. A Chinese source confirms that Braun raised such objections; it places the event at the end of December 1935, however (Wayaobao session). See Zhou and Chu, *Gongchan guoji*, p. 278.

140. On this, see Yang and Yang, *Gongchan guoji*, pp. 347, 375–76.

141. On this, see Garver, "Origins," pp. 47–53. Garver's thesis that the Eastern Expedition conflicted with Soviet interests is not convincing and is refuted by Sheng, who relies on stronger sources; cf. Sheng, "Formation of the Anti-Japanese United Front," pp. 158–63 passim.

142. Braun, *Comintern Agent in China*, p. 152.

143. On this paragraph, see telegram from Luo Fu, Bo Gu, Zhou Enlai, and Mao Zedong to Wang Ming, August 25, 1936, cited in Yang Kuisong, *Zhongjian didai de geming*, p. 293, Yang and Yang, *Gongchan guoji*, p. 380; and *Zhou Enlai nianpu, 1949–1976*, vol. 1, p. 318. See also telegram from the Comintern Secretariat to the CCP Secretariat, September 11, 1936, cited in Yang Kuisong, *Zhongjian didai de geming*, p. 294; telegram from Zhang Wentian, Zhou Enlai, Bo Gu, and Lin Biao to Zhu De, Zhang Guotao, He Long, and Ren Bishi, November 8, 1936, cited in Yang and Yang, *Gongchan guoji*, p. 384; Yang Kuisong, *Zhongjian didai de geming*, pp. 286, 293–94; *Zhou Enlai nianpu, 1949–1976*, vol. 1, pp. 321–22; Yang and Yang, *Gongchan guoji*, pp. 380, 384–85; Sheng, "Formation of the Anti-Japanese United Front," pp. 162–63.

144. On this section, see Chang Kuo-t'ao, *Rise*, vol. 2, p. 566.

145. Braun, *Comintern Agent in China*, p. 209.

146. A.I. Cherepanov, *Zapiski voennogo sovetnika v Kitae*, p. 603.

147. Ambassador Luganets-Orelskii to Chief Military Adviser Cherepanov; see Cherepanov, *Zapiski*

voennogo sovetnika, pp. 602–3. On this paragraph, see Arlen Meliksetov, "1940s . . . ," p. 26; Liu Zhiqing, "Shi lun 'Zhong-Su youhao tongmen tiaoyue' di qianding," p. 73; and P.P. Vladimirov, *Osobyi raion Kitaia*, p. 27.

148. Zhou and Chu, *Gongchan guoji*, p. 410; P.P. Vladimirov, *Osobyi raion Kitaia*, pp. 171–72.

149. On this paragraph, cf. Eva-Maria Stolberg, *Stalin und die chinesischen Kommunisten, 1945–1953*, pp. 28, 39.

150. On this paragraph, see A.Ia. Kaliagin, *Po neznakomym dorogam*, passim; *Na kitaiskoi zemle*, passim; Cherepanov, *Zapiski voennogo sovetnika*, pp. 599–647; B.A. Borodin, *Pomoshch' SSSR kitaiskomu narodu v antiiaponskoi voine, 1937–1941*, passim; Yang Kuisong, "Soviet Factor," pp. 18–19; Meliksetov, "1940 . . . ," p. 26; Ledovskii, "Zapiski diplomata," p. 109; Garver, *Chinese-Soviet Relations*, pp. 18–21, 37–52, 104–8; Nancy Bernkopf Tucker, *Patterns in the Dust*, p. 16, with references to secondary literature; *Voennaia pomoshch' SSSR v osvoboditel'noi bor'be kitaiskogo naroda*, pp. 45–84; Arthur N. Young, *China and the Helping Hand, 1937–1945*, 1963, p. 254.

151. Mao Zedong, "On New Democracy," *Zhongguo wenhua* 1 (1940), cited in *Selected Works of Mao Tse-tung*, vol. 2, p. 355.

152. Telegram from Mao Zedong to Zhou Enlai in January 1941 (with no precise date), and telegram from Zhou Enlai to Mao Zedong, February 9, 1941, both cited in Zhou and Chu, *Gongchan guoji*, pp. 373, 377.

153. Yang Kuisong, *Zhongjian didai de geming*, p. 361; Garver, *Chinese-Soviet Relations*, pp. 238–39. On the context of the incidents, see this chapter below.

154. On this and the following paragraph, see Kuo, *Analytical History*, vol. 3, pp. 286–89; Kuo Heng-yü, *Maos Weg zur Macht*, pp. 93–95; Sheng, "Formation of the Anti-Japanese United Front," p. 169.

155. On this paragraph, see Chang Kuo-t'ao, *Rise*, vol. 2, pp. 566–73; Kuo, *Analytical History*, vol. 3, pp. 333–337; Kuo Heng-yü, *Maos Weg zur Macht*, pp. 97–99; Braun, *Comintern Agent in China*, pp. 217–23; Garver, *Chinese-Soviet Relations*, pp. 65–67. For further details on the disputes between Wang Ming and Mao Zedong within the Party, see this chapter, above (section "Mao Zedong's Rise to Party Leader").

156. CCP manifesto, December 25. 1937, cited in Kuo, *Analytical History*, vol. 3, pp. 342, 368–71; see also pp. 342–46.

157. Internal document made accessible by Zhang Guotao, cited in Liu, *Military History*, pp. 205–6, with source reference. See also *A Brief History of the Chinese National Revolutionary Forces*, p. 133.

158. Chang Kuo-t'ao, *Rise*, vol. 2, p. 568; Liu, *Military History*, p. 205.

159. Chang Kuo-t'ao, *Rise*, vol. 2, p. 568.

160. Liu, *Military History*, p. 205.

161. Lyman P. Van Slyke, ed., *The Chinese Communist Movement*, p. 67.

162. Chuikov, *Missiia v Kitae*, pp. 141–50, passim.

163. Yang and Yang, *Gongchan guoji*, p. 523.

164. Huang Xiurong, *Gongchan guoji*, vol. 2, p. 282.

165. Garver, *Chinese-Soviet Relations*, pp. 141–42, with source references. Yang Yunruo and Yang Kuisong do not mention any Soviet pressure, see Yang and Yang, *Gongchan guoji*, p. 515.

166. Chuikov, *Missiia v Kitae*, p. 144; Reardon-Anderson, *Yenan and the Great Powers*, p. 14.

167. Chuikov, *Missiia v Kitae*, p. 145.

168. Telegram from Mao Zedong and Wang Jiaxiang to Peng Dehuai, November 3, 1940, cited in Yang and Yang, *Gongchan guoji*, p. 516; Michael M. Sheng, "America's Lost Chance in China?" pp. 145–46, with source references.

169. Yang and Yang, *Gongchan guoji*, p. 518.

170. Ibid., with source reference.

171. Mao Zedong to Li Kenong, November 6, 1940, cited in Sheng, "America's Lost Chance in China?" p. 146, with source reference.

172. Mao Zedong, November 1, 1940, cited in Yang and Yang, *Gongchan guoji*, p. 516.

173. On this and the following sentence, see telegram from Mao Zedong to Zhou Enlai, November 6, 1940, cited in ibid., p. 517.

174. Telegram from Mao Zedong to Zhou Enlai and Dong Biwu, January 15, 1941, cited in ibid., p. 521.

175. On this paragraph, see Chalmers A. Johnson, *Peasant Nationalism and Communist Power*, pp. 136–40; Kuo, *Analytical History*, vol. 4, pp. 275–94; Yang and Yang, *Gongchan guoji*, pp. 519–21. The causalities in the Wannan incident appear not yet to be entirely clear in some points.

176. On this paragraph, see Deputy Minister of Foreign Affairs Lozovskii to Shao Lizi, January 17, 1941; Chuikov to Zhou Enlai and Ye Jianying on January 14, 1941; telegrams from Dimitrov to the CCP leadership, January 20 and February 15, 1941; Paniushkin to Chiang Kai-shek, January 25, 1941, all cited in Zhou

and Chu, *Gongchan guoji*, pp. 374–75, and Yang and Yang, *Gongchan guoji*, pp. 520–21. See also Zhou and Chu, *Gongchan guoji*, pp. 374–76; Yang and Yang, *Gongchan guoji*, pp. 519–22.

177. On this paragraph, see Mao Zedong, January 20, 1941 and telegram from Mao Zedong to Zhou Enlai, January 29, 1941, both cited in Zhou and Chu, *Gongchan guoji*, pp. 375–76, and Yang and Yang, *Gongchan guoji*, p. 523; and resolution of the CC of the CCP, January 29, 1941, in *Zhonggong zhongyang wenjian xuanji*, vol. 13, pp. 26–30. See also Zhou and Chu, *Gongchan guoji*, pp. 375, 377; Yang and Yang, *Gongchan guoji*, p. 522.

178. See Yang and Yang, *Gongchan guoji*, p. 523.

179. Ibid., p. 524.

180. Telegram from Mao Zedong to Zhou Enlai, February 14, 1941, cited in ibid., p. 524, and *Zhou Enlai zhuan*, p. 491.

181. Yang and Yang, *Gongchan guoji*, p. 526. On this paragraph, see Johnson, *Peasant Nationalism*, pp. 139–40; Yang and Yang, *Gongchan guoji*, pp. 524–25.

182. Text of the pact and declaration in *Pravda*, April 14, 1941. On this paragraph, see also Ulam, *Expansion*, p. 308; Yang and Yang, *Gongchan guoji*, pp. 530–31; Garver, *Chinese-Soviet Relations*, pp. 114–17.

183. Telegram from Mao Zedong to Zhou Enlai, April 17, 1941, cited in Zhou and Chu, *Gongchan guoji*, p. 379. On this paragraph, see telegram from Mao Zedong to Zhou Enlai, April 14, 1941, cited in ibid.

184. Liu, *Military History*, p. 205.

185. On this paragraph, see Kuo, *Analytical History*, vol. 4, pp. 502–3, 547–49, with source references; Zhou and Chu, *Gongchan guoji*, pp. 38–39, with source references.

186. The secondary literature in the People's Republic of China also refers to this contrast explicitly if also uncritically; see Yang and Yang, *Gongchan guoji*, pp. 530–31.

187. Zhou and Chu, *Gongchan guoji*, p. 379.

188. Steven I. Levine, *Anvil of Victory*, pp. 19–20, 103; Odd Arne Westad, *Cold War and Revolution*, p. 85; M.F. Iur'ev, *Vooruzhennye sily KPK v osvoboditel'noi bor'be kitaiskogo naroda*, p. 225.

189. On this and the rest of the paragraph, see Zhou Baozhong in a letter to the "Soviet representative" Wang Xinlin (cover name), May 10, 1941, cited in Zhou and Chu, *Gongchan guoji*, pp. 381–82.

190. Chuikov, *Missiia v Kitae*, pp. 146–47; Huang Xiurong, *Gongchan guoji*, vol. 2, pp. 292–93; Yang and Yang, *Gongchan guoji*, pp. 533–34; Zhou and Chu, *Gongchan guoji*, p. 383.

191. Telegram from Mao Zedong, Zhu De, and Wang Jiaxiang to Peng Dehuai, May 18, 1941, cited in Yang and Yang, *Gongchan guoji*, p. 534, and Yang Kuisong, *Zhongjian didai de geming*, p. 356. See also Zhou and Chu, *Gongchan guoji*, p. 387. These sources mistakenly put off the event to July 1941.

192. Yang and Yang, *Gongchan guoji*, p. 534; Zhou and Chu, *Gongchan guoji*, p. 387.

193. On this and the following sentence, see Mao Zedong's instruction to Zhou Enlai, May 14, 1941, cited in Yang and Yang, *Gongchan guoji*, p. 534. See also Zhou and Chu, *Gongchan guoji*, p. 387. This source mistakenly puts off the event to July 1941.

194. Mao Zedong to Zhou Enlai, undated, cited in Huang Xiurong, *Gongchan guoji*, vol. 2, p. 293; Mao Zedong to Zhou Enlai, dated May 1941, cited in Zhou and Chu, *Gongchan guoji*, p. 383.

195. Mao Zedong to Zhou Enlai, dated May 1941, cited in Zhou and Chu, *Gongchan guoji*, p. 383.

196. Chuikov, *Missiia v Kitae*, pp. 144, 147.

197. Mao Zedong, Zhu De, and Wang Jiaxiang in a telegram to Peng Dehuai, May 18, 1941, cited in Yang and Yang, *Gongchan guoji*, p. 534.

198. See Yang Kuisong, "Soviet Factor," p. 19, with source references.

199. On this paragraph, see *Jiefang ribao*, June 23, 1941, cited in Garver, *Chinese-Soviet Relations*, p. 238; *Jiefang*, no. 85 (September 30, 1939), cited in Sheng, "America's Lost Chance in China?" p. 142.

200. Resolution and appeal of the ECCI, June 22, 1941 cited in Zhou and Chu, *Gongchan guoji*, pp. 383–85; speech by Stalin, July 3, 1941, printed in *Pravda*, July 3, 1941, p. 1; instruction of the CC of the CCP, July 12, 1941, in *Zhonggong zhongyang wenjian xuanji*, vol. 13, p. 164.

201. Mao Zedong's instruction, June 23, 1941, cited in Zhou and Chu, *Gongchan guoji*, p. 385. See also the resolution of the CC of the CCP, June 23, 1941, in *Jiefang*, no. 131–32, cited in Yang and Yang, *Gongchan guoji*, p. 532; Yang and Yang, *Gongchan guoji*, p. 535.

202. Yang and Yang, *Gongchan guoji*, pp. 532–33, 535.

203. Ribbentrop to Matsuoka, July 1, 1941; cited in Ulam, *Expansion*, p. 323.

204. *TsGAOR*, f. 7867, op. 1, d. 275, l. 266, cited in A.A. Arkad'ev, "Politika Iaponii v otnoshenii SSSR posle nachala vtoroi mirovoi voiny," p. 38.

205. On this paragraph, see Li Haiwen, "Mao Zedong yu Si-da-lin jiaowang," p. 12, with no source

reference. Li has access to the Party archive. Her statements are confirmed in part by archival documents cited by other authors; see the following paragraphs. These statements are confirmed in part by O. Vladimirov, *Sovetsko-kitaiskie otnosheniia v sorokovykh-vos'midesiatykh godach*, pp. 64–66.

206. Although the publication was created by the anti-Chinese propaganda machine of the notorious Rakhmanin, deputy department head of the CC of the CPSU (Vladimirov was one of his pseudonyms), the statements cited here and in the following are apparently based on archival documents and can in principle be considered authentic.

207. Yang and Yang, *Gongchan guoji*, p. 535; Zhou and Chu, *Gongchan guoji*, p. 387; O. Vladimirov and V. Riazantsev, *Stranitsy politicheskoi biografii Mao Tszė-duna*, p. 64; Garver, *Chinese-Soviet Relations*, p. 238, with source references.

208. Zhou and Chu, *Gongchan guoji*, p. 387. In early September, Moscow repeated its request to Yenan to send the Eighth Route Army in the case of an attack on the USSR; see ibid., p. 389.

209. Yang and Yang, *Gongchan guoji*, p. 536. Tokyo finally decided in September to expand to the south and not to the west. On the Japanese plans of the time, see also Ulam, *Expansion*, pp. 321–22; Yang and Yang, *Gongchan guoji*, p. 536.

210. On this and the following sentence, see Vladimirov and Riazantsev, *Stranitsy politicheskoi biografii Mao Tszė-duna*, p. 64. Cf. Niu Jun, "The Origins of the Sino-Soviet Alliance," p. 48.

211. On this and the following sentence, see Yang and Yang, *Gongchan guoji*, p. 535, with source references; Yang Kuisong, *Zhongjian didai de geming*, p. 361.

212. Telegram from Mao Zedong, Zhu De, Wang Jiaxiang, and Ye Jianying to Peng Dehuai, July 2, 1941, cited in Zhou and Chu, *Gongchan guoji*, p. 385; telegram from Mao Zedong to Zhou Enlai, July 6, 1941, cited in Yang and Yang, *Gongchan guoji*, pp. 535–36.

213. On this and the following five sentences, see telegram from Mao Zedong to Zhou Enlai, July 15, 1941, cited in Yang and Yang, *Gongchan guoji*, p. 536; Yang Kuisong, *Zhongjian didai de geming*, p. 361; Zhou and Chu, *Gongchan guoji*, pp. 387–88. According to Soviet accounts, the CCP leadership did not follow the Soviet request to disrupt rail traffic. See Vladimirov and Riazantsev, *Stranitsy politicheskoi biografii Mao Tsze-duna*, p. 64.

214. Yang Kuisong, *Zhongjian didai de geming*, p. 361.

215. See Garver, *Chinese-Soviet Relations*, pp. 238–39.

216. See Yang Kuisong, *Zhongjian didai de geming*, p. 361. Cf. *Voennaia pomoshch' SSSR*, p. 83.

217. On this paragraph, see Vladimirov and Riazantsev, *Stranitsy politicheskoi biografii Mao Tsze-duna*, p. 64.

218. Telegram from Dimitrov to the CC of the CCP, October 7, 1941, contents reproduced in Yang and Yang, *Gongchan guoji*, p. 537, and Zhou and Chu, *Gongchan guoji*, pp. 389–90. The full text of the telegram has apparently never been published up to now. It is also missing from a recent publication of archival documents on the relations between the Comintern and the CCP; see Iuri Ovchinnikov, ed. and trans., "Comintern-CCP Relations," parts 1–2.

219. Yang and Yang, *Gongchan guoji*, p. 537.

220. Ibid., p. 538. It is possible that these demands were also included in Dimitrov's fifteen-point list of October 7, although Yang Yunruo and Yang Kuisong do not state this explicitly.

221. Instruction of the Military Commission of the Central Committee, November 7, 1941, cited in ibid. Cf. *Voennaia pomoshch' SSSR*, pp. 82–83.

222. Wang Ming at a working session of the Secretariat of the Central Committee, October 8, 1941, cited in Yang and Yang, *Gongchan guoji*, p. 541, and Zhou and Chu, *Gongchan guoji*, p. 390 (Wang Ming's speech was mistakenly dated October 18); Yang Kuisong, *Zhongjian didai de geming*, p. 362; Van Min, *Polveka KPK*, p. 39. Wang provides in part inaccurate data.

223. Yang and Yang, *Gongchan guoji*, p. 539. On this paragraph, see ibid., pp. 537–42, and Van Min, *Polveka KPK*, p. 39. The presentation by Yang and Yang is unclear at times, as it apparently refers to different time periods and is based on sources of differing reliability. In one case, it is illogical, as the data does not support the causal connection; see Yang and Yang, *Gongchan guoji*, p. 541, nn. 3, 4, p. 542 n. 1.

224. Yang and Yang, *Gongchan guoji*, p. 540.

225. In this wording in ibid., p. 542, but Zhou and Chu, *Gongchan guoji*, p. 390, say "unanimous."

226. Yang and Yang, *Gongchan guoji*, p. 542; Zhou and Chu, *Gongchan guoji*, p. 390.

227. Yang and Yang, *Gongchan guoji*, p. 542.

228. Yang Kuisong, *Zhongjian didai de geming*, p. 362.

229. On this paragraph, see Zhou and Chu, *Gongchan guoji*, p. 390; Yang and Yang, *Gongchan guoji*, p. 542; Yang Kuisong, *Zhongjian didai de geming*, p. 362; Wan Min, *Polveka KPK*, pp. 39–54; P.P. Vladimirov,

Osobyi raion Kitaia, passim, esp. p. 179; Wang Ming's son to the Moscow sinologist Konstantin Shevelev, oral communication from Shevelev to the author in 1995.

230. Instruction of the Military Commission of the CC of the CCP on the military development of bases for the struggle against the Japanese, November 7, 1941, cited in Yang Kuisong, *Zhongjian didai de geming*, p. 357.

231. On this and the following sentence, see Zhou and Chu, *Gongchan guoji*, pp. 390–91.

232. Resolution of the ECCI Secretariat, February 27, 1942, cited in ibid., pp. 393–94.

233. Cited in Vladimirov and Riazantsev, *Stranitsy politicheskoi biografii Mao Tsze-duna*, p. 65 (Mao quote undated).

234. On this paragraph, see Allen S. Whiting and General Sheng Shi-ts'ai, *Sinkiang*, passim; O. Edmund Clubb, *China and Russia*, pp. 323–29; Han-jung Ziemann, *Die Beziehungen Sinkiangs (Ostturkestan) zu China und der UdSSR, 1917–1945*, passim; Zhou and Chu, *Gongchan guoji*, p. 400.

235. The Western literature does not rule out the possibility that this version could be accurate, either. See also Whiting and Sheng, *Sinkiang*, pp. 85–86; Clubb, *China and Russia*, p. 323.

236. Clubb, *China and Russia*, p. 323.

237. On this paragraph, see Zhou and Chu, *Gongchan guoji*, pp. 394–99, 401, passim.

238. Whiting and Sheng, *Sinkiang*, pp. 201, 205–6.

239. Zhou and Chu, *Gongchan guoji*, p. 395.

240. Letter from Dimitrov to Mao Zedong, June 15, 1942, cited in Zhou and Chu, *Gongchan guoji*, p. 396.

241. On this paragraph, see P.P. Vladimirov, *Osobyi raion Kitaia*, pp. 53, 56, 58, 94, 99.

242. On this paragraph, see Garver, *Chinese-Soviet Relations*, pp. 240–41.

243. Chiang Kai-shek, *Soviet Russia in China*, p. 114.

244. P.P. Vladimirov, *Osobyi raion Kitaia*, pp. 640–41.

245. Resolution of the CC of the CCP on the resolution of the ECCI Presidium on the dissolution of the Comintern, *Jiefang ribao*, May 27, 1943, cited in Kukushkin, "Komintern i edinyi natsional'nyi antiiaponskii front," p. 377.

246. Resolution of the Politburo of the CC of the CCP, undated, cited in Garver, *Chinese-Soviet Relations*, pp. 247–48.

247. On this and the following sentence, see Mao Zedong, May 26, 1943, in *Jiefang ribao*, May 28, 1943, cited in Schram, *Political Thought*, p. 422; see also Zhou and Chu, *Gongchan guoji*, pp. 403–4.

248. P.P. Vladimirov, *Osobyi raion Kitaia*, pp. 151–53.

249. On this paragraph, see Zhou and Chu, *Gongchan guoji*, pp. 401–6; Garver, *Chinese-Soviet Relations*, pp. 247–49.

250. On this paragraph, see Zhou and Chu, *Gongchan guoji*, p. 410; P.P. Vladimirov, *Osobyi raion Kitaia*, pp. 171–72.

251. On this paragraph, see Zhou and Chu, *Gongchan guoji*, pp. 249, 424.

252. On this paragraph, see ibid., p. 417; Yang Kuisong, "Soviet Factor," p. 21; Sheng, "America's Lost Chance in China?" p. 152; Tuchman, *Sand Against the Wind*, pp. 477–78; Bagby, *Eagle-Dragon Alliance*, pp. 112–15; Michael H. Hunt, *The Genesis of Chinese Communist Foreign Policy*, pp. 154–55. On the CCP's recognizable (and successful) efforts to provide foreign visitors in Yenan a favorable impression of the Chinese Communists since 1936, see Hunt, *The Genesis of Chinese Communist Foreign Policy*, pp. 151–52.

253. Directive of the CC of the CCP "on diplomatic work" (*Guanyu waijiao gongzuo zhishi*), August 18, 1944, in *Zhonggong zhongyang wenjian xuanji*, vol. 14, pp. 314–18. The passages cited in this paragraph are from pages 314–15. See also Zhou and Chu, *Gongchan guoji*, p. 417, and the Russian version —inaccurately translated—in *PDV* 1 (1972), pp. 185–87. The directive was sent to the CC bureaus and brought to the attention of the Party committees in the "liberated regions"; see PDV 1 (1972), p. 185.

254. Mao Zedong to John Service, July 26, 1944, in Esherick, *Lost Chance in China*, p. 290.

255. *Zhonggong zhongyang wenjian xuanji*, vol. 14, pp. 316–17.

256. See Reardon-Anderson, *Yenan and the Great Powers*, pp. 41–42; John W. Garver, *Foreign Relations of the People's Republic of China*, pp. 35–36.

257. *Zhonggong zhongyang wenjian xuanji*, vol. 14, pp. 317–18.

258. Mao Zedong to John Service, in Esherick, *Lost Chance in China*, pp. 301–2.

259. Mao Zedong to David D. Barrett, December 8, 1944, in *FRUS 1944*, vol. 6, *China*, p. 731, cited in Reardon-Anderson, *Yenan and the Great Powers*, p. 56.

260. Mao Zedong to John Service, August 23, 1944, cited in Esherick, *Lost Chance in China*, p. 307.

261. Ibid., pp. 301, 304.

262. Zhou and Chu, *Gongchan guoji*, p. 418. On this paragraph, see Zhou and Chu, *Gongchan guoji*, pp.

417–18; Tuchman, *Sand Against the Wind*, pp. 476–78; Tucker, *Patterns in the Dust*, p. 44; Reardon-Anderson, *Yenan and the Great Powers*, pp. 36–50; Bagby, *Eagle-Dragon Alliance*, pp. 104–22.

263. For further details on the Soviet go-betweens, see this chapter above.

264. On this paragraph, see the report by John Service, March 23, 1945, in Esherick, *Lost Chance in China*, pp. 350–53. See also Iurii Vlasov, "Povest' ob ottse," pt. 1, p. 125; P.P. Vladimirov, *Osobyi raion Kitaia*, p. 420.

265. P.P. Vladimirov, *Osobyi raion Kitaia*, pp. 421–22. Vlasov's commentary appears credible in the context of information available from other sources. This is true of most of his judgments of relations between the CCP leadership and the Americans in this period. See P.P. Vladimirov, *Osobyi raion Kitaia*, pp. 306–78 passim.

266. On this paragraph, see Tien-fong Cheng, *A History of Sino-Russian Relations*, p. 252; Wei, *China and Soviet Russia*, p. 170; Tuchman, *Sand Against the Wind*, pp. 476–78.

267. Letter from Zhou Enlai to Stilwell, written as the result of a CC session, September 22, 1944, in *Zhou Enlai nianpu, 1898–1949*, p. 583.

268. Liang Chin-tung, *General Stilwell in China, 1942–1944*, p. 235.

269. Peng Dehuai said this in Yenan to the American journalist Theodore H. White, see Theodore H. White, *In Search of History*, p. 189. Peng was clearly exaggerating when he provided the information on troop strengths. In a letter to Dimitrov, December 11, 1944, Mao Zedong speaks of up to 650,000 men available to the CCP; RTsKhIDNI, f. 17, op. 128, d. 10, l. 9.

270. Theodore H. White, in *Life*, December 18, 1944, cited in Zhou and Chu, *Gongchan guoji*, p. 419.

271. On this paragraph, see Bagby, *Eagle-Dragon Alliance*, pp. 157–63; Reardon-Anderson, *Yenan and the Great Powers*, pp. 51–57; P.P. Vladimirov, *Osobyi raion Kitaia*, pp. 403–5, esp. p. 403.

272. On this paragraph, see Zhou and Chu, *Gongchan guoji*, pp. 417–18; Tucker, *Patterns in the Dust*, p. 44.

273. Mao Zedong to David D. Barret, December 10, 1944, *FRUS 1944*, vol. 6, *China*, pp. 61–62, cited in Sheng, "America's Lost Chance in China?" p. 154.

274. Reardon-Anderson, *Yenan and the Great Powers*, p. 61.

275. On this paragraph, see ibid., pp. 61–63; Bagby, *Eagle-Dragon Alliance*, pp. 163–65, 168–69, 182.

276. On this paragraph, see Reardon-Anderson, *Yenan and the Great Powers*, pp. 63–64; Bagby, *Eagle-Dragon Alliance*, p. 165. See also P.P. Vladimirov, *Osobyi raion Kitaia*, pp. 421–23.

277. On this paragraph, see Bagby, *Eagle-Dragon Alliance*, pp. 164–66; Reardon-Anderson, *Yenan and the Great Powers*, p. 64; Tucker, *Patterns in the Dust*, pp. 44–45; White, *In Search of History*, p. 190; Furuya, *Chiang Kai-shek*, p. 844.

278. Bagby, *Eagle-Dragon Alliance*, p. 166; Reardon-Anderson, *Yenan and the Great Powers*, p. 64.

279. P.P. Vladimirov, *Osobyi raion Kitaia*, p. 421.

280. Reardon-Anderson, *Yenan and the Great Powers*, p. 65; Bagby, *Eagle-Dragon Alliance*, p. 166; P.P. Vladimirov, *Osobyi raion Kitaia*, pp. 421–22, 431.

281. On this paragraph, see Yang Kuisong, "Soviet Factor," p. 21; and Reardon-Anderson, *Yenan and the Great Powers*, p. 66.

282. *Zhou Enlai nianpu, 1898–1949*, p. 594.

283. Telegram from Mao Zedong, January 25, 1945, cited in Sheng, "America's Lost Chance in China?" p. 154.

284. On this and the following sentence, see Reardon-Anderson, *Yenan and the Great Powers*, p. 66.

285. On this and the following sentence, see telegram from Mao Zedong to Zhou Enlai, February 3, 1945, cited in Sheng, "America's Lost Chance in China?" p. 155.

286. Mao Zedong at the Seventh Plenary Session of the Sixth CC of the CCP, March 31, 1945, cited in Yang Kuisong, "Soviet Factor," p. 21.

287. On this paragraph, see Reardon-Anderson, *Yenan and the Great Powers*, pp. 66–67, 69.

288. On this paragraph, see P.P. Vladimirov, *Osobyi raion Kitaia*, pp. 475–76.

289. See also Reardon-Anderson, *Yenan and the Great Powers*, p. 164.

290. Mao Zedong at the Seventh Plenary Session of the Sixth CC of the CCP, March 31, 1945, cited in Yang Kuisong, "Soviet Factor," p. 21.

291. Mao Zedong to John Service, March 13, 1945, cited in Esherick, *Lost Chance in China*, p. 373.

292. *Xinhua ribao*, March 30, 1945, cited in Reardon-Anderson, *Yenan and the Great Powers*, p. 88.

293. On this and the two following sentences, see Mao Zedong to John Service, April 1, 1945, cited in Esherick, *Lost Chance in China*, p. 383.

294. Bagby, *Eagle-Dragon Alliance*, p. 176; Reardon-Anderson, *Yenan and the Great Powers*, p. 86;

Sheng, "America's Lost Chance in China?" p. 156. The report of Hurley's press conference appeared in the newspapers under such headlines as "U.S. Bars Arms Aid to Reds in China"; see Bagby, *Eagle-Dragon Alliance*, p. 156.

295. For this and the following two sentences, see P.P. Vladimirov, *Osobyi raion Kitaia*, pp. 494–95.

296. *Xinhua ribao*, April 5, 1945, cited in Reardon-Anderson, *Yenan and the Great Powers*, p. 86.

297. See also ibid., p. 74.

298. S.T. Mazorov, "Sovetsko-iaponskie otnosheniia v khode vtoroi mirovoi voiny." In *SSSR—Iaponiia: K 50-letiiu ustanovleniia sovetsko-iaponskikh diplomaticheskikh otnoshenii (1925–1975)*. Moscow: Nauka, 1978, p. 62; Wei, *China and Soviet Russia*, p. 174.

299. On this and the following three sentences, see P.P. Vladimirov, *Osobyi raion Kitaia*, pp. 501–2. John Service reproduces a more cautious assessment by the CCP of the Soviet Army's willingness to cooperate in a report dated March 14, 1945; see Esherick, *Lost Chance in China*, pp. 345–50.

300. On this, see the instruction of the CCP leadership in Pucha (Hebei Province), April 12, 1945, which was confirmed by Mao Zedong, cited in Zhou and Chu, *Gongchan guoji*, p. 422.

301. On this paragraph, see Reardon-Anderson, *Yenan and the Great Powers*, pp. 86–90, with source references. Sheng is not convincing when he draws the conclusion that Yenan had already made the shift in its policy toward the United States from an article in *Jiefang ribao*, April 9 and 10, criticizing Washington's policy toward South America; see Sheng, "America's Lost Chance in China?" pp. 156–57.

302. On this paragraph, see Reardon-Anderson, *Yenan and the Great Powers*, pp. 87–94, passim, esp. the report of the Dixie Mission, June 6, 1945, on p. 94.

303. On this and the two following sentences, see *Xinhua ribao*, June 1, 1945, cited in Reardon-Anderson, *Yenan and the Great Powers*, p. 92.

304. See also Esherick, *Lost Chance in China*, pp. xii–xxiii; John S. Service, *The Amerasia Papers*.

305. On this paragraph, see Mao Zedong in a summarizing report, May 31, 1945, cited in Yang Kuisong, "Soviet Factor," p. 22. See also P.P. Vladimirov, *Osobyi raion Kitaia*, p. 576.

306. On this paragraph, see Mao Zedong in a final report, May 30, 1945, cited in Westad, *Cold War and Revolution*, pp. 72–73; see also ibid., pp. 73–75.

307. *Jiefang ribao*, June 14 and 17, 1945, cited in Reardon-Anderson, *Yenan and the Great Powers*, p. 95; *China White Paper*, pp. 101–2, 105; Richard C. Thornton, *China: A Political History*, pp. 170–71.

308. On this and the following sentence, see Reardon-Anderson, *Yenan and the Great Powers*, p. 95.

309. Ibid., p. 75; P.P. Vladimirov, *Osobyi raion Kitaia*, pp. 455–56, 569.

310. P.P. Vladimirov, *Osobyi raion Kitaia*, p. 569.

311. Ibid., pp. 453–56 passim.

312. Ibid., pp. 569–71.

313. *Jiefang ribao*, June 25, 1945, cited in Reardon-Anderson, *Yenan and the Great Powers*, p. 96.

314. Mao Zedong in commentaries for the Xinhua News Agency, July 10 and 12, 1945, in *Mao Zedong xuanji*, vol. 3, pp. 235–40.

315. Cited in Schram, *Mao Tse-tung*, p. 232. On this paragraph, see also Sheng, "America's Lost Chance in China?" p. 157.

Notes to Chapter 2

1. For the corresponding considerations in Washington, see John Hansen Feaver, *The Truman Administration and China, 1945–1950*, pp. 54–55.

2. On this and the following paragraph, see Warren F. Kimball, ed., *Churchill and Roosevelt*, vol. 1, p. 502; ibid., vol. 2, pp. 213, 393–94, 608–9; vol. 3, p. 237; Tien-fong Cheng, *A History of Sino-Russian Relations*, p. 252; Brian Joseph Murray, *Western versus Chinese Realism*, pp. 122–23.

3. On the American and Soviet interest in convincing the opponents in the Chinese civil war to cooperate, see Murray, *Western versus Chinese Realism*, pp. 122–27.

4. Kimball, *Churchill and Roosevelt*, vol. 2, p. 608.

5. On this, see Raymond L. Garthoff, "The Soviet Intervention in Manchuria, 1945–46," p. 70.

6. *The China White Paper*, p. 72.

7. On this and the following sentence, see Deputy Minister of Foreign Affairs Lozovskii in a report for Stalin and Molotov, December 29, 1945, AP RF, f. 3, op. 86, d. 146, ll. 20–21, 24 (copy of the original).

8. Ibid., ll. 24–25.

9. A.M. Ledovskii, *Kitaiskaia politika SShA i sovetskaia diplomatiia, 1942–1954*, pp. 121–22.

10. See Murray, *Western versus Chinese Realism*, p. 129. Deputy Minister of Foreign Affairs Lozovskii's

report for Stalin and Molotov, December 29, 1945, shows that Moscow was aware of this fact; AP RF, f. 3, op. 86, d. 146, ll. 21–22 (copy of the original).

11. Instruction of the CC of the CCP, January 3, 1946, cited in Yang Kuisong, "The Soviet Factor and the CCP's Policy toward the United States in the 1940s," p. 25.

12. Mao Zedong to Edgar Snow, July 16, 1936, cited in Edgar Snow, *Red Star over China*, p. 96.

13. Undated interview of Mao Zedong by Edgar Snow that took place between June and October 1936, cited in ibid.

14. P.P. Vladimirov, *Osobyi raion Kitaia, 1942–1945*, p. 399.

15. Directive of the CC of the CCP, November 28, 1945, in *Zhonggong zhongyang wenjian xuanji*, vol. 15, p. 455.

16. On this paragraph, see Mao Zedong's speeches to the Seventh Party Congress, May 24, May 25, and May 31, and June 10, 1945, cited in Yang Kuisong, "Soviet Factor," p. 22; P.P. Vladimirov, *Osobyi raion Kitaia*, p. 572.

17. On this paragraph, see James Reardon-Anderson, *Yenan and the Great Powers*, pp. 96–99.

18. For details, see further below in this section.

19. See the treaty and exchange of notes, August 14, 1945; Chinese texts in *Treaties Between the Republic of China and Foreign States (1927–1957)*, pp. 505–10; Russian texts in *Sovetsko-kitaiskie otnosheniia, 1917–1957*, pp. 196–98, 205–6. See also O. Edmund Clubb, *China and Russia*, pp. 344–47; Theodore H. White, *In Search of History*, p. 238.

20. Texts of the agreements of August 14, 1945, in *Treaties Between the Republic of China and Foreign States*, pp. 512–22; *Sovetsko-kitaiskie otnosheniia*, pp. 198–205.

21. Text in *Treaties Between the Republic of China and Foreign States*, p. 522; *Sovetsko-kitaiskie otnosheniia*, p. 203.

22. Protocol of the talk between Stalin and T.V. Soong, July 11, 1945, in *Treaties Between the Republic of China and Foreign States*, p. 523.

23. Agreement Regarding Entry of the Soviet Union into the War Against Japan, February 11, 1945, signed by Stalin, Roosevelt, and Churchill, text in *The Conferences at Malta and Yalta, 1945*, p. 984. The discussion of the Yalta agreement took place on February 10, 1945, and lasted only a few hours; see also the American protocol (with Stalin's draft and a proposed modification by Harriman) in ibid., pp. 894–97.

24. Barbara W. Tuchman, *Sand Against the Wind*, p. 353.

25. By Chinese Changchun Railroad, the combined Chinese Eastern Railroad and Southern Manchurian Railroad are meant.

26. Tuchman, *Sand Against the Wind*, p. 515.

27. Ibid.

28. Churchill, March 5, 1946, cited in Henry Wei, *China and Soviet Russia*, p. 175.

29. On this and the following paragraph, see Cheng, *Sino-Russian Relations*, pp. 51–62; Aitchen K. Wu, *China and the Soviet Union*, pp. 73–89; Gottfried-Karl Kindermann, *Der Ferne Osten in der Weltpolitik des industriellen Zeitalters*, pp. 84–99.

30. Text in *Russko-kitaiskie otnosheniia, 1698–1916*, pp. 73–74. The treaty on the CER was signed on August 27, 1896; text in ibid., pp. 74–77.

31. Ibid., p. 77.

32. See the Sino-Russian Convention of March 15, 1898, in ibid., pp. 78–80.

33. Kindermann, *Der Ferne Osten*, p. 134.

34. For further details, see Dieter Heinzig, *Sowjetische Militärberater bei der Kuomintang, 1923–1927*, pp. 25–27, with references to sources and secondary literature.

35. Text of the agreement in *Sovetsko-kitaiskie otnosheniia, 1917–1957*, pp. 82–85.

36. Ibid., pp. 90–92.

37. Protocol of December 22, 1929, in ibid., pp. 145–47. On this paragraph, see Cheng, *Sino-Russian Relations*, pp. 149–59; Wu, *China and the Soviet Union*, pp. 198–221.

38. On this paragraph, see Wei, *China and Soviet Russia*, pp. 103–11; Kindermann, *Der Ferne Osten*, pp. 295–309.

39. On this paragraph, see Kimball, ed., *Churchill and Roosevelt*, vol. 3, p. 527; Shi Yinhong, "Cong Ya-er-ta mimi xieding dao 1945 nian Zhong Su youhao tongmeng tiaoyue," pp. 61–70. A historian from the People's Republic of China described the agreement as a "dirty business" that was concluded behind the back of China, America and Great Britain's comrade in arms of many years, in which Chinese sovereignty was set as a prize; see Lin Jun, *Zhong Su guanxi, 1689–1989*, p. 142.

40. Xu Yan, *Sulian chubing Zhongguo Dongbei jishi*, p. 3. Xu had access to Party and military archives in the People's Republic of China.

41. On this and the following sentence, see Mao Zedong to the Yugoslav ambassador in September 1956, cited in *Mao Zedong waijiao wenxuan*, pp. 253–54.

42. Ambassador Tao Ming-wei's telegram to Chiang Kai-shek, March 12, 1945, cited in Liang Chin-tung, "The Sino-Soviet Treaty," p. 374.

43. Entry of March 15, 1945, in Chiang's diary, cited in Furuya Keiji, *Chiang Kai-shek*, p. 822. Other authors report that the president's office of the KMT government received the complete text of the agreement on the day it was concluded. See, for example, Lin, *Zhong Su guanxi*, p. 144, with source references.

44. On this and the following sentence, see Wei, *China and Soviet Russia*, pp. 174–75.

45. *China White Paper*, p. 116; Chiang Kai-shek's telegram to T.V. Soong, May 22, 1945, cited in Liang, "Sino-Soviet Treaty," p. 375.

46. Liang, "Sino-Soviet Treaty," p. 375.

47. Zhou Wenqi and Chu Liangru, eds., *Gongchan guoji, Sulian he Zhongguo gongchandang guanxi biannianshi, 1919–1991*, p. 447; John W. Garver, *Chinese-Soviet Relations, 1937–1945*, p. 229. Garver mistakenly gives the date of publication as January, which Goncharov, Lewis, and Xue copied as such. See Sergei Goncharov, John W. Lewis, and Xue Litai, *Uncertain Partners*, p. 3.

48. To avoid confusion with the armed forces of the CCP, the Soviet armed forces will usually be called Soviet Army in this book. For further information, see chapter 1 n. 75.

49. Tai Wan-chin, "The Conclusion of the 1945 Sino-Soviet Treaty of Friendship and Alliance," p. 132 n. 31, with reference to the secondary literature.

50. On this paragraph, see Clubb, *China and Russia*, p. 349.

51. *China White Paper*, p. 99.

52. Andrei Ledovskii recently made the Soviet protocol of the negotiations stored in the Russian presidential archives accessible for the first time in an unpublished manuscript: see A.M. Ledovskii, "Peregovory I.V. Stalina s Sun Tszywenem o sovetsko-kitaiskom dogovore, iiun'–avgust 1945 g." (henceforth: Ledovskii, Peregovory I.V. Stalina s Sun Tszywenem"). The author bases his account of the negotiations primarily on this source, which is described as "Soviet protocol." Garver assessed the official Chinese protocols of the talks in the form in which they were published in *Zhonghua minguo zhongyao shiliao chubian, Dui Ri kangzhan shiqi*, vol. 3, bk. 2 (Taipei; n.p. 1981), pp. 572–651; see Garver, *Chinese-Soviet Relations*, pp. 214–28. Passages from the Chinese protocols are cited by Liang Chin-tung, while other parts of the negotiations are reproduced indirectly on the basis of unpublished papers from Chiang Kai-shek's private archives; see Liang, "Sino-Soviet Treaty," pp. 377–88. Murray sporadically cites Chinese protocols from the Archives of the Foreign Ministry of the Republic of China (*Waijiaobu dang'an zixun chu*), henceforth cited as AFMROC; see Murray, *Western versus Chinese Realism*, pp. 127–45, passim. Furuya, *Chiang Kai-shek*, pp. xxxiii–xxxviii, cites excerpts from Chiang Ching-kuo's notes. Elleman quotes from notes taken by Victor Hoo during the talks, which are stored in the Victor Hoo Collection, Hoover Institution Archives; see Bruce A. Elleman, *Diplomacy and Deception*, pp. 232–44, passim.

53. For further details, see this chapter below.

54. On this and the following sentence, see Harry S. Truman, *Year of Decisions*, pp. 239–44, 354–55. On this paragraph, see Garver, *Chinese-Soviet Relations*, p. 214.

55. On Mao Zedong's cool reception, see also chapter 4 below.

56. AP RF, f. 45, op. 1, d. 322, l. 3, cited in Ledovskii, "Peregovory I.V. Stalina s Sun Tszywenem," p. 3.

57. Murray, *Western versus Chinese Realism*, p. 128 (Chinese protocol, cited in a protocol in the AFMROC).

58. On this paragraph, see Garver, *Chinese-Soviet Relations*, pp. 214–15; Murray, *Western versus Chinese Realism*, p. 128.

59. Chiang Ching-kuo later claimed that Stalin had opened the talk on July 2 "by throwing a copy of the Yalta Far Eastern agreement before Soong and rudely asking him if he had previously seen the agreement" and had added: "This is our position." See Jiang Jingguo, *Fuzhong zhi yuan*, p. 63, cited in Garver, *Chinese-Soviet Relations*, p. 215. Chiang clearly made this up out of whole cloth, as there is no corresponding reference at this point in either the Chinese protocols or the Soviet protocol. See also Murray, *Western versus Chinese Realism*, p. 128; AP RF, f. 45, op. 1, d. 322, l. 3, cited in Ledovskii, "Peregovory I.V. Stalina s Sun Tszywenem," p. 5. It is unclear why Garver, who knew the Chinese protocol, assumed that Chiang's information was accurate. See also Garver, *Chinese-Soviet Relations*, p. 215. Goncharov, Lewis, and Xue accepted Garver's account—clearly without checking it. See Goncharov, Lewis, and Xue, *Uncertain Partners*, p. 3.

60. On this paragraph, see the Soviet protocol, AP RF, f. 45, op. 1, d. 322, l. 3, cited in Ledovskii, "Peregovory I.V. Stalina s Sun Tszywenem," p. 5.

61. For further details on this paragraph, see Heinzig, *Sowjetische Militärberater*, pp. 33–34, 36, 39. On the status of Outer Mongolia after the 1924 treaty, see Elleman, *Diplomacy and Deception*, pp. 85–113.

62. Soviet protocol, AP RF, f. 45, op. 1, d. 322, l. 4, cited in Ledovskii, "Peregovory I.V. Stalina s Sun Tszywenem," pp. 5–6.

63. On this and the rest of the paragraph, see Chiang Kai-shek to Petrov June 26, 1945, cited in Garver, *Chinese-Soviet Relations*, p. 216.

64. Soviet protocol, copy of the original, AP RF, f. 45, op. 1, d. 322, l. 4, in Ledovskii, "Peregovory I.V. Stalina s Sun Tszywenem," pp. 6–7. Cf. Victor Hoo's notes; see Elleman, *Diplomacy and Deception*, p. 233.

65. Soviet protocol, copy of the original, AP RF, f. 45, op. 1, d. 322, l. 7, cited in Ledovskii, "Peregovory I.V. Stalina s Sun Tszywenem," p. 7.

66. Roosevelt informed Ambassador Wei Tao-ming in March 1945, and Truman had Ambassador Harriman inform T.V. Soong "privately" and "confidentially" on July 8. On July 4, Truman interpreted the term "status quo" to Harriman to mean that the sovereignty over Outer Mongolia remained with China de jure but was not exercised de facto. See telegram from Ambassador Wei Tao-ming to Chiang Kai-shek, March 12, 1945, in Liang, "Sino-Soviet Treaty," p. 374; Truman, *Year of Decisions*, p. 241; Garver, *Chinese-Soviet Relations*, pp. 218, 220.

67. See also Garver, *Chinese-Soviet Relations*, pp. 215–16.

68. Soviet protocol, AP RF, f. 45, op. 1, d. 322, l. 5, cited in Ledovskii, "Peregovory I.V. Stalina s Sun Tszywenem," pp. 7–8.

69. Italics in the text of the agreement.

70. Soviet protocol of the session on July 8, 1945, copy of the original, AP RF, f. 45, op. 1, d. 322, l. 20, cited in Ledovskii, "Peregovory I.V. Stalina s Sun Tszywenem," p. 28. See also the survey of the session in Garver, *Chinese-Soviet Relations*, pp. 218–19.

71. On this and the following sentence, see AP RF, f. 45, op. 1, d. 322, l. 21, cited in Ledovskii, "Peregovory I.V. Stalina s Sun Tszywenem," p. 29.

72. On this and the rest of the paragraph, see AP RF, f. 45, op. 1, d. 322, ll. 22, 24, 25, 27, cited in ibid. pp. 30, 32–33, 35.

73. Petrov's record of the talk with Chiang Ching-kuo on July 8, 1945, copy of the original, AVP RF (without shelf mark), cited in ibid. pp. 36–37.

74. On this and the rest of the paragraph, see Jiang Jingguo, *Fengyu zhong de ningjing*, pp. 67–70, cited in Furuya, *Chiang Kai-shek*, pp. xxx–xxxii. See also Garver, *Chinese-Soviet Relations*, p. 217; Liang, "Sino-Soviet Treaty," pp. 380. Garver refers to the cited publication by Chiang Ching-kuo. Liang cited from Jiang Jingguo, *Fuzhong zhi yuan*, p. 67. Ledovskii, who could consult the protocols of Stalin's talks in the Russian presidential archives, did find the protocol of Chiang Ching-kuo's talk with Petrov on July 8, 1945, but none of a talk between Chiang and Stalin. It is conceivable, however, that this protocol, if it exists, is still being censored for the brutal openness with which Stalin discussed the allegedly sovereign Outer Mongolia. It can be shown that Chiang made inaccurate statements about his talks with Stalin in late 1945 and early 1946; see this chapter below.

75. The discussion of Chiang's telegram in this and the following paragraphs is based on the text of the telegram contained in the copy of the original Soviet protocol of the session on July 9, 1945, AP RF, f. 45, op. 1, d. 322, ll. 30–31, cited in Ledovskii, "Peregovory I.V. Stalina s Sun Tszywenem," pp. 40–42. It is apparently a telegram that Chiang had sent to Soong on July 6, 1945. See also the literal citations from it in Tao Wenzhao, Yang Kuisong, and Wang Jianlang, *Kang Ri zhanzheng shiqi zhong Zhongguo duiwai guanxi*, p. 487. Garver does not mention Chiang's telegram in his account of the session; see Garver, *Chinese-Soviet Relations*, pp. 220–21.

76. On this paragraph, see the Soviet protocol of the session on July 9, 1945, copy of the original, AP RF, f. 45, op. 1, d. 322, ll. 31–34, cited in Ledovskii, "Peregovory I.V. Stalina s Sun Tszywenem," pp. 42–45.

77. Soviet protocol of the session on July 9, 1945, copy of the original, AP RF, f. 45, op. 1, d. 322, l. 32, cited in ibid., p. 43. See also the text of the Chinese protocol with essentially identical contents, cited in Garver, *Chinese-Soviet Relations*, pp. 221–22.

78. The Soviet and Chinese protocols, to the extent that these are reproduced by Ledovskii and Garver, do not include this passage. It can be found in Murray, *Western versus Chinese Realism*, p. 130. Murray refers to a Chinese protocol, which he cites from documents in the AFMROC.

79. The Chinese protocol notes in brackets after this sentence: "Stalin smiles." See Garver, *Chinese-Soviet Relations*, p. 221.

80. Soviet protocol of the session on July 9, 1945, copy of the original, AP RF, f. 45, op. 1, d. 322, l. 38, cited in Ledovskii, "Peregovory I.V. Stalina s Sun Tszywenem," pp. 48–49. The Chinese protocol Garver

assessed has essentially identical contents. It includes two sentences at the end that are missing in the Soviet protocol. According to these, Soong said that his government was prepared to include the Communists in the war cabinet that had already been proposed and in the military commission; they could not allow them to turn against the government. Stalin replied evasively that the current situation really did not make a good impression. See Garver, *Chinese-Soviet Relations*, p. 221.

81. AP RF, f. 45, op. 1, d. 322, l. 37, cited in Ledovskii, "Peregovory I.V. Stalina s Sun Tszywenem," pp. 47–48.

82. Garver, *Chinese-Soviet Relations*, p. 220.

83. On this paragraph, see the Soviet protocol of the session on July 9, 1945, copy of the original, AP RF, f. 45, op. 1, d. 322, ll. 32–33, cited in Ledovskii, "Peregovory I.V. Stalina s Sun Tszywenem," p. 43–44.

84. Soviet protocol of the session on July 10, 1945, copy of the original, AP RF, f. 45, op. 1, d. 322, l. 41, cited in ibid., p. 52.

85. Soviet protocol of the session on July 11, 1945, copy of the original, AP RF, f. 45, op. 1, d. 322, l. 55, cited in ibid., p. 66.

86. On this paragraph, see the Soviet protocol of the talk between Molotov and T.V. Soong on July 10, 1945, AP RF, f. 45, op. 1, d. 322, ll. 39–53, cited in ibid., pp. 49–53; Soviet protocol of the sessions on July 11 and 12, 1945, AP RF, f. 45, op. 1, d. 322, ll. 54–64, 66–69, cited in ibid., pp. 64–78.

87. Truman's telegram to Hurley, July 23, 1945, cited in *The Conference of Berlin (The Potsdam Conference), 1945*, vol. 2, p. 1241.

88. On this paragraph, see *Conference of Berlin*, vol. 2, p. 476; Truman, *Year of Decisions*, pp. 239–44; Soviet protocol of the talk between Hurley and Petrov on July 27, 1945, copy of the original, AVP RF (without shelf mark), cited in Ledovskii, "Peregovory I.V. Stalina s Sun Tszywenem," pp. 82–84; Garver, *Chinese-Soviet Relations*, pp. 222–24.

89. On this and the following sentence, see Bevin Alexander, *The Strange Connection*, p. 33. In his memoirs, Truman tries to give the impression, however, that he was still interested in having the USSR enter the war as soon as possible after the first atom bomb had been dropped. See also Truman, *Year of Decisions*, p. 355.

90. Ibid., p. 354. See also Warren I. Cohen, *America's Response to China*, p. 181.

91. Soviet protocol, AP RF, f. 45, op. 1, d. 322, ll. 83–91, cited in Ledovskii, "Peregovory I.V. Stalina s Sun Tszywenem," pp. 95–103.

92. AP RF, f. 45, op. 1, d. 322, l. 84, in ibid., p. 96.

93. Murray, *Western versus Chinese Realism*, p. 140 (Chinese protocol, cited in the AFMROC files).

94. On this and the following sentence, see AP RF, f. 45, op. 1, d. 322, ll. 87, 89, cited in Ledovskii, "Peregovory I.V. Stalina s Sun Tszywenem," pp. 99–101.

95. For further details, see Han-jung Ziemann, *Die Beziehungen Sinkiangs (Ostturkestan) zu China und der UdSSR, 1917–1945*, pp. 246–52.

96. On the Altai region, see also Garver, *Chinese-Soviet Relations*, p. 226.

97. On this and the rest of the paragraph, see the Soviet protocol of the session on August 10, 1945, copy of the original, AP RF, f. 45, op. 1, d. 322, ll. 89–90, cited in Ledovskii, "Peregovory I.V. Stalina s Sun Tszywenem," p. 102.

98. AP RF, f. 45, op. 1, d. 322, l. 87, cited in ibid., p. 100.

99. Exchange of notes of August 14, 1945 in *Treaties Between the Republic of China and Foreign States*, pp. 508–9; *Sovetsko-kitaiskie otnosheniia*, pp. 205–6. The statement originally proposed was made in the form of an exchange of notes on August 14, 1945.

100. On this paragraph, see the Soviet protocol of the session on August 10, 1945, copy of the original, AP RF, f. 45, op. 1, d. 322, l. 88, cited in Ledovskii, "Peregovory I.V. Stalina s Sun Tszywenem," pp. 99–100.

101. Soviet protocol of the talk between Molotov and T.V. Soong on August 12, 1945, AVP RF, f. 06, op. 7, p. 36, d. 507, ll. 108–13, contents reproduced in ibid., pp. 104–6.

102. For further details, see Garver, *Chinese-Soviet Relations*, p. 225.

103. On this and the following three paragraphs, see the Soviet protocol of the session on August 13, 1945, copy of the original, AP RF, f. 45, op. 1, d. 322, ll. 93–97, cited in Ledovskii, "Peregovory I.V. Stalina s Sun Tszywenem," pp. 106–10.

104. On this and the rest of the paragraph, see Truman, *Year of Decisions*, p. 355; Garver, *Chinese-Soviet Relations*, p. 228. The Soviet protocol contains only the following exchange on the topic of war booty: "Furthermore, Soong addresses the question of handing over to China all military goods that will be taken from the Japanese. Comrade Stalin said that they could discuss the question of the military goods in Manchuria." AP RF, f. 45, op. 1, d. 322, l. 79, cited in Ledovskii, "Peregovory I.V. Stalina s Sun Tszywenem," p. 94. If, in fact, nothing more was said, Harriman interpreted Stalin's terse sentence correctly.

105. For further details, see this chapter below.

106. "Meeting between Dr. Soong and Mr. Molotov," Hoover Institution Archives, Victor Hoo Collection, Box 7, August 14, 1945, cited in Bruce A. Elleman, "End of Extraterritoriality," p. 81.

107. "Treaty of Alliance," Hoover Institution Archives, T.V. Soong Collection, Schedule A, Box 25, August 14, 1945, cited in ibid. The sentence derives from a draft, which, to judge from the use of the term "Russian" (instead of "Soviet"), the Chinese side presented. The final text has apparently not become known to date.

108. Elleman, "End of Extraterritoriality," p. 75.

109. Vladislav Zubok and Constantine Pleshakov were of the opinion that Stalin used the invasion of Manchuria, control of northern Sinkiang, and the CCP as trump cards in order "to obtain all the concessions he wanted." See Vladislav Zubok and Constantine Pleshakov, *Inside the Kremlin's Cold War*, p. 56.

110. On this and the following sentence, see "Protokol o rezul'tatach provedeniia plebistsita po voprosu o nezavizimosti Mongol'skoj Narodnoj Respubliki," AVP RF, f. 07, op. 23a, p. 18, d. 235, ll. 129–33.

111. See ibid., ll. 124–27.

112. S. Luzianin, "Ialtinskaia konferentsiia i problemy mezhdunarodno-pravovogo oformleniia MNR nakanune i v gody vtoroi mirovoi voiny," p. 59. Luzianin refers to unpublished Russian archival documents.

113. AVP RF, f. 07, op. 23a, p. 18, d. 235, ll. 126–27.

114. *Sto sorok besed s Molotovym*, p. 14.

115. On this paragraph, see also Iriye Akira, *The Cold War in Asia*, p. 122; Cohen, *America's Response to China*, pp. 181–82. Soong gave Harriman the impression that he was satisfied with the results of the negotiations. See also Harriman's report to Truman, August 14, 1945, *FRUS 1945*, vol. 7, *The Far East: China*, p. 972, cited in Wesley M. Bagby, *The Eagle-Dragon Alliance*, pp. 216, 263 n. 44.

116. See A.M. Ledovskii, "The World Reaction to the Soviet-Chinese Treaty," pp. 153–54, passim.

117. See ibid.; Garver, *Chinese-Soviet Relations*, p. 229.

118. The report of September 10, 1945, is reproduced in Murray, *Western versus Chinese Realism*, pp. 141–44.

119. Telegram from T.V. Soong to Chiang Kai-shek, August 18, 1945, cited in Goncharov, Lewis, and Xue, *Uncertain Partners*, p. 6.

120. On this and the following sentence, see Wei, *China and Soviet Russia*, p. 184; Cheng, *Sino-Russian Relations*, p. 274.

121. On this and the following sentence, see *China White Paper*, pp. 120–21.

122. Odd Arne Westad, *Cold War and Revolution*, p. 94.

123. Ibid.

124. See also, for example, Shi Yuanhua, *Zhonghua minguo waijiao shi*, p. 625.

125. For further details, see Bagby, *Eagle-Dragon Alliance*, p. 217, with source references; Cohen, *America's Response to China*, p. 182.

126. Xu, *Sulian chubing Zhongguo Dongbei jishi*, p. 3 (Xu had access to Party and military archives in the People's Republic of China); report of the head of the American observer mission, Ivan Yeaton, to General Albert C. Wedemeyer, August 15, 1949, on a talk with Zhou Enlai, National Archives, Washington, DC, cited in Westad, *Cold War and Revolution*, p. 80; Reardon-Anderson, *Yenan and the Great Powers*, p. 72; Liu Zhiqing, "Shi lun 'Zhong-Su youhao tongmeng tiaoyue' de qianding," p. 73.

127. Zhou Enlai to Ivan Yeaton, August 15, 1945, cited in Westad, *Cold War and Revolution*, p. 80. According to a Soviet author, Zhou Enlai said in the 1970s that the treaty had made Chiang Kai-shek "so confident about his power that he unleashed the civil war against the Communists." There is, however, no statement of this kind by Zhou Enlai in the source cited (*New York Times*, October 27, 1973) or in the interview of Zhou reproduced in the *New York Times*, October 29, 1973. See E.P. Bazhanov, *Kitai i vneshnii mir*, pp. 36, 345 n. 20.

128. Stalin to Chiang Ching-kuo, December 30, 1945, cited in the Soviet protocol, AP RF, f. 45, op. 1, l. 102 (copy of the original); see also the published protocol in A.M. Ledovskii, "Stalin i Chan Kaishi," p. 112.

129. On this and the two following sentences, see Chen Jian, *China's Road to the Korean War*, p. 17. See also Yang Kuisong, "The Soviet Factor and the CCP's Policy Toward the United States in the 1940s," p. 23, which presents a slightly different opinion.

130. Hurley on September 6, 1945, cited in Bagby, *Eagle-Dragon Alliance*, p. 217.

131. Garver, *Chinese-Soviet Relations*, p. 262, with source references.

132. John Service in a report on March 14, 1945, on the basis of talks with CCP leaders, in Joseph W. Esherick, ed., *Lost Chance in China*, pp. 345, 349–50. On this paragraph, see also Westad, *Cold War and Revolution*, p. 80.

133. Xu, *Sulian chubing Zhongguo dongbei jishi*, p. 3.
134. On this paragraph, see Clubb, *China and Russia*, p. 349.
135. Cited in Westad, *Cold War and Revolution*, 78.
136. Instruction of the CC of the CCP of August 10, 1945, in *Zhonggong zhongyang wenjian xuanji*, vol. 15, pp. 215–16.
137. Telegram from the CC of the CCP to the Central Chinese Bureau, August 10, 1945, in *Mao Zedong nianpu, 1893–1949*, vol. 3, p. 1.
138. On this paragraph, see orders nos. 1–7, August 10 and 11, 1945, reproduced in *Zhonggong zhongyang wenjian xuanji*, vol. 15, pp. 217–25. See also *Mao Zedong nianpu*, vol. 3, p. 9; *Zhou Enlai nianpu, 1898–1949*, p. 613; Yang Kuisong, *Zhongjian didai de geming*, p. 404; Yang Kuisong, "Zhongguo gongchandang duoquan Dongbei zhanlüe yanbian yu Sulian," p. 60; Reardon-Anderson, *Yenan and the Great Powers*, p. 100; *History of the Sino-Japanese War (1937–1945)*, pp. 547–48; telegram from the CC, August 12, 1945, cited in Zhou and Chu, *Gongchan guoji*, p. 430; Westad, *Cold War and Revolution*, pp. 78–79; *Jiefang ribao*, August 13, 1945, cited in Garver, *Chinese-Soviet Relations*, pp. 262–63; Michael M. Sheng, "Response," p. 181; Michael M. Sheng, *Maoist Dualism and Chinese Communist Foreign Relations*, pp. 199–204. The advance into Manchuria was intended only partially seriously, however. See this chapter below (section "Moscow's Double Game in Manchuria").
139. Politburo resolution at a session in Wayaobao (northern Shaanxi), December 17–25, 1935, cited in John W. Garver, "The Origins of the Second United Front, p. 41, with source references.
140. Westad, *Cold War and Revolution*, pp. 81–82; Reardon-Anderson, *Yenan and the Great Powers*, p. 100.
141. On this paragraph, see Westad, *Cold War and Revolution*, p. 80.
142. Garver, *Chinese-Soviet Relations*, p. 263.
143. For the two telegrams, see *Mao Zedong nianpu*, vol. 3, p. 13; *Zhou Enlai nianpu*, p. 615; Xu, *Sulian chubing Zhongguo Dongbei jishi*, p. 103; Shi Zhe, *Zai lishi juren shenbian*, p. 415; P.P. Vladimirov, *Osobyi raion Kitaia*, pp. 645–47; Yang Kuisong, *Zhongjian didai de geming*, p. 399; Sheng, *Maoist Dualism*, pp. 204–5. The text of the telegrams is not yet accessible. They are supposedly not in the CCP archives, as they were among the documents that were burned in 1947 before the Party leadership withdrew from Yenan. See Xu, *Sulian chubing Zhongguo Dongbei jishi*, p. 103. They have not yet surfaced in the Soviet archives. Michael Sheng's assessment that Stalin's telegrams and the Chinese reaction were an example of the harmonious relations between Moscow and the CCP leadership sounds bizarre. See Sheng, "Response," p. 181.
144. Li Haiwen, "Mao Zedong yu Si-da-lin jiaowang," pp. 12–13. The author did not provide source references, which is common in the People's Republic of China when Party archives are used. She is employed in the Party archives.
145. On this and the following sentence, see Reardon-Anderson, *Yenan and the Great Powers*, p. 103.
146. On this and the rest of the paragraph, see Stalin to Chiang Ching-kuo, December 30, 1945, cited in the Soviet protocol of the talk, AP RF, f. 45, op. 1, l. 102 (copy of the original). See also the published protocol in Ledovskii, "Stalin i Chan Kaishi," p. 111.
147. Mao to Iudin, March 31, 1956, Soviet record of the talk, in *PDV* 5 (1994), p. 105.
148. Mao Zedong to Iudin, July 22, 1958, Chinese record of the talk, in *Dang de wenxian* 1 (1994), pp. 16–17. See also Yu Zhan and Zhang Guangyou, "Guanyu Si-da-lin ceng fou quanzu wo guo Changjiang de tantao," pp. 20–21.
149. Mao to Iudin, July 22, 1958, Chinese record of the talk, in *Dang de wenxian*, 1 (1994), p. 17.
150. Mao, September 24, 1962, at the Tenth Plenary Session of the Eighth CC of the CCP, in *Mao Zedong tongzhi shi dangdai zui weida de Ma-ke-si Lie-ning zhuyi zhe*, 1969, p. 432.
151. Stalin, February 8, 1948 in Moscow; see Vladimir Dedijer, *Tito Speaks*, p. 331. It is not clear what Stalin is referring to when he speaks of discussions with Chinese comrades in Moscow. It is possible that these Chinese were Liu Shaoqi and Gao Gang, who met with Soviet leaders in 1945 in the USSR, according to Soviet reports. See O.B. Borisov and B.T. Koloskov, *Sovetsko-kitaiskie otnosheniia, 1945–1970*, p. 28.
152. Milovan Djilas, *Conversations with Stalin*, pp. 164–65.
153. See Mao Zedong to Ambassador Iudin, July 22, 1958, Chinese record of the talk, in *Dang de wenxian* 1 (1994), p. 17.
154. See also Yang Kuisong, *Zhongjian didai de geming*, p. 439.
155. Telegrams from the CC of the CCP to the Central Chinese Bureau, August 20 and 21, 1945, in *Mao Zedong nianpu*, vol. 3, p. 9; Sheng, *Maoist Dualism*, pp. 302–3.
156. Directives of the CC and Military Commission of the CCP, August 22, 1945, in *Zhonggong zhongyang wenjian xuanji*, vol. 15, p. 243–44. See also the summary of the contents of the directives in Zhou and Chu,

Gongchan guoji, p. 433; Mao's speech to an extended Politburo session on August 23, 1945 in *Mao Zedong nianpu*, vol. 3, p. 10; the internal Party circular of August 26, 1945, in *Selected Works of Mao Tse-tung*, vol. 4, pp. 47–50; and the CC directive of August 29, 1945, in *Zhonggong zhongyang wenjian xuanji*, vol. 15, pp. 257–58.

157. P.P. Vladimirov, *Osobyi raion Kitaia*, pp. 636–42. According to the entry of August 16, 1945, Yenan was of the opinion that "in establishing the theaters of war, the USSR had merely taken over Manchuria and forgotten the interests of the CCP." Ibid., p. 639.

158. Ibid., pp. 645–47.

159. On this and the rest of the paragraph, see Secretariat of the North China Bureau of the CCP, August 30, 1945, cited in Garver, *Chinese-Soviet Relations*, p. 263.

160. On this paragraph, see Yang Kuisong, "Soviet Factor," pp. 23–24; Westad, *Cold War and Revolution*, p. 82.

161. Mao Zedong in a talk with a few comrades, August 23, 1945, cited in Yang Kuisong, "Soviet Factor," p. 23

162. On this and the two following sentences, see Mao Zedong at an extended Politburo session, August 23, 1945, in *Mao Zedong nianpu*, vol. 3, p. 10.

163. Cited in Yang Kuisong, *Zhongjian didai de geming*, p. 423.

164. On this and the following sentence, see Mao Zedong in Chungking, August 29, 1945, cited in Westad, *Cold War and Revolution*, pp. 80–81. See also P.P. Vladimirov, *Osobyi raion Kitaia*, pp. 645–47. The assumption made in this source does appear plausible in the overall historical context. See also Garver, *Chinese-Soviet Relations*, p. 263; Yang Kuisong, "Soviet Factor," pp. 23, 25.

165. On this and the following sentence, see Reardon-Anderson, *Yenan and the Great Powers*, p. 103.

166. *Jiefang ribao*, August 27, 1945, cited in O. Borisov, *Sovetskii Soiuz i mandzhurskaia revoliutsionnaia baza, 1945–1949*, p. 91; *Jiefang ribao*, August 29, 1945, cited in Liu Zhiqing, "Shi lun 'Zhong-Su youhao tongmeng tiaoyue' de qianding," p. 73.

167. See Mao Zedong, August 29, 1945, in Chungking, cited in Westad, *Cold War and Revolution*, p. 81; Yang Kuisong, "Soviet Factor," p. 23; Garver, *Chinese-Soviet Relations*, p. 263 (Garver mistakenly gives August 27 as the date of Mao's arrival); Furuya, *Chiang Kai-shek*, p. 851.

168. AVP RF, f. 56, op. 6, p. 47, d. 67, cited in A.M. Ledovskii, "Na diplomaticheskoi rabote v Kitae v 1942–1952 gg," p. 117. Petr Vlasov apparently used the same archival document in compiling the "diary," as we find the following blunt statement in it: "He went as if to his excution." See P.P. Vladimirov, *Osobyi raion Kitaia*, p. 648.

169. *China White Paper*, p. 99; Bagby, *Eagle-Dragon Alliance*, p. 217; P.P. Vladimirov, *Osobyi raion Kitaia*, p. 647.

170. Ledovskii, "Na diplomaticheskoi rabote v Kitae," pp. 116–17.

171. P.P. Vladimirov, *Osobyi raion Kitaia*, p. 648.

172. On this paragraph, see M.S. Kapitsa, "Na raznych paralleliakh," pp. 13–14.

173. *China White Paper*, p. 109.

174. Mao Zedong, Zhou Enlai and Wang Ruofei to Petrov, October 10, 1945, AVP RF, f. 0100, op. 40, d. 248, cited in Ledovskii, "Stalin i Chan Kaishi," p. 110 n. 21.

175. *China White Paper*, p. 108.

176. Reardon-Anderson, *Yenan and the Great Powers*, pp. 114, 144.

177. This is assumed by Reardon-Anderson. See ibid., pp. 114–15.

178. Andrei M. Ledovskii to the author, March 23, 1995. Ledovskii was employed as a diplomat at the Soviet embassy in China from 1942 to 1949. See also Ledovskii, "Stalin i Chan Kaishi," p. 102.

179. See also the Russian and Chinese texts of the treaty of alliance of August 14, 1945, and of the exchange of notes on the same day, in *Sovetsko-kitaiskie otnosheniia*, pp. 196–98, 205–6; *Treaties Between the Republic of China and Foreign States*, pp. 505–9.

180. Russian text in *Sovetsko-kitaiskie otnosheniia*, pp. 204–5.

181. See this chapter above (section "The Sino-Soviet Treaty of Alliance of August 14, 1945: Moscow's Grab for Czarist Privileges").

182. See also Steven I. Levine, "Soviet-American Rivalry in Manchuria and the Cold War," pp. 15–16; Westad, *Cold War and Revolution*, p. 120.

183. Cited in Yang Kuisong, "Zhongguo gongchandang duoquan Dongbei," p. 60. The citation is not dated.

184. Cited in ibid. The citation is not dated. On this paragraph, see ibid.

185. On this paragraph, see Zhou and Chu, *Gongchan guoji*, pp. 397–98, 401, 426–27; Borisov, *Sovetskii*

Soiuz i mandzhurskaia revoliutsionnaia baza, p. 71; A.M. Ledovskii, *SSSR, SShA i narodnaia revoliutsiia v Kitae*, p.109; M.F. Iur'ev, *Vooruzhennye sily KPK v osvoboditel'noi bor'be kitaiskogo naroda*, pp. 266–67; Reardon-Anderson, *Yenan and the Great Powers*, pp. 105, 120–21.

186. Order no. 2 of August 11, 1945, in *Jiefang ribao*, August 12, 1945, cited in *Zhonggong zhongyang wenjian xuanji*, vol. 15, p. 219. Order no. 6 of August 11, 1945, also calls for military activity to be coordinated with the Soviet Army; see ibid., p. 223.

187. Yang Kuisong, "Zhongguo gongchandang duoquan Dongbei," p. 60.

188. On this paragraph, see ibid.; Yang Kuisong, *Zhongjian didai de geming*, p. 404; Reardon-Anderson, *Yenan and the Great Powers*, pp. 105, 107.

189. See Chiang Kai-shek, *Soviet Russia in China*, p. 142.

190. For further details, see this chapter above.

191. On this paragraph, see Yang Kuisong, "Zhongguo gongchandang duoquan Dongbei," p. 60.

192. On this paragraph, see instructions of the CC of the CCP, August 26 and 29, 1945, cited in ibid., p. 61. The instruction of August 29, 1945, is reproduced in *Zhonggong zhongyang wenjian xuanji*, vol. 15, p. 258.

193. CC directive of August 29, 1945, reproduced in *Zhonggong zhongyang wenjian xuanji*, vol. 15, pp. 257–58.

194. On this paragraph, see Reardon-Anderson, *Yenan and the Great Powers*, pp. 107–8; Westad, *Cold War and Revolution*, p. 83.

195. On this paragraph, see the CC of the CCP directive to the branch bureau in Shandong, September 11, 1945, in *Zhonggong zhongyang wenjian xuanji*, vol. 15, p. 274.

196. This was reported to an official in the Soviet consulate general in Harbin on December 27, 1949; AVP RF, f. 07, op. 43, d. 13, p. 302, l. 13. On this paragraph, see Xu Yan, *Sulian chubing Zhongguo Dongbei jishi*, pp. 171–74; Westad, *Cold War and Revolution*, pp. 83, 87; Mrs. Zhao Yan as contemporary witness to the author, July 23, 1994.

197. On this paragraph, see Westad, *Cold War and Revolution*, p. 84.

198. On this and the following paragraph, see *Zhongguo gongchandang chuangye sanshi nian (1919–1949)*, p. 587; *Liu Shaoqi nianpu, 1898–1969*, vol. 1, p. 490; Huang Zheng, ed., *Liu Shaoqi yi sheng*, pp. 218–19; *Zhu De zhuan*, p. 547; Xu, *Sulian chubing Zhongguo Dongbei jishi*, p. 109; Shi Zhe, *Zai lishi juren shenbian*, p. 310; Yang Kuisong, "Zhongguo gongchandang duoquan Dongbei," p. 62; Yang Kuisong, *Zhongjian didai de geming*, p. 407; Zhou and Chu, *Gongchan guoji*, pp. 438–39; Bi Xiantian, "Zhanhou Dongbei wenti yu Zhong-Su guanxi zouxiang," p. 62; Westad, *Cold War and Revolution*, p. 84. The author learned Belorussov's real name from V. Ia. Sidikhmenov, who accompanied Belorussov to Yenan as interpreter. Belorussov's name was for unknown reasons reproduced as Wei-si-bie-fu in the CC of the CCP files, and transcribed Bei-lu-nuo-suo-fu in the Chinese secondary literature. See also Wu Xiuquan, *Huiyi yu huainian*, p. 199. Here and in Zhou and Chu, *Gongchan guoji*, as well as by Sidikhmenov, the arrival of the aircraft is dated September 15, whereas all the other cited publications place it on the previous day. The author thanks his Russian colleague Konstantin Shevelev for providing the information from Sidikhmenov cited here and further below on April 29 and July 7, 1995. On this and the following paragraphs, see Zeng Kelin's report in Zeng Kelin, *Rongma shengya de huiyi*, pp. 226–39. Although Zeng reported to the Politburo, he learned nothing about the talks with Belorussov at that time. Ibid., p. 234.

199. Yang Kuisong, *Zhongjian didai de geming*, p. 407.

200. On this paragraph, see *Liu Shaoqi nianpu*, vol. 1, pp. 490–91; Huang Zheng, ed., *Liu Shaoqi yi sheng*, pp. 219–20. In a letter addressed to Marshal Rodion Malinovskii on September 15, 1945, Liu Shaoqi, Zhu De, and Ren Bishi explicitly asked for approval for the Eighth Route Army to stay in Jehol and Liaoning; *Liu Shaoqi nianpu*, vol. 1, p. 491.

201. Westad, *Cold War and Revolution*, p. 84.

202. Zhou and Chu, *Gongchan guoji*, p. 439.

203. Ibid.

204. Ibid.

205. Xu, *Sulian chubing Zhongguo Dongbei jishi*, p. 109; Yang Kuisong, *Zhongjian didai de geming*, p. 407; Yang Kuisong, "Zhongguo gongchandang duoquan Dongbei," p. 62; Wu Xiuquan, *Huiyi yu huainian*, p. 199.

206. Yang Kuisong, "Zhongguo gongchandang duoquan Dongbei," p. 65; Yang Kuisong, *Zhongjian didai de geming*, p. 413. This citation is neither dated nor indentified further in any other way. The attribution in terms of date and subject matter that the author of the present book provides arises from the context.

207. On this paragraph, see Sidikhmenov's written communication to the author, April 16, 1995. Zeng Kelin indicates that the two Soviet officers were sent to Yenan in Molotov's service (in Dieter Heinzig

archives, Erftstadt, Germany). See Zeng Kelin, "Xinhuo," pp. 20–32; the author thanks his colleague Konstantin Shevelev for this reference. The author was unable to convince Belorussov to comment on the events of that time in a telephone conversation on October 17, 1995. Belorussov, who did not write his memoirs, is apparently determined to take his secrets with him to the grave. The conversation ended when he observed categorically that he thought that it is necessary to remain silent (*Dumaiu, chto nado molchat'*).

208. Westad, *Cold War and Revolution*, p. 85. On this and the following sentence, see also Huang Zheng, *Liu Shaoqi yi sheng*, p. 266; *Zhu De zhuan*, pp. 547–48.

209. According to *Liu Shaoqi nianpu*, vol. 1, p. 491, the aircraft had already flown back on September 15.

210. On this and the following sentence, see Wu Xiuquan, *Huiyi yu huainian*, pp. 199–200; Peng Zhen, "Dongbei jiefang zhanzheng de tou jiu yue," p. 349; Sidikhmenov's communication to the author, July 19, 1995.

211. On this and the following sentence, see *Zhu De zhuan*, p. 584; Reardon-Anderson, *Yenan and the Great Powers*, pp. 120–21; Steven I. Levine, *Anvil of Victory*, pp. 108–9.

212. Yang Kuisong, *Zhongjian didai de geming*, pp. 407–8; Yang Kuisong, "Zhongguo gongchandang duoquan Dongbei," p. 62.

213. Sidikhmenov's communications to the author, April 16, 1995, and July 19, 1995.

214. Andrei Ledovskii to the author, March 23, 1995.

215. On this paragraph, see protocol of the Politburo session on September 14, 1945, cited in *Zhu De zhuan*, pp. 547–48; Yang Kuisong, "Zhongguo gongchandang duoquan Dongbei," p. 62.

216. On this paragraph, see Yang Kuisong, "Zhongguo gongchandang duoquan Dongbei," p. 62. See also Yang Kuisong, *Zhongjian didai de geming*, p. 407.

217. See, for example, Reardon-Anderson, *Yenan and the Great Powers*, p. 119.

218. Ibid., pp. 120–22.

219. On this paragraph, see *Zhu De zhuan*, p. 548. See also Yang Kuisong, "Zhongguo gongchandang duoquan Dongbei," p. 63.

220. Yang Kuisong, "Zhongguo gongchandang duoquan Dongbei," p. 63; *Zhu De zhuan*, p. 548.

221. Protocol of the Politburo session on September 19, 1945, cited in *Zhu De zhuan*, p. 548.

222. Yang Kuisong, "Zhongguo gongchandang duoquan Dongbei," p. 62.

223. Instructions of the CC of the CCP to the Northeast Bureau of November 19 and 20, 1945, cited in Yang Kuisong, "Soviet Factor," p. 24.

224. Yang Kuisong, *Zhongjian didai de geming*, p. 409.

225. *Zhu De zhuan*, p. 548.

226. Garthoff, "Soviet Intervention in Manchuria," p. 72; *Selected Works of Mao Tse-tung*, vol. 4, p. 84 ; Borisov, *Sovetskii Soiuz i mandzhurskaia revoliutsionnaia baza*, p. 139.

227. On this paragraph, see *Zhu De zhuan*, p. 548.

228. Eva-Maria Stolberg, *Stalin und die chinesischen Kommunisten, 1945–1953*, p. 99.

229. On this paragraph, see Garthoff, "Soviet Intervention in Manchuria," p. 72; Westad, *Cold War and Revolution*, pp. 83–87, 89–90, 121; Stolberg, *Stalin und die chinesischen Kommunisten*, pp. 74–75.

230. On this paragraph, see Yang Kuisong, "Zhongguo gongchandang duoquan Dongbei," p. 63.

231. On this paragraph, see William Pace Head, *America's China Sojourn*, p. 239; Westad, *Cold War and Revolution*, pp. 103–4, 113–17, 121; Reardon-Anderson, *Yenan and the Great Powers*, pp. 110–12.

232. On this and the following sentence, see Westad, *Cold War and Revolution*, p. 95.

233. *Cong Yan'an dao Beijing*, p. 34.

234. See protocol of the meeting between Stalin and T.V. Soong, August 14, 1945, in *Treaties Between the Republic of China and Foreign States*, p. 523.

235. On this and the following sentence, see *History of the Sino-Japanese War*, p. 559; Clubb, *China and Russia*, pp. 353, 356–57.

236. On this paragraph, see Westad, *Cold War and Revolution*, p. 121; Yang Kuisong, *Zhongjian didai de geming*, p. 416.

237. Yang Kuisong, *Zhongjian didai de geming*, p. 412; Yang Kuisong, "Zhongguo gongchandang duoquan Dongbei," p. 43. The Soviet recommendation is neither dated nor supported by source references, but seems authentic. The author had access to the CCP archives. From the context, the recommendation was made between the beginning of October and the middle of November. On this paragraph, see Yang Yang Kuisong, *Zhongjian didai de geming*, pp. 408–9, 412.

238. Instruction of the Military Commission of the CC of the CCP, September 28, 1945, cited in Yang Kuisong, *Zhongjian didai de geming*, p. 409.

239. Furuya, *Chiang Kai-shek*, p. 857. On this paragraph, see Yang Kuisong, *Zhongjian didai de geming*, p. 409; Yang Kuisong, "Zhongguo gongchandang duoquan Dongbei," p. 63.

240. On this paragraph, see Clubb, *China and Russia*, pp. 349–50; Westad, *Cold War and Revolution*, p. 95.

241. Agreement on Dairen of August 14, 1945, in *Sovetsko-kitaiskie otnosheniia*, p. 203.

242. Ledovskii, "Na diplomaticheskoi rabote v Kitae," p. 127.

243. Talks between Petrov and Chiang Kai-shek, October 18 and 23, 1945, in AVP RF, f. 0100, op. 40, p. 248, d. 7, cited in ibid., p. 127; Yang Kuisong, *Zhongjian didai de geming*, p. 416.

244. Yang Kuisong, "Zhongguo gongchandang duoquan Dongbei," p. 64.

245. Yang Kuisong, *Zhongjian didai de geming*, p. 411.

246. Communication of October 25, 1945, cited in ibid.

247. Telegram from the Northeast Bureau to the CC of the CCP, October 25, 1945, cited in ibid.

248. On this paragraph, see Clubb, *China and Russia*, p. 352; Westad, *Cold War and Revolution*, pp. 95–96, 112; Yang Kuisong, "Zhongguo gongchandang duoquan Dongbei," p. 64; Yang Kuisong, *Zhongjian didai de geming*, pp. 410–11, 416; Reardon-Anderson, *Yenan and the Great Powers*, p. 123.

249. On this, see Westad, *Cold War and Revolution*, pp. 115–16.

250. On this paragraph, see Yang Kuisong, *Zhongjian didai de geming*, pp. 412–13, 416; Yang Kuisong, "Zhongguo gongchandang duoquan Dongbei," p. 65.

251. On this paragraph, see Westad, *Cold War and Revolution*, pp. 121–22.

252. On this paragraph, see Chiang, *Soviet Russia in China*, p. 143; Reardon-Anderson, *Yenan and the Great Powers*, pp. 117–18, 123; Yang Kuisong, *Zhongjian didai de geming*, p. 412; Yang Kuisong, "Zhongguo gongchandang duoquan Dongbei," p. 64; *Military Campaigns in China*, pp. 101, 104; Trevor N. Dupuy, *The Military History of the Chinese Civil War*, p. 71. Yang Kuisong provides an inaccurate date for the landing.

253. Reardon-Anderson, *Yenan and the Great Powers*, p. 117.

254. Telegram from the CC of the CCP, November 23, 1945, cited in Yang Kuisong, *Zhongjian didai de geming*, p. 415.

255. On this paragraph, see Levine, *Anvil of Victory*, p. 58; *Military Campaigns in China*, pp. 101–2, 104; Dupuy, *Military History*, p. 71; Westad, *Cold War and Revolution*, p. 125; Yang Kuisong, "Zhongguo gongchandang duoquan Dongbei," p. 64; Yang Kuisong, *Zhongjian didai de geming*, p. 416; Chiang, *Soviet Russia in China*, pp. 146–47.

256. On this and the following sentence, see Reardon-Anderson, *Yenan and the Great Powers*, pp. 123–24.

257. On this paragraph, see Chiang, *Soviet Russia in China*, pp. 146–47; Yang Kuisong, "Zhongguo gongchandang duoquan Dongbei," p. 65; Reardon-Anderson, *Yenan and the Great Powers*, p. 124.

258. On this paragraph, see Yang Kuisong, *Zhongjian didai de geming*, p. 413; Westad, *Cold War and Revolution*, p. 123; Reardon-Anderson, *Yenan and the Great Powers*, pp. 124–25.

259. On this paragraph, see Westad, *Cold War and Revolution*, p. 122; Clubb, *China and Russia*, p. 353.

260. On this paragraph, see M.I. Sladkovskii, *Istoriia torgovo-ekonomicheskikh otnoshenii SSSR s Kitaem (1917–1974)*, pp. 162–64; Levine, *Anvil of Victory*, pp. 49, 51; Westad, *Cold War and Revolution*, p. 124; Clubb, *China and Russia*, p. 356; Yang Kuisong, *Zhongjian didai de geming*, pp. 417–18.

261. Instruction of the CC of the CCP to the Northeast Bureau, October 6, 1945, cited in Yang Kuisong, *Zhongjian didai de geming*, pp. 409–10.

262. Westad, *Cold War and Revolution*, p. 88.

263. On this paragraph, see instructions of the CC of the CCP to the Northeast Bureau, October 28, 1945, cited in Yang Kuisong, *Zhongjian didai de geming*, pp. 410–11.

264. Westad, *Cold War and Revolution*, pp. 123–24; Garthoff, "Soviet Intervention in Manchuria," p. 73; Yang Kuisong, *Zhongjian didai de geming*, p. 412.

265. Westad, *Cold War and Revolution*, pp. 125–26.

266. Coded telegram from Nikolai Fedorenko to Stalin, August 4, 1951, about Ambassador Iudin's talk with Gao Gang in Mukden, AVP RF, f. 059A, op. 3, d. 4, p. 1, unpaginated. On this paragraph, see Westad, *Cold War and Revolution*, pp. 121–23, 125–27; Garthoff, "Soviet Intervention in Manchuria," p. 73.

267. On this paragraph, see Yang Kuisong, *Zhongjian didai de geming*, p. 414; Yang Kuisong, "Zhongguo gongchandang duoquan Dongbei," pp. 64, 66; Westad, *Cold War and Revolution*, p. 125.

268. On this paragraph, see instructions of the CC of the CCP to the Northeast Bureau, November 19, 23, and 28, 1945, and other sources, cited in Westad, *Cold War and Revolution*, p. 126; Sheng, "Response," pp. 182–83, with source references.

269. On this paragraph, see instructions of the CC of the CCP, November 19 and 20, 1945, cited in Yang Kuisong, *Zhongjian didai de geming*, p. 414; Yang Kuisong, "Zhongguo gongchandang duoquan Dongbei," p. 66.

270. On this paragraph, see Dupuy, *Military History*, pp. 71–72; Yang Kuisong, *Zhongjian didai de geming*, p. 415.

271. On this paragraph, see Yang Kuisong, *Zhongjian didai de geming*, pp. 414–15; Yang Kuisong, "Zhongguo gongchandang duoquan Dongbei," p. 66. On the cooperation between the Soviet Army and Chinese Communists in Port Arthur and Dairen, see Xia Zhenduo, "Sulian hongjun jinzhu 10 nian," pp. 131–32.

272. See Levine, *Anvil of Victory*, p. 51; Clubb, *China and Russia*, p. 353.

273. On this paragraph, see Levine, *Anvil of Victory*, p. 51; Yang Kuisong, *Zhongjian didai de geming*, p. 416; Reardon-Anderson, *Yenan and the Great Powers*, pp. 125–26.

274. On this paragraph, see Yang Kuisong, "Zhongguo gongchandang duoquan Dongbei," p. 67; Yang Kuisong, "Soviet Factor," pp. 25–27; Furuya, *Chiang Kai-shek*, pp. 862–65; *China White Paper*, pp. 140–43, 622–26; Westad, *Cold War and Revolution*, pp. 143–44, 155; Suzanne Pepper, "The KMT-CCP Conflict, 1945–1949," p. 729; Levine, *Anvil of Victory*, p. 125; Reardon-Anderson, *Yenan and the Great Powers*, pp. 129–30, 134. On the resumption of negotiations, see also this chapter below.

275. Reardon-Anderson, *Yenan and the Great Powers*, p. 133.

276. *Treaties Between the Republic of China and Foreign States*, pp. 521–22. The Russian text has "*Tri Vostochnye provintsii Kitaia*"; see *Sovetsko-kitaiskie otnosheniia*, p. 204.

277. On this paragraph, see Yang Kuisong, "Zhongguo gongchandang duoquan Dongbei," p. 67; Yang Kuisong, *Zhongjian didai de geming*, p. 416; Reardon-Anderson, *Yenan and the Great Powers*, p. 133.

278. Reardon-Anderson, *Yenan and the Great Powers*, p.134.

279. On this paragraph, see instructions of the CC of the CCP to the Northeast Bureau, January 27, 1946, cited in Yang Kuisong, "Zhongguo gongchandang duoquan Dongbei," p. 68; Reardon-Anderson, *Yenan and the Great Powers*, p. 134.

280. On this paragraph, see Westad, *Cold War and Revolution*, pp. 147–48, with source references.

281. Instructions of the CC of the CCP to the Northeast Bureau, January 27, 1946, cited in Yang Kuisong, "Zhongguo gongchandang duoquan Dongbei," p. 68.

282. Instructions of the CC of the CCP to the delegation in Chungking, January 21, 1946, and *Xinhua ribao*, February 16, 1946, both cited in ibid.

283. Clubb, *China and Russia*, pp. 356–57.

284. On this paragraph, see Stalin to Chiang Ching-kuo, December 30, 1949, Soviet protocol, cited in Ledovskii, "Stalin i Chan Kaishi," pp. 118–19; Westad, *Cold War and Revolution*, pp. 118–19, 135; Levine, *Anvil of Victory*, p. 239; Chiang, *Soviet Russia in China*, p. 147; Furuya, *Chiang Kai-shek*, p. 869.

285. On this paragraph, see Yang Kuisong, *Zhongjian didai de geming*, p. 430; Yang Kuisong, "Zhongguo gongchandang duoquan Dongbei," p. 69; Yang Kuisong, "Soviet Factor," pp. 26, 28–29; Helmut Handzik, *Politische Bedingungen sowjetischer Truppenabzüge, Drei Fallstudien: Asien 1925, 1946 und 1948*, p. 130.

286. On this paragraph, see Reardon-Anderson, *Yenan and the Great Powers*, pp. 139–41; telegram from the delegation in Chungking to the CC of the CCP, March 11, 1946, cited in Yang Kuisong, "Zhongguo gongchandang duoquan Dongbei," p. 69; Westad, *Cold War and Revolution*, p. 156.

287. On this paragraph, see Reardon-Anderson, *Yenan and the Great Powers*, pp. 138–41.

288. Yang Kuisong, "Zhongguo gongchandang duoquan Dongbei," p. 69.

289. Ibid.

290. On this paragraph, see Garthoff, "Soviet Intervention in Manchuria," p. 76; Westad, *Cold War and Revolution*, pp. 152, 157, 159; Chiang, *Soviet Russia in China*, pp. 164–65; Levine, *Anvil of Victory*, pp. 78–79; Reardon-Anderson, *Yenan and the Great Powers*, pp. 146, 149, 152; *Liu Shaoqi nianpu*, vol. 2, pp. 20–24.

291. *Liu Shaoqi nianpu*, vol. 2, pp. 114, 142, 144–45, 150.

292. Yang Kuisong explicitly stresses the causal relationship between the Soviet recommendations and the CCP resolution. See Yang Kuisong, "Soviet Factor," p. 27.

293. On this paragraph, see Westad, *Cold War and Revolution*, p. 156; Yang Kuisong, "Soviet Factor," p. 27, with reference to documents from the CCP archives.

294. On this paragraph, see Yang Kuisong, "Zhongguo gongchandang duoquan Dongbei," pp. 69–70; Reardon-Anderson, *Yenan and the Great Powers*, pp. 148–50, 152. A senior Communist officer stationed in Manchuria at the time reports, however, that his comrades and he repeatedly heard from Yenan between March 15 and March 20 that there was still hope for peace. Only on March 21 had the Party leadership telegraphed that it was finally gone. See Reardon-Anderson, *Yenan and the Great Powers*, p. 151.

295. Levine, *Anvil of Victory*, pp. 79–80. According to the report of an American observer in Yenan, Zhou was bitter about the Party leadership's criticism. See *FRUS 1946*, vol. 9, *The Far East: China*, pp. 596, 742, cited in Reardon-Anderson, *Yenan and the Great Powers*, p. 153.

296. On this paragraph, see Reardon-Anderson, *Yenan and the Great Powers*, pp. 152–54.
297. On this paragraph, see ibid., pp. 154–55.
298. On this paragraph, see ibid., pp. 99–100; Westad, *Cold War and Revolution*, pp. 159, 165; Garthoff, "Soviet Intervention in Manchuria," p. 77; Reardon-Anderson, *Yenan and the Great Powers*, pp. 153–56.
299. Westad, *Cold War and Revolution*, p. 162.
300. See Chiang Kai-shek's diary, entry of March 13, 1946, cited in Furuya, *Chiang Kai-shek*, pp. 874–75. Suzanne Pepper puts it in similar terms but somewhat more cautiously; see Pepper, "KMT-CCP Conflict," p. 732. See also Levine, *Anvil of Victory*, p. 149.
301. On this paragraph, see Garthoff, "Soviet Intervention in Manchuria," pp. 76–77; Levine, *Anvil of Victory*, pp. 80–81, 83; Reardon-Anderson, *Yenan and the Great Powers*, pp. 148–49; Westad, *Cold War and Revolution*, p. 162; Handzik, *Drei Fallstudien: Asien 1925, 1946 und 1948*, p. 136.
302. Garthoff represents this opinion. See Garthoff, "Soviet Intervention in Manchuria," p. 77.
303. For further details regarding the Soviet war booty see Chiang, *Soviet Russia in China*, pp. 168–71; Clubb, *China and Russia*, pp. 353–57; Westad, *Cold War and Revolution*, pp. 100–104; Furuya, *Chiang Kai-shek*, pp. 870–73; Garthoff, "Soviet Intervention in Manchuria," p. 74.
304. Garthoff, "Soviet Intervention in Manchuria," pp. 77–78.
305. Lloyd E. Eastman, *Seeds of Destruction*, pp. 223–24, cited in Levine, *Anvil of Victory*, p. 239.
306. On Soviet assistance, see Borisov, *Sovetskii Soiuz i mandzhurskaia revoliutsionnaia baza*, pp. 181–204.
307. Ibid., p. 185.
308. Mao Zedong to Mikoyan, January 31 and February 6 or 7, 1949, AP RF, f. 39, op. 1, d. 39, l. 2, 86, cited in A.M. Ledovskii, "Sekretnaia missiia A.I. Mikoiana v Kitai (ianvar'–fevral' 1949 g.)," pt. 2, pp. 97, 100.
309. Mao Zedong to Mikoyan, February 5, 1949, AP RF, f. 39, op. 1, d. 39, l. 37, cited in ibid., p. 97. On this paragraph, see Levine, *Anvil of Victory*, pp. 239–40.
310. On this paragraph, see Sladkovskii, *Istoriia torgovo-ekonomicheskikh otnoshenii*, pp. 169–83; Stolberg, *Stalin und die chinesischen Kommunisten*, pp. 104–7, with further numbers and source references.
311. Li Lisan, for example, observed that the war booty from Manchuria was not worth much in comparison with the war losses suffered by the Soviet Union (in Europe); see Garthoff, "Soviet Intervention in Manchuria," p. 74.
312. On this paragraph, see Clubb, *China and Russia*, pp. 355–56; Stolberg, *Stalin und die chinesischen Kommunisten*, pp. 88–92, with further numbers and source references.
313. For further details, see this chapter above.
314. On this paragraph, see *China White Paper*, pp. 105–12; Bagby, *Eagle-Dragon Alliance*, p. 218.
315. For further details, see this chapter above.
316. On this paragraph, see instruction of the CC of the CCP, November 28, 1945, cited in Westad, *Cold War and Revolution*, p. 129.
317. On this paragraph, see Mao Zedong to Zhou Enlai and Wang Ruofei, November 7, 1945, and CC of the CCP, November 28, 1945, both cited in Westad, *Cold War and Revolution*, pp. 128–29.
318. For further details, see this chapter above.
319. On this paragraph, see Tuchman, *Sand Against the Wind*, pp. 526–27; Westad, *Cold War and Revolution*, pp. 131–33.
320. Ivan Yeaton to Albert C. Wedemeyer, December 20, 1945, cited in Reardon-Anderson, *Yenan and the Great Powers*, p. 128.
321. Yang Kuisong, *Zhongjian didai de geming*, pp. 417–18. On this paragraph, see Westad, *Cold War and Revolution*, pp. 137–38.
322. On this paragraph, see Yang Kuisong, *Zhongjian didai de geming*, p. 418; Westad, *Cold War and Revolution*, pp. 137–38. On the resumption of negotiations, see also this chapter above.
323. On this paragraph, see Clubb, *China and Russia*, pp. 354–55; Westad, *Cold War and Revolution*, 134–35, 141–42; Murray, *Western versus Chinese Realism*, pp. 160–68.
324. This is the Soviet protocol, which is stored in the Russian presidential archives (AP RF, f. 45, op. 1, ll. 98–121, 123–40), published in Ledovskii, "Stalin i Chan Kaishi," pp. 109–28. The author possesses a copy of the original. The Chinese protocol apparently is not yet accessible, but Chiang Ching-kuo often discussed the talks. See Murray, *Western versus Chinese Realism*, p. 171 (with references to the Chinese sources), and Furuya, *Chiang Kai-shek*, p. 869. For speculation about why the Chinese protocols are not accessible, see the discussion later in this section.
325. On this paragraph, see Murray, *Western versus Chinese Realism*, pp. 169–71, where the contents of

the instructions are reproduced from the files of the AFMROC. See also Westad, *Cold War and Revolution*, pp. 140–41.

326. Ledovskii, "Stalin i Chan Kaishi," p. 104. Ledovskii was employed as a diplomat at the Soviet embassy in Chungking at the time.

327. On this paragraph, see Howard L. Boorman and Richard C. Howard, eds., *Biographical Dictionary of Republican China*, vol. 1, pp. 305–9; Valentin Zaitsev, "Liubimaia mnogotirazhka prezidenta Taivania," pp. 42–45. According to a document from the NKVD archives, Chiang informed on a Soviet citizen for being a "Trotskyite" and "enemy of the people"; ibid., p. 44. Unlike what Zaitsev assumes, this could, however, be seen as something other than evidence that Chiang left the USSR as a "firmly convinced Bolshevik." It is more probable that he behaved this way after he had himself been denounced as a Trotskyite in order to demonstrate his loyalty to Soviet power and not to endanger his exit visa at the last moment in this period of gruesome political persecution.

328. On this paragraph, see Ledovskii, "Stalin i Chan Kaishi," pp. 103, 109; Chiang, *Soviet Russia in China*, p. 147.

329. The report has only recently been made accessible and is in the Russian presidential archives (AP RF, f. 3, op. 86, d. 146, ll. 20–26). The author possesses a copy of the original. The report was published in Ledovskii, "Stalin i Chan Kaishi," pp. 105–8.

330. On this and three following paragraphs, see Ledovskii, "Stalin i Chan Kaishi," pp. 105–8, passim.

331. This is also admitted by Ledovskii; see ibid., p. 107 n. 14.

332. The contents of the letter are reproduced from documents in the AFMROC in Murray, *Western versus Chinese Realism*, p. 169.

333. On this paragraph, see the Soviet protocol of the talk on December 30, 1945, cited in Ledovskii, "Stalin i Chan Kaishi," p. 109 n. 3.

334. On this paragraph, see ibid., pp. 110–11.

335. On this paragraph, see ibid., p. 111.

336. On this paragraph, see ibid., pp. 111–14.

337. On military aid during the 1920s, see Heinzig, *Sowjetische Militärberater*, pp. 99–287.

338. On this paragraph, see Ledovskii, "Stalin i Chan Kaishi," pp. 114–16.

339. Ledovskii assumes that this was a misunderstanding by the translator V. Pavlov, who drew up the protocol, because the Soviet Army only helped the KMT "in words," but actually supported the Communists; see Ledovskii, "Stalin i Chan Kaishi," p. 116 n. 28. Ledovskii ignores here that Moscow had changed course temporarily in mid-November and handed over at least the largest cities to the KMT; see this chapter above (section "Moscow's Double Game in Manchuria").

340. On this paragraph, see ibid., pp. 116–17.

341. On this paragraph, see ibid., pp. 117–18.

342. On this paragraph, see ibid., pp. 118–19.

343. On this and the following paragraph, see ibid., p. 120.

344. On this and the following paragraph, see ibid., pp. 121–22.

345. On this paragraph, see ibid., pp. 123–24.

346. On this paragraph, see ibid., p. 125.

347. On this paragraph, see ibid., p. 126.

348. On this paragraph, see ibid., pp. 125–28.

349. On this paragraph, see ibid., p. 129. The actual text of the letter can also be found here.

350. See chapter 1 above.

351. See Yang Kuisong, "Zhongguo gongchandang duoquan Dongbei," p. 67.

352. Ibid.

353. Furuya, *Chiang Kai-shek*, p. 869, with no source references.

354. On this and the two following sentences, see Chiang, *Soviet Russia in China*, pp. 147–48.

355. Text in Ledovskii, "Stalin i Chan Kaishi," p. 129.

356. Instruction of the CC of the CCP, January 3, 1946, cited in Yang Kuisong, "Soviet Factor," p. 25.

357. On this paragraph, see ibid., with source references.

358. Zhou Enlai to the CC of the CCP, January 28, 1946, cited in ibid., p. 26.

359. See Levine, *Anvil of Victory*, p. 239.

360. On this paragraph, see Furuya, *Chiang Kai-shek*, pp. 862–64; Pepper, "KMT-CCP Conflict," pp. 729–30; Westad, *Cold War and Revolution*, pp. 143–44; Reardon-Anderson, *Yenan and the Great Powers*, p. 135.

361. Yang Kuisong, *Zhongjian didai de geming*, p. 422.

362. Ibid.
363. Ibid., p. 419.
364. Instruction of the CC of the CCP on the cease-fire, January 12, 1946, cited in ibid., p. 420. On this paragraph, see Westad, *Cold War and Revolution*, p. 147, with source references.
365. On this paragraph, see Yang Kuisong, *Zhongjian didai de geming*, p. 422.
366. Ibid., p. 423.
367. Instruction of the propaganda department of the CC of the CCP, January 13, 1946, cited in ibid., p. 420.
368. Instructions of the CC of the CCP, February 1, 1946, cited in ibid., p. 424.
369. On this paragraph, see *China White Paper*, pp. 140–43, 622–26; Furuya, *Chiang Kai-shek*, pp. 864–65; Westad, *Cold War and Revolution*, p. 155; Yang Kuisong, "Soviet Factor," p. 27.
370. On this paragraph, see Chiang, *Soviet Russia in China*, p. 162; Westad, *Cold War and Revolution*, p. 155.
371. On this paragraph, see Yang Kuisong, *Zhongjian didai de geming*, pp. 426–27; Yang Kuisong, "Soviet Factor," pp. 26–27.
372. On this paragraph, see Yang Kuisong, *Zhongjian didai de geming*, p. 427.
373. On this and the following sentence, see message from the CC of the CCP to the Northeast Bureau of the CCP, February 12, 1946, cited in Yang Kuisong, "Soviet Factor," p. 27.
374. On this paragraph, see Nancy Bernkopf Tucker, *Patterns in the Dust*, pp. 45–46, with source references; Chiang Kai-shek, diary entry of February 1, 1946, cited in Furuya, *Chiang Kai-shek*, p. 865; Reardon-Anderson, *Yenan and the Great Powers*, p. 135.
375. On this paragraph, see Levine, *Anvil of Victory*, pp. 78–79.
376. On this paragraph, see Yang Kuisong, *Zhongjian didai de geming*, p. 439, and this chapter above (section "Moscow's Double Game in Manchuria").
377 *Mao Zedong xuanji*, vol. 4, pp. 1181–82; *Selected Works of Mao Tse-tung*, vol. 4, pp. 87–88.
378. Instruction of the CC of the CCP to Zhou Enlai, June 30, 1946, cited in Yang Kuisong, *Zhongjian didai de geming*, p. 437.
379. See ibid., pp. 440–41.
380. On this paragraph, see Reardon-Anderson, *Yenan and the Great Powers*, pp. 158–59; Garthoff, "Soviet Intervention in Manchuria," p. 77.
381. On this and the following sentence, see CC of the CCP to Ruo Ronghuan, June 30, 1946, cited in Yang Kuisong, "Soviet Factor," p. 29.
382. See also Clubb, *China and Russia*, p. 357.
383. *Jiefang ribao*, September 12, 1946, cited in Yang Kuisong, "Soviet Factor," p. 28.
384. Mao Zedong's talk with Anna Louise Strong in August 1946, in *Selected Works of Mao Tse-tung*, vol. 4, pp. 99.
385. Mao Zedong's talk with Liu Shaoqi and Zhou Enlai on November 21, 1946, cited in Yang Kuisong, *Zhongjian didai de geming*, p. 438. On this paragraph, see also pp. 442–43.
386. On this paragraph, see Mao Zedong, October 1, 1946, in *Selected Works of Mao Tse-tung*, vol. 4, pp. 113–14; Yang Kuisong, *Zhongjian didai de geming*, pp. 443–44; Mao Zedong's talk with Liu Shaoqi and Zhou Enlai, November 21, 1946, cited in Yang Kuisong, *Zhongjian didai de geming*, p. 444.
387. On this paragraph, see Yang Kuisong, *Zhongjian didai de geming*, p. 444. Mao had already said something similar on September 29, 1946; see *Selected Works of Mao Tse-tung*, vol. 4, pp. 109–10.
388. Yang Kuisong, *Zhongjian didai de geming*, p. 444.
389. On this paragraph, see *FRUS 1947*, vol. 7, *The Far East: China*, passim, cited in Alexander, *Strange Connection*, pp. 72, 75 n. 7; *FRUS 1948*, vol. 7, *The Far East: China*, passim, cited in Alexander, *Strange Connection*, pp. 78, 87 n. 1; *FRUS 1949*, vol. 8, *The Far East: China*, pp. 264, 269; *China White Paper*, pp. 305, 307; Garthoff, "Soviet Intervention in Manchuria," pp. 77–80; Clubb, *China and Russia*, pp. 357–60; Sladkovskii, *Istoriia torgovo-ekonomicheskikh otnoshenii*, p. 168; S.L. Tikhvinskii, "Perepiska I.V. Stalina s Mao Tszedunom v ianvare 1949 g," p. 133.
390. Yang Kuisong, "Soviet Factor," p. 29.
391. See also He Di, "The Evolution of the Chinese Communist Party's Policy Toward the United States, 1944–1949," p. 45. Hunt assumes unconvincingly that intensive efforts to achieve closer cooperation with Moscow only began in the summer of 1948. See Michael H. Hunt, *The Genesis of Chinese Communist Foreign Policy*, p. 178.
392. PLA manifesto, October 10, 1947, in *Mao Zedong xuanji*, vol. 4, p. 1235; *Selected Works of Mao Tse-tung*, vol. 4, p. 147.

393. On this paragraph, see also Bo Yibo, *Ruogan zhongda juece yu shibian de huigu*, vol. 1, p. 36; Yang Kuisong, "Soviet Factor," p. 32.

394. On this paragraph, see Günther Nollau, *International Communism and World Revolution*, pp. 216–25; Hans Wassmund, *Grundzüge der Weltpolitik: Daten und Tendenzen von 1945 bis zur Gegenwart* (Munich: 1982), pp. 34–36.

395. Yang Kuisong, *Zhongjian didai de geming*, p. 446.

396. I.V. Kovalev, "Dialog Stalina s Mao Tszedunom," pt. 1, pp. 79–80. For further details, see chapter 3 below (section "The Second Top Level Meeting"). On this paragraph, see Dedijer, *Tito Speaks*, pp. 300–301.

397. On this and the two following sentences, see Yang Kuisong, *Zhongjian didai de geming*, p. 445.

398. Manifesto of the Chinese People's Liberation Army, October 10, 1947, in *Mao Zedong xuanji*, vol. 4, p. 1235; *Selected Works of Mao Tse-tung*, vol. 4, p. 147.

399. Instruction of the CC of the CCP, October 27, 1947, cited in Yang Kuisong, *Zhongjian didai de geming*, p. 445.

400. On this paragraph, see Yang Kuisong, "Soviet Factor," pp. 28–29; Yang Kuisong, *Zhongjian didai de geming*, pp. 444–46.

401. He Di, "Evolution," p. 45; Liu Shaoqi to the *Pravda* journalist O. Chechetkina, January 27, 1950, AVP RF, f. 0100, op. 43, d. 11, l. 22–23.

402. Radio Moscow, December 1, 1947, reproduced in *Renmin ribao*, December 7, 1947, cited in Xiang Qing, You Zhifu, and Liu Dexi, eds., *Sulian yu Zhongguo geming*, p. 555. See also pp. 555–56.

403. *Renmin ribao*, January 1, 1948, cited in *Mao Zedong ji*, vol. 10, pp. 97–115 passim, esp. pp. 97–98, 114–16. The article reproduces a report that Mao delivered at a meeting of the CC of the CCP on December 25, 1947; see ibid., p. 97. It was not included in Mao's *Selected Works*. Cf. *Mao Zedong xuanji*, vol. 4, pp. 1243–60; *Selected Works of Mao Tse-tung*, vol. 4, pp. 157–73. On this and Qiao Guanhua's previously published articles, see Nagai Yonosuke and Iriye Akira, *The Origins of the Cold War in Asia*, pp. 238–39.

404. Peter M. Kuhfus, "'Ein Held, der mit der Zeit geht,'" pp. 232–33.

405. Mao Zedong to Ambassador Iudin, March 31, 1956, Soviet record of the talk, in *PDV* 5 (1994), p. 105.

406. On this paragraph, see *Mao Zedong tongzhi shi dangdai zui weida de Ma-ke-si Lie-ning zhuyi zhe*, p. 432; Wu Xiuquan, *Huiyi yu huainian*, p. 235; Ronald L. McGlothlen, *Controlling the Waves*, pp. 136–37; Yugoslav deputy foreign minister Alois Bebler to a Western diplomat, cited in Elisabeth Barker, *The British Between the Superpowers, 1945–50*, p. 172.

407. Resolution of the CC of the CCP, July 10, 1948, cited in Xiang, You, and Liu, *Sulian yu Zhongguo geming*, p. 560.

408. Instructions of the CC of the CCP on the problem of the Yugoslav Communist Party, August 4, 1948, cited in Yang Kuisong, "Soviet Factor," p. 29, based on archival documents. On the effects of the conflict with Yugoslavia on CCP policy, see also Wang Jin, "Lüe lun xin Zhongguo chengli chuqi de dui Su 'yi bian dao' zhengce," pp. 110–11.

409. See also Xiang, You, and Liu, *Sulian yu Zhongguo geming*, pp. 561–62.

410. Mao Zedong, "Quan shijie geming liliang tuanjie qilai, fandui zhuyi de qinlüe," in *Renmin ribao*, November 7, 1948, cited in *Mao Zedong ji*, vol. 10, p. 186. The article appeared in the Cominform newspaper *Für dauerhaften Frieden, für Volksdemokratie* 24, no. 21 (November 1, 1948). The cited passage is included in loose translation there. It was left out of vol. 4 of Mao's *Selected Works*, which appeared in 1960—apparently because the CCP leadership in the meantime wanted more than before to follow its own path independent of the USSR.

411. Liu Shaoqi, "Lun guoji zhuyi yu minzu zhuyi," dated November 1, 1948, cited in *Liu Shaoqi xuanji*, pp. 244–45. According to a report by the American embassy in Moscow, June 27, 1949, the article had been reproduced "recently" in *Pravda*; see *FRUS 1949*, vol. 8. *The Far East: China*, p. 400. But it cannot be found in *Pravda*.

412. These questions were still raised in China at the end of 1949. A publication appearing at the time in Shanghai stated that the factories had been dismantled so that they would not fall into the hands of the KMT. Oral communication of a Chinese interpreter with the Harbin Association for Chinese Soviet Friendship to an official of the Soviet Consulate General in Harbin, December 28, 1949, AVP RF, f. 07, op. 43, d. 13, p. 302, l. 15.

413. Liu Shaoqi to the *Pravda* journalist Chechetkina, January 27, 1950, AVP RF, f. 0100, op. 43, d. 11, ll. 22–23; see also Kovalev, "Dialog Stalina s Mao Tszedunom," pt. 2, p. 88.

414. Radio North Shaanxi, March 18, 1949, cited in *FRUS 1949*, vol. 8, *The Far East: China*, pp. 192–93.

415. Telegram from Ambassador Stuart to the American Secretary of State, March 22, 1949, in ibid., pp.

192–93. The Truman administration ceased to anticipate a "Titoist China" after March 1949. See McGlothlen, *Controlling the Waves*, pp. 138–39, 158.

416. Kovalev, "Dialog Stalina s Mao Tszedunom," pt. 2, p. 86.

417. Stalin to Mao Zedong, June 15 and July 1, 1947, cited in Ledovskii, "Sekretnaia missiia A.I. Mikoiana v Kitai," pt. 1, p. 98; A.M. Ledovskii, "Vizit Mao Tszeduna v Sovetskii Soiuz," v dekabre 1949–fevral' 1950 g.," pt. 1, p. 1.

418. See Shi Zhe, *Zai lishi juren shenbian*, p. 315; Shi Zhe and Chen Jian, "With Mao and Stalin," chapter 6, p. 5. For further details about Orlov, see chapter 1 above.

419. Mao Zedong to Stalin, April 26 and May 10, 1948, Stalin to Mao Zedong, April 29 and May 10, 1948, cited in Ledovskii, "Sekretnaia missiia A.I. Mikoiana v Kitai," pt. 1, p. 98. See also *Zhonghua renmin gongheguo waijiao shi, 1949–1956*, p. 10. Shi Zhe reports that the CCP leadership resolved at Mao's suggestion in mid-April 1948 that Mao should visit the Soviet Union and that Stalin requested a postponement in May. See Shi Zhe, *Zai lishi juren shenbian*, pp. 366–67, 372, 374; Shi Zhe, "Peitong Mao zhuxi fang Su," *Renwu* 5 (1988), p. 6.

420. Kovalev, "Dialog Stalina s Mao Tszedunom," pt. 2, p. 77.

421. Mao Zedong to Stalin, July 4, 1948, and July 14, 1948, and Stalin to Mao Zedong, July 14, 1948, cited in Ledovskii, "Sekretnaia missiia A.I. Mikoiana v Kitai," pt. 1, pp. 98–99.

422. Orlov's report on Mao's reactions is in his telegram to Stalin, July 14, 1948, AP RF, f. 39, d. 31, l. 40, cited in ibid., p. 99; *Wang Dongxing riji*, p. 151. Wang Dongxing gives the mistaken impression that Stalin invited Mao in April 1948.

423. On this paragraph, see telegram from Orlov to Stalin, August 28, 1948, AP RF, f. 39, op. 1, d. 31, l. 41, cited in Ledovskii, "Sekretnaia missiia A.I. Mikoiana v Kitai," pt. 1, pp. 99–100.

424. The Russian text used the term *ukazaniia.* The Chinese original probably read *zhish.*

425. Telegrams from Mao Zedong to Stalin, September 28, October 16, November 21, and December 30, 1948, cited in Ledovskii, "Sekretnaia missiia A.I. Mikoiana v Kitai," pt. 1, p. 100, and according to Bo, *Ruogan zhongda juece*, vol. 1, pp. 36–37.

Notes to Chapter 3

1. *Military Campaigns in China*, pp. 165–66.

2. In January 1955, Zhou Enlai explained to the Chinese ambassador-designate in Moscow, Liu Xiao: "The Soviet Union's opinion of our ability to liberate all of China remained very reserved for a long time." See Liu Xiao, *Chushi Sulian ba nian*, pp. 4–5.

3. For a detailed discussion, see chapter 2 above.

4. I.V. Kovalev, "Dialog Stalina s Mao Tszedunom," pt. 2, p. 86. An English version appeared in *Far Eastern Affairs* 2 (1992), pp. 94–111. Cf. Chen Jian, China's Road to the Korean War, p. 9.

5. Telegram from Stalin (under code name: Filippov) to Mao Zedong, January 10, 1949, AP RF, cited in S.L. Tikhvinskii, "Perepiska I.V. Stalina s Mao Tszedunom v ianvare 1949 g.," p. 134.

6. On this paragraph, see *Zapiska A.I. Mikoiana v Prezidium TsK KPSS o poezdke v Kitai v ianvare–fevrale 1949 g.*, cited in A.M. Ledovskii, "Sekretnaia missiia A.I. Mikoiana v Kitai (ianvar'–fevral' 1949 g.)," pt. 1, p. 101.

7. Telegram from Stalin (under code name: Filippov) to Mao Zedong, January 14, 1949, AP RF, f. 45, op. 1, d. 330, l. 113, cited in Tikhvinskii, "Perepiska I.V. Stalina s Mao Tszedunom," p. 138. A similar text is also presented in *Zhonghua renmin gongheguo waijiao shi, 1949–1956*, p. 10. The Chinese text is obviously reproduced falsely for prestige reasons and states that Stalin said "agree" (instead of "insist"). In a version of the final paragraph of the telegram also found in the presidential archives, but in another source, we find "the questions that interest us" rather than "the questions that interest you" AP RF, f. 39, op. 1, d. 31, l. 72, cited in A.M. Ledovskii, "Vizit Mao Tszeduna v Sovetskii Soiuz v dekabre 1949–fevral' 1950 g.," pt. 1, p. 3. Bo Yibo made a claim—one that is not very credible in the light of the two telegrams from Stalin—that transport problems and the need to command two military campaigns were the reasons Mao's trip did not take place at that time. See Bo Yibo, *Ruogan zhongda juece yu shibian de huigu*, vol. 1, p. 37.

8. Zhou Enlai to Roshchin, November 10, 1949, AVP RF, f. 07, op. 22, d. 220, l. 53.

9. For a detailed discussion, see chapter 4 below.

10. *Mao Zedong nianpu, 1893–1949*, vol. 3, p. 439. Cf. *Zhonghua renmin gongheguo waijiao shi*, p. 10.

11. For details of the Mikoyan visit, see this chapter below.

12. Minister-Counselor Clark in Canton to Acheson, February 5, 1949, in *FRUS 1949*, vol. 8, *The Far East: China*, p. 107; A.M. Ledovskii, "Sekretnaia missiia A.I. Mikoiana v Kitai," pt. 2, pp. 95–96.

13. Fedorenko's record of a discussion with embassy counselor Chen Ting, January 15, 1949, AVP RF, f. 07, op. 22, d. 220.

14. Ledovskii, "Sekretnaia missiia A.I. Mikoiana v Kitai," pt. 2, p. 95.

15. On this and the following sentence, see A.M. Ledovskii, "Nankinskie zapiski," p. 42.

16. Clark to Acheson on April 15, 1949, in *FRUS 1949*, vol. 8, *The Far East: China*, p. 240.

17. According to the then Soviet consul general in Peiping, Sergei Tikhvinskii, on March 24, 1995, to the author; see also the then head of the Nanking branch of the Soviet embassy, Andrei Ledovskii, in Ledovskii, "Nankinskie zapiski," p. 41.

18. Stuart to Acheson, February 3, 1949, in *FRUS 1949*, vol. 8, *The Far East: China*, p. 104.

19. For further details, see this chapter below.

20. Telegram from Kovalev to Stalin, May 23, 1949, AP RF, cited in S.L. Tikhvinskii, "O 'sekretnom demarshe' Chzhou En'laia i neofitsial'nykh peregovorakh KPK s amerikantsami v iiune 1949 g.," p. 136.

21. Telegram from Stalin to Kovalev, May 26, 1949, in ibid. Unfortunately—for whatever reasons—Tikhvinskii reproduces only this one sentence, however significant. On this paragraph, cf. *The China White Paper*, p. 307.

22. Telegram from the CC of the CCP to the Nanking Municipal Committee, June 3, 1949, cited in *Zhonggong zhongyang zai Xiangshan*, p. 433.

23. See this chapter above (section "Rethinking in Moscow").

24. Zhou's discussion with Ambassador Liu Xiao took place in late January 1955. See Liu Xiao, *Chushi Sulian ba nian*, pp. 4–5.

25. Ledovskii, "Sekretnaia missiia A.I. Mikoiana v Kitai," pt. 2, p. 96.

26. Mao Zedong to Ambassador Iudin, July 22, 1958, Chinese discussion notes, in *Dang de wenxian*, 1 (1994), p. 16.

27. On this statement and the next sentence, see Ledovskii, "Nankinskie zapiski," p. 41.

28. Ibid., p. 13; Ledovskii to the author, March 27, 1995.

29. Stuart to the Secretary of State, April 12, 1949, in *FRUS 1949*, vol. 8, *The Far East: China*, p. 678, and in Kenneth W. Rea and John C. Brewer, ed., *The Forgotten Ambassador*, p. 316. The garbled language reflects the telegraphic style of the original document.

30. Shi Zhe, *Zai lishi juren shenbian*, p. 418.

31. Shi Zhe to the author, May 1993.

32. Ledovskii to the author, March 27, 1995.

33. Ibid.

34. Tikhvinskii to the author, March 24, 1995.

35. For further details, see this chapter below.

36. Kovalev, "Dialog Stalina s Mao Tszedunom," pt. 1, pp. 87–88. An English version appeared in *Far Eastern Affairs* 1 (1992), pp. 100–116.

37. Melvyn P. Leffler, *A Preponderance of Power*, pp. 222–24.

38. Kovalev, "Dialog Stalina s Mao Tszedunom," pt. 1, p. 88. Kovalev, who, as far as one can tell, identified with Stalin's China policy until his death, notes at this point that despite his otherwise accurate analysis of the situation, Stalin's observation about the Russian people had a demogogical aftertaste. We can only agree with this.

39. See Ledovskii, "Nankinskie zapiski," pp. 14, 16–92 passim. For further details, see this chapter below.

40. The Russian language texts of Stalin's telegrams to Mao Zedong, January 10 and 11, 1949, cited in Tikhvinskii, "Perepiska I.V. Stalina s Mao Tszedunom," pp. 133–35, and Ledovskii, "Sekretnaia missiia A.I. Mikoiana v Kitai," pt. 1, pp. 102–3; (incomplete) Chinese texts of the telegrams in Yu Zhan and Zhang Guangyou, "Guanyu Si-da-lin ceng fou quanzu wo guo Changjiang de tantao," p. 19. An English-language version of the exchange of telegrams between Stalin and Mao appeared in *CWIHPB*, 6–7 (Winter 1995–1996), pp. 27–29. Cf. Odd Arne Westad, "Rivals and Allies: Stalin, Mao, and the Chinese Civil War, January 1949," ibid., pp. 7, 27. Cf. also the presentation of the efforts to arrange peace negotiations from the perspective of the Soviet embassy in China in *Posol'stvo SSSR v Kitaiskoi Narodnoi Respublike, Otchetnyi doklad o polozhenii v Kitae za 1949 god*, with an accompanying message by P.A. Shibaev to the Soviet Foreign Ministry, March 8, 1959, AVP RF, f. 0100, op. 43, p. 103, d. 18, ll. 10–17; the annual report was 220 pages long.

41. On this and the following paragraph, see Stalin's telegram to Mao Zedong, January 10, 1949, AP RF, cited in Tikhvinskii, "Perepiska I.V. Stalina s Mao Tszedunom," p. 134, and Ledovskii, "Sekretnaia missiia A.I. Mikoiana v Kitai," pt. 1, p. 103.

42. On this and the following paragraph, see Stalin's telegram to Mao Zedong, January 11, 1949, cited in Tikhvinskii, "Perepiska I.V. Stalina s Mao Tszedunom," pp. 134–35, and Ledovskii, "Sekretnaia missiia A.I. Mikoiana v Kitai," pt. 1, p. 103.

43. On this and the following quotation, telegram from Mao Zedong to Stalin, January 11, 1949 (Chinese version), respectively January 12 or 13, 1949 (Russian versions). Chinese text in Yu and Zhang, "Guanyu Si-da-lin," p. 20; Russian text in Tikhvinskii, "Perepiska I.V. Stalina s Mao Tszedunom," pp. 135–36, and Ledovskii, "Sekretnaia missiia A.I. Mikoiana v Kitai," pt. 1, p. 103. The fact that different dates are given is usually explained in terms of the different time zones in which Moscow and Peking are located.

44. Telegrams from Stalin to Mao Zedong, January 14 and January 15, 1949, and telegram from Mao Zedong to Stalin, January 14, 1949, cited in Tikhvinskii, "Perepiska I.V. Stalina s Mao Tszedunom," pp. 136–39, and Ledovskii, "Sekretnaia missiia A.I. Mikoiana v Kitai," pt. 1, pp. 103–4.

45. Cf. also *FRUS 1949*, vol. 8, *The Far East: China*, pp. 41–42; *Selected Works of Mao Tse-tung*, vol. 4, pp. 315–19.

46. Cited in Tikhvinskii, "Perepiska I.V. Stalina s Mao Tszedunom," p. 138.

47. Ibid., p. 139.

48. See Vladislav Zubok and Constantine Pleshakov, *Inside the Kremlin's Cold War*, p. 57.

49. Yu and Zhang, "Guanyu Si-da-lin," p. 19. A similar interpretation can be found in Yang Kuisong, *Zhongjian didai de geming*, p. 447.

50. Tikhvinskii, "Perepiska I.V. Stalina s Mao Tszedunom," p. 138.

51. This is reported by Ledovskii, who was first secretary of the Soviet embassy in Nanking at the time. See Ledovskii, "Sekretnaia missiia A.I. Mikoiana v Kitai," pt. 2, pp. 94–95.

52. *Sovetsko-kitaiskie otnosheniia, 1917–1957*, p. 209.

53. Text of the U.S. government's aide-mémoire in *FRUS 1949*, vol. 8, *The Far East: China*, pp. 41–42; see also ibid., pp. 38–39, 47–48.

54. Ibid., pp. 25, 38–39.

55. AP RF, f. 39, op. 1, d. 39, p. 47, cited in Ledovskii, "Sekretnaia missiia A.I. Mikoiana v Kitai," pt. 2, p. 95.

56. According to a Soviet publication, Liu Shaoqi and Gao Gang traveled to the USSR at the head of a delegation from the "people's democratic areas in Manchuria" "in 1945," where they met with the Soviet leadership. No details or sources are indicated. The trip is not confirmed by any other source. See O.B. Borisov and B.T. Koloskov, *Sovetsko-kitaiskie otnosheniia, 1945–1970*, p. 28. It is possible that this reflects a confusion with the trip to Moscow by a delegation led by Liu and Gao in the summer of 1949; see further details in this chapter below.

57. For further details, see chapter 2 and this chapter above.

58. "Memuary Nikity Sergeevicha Khrushcheva," p. 72. Khrushchev mistakenly indicates that Mikoyan's trip took place during the war.

59. Shi Zhe, *Zai lishi juren shenbian*, pp. 372–88; Kovalev, "Dialog Stalina s Mao Tszedunom," pt. 1, pp. 85–86.

60. Ledovskii, "Sekretnaia missiia A.I. Mikoiana v Kitai," pp. 100–10; pt. 1, ibid., pt. 2, pp. 94–105. An English-language version appeared in *Far Eastern Affairs* 2 (1995), pp. 72–94, and 3 (1995), pp. 74–90. It is not always clear from Ledovskii's article whether he based his description of the talks in Xibaipo on the telegrams Mikoyan sent to Stalin or on other contemporary documents—such as, for example, minutes of the talks. He also only cites from a small portion of the documents available to him, which consisted of more than seventy pages.

61. Shi Zhe, *Zai lishi juren shenbian*, p. 372; *Wang Dongxing riji*, p. 153; Ledovskii, "Sekretnaia missiia A.I. Mikoiana v Kitai," pt. 1, p. 104. Shi Zhe and Wang Dongxing both agree in saying that they were the only ones to pick up Mikoyan. Mikoyan claims—possibly in an attempt to increase his status in Moscow—that he was picked up by "commander in chief Zhu De" and "Politburo member Ren Bishi" in addition to the interpreter, Shi Zhe.

62. On this paragraph, see Ledovskii, "Sekretnaia missiia A.I. Mikoiana v Kitai," pt. 1, p. 104; Shi Zhe, *Zai lishi juren shenbian*, pp. 372–75, 388; Kovalev, "Dialog Stalina s Mao Tszedunom," pt. 1, pp. 85–86; *Zhonghua renmin gongheguo waijiao shi*, p. 11; *Mao Zedong nianpu*, vol. 3, p. 448; *Zhou Enlai nianpu, 1898–1949*, p. 810. See also Chen Jian, "The Sino-Soviet Alliance and China's Entry into the Korean War," pp. 6–7, with further references. See further details on Kovalev's activity in this chapter below. Shi Zhe claims that he interpreted alone because of Evgenii Kovalev's poor command of Chinese and that Mikoyan was so angry about this that he threatened to beat Kovalev. See Shi Zhe, "Peitong Mao zhuxi fang Su," *Renwu* 5 (1988), p. 6. According to Shi Zhe's information, Mikoyan arrived in Xibaipo on January 31. This

cannot be correct; according to Mikoyan, who based his account on archival documents, his first talk with Mao had already taken place on January 30. See Ledovskii, "Sekretnaia missiia A.I. Mikoiana v Kitai," pt. 2, p. 96.

63. Arlen Meliksetov, "1940s . . . ," pp. 32–33. Meliksetov even believed this to be "one of the main reasons."

64. Ledovskii, "Sekretnaia missiia A.I. Mikoiana v Kitai," pt. 1, p. 109.

65. For further details, see chapter 2 and this chapter above.

66. On this paragraph, see Shi Zhe, *Zai lishi juren shenbian*, p. 374; *Zhonghua renmin gongheguo waijiao shi*, pp. 10–11; *Zhou Enlai zhuan, 1898–1949*, pp. 742–43. Shi Zhe mistakenly assumed that the first meeting took place on January 31. Cf. Ledovskii, "Sekretnaia missiia A.I. Mikoiana v Kitai," pt. 2, p. 96.

67. For further details, see Ledovskii, "Sekretnaia missiia A.I. Mikoiana v Kitai," pt. 2, pp. 96–97.

68. Ibid., pt. 1, p. 109; ibid., pt. 2, p. 95; *Zhou Enlai nianpu, 1898–1949*, p. 743; *Zhou Enlai zhuan*, p. 743; Zhu Yuanshi, "Liu Shaoqi 1949 nian mimi fang Su," p. 75, with no indication of sources; Kovalev, "Dialog Stalina s Mao Tszedunom," pt. 2, pp. 95–96; *Zhonghua renmin gongheguo waijiao shi*, p. 11; Shi Zhe, *Zai lishi juren shenbian*, pp. 386–87.

69. On this paragraph, see Shi Zhe, *Zai lishi juren shenbian*, pp. 375, 385; Shi Zhe, quoted in Li Haiwen, "Mao Zedong Liu Shaoqi fang Su ji Zhong-Su liang dang guanxi zhong de yixie wenti," p. 93; Zhu Yuanshi, "Liu Shaoqi 1949 nian mimi fang Su," p. 75, with no indication of sources; *Zhonghua renmin gongheguo waijiao shi*, pp. 10–11.

70. See more on this in this same section below.

71. On this paragraph, see Kovalev, "Dialog Stalina s Mao Tszedunom," pt. 1, p. 86.

72. For the details of the circumstances of Roshchin's move to Canton, see this chapter above.

73. On this section, see Ledovskii, "Sekretnaia missiia A.I. Mikoiana v Kitai," pt. 2, pp. 95–96. For further details, see this chapter above (section "Affront to the CCP").

74. Ibid., p. 95. For further details, see this chapter above (section "Moscow and the CCP at the Beginning of 1949").

75. AP RF, f. 3, op. 65, d. 606, ll. 1–17, cited in ibid., pt. 1, pp. 100–110 (henceforth: Mikoyan report, September 22, 1960).

76. On this paragraph, see Mikoyan report, September 22, 1960, in ibid., p. 105; AP FR, f. 39, op. 1, d. 39, l. 69, cited in ibid., p. 102.

77. AP RF, f. 39, op. 1, d. 39, l. 92, cited in ibid., pt. 2, p. 102.

78. Ibid.

79. On this and the following sentence, see Mikoyan report, September 22, 1960, in ibid., pt. 1, p. 105.

80. Mao to Mikoyan, January 30, 1949, cited in ibid., pt. 2, p. 96.

81. On this and the following sentence, see *Mao Zedong nianpu*, vol. 3, p. 448; Shi Zhe, *Zai lishi juren shenbian*, p. 375.

82. On this paragraph, see Shi Zhe's communication to the author, March 16, 1996.

83. "Memuary Nikity Sergeevicha Khrushcheva," p. 73.

84. Ibid. Khrushchev mistakenly places Mikoyan's visit during wartime.

85. Stalin to Chiang Ching-kuo, January 3, 1946; see A.M. Ledovskii, "Stalin i Chan Kaishi," p. 121.

86. Telegram from Mikoyan to Stalin, February 5, 1949, cited in Ledovskii, "Sekretnaia missiia A.I. Mikoiana v Kitai," pt. 2, p. 102.

87. On this paragraph, see ibid.

88. On this and the following paragraph, see Mikoyan report, September 22, 1960, cited in ibid., pt. 1, p. 105.

89. Mikoyan report, September 22, 1960, cited in ibid. and A.M. Ledovskii, "Sekretnaia missiia delegatsii TsK Kommunisticheskoi Partii Kitaia v Moskvu v iiune–avguste 1949 g.," p. 8.

90. Andrei Ledovskii, who was employed as a diplomat in the Soviet embassy in Nanking at the time, also made the same argument. See Ledovskii, "Sekretnaia missiia delegatsii TsK," p. 8.

91. Ibid., p. 7. No date is provided for the relevant telegram, and no source is given.

92. On this paragraph, see the Mikoyan report, September 22, 1960, cited in Ledovskii, "Sekretnaia missiia A.I. Mikoiana v Kitai," pt. 1, pp. 105–6, and Ledovskii, "Sekretnaia missiia delegatsii TsK," pp. 8–9.

93. Shi Zhe, *Zai lishi juren shenbian*, p. 419.

94. Telegram from Mao Zedong to the Soviet Party leadership, November 30, 1947, in Mikoyan report, September 22, 1960, cited in Ledovskii, "Sekretnaia missiia A.I. Mikoiana v Kitai," pt. 1, p. 102.

95. Mikoyan report, September 22, 1960, cited in ibid.

96. Telegram from Stalin to Mao Zedong, April 20, 1948, AP RF, f. 39, op. 1, d. 31, l. 28–29, cited in ibid., pt. 2, pp. 101–2, and Ledovskii, "Sekretnaia missiia delegatsii TsK," p. 8.

97. Wolfgang Leonhard, *Child of the Revolution*, p. 303. On this paragraph, see Ledovskii, "Sekretnaia missiia A.I. Mikoiana v Kitai," pt. 1, p. 102.

98. Mikoyan report, September 22, 1960, cited in Ledovskii, "Sekretnaia missiia A.I. Mikoiana v Kitai," pt. 1, pp. 102, 105. The same can be seen in *Mao Zedong nianpu*, vol. 3, p. 449.

99. *Mao Zedong nianpu*, vol. 3, p. 449.

100. Ibid.; *Zhou Enlai nianpu, 1898–1949*, p. 811; *Zhonghua renmin gongheguo waijiao shi*, p. 11; Shi Zhe, *Zai lishi juren shenbian*, p. 376. None of these works give any indication of sources.

101. *Zhonghua renmin gongheguo waijiao shi*, p. 11.

102. "An Lie-duo-fu-si-ji, Mi-gao-yang yu Mao Zedong de mimi tanpan (1949 nian 1–2 yue) (zhong)," *Dang de wenxian* 1 (1996), pp. 90–96.

103. On this and the rest of the paragraph, see ibid., p. 91.

104. *Mao Zedong xuanji*, vol. 4, p. 1237; English version in *Selected Works of Mao Tse-tung*, vol. 4, p. 150.

105. On this and the rest of the paragraph, see information provided by Shi Zhe to the author, March 16, 1996, and April 7, 1996.

106. Xu Yan, *Sulian chubing Zhongguo Dongbei jishi*, p. 103.

107. On this paragraph, see Ledovskii, "Vizit Mao Tszeduna v Sovetskii Soiuz," pt. 1, p. 10. Ledovskii did not indicate his source. From the context, however, it would appear that he was referring to a telegram to Stalin in which Mikoyan reported on the February 4 talks.

108. Telegram from Mikoyan to Stalin, AP RF, f. 39, op. 1, d. 39, ll. 62–63, reproduced by the author from a copy of the original, cited in ibid., p. 11. Ledovskii did not date the telegram.

109. Mikoyan report, September 22, 1960, cited in Ledovskii, "Sekretnaia missiia A.I. Mikoiana v Kitai," pt. 1, p. 106.

110. Ledovskii, "Vizit Mao Tszeduna v Sovetskii Soiuz," pt. 1, p. 11.

111. Soviet minutes of the negotiations of July 9, August 12, and August 13, 1945; see AP RF, f. 45, op. 1, d. 322, ll. 30–33, 93–97, and AVP RF, f. 06, op. 7, p. 36, d. 507, ll. 108–13, both cited in A.M. Ledovskii, "Peregovory I.V. Stalina s Sun Tszywenem o sovetsko-kitaiskom dogovore, iiun'–avgust 1945 g.," pp. 40–44, 104–10. For further details, see chapter 2 above.

112. On this and the following sentence, Ledovskii, "Vizit Mao Tszeduna v Sovetskii Soiuz," pt. 1, p. 12.

113. On this paragraph, see Mikoyan report, September 22, 1960, cited in Ledovskii, "Sekretnaia missiia A.I. Mikoiana v Kita," pt. 1, p. 106.

114. Telegram from Mikoyan to Stalin, February 6, 1949, AP RF, f. 39, op. 1, d. 39, ll. 78–79, reproduced by the author from a copy of the original, cited in Ledovskii, "Vizit Mao Tszeduna v Sovetskii Soiuz," pt. 1, p. 13; cf. Ledovskii, "Sekretnaia missiia A.I. Mikoiana v Kitai," pt. 2, pp. 100–101.

115. Telegram from Mikoyan to Stalin, February 6, 1949, AP RF f. 39, op. 1, d. 39, ll. 78–79, reproduced by the author from a copy of the original, cited in Ledovskii, "Vizit Mao Tszeduna v Sovetskii Soiuz," pt. 1, p. 13; cf. Ledovskii, "Sekretnaia missiia A.I. Mikoiana v Kitai," pt. 2, p. 101.

116. Mikoyan report, September 22, 1960, cited in Ledovskii, "Sekretnaia missiia A.I. Mikoiana v Kitai," pt. 1, p. 106.

117. Telegram from Mikoyan to Stalin, AP RF, f. 39, op. 1, d. 39, ll. 78–79, cited in Ledovskii, "Vizit Mao Tszeduna v Sovetskii Soiuz," pt. 1, p. 13; cf. Ledovskii, "Sekretnaia missiia A. I. Mikoiana v Kitai," pt. 2, p. 101.

118. Mikoyan report, September 22, 1960, cited in Ledovskii, "Sekretnaia missiia A.I. Mikoiana v Kitai," pt. 1, p. 106.

119. For the prior history of the Mongolian question and the role it played in the negotiations leading up to the treaty of August 14, 1945, see chapter 2 above.

120. Mikoyan report, September 22, 1960, cited in Ledovskii, "Sekretnaia missiia A.I. Mikoiana v Kitai," pt. 1, p. 107. Reports by the contemporary witnesses Kovalev and Shi Zhe on the Mongolian controversy are unreliable. Kovalev remembered sharp differences of opinion on this question, but mistakenly believed that Mao had not presented his wishes in his own name and had attributed them to the left wing of the KMT government. Kovalev apparently confused the situation in Xibaipo with the one that arose during Liu Shaoqi's visit to Moscow. Liu did, in fact, refer then to powers outside the CCP that wanted Mongolia back. Shi Zhe denied that Mao even raised the Mongolian question with Mikoyan. See Kovalev, "Dialog Stalina s Mao Tszedunom," pt. 1, p. 86; Shi Zhe quoted in Li Haiwen, "Mao Zedong Liu Shaoqi fang Su," p. 93.

121. Mikoyan report, September 22, 1960, cited in Ledovskii, "Sekretnaia missiia A.I. Mikoiana v Kitai," pt. 1, p. 107.

122. Ibid.

123. Cf. this chapter below.

124. Edgar Snow's undated interview with Mao Zedong, which took place sometime between June and October 1936, cited in Edgar Snow, *Red Star over China*, p. 96.

125. For details on this question, see this chapter below (section "The CCP's Policy Toward the West"). On this paragraph, see Mikoyan report, September 22, 1960, cited in Ledovskii, "Sekretnaia missiia A.I. Mikoiana v Kitai," pt. 1, p. 107.

126. Undated report from Xibaipo by Mikoyan, AP RF, f. 39, op. 1, d. 39, ll. 20–21, cited in Ledovskii, "Sekretnaia missiia delegatsii TsK," p. 83 n. 10.

127. For further details, see chapter 1 above.

128. For further details, see chapter 1 above.

129. On this paragraph, see Mikoyan report, September 22, 1960, cited in Ledovskii, "Sekretnaia missiia A.I. Mikoiana v Kitai," pt. 1, pp. 104–5.

130. On this paragraph, see ibid.

131. For further details, see chapter 1 above.

132. CCP Statutes of June 11, 1945, General Program, cited in *Zhongguo gongchandang dangzhang*, 4th ed., 1949, n.p. On this paragraph, see Mikoyan report, September 22, 1960, cited in Ledovskii, "Sekretnaia missiia A.I. Mikoiana v Kitai," pt. 1, pp. 107–9.

133. Mikoyan report, September 22, 1960, cited in Ledovskii, "Sekretnaia missiia A.I. Mikoiana v Kitai," pt. 1, pp. 108–9.

134. *Mao Zedong nianpu*, vol. 3, p. 449. Cf. *Zhou Enlai nianpu, 1898–1949*, p. 811; Shi Zhe, *Zai lishi juren shenbian*, p. 375.

135. On this paragraph, see Shi Zhe, *Za lishi juren shenbian*, pp. 375–78; Shi Zhe, "Peitong Mao zhuxi fang Su," *Renwu* 5 (1988), p. 6. Cf. also *Zhonghua renmin gongheguo waijiao shi*, p. 11.

136. Liu Shaoqi to Mikoyan, February 3, 1949, AP RF, f. 39, op. 1, d. 39, l. 41, cited in Ledovskii, "Sekretnaia missiia A.I. Mikoiana v Kitai," pt. 2, p. 102.

137. On this paragraph, see AP RF, f. 39, op. 1, d. 39, ll. 42–43, 46, cited in ibid., 102–3.

138. AP RF, f. 39, op. 1, d. 39, l. 46, cited in ibid., p. 103.

139. On this paragraph, see Mao Zedong to Mikoyan, February 7, 1949, AP RF, f. 39, op. 1, d. 39, ll. 94–95, cited in ibid., pp. 103–4.

140. On this paragraph, see AP RF, f. 39, op. 1, d. 39, ll. 89–90, cited in ibid., p. 104.

141. Telegram from Mikoyan to Stalin about a talk with Mao Zedong on February 4, 1949, AP RF, f. 39, op. 1, d. 39, l. 54, cited in ibid., p. 103.

142. On this paragraph, see Shi Zhe, *Zai lishi juren shenbian*, pp. 379–81; *Mao Zedong nianpu*, vol. 3, p. 448. Cf. *Zhonghua renmin gongheguo waijiao shi*, p. 11.

143. On this paragraph, see Shi Zhe, *Zai lishi juren shenbian*, p. 382.

144. On this paragraph, see AP RF, f. 39, op. 1, d. 39, l. 69, cited in Ledovskii, "Sekretnaia missiia A.I. Mikoiana v Kitai," pt. 2, p. 102.

145. On this paragraph, see Shi Zhe, *Zai lishi juren shenbian*, p. 385.

146. On this paragraph, see AP RF, f. 39, op. 1, d. 39, l. 29, cited in Ledovskii, "Sekretnaia missiia A.I. Mikoiana v Kitai," pt. 2, p. 99.

147. Ibid.

148. Mikoyan incorrectly mentions the name Luo Yonghan; see ibid., pp. 97, 105 n. 12.

149. On this paragraph, see AV RF, f. 39, op. 1, d. 39, l. 58, cited in ibid., p. 97.

150. On this paragraph, see Liu Shaoqi to Mikoyan, February 3, 1949, AP RF, f. 39, op. 1, d. 39, l. 44, cited in ibid., pp. 99–100.

151. On this paragraph, see Ren Bishi and Zhu De to Mikoyan, February 2, 1949, AP RF, f. 39, op. 1, d. 39, l. 37, cited in ibid., p. 99.

152. Telegram from Mao Zedong to Stalin, January 8, 1949, AP RF, f. 39, op. 1, d. 37, l. 1, cited in ibid., p. 98.

153. On this and the following sentence, see ibid., pp. 106–7.

154. On this paragraph, see Mao Zedong and Zhou Enlai to Mikoyan, February 4, 1949, AP RF, f. 39, op. 1, d. 39, ll. 57–58, cited in ibid., pt. 2, pp. 97–98.

155. On this paragraph, see Ren Bishi to Mikoyan, February 2, 1949, AP RF, f. 39, op. 1, d. 39, l. 37, cited in ibid., p. 99.

156. Mikoyan in a telegram to Stalin, AP RF, f. 39, op. 1, d. 39, ll. 85–86, cited in Ledovskii, "Vizit Mao Tszeduna v Sovetskii Soiuz," pt. 1, p. 17; Ledovskii, "Sekretnaia missiia A.I. Mikoiana v Kitai," pt. 2, p. 100. Ledovskii did not date the telegram.

157. Mikoyan in a telegram to Stalin, AP RF, f. 39, op. 1, d. 39, ll. 84–85, cited in Ledovskii, "Vizit Mao Tszeduna v Sovetskii Soiuz," pt. 1, p. 17. Ledovskii did not date the telegram.

158. On this paragraph, see Mikoyan in a telegram to Stalin, AP RF, f. 39, op. 1, d. 39, ll. 84–86, cited in ibid.; Ledovskii, "Sekretnaia missiia A.I. Mikoiana v Kitai," pt. 2, p. 100. Ledovskii did not date the telegram.

159. On this paragraph, see the telegram from Orlov to Stalin, August 28, 1948, AP RF, f. 39, op. 1, d. 31, l. 41, cited in Ledovskii, "Sekretnaia missiia A.I. Mikoiana v Kitai," pt. 1, p. 99.

160. On this paragraph, see Mao Zedong to Mikoyan, February 7, 1949, AP RF, f. 39, op. 1, d. 39, l. 95, cited in ibid., pt. 2, pp. 103–4.

161. *Zhonghua renmin gongheguo waijiao shi*, p. 11.

162. Kovalev, "Dialog Stalina s Mao Tszedunom," pt. 2, p. 88.

163. *Zhonghua renmin gongheguo waijiao shi*, p. 11, with no indication of sources.

164. Shi Zhe, *Zai lishi juren shenbian*, p. 386.

165. Kovalev, "Dialog Stalina s Mao Tszedunom," pt. 1, p. 86.

166. On this paragraph, see Ledovskii, "Sekretnaia missiia A.I. Mikoiana v Kitai," pt. 2, pp. 104–5.

167. On this and on the following three sentences, see Kovalev, "Dialog Stalina s Mao Tszedunom," pt. 2, p. 87.

168. For further details, see this chapter below.

169. For further details, see chapter 2 above.

170. On this paragraph, see Ledovskii, "Sekretnaia missiia A.I. Mikoiana v Kitai," pt. 1, pp. 109–10.

171. Mao Zedong to Iudin, July 22, 1958, Chinese discussion notes, in *Dang de wenxian* 1 (1994), p. 16. Cf. Yu and Zhang, "Guanyu Si-da-lin," p. 21.

172. Mao Zedong's talks with Iudin, March 31, 1956, in *PDV* 5 (1994), p. 109 n. 18.

173. Kovalev, "Dialog Stalina s Mao Tszedunom," pt. 1, p. 85. With the exception of a passage in Khrushchev's memoirs, Western researchers had no knowledge of Kovalev's mission until 1991. Khrushchev mentioned that Stalin had nominated a former wartime people's commissar and railroad engineer, whom he considered a personal confidant, to become the first Soviet ambassador to the People's Republic of China. He had reported intensively on anti-Soviet tendencies in the CCP leadership. Khrushchev could not remember his name. He thought it was Kovalev, but was mistaken in that it was Roshchin who was nominated as the first ambassador. This confusion was increased when the publisher of the memoirs mistakenly identified the person Khrushchev described as the "first ambassador"—that is, Kovalev—with Paniushkin, who went to Peking as the second ambassador. See *Khrushchev Remembers: The Last Testament*, pp. 242–43.

174. Tikhvinskii to the author, September 15, 1992.

175. Shi Zhe, "Gao Gang ceng jianyi jiang Dongbei huagui Sulian ma?" pp. 84–85.

176. This is confirmed by two contemporary witnesses. Kovalev stated that with regard to China, Stalin "kept everything in his own hands." Shi Zhe supported this, saying that all questions involving China were "handled and decided by Stalin himself." See Kovalev, "Dialog Stalina s Mao Tszedunom," pt. 1, p. 84; Shi Zhe, *Zai lishi juren shenbian*, p. 418.

177. On this paragraph, see Kovalev, "Dialog Stalina s Mao Tszedunom," pt. 1, pp. 85, 90–91, 92–93 n. 11; *Khrushchev Remembers: The Last Testament*, p. 243.

178. Li Yueran, *Waijiao wutai shang de xin Zhongguo lingxiu*, pp. 11–14; a contemporary Soviet witness, who chooses to remain anonymous, to the author.

179. On this paragraph, see Kovalev, "Dialog Stalina s Mao Tszedunom," pt. 1, pp. 84–85; Shi Zhe, *Zai lishi juren shenbian*, p. 390; telegram from Mao Zedong to Stalin, January 8, 1949, AP RF, f. 30, op. 1, d. 37, l. 1, cited in Ledovskii, "Sekretnaia missiia A.I. Mikoiana v Kitai," pt. 2, p. 98; Ledovskii, "Sekretnaia missiia delegatsii TsK," pp. 61–62.

180. Kovalev, "Dialog Stalina s Mao Tszedunom," pt. 1, p. 85. Exactly when this transition occurred remains unclear. There was still a direct correspondence in mid-January; in May, it already ran via Kovalev. On this, see Tikhvinskii, "Perepiska I.V. Stalina s Mao Tszedunom," pp. 133–39; Tikhvinskii, "O 'sekretnom demarshe,'" pp. 134–36.

181. On this, see telegram to Stalin, May 18, 1949, AP RF, f. 45, op. 1, d. 331, ll. 59–61; Shi Zhe, *Zai lishi juren shenbian*, pp. 315, 390; Iurii Vlasov, "Povest' ob ottse," pt. 1, p. 128. For information about Orlov, see chapter 1 above (section "Soviet Go-Betweens in Yenan").

182. Shi Zhe, quoted in Li Haiwen, "Mao Zedong Liu Shaoqi fang Su," p. 95.

183. Telegram from the CC of the CCP to Wang Jiaxiang, November 9, 1949, in *Jianguo yilai Mao Zedong wengao*, vol. 1, p. 131.

184. For further details, see this chapter below.
185. This can be seen in the telegram from the CC of the CCP to Wang Jiaxiang, November 9, 1949, in *Jianguo yilai Mao Zedong wengao*, vol. 1, p. 131.
186. On this paragraph, see Kovalev, "Dialog Stalina s Mao Tszedunom," pt. 1, p. 85; ibid., pt. 2, pp. 89–90; telegram from Mao Zedong to Stalin, November 12, 1949, in *Jianguo yilai Mao Zedong wengao*, vol. 1, p. 135; Shi Zhe, *Zai lishi juren shenbian*, pp. 392–93; Li Yueran, *Waijiao wutai*, pp. 4, 6–11; Ledovskii, "Sekretnaia missiia A.I. Mikoiana v Kitai," pt. 2, p. 105 n. 19. For further details of Kovalev's activities during Mao's stay in Moscow, see chapter 4 below.
187. For further details on this paragraph, see chapter 4 below.
188. Kovalev, "Dialog Stalina s Mao Tszedunom," pt. 2, p. 90.
189. Message from Vyshinskii to Stalin, February 3, 1950, AP RF, f. 3, op. 65, d. 369, l. 46.
190. Conversation between Gromyko and Wang Jiaxiang, February 22, 1950, AVP RF, f. 0100, op. 43, p. 302, d. 8, l. 26.
191. For further information on this and the following sentence, see *General'noe Konsul'stvo SSSR v gor. Pekine, Konsul'skii otchet za 1949 god*, AVP RF, f. 0100, op. 43, d. 20, l. 6.
192. For further information on this and the following sentence, see Xu, *Sulian chubing Zhongguo Dongbei jishi*, p. 130.
193. Zhou Enlai stated this explicitly to Mikoyan on February 4, 1949. See AP RF, f. 39, op. 1, d. 39, l. 57, cited in Ledovskii, "Sekretnaia missiia A.I. Mikoiana v Kitai," pt. 2, p. 97.
194. On this paragraph, see Ledovskii to the author, March 27, 1995. For further details on Kovalev's mission, see this chapter above.
195. Ledovskii, "Nankinskie zapiski," p. 42.
196. For further details on the meetings, see this section below and this chapter below.
197. See chapter 2 above.
198. Cf. Steven M. Goldstein, "Sino-American Relations, 1948–1950," pp. 132, 137; Nancy Bernkopf Tucker, *Patterns in the Dust*, pp. 15–16.
199. See chapter 1 above ("Beginnings of a CCP Foreign Policy: Flirting with the United States at the End of the War").
200. Telegram from the CC of the CCP to the Northeast Bureau, November 1, 1948, in *Zhou Enlai nianpu, 1898–1949*, p. 794.
201. On this paragraph, see Chen Jian, "The Ward Case and the Emergence of Sino-American Confrontation, 1948–1950," p. 151; Yang Kuisong, "The Soviet Factor and the CCP's Policy Toward the United States in the 1940s," p. 30; Yang Kuisong, "Hua-de shijian yu xin Zhongguo dui Mei zhengce de queding," p. 105.
202. Telegram from Mao Zedong to Gao Gang, November 17, 1948, cited in Yang Kuisong, "Hua-de shijian," pp. 108–9.
203. Shi Yinhong, *Du-lu-men zhengfu dui xin Zhongguo de zhengce cong dishi dao zhanzheng de lishi huigu*, Dissertation. The University of Nanjing, 1987, pp. 62–63, 72–74, cited in Chen Jian, "Ward Case," p. 164. Yang Kuisong appears to assume that the CCP leadership was convinced that the Americans were engaged in espionage. See Yang Kuisong, "The Soviet Factor," p. 31.
204. On this paragraph, see the telegram from the CC of the CCP to the Northeast Bureau, November 10, 1948, cited in *Zhou Enlai nianpu, 1898–1949*, p. 796; telegram from the Northeast Bureau to the CC of the CCP, November 11, 1948, cited in Yang Kuisong, "Hua-de shijian," p. 106; telegram from Gao Gang to Mao Zedong, November 16, 1948, cited in Yang Kuisong, "Soviet Factor," pp. 30–31; telegram from Mao Zedong to Gao Gang and the Northeast Bureau of the CCP, November 17, 1948, cited in ibid., p. 31; telegram from Mao Zedong to the Northeast Bureau, November 18, 1948, cited in Chen Jian, "Ward Case," pp. 153–56; *FRUS 1949*, vol. 8, *The Far East: China*, pp. 1044–50; Ronald L. McGlothlen, *Controlling the Waves*, pp. 148–52.
205. On this paragraph, see Yang Kuisong, "Hua-de shijian," p. 107.
206. On this paragraph, see Kovalev, "Dialog Stalina s Mao Tszedunom," pt. 2, pp. 83–84. Kovalev's contacts with the CC of the CCP passed through Gao Gang.
207. On this and the following sentence, see the telegram from the CC of the CCP of the Northeast Bureau, November 23, 1948, cited in Yang Kuisong, "Soviet Factor," p. 31.
208. On this paragraph, see Yang Kuisong, "Hua-de shijian," pp. 107–9; Yang Kuisong, "Soviet Factor," pp. 30–31; Mao Zedong's notes for Zhou Enlai, November 18 and 21, 1948, cited in Yang Kuisong, "Hua-de shijian," p. 109; telegram from Zhou Enlai to the Northeast Bureau, November 23, 1948, cited in ibid.; and Yang Kuisong, "Soviet Factor," p. 31.

209. Telegram from Ambassador Stuart to the American Secretary of State, March 10, 1949, in *FRUS 1949*, vol. 8, *The Far East: China*, p. 174.

210. Radio Nord-Shaanxi, March 18, 1949, cited in ibid., pp. 192–93.

211. Telegram from Ambassador Stuart to the American Secretary of State, March 22, 1949, in ibid., pp. 192–94.

212. Speech by Mao Zedong during a Politburo session, January 1949, cited in Yang Kuisong, "Soviet Factor," p. 31. On this paragraph, see Report of the Northeast Bureau of the CCP, November 24, 1948 and *Renmin ribao*, June 19, 1949, both cited in ibid.

213. "Central Committee instructions on diplomatic activity," January 19, 1949, drafted by Mao Zedong and Zhou Enlai for the CC of the CCP, cited in *Zhonggong zhongyang wenjian xuanji*, vol. 18, pp. 44–49; *Zhou Enlai zhuan*, p. 742.

214. On this paragraph, see *Dangdai zhongguo waijiao*, pp. 3–5; *Mao Zedong xuanji*, vol. 4, pp. 1435–36; Qiang Zhai, *The Dragon, the Lion, and the Eagle*, p. 25, with further references.

215. "Central Committee instructions on diplomatic activity," January 19, 1949, drafted by Mao Zedong and Zhou Enlai for the CC of the CCP, cited in *Zhonggong zhongyang wenjian xuanji*, vol. 18, p. 45.

216. Yang Kuisong, "Soviet Factor," pp. 31–32. Yang does not indicate his sources, but the text is obviously based on archival documents.

217. For further details, see this and the following paragraph in this section and this chapter below (section "The CCP's Tactical Explanatory Contacts").

218. On this paragraph, see Ledovskii, "Nankinskie zapiski," pp. 8, 12–92 passim; Huang Hua, "Nanjing jiefang chiqi wo tong Si-tu-lei-deng de jici jiechu," pp. 24–25; Ledovskii to the author, March 27, 1995.

219. On this paragraph, see Ledovskii, "Nankinskie zapiski," pp. 12–13, 14, 16.

220. For further details, see this chapter above.

221. Ledovskii, "Nankinskie zapiski," p. 13.

222. On this paragraph, see Huang Hua to Ledovskii, May 4, 1949, in ibid., pp. 15, 17.

223. Conversation between Huang Hua and Ledovskii, May 15, 1949, in ibid., p. 28.

224. Mao Zedong's telegram to the PLA's General Front Municipal Committee, April 28, 1949, in *Dang de wenxian* 4 (1989), p. 43, cited in Qiang, *Dragon*, p. 14.

225. On this paragraph, see ibid., pp. 14–15.

226. Huang Hua to Ledovskii, May 10, 1949, in Ledovskii, "Nankinskie zapiski," pp. 22–23.

227. Ledovskii to Huang Hua, May 10, 1949, in ibid., p. 22.

228. Discussion between Huang Hua and Ledovskii, May 10, 1949, in ibid., p. 23.

229. Telegram drafted by Mao for the Military Commission of the CC of the CCP to the Main Front Municipal Committee, April 28, 1949, in *Mao Zedong nianpu*, vol. 3, p. 490. On this paragraph, see the discussion between Ledovskii and Huang Hua on May 10, 1949, in Ledovskii, "Nankinskie zapiski," pp. 23–24.

230. Discussion between Ledovskii and Huang Hua, May 10, 1949, in Ledovskii, "Nankinskie zapiski," p. 25.

231. Discussion between Ledovskii and Huang Hua, May 10, 1949, in ibid., pp. 25–26.

232. Ibid., p. 26.

233. Ibid., pp. 26–27.

234. On this paragraph, see ibid., pp. 17–18.

235. Chen Jian, "Ward Case," p. 158.

236. McGlothlen, *Controlling the Waves*, pp. 148–49.

237. On this paragraph, see ibid., p. 151; Tucker, *Patterns in the Dust*, pp. 14–16; Chen Jian, "Ward Case," pp. 150–70 passim, in particular pp. 161–63, 165–66.

238. On this paragraph, see Yang Kuisong, "Soviet Factor," pp. 31–32. For further details about Huang Hua's talks with Stuart, see this chapter below.

239. Stalin to Harriman and Hopkins, June 1, 1944, cited in W. Averell Harriman and Elie Abel, *Special Envoy to Churchill and Stalin, 1941–1946*, p. 472.

240. Cf. E.P. Bazhanov, "Sovetsko-kitaiskie otnosheniia," p. 8, with no indication of sources.

241. The head of the Indian mission in Tokyo, Chakravarty, to the American diplomat Cloyce K. Huston, early January 1949, in *FRUS 1949*, vol. 8, *The Far East: China*, p. 24.

242. The U.S. chargé d'affaires in the Soviet Union, Kohler, to the Secretary of State, June 27, 1949, in ibid., p. 398 (diplomatic telegraphic style used in quote here and elsewhere).

243. Stalin in a telegram, April 1949, in Kovalev, "Dialog Stalina s Mao Tszedunom," pt. 2, p. 87; the precise date is not indicated.

244. Telegram from Stalin to Mao Zedong, April 1949, cited in ibid., p. 82; the precise date is not indicated.

245. Message from Stalin to the Minister of Economic Affairs of Manchuria, Chen Yun, March 15, 1949, cited in ibid., p. 81.

246. Stalin to Liu Shaoqi in Moscow, July 11, 1949, cited in ibid., p. 82. Goncharov, Lewis, and Xue later corrected the date mistakenly given for the talk here (July 27); cf. Sergei N. Goncharov, John W. Lewis, and Xue Litai, *Uncertain Partners*, p. 69.

247. In August 1948, Mao had requested a credit of US$30 million; in February 1949, he asked for US$300 million. See telegram from Orlov to Stalin, August 28, 1948, AP RF, f. 39, op. 1, d. 31, l. 40, cited in Ledovskii, "Sekretnaia missiia A.I. Mikoiana v Kitai," pt. 1, p. 99; Mao to Mikoyan, February 6, 1949, AP RF, f. 39, op. 1, d. 39, ll. 84–85, cited in Ledovskii, "Vizit Mao Tszeduna v Sovetskii Soiuz," pt. 1, p. 17; Ledovskii, "Sekretnaia missiia A.I. Mikoiana v Kitai," pt. 2, p. 100. For further details, see chapter 2 above and this chapter (sections "Rethinking in Moshan" and "The First Top-Level Soviet Meeting").

248. On this and the following sentence, see this chapter below; telegram from the CC of the CCP to Liu Shaoqi and Wang Jiaxiang, August 10, 1949, cited in *Zhonggong zhongyang zai Xiangshan*, p. 458; Kovalev, "Dialog Stalina s Mao Tszedunom," pt. 1, pp. 84–85; Shi Zhe, *Zai lishi juren shenbian*, pp. 407–8, 415.

249. For further information on this, see this chapter below.

250. Telegram from Stalin to Mao Zedong, April 1949, extracts of which are printed in Kovalev, "Dialog Stalina s Mao Tszedunom," pt. 2, p. 82; Kovalev does not provide the exact date. See also Kovalev's December 1949 report for Stalin, extracts of which are in ibid.

251. Stalin to Liu Shaoqi, July 11, 1949, cited in ibid.; Kovalev, in his December 1949 report for Stalin, ibid.

252. For further details on the contacts described in this and the following sentence, see this chapter below.

253. On this paragraph, see Kovalev, "Dialog Stalina s Mao Tszedunom," pt. 2, pp. 82, 84–85.

254. On the individual initiatives, cf. Tucker, *Patterns in the Dust*, pp. 47–48; Mao, March 5, 1940, in *Selected Works of Mao Tse-tung*, vol. 4, p. 371.

255. *Selected Works of Mao Tse-tung*, vol. 4, p. 371.

256. Telegram from the Military Commission of the CC of the CCP to the Main Front Municipal Committee, April 28, 1949, in *Mao Zedong nianpu*, vol. 3, pp. 489–90.

257. Mao Zedong, April 30, 1949, in *Selected Works of Mao Tse-tung*, vol. 4, p. 402.

258. Mao Zedong, June 15, 1949, in ibid., p. 408.

259. Stuart to Huang Hua, June 6, 1949, as reported by Stuart to the Secretary of State, June 8, 1949, cited in Rea and Brewer, *Forgotten Ambassador*, pp. 325–26.

260. Chen Jian, "Ward Case," pp. 159, 165, 169.

261. *FRUS 1949*, vol. 8, *The Far East: China*, p. 776. For further details, see this chapter below (section "The CCP's Tactical Explanatory Contacts").

262. Dean Acheson, *Present at the Creation*, p. 340, cited in Chen Jian, "Ward Case," p. 149.

263. Unaware of Yang Kuisong's and Chen Jian's conclusions and of the materials presented by Kovalev, Michael M. Sheng mistakenly describes the CCP's new approach as a "self-imposed isolationist policy"; see Michael M. Sheng, *Maoist Dualism and Chinese Communist Foreign Relations, 1935–1949*, p. 371. This is no criticism of him, because these publications were not available to him at the time. But it is less comprehensible why he does not even suggest the possibility of a "Soviet-imposed isolationist policy."

264. Chen Jian, "Ward Case," p. 160.

265. To support the claim that Moscow based its policy in the spring of 1949 on the hope that the CCP would not be able to control all of China, one could point to a contemporary statement by Soviet ambassador Roshchin to his American colleague in Canton that if the Communists did finally seize power, they would be "riding a tiger" and be confronted with situations impossible of solution. Washington should not worry too much, he said. History had shown that no one could control the clannish Chinese; the Communists could not succeed any more than had the Kuomintang. Roshchin to Minister-Counselor Clark, March 31, 1949, *FRUS 1949*, vol. 8, *The Far East: China*, pp. 217–18.

266. Tucker, *Patterns in the Dust*, p. 16.

267. *Zhonggong dangshi dashi nianbao shuoming*, p. 133, cited in Goncharov, Lewis, and Xue, *Uncertain Partners*, p. 306 n. 25.

268. Mao, April 11, 1957, cited in Wang Fangming, "'Shishi jiu shi, duli sikao'—huiyi Mao zhuxi 1957 nian yici qinqie jianghua."

269. Shi Zhe, *Zai lishi juren shenbian*, p. 435.

270. Ibid., p. 414.

271. Liu Xiao, *Chushi Sulian ban ian*, pp. 4–5. Cf. *Shijie zhishi* 3 (1987), p. 15, cited in E.P. Bazhanov, *Kitai i vneshnii mir*, p. 35.

272. The authors who assume that it was Mikoyan who passed on the advice include Nie Rongzhen and Qu Xing, cited in Kovalev, "Dialog Stalina s Mao Tszedunom," pt. 1, p. 91 n. 4. Cf. this and the following sentence in Goncharov, Lewis, and Xue, *Uncertain Partners*, pp. 42–43, 306 n. 28.

273. Cf. Yang Kuisong, *Zhongjian didai de geming*, p. 447; Xiang Qing, "Guanyu Si-da-lin ceng fou quanzu jiafang dajun guo jiang zhi wojian." *Dang de wenxian* 6 (1989) Liao Gailong and Chen Guangxiang, cited in Chen Jian, "Sino-Soviet Alliance," p. 5. Similar suggestions can be found in a commentary by Guo Moruo on a poem composed by Mao Zedong after crossing the Yangtze; see *Hongqi* 1 (1964), cited in John Gittings, *Survey of the Sino-Soviet Dispute*, p. 153. Cf. also Stuart R. Schram, *Mao Tse-tung*, p. 245; Gordon Chang, *Friends and Enemies*, p. 28; Gittings, *Survey of the Sino-Soviet Dispute*, pp. 13–14.

274. See Goncharov, Lewis, and Xue, *Uncertain Partners*, p. 42.

275. Kovalev, "Dialog Stalina s Mao Tszedunom," pt. 1, p. 86.

276. Shi Zhe, *Zai lishi juren shenbian*, pp. 371, 374.

277. On this and the following sentence, see Ledovskii, "Sekretnaia missiia A.I. Mikoiana v Kitai," pt. 1, pp. 100–110, esp. 105.

278. Ibid., pt. 2, pp. 94–105.

279. See, for example, Yu and Zhang, "Guanyu Si-da-lin," pp. 15–21. Other authors from the People's Republic of China refrain from a clear position. See, for example, Zhou Wenqi, "Guanyu Si-da-lin ceng fou quanzu wo jun daguo Changjiang wenti de jiutao," pp. 9–10.

280. On this and the rest of the paragraph, see Mao Zedong's talks with Iudin, March 31, 1956, in *PDV* 5 (1994), pp. 103–9, and July 22, 1958, in *Dang de wenxian* 1 (1994), pp. 16–20.

281. Wang Qi to the author, May 23, 1993.

282. Li Zhisui, *Mao Zedong siren yisheng*, p. 112; Li Zhisui, *The Private Life of Chairman Mao*, p. 117.

283. *Wang Dongxing riji*, p. 167.

284. Kovalev, "Dialog Stalina s Mao Tszedunom," pt. 2, p. 82. The telegram is not dated precisely.

285. Cf. Tikhvinskii, "O 'sekretnom demarshe,'" pp. 133–38 passim; Tikhvinskii, "Perepiska I.V. Stalina s Mao Tszedunom," pp. 133–39; Yu and Zhang, "Guanyu Si-da-lin," pp. 19–20.

286. Telegram from Mao Zedong to Stalin, January 13, 1949, cited in Tikhvinskii, "Perepiska I.V. Stalina s Mao Tszedunom," p. 136. Mao had also previously published articles that were sent to the Soviet leadership in which he announced the decision to cross the Yangtze. Cf., for example, the 1949 New Year's message, written on December 30, 1948, in *Selected Works of Mao Tse-tung*, vol. 4, p. 301.

287. Telegram from Mao Zedong to Stalin, January 13, 1949, cited in Tikhvinskii, "Perepiska I.V. Stalina s Mao Tszedunom," pp. 136–38.

288. Ibid., p. 132.

289. Shi Zhe, *Zai lishi juren shenbian*, pp. 370–71.

290. Huang Hua, "Nanjing jiefang," p. 23.

291. On this paragraph, see Yu Zhanbang,"Mao Zedong yu Zhang Zhizhong de yi ci zhongyao tanhua," pp. 148–55, esp. p. 151.

292. Zhang Zhizhong to the assistant to U.S. ambassador Stuart, Fu Jingpo, cited in Zhou Wenqi, "Guanyu Si-da-lin," p. 9, without indication of date or source.

293. Wu Tiecheng to Lewis Clark, April 2, 1949, *FRUS 1949*, vol. 8, *The Far East: China*, pp. 222–23.

294. On this paragraph, see Zhu Yuanshi, "Liu Shaoqi 1949 nian mimi fang Su," p. 75, with no indication of sources.

295. On this paragraph, see Xiang Qing, You Zhitu, and Liu Dexi, eds., *Sulian yu Zhongguo geming*, p. 568. Wang Yongxiang provides a similar interpretation of the telegram in *Jin wan bao*, April 3, 1997, cited in *Baokan wenzhai*, April 10, 1997, p. 3. Wang Yongxiang actually describes Mao Zedong's and Zhou Enlai's above-mentioned statements as untrustworthy (*bu kekao de*).

296. Cf. Goncharov, Lewis, and Xue, *Uncertain Partners*, p. 44.

297. For further details, see chapter 2 above (section "Moscow Holds Back") and this chapter above (section "Rethinking in Moscow").

298. For further details on this and the following sentence, see this chapter above (section "The First Top Level Secret Meeting").

299. For further details on this and the following paragraph, see Kovalev, "Dialog Stalina s Mao Tszedunom," pt. 2, p. 88. Neither telegram is dated precisely.

300. See this chapter above (section "Rethinking in Moscow").

301. Zhou Enlai to Ambassador Roshchin, November 10, 1949, AVP RF, f. 07, op. 22, d. 220, l. 52.

302. On this paragraph, see Bo, *Ruogan zhongda juece*, vol. 1, p. 37; Shi Zhe, *Zai lishi juren shenbian*, p. 395; Zhu Yuanshi, "Liu Shaoqi 1949 nian mimi fang Su," p. 76; *Zhonghua renmin gongheguo waijiao shi*, p. 12.

303. On this paragraph, see *Zhonghua renmin gongheguo waijiao shi*, p. 12; Shi Zhe, *Zai lishi juren shenbian*, p. 395; Zhu Yuanshi, "Liu Shaoqi 1949 nian mimi fang Su," pp. 75–76.

304. On this paragraph, see Zhu Yuanshi, "Liu Shaoqi 1949 nian mimi fang Su," pp. 75–76. Cf. *Zhonghua renmin gongheguo waijiao shi*, p. 12. For details about Roshchin's actions, see this chapter above.

305. *Zhonghua renmin gongheguo waijiao shi*, p. 11. For further details on the Mikoyan visit, see this chapter above.

306. Mao Zedong to Mikoyan, February 7, 1949, AP RF, f. 39, op. 1, d. 39, 1.95, cited in Ledovskii, "Sekretnaia missiia A.I. Mikoiana v Kitai," pt. 2, p. 104.

307. Ibid., pp. 77–100, 103–105.

308. Kovalev, "Dialog Stalina s Mao Tszedunom," pt. 2, p. 87.

309. On this and the remainder of the paragraph, see the telegram from Mao Zedong to Stalin, May 3, 1949, in *Mao Zedong nianpu*, vol. 3, p. 496.

310. For further details, see Chen Jian, "Sino-Soviet Alliance," p. 3. Reliable source material became available only in 1996, however, when some of the documents dealing with the Liu mission in the Russian presidential archives were published. See A.M. Ledovskii, "Vizit v Moskvu delegatsii Kommunisticheskoi partii Kitaia v iiune—avguste 1949 g.," pt. 1, pp. 66–83, and pt. 2, pp. 84–94. An English-language version appeared in *Far Eastern Affairs* 4 (1996), pp. 65–86, and 5 (1996), pp. 85–97. The author possesses copies of some of the original archival documents, for which he thanks Andrei Ledovskii, who made them available to him in the context of an unpublished manuscript (Ledovskii, "Sekretnaia missiia delegatsii TsK").

311. Zhou Enlai to Roshchin, November 10, 1949, AVP RF, f. 07, op. 22, d. 220, l. 52.

312. On this and the following sentence, see *Pravda*, July 31, 1949, p. 3.

313. Note from the acting foreign minister, Ye Gongchao, to the Soviet chargé d'affaires, V.V. Vaskov, delivered on August 6, 1949, AVP RF, f. 07, op. 22a, d. 198, ll. 13–14.

314. Draft of the response with Vyshinskii's letter to Stalin, August 25, 1949, including prior drafts, AVP RF, f. 07, op. 22a, d. 198, ll. 11–12, 16–21, 23–26. The letter bears a handwritten note that appears to be by Vyshinskii, dated August 29, 1949: "Do not answer until there is another query."

315. Cf. Colin Mackerras and Robert Chan, *Modern China*, p. 437.

316. The discussion described here took place during the June 27 session. See below in this same section.

317. On this paragraph, see Ledovskii, "Sekretnaia missiia delegatsii TsK," pp. 16–17. Ledovskii was first secretary of the Soviet embassy in Nanking at the time.

318. A publication from the People's Republic of China indicates that Stalin had insisted on maintaining strict secrecy about the visit. See *Zhonghua renmin gongheguo waijiao shi*, p. 14.

319. On this point, see this chapter below.

320. The travel dates are based on the following sources: *Liu Shaoqi nianpu, 1898–1969*, vol. 2, p. 217; *Zhonghua renmin gongheguo waijiao shi*, p. 12; *Zhonggong zhongyang zai Xiangshan*, pp. 439, 441, 460, 465; *Mao Zedong nianpu*, vol. 3, p. 542; Xiang, You, and Liu, *Sulian yu Zhongguo geming*, p. 570; Bo, *Ruogan zhongda juece*, vol. 1, p. 37. Shi Ze originally claimed that the delegation left Peiping on July 2 and arrived in Moscow "on the sixth day"; see Shi Zhe, *Zai lishi juren shenbian*, pp. 396–97. This version was repeated in Zhu Yuanshi, "Liu Shaoqi 1949 nian mimi fang Su," p. 76. Shi Zhe corrected himself in the meantime and gave June 21 as the date of departure and—once again inaccurately—June 28 as the date of arrival; see Shi Zhe and Chen Jian, "With Mao and Stalin," p. 26. An inaccurate arrival date (July 10) is also provided in Michael H. Hunt, *The Genesis of Chinese Communist Foreign Policy*, p. 180. When they wrote their book, Goncharov, Lewis, and Xue could only base their account on the original version of Shi Zhe's memoirs and mistakenly assumed that the delegation arrived in Moscow around July 9. See Goncharov, Lewis, and Xue, *Uncertain Partners*, p. 64.

321. Bo, *Ruogan zhongda juece*, vol. 1, p. 37; *Zhonggong zhongyang zai Xiangshan*, p. 455. For further details on the reasons for Gao Gang's premature departure, see below in this same section.

322. Xiang, You, and Liu, *Sulian yu Zhongguo geming*, p. 572; *Zhonggong zhongyang zai Xiangshan*, p. 460.

323. *Zhonggong zhongyang zai Xiangshan*, p. 460.

324. On this paragraph, see Shi Zhe, *Zai lishi juren shenbian*, pp. 392, 396; Shi Zhe, "Peitong Mao zhuxi fang Su," Renwu 5 (1988), p. 7; Shi Zhe, in Li Haiwen, "Mao Zedong Liu Shaoqi fang Su," p. 91; Zhu Yuanshi, "Liu Shaoqi 1949 nian mimi fang Su," pp. 76, 80; *Zhonghua renmin gongheguo waijiao shi*, p. 14; *Jianguo yilai Mao Zedong wengao*, vol. 1, p. 81.

325. Wolfgang Bartke, *Biographical Dictionary and Analysis of China's Party Leadership, 1922–1988*, p. 44.
326. Ledovskii, "Vizit v Moskvu delegatsii," pt. 1, p. 68.
327. Bartke, *Biographical Dictionary*, pp. 216–17; Donald W. Klein and Anne B. Clark, *Biographic Dictionary of Chinese Communism*, vol. 2, pp. 896–98. On this paragraph, see Shi Zhe, *Zai lishi juren shenbian*, pp. 395–97, 425–26.
328. Kovalev, "Dialog Stalina s Mao Tszedunom," pt. 2, p. 78.
329. Following publication of Kovalev's memoirs, Shi Zhe made an attempt to destroy Kovalev's credibility in polemical publications that was not very convincing. In doing so, he went so far as to claim that Kovalev attended only the first session—and even then only at the beginning, when Stalin had him summoned to report to him on his trip and then sent him out again. He thus claimed that Kovalev could say nothing on the basis of his own observations about the progress of the session. This is false, as can be seen from the list of visitors to Stalin's office. Cf. Li Haiwen, "Mao Zedong Liu Shaoqi fang Su," p. 91; "Posetiteli kremlevskogo (kabineta, I. V. Stalina)," pt. 1, pp. 56–57.
330. Shi Zhe, *Zai lishi juren shenbian*, pp. 397–425 passim.
331. *Zhonghua renmin gongheguo waijiao shi*, p. 12.
332. AP RF, f. 45, op. 1, d. 329, ll. 1–7, cited in Ledovskii, "Sekretnaia missiia delegatsii TsK," pp. 18–23, 27; Kovalev, "Dialog Stalina s Mao Tszedunom," pt. 1, p. 88; ibid., pt. 2, pp. 78, 82; "Posetiteli kremlevskogo kabineta," pt. 1, pp. 56–58.
333. "Posetiteli kremleuskogo kabineta," pt. 1, pp. 56–58. On August 5, a date that cannot be documented from other sources, Liu was accompanied by Wang Jiaxiang and Shi Zhe (Karskii); on the Soviet side, Stalin, Beriia, Bulganin, Malenkov, Mikoyan, Kaganovich, and Kovalev were present. Ibid., pp. 57–58.
334. On this and the following sentence, see Shi Zhe, *Zai lishi juren shenbian*, pp. 420–23; Bo, *Ruogan zhongda juece*, vol. 1, p. 37; *Zhonggong zhongyang zai Xiangshan*, p. 448.
335. Shi Zhe, *Zai lishi juren shenbian*, p. 397.
336. For further details on this and the following sentence, see further below in this section.
337. On this paragraph, see Shi Zhe, *Zai lishi juren shenbian*, p. 398. It is not clear when this banquet took place or what the important political questions were. Another Chinese publication mentions a reception at Stalin's dacha in Kuntsevo, which took place on July 10. See *Liu Shaoqi nianpu*, vol. 1, p. 218. In a later edition of his memoirs, Shi Zhe claimed that Stalin had offered a credit of US$300 million to his Chinese guests great pleasure; see Shi and Chen, "With Mao and Stalin," p. 27. It can be proved that Stalin made this proposal in the official working session on June 27. On this, see below in this section. Cf. Bo, *Ruogan zhongda juece*, vol. 1, p. 37, which mentions a reception by Stalin, Molotov, Malenkov, and Mikoyan on June 28. This information can be found in the same terms in *Zhonggong zhongyang zai Xiangshan*, p. 442.
338. The Soviet record of the talk, which is classified "top secret [*sovershenno sekretno*]" is in the Russian presidential archives; see AP RF, f. 45, op. 1, d. 329, ll. 1–7. It was drawn up by Kovalev and corrected on the basis of instructions from Stalin. Stalin added underlining and numbered the twelve points. (The underlined words are italicized in this book.) The note "to comrade Molotov V.M. in person" is written on the record. The record has since been published; see Ledovskii, "Vizit v Moskvu delegatsii," pt. 1, pp. 68–70, 67 n. Neither Shi Zhe nor Kovalev report on this talk. It is therefore also not mentioned in Goncharov, Lewis, and Xue, *Uncertain Partners*. See "Posetiteli kremlevskogo kabineta," pt. 1, p. 56. *Liu Shaoqi nianpu*, vol. 2, p. 217, dated the talk by Peking time, thus on June 28.
339. Ledovskii, "Vizit v Moskvu delegatsii," pt. 1, p. 68. Shi Zhe's name is not included on the list of visitors to Stalin's office. See "Posetiteli kremlevskogo kabineta," pt. 1, p. 56.
340. On this paragraph, see "Posetiteli kremlevskogo kabineta," pt. 1, p. 56.
341. This is an unusual term. It no doubt means the USSR's Eastern European satellite states, which are usually called "people's democracies" in Communist linguistic usage. In Ledovskii, "Sekretnaia missiia delegatsii TsK," p. 19, where the text is paraphrased, we accordingly find the phrase "Eastern European people's democracies" used at this point. It would appear that it was the Communist victory in China that led Stalin to distinguish between "Western democratic countries" and "Eastern democratic countries." This terminology did not become established. "Western" is also capitalized in the record of the talk.
342. On this paragraph, see Ledovskii, "Vizit v Moskvu delegatsii," pt. 1, p. 68.
343. On this paragraph, see ibid. Stalin drew two vertical lines in the margin next to the paragraph through "the government of Manchuria that already existed." Copy of the original.
344. Ledovskii, "Vizit Mao Tszeduna v Sovetskii Soiuz," pt. 1, p. 18, with no indication of sources.
345. On this and the following sentence, see Shi Zhe, "With Mao and Stalin," pt. 2, p. 79.

346. Letter from Liu Shaoqi to Stalin, July 25, 1949, AP RF, f. 39, op. 1, d. 38 l. 168.
347. Telegram from Mao Zedong to Liu Shaoqi, Gao Gang, and Wang Jiaxiang, July 25, 1949, cited in *Mao Zedong nianpu*, vol. 3, p. 538. The text of the trade agreement has not yet been made accessible.
348. *Zhonghua renmin gongheguo waijiao shi*, p. 14.
349. *Zhonggong zhongyang zai Xiangshan*, p. 455; *Mao Zedong nianpu*, vol. 3, p. 541.
350. Talk between Stalin and Mao, January 22, 1950, AP RF, f. 45, op. 1, d. 329, l. 35.
351. *Pravda*, July 31, 1949, p. 3.
352. Letter from Liu Shaoqi to Mikoyan, July 25, 1949, AP RF, f. 39, op. 1, d. 38, l. 16.
353. On this paragraph, see Ledovskii, "Vizit v Moskvu delegatsii," pt. 1, p. 68.
354. Ibid., pp. 68–69.
355. On this and the following three paragraphs, see ibid., p. 69.
356. On this paragraph, see ibid., pp. 69–70.
357. On this and the following three paragraphs, see ibid., p. 70.
358. Stalin underlined the words "most modern" and marked them with a question mark.
359. Mao Zedong, "Lun renmin minzhu zhuanzheng," in *Renmin ribao*, July 1, 1949, cited in *Mao Zedong ji*, vol. 10, pp. 291–307. There is a slightly modified version of the article in *Mao Zedong xuanji*, vol. 4, pp. 1473–86; an English-language version appears in *Selected Works of Mao Tse-tung*, vol. 4, pp. 411–24.
360. For further details, see chapter 2 above.
361. Record of the talk with Zhang Zhizhong on April 8,1949, in Yu Zhanbang, "Mao Zedong yu Zhang Zhizhong," pp. 152–53. According to information from the publisher of the record of the talk, Yu Zhanbang, who was Zhang Zhizhong's secretary, the record is based on minutes that were revised while Zhang was alive.
362. Ibid., pp. 153, 155.
363. He Di, "The Evolution of the Chinese Communist Party's Policy Toward the United States, 1944–1949," p. 39, with no indication of sources.
364. *Renmin Ribao*, July 1, 1949, cited in *Mao Zedong ji*, vol. 10, pp. 296–98; English-language text from *Selected Works of Mao Tse-tung*, vol. 4, pp. 415–17. This English-language edition is based on *Mao Zedong xuan ji*, 1st ed., vol. 4 (Peking: Renmin chubanshe, 1960). In the text of the article published in that edition on pp. 1473–93, the two sentences printed in italics in the quotation that appears in this book are missing; cf. p. 1478. They are therefore also missing in the English-language edition; cf. p. 415.
365. Cf. Kovalev, "Dialog Stalina s Mao Tszedunom," pt. 2, p. 88.
366. Ibid., p. 86.
367. *Pravda*, July 6, 1949, p. 3.
368. The author bases his views of the additional motives, among other things, on information from a historian from the People's Republic of China, who had broad access to unpublished contemporary documents in the CCP archives. He wishes to remain anonymous.
369. He Di also takes this position. See He Di, "Evolution," pp. 45–46.
370. Mao Zedong's speech of June 15, 1949; see *Selected Works of Mao Tse-tung*, vol. 4, p. 407.
371. Deng Xiaoping, July 19, 1949, in *Deng Xiaoping wenxuan*, vol. 1, p. 134.
372. Ibid.
373. Kovalev supposedly received this information from Liu Xiao, who later became ambassador in Moscow but at that time was a member of CCP military espionage. See Kovalev, "Dialog Stalina s Mao Tszedunom," pt. 1, p. 87.
374. Ibid., p. 88. Shi Zhe challenged these facts. See Li Haiwen, "Mao Zedong Liu Shaoqi fang Su," p. 92.
375. This fear can be seen in Mao's public speech of June 15, 1949; see *Selected Works of Mao Tse-tung*, vol. 4, p. 407.
376. On this paragraph, see Liu Shaoqi to Shibaev, December 25, 1949, AVP RF, f. 0100, op. 43, p. 302, d. 10, l. 27.
377. The term "propaganda" had positive connotations in Communist linguistic usage.
378. Liu Shaoqi to the journalist O. Chechetkina, January 27, 1950, AVP RF, f. 0100, op. 43, d. 11, l. 23–24.
379. Stuart to the Secretary of State, July 6, 1949, in Rea and Brewer, *Forgotten Ambassador*, pp. 335–36.
380. The reference is to Mao's speech on June 15, 1949, which the English service of the Xinhua News Agency published on June 20, 1949. It is reprinted in *Selected Works of Mao Tse-tung*, vol. 4, pp. 405–8. The passage Mao referred to is on p. 408. On the dating of the speech, cf. *FRUS 1949*, vol. 8, *The Far East: China*, pp. 392, 775 n. 82.
381. On this paragraph, see *FRUS 1949*, vol. 8, *The Far East: China*, pp. 774–76. For further details on the Chen Mingshu message of July 11, 1949, see this chapter below.

382. Stuart to Acheson, July 13, 1949, in ibid., vol. 8, pp. 782–83.

383. See Tucker, *Patterns in the Dust*, p. 47; Manfred Vasold, "Versäumte Gelegenheiten?" p. 251, with references to source. For further details, see this chapter below.

384. Ledovskii, "Vizit v Moskvu delegatsii," pt. 1, p. 70.

385. Shi Zhe, *Zai lishi juren shenbian*, p. 398.

386. *Liu Shaoqi nianpu*, vol. 2, p. 217; *Zhonggong zhongyang zai Xiangshan*, p. 443.

387. Russian-language letter, cited in a copy of the original in the author's possession; AP RF, f. 45, op. 1, d. 328, l. 32. An eighteen-page Russian-language version of the report, which was stamped "*sovershenno sekretno* [top secret]" by Stalin's office, is in the Russian presidential archives (AP RF, f. 45, op. 1, d. 328, ll. 33–50). It has since been published; see Ledovskii, "Vizit v Moskvu delegatsii," pt. 1, pp. 71–82 (the archival pagination noted there is inaccurate). The author is in possession of copies of two pages of the report with handwritten comments by Stalin. S.P. Andreev translated the report. The Russian style at times sounds wooden and stilted, which may be explained by the shortness of the deadline—within the same day—with which the report had to be translated. See copy of the original of the final page of the report, AP RF, f. 45. op. 1, d. 328, l. 50 (in the author's possession). The Russian version of the report, evaluated here was not available to Goncharov, Lewis, and Xue. They assumed that the report had "apparently not survived" and based their account solely on Shi Zhe's version, which is incomplete. See Goncharov, Lewis, and Xue, *Uncertain Partners*, pp. 64–66.

388. The substance of the report is in part reproduced literally and in part paraphrased below. Stalin is responsible for all of the underlining included in the following translations of excerpts of the report. (The underlined words are italicized in this book.) Stalin's underlining and comments are not always the same in the available sources (the two original pages; Ledovskii, "Sekretnaia missiia delegatsii TsK"; Ledovskii, "Vizit Mao Tszeduna v Sovetskii Soiuz," pt. 1; and Ledovskii, "Vizit v Moskvu delegatsii," pt. 1).

389. Cf. Shi Zhe, *Zai lishi juren shenbian*, pp. 398–403; Zhu Yuanshi, "Liu Shaoqi 1949 nian mimi fang Su," pp. 77–79; *Zhonggong zhongyang zai Xiangshan*, p. 444. One gets the impression that the latter two publications took their versions from Shi Zhe.

390. Stalin added the following comment questioning this information: "1) Are there any *pilots?* Are there any *sailors?*" On this paragraph, see Ledovskii, "Vizit v Moskvu delegatsii," pt. 1, p. 71.

391. On this paragraph, see ibid., pp. 71–72.

392. On this and the two following paragraphs, see ibid., p. 72.

393. For further information, see chapter 1 above.

394. Stalin drew a vertical line alongside this and the following sentence. See Ledovskii, "Vizit v Moskvu delegatsii," pt. 1, p. 73.

395. On this and the following sentence, see ibid., p. 73. In the version of these two paragraphs in the report presented by Shi Zhe, the information on the Communists' secret dominating role in establishing the PCC is missing for obvious reasons. See Shi Zhe, *Zai lishi juren shenbian*, p. 400.

396. Comment in the margin by Stalin: "2) And will there *also be provincial governments?"*

397. Comment in the margin by Stalin: "3) What is this: de facto the *president?*"

398. On this paragraph, see Ledovskii, "Vizit v Moskvu delegatsii," pt. 1, pp. 73–74.

399. Comment in the margin by Stalin: "4) What is meant by 'bureaucratic' capital?"

400. Comment in the margin by Stalin: "Yes!"

401. Shi Zhe's version differs by first mentioning an "alliance of workers and peasants [*gongnong lianmeng*]," but then an "alliance of workers, peasants and revolutionary intellectuals [*gongren, nongmin yu geming zhishi fenzi de lianmeng*]." See Shi Zhe, *Zai lishi juren shenbian*, p. 400.

402. Comment in the margin by Stalin: "5) Yes!"

403. Comment in the margin by Stalin: "Yes!"

404. On this and the following paragraph, see Ledovskii, "Vizit v Moskvu delegatsii," pt. 1, p. 74.

405. The Chinese original says "Soviet system [*suweiai zhi*]"; see Shi Zhe, *Zai lishi juren shenbian*, p. 401.

406. Stalin drew a vertical line in the margin beside this paragraph and noted: "Yes!" See Ledovskii, "Sekretnaia missiia delegatsii TsK," p. 32. Cf. Ledovskii, "Vizit v Moskvu delegatsii," pt. 1, p. 74, where there is no reference to the vertical line.

407. Comment in the margin by Stalin: "Yes!"

408. Code name of Anastas Mikoyan. This is a reference to Mikoyan's talks with the CCP leadership in late January and early February, 1949, in Xibaipo. On this, see this chapter above.

409. On this and the following three paragraphs, see Ledovskii, "Vizit v Moskvu delegatsii," pt. 1, p. 75.

410. Stalin marked this sentence with a vertical line in the margin and noted: "Collective contracts, which exclude strikes."

411. Comment in the margin by Stalin referring to the entire paragraph: "7) Will there be *local* (provincial) governments?" Cf. also Ledovskii, "Sekretnaia missiia delegatsii TsK," p. 34.

412. Comment by Stalin: "Yes!"

413. On this paragraph, see Ledovskii, "Vizit v Moskvu delegatsii," pt. 1, pp. 75–76.

414. The literal translation is "main leader [*osnovnoi rukovoditel'*]."

415. Comment in the margin by Stalin: "8) and the comprador group?" On this and the following paragraph, see Ledovskii, "Vizit v Moskvu delegatsii," pt. 1, p. 76.

416. On the debate over the concept of a coalition government, see this chapter above.

417. The Russian text used *provotsirovat'* [provoke]" for "obstruct"—an example of the poor Russian that had by then become commonplace among Communist cadres.

418. On this and the following paragraph, see Ledovskii, "Vizit v Moskvu delegatsii," pt. 1, p. 77.

419. Comment by Stalin: "9) Yes."

420. AP RF, f. 45, op. 1, d. 328, ll. 43–44, cited in Ledovskii, "Sekretnaia missiia delegatsii TsK," p. 36 and Ledovskii, "Vizit v Moskvu delegatsii," pt. 1, pp. 77–78. Stalin drew a thick vertical line in the margin alongside the four points.

421. Stalin circled the word "confiscated," drew two vertical lines in the margin alongside this point, and noted: "10) By whom?" See Ledovskii, *Sekretnaia missiia delegatsii TsK*, p. 37, and Ledovskii, "Vizit v Moskvu delegatsii," pt. 1, p. 78. Ledovskii, "Vizit v Moskvu delegatsii" does not indicate the circled word or the lines in the margin.

422. AP RF, f. 45, op. 1, d. 328, l. 44, cited in Ledovskii, "Sekretnaia missiia delegatsii TsK," p. 37, and Ledovskii, "Vizit v Moskvu delegatsii," pt. 1, p. 78. In Shi Zhe's version, there is only the observation that there had not been very significant foreign investment during the Second World War in the place of this paragraph. See Shi Zhe, *Zai lishi juren shenbian*, p. 402.

423. Comment in the margin by Stalin: "11) What restrictions?"

424. AP RF, f. 45, op. 1, d. 328, l. 44–45, cited in Ledovskii, "Sekretnaia missiia delegatsii TsK," p. 3, and Ledovskii, "Vizit v Moskvu delegatsii," pt. 1, p. 78.

425. AP RF, f. 45, op. 1, d. 328, l. 45, cited in Ledovskii, "Sekretnaia missiia delegatsii TsK," p. 38, and Ledovskii, "Vizit v Moskvu delegatsii," pt. 1, pp. 78–79.

426. Two vertical lines and a comment in the margin by Stalin: "12) Yes!"

427. AP RF, f. 45, op. 1, d. 328, l. 45, cited in Ledovskii, "Sekretnaia missiia delegatsii TsK," p. 38, and Ledovskii, "Vizit v Moskvu delegatsii," pt. 1, p. 79.

428. AP RF, f. 45, op. 1, d. 328, ll. 45–46, cited in Ledovskii, "Sekretnaia missiia delegatsii TsK," pp. 38–39, and Ledovskii, "Vizit v Moskvu delegatsii," pt. 1, p. 79.

429. Stalin drew two vertical lines in the margin from here to the end of the paragraph.

430. AP RF, f. 45, op. 1, d. 328, l. 45–46, cited in Ledovskii, "Sekretnaia missiia delegatsii TsK," p. 39; reproduced in part from a copy of the original page 14 of the report. Cf. Ledovskii, "Vizit v Moskvu delegatsii," pt. 1, p. 79. Stalin's additions reported in the previous footnote are not indicated in ibid., p. 79.

431. This paragraph is reproduced on the basis of a copy of the original page 14 of the report (AP RF, f. 45, op. 1, d. 328, l. 46). Cf. Ledovskii, "Vizit v Moskvu delegatsii," pt. 1, p. 79, which does not indicate the underlinings by Stalin.

432. Stalin wrote a "P" (for *pravil'no*, which means "right") in the margin next to this sentence. Copy of the original page 14 of the report. This indication is missing in Ledovskii, "Vizit v Moskvu delegatsii," pt. 1, p. 79.

433. Stalin provided the following comment in the margin on the underlined passage: "13) Differentiated." He wrote diagonally across the text: "Anyone who does not recognize China should be given no commercial privileges with China (the crisis in the USA will force the USA to value trade with China highly). Use Chinese businessmen for this." Stalin drew a vertical line in the margin alongside the sentences in this paragraph beginning with "It is possible that . . ." through ". . . not hurry" and wrote a "P" (for "right"). Copy of the original of page 14 of the report and Ledovskii, "Vizit v Moskvu delegatsii," pt. 1, p. 79.

434. Stalin wrote the word "Yes" diagonally across this sentence and noted in the margin: "14) It is better not to hurry." He also wrote the comment "P" ("right") in the margin. See copy of the original of page 14 of the report. There is no reference to the comment "P" in the margin in Ledovskii, "Vizit v Moskvu delegatsii," pt. 1, p. 80.

435. Comment in the margin by Stalin: "Yes!"

436. Comment in the margin by Stalin: "15) Yes!" See Ledovskii, "Vizit v Moskvu delegatsii," pt. 1, p. 80; Ledovskii, "Sekretnaia missiia delegatsii TsK," p. 40. The underlining is indicated only in Ledovskii's unpublished manuscript.

437. Comment in the margin by Stalin: "Yes." This paragraph in AP RF, f. 45, op. 1, d. 328, ll. 46–47, is found in Ledovskii, "Sekretnaia missiia delegatsii TsK," pp. 39–40; reproduced in part from a copy of the original of page 14 of the report. Cf. Ledovskii, "Vizit v Moskvu delegatsii," pt. 1, pp. 79–80, which does not indicate all the underlinings and the final comment, "Yes."

438. AP RF, f. 45, op. 1, d. 328, l. 47, cited in Ledovskii, "Sekretnaia missiia delegatsii TsK," p. 40; Ledovskii, "Vizit v Moskvu delegatsii," pt. 1, p. 80. Stalin added two vertical lines in the margin beginning where it reads "We are prepared" through the end of the paragraph and the comment "Yes!" These and the other statements in the third section are reproduced in summary form in Shi Zhe's version, but their contents are essentially equivalent. See Shi Zhe, *Zai lishi juren shenbian*, pp. 402–3. Similarly in *Zhonghua renmin gongheguo waijiao shi*, p. 12; Zhu Yuanshi, "Liu Shaoqi 1949 nian mimi fang Su," pp. 78–79.

439. Comment by Stalin: "16) Yes." See Ledovskii, "Vizit v Moskvu delegatsii," pt. 1, p. 80. In Ledovskii, "Sekretnaia missiia delegatsii TsK," p. 40, there is no reference to Stalin's comment "Yes."

440. This and the following paragraph from AP RF, f. 45, op. 1, d. 328, l. 47, cited in Ledovskii, "Sekretnaia missiia delegatsii TsK," pp. 40–41, and Ledovskii, "Vizit v Moskvu delegatsii," pt. 1, p. 80. The declaration mentioned was made on April 3, 1949. The text was printed in *Pravda*, April 10, 1949, p. 4.

441. Comment by Stalin on this sentence: "Yes indeed" ("*Tak*"). See Ledovskii, "Vizit v Moskvu delegatsii," pt. 1, p. 80. Ledovskii, "Sekretnaia missiia delegatsii TsK," p. 41, has no reference to this comment.

442. See below, section entitled "Questions of Soviet-Chinese Relations."

443. Ledovskii, "Vizit v Moskvu delegatsii," pt. 1, p. 80.

444. For further details, see chapter 4 above.

445. AP RF, f. 45, op. 1, d. 328, l. 47, cited in Ledovskii, "Sekretnaia missiia delegatsii TsK," p. 40, and Ledovskii, "Vizit v Moskvu delegatsii," pt. 1, p. 80.

446. AP RF, f. 45, op. 1, d. 328, l. 48, cited in Ledovskii, "Vizit Mao Tszeduna v Sovetskii Soiuz," pt. 1, pp. 13–14; Ledovskii, "Sekretnaia missiia delegatsii TsK," pp. 41, 74; Ledovskii, "Vizit v Moskvu delegatsii," pt. 1, p. 80. At this point, Stalin made the comment "P" ("right") in the margin. In Ledovskii's version, there is no reference to the underlining and the comment "P." As in the case cited in the previous paragraph, ("continue to observe") "observe" here is translated awkwardly as *priniat'* instead of the usual *sobliudat'* or *priderzhivat'sia*. But the meaning is clear.

447. For further details, see chapter 2 above. Cf. Dieter Heinzig, "Verriet Stalin im Jahre 1945 die kommunistische Revolution in China?" p. 9.

448. Ledovskii, "Vizit Mao Tszeduna v Sovetskii Soiuz," pt. 1, p. 10. For further details, see this chapter above.

449. AP RF, f. 39, op. 1, d. 328, l. 48, cited in Ledovskii, "Vizit Mao Tszeduna v Sovetskii Soiuz," pt. p. 14, and Ledovskii, "Sekretnaia missiia delegatsii TsK," p. 41. The contents of the three alternatives are reported identically in *Zhonghua renmin gongheguo waijiao shi*, p. 12. Shi Zhe did not mention them; cf. Shi Zhe, *Zai lishi juren shenbian*, p. 403. Similarly, see Zhu Yuanshi, "Liu Shaoqi 1949 nian mimi fang Su," p. 79.

450. AP RF, f. 45, op. 1, d. 328, l. 48, cited in Ledovskii, "Vizit Mao Tszeduna v Sovetskii Soiuz," pt. 1, p. 14; Ledovskii, "Sekretnaia missiia delegatsii TsK," p. 74; and Ledovskii, "Vizit v Moskvu delegatsii," pt. 1, p. 81.

451. AP RF, f. 45, op. 1, d. 328, l. 48, cited in Ledovskii, "Sekretnaia missiia delegatsii TsK," p. 42, and Ledovskii, "Vizit v Moskvu delegatsii," pt. 1, p. 81.

452. AP RF, f. 45, op. 1, d. 328, l. 49, cited in Ledovskii, "Sekretnaia missiia delegatsii TsK," p. 42; Ledovskii, "Vizit v Moskvu delegatsii," pt. 1, p. 81. The following discussions of this controversial topic are not reproduced in the Chinese literature on Liu's report that are cited above. They are not even mentioned except in Shi Zhe, who stated simply that something was said about the Soviet troops stationed in Port Arthur and Mongolian independence (*menggu duli*). See Shi Zhe, *Zai lishi juren shenbian*, p. 403.

453. The abbreviation of Mongolian People's Republic.

454. See also Shi Zhe, *Zai lishi juren shenbian*, p. 449.

455. Comment by Stalin: "18) Yes." Stalin drew two vertical lines in the margin next to this paragraph. See Ledovskii, "Vizit v Moskvu delegatsii," pt. 1, p. 81, and Ledovskii, "Sekretnaia missiia delegatsii TsK," p. 42.

456. Ledovskii, "Sekretnaia missiia delegatsii TsK," p. 42. Ledovskii, "Vizit v Moskvu delegatsii," pt. 1, p. 81, has no reference to this comment.

457. Ledovskii, "Sekretnaia missiia delegatsii TsK," p. 42. Ledovskii, "Vizit v Moskvu delegatsii," pt. 1, p. 81, has no reference to any underlining.

458. Cf. Liu Shaoqi to O. Chechetkina, January 27, 1950, AVP RF, f.100, op. 43, d. 11, ll. 22–23; Kovalev, "Dialog Stalina s Mao Tszedunom," pt. 2, p. 88.

459. An official in the Soviet consulate general in Harbin learned this on December 28, 1949; see AVP RF, f. 07, op. 43, d. 13, p. 302, l. 15.

460. Telegram from the CC of the CCP to the Northeast Bureau, February 7, 1949, cited in Zhou Wenqi and Zhu Liangru, eds., *Gongchan guoji, Sulian he Zhongguo gongchandang guanxi biannianshi, 1919–1991*, p. 459. At the end of the telegram, it said that the news should be coordinated with the Soviet side.

461. On this and the following sentence, see Mikoyan's report, September 22, 1960, in Ledovskii, "Sekretnaia missiia A.I. Mikoiana v Kitai," pt. 1, p. 107. See this chapter above.

462. For further details, see ibid.

463. Undated telegram from Mikoyan to Stalin, cited in Ledovskii, "Sekretnaia missiia A.I. Mikoiana v Kitai," pt. 2, pp. 100–101; ibid., pt. 1, p. 106. For further details, see this chapter above.

464. Kovalev, "Dialog Stalina s Mao Tszedunom," pt. 2, p. 86.

465. Stalin drew two vertical lines in the margin next to this paragraph and commented "19) Yes! We will help." See AP RF, f. 45, op. 1, d. 328, l. 49, cited in Ledovskii, "Sekretnaia missiia delegatsii TsK," pp. 42–43, and Ledovskii, "Vizit v Moskvu delegatsii," pt. 1, p. 81.

466. AP RF, f. 45, op. 1, d. 328, l. 49, cited in Ledovskii, "Sekretnaia missiia delegatsii TsK," p. 43. Ledovskii, "Vizit v Moskvu delegatsii," pt. 1, p. 81, makes no reference to the underlining or the willingness to help.

467. In the margin next to the last part of the paragraph, which mentioned sending out specialists, teachers, delegations and students, Stalin put two vertical lines and the comment "20) Yes!" See AP RF, f. 45, op. 1, d. 328, l. 49, cited in Ledovskii, "Sekretnaia missiia delegatsii TsK," p. 43, and Ledovskii, "Vizit v Moskvu delegatsii," pt. 1, pp. 81–82.

468. Comment in the margin by Stalin: "21) No!" See Ledovskii, "Sekretnaia missiia delegatsii TsK," p. 44; Ledovskii, "Vizit v Moskvu delegatsii," pt. 1, p. 82. The CCP's willingness to play a subordinate role announced in this paragraph is not mentioned in the Chinese literature cited on Liu's report.

469. Comment in the margin by Stalin: "No!"

470. Comment in the margin next to the underlined passage by Stalin: "22) Yes!"

471. AP RF, f. 45, op. 1, d. 328. l. 50, cited in Ledovskii, "Sekretnaia missiia delegatsii TsK," p. 44, and Ledovskii, "Vizit v Moskvu delegatsii," pt. 1, p. 82.

472. Comment in the margin and diagonally across the text of this paragraph by Stalin: "23) Yes. In late 1949 after est. of dipl. Relations." AP RF, f. 45, op. 1, d. 328, l. 50, cited in Ledovskii, "Sekretnaia missiia delegatsii TsK," p. 44, and Ledovskii, "Vizit v Moskvu delegatsii," pt. 1, p. 82.

473. In the margin next to this sentence, Stalin wrote "and the silver?" AP RF, f. 45, op. 1, d. 238, l. 50, cited in Ledovskii, "Sekretnaia missiia delegatsii TsK," p. 44, and Ledovskii, "Vizit v Moskvu delegatsii," pt. 1, p. 82. By this he meant the delivery of silver that the Chinese side had already requested from Mikoyan in Xibaipo in connection with the credit. The silver was to be used to issue a stable Chinese currency.

474. On this and the following sentence, see AP RF, f. 45, op. 1, d. 238, l. 50, copy of the original; cf. Ledovskii, "Vizit v Moskvu delegatsii," pt. 1, p. 82.

475. This phrase, for example, was used in a telegram that Liu Shaoqi, Zhu De, and Zhou Enlai sent to Mao Zedong on December 21, 1949. See *Zhonghua renmin gongheguo waijiao shi*, p. 18. The long-serving Soviet China diplomat M.S. Kapitsa noted that Mao used the phrase *qing zhi* when he asked Stalin for "instructions." See M.S. Kapitsa, *Na raznykh paralleliakh*, cited in A. Khazanov, "Informatsiia iz pervykh ruk," p. 62. Kapitsa's book, which was not available to the author, is cited here without page numbers.

476. For further details, see below in this chapter.

477. See this chapter below (section "Discussion of Liu's Report and Wish List." There was another discussion at the banquet on July 30. See this chapter below (section "The Elder Brother–Younger Brother" Debate).

478. Talk between Stalin and Zhou Enlai, September 19, 1952, AP RF, f. 45, op. 1, d. 329, l. 99.

479. Cf. Yang Kuisong, *Zhongjian didai de geming*, p. 333 n. 1. The CCP also called internal Party instructions (for example by the Central Committee) "*zhishi*"; see ibid., p. 347 n.2.

480. The letter was addressed as follows: "To the CC of the CPSU(B). To Comrade Stalin I.V." It was signed by Liu in his capacity of "head of the delegation of the CC of the CCP" in Chinese characters. S.P. Andreev provided the translation into Russian—as was the case with the July 4 report to Stalin. The original of the Russian-language version is in the Russian presidential archives: AP RF, f. 45, op. 1, d. 328, ll. 351–55. The author thanks Andrei Ledovskii for making the document available almost in its entirety and for the most part as a copy of the original; see Ledovskii, "Sekretnaia missiia delegatsii TsK," pp. 45–49. It has been published in full in the meantime in Ledovskii, "Vizit v Moskvu delegatsii," pt. 2, pp. 84–88, where the date is mistakenly given as June 6. Before, this letter had been mentioned only by Shi Zhe, who summarized its contents in two sentences; see Shi Zhe, *Zai lishi juren shenbian*, pp. 403–4.

481. Stalin drew a cross and a vertical line in the margin next to this paragraph; AP RF, f. 45, op. 1, 328, l. 51, reproduced from a copy of the original. See Ledovskii, "Sekretnaia missiia delegatsii TsK," p. 45, and Ledovskii, "Vizit v Moskvu delegatsii," pt. 2, p. 84.

482. AP RF, f. 45, op. 1, d. 328, l. 51, reproduced from the copy of the original, presented in Ledovskii, "Sekretnaia missiia delegatsii TsK," p. 45, and Ledovskii, "Vizit v Moskvu delegatsii," pt. 2, pp. 84–85.

483. Stalin marked this and the following sentence with a vertical line in the margin. See Ledovskii, "Vizit v Moskvu delegatsii," pt. 2, p. 85.

484. AP RF, f. 45, op. 1, d. 328, ll. 51–52, cited in Ledovskii, "Sekretnaia missiia delegatsii TsK," pp. 45–46, and Ledovskii, "Vizit v Moskvu delegatsii," pt. 2, pp. 85–86.

485. AP RF, f. 45, op. 1, d. 328, ll. 52–53, cited in Ledovskii, "Sekretnaia missiia delegatsii TsK," p. 46, and Ledovskii, "Vizit v Moskvu delegatsii," pt. 2, p. 86.

486. Stalin drew a vertical line in the margin alongside the beginning of this paragraph through the word "departments." AP RF, f. 45, op. 1, d. 328, l. 53, reproduced from a copy of the original, cited in Ledovskii, "Sekretnaia missiia delegatsii TsK," p. 47; cf. Ledovskii, "Vizit v Moskvu delegatsii," pt. 2, p. 86.

487. Stalin marked a vertical line in the margin next to the first sentence of this paragraph. AP RF, f. 45, op. 1, d. 328, ll. 53–54, reproduced from a copy of the original, cited in Ledovskii, "Sekretnaia missiia delegatsii TsK," p. 47; cf. Ledovskii, "Vizit v Moskvu delegatsii," pt. 2, p. 86.

488. Underlining by Stalin. (The underlined words are italicized in this book.) Stalin drew a vertical line in the margin from the beginning of this paragraph through the words "Ulan Bator to the USSR," and four intersecting lines next to the rest of the paragraph. This paragraph and the two following paragraphs are reproduced from a copy of the original, cited in Ledovskii, "Sekretnaia missiia delegatsii TsK," pp. 47–48; cf. Ledovskii, "Vizit v Moskvu delegatsii," pt. 2, p. 87.

489. Stalin drew four intersecting lines in the margin next to this paragraph. Ledovskii, "Vizit v Moskvu delegatsii," pt. 2, p. 87.

490. Underlining by Stalin. He drew a vertical line and two additional crosses in the margin next to this paragraph. Ledovskii, "Vizit v Moskvu delegatsii," pt. 2, p. 87.

491. Underlining by Stalin. Stalin drew a vertical line in the margin beside the beginning of the paragraph through the words "reconstruct industry in Dairen even better," indicated the sentence about provisions with a cross, highlighted the questions on the opening of Dairen harbor for the export of coal, among other things, and about the ability of American and English ships to use the harbor with two crosses, and marked the paragraph about the hydroelectric power plant on the Yalu with four intersecting lines. AP RF, f. 45, op. 1, d. 328, ll. 54–55, reproduced from a copy of the original, cited in Ledovskii, "Sekretnaia missiia delegatsii TsK," p. 48; cf. Ledovskii, "Vizit v Moskvu delegatsii," pt. 2, p. 87.

492. Stalin drew a vertical line and a cross in the margin next to this paragraph. AP RF, f. 45, op. 1, d. 328, l. 55, reproduced from a copy of the original, cited in Ledovskii, "Sekretnaia missiia delegatsii TsK," pp. 48–49; cf. Ledovskii, "Vizit v Moskvu delegatsii," pt. 2, p. 87.

493. Reproduced from a copy of the original, cited in Ledovskii, "Sekretnaia missiia delegatsii TsK," p. 49; cf. Ledovskii, "Vizit v Moskvu delegatsii," pt. 2, p. 88.

494. On this paragraph, see Shi Zhe, *Zai lishi juren shenbian*, p. 404; *Zhonggong zhongyang zai Xiangshan*, p. 447. Access has not yet been permitted to minutes of this session. The following reconstruction of the session is based primarily on information provided by Shi Zhe and documents from Kovalev's personal archive, in addition to one Chinese book whose author had access to the CCP archives; on the latter, see *Zhonghua renmin gongheguo waijiao shi*, pp. 12–13. According to Shi Zhe, the July 11 session was the third, while Kovalev incorrectly spoke of it as the first session. See Shi Zhe, *Zai lishi juren shenbian*, p. 398; Kovalev, "Dialog Stalina s Mao Tszedunom," pt. 1, p. 88. On the basis of all the circumstances that have become known, however, it must be the second session. This can be seen from *Liu Shaoqi nianpu*, vol. 2, p. 218. This is also suggested by the fact that Ledovskii found out from Soviet archival documents that a second session had been planned for July 11 after the first session on June 27; see Ledovskii, "Sekretnaia missiia delegatsii TsK," p. 45.

495. On this paragraph, see Shi Zhe, *Zai lishi juren shenbian*, p. 406; Shi and Chen, "With Mao and Stalin," p. 34.

496. On this paragraph, see Liu Jiecheng, *Mao Zedong yu Si-da-lin*, p. 543. Liu makes some literal citations, but provides no indication of his sources.

497. Shi Zhe, *Zai lishi juren shenbian*, pp. 404–5.

498. Ibid., p. 405.

499. The reference appears to be to the July 27 session.

500. Shi Zhe, *Zai lishi juren shenbian*, pp. 408–9; Shi and Chen, "With Mao and Stalin," pp. 36–37.

501. According to a record drawn up by Kovalev, reproduced in Kovalev, "Dialog Stalina s Mao Tszedunom," pt. 2, p. 82. The citation is incorrectly dated July 27 here. Goncharov, Lewis, and Xue corrected this; see Goncharov, Lewis, and Xue, *Uncertain Partners*, p. 69.

502. On this paragraph, see Shi Zhe, *Zai lishi juren shenbian*, p. 405; *Zhonghua renmin gonghegua waijiao shi*, p. 13.

503. Kovalev, "Dialog Stalina s Mao Tszedunom," pt. 2, p. 82.

504. Ibid., pp. 82–83. For further details, see chapter 4 below. Cf. also Goncharov, Lewis, and Xue, *Uncertain Partners*, p. 69.

505. *Zhonghua renmin gongheguo waijiao shi*, p. 13.

506. Ibid.; in almost the same terms, Shi Zhe, *Zai lishi juren shenbian*, p. 405.

507. Hu Qiaomu, *Hu Qiaomu huiyi Mao Zedong*, p. 551.

508. On this paragraph, see *Zhonghua renmin gongheguo waijiao shi*, p. 13, with no indication of sources and yet clearly based on archival documents; Shi Zhe, *Zai lishi juren shenbian*, p. 405. Kovalev said nothing about this topic.

509. AP RF, f. 39, op. 1, d. 39, ll. 78–79, in Ledovskii, "Sekretnaia missiia A.I. Mikoiana v Kitai," pt. 2, pp. 100–101; ibid., pt. 1, p. 106. For further details, see this chapter above.

510. *Zhonghua renmin gongheguo waijiao shi*, p. 13, with no indication of sources and yet clearly based on archival documents; Shi Zhe, *Zai lishi juren shenbian*, p. 405. Cf. Shi Zhe quoted in Li Haiwen, "Mao Zedong Liu Shaoqi fang Su," pp. 92–93.

511. See Kovalev's notes, which he took during the July 11 talk, in Kovalev, "Dialog Stalina s Mao Tszedunom," pt. 2, p. 86.

512. On this and the following two sentences, see Stalin's telegram to Mikoyan, February 6, 1949, AP RF, f. 39, op. 1, d. 39, ll. 78–79, reproduced from a copy of the original, cited in Ledovskii, "Vizit Mao Tszeduna v Sovetskii Soiuz," pt. 1, pp. 12–13. For further details, see this chapter above. On this paragraph, cf. Goncharov, Lewis, and Xue, *Uncertain Partners*, pp. 67–68.

513. Goncharov, Lewis, and Xue, *Uncertain Partners*, pp. 68, 311 n. 124, based on Kovalev's unpublished memoirs and Kovalev, "Dialog Stalina s Mao Tszedunom," pt. 2, p. 86.

514. AP RF, f. 45, op. 1, d. 328, l. 48, cited in Ledovskii, "Sekretnaia missiia delegatsii TsK," p. 42. For further details, see above in this section.

515. See Kovalev's notes, which he took during the talk, in Kovalev, "Dialog Stalina s Mao Tszedunom," pt. 2, p. 78. Kovalev did not date the quotation from Stalin. As Stalin made explicit reference to Liu's report and the report was discussed essentially in the July 11 session, one can assume that Stalin made his observations in this session.

516. On this paragraph Shi Zhe, *Zai lishi juren shenbian*, p. 406.

517. On this paragraph, see ibid., pp. 406–7; *Zhonghua renmin gongheguo waijiao shi*, p. 13.

518. Shi Zhe, *Zai lishi juren shenbian*, p. 406.

519. Kovalev, "Dialog Stalina s Mao Tszedunom," pt. 2, p. 79. Kovalev moved Liu's question and Stalin's answer to "July 1949"; no precise dating is given. See Goncharov, Lewis, and Xue, *Uncertain Partners*, pp. 232–33. The text can be found here in an English translation.

520. Cf. recent discoveries about the creation and duties of the Cominform based on hitherto unpublished archival documents in L.Ia. Gibanskii, "Kominform v deistvii, 1947–1948 gg," pt. 1, pp. 149–70, and pt. 2, pp. 157–72. On the founding of the Cominform, see chapter 2 above.

521. Kovalev, "Dialog Stalina s Mao Tszedunom," pt. 2, p. 79. Shi Zhe provided only a brief description of this statement by Stalin, but its meaning was the same. In his presentation, Stalin also claimed that all parties had equal rights in the Cominform. No party could force its will on any other. Shi Zhe, *Zai lishi juren shenbian*, p. 413.

522. On this and the following paragraph, see Kovalev, "Dialog Stalina s Mao Tszedunom," pt. 2, pp. 79–80. This part of Stalin's discussion is not included in Shi Zhe, *Zai lishi juren shenbian*, p. 413. Cf. Li Haiwen, "Mao Zedong Liu Shaoqi fang Su," p. 90.

523. Telegram from Kovalev to Stalin, delivered through the GRU Administration of the Soviet General Staff, May 18, 1949, AP RF, f. 45, op. 1, d. 329, ll. 59–60.

524. Cf. the different interpretation in Goncharov, Lewis, and Xue, *Uncertain Partners*, pp. 71–73.

525. For further details, see this chapter below.

526. Kovalev's unpublished memoirs, cited in Goncharov, Lewis, and Xue, *Uncertain Partners*, pp. 71–72.

527. Shi Zhe did not date the following discussion, but suggested that it had taken place after the discussion of the question of the CCP's membership in the Cominform. See Shi Zhe, *Zai lishi juren shenbian*, pp. 413–14.

528. This sentence is not included in the original version of Shi Zhe's memoirs; cf. Shi Zhe, *Zai lishi juren shenbian*, p. 414. He added it in a revised version; see Shi Zhe, "With Mao and Stalin," pt. 2, p. 82. On the events of August 1945, see chapter 2 above.

529. The reference was to the Marshall mission (1945–46).

530. On this, see Shi Zhe, *Zai lishi juren shenbian*, p. 414; Shi Zhe, "Peitong Mao zhuxi fang Su," Renwu 5 (1988), pp. 7–8, Shi Zhe, "With Mao and Stalin," pt. 2, p. 83; Zhu Yuanshi, "Liu Shaoqi 1949 nian mimi fang Su," p. 77, with source references; *Zhonghua renmin gongheguo waijiao shi*, p. 12. Kovalev did not mention this episode; supposedly no reference to it could be found in his archives. See Kovalev, "Dialog Stalina s Mao Tszedunom," pt. 2, p. 90 n. 1.

531. Cited in the revised version of Shi Zhe's memoirs; see Shi Zhe, "With Mao and Stalin," pt. 2, p. 83. Cf. the compressed statement in the original version, Shi Zhe, *Zai lishi juren shenbian*, p. 414.

532. Shi Zhe, *Zai lishi juren shenbian*, p. 415; Shi and Chen, "With Mao and Stalin," p. 39.

533. Goncharov, Lewis, and Xue, to whom the Soviet minutes of the June 27 session were not available, mistakenly assumed, on the basis of Mikoyan's undated transcript, that Stalin made this promise on July 11. Kovalev mentioned "40 to 50 aircraft." See Goncharov, Lewis, and Xue, *Uncertain Partners*, p. 312 n. 136.

534. On this and the following sentence, see Shi Zhe, *Zai lishi juren shenbian*, p. 407; Goncharov, Lewis, and Xue, *Uncertain Partners*, p. 70, with source references.

535. On this and the following sentence, see Shi Zhe, *Zai lishi juren shenbian*, p. 408.

536. Cf. Chen Jian, *China's Road*, p. 89.

537. On this paragraph, see He Di, "The Last Campaign to Unify China," pp. 1–3.

538. Kovalev, "Dialog Stalina s Mao Tszedunom," pt. 1, 88; Kovalev's personal archive and Kovalev's unpublished memoirs, cited in Goncharov, Lewis, and Xue, *Uncertain Partners*, p. 69.

539. Kovalev, "Dialog Stalina s Mao Tszedunom," pt. 1, p. 88.

540. *Khrushchev Remembers: The Glasnost Tapes*, pp. 100–101.

541. Kovalev, "Dialog Stalina s Mao Tszedunom," pt. 1, p. 88; Goncharov, Lewis, and Xue, *Uncertain Partners*, pp. 69–70.

542. Li Haiwen, "Mao Zedong Liu Shaoqi fang Su," p. 92. Shi Zhe did not address this topic in his memoirs. Cf. Shi Zhe, "Peitong Mao zhuxi fang Su," Renwu 5 (1988), pp. 7–11; Shi Zhe, *Zai lishi juren shenbian*, pp. 397–24.

543. On this and the following three sentences, see Shi Zhe, *Zai lishi juren shenbian*, pp. 405–6.

544. Ibid., p. 408.

545. For further details, see this chapter above in this same section (section "Liu's Wish List").

546. Letter from Liu Shaoqi to Mikoyan, July 22, 1949, AP RF, f. 39, op. 1, d. 38, l. 166. For further details on this paragraph, see Bo, *Ruogan zhongda juece*, vol. 1, p. 37; Shi Zhe, *Zai lishi juren shenbian*, pp. 419–23.

547. Liu Shaoqi's letter to Stalin, July 6, 1949, AP RF, f. 45, op. 1, d. 328, l. 1, reproduced from a copy of the original, cited in Ledovskii, "Sekretnaia missiia delegatsii TsK," p. 46.

548. Vyshinskii's conversation notes, AP RF, f. 3, op. 65, d. 500, l. 172, cited in Ledovskii, Sekretnaia missiia delegatsii TsK," p. 58; ibid., pp. 53–54; Ledovskii, "Vizit v Moskvu delegatsii," pt. 2, p. 90.

549. Vyshinskii's conversation notes, AP RF, f. 3, op. 65, d. 500, ll. 167–70, copy of the original, cited in Ledovskii, "Sekretnaia missiia delegatsii TsK," pp. 54–56; cf. Ledovskii, "Vizit v Moskvu delegatsii," pt. 2, pp. 91–92.

550. The Council of Foreign Ministers was founded by the foreign ministers of the United States, Great Britain, France, and China on the basis of resolution at the Potsdam Conference.

551. Vyshinskii's conversation notes, AP RF, f. 3, op. 65, d. 500, ll. 170–71, copy of the original, cited in Ledovskii, "Sekretnaia missiia delegatsii TsK," pp. 56–57; cf. Ledovskii, "Vizit v Moskvu delegatsii," pt. 2, p. 92.

552. Vyshinskii's conversation notes, AP RF, f. 3, op. 65, d. 500, ll. 171–72, copy of the original, cited in Ledovskii, "Sekretnaia missiia delegatsii TsK," pp. 57–58; cf. Ledovskii, "Vizit v Moskvu delegatsii," pt. 2, p. 92.

553. Kovalev, "Dialog Stalina s Mao Tszedunom," pt. 1, p. 88; Bo, *Ruogan zhongda juece*, vol. 1, p. 37; *Zhonggong zhongyang zai Xiangshan*, p. 454. Shi Zhe did not mention that a session was held on July 27.

554. Kovalev, "Dialog Stalina s Mao Tszedunom," pt. 1, p. 88.

555. It is unclear to what the mention of Hong Kong was a reference.

556. Kovalev, "Dialog Stalina s Mao Tszedunom," pt. 1, p. 88; Goncharov, Lewis, and Xue, *Uncertain Partners*, p. 70.

557. In the Russian version of the telegram available to us, the term *aviiatsiia* (neutrally, "air transport")

was used here. From the context, however, we can deduce that *voennaia aviiatsiia* ("air force") was clearly meant.

558. The telegram is available to us only in the Russian translation given to Stalin. AP RF, f. 45, op. 1, d. 328, ll. 137–40, reproduced from a copy of the original, cited in Ledovskii, "Sekretnaia missiia delegatsii TsK," p. 52; cf. Ledovskii, "Vizit v Moskvu delegatsii," pt. 2, p. 90.

559. See AP RF, f. 45, op. 1, d. 328, l. 140, reproduced from a copy of the original, cited in Ledovskii, "Sekretnaia missiia delegatsii TsK," p. 52; cf. Ledovskii, "Vizit v Moskvu delegatsii," pt. 2, p. 90.

560. Telegram from Mao Zedong to Zhou Enlai, July 10, 1949, cited in *Mao Zedong nianpu*, vol. 3, p. 529, and *Zhonggong zhongyang zai Xiangshan*, p. 447.

561. *Mao Zedong nianpu*, vol. 3, p. 529.

562. See telegram from the Secretariat of the CC of the CCP to Liu Shaoqi, July 26, 1995, cited in *Zhonggong zhongyang zai Xiangshan*, p. 454.

563. Ibid.

564. Telegram from Mao Zedong to Liu Shaoqi, July 25, 1949, AP RF, f. 45, op.1, d. 328, l. 139, reproduced from a copy of the original, cited in Ledovskii, "Sekretnaia missiia delegatsii TsK," pp. 51–52; cf. Ledovskii, "Vizit v·Moskvu delegatsii," pt. 2, pp. 89–90.

565. On this paragraph, see *Zhonggong zhongyang zai Xiangshan*, p. 454. It says here that Stalin expressed his support immediately (*dangji biaoshi zhichi*). See also Goncharov, Lewis, and Xue, *Uncertain Partners*, pp. 73–74, with source references.

566. For further details and sources for this and the following paragraph, see below in this section.

567. For further details on the events presented in this paragraph and the sources, see the following paragraphs.

568. On this paragraph, see Howard L. Boorman and Richard C. Howard, *Biographical Dictionary of Republican China*, vol. 2, pp. 234–35; Bartke, *Biographical Dictionary*, p. 44.

569. Cf., for example, reports of American diplomats in August 1949, in *FRUS 1949*, vol. 9, *The Far East: China*, pp. 957, 959–60, 962–63.

570. *Zhonghua renmin gongheguo waijiao shi*, p. 14.

571. On this paragraph, see Kovalev, "Dialog Stalina s Mao Tszedunom," pt. 1, p. 89. At that time, there were sixteen Soviet republics, as the Karelo-Finnish ASSR still existed. On another occasion, Kovalev reported that Gao Gang had advocated an autonomous status for Manchuria or a federal link to China. Stalin had rejected the proposal and—seriously or as a joke—had added that Gang must want to become a second Zhang Tso-lin. There is no evidence of when or on what occasion this dialogue was supposed to have taken place. See A.M. Ledovskii, *Delo Gao Gan—Zhao Shushi*, p. 30.

572. After his return, Liu criticized Gao in Peiping. *Zhonghua renmin gongheguo waijiao shi*, p. 14.

573. On this paragraph, see Kovalev, "Dialog Stalina s Mao Tszedunom," pt. 1, p. 89.

574. Shi Zhe, *Zai lishi juren shenbian*, pp. 397–419.

575. Shi Zhe to the historian Zhu Ruizhen, February 1992. The author thanks Zhu Ruizhen for this information. Cf. Li Haiwen, "Mao Zedong Liu Shaoqi fang Su," p. 90.

576. On this and the two following sentences, see Shi Zhe, quoted in ibid.

577. Shi Zhe, "Gao Gang ceng jianyi jiang Dongbei huagui Sulian ma?" p. 85. The list of visitors to Stalin's office for example shows that Kovalev was present at the meetings of the two delegations on July 27 and August 5. See "Posetiteli kremlevskogo kabineta," pt. 1, p. 56–57.

578. Communication to the author from Zhu Ruizhen. Zhu received this information "from a Soviet source" in 1956.

579. See further above in this section.

580. On this paragraph, see Kovalev, "Dialog Stalina s Mao Tszedunom," pt. 1, p. 89; Bo, *Ruogan zhongda juece*, vol. 1, p. 37; *Zhonggong zhongyang zai Xiangshan*, p. 455.

581. On this paragraph, see Shi Zhe, quoted in Li Haiwen, "Mao Zedong Liu Shaoqi fang Su," p. 90.

582. Telegram from Stalin to Kovalev, October 11, 1949, cited in A. M. Ledovskii, "Vizit Mao Tszeduna v Sovetskii Soiuz," pt. 2, p. 3, with no indication of the archival reference.

583. On this and the following two paragraphs, see Kovalev, "Dialog Stalina s Mao Tszedunom," pt. 1, pp. 89–90; Ledovskii, "Vizit Mao Tszeduna v Sovetskii Soiuz," pt. 2, pp. 2–6. Ledovskii was employed as first secretary of the Soviet embassy in Peking at the time. Kovalev's report on Gao Gang's secret ties to Moscow and Stalin's betrayal of Gao Gang is confirmed in essential points by Khrushchev's and Bo Yibo's memoirs; see Kovalev, "Dialog Stalina s Mao Tszedunom," pt. 1, pp. 92–93 n. 11, with literal citations. Shi Zhe did not mention these events; cf. Shi Zhe, "Peitong Mao zhuxi fang Su," Renwu 5 (1988), p. 11.

584. For details on the events described in this paragraph and for source references, see chapter 4 below.

585. Kovalev, "Dialog Stalina s Mao Tszedunom," pt. 1, p. 91; Ledovskii, "Vizit Mao Tszeduna v Sovetskii Soiuz, pt. 2, pp. 2–6; Ledovskii, "Sekretnaia missiia delegatsii TsK," pp. 23–26 passim. Ledovskii was Soviet consul general from 1950 to 1952 in Mukden, where he cultivated close contacts with Gao Gang.

586. On this paragraph, see Frederick C. Teiwes, *Politics at Mao's Court*, passim.

587. On this paragraph, see Gao Gang to Iudin in early August 1951 in Mukden, see Fedorenko's telegram to Stalin (Filippov), August 4, 1951 ("top secret"), AVP RF, f. 059A, op. 3, p. 1, d. 4, ll. 9–10; Iudin's telegram to Malenkov and Khrushchev, February 3, 1954 ("top secret"), AVP RF, f. 059A, op. 3, p. 1, d. 4, ll. 13–17.

588. It can be assumed that the banquet that Shi Zhe dates "more or less July 27" took place on July 30. See Goncharov, Lewis, and Xue, *Uncertain Partners*, p. 312 n. 132.

589. On this paragraph, see Shi Zhe, *Zai lishi juren shenbian*, pp. 410–11; Shi Zhe, "Peitong Mao zhuxi fang Su," *Renwu* 5 (1988), p. 10.

590. On this paragraph, see Shi Zhe, *Zai lishi juren shenbian*, p. 412; Shi Zhe, "Peitong Mao zhuxi fang Su," p. 10; Zhu Yuanshi, "Liu Shaoqi 1949 nian mimi fang Su," *Renwu* 5 (1988), pp. 76–77.

591. On this paragraph, see Shi Zhe, *Zai lishi juren shenbian*, p. 414; Shi Zhe, "Peitong Mao zhuxi fang Su," *Renwu* 5 (1988), p. 10. Cf. also Zhu Yuanshi, "Liu Shaoqi 1949 nian mimi fang Su," p. 77; *Zhonghua renmin gongheguo waijiao shi*, p. 12.

592. On this paragraph, see Kovalev, "Dialog Stalina s Mao Tszedunom," pt. 2, p. 78.

593. Ibid.

594. For further details, see chapter 1 above.

595. Mao Zedong to Iudin, July 22, 1958, Chinese notes of the talk, in *Dang de wenxian* 1 (1994), p. 16.

596. Cf. further above in this same section.

597. "Posetiteli kremlevskogo kabineta," pt. 1, pp. 57–58. According to Shi Zhe, this session was held in "mid-August" in a smaller circle and was shorter than usual. See Shi Zhe, *Zai lishi juren shenbian*, p. 415; Shi and Chen, "With Mao and Stalin," p. 42. Kovalev did not mention the session.

598. For further details, see further below in this section.

599. On this paragraph, see Shi Zhe, *Zai lishi juren shenbian*, pp. 416–17.

600. On this paragraph, see ibid., pp. 417–18. Shi Zhe added the last half of the sentence paraphrased here as "so that they. . . ," which can be understood as a warning, in the revised version. See Shi Zhe, "With Mao and Stalin," p. 87.

601. On this paragraph, see Shi Zhe, *Zai lishi juren shenbian*, p. 416; Shi Zhe, "With Mao and Stalin," pt. 2, p. 88; Ledovskii, "Sekretnaia missiia A.I. Mikoiana v Kitai," pt. 2, p. 96. For further details on Mikoyan's response, see this chapter above.

602. On this paragraph, see Shi Zhe, *Zai lishi juren shenbian*, pp. 418–19; Shi Zhe, "With Mao and Stalin," pt. 2, pp. 88. Stalin's intervention is described in greater detail in the revised version. The literal citations are marked in this publication.

603. On this and the following two sentences, see Shi Zhe, *Zai lishi juren shenbian*, p. 418.

604. Letter from Liu Shaoqi to Stalin, August 2, 1949, AP RF, f. 3, op. 65, d. 364, ll. 1–2.

605. On this paragraph, see Shi Zhe, *Zai lishi juren shenbian*, pp. 415–16.

606. Mao's telegram to Liu Shaoqi arrived on July 31, 1949. See Zhu Yuanshi, "Liu Shaoqi 1949 nian mimi fang Su," p. 79.

607. On this paragraph, see letter from Liu Shaoqi to Stalin, August 2, 1949, delivered by Kovalev on August 3, AP RF, f. 3, op. 65, d. 364, ll. 1–2.

608. On this paragraph, see Zhu Yuanshi, "Liu Shaoqi 1949 nian mimi fang Su," p. 80; *Zhonggong zhongyang zai Xiangshan*, p. 461.

609. On this and the following sentence, see AP RF, op. 1, d. 39, l. 29, cited in Ledovskii, "Sekretnaia missiia A.I. Mikoiana v Kitai," pt. 2, p. 99. For further details, see this chapter above.

610. For further details on this paragraph, see in this section above.

611. On this paragraph, see telegram from Mao Zedong to Liu Shaoqi, July 25, 1949, forwarded to Stalin, AP RF, f. 45, op. 1, d. 328, l. 138, reproduced from a copy of the original, cited in Ledovskii, "Sekretnaia missiia delegatsii TsK," p. 50; Ledovskii "Vizit v Moskvu delegatsii," pt. 2, pp. 88–89; telegram from Mao Zedong to Liu Shaoqi and Wang Jiaxiang, August 4, 1940, cited in *Mao Zedong nianpu*, vol. 3, p. 541.

612. On this paragraph, see *Zhonggong zhongyang zai Xiangshan*, p. 460; Shi Zhe, *Zai lishi juren shenbian*, pp. 407–8; Goncharov, Lewis, and Xue, *Uncertain Partners*, p. 70, with source references.

613. On this paragraph, see Xiang, You, and Liu, *Sulian yu Zhongguo geming*, p. 572.

614. AP RF, f. 45, op. 1, d. 329, l. 38. For further details, see chapter 4 below.

615. Telegram from Mao Zedong to Liu Shaoqi and Wang Jiaxiang, August 4, 1949, in *Mao Zedong nianpu*, vol. 3, p. 541.

616. *Zhonggong zhongyang zai Xiangshan*, p. 455; *Mao Zedong nianpu*, vol. 3, p. 529.

617. On this paragraph, see telegram from Mao Zedong to Liu Shaoqi and Wang Jiaxiang, July 31, 1949, and telegram from Zhou Enlai to Liu Shaoqi and Wang Jiaxiang, July 31, 1949, both cited in *Zhonggong zhongyang zai Xiangshan*, p. 455, 459–60; Shi Zhe, *Zai lishi juren shenbian*, pp. 407–8.

618. On this paragraph, see telegrams from the CC of the CCP to Liu Shaoqi and Wang Jiaxiang, August 6. 1949, and August 11, 1949, and to Gao Gang, August 11, 1949, cited in *Zhonggong zhongyang zai Xiangshan*, pp. 458–59; telegram from the CC of the CCP to Gao Gang, August 11, 1949, in *Zhou Enlai nianpu*, p. 836; telegram from Mao Zedong to Liu Shaoqi, July 25, 1949, AP RF, f. 45, op. 1, d. 328, ll. 138–39, copy of the original, cited in Ledovskii, "Sekretnaia missiia delegatsii TsK," pp. 50–51.; Ledovskii, "Vizit v Moskvu delegatsii," pt. 2, p. 89; B.T. Kulik, "Kitaiskaia Narodnaia Respublika v period stanovleniia (1949–1952 gg.)," pt. 2, p. 75, based on materials from the AVP RF; Shi Zhe, *Zai lishi juren shenbian*, p. 408; AVP RF, f. 07, op. 23a, p. 18, d. 237, l. 2.

619. AVP RF, f. 07, op. 23a, d. 237, l. 125.

620. On this paragraph, Stalin in the June 27 session, 1949, see this chapter above (section "The First Session: Questions of Soviet Aid").

621. Kovalev's unpublished memoirs, cited in Goncharov, Lewis, and Xue, *Uncertain Partners*, p. 71.

622. On this paragraph, see Shi Zhe, *Zai lishi juren shenbian*, pp. 409–10; Thomas B. Cochran, Robert S. Norris, and Oleg A. Bukharin, *Making the Russian Bomb*, p. 49.

623. On this paragraph, see Kovalev's unpublished memoirs, cited in Goncharov, Lewis, and Xue, *Uncertain Partners*, pp. 74, 233–34 (doc. 8).

624. Document from Kovalev's archive and Kovalev's unpublished memoirs, both cited in Goncharov, Lewis, and Xue, *Uncertain Partners*, p. 74.

625. On this and the following sentence, see Zubok and Pleshakov, *Inside the Kremlin's Cold War*, p. 61, with source references.

626. On this and the following sentence, see Goncharov, Lewis, and Xue, *Uncertain Partners*, p. 74, with source references.

627. For further details on this paragraph, see chapter 4 below.

628. On this and the following sentence, see further above in this chapter (June 27 session).

629. *Pravda*, July 31, 1949, p. 3.

630. On this paragraph, see telegram from Mao Zedong to Liu Shaoqi, Gao Gang, and Wang Jiaxiang, August 4, 1949, cited in *Mao Zedong nianpu*, vol. 3, p. 541.

631. On this and the following sentence, see S. Shtemenko and Gromyko to Malenkov, November 26, 1949, AP RF, f. 007, op. 22a, p. 13, d. 198, ll. 34–35.

632. Mao Zedong to Stalin, December 16, 1949, AP RF, f. 45, op. 1, d. 329, l. 13.

633. Letter from Liu Shaoqi to Mikoyan, August 5, 1949, AP RF, f. 3, op. 65, d. 500, ll. 3–4.

634. For further details, see chapter 4 below. On this paragraph, see Shi Zhe, *Zai lishi juren shenbian*, p. 409; Shi Zhe, "With Mao and Stalin," pt. 2, pp. 79–81; Zhu Yuanshi, "Liu Shaoqi 1949 nian mimi fang Su," p. 79.

635. See further below in this section.

636. Kulik, "Kitaiskaia Narodnaia Respublika," pt. 2, p. 75, based on materials in the AVP RF.

637. On this and the following sentence, Stalin wrote the comment "Yes!" in the margin next to the corresponding passage in Liu Shaoqi's report of July 4, 1949. Cf. further above in this chapter (section "Liu's Written Report to Stalin").

638. Letter from Liu Shaoqi to Stalin, July 6, 1949, AP RF, f. 45, op. 1, d. 328, l. 53, cited in Ledovskii, "Sekretnaia missiia delegatsii TsK," p. 47.

639. On this paragraph, see telegram from Mao Zedong to Liu Shaoqi, Gao Gang, and Wang Jiaxiang, July 25, 1949, cited in *Mao Zedong nianpu*, vol. 3, p. 538; telegram from Mao Zedong to Liu Shaoqi and Wang Jiaxiang, August 7, 1949, cited in ibid., vol. 3, pp. 541–42. Cf. Liu Shaoqi's letter to Stalin, July 6, 1949, AP RF, f. 45, op. 1, d. 328, ll. 54–55, cited in Ledovskii, "Sekretnaia missiia delegatsii TsK," p. 48.

640. On this paragraph, see Shi Zhe, *Zai lishi juren shenbian*, pp. 426–28.

641. *Pravda*, November 11, 1949. On this paragraph, see Shi Zhe, *Zai lishi juren shenbian*, pp. 428–29.

642. On this and the following paragraph, see Shi Zhe, *Zai lishi juren shenbian*, pp. 428–29.

643. Li Kenong to Roshchin, November 17, 1949, AVP RF, f. 07, op. 22, d. 220, l. 70.

644. Zhu Yuanshi, "Liu Shaoqi 1949 nian mimi fang Su," p. 80.

645. On this paragraph, see AP RF, f. 39, op. 1, d. 39, ll. 29, 37, 57, 58, cited in Ledovskii, "Sekretnaia missiia A.I. Mikoiana v Kitai," pt. 2, pp. 97–100; Ledovskii, "Sekretnaia missiia delegatsii TsK," p. 20. For further details, see chapter 4 above.

646. On this paragraph, *see* further above in this section.

647. On this and the following sentence, see Mao Zedong to Liu Shaoqi, Gao Gang and Wang Jiaxiang, August 4, 1949, cited in *Mao Zedong nianpu*, vol. 3, p. 541.

648. On this and the following sentence, see Liu's letter to Stalin, August 2, 1949, AP RF, f. 3, op. 65, d. 364, l. 2, cited in Ledovskii, "Sekretnaia missiia delegatsii TsK," pp. 58–59 (in part from a copy of the original); cf. Ledovskii, "Vizit v Moskvu delegatsii," pt. 2, p. 93.

649. "Instruktsiia dlia sovetskich spetsialistov, rabotaiushchich v Kitae," August 9, 1949, AVP RF f. 07, op. 23a, d. 221, l. 2.

650. The fact that an agreement with this title was concluded is mentioned in Wang Jiaxiang to Gromyko, February 21, 1950, AVP RF, f. 0100, op. 43, p. 302, d. 8, l. 24. The text is not yet available. One month later, Wang claimed that it had only been an "arrangement" (in Russian: *dogovorennost'*), that was not translated into a formal agreement at that time; Wang Jiaxiang to Gromyko, March 22, 1950, AVP RF, f. 0100, op. 43, p. 302, d. 8, l. 50.

651. Brian Joseph Murray, *Western versus Chinese Realism*, pp. 461–62, with reference to archival documents. On this paragraph, see telegram from the CC of the CCP to Liu Shaoqi and Wang Jiaxiang, August 10, 1949, cited in *Zhonggong zhongyang zai Xiangshan*, p. 458; Shi Zhe, *Zai lishi juren shenbian*, p. 415.

652. *Mao Zedong nianpu*, vol. 3, p. 541. This information appears to be based on archival documents. Shi Zhe also spoke of more than two hundred experts; see Shi Zhe, *Zai lishi juren shenbian*, p. 425.

653. On this and the following sentence, see telegram from Liu Shaoqi to the CC of the CCP, August 26, 1949, in *Zhonggong zhongyang zai Xiangshan*, p. 463. This is the first number of accompanying experts that is supported by documents. Other numbers mentioned range from 80 to 250; see n. 655 below.

654. On this paragraph, see Zhu Yuanshi, "Liu Shaoqi 1949 nian mimi fang Su," pp. 79–80; Kovalev, "Dialog Stalina s Mao Tszedunom," pt. 1, pp. 84–85; Shi Zhe, *Zai lishi juren shenbian*, p. 425; Li Yueran, *Waijiao wutai*, pp. 4–13, passim. Li Yueran acted as interpreter for Kovalev and the group of Soviet experts at that time. The information provided by Shi Zhe (80) and Kovalev (250) on the number of accompanying experts is not accurate, as already discussed in this section; cf. Shi Zhe, *Zai lishi juren shenbian*, p. 425, and Kovalev, "Dialog Stalina s Mao Tszedunom," pt. 1, p. 85. Equally inaccurate are earlier numbers cited by Shi Zhe, who stated that 150 experts were requested and that 96 traveled with Kovalev; see Shi Zhe, "Peitong Mao zhuxi fang Su," Renwu 5 (1988), p. 11.

655. On this paragraph, see Kovalev, "Dialog Stalina s Mao Tszedunom," pt. 1, p. 85; Ledovskii, "Sekretnaia missiia delegatsii TsK," pp. 61–62; Kulik, "Kitaiskaia Narodnaia Respublika," pt. 2, p. 76, based on documents from the AVP RF.

656. Shi Zhe, *Zai lishi juren shenbian*, pp. 425–26; Li Yueran, *Waijiao wutai*, pp. 10–11.

657. *Mao Zedong nianpu*, vol. 3, pp. 542–43; *Zhonggong zhongyang zai Xiangshan*, p. 465.

658. On this paragraph, see Zhu Yuanshi, "Liu Shaoqi 1949 nian mimi fang Su," p. 80; Shi Zhe, *Zai lishi juren shenbian*, p. 426.

659. On this paragraph, see Li Yueran, *Waijiao wutai*, pp. 14–20 passim; *Zhonggong zhongyang zai Xiangshan*, p. 465.

660. On this paragraph, see Li Yueran, *Waijiao wutai*, p. 15.

661. On this paragraph, see ibid., p. 22; Klein and Clark, *Biographic Dictionary*, vol. 1, p. 150.

662. Zhu Yuanshi, "Liu Shaoqi 1949 nian mimi fang Su," p. 80; cf. Shi Zhe, *Zai lishi juren shenbian*, p. 395.

663. For further details, see McGlothlen, *Controlling the Waves*, pp. 135–43. A discussion of the topic presented in this section that is convincing in its broad outline and based firmly on the sources can be found in Vasold, "Versäumte Gelegenheiten?" pp. 241–71 passim.

664. Cf. Tucker, *Patterns in the Dust*, p. 48.

665. For further information on the lost chance school, see Goldstein, "Sino-American Relations," pp. 120–21; Michael M. Sheng, "America's Lost Chance in China?" pp. 135–36, with references to the literature on the creation of the school; Warren I. Cohen, "Introduction: 'Was There a 'Lost Chance' in China?" pp. 71–75. By now most historians support the thesis that there was no lost chance. See the most decisive and convincing statement in Chen Jian, "The Myth of America's 'Lost Chance' in China," pp. 77–86. The following authors present similar arguments: John W. Garver, "Little Chance," pp. 87–94; Michael M. Sheng, "The Triumph of Internationalism," pp. 95–104; Odd Arne Westad, "Losses, Chances, and Myths," pp. 105–15.

666. See *FRUS 1949*, vol. 8, pp. 741–802; Huang Hua, "Nanjing jiefang," pp. 23–32.

667. On this paragraph, see Kovalev's telegram to Stalin, May 23, 1949, AP RF, f. 45, op. 1, d. 331, l. 66–69, passim, cited in Tikhvinskii, "O 'sekretnom demarshe,'" pp. 135–36; Stalin's telegram to Mao Zedong (via Kovalev), May 26, 1949, AP RF, f. 45, op. 1, d. 131, ll. 73–75, passim, cited ibid., p. 136.

668. Ledovskii, "Nankinskie zapiski," pp. 12–13, 19–71 passim. For further details on the first meeting, see this chapter above.
669. For further details on the Ward case, see this chapter above.
670. On this paragraph, see Yang Kuisong, "Soviet Factor," p. 32.
671. *FRUS 1949*, vol. 8, *The Far East: China*, pp. 173–74, 176–77.
672. Ibid., pp. 230–31.
673. Huang Hua to Ledovskii, May 10, 1949, in Ledovskii, "Nankinskie zapiski," p. 19.
674. There is no reference at least for this time period in the official collection of State Department documents. Cf. *FRUS 1949*, vol. 8, *The Far East: China*, pp. 271–78, 723–42.
675. On this and the following sentence, see Huang Hua, *Nanjing jiefang*, p. 24.
676. Ledovskii, "Nankinskie zapiski," p. 8.
677. On this paragraph, see Stuart to Acheson, May 11, 1949, in Rea and Brewer, *Forgotten Ambassador*, pp. 322–23; Huang Hua, "Nanjing jiefang," pp. 25–26.
678. Stuart to Acheson, May 11, 1949, *FRUS 1949*, vol. 8, *The Far East: China*, pp. 741–42.
679. Ibid., p. 742.
680. Ledovskii, "Nankinskie zapiski," p. 19.
681. Ibid., pp. 19–20.
682. Ibid., p. 20.
683. Ibid., pp. 20–21.
684. *FRUS 1949*, vol. 8, *The Far East: China*, pp. 173–77, 230–31.
685. See Ledovskii, "Nankinskie zapiski," p. 19.
686. Telegram from the CC of the CCP the Nanking Municipal Committee to be forwarded to the East China Bureau, May 10, 1949, cited in *Mao Zedong nianpu*, vol. 3, pp. 499–500.
687. Stuart to Acheson, May 14, 1949, in Rea and Brewer, *Forgotten Ambassador*, pp. 324–25.
688. Huang Hua, "Nanjing jiefang," p. 26.
689. Ibid., pp. 26–27.
690. On this and the following paragraph, see Kovalev's telegram to Stalin, May 23, 1949, in AP RF, f. 45, op. 1, d. 331, ll. 66–69, cited in Tikhvinskii, "O 'sekretnom demarshe,'" pp. 135–36.
691. Telegram from Stalin to Kovalev, May 26, 1949, in AP RF, f. 45, op. 1, d. 331, ll. 66–69, 73–75, cited in ibid.
692. On this and the following sentence, see Tikhvinskii in a talk with Yang Kuisong and in his lecture to the Institute of Contemporary History of the Chinese Academy for Social Sciences in Peking; both the talk and the lecture took place in April 1994. See Yang Kuisong, "Hua-de shijian," p. 115.
693. For further details on the two telegrams, see the three previous paragraphs.
694. On this paragraph, see Kovalev, "Dialog Stalina s Mao Tszedunom," pt. 2, pp. 84–85.
695. *FRUS 1949*, vol. 8, *The Far East: China*, pp. 745–46; Stuart to Acheson, May 14, 1949, in Rea and Brewer, *Forgotten Ambassador*, pp. 325–25. Cf. also Yang Kuisong, "Soviet Factor," pp. 32–33; Tucker, *Patterns in the Dust*, pp. 47, 232 n. 27.
696. Ledovskii, "Nankinskie zapiski," pp. 28–29.
697. Telegram from the CC of the CCP to the Nanking Municipal Committee, June 3, 1949, in *Mao Zedong nianpu*, vol. 3, p. 514, and *Zhonggong zhongyang zai Xiangshan*, p. 433.
698. On this paragraph, see Huang Hua, "Nanjing jiefang," p. 28.
699. On this paragraph, see ibid., pp. 28–29.
700. On this paragraph, see ibid., p. 29.
701. *FRUS 1949*, vol. 8, *The Far East: China*, pp. 752–53.
702. Ibid., p. 753.
703. Ibid., pp. 377–78.
704. Yang Kuisong, "Soviet Factor," p. 33.
705. On this paragraph, see Huang Hua, "Nanjing jiefang," pp. 29–30; cf. Yang Kuisong, "Soviet Factor," p. 33.
706. Yuan Ming, "The Failure of Conception," p. 144.
707. Talk between Ledovskii and Huang Hua, June 11, 1949, reproduced in Ledovskii, "Nankinskie zapiski," p. 47. In his notes, Ledovskii mentioned only one supposed meeting between Huang and Stuart and Fu Jingpo on June 8. His report nonetheless referred to the contents of both Huang's meeting with Stuart and Fu on June 6 and the one with Fu alone on June 8.
708. For further details, see below in this section.
709. For further details, see below in this section.

710. On this paragraph, see Stuart to Acheson, June 30, 1949, *FRUS 1949*, vol. 8, *The Far East: China*, p. 766.

711. On this paragraph, see ibid., pp. 766–67.

712. Huang Hua, "Nanjing jiefang," p. 30.

713. Talk between Huang Hua, Ledovskii, and Shibaev, July 1, 1949, reproduced in Ledovskii, "Nankinskie zapiski," pp. 52–53.

714. For further details, see below in this section.

715. On this paragraph, see the talks between Huang Hua and Ledovskii, July 11 and 25, 1949, in Ledovskii, "Nankinskie zapiski," pp. 54–55, 67–68.

716. On this paragraph, see talk between Huang and Ledovskii, June 19, 1949, in ibid., pp. 49–50.

717. Clubb to Acheson, July 11, 1949, in *FRUS 1949*, vol. 8, *The Far East: China*, p. 779. At this time, Clubb apparently did not yet know anything about the decision Acheson had taken on July 1 not to allow Stuart to travel to Peiping.

718. For further details about this démarche, see below in this section.

719. The reference is to Stuart's telegram to Acheson, June 30, 1949, in *FRUS 1949*, vol. 8, *The Far East: China*, pp. 766–67.

720. Memorandum to the Director of the Policy Planning Staff of the State Department, George F. Kennan, drawn up by John P. Davies, Office of Far Eastern Affairs, June 30, 1949. It was presented to Secretary of State Acheson the following day. See ibid., pp. 768–69.

721. Cf. Yuan, "Failure of Conception," p. 144. The speech was first published on July 1, 1949, in *Renmin ribao* and July 7, 1949, in *Qunzhong* (Hong Kong); cf. *Mao Zedong ji*, vol. 10, p. 307.

722. On this paragraph, see *FRUS 1949*, vol. 8, *The Far East: China*, p. 769 (the telegraphic style has been retained in all American diplomatic cables quoted in this book); Huang Hua, "Nanjing jiefang," p. 31; McGlothlen, *Controlling the Waves*, p. 145.

723. Stuart to Acheson, July 14, 1949, in *FRUS 1949*, vol. 8, *The Far East: China*, p. 784.

724. Stuart to Acheson, July 6, 1949, in Rea and Brewer, *Forgotten Ambassador*, pp. 335–36. For further details on the July 1 article, see this chapter above.

725. CC of the CCP to the Nanking Municipal Committee of the CCP, June 30, 1949, cited in Yang Kuisong, "Soviet Factor," p. 33.

726. Ibid., with no indication of sources.

727. According to Huang Hua, cited in John W. Garver, *Foreign Relations of the People's Republic of China*, p. 41, with source references.

728. *FRUS 1949*, vol. 8, *The Far East: China*, pp. 776–79. Zhou's message clearly reached Stuart through Chen Mingshu; see ibid., p. 774. To this extent, the statement by Dean Rusk, then deputy undersecretary of state, that Zhou presented his message to Stuart in a direct meeting is inaccurate. The same is therefore also true of the corresponding hypothesis in a publication on Acheson's China policy, see McGlothlen, *Controlling the Waves*, pp. 145–46 n. 260. The statement found there that none of the telegrams Stuart sent between July 19 and August 2 are available is inaccurate; so are the resulting fanciful commentaries. Stuart sent the State Department several telegrams during this period, see *FRUS 1949*, vol. 8, *The Far East: China*, pp. 449–66, 795–809 passim.

729. Acheson in a speech in San Francisco, March 1950, cited in McGlothlen, *Controlling the Waves*, p. 159.

730. Nitze to McGlothlen, cited in ibid., pp. 161–62.

731. *FRUS 1949*, vol. 8, *The Far East: China*, p. 705.

732. *Renmin ribao*, August 19, 1949, cited in *Mao Zedong ji*, vol. 10, p. 331, and *Mao Zedong xuanji*, vol. 4, pp. 1500–150; for a version in English, see *Selected Works of Mao Tse-tung*, vol. 4, pp. 438–39.

733. *Selected Works of Mao Tse-tung*, vol. 4, pp. 411–42. For further details, see this chapter above.

734. Ibid., pp. 405–9. The speech was given on June 15, but published in English on June 20; cf. *FRUS 1949*, vol. 8, *The Far East: China*, p. 392.

735. *FRUS 1949*, vol. 8, pp. 774–76.

736. On this and the following sentence, see Stuart to Acheson, July 13, 1949, in ibid., pp. 782–83.

737. Ibid., p. 771 n. 78.

738. On this paragraph, see ibid., pp. 357, 373, 397, 497. On the Zhou démarche, see Tucker, *Patterns in the Dust*, pp. 48, 174–75, 233 n. 30.

739. On this paragraph, see Clubb to Acheson, June 24, 1949, in *FRUS 1949*, vol. 8, *The Far East: China*, pp. 397–98; Clubb to Acheson, August 18, 1949, in ibid., pp. 496–98.

740. On this paragraph, see Consul General in Peiping, Clubb, to the Secretary of State, June 1, 1949, in ibid., pp. 357–60.

741. On this paragraph, see Elisabeth Barker, *The British Between the Superpowers, 1945–50*, p. 172, which is based on Foreign Office records; Acting Secretary of State, Webb, to Clubb, June 14, 1949, in *FRUS 1949*, vol. 8, *The Far East: China*, p. 385.

742. Clubb to Acheson, June 2, 1949, in ibid., pp. 363–64.

743. Clark to Acheson, June 6, 1949, in ibid., p. 370.

744. Stuart to Acheson, June 7, 1949, in ibid., pp. 372–72.

745. Memorandum of a conversation with the President by Acting Secretary of State, Webb, June 16, 1949, in ibid., p. 388.

746. Webb, Acting Secretary of State, to Clubb, June 14, 1949, in ibid., pp. 384–85.

747. Clubb to Acheson, June 27, 1949, in ibid., pp. 398–99.

748. Kohler to Acheson, June 27, 1949, in ibid., pp. 399–400.

749. Cf. Dieter Heinzig, "From the 'Middle of the Road' to 'Yibian dao,'" pp. 59–72. This position is often taken by Western researchers; cf. for example Tucker, *Patterns in the Dust*, pp. 46–48, and Dick Wilson, *Zhou Enlai*, pp. 177–78.

750. All of the historians the author asked in the People's Republic of China assumed that the Zhou démarche was authentic in the sense that the CCP leadership did transmit such a message. They equally unanimously take the position that the two factions claimed in the message did not exist. Shi Zhe thus claimed to the author that Zhou Enlai and Liu Shaoqi supported Mao on the questions under discussion. The author thanks Zhu Ruizhen, who gave him Shi Zhe's opinion.

751. See chapter 2 above.

752. Ledovskii to the author, March 27, 1995.

753. Ibid.; Tikhvinskii to Ledovskii, September 1995.

754. Tikhvinskii to the author, September 15, 1992; Tikhvinskii to Ledovskii, September 1995. Accordingly, there is no reference to this in Tikhvinskii's China memoirs; see S.L. Tikhvinskii, *Kitai v moei zhizni*, pp. 75–79.

755. Tikhvinskii in April 1994 in a talk with Yang Kuisong and in his lecture to the Institute for Contemporary History of the Chinese Academy for Social Studies in Peking. See Yang Kuisong, "Hua-de shijian," p. 115.

756. See the beginning of this section and this chapter above.

757. See this chapter above.

758. See this chapter above.

759. *FRUS 1949*, vol. 8, *The Far East: China*, pp. 377–78.

760. Interview with Chiang Kai-shek, July 4, 1949, in *FRUS 1949*, vol. 8, *The Far East: China*, p. 413.

761. Tikhvinskii, "O 'sekretnom demarshe,'" p. 137. The historical literature of the People's Republic of China does not mention the Zhou demarche; see, for example, *Zhou Enlai nianpu, 1898–1949*, p. 828. There is supposedly no reference to the Zhou démarche on Zhou Enlai's work schedule, which is preserved in the party archives; Chen Jian to the author, on July 14, 1997. If this is true, it could lead us to support Tikhvinskii's hypothesis that this was an action carried out by the CCP secret service, albeit hardly without Zhou's knowledge.

762. Shi Yinhong, a Nanking-based contemporary historian and a specialist on CCP foreign policy in this period, said the same to the author, February 10, 1995.

763. A comparable uncertainty can be seen, for example, in a public speech by Mao Zedong, June 15, 1994. See *Selected Works of Mao Tse-tung*, vol. 4, p. 407. A Chinese historian with access to the Party archives who wishes to remain anonymous confirmed to the author, on the basis of documents found in the archives, that this was one of the reasons for the Zhou démarche.

764. See Kovalev, "Dialog Stalina s Mao Tszedunom," pt. 2, p. 84.

765. CC of the CCP to the Nanking Municipal Committee of the CCP on June 30, 1949, cited in Yang Kuisong, "Soviet Factor," p. 33.

766. *FRUS 1949*, vol. 8, *The Far East: China*, pp. 774–76.

767. On this paragraph, see Stuart to Acheson, June 8, 1949, in ibid., p. 753; John Wesley Jones, Counselor of the Embassy, to Acheson, August 15, 1949, in ibid., pp. 814–15; McGlothlen, *Controlling the Waves*, pp. 147–52; Chen Jian, "Ward Case," pp. 161–62. For further details on the Ward case, see this chapter above.

768. See Chen Jian, "Ward Case," pp. 161–62.

769. Jones, Counselor of the U.S. Embassy in Nanking, to the Secretary of State, September 5, 1949, in *FRUS 1949*, vol. 8, *The Far East: China*, p. 520. On this paragraph, see Vasold, "Versäumte Gelegenheiten?" pp. 257–59, with further source references. Secretary of State Dean Acheson was apparently influenced

primarily by the acting ambassador in Nanking, John W. Jones, and the consul general in Shanghai, John Cabot; see ibid., pp. 258–59.

770. On this paragraph, see McGlothlen, *Controlling the Waves*, p. 153.

771. Ibid., p. 154.

772. On this, see Chen Jian, "Ward Case," p. 167.

773. *FRUS 1949*, vol. 8, *The Far East: China*, p. 776.

774. See Vasold, "Versäumte Gelegenheiten?" pp. 243–44, 259, with source references; McGlothlen, *Controlling the Waves*, pp. 138–39, with source references.

775. Shi Zhe, *Zai lishi juren shenbian*, p. 419. For further details, see this chapter above.

776. Telegram from Gromyko to Zhou Enlai, October 2, 1949, *Pravda*, October 3, 1949.

777. Telegram from Zhou Enlai to Gromyko, October 3, 1949, AVP RF, f. 07, op. 22a, d. 198, l. 5; *Renmin ribao*, October 4,1949, cited in *Sovetsko-kitaiskie otnosheniia, 1917–1957*, p. 216.

778. Gromyko to Stalin, October 2, 1949, AVP RF, f. 07, op. 22a, d. 198, ll. 1–2.

779. Aitchen K. Wu, *China and the Soviet Union*, p. 330.

780. *Sovetsko-kitaiskie otnosheniia*, p. 215.

781. Gromyko to Stalin, October 3, 1949, AVP RF, f. 07, op. 22a, d. 198, ll. 27–28.

782. On this and the following sentence, see Mao Zedong to Roshchin, October 16, 1949, AVP RF, f. 07, op. 22, delo 220, l. 50.

783. Kovalev's memoirs, cited in Goncharov, Lewis, and Xue, *Uncertain Partners*, pp. 313–14 n. 3.

784. Gromyko to Stalin, October 3, 1949, AVP RF, f. 07, op. 22a, d. 198, ll. 1–2.

785. On this and the following sentence, see AVP RF, op. 22a, d. 198, ll. 6–10.

786. On this paragraph, see *Pravda*, October 2–November 25, 1945. China first appeared in *Pravda* on November 23 in a somewhat obscure report on the arrival of the Korean delegates at the Conference of Trade Unions from the Countries of Asia and Oceania and the Conference of Women of the Asian Countries in Peking.

787. On this paragraph, see the statement by the State Department on October 4, 1949, cited in Tang Tsou, *America's Failure in China, 1941–1950*, vol. 2, p. 515; Yuan, "Failure of Conception," p. 145.

788. On this paragraph, see *Pravda*, October 7, 1949; *Liu Shaoqi nianpu*, vol. 2, pp. 228–29; *Zhonghua renmin gongheguo shilu*, vol. 1, pt. 1, pp. 60–61; Tikhvinskii, *Kitai v moei zhizni*, pp. 85–86; *Nash drug Kitai*, p. 212.

789. Wu, *China and the Soviet Union*, p. 330.

790. Tikhvinskii, *Kitai v moei zhizni*, p. 86.

791. On this paragraph, see Tikhvinskii to the author, March 24, 1995; Tikhvinskii, *Kitai v moei zhizni*, p. 86.

792. For further details, see this chapter above.

793. On this paragraph, see Tikhvinskii, *Kitai v moei zhizni*, p. 87; *FRUS 1949*, vol. 8, *The Far East: China*, p. 24; statements by a Russian contemporary witness who wishes to remain anonymous.

794. Tikhvinskii, *Kitai v moei zhizni*, p. 87; Ledovskii in interviews with Sergei N. Goncharov, November 1990 and March 1991, Kovalev in his unpublished memoirs, both cited in Goncharov, Lewis, and Xue, *Uncertain Partners*, p. 314 n. 22.

795. AVP RF, f. 47, op. 22, d. 202, l. 49.

796. For further details, see chapter 2, above.

797. On this paragraph, see Roshchin's notes of December 16, 1949, AVP RF, f. 47, op. 22, d. 220, ll. 48–51; Mao's speech from *Jianguo yilai Mao Zedong wengao*, vol. 1, p. 71; Shi Zhe, *Zai lishi juren shenbian*, p. 431; Tikhvinskii, *Kitai v moei zhizni*, pp. 87–88.

798. Note from Mao Zedong to Stalin, October 20, 1995, in *Jianguo yilai Mao Zedong wengao*, vol. 1, p. 81.

799. Mao Zedong to Mikoyan, February 4, 1949, AP RF, f. 39, op. 1, d. 39, ll. 57–58, cited in Ledovskii, "Sekretnaia missiia A.I. Mikoiana v Kitai," pt. 2, pp. 97–98.

800. See, for example, Gromyko to Stalin, November 12, 1949, AVP RF, f. 07, op. 22a, d. 198, ll. 29–31.

801. Telegrams from Mao Zedong to Wang Jiaxiang, November 9, 1949, and from Mao Zedong to Stalin, November 12, 1949, and November 21, 1949, in *Jianguo yilai Mao Zedong wengao*, vol. 1, pp. 131, 135, 148.

802. On this paragraph, see ibid., p. 81; Klein and Clark, *Biographic Dictionary*, vol. 2, p. 898; Chang Kuo-t'ao, *The Rise of the Chinese Communist Party, vol. 2, 1928–1938*, p. 592; Bartke, *Biographical Dictionary*, p. 357. Tikhvinskii mistakenly assumed that Wang was a member of the Politburo then. See Tikhvinskii, *Kitai v moei zhizni*, p. 87.

803. On this paragraph, see AP RF, f. 39, op. 1, d. 39, ll. 57–58, cited in Ledovskii, "Sekretnaia missiia A.I. Mikoiana v Kitai," pt. 2, p. 97.
804. Tikhvinskii, *Kitai v moei zhizni*, p. 88.
805. Liu Shaoqi to Roshchin, October 25, 1949, AVP RF, f. 07, op. 22, d. 220, l. 16.
806. Gromyko to Stalin, November 12, 1949, AVP RF, f. 07, op. 22a, d. 198, ll. 29–31.
807. See this chapter above.
808. *Für dauerhaften Frieden, für Volksdemokratie*, no. 29 (November 29, 1949).
809. Liu Shaoqi to the Soviet chargé d'affaires, Shibaev, December 25, 1949, AVP RF, f. 0100, op. 43, p. 302, d. 10, ll. 18–19.
810. On this and the following sentence, see Kovalev, "Dialog Stalina s Mao Tszedunom," pt. 2, p. 83.
811. See the text of the speech in *Pravda*, January 4, 1950.
812. On this paragraph, see Ledovskii, "Vizit Mao Tszeduna v Sovetskii Soiuz," pt. 2, pp. 7–8.
813. Liu Shaoqi to the Soviet chargé d'affaires, Shibaev, December 25, 1949, AVP RF, f. 0100, op. 43, p. 302, d. 10, ll. 19–21.
814. Ibid., ll. 19–20.
815. On this paragraph, see Kovalev, "Dialog Stalina s Mao Tszedunom," pt. 2, pp. 80, 90 n. 3; Ledovskii, "Vizit Mao Tszeduna v Sovetskii Soiuz," pt. 2, p. 9; Goncharov, Lewis, and Xue, *Uncertain Partners*, p. 78.
816. According to Ledovskii, "Vizit Mao Tszeduna v Sovetskii Soiuz," pt. 2, p. 9.
817. On this paragraph, see Liu Shaoqi to the Soviet chargé d'affaires, Shibaev, December 25, 1949, AVP RF, f. 0100, op. 43, p. 302, d. 10, ll. 18–21; Kovalev, "Dialog Stalina s Mao Tszedunom," pt. 2, pp. 80, 90–91 n. 3; Ledovskii, "Vizit Mao Tszeduna v Sovetskii Soiuz," pt. 2, pp. 9–10; *Pravda*, January 4, 1950.
818. Liu Shaoqi to the Soviet chargé d'affaires, Shibaev, December 25, 1949, AVP RF, f. 0100, op. 43, p. 302, d. 10, ll. 21–23.
819. Ibid., l. 23.
820. Ibid.
821. Ibid., ll. 23–26.
822. Ibid., ll. 26–27.
823. Ibid., ll. 28–29.
824. Ibid., l. 23; cf. also ll. 24, 26, and 28.
825. Liu Shaoqi to O. Chechetkina, January 27, 1950; see Kulik, "Kitaiskaia Narodnaia Respublika," pt. 2, p. 82.

Notes to Chapter 4

1. For further details, see chapters 2 and 3 of this book.
2. Shi Zhe, *Zai lishi juren shenbian. Shi Zhe huiyi lu*, p. 405.
3. Zhou Enlai to Ambassador Iudin, November 10, 1949, AVP RF, f. 07, op. 22, p. 36, d. 220, l. 52.
4. On this and the following two sentences, see Zhou Enlai to Roshchin, November 10, 1949, AVP RF, f. 07, op. 22, d. 220, ll. 52–53. It is not clear from Zhou whether Mao presented his request to travel to Moscow or Kovalev related Stalin's invitation first. Ledovskii, who reports that he was able to see Kovalev's report of the talk in the Russian presidential archives, states that the initiative came from Mao. See A.M. Ledovskii, "Vizit Mao Tszeduna v Sovetskii Soiuz v dekabre 1949–fevral' 1950 g.," pt. 1, p. 4.
5. Ledovskii, "Vizit Mao Tszeduna v Sovetskii Soiuz," pt. 1, p. 5b, with reference to Kovalev's report on his talk with Mao Zedong on November 5, 1949.
6. The author is assuming that the November 8, 1949, telegram Zhou Enlai mentioned to Roshchin is the same as the one that is summarized in the telegram from the CC of the CCP to Wang Jiaxiang, November 9, 1949. See AVP RF, f. 07, op. 22, d. 220, l. 53; *Jianguo yilai Mao Zedong wengao*, vol. 1, p. 131. Most of the documents from *Jianguo yilai Mao Zedong wengao* cited below are available in English in Sergei Goncharov, John W. Lewis, and Litai Xue, *Uncertain Partners*, app.
7. On this and the following two sentences, see the telegram from the CC of the CCP to Ambassador Wang Jiaxiang, November 9, 1949, in *Jianguo yilai Mao Zedong wengao*, vol. 1, p. 131.
8. Goncharov, Lewis, and Xue give the impression that Mao only proposed that Zhou should come later and thus contradict the contents of the telegram from the CC of the CCP to Wang Jiaxiang on November 9, 1949, which they print in their book. See Goncharov, Lewis, and Xue, *Uncertain Partners*, pp. 83, 236 (document 13).
9. On this paragraph, see Nikolai Roshchih's talk with Zhou Enlai, November 10, 1949, AVP RF, f. 07, op. 22, d. 220, ll. 53–54.

10. On this and the following two sentences, see the telegram from Mao Zedong to Stalin, November 12, 1949, in *Jianguo yilai Mao Zedong wengao*, vol. 1, p. 135; Shi Zhe, *Zai lishi juren shenbian*, p. 431.

11. Gromyko to Stalin, November 26, 1949, AV P RF, f. 07, op. 22a, d. 198, l. 31.

12. Goncharov, Lewis, and Xue make this claim on the basis of Chinese sources; see *Uncertain Partners*, p. 84.

13. On this and the following sentence, see Gromyko to Stalin, November 12, 1949, AVP RF, f. 07, op. 22a, d. 198, ll. 29–31.

14. Cf. Goncharov, Lewis, and Xue, *Uncertain Partners*, p. 84, with source references.

15. The reference is to the harbors of Port Arthur and Dairen.

16. Li Kenong to Roshchin, November 17, 1949, AVP RF, f. 07, op. 22, p. 36, d. 220, l. 68.

17. *Zhou Enlai nianpu, 1949–1976*, vol. 1, p. 14.

18. On this and the following two sentences, see I.V. Kovalev, "Dialog Stalina s Mao Tszedunom," pt. 2, pp. 88–89.

19. Mao Zedong to Roshchin, January 1, 1950, AVP RF, f. 0100, op. 43, p. 302, d. 10, l. 3.

20. Shi Zhe, "Peitong Mao zhuxi fang Su," *Renwu* 5 (1988), p. 11. Shi Zhe did not discuss the planning for the trip again in the versions of his memoirs that were published later.

21. Mao Zedong's talk with Ambassador Pavel Iudin, March 31, 1956, Soviet record of the talk, in *PDV* 5 (1994), p. 105.

22. Wu Xiuquan, *Huiyi yu huainian*, p. 236.

23. On this and the following sentence, see *Zhonghua renmin gongheguo waijiao shi*, p. 16; Wei Shiyan, "Diyi ci fangjian Sulian," p. 302; Shi Zhe and Chen Jian, "With Mao and Stalin," p. 46.

24. On the dichotomy between ideological bonds and national interests in Stalin and Mao, see Michael H. Hunt, *The Genesis of Chinese Communist Foreign Policy*, p. 178.

25. Kovalev, "Dialog Stalina s Mao Tszedunom," pt. 2, p. 86.

26. *Khrushchev Remembers: The Glasnost Tapes*, p. 155. Khrushchev made this observation in connection with Stalin's behavior toward Ho Chi Minh in February 1950.

27. *Khrushchev Remembers*, ed. Edward Crankshaw, p. 128.

28. Mao Zedong in his speech "On the Ten Great Relationships [*Lun shi da guanxi*]," given at a meeting of the Politburo of the CC of the CCP, April 25, 1956, in *Mao Zedong xuanji*, vol. 5, p. 286. This passage is not included in some earlier editions of this speech; cf. John K. Leung and Michael Y.M. Kao, ed., *The Writings of Mao Zedong, 1949–1976*, vol. 2, pp. 43–44, 59–61.

29. Dmitri Volkogonov, *Stalin*, p. 244.

30. This can be confirmed in the archival documents. After Stalin, Molotov stood at the head of the list of Politburo members responsible for China to whom copies of all important documents regarding that country were sent. See, for example, AP RF, f. 3, op. 65, d. 369, l. 6; AVP RF, f. 0100, op. 23a, p. 18, d. 235, l. 43. Khrushchev confirmed that Stalin decided all fundamental issues involving China in conjunction with Molotov at that time; see "Memuary Nikity Sergeevicha Khrushcheva," p. 73.

31. On this and the rest of the paragraph, Mao Zedong's talk with Iudin, October 29, 1950, Iudin's record of the talk; see "Telegramma posla SSSR P.F. Iudina Pervomu sekretariu TsK KPSS N.S. Khrushchevu 30 oktiabria 1957 goda" (AP RF, f. 3, op. 65, d. 589, ll. 173–79), p. 112.

32. On this paragraph, see Kovalev, "Dialog Stalina s Mao Tszedunom," pt. 2, p. 88.

33. On this paragraph, see ibid.

34. On this paragraph, see *Wang Dongxing riji*, pp. 154–56; Shi Zhe, *Zai lishi juren shenbian*, pp. 432, 434; Kovalev, "Dialog Stalina s Mao Tszedunom," pt. 2, p. 88; *Jianguo yilai Mao Zedong wengao*, vol. 1, p. 135; *Pravda*, December 17, 1949; N.T. Fedorenko, "Stalin i Mao," p. 150; Peter M. Kuhfus, "'Ein Held, der mit der Zeit geht,'" pp. 216–20; Wolfgang Bartke, *Biographical Dictionary and Analysis of China's Party Leadership, 1922–1988*, pp. 10, 356–57; AP RF, f. 45, op. 1, d. 329, ll. 17, 38. Wang Dongxing gave the following as the names of the other Chinese who accompanied Mao: Teng Daiyuan, commander of the PLA railroad corps, and Li Jiaji, Luo Duanliao, Chen Jiancheng, Shen Jianxin, and Tian Shubin, who held lower political ranks. See *Wang Dongxing riji*, pp. 154–55.

35. On this paragraph, see Kovalev, "Dialog Stalina s Mao Tszedunom," pt. 2, p. 88; Kovalev, unpublished memoirs, cited in Goncharov, Lewis, and Xue, *Uncertain Partners*, p. 84. Cf. E.P. Bazhanov, *Kitai i vneshnii mir*, p. 47.

36. The bodyguard accompanying Mao, however, claimed that Mao refused Gao Gang's request to accompany him as far as Manzhouli. See *Wang Dongxing riji*, p. 156.

37. On this paragraph, see Shi Zhe, *Zai lishi juren shenbian*, pp. 432–33; Shi Zhe, "Peitong Mao zhuxi fang Su," *Renwu* 5 (1988), pp. 11–12; Shi and Chen, "With Mao and Stalin," pp. 46–47; *Pravda*, December

17, 1949; *Wang Dongxing riji*, pp. 154–56; Goncharov, Lewis, and Xue, *Uncertain Partners*, pp. 84–85; Wei Shiyan, "Diyi ci fangjian Sulian," p. 302.

38. On this paragraph, see *Pravda*, December 17, 1949; *Renmin ribao*, December 18, 1949; Kovalev, "Dialog Stalina s Mao Tszedunom," pt. 2, p. 89; Wei Shiyan, "Diyi ci fangjian Sulian," p. 302; Shi Zhe, *Zai lishi juren shenbian*, p. 433; Shi Zhe, "Peitong Mao zhuxi fang Su," *Renwu* 5 (1988), p. 12. Fedorenko's account of the reception committee and Mao's speech is for the most part inaccurate; see Fedorenko, "Stalin i Mao," p. 151.

39. John W. Garver, *Chinese-Soviet Relations, 1937–1945*, p. 214; Brian Joseph Murray, *Western versus Chinese Realism*, p. 128. On this paragraph, see Kovalev, "Dialog Stalina s Mao Tszedunom," pt. 2, p. 89; *Pravda*, December 17, 1949. Cf. chapter 2 above.

40. Kovalev, "Dialog Stalina s Mao Tszedunom," pt. 2, p. 89.

41. As Molotov put it explicitly; see *Sto sorok besed s Molotovym*, p. 114. Molotov's information was confirmed to the author in a conversation on June 6, 1995, by Liparit Kiuzadzhian. The latter often stayed at the Blizhniaia dacha as adviser on China to the CC of the CPSU between 1965 and 1970 and had the dacha staff, which had been working there in 1949–50, show him the rooms Mao Zedong had used. Wang Dongxing—as did Shi Zhe—mentions a three-story dacha called Jiemei He (literally, "Sisters' River") that was twenty-seven kilometers southwest of Moscow. See *Wang Dongxing riji*, p. 157; Shi Zhe, *Zai lishi juren shenbian*, p. 434. According to Kovalev, the dacha was in Usovo; see Kovalev, "Dialog Stalina s Mao Tszedunom," pt. 2, p. 89. This is also about twenty-five kilometers southwest of Moscow as the crow flies. Goncharov, Lewis, and Xue, on the one hand, cite Kovalev (ibid.), and thus assume that the dacha was in Usovo, and at the same time (without citing sources) speak of a dacha in Usachevka; see Goncharov, Lewis, and Xue, *Uncertain Partners*, p. 316 n. 54.

42. Volkogonov, *Stalin*, p. 452.

43. Shi Zhe, *Zai lishi juren shenbian*, p. 434. Wang Dongxing confirmed that Mao, Ye, and Shi were housed on the same floor. See *Wang Dongxing riji*, p. 157.

44. Goncharov, Lewis, and Xue, *Uncertain Partners*, p. 316 n. 54.

45. *Izvestiia*, December 17, 1949, cited in *Sovetsko-kitaiskie otnosheniia, 1917–1957*, pp. 216–17.

46. *Renmin ribao*, December 18, 1949.

47. On this paragraph, see *Pravda*, December 17, 1949, p. 1, and January 21, 1950, p. 1.

48. *Pravda*, December 22, 1949, p. 3.

49. On this and the following three sentences, see Liu Jiecheng, *Mao Zedong yu Si-da-lin*, pp. 20, 217–19; Liu Jiecheng, *Mao Zedong yu Si-da-lin huiwu jishi*, p. 204. Liu did not indicate his sources. He probably based his statement on interviews with Shi Zhe. Shi Zhe's memoirs, however, make no such claim; see Shi Zhe, *Zai lishi juren shenbian*, pp. 433–67. The Soviet minutes of the talk on January 22, 1950, did not make any reference to any such exchange between Shi Zhe and Stalin, either.

50. *Renmin ribao*, December 18, 1949; *Pravda*, December 22, 1949, p. 6. The title "generalissimo" is used here in connection with Stalin's seventieth birthday.

51. Liu Jiecheng, *Mao Zedong yu Si-da-lin*, p. 39.

52. Cf., for example, the report on Gromyko's reception of the Hungarian ambassador, *Pravda*, October 9, 1949, p. 4. Other examples can be found in *Pravda*, October 14, 1949, p. 1; *Pravda*, November 5, 1949, p. 4; and *Pravda*, November 25, 1949, p. 4.

53. Zhu Zhongli, Wang Jiaxiang's wife, who accompanied her husband to Moscow at the time, confirmed that Fedorenko spoke a very correct (*hen biaozhun*) Chinese and had a very profound knowledge of Chinese history; see Zhu Zhongli, *Mao Zedong Wang Jiaxiang zai wo de shenghuo zhong*, p. 185.

54. On this paragraph, see *Pravda*, December 17, 1949; "Zapis' besedy tovarishcha Stalina I.V. s predsedatelem Tsentral'nogo Narodnogo Pravitel'stva Kitaiskoi Narodnoi Respubliki Mao Tsze-dunom 16 dekabria 1949 g," (henceforth: "Zapis' besedy Stalina s Mao 16 dekabria 1949 g.") AP RF, f. 45, op. 1, d. 329, ll. 9–17; "Posetiteli kremlevskogo kabineta I.V. Stalina," pt. 1, p. 60; telegram from Mao Zedong to Liu Shaoqi, December 14, 1949, in *Zhonghua renmin gongheguo waijiao shi*, p. 17, and Wei Shiyan, "Diyi ci fangjian Sulian," p. 303; Nikolai Fedorenko's talk with Chinese embassy counselor Cheng Dingsi, January 15, 1949, AVP RF, f. 07, op. 22, d. 220, l. 1.

55. See N.T. Fedorenko, "Nochnye besedy"; Fedorenko, "Stalin i Mao," pp. 149–64; N.T. Fedorenko, "Stalin i Mao Tszedun," pt. 1, pp. 98–113, and pt. 2, pp. 83–95. For example, Fedorenko took an entire printed page to describe how a telegram was written to Mao Zedong "on an important matter" without wasting a single word on its contents. See ibid., p. 104.

56. In March 1995, Fedorenko—then eighty-two years old and, by his own accounts, seriously ill—informed the author that he would be willing to speak about politically relevant details only in return for a

large honorarium paid in dollars. He said he would soon publish his memoirs in the People's Republic of China. Chen Jian reported to the author that these have since appeared, but that they have provided as little political information as Fedorenko's various other previous publications.

57. Shi Zhe, "Peitong Mao zhuxi fang Su," *Renwu* 5 (1988), pp. 12–14; Shi Zhe, "Peitong Mao zhuxi fang Su," in *Feng yu gu*, pp. 54–58; Shi Zhe, *Zai lishi juren shenbian*, pp. 434–36. Shi Zhe has also published numerous articles and interviews in journals and newspapers.

58. In a discussion with the author in May 1993, the then eighty-seven-year-old Shi Zhe could not or did not want to remember important events in the field of Sino-Soviet relations in 1949. He was also no longer able to speak Russian well. Party censors blocked publication of an earlier version of his memoirs.

59. Shi Zhe, *Zai lishi juren shenbian*, p. 434.

60. Fedorenko, "Stalin i Mao," pp. 151–52. In May 1995, Fedorenko insisted on this version to the author. It is contradicted by his own report on the talk that has since been made accessible; see the following note.

61. "Zapis' besedy Stalina s Mao 16 dekabria 1949 g.," AP RF, f. 45, op. 1, d. 329, ll. 9–17. At the end of the minutes, there is a typed note that reads "Drawn up by comrade Fedorenko." Vyshinskii signed the minutes by hand on December 31, 1949; an English-language version appeared in *CWIHPB* 6–7 (Winter 1995–1996), pp. 5–7. Conspectus of the talk in A.M. Ledovskii, "Peregovory I.V. Stalina s Mao Tszedunom v dekabre 1949–fevrale 1959 g.," pp. 25–32. Cf. Chen Jian, "Comparing Russian and Chinese Sources," pp. 20–21; Odd Arne Westad, "Unwrapping the Stalin-Mao Talks," pp. 23–24.

62. Telegram from Mao Zedong to Liu Shaoqi, December 18, 1949 (henceforth: Mao to Liu, December 18, 1949), contents reproduced in *Zhonghua renmin gongheguo waijiao shi*, pp. 17–18, and Wei Shiyan, "Diyi ci fangjian Sulian," p. 303. The U.S. intelligence service may have intercepted this telegram; see Richard C. Thornton, *Odd Man Out*, pp. 44–45, 122.

63. Shi Zhe has since supplemented his memoirs with the help of historian Chen Jian. The new text will be published in English under the title *With Mao and Stalin: Shi Zhe's Memoirs.* The author thanks Chen Jian for allowing him to look at the unpublished manuscript, which will be cited as follows: Shi Zhe and Chen Jian, "With Mao and Stalin." On the December 16, 1949 talk, see also Goncharov, Lewis, and Xue, *Uncertain Partners*, pp. 85–88. Mao's telegram to Liu Shaoqi was not available to these authors. They mention the "Soviet minutes," but do not evaluate them in detail. They correctly observe that according to these minutes, Mao had made clear his goal of replacing the old treaty with a new one, which Stalin immediately refused to do with a reference to the Yalta agreement. Goncharov, who had privileged access to the MID archive through his activities in the Soviet Foreign Ministry, clearly saw the minutes but could only make limited use of them.

64. On this and the rest of the paragraph, see Fedorenko, "Stalin i Mao," pp. 154–55.

65. Mao to Liu, December 18, 1949, cited in *Zhonghua renmin gongheguo waijiao shi*, p. 17.

66. Shi Zhe, *Zai lishi juren shenbian*, p. 435. The incident reported by Shi Zhe has been echoed repeatedly without confirmation in the relevant literature in the People's Republic of China. See, for example, Wei Shiyan, "Diyi ci fangjian Sulian," p. 303; Chen Dunde, *Mao Zedong yu Jiang Jieshi, 1949–1976*, pp. 71–72; Liu Jiecheng, *Mao Zedong yu Si-da-lin*, pp. 16–17. According to Shi Zhe, Stalin had said something similar to Liu Shaoqi five months earlier; see Shi Zhe, *Zai lishi juren shenbian*, p. 414.

67. On this paragraph, see AP RF, f. 45, op. 1, d. 329, ll. 9–10.

68. Mao to Liu, December 18, 1949, cited in *Zhonghua renmin gongheguo waijiao shi*, p. 17.

69. Mao Zedong's talk with Iudin, March 31, 1956, Soviet record of the talk, *PDV* 5 (1994), p. 105.

70. AP RF, f. 39, op. 1, d. 328, l. 48, cited in Ledovskii, "Vizit Mao Tszeduna v Sovetskii Soiuz," pt. 1, p. 14. For further details, see chapter 3 above.

71. On this and the rest of the paragraph, see *Zhonghua renmin gongheguo waijiao shi*, p. 13; Shi Zhe, *Zai lishi juren shenbian*, pp. 405–6. For further details, see chapter 3 above.

72. "Zapis' besedy Stalina s Mao 16 dekabria 1949 g.,"AP RF, f. 45, op. 1, d. 329, l. 10.

73. On this and the following sentence, see ibid.

74. Ibid., ll. 10–11.

75. Mao to Liu, December 18, 1949, cited in *Zhonghua renmin gongheguo waijiao shi*, p. 17; Wei Shiyan, "Diyi ci fangjian Sulian," p. 303. Mao's report did not mention southern Sakhalin or the CCR problem.

76. On this paragraph, see the Agreement Regarding Entry of the Soviet Union into the War Against Japan, February 11, 1945, *The Conferences at Malta and Yalta, 1945*, p. 984; the Treaty of Friendship and Alliance Between the Republic of China and the USSR of August 14, 1945, and the agreements that were signed on the same day concerning the Changchun Railroad, Port Arthur and Dairen, in *Treaties Between the*

Republic of China and Foreign States, pp. 505–7, 516–22 (Chinese and English text), and *Sovetsko-kitaiskie otnosheniia*, pp. 196–203 (Russian text). For further details, see chapter 2 above.

77. *Zhonghua renmin gongheguo waijiao shi*, p. 13; Shi Zhe, *Zai lishi juren shenbian*, p. 405. For further details, see chapter 3 above.

78. On this and the following two sentences, see A.M. Ledovskii, "Sekretnaia missiia A.I. Mikoiana v Kitai (ianvar′–fevral′ 1949 g.)," pt. 2, p. 100; ibid., pt. 1, p. 106. For further details, see chapter 3 above.

79. For further details, see chapter 3 above.

80. On this paragraph, see AP RF, f. 45, op. 1, d. 329, l. 11; Mao to Liu, December 18, 1949, cited in *Zhonghua renmin gongheguo waijiao shi*, p. 17. The reproduction of the telegram in Wei Shiyan, "Diyi ci fangjian Sulian," p. 303, is missing any reference to Mao's reply.

81. Cf. Chen Jian, *China's Road to the Korean War*, p. 89.

82. See Wu Xiuquan, *Huiyi yu huainian*, p. 241.

83. See this chapter below.

84. AP RF, f. 45, op. 1, d. 329, ll. 11–12.

85. Mao to Liu, December 18, 1949, cited in *Zhonghua renmin gongheguo waijiao shi*, p. 17.

86. On this paragraph, see ibid.

87. On this and the following two sentences, see AP RF, f. 45, op. 1, d. 329, l. 12.

88. See Stalin's talk with Mao, January 22, 1950, AP RF, f. 45, op. 1, d. 239, l. 32. Cf. this chapter below.

89. Mao to Liu, December 18, 1949, in *Zhonghua renmin gongheguo waijiao shi*, p. 18; Wei Shiyan, "Diyi ci fangjian Sulian," p. 303.

90. On this paragraph, see AP RF, f. 45, op. 1, d. 329, l. 12; Mao to Liu, December 18, 1949, cited in *Zhonghua renmin gongheguo waijiao shi*, p. 18.

91. On this paragraph, see Mao to Liu, December 18, 1949, cited in ibid.; telegram from Mao Zedong to Liu Shaoqi and Zhou Enlai, December 19, 1949, in *Jianguo yilai Mao Zedong wengao*, vol. 1, p. 193; telegram from Liu Shaoqi, Zhu De, and Zhou Enlai to Mao Zedong, December 21, 1949, cited in *Zhonghua renmin gongheguo waijiao shi*, p. 18.

92. Shi Zhe, *Zai lishi juren shenbian*, p. 436. Numerous authors in the People's Republic of China have repeated Shi Zhe's report about "a thing that both looked good and also tasted good." Cf., for example, Chen Dunde, *Mao Zedong yu Jiang Jieshi, 1949–1976*, p. 72; Liu Jiecheng, *Mao Zedong yu Si-da-lin*, pp. 18–19.

93. Shi Zhe, *Zai lishi juren shenbian*, p. 436.

94. For further details on this and the following sentence, see this chapter above.

95. Shi Zhe, *Zai lishi juren shenbian*, p. 437.

96. On this and the following sentence, cf. Goncharov, Lewis, and Xue, *Uncertain Partners*, pp. 86–87.

97. For further details, see chapter 3 above.

98. On this and the following sentence, see AP RF, f. 45, op. 1, d. 329, l. 12.

99. On this paragraph, see ibid.

100. AVP RF, f. 07, op. 23a, p. 18, d. 235, l. 44. The text of this draft has not yet been made accessible.

101. On this paragraph, see the telegram from Peng Dehuai to Mao Zedong, AP RF, f. 45, op. 1, d. 334, ll. 4–7, cited in Ledovskii, "Vizit Mao Tszeduna v Sovetskii Soiuz," pt. 1, pp. 33–35. Ledovskii did not date the telegram. From the context, it was sent between December 29, 1949, and January 2, 1950.

102. AP RF, f. 45, op. 1, d. 329, ll. 12–13.

103. On this and the following sentences, see Liu Shaoqi's report of July 4, 1949, AP RF, f. 45, op. 1, d. 328, l. 34, cited in Ledovskii, "Vizit Mao Tszeduna v Sovetskii Soiuz," pt. 1, p. 20. For further details, see chapter 3 above.

104. On this and the rest of the paragraph, see the telegram from Mao Zedong to Liu Shaoqi, July 25, 1949, AP RF, f. 45, op. 1, d. 328, ll. 139–40, reproduced from a copy of the original, cited in A.M. Ledovskii, "Sekretnaia missiia delegatsii TsK Kommunisticheskoi Partii Kitaia v Moskvu v iiune–avguste 1949 g.," pp. 51–52.

105. On this and the following sentence, see Kovalev, "Dialog Stalina s Mao Tszedunom," pt. 1, p. 88. Shi Zhe denied—not very convincingly—that Stalin and Liu had ever discussed the possibility of Soviet support for an invasion of Taiwan in Moscow; see Shi Zhe quoted in Li Haiwen, in Li Haiwen, "Mao Zedong Liu Shaoqi fang Su ji Zhong-Su liang dang guanxi zhong de yixie wenti," p. 92. Without citing any sources, Ledovskii claimed that Stalin had avoided Liu's request; Ledovskii, "Vizit Mao Tszeduna v Sovetskii Soiuz," pt. 1, p. 21.

106. *Zhonggong zhongyang zai Xiangshan*, p. 454; *Mao Zedong nianpu, 1893–1949*, vol. 3, pp. 458, 541. For further details on this paragraph, see chapter 3 above.

107. AP RF, f. 45, op. 1, d. 329, ll. 13–14.
108. Ibid., l. 14.
109. Mao Zedong to Kovalev, January 12, 1950, AP RF, f. 3, op. 65, d. 533, l. 73, cited in Ledovskii, "Vizit Mao Tszeduna v Sovetskii Soiuz," pt. 1, p. 47.
110. AP RF, f. 45, op. 1, d. 329, l. 14.
111. Ibid., ll. 14–15.
112. Ibid., ll. 15–16.
113. For more on this subject, see this chapter below.
114. For further details, see this chapter below.
115. AP RF, f. 45, op. 1, d. 329, ll. 16–17.
116. Ibid., l. 17. Shi Zhe mistakenly indicated that this talk took place before Mao's return to China (i.e., in January or February 1950); Shi Zhe, *Zai lishi juren shenbian*, p. 459.
117. On this paragraph, see "Memuary Nikity Sergeevicha Khrushcheva," pp. 75–76.
118. On this and the following two sentences, see telegram from Mao Zedong to Stalin, March 30, 1950, forwarded to Stalin by Roshchin, cited in Ledovskii, "Vizit Mao Tsze v Sovetskii Soiuz," pt. 2, p. 48.
119. On this paragraph, see Kuhfus, "'Ein Held, der mit der Zeit geht,'" pp. 238–303 passim; Shi Zhe, *Zai lishi juren shenbian*, p. 459; letter from Wang Jiaxiang to Malenkov, February 27, 1950, AVP RF, f. 0100, op. 43, p. 3102, d. 9, l. 107. On Iudin, see *Bol'shaia Sovetskaia Entsiklopediia*, 3d ed., vol. 30, p. 364; *Filosofskaia Entsiklopediia*, vol. 5, p. 596.
120. On this paragraph, see Mao Zedong's talk with Iudin, July 22, 1958, Chinese record of the talk, *Dang de wenxian* 1 (1994), pp. 16–17. Iudin was Soviet ambassador in China from 1953 to 1959.
121. *Zhonghua renmin gongheguo waijiao shi*, p. 17.
122. On this paragraph, see Shi and Chen, "With Mao and Stalin," p. 51 (Shi Zhe's published memoirs do not include this passage; cf. Shi Zhe, *Zai lishi juren shenbian*, p. 437); Kovalev, "Dialog Stalina s Mao Tszedunom," pt. 2, p. 89.
123. Mao Zedong's talk with Iudin, March 31, 1956, Soviet record of the talk, in *PDV* 5 (1994), p. 105.
124. *Wang Dongxing riji*, p. 159.
125. Ibid., p. 160.
126. Kovalev, "Dialog Stalina s Mao Tszedunom," pt. 2, p. 89.
127. Zhu Zhongli, *Mao Zedong Wang Jiaxiang*, p. 185.
128. On this paragraph, see Kovalev, "Dialog Stalina s Mao Tszedunom," pt. 2, p. 89.
129. On this paragraph, see Zhu Zhongli, *Mao Zedong Wang Jiaxiang*, p. 181.
130. On this and the following two sentences, see *Pravda*, December 24, 1949, p. 1.
131. On this and the following paragraph, see Kovalev's report to Stalin on a talk with Mao Zedong, December 20, 1949, in Goncharov, Lewis, and Xue, *Uncertain Partners*, pp. 238–39, and Kovalev, "Dialog Stalina s Mao Tszedunom," pt. 2, pp. 89–90, where it is excerpted and incorrectly dated.
132. Kovalev, "Dialog Stalina s Mao Tszedunom," pt. 2, p. 89; cf. also Zhu Zhongli, *Mao Zedong Wang Jiaxiang*, p. 184.
133. On this and the following sentence, see "Memuary Nikity Sergeevicha Khrushcheva," p. 73. Cf. *Khrushchev Remembers: The Last Testament*, p. 240.
134. Mao Zedong's telegram to Liu Shaoqi and Zhou Enlai, December 19, 1949, in *Jianguo yilai Mao Zedong wengao*, vol. 1, p. 193.
135. On this paragraph, see Kovalev, "Dialog Stalina s Mao Tszedunom," pt. 2, p. 90.
136. See the photograph published on the front page of *Pravda*, December 22, 1949. A comparison with a photograph published by the Chinese reveals that a few less-important personalities were covered over when the photograph was touched up in the *Pravda* version, including, among others, Shi Zhe, who was standing diagonally behind Mao Zedong; see Shi Zhe, Shi Zhe, *Zai lishi juren shenbian*, third photo after the title page. Cf. *Wang Dongxing riji*, p. 161. Shi Zhe was mistaken when he claimed that the general secretary of the Hungarian CP, Mátyás Rákosi, had stood on Mao's right; see Shi Zhe, *Zai lishi juren shenbian*, p. 441.
137. On this paragraph, see *Pravda*, December 22, 1949, pp. 1–4; *Renmin ribao*, January 23, 1950, p. 1; Shi Zhe, *Zai lishi juren shenbian*, p. 441; Fedorenko, "Stalin i Mao Tszedun," pt. 1, p. 98. The Chinese text of Mao Zedong's speech appeared in *Renmin ribao*, December 23, 1949; the Russian text in *Pravda*, December 22, 1949, p. 3. Mao was the only foreign participant for whom *Pravda* noted that the entire audience had risen to its feet at the end of his speech. Mao reported to Peking that his speech had been received "with great enthusiasm." Everyone had risen to their feet three times, he noted, and they had applauded for a long time. On Mao Zedong's instructions, the text of the speech provided for publication in China was based on the one published in the USSR. See *Jianguo yilai Mao Zedong wengao*, vol. 1, p. 196.

138. On this paragraph, see Fedorenko, "Stalin i Mao Tszedun," pt. 1, pp. 98–99.

139. Shi Zhe, "Peitong Mao zhuxi fang Su," *Renwu* 5 (1988), p. 14; Shi Zhe, *Zai lishi juren shenbian*, p. 441; *Zhonghua renmin gongheguo waijiao shi*, p. 18. The last-mentioned publication stated that the ballet *Swan Lake* was performed. This was clearly a confusion with the performance of *Swan Lake* that Mao attended on February 13, 1950; see this chapter below.

140. *Wang Dongxing riji*, p. 163.

141. *Pravda*, December 23, 1949, p. 1.

142. *Wang Dongxing riji*, p. 164.

143. As the American ambassador in Moscow, Alan G. Kirk, put it in a report on Mao's arrival that he cabled to the State Department, December 18, 1949; see *FRUS 1949*, vol. 8, *The Far East: China*, p. 637.

144. See also Goncharov, Lewis, and Xue, *Uncertain Partners*, p. 90.

145. TASS report from Peking, in *Pravda*, December 22, 1949, p. 6.

146. On this paragraph, see telegram from Mao Zedong to the CC of the CCP, December 22, 1949, in *Jianguo yilai Mao Zedong wengao*, vol. 1, p. 197; *Zhonghua renmin gongheguo waijiao shi*, p. 18. Cf. Shi and Chen, "With Mao and Stalin," p. 51.

147. *Wang Dongxing riji*, p. 164.

148. *Zhonghua renmin gongheguo waijiao shi*, pp. 18–19; *Wang Dongxing riji*, pp. 164–65; Wei Shiyan, "Diyi ci fangjian Sulian," p. 304; Qiu Jing, "Mao Zedong yu Si-da-lin de huiwu," p. 53; Shi and Chen, "With Mao and Stalin," p. 52. No minutes are yet available for this talk. It is probable that the minutes can be found in the Russian presidential archives under the call number AP RF, f. 45, op. 1, d. 239, ll. 18–29, but the author's attempts to discover whether this is true have been unsuccessful. The list of visitors to Stalin's office in the Kremlin does not indicate any meeting on December 24; see "Posetiteli kremlevskogo kabineta," pt. 1, p. 60. Thus, the talk must have taken place somewhere else.

149. Mao Zedong's talk with Iudin, March 31, 1956, Soviet record of the talk, in *PDV* 5 (1994), p. 105.

150. Bo Yibo, who was apparently familiar with the relevant archival materials, drew a causal connection between Stalin's becoming aware of the Kovalev report and his dilatory treatment of Mao Zedong, and added euphemistically that the Soviet side was "not sufficiently active for a while"; see Bo Yibo, *Ruogan zhongda juece yu shibian de huigu*, vol. 1, p. 40.

151. On this paragraph, see Kovalev, "Dialog Stalina s Mao Tszedunom," pt. 1, p. 90; ibid., pt. 2, p. 82; Bo, *Ruogan zhongda juece*, vol. 1, p. 40; Kovalev's diary and memoirs (both unpublished), cited in Goncharov, Lewis, and Xue, *Uncertain Partners*, pp. 91, 240; Liu Jiecheng, *Mao Zedong yu Si-da-lin*, p. 130. Cf. *Khrushchev Remembers. The Last Testament*, pp. 242–43. Khrushchev could not remember Kovalev's name. The memoir's publisher mistakenly assumed that it was Aleksandr Paniushkin, later the Soviet ambassador in Peking, and not Kovalev; see *Khrushchev Remembers*, p. 243 n. 24.

152. On this paragraph, see Kovalev, "Dialog Stalina s Mao Tszedunom," pt. 1, pp. 90–91; Shi Zhe, *Zai lishi juren shenbian*, p. 438; Shi and Chen, "With Mao and Stalin," p. 52; *Khrushchev Remembers: The Last Testament*, pp. 242–43; Bo, *Ruogan zhongda juece*, vol. 1, p. 40.

153. The text of Kovalev's report, which was apparently very voluminous and was subdivided into at least twelve sections, has not yet been published in its entirety. The information on its contents provided in this and the following paragraph is based on both of the following sources: a literal excerpt from the foreign policy section in Kovalev, "Dialog Stalina s Mao Tszedunom," pt. 2, pp. 82–83, and Bo, *Ruogan zhongda juece*, vol. 1, p. 40. The text of the report was apparently available to Bo.

154. Kovalev, "Dialog Stalina s Mao Tszedunom," pt. 1, p. 91.

155. According to Khrushchev, this was true at least of the information Kovalev had sent to the Soviet leadership in the previous months. On this and the rest of the paragraph, see *Khrushchev Remembers: The Last Testament*, p. 243.

156. Khrushchev observed this explicitly about the information Kovalev had previously sent to Moscow. And yet Mao had done nothing about the people who had supposedly stirred up anti-Soviet sentiment, he added. See ibid.

157. Mao Zedong's talk with Iudin, March 31, 1956, Soviet record of the talk, in *PDV* 5 (1994), p. 106.

158. *Wang Dongxing riji*, pp. 167–68. The information provided in Goncharov, Lewis, and Xue, according to which Mao held three talks with Stalin between December 23 and December 30, is apparently also not accurate. The report of a meeting on December 25 clearly refers to the meeting on December 24, which lasted until after midnight, or—as in the case of the report of a meeting on December 30—to the fact that Mao did not ask Stalin a question or did not deliver a telegram to him directly, but on the telephone or by messenger. Cf. Goncharov, Lewis, and Xue, *Uncertain Partners*, p. 91.

159. On this paragraph, see *Wang Dongxing riji*, pp. 160, 162–63, 165, 169–70; Shi Zhe, *Zai lishi juren*

shenbian, p. 441; Wei Shiyan, "Diyi ci fangjian Sulian," p. 304; telegram from Mao Zedong to the CC of the CCP, December 22, 1949, in *Jianguo yilai Mao Zedong wengao*, vol. 1, p. 197; Zhu Zhongli, *Mao Zedong Wang Jiaxiang*, p. 184.

160. Mao Zedong's talk with Iudin, March 31, 1956, Soviet record of the talk, in *PDV* 5 (1994), pp. 105–6.

161. On this paragraph, see *Wang Dongxing riji*, p. 165; Shi Zhe, *Zai lishi juren shenbian*, p. 437; *Zhonghua renmin gongheguo waijiao shi*, p. 19.

162. On this paragraph, see Zhu Zhongli, *Mao Zedong Wang Jiaxiang*, p. 187.

163. Telegram from Mao Zedong to the CC of the CCP, January 2, 1950, in *Jianguo yilai Mao Zedong wengao*, vol. 1, p. 212.

164. Kovalev's unpublished memoirs, cited in Goncharov, Lewis, and Xue, *Uncertain Partners*, p. 92.

165. On this paragraph, see Shi Zhe, *Zai lishi juren shenbian*, pp. 437–38; *Zhonghua renmin gongheguo waijiao shi*, p. 19; Bo, *Ruogan zhongda juece*, vol. 1, p. 41; Wei Shiyan, "Diyi ci fangjian Sulian," p. 304.

166. On this and the following three sentences, see Mao Zedong's talk with Iudin, July 22, 1958, Chinese record of the talk, in *Dang de wenxian* 1 (1994), p. 17.

167. On this and the following two sentences, see Mao Zedong's talk with Iudin, March 31, 1956, Soviet record of the talk, in *PDV* 5 (1994), p. 106. Iudin put the phrase "to rest up at the dacha *[otsypat'sia na dache]*" in quotation marks.

168. On this and the following two sentences, Anastas Mikoyan to his son Sergo. The author thanks Sergo Mikoyan for his oral communication in May 1995.

169. "Memuary Nikity Sergeevicha Khrushcheva," p. 73. The translation on p. 240 of *Khrushchev Remembers: The Last Testament*, is inaccurate.

170. Anastas Mikoyan reported to his son Sergo that Molotov's visit was a reaction to Mao's threat to leave. The author thanks Sergo Mikoyan for his oral communication in May 1995.

171. On this paragraph, see *Sto sorok besed s Molotovym*, p. 114; Shi Zhe, *Zai lishi juren shenbian*, p. 437. It has not yet been possible to determine the exact date of the visit.

172. *Zhonghua renmin gongheguo waijiao shi*, p. 19.

173. Shi and Chen, "With Mao and Stalin," p. 52.

174. On this paragraph, see Roshchin's record of the talk, AVP RF, f. 0100, op. 43, p. 302, d. 10, ll. 1–4. The first copy was for Foreign Minister Vyshinskii, the second for the First Far Eastern Department of the Foreign Ministry.

175. Ibid., l. 3.

176. Ibid., ll. 1–4.

177. Ibid., l. 4.

178. Only the trip to Leningrad took place. See this chapter.

179. On this, see this chapter below.

180. On this, see chapter 3.

181. It has not been possible to determine which press agency published the report, when it occurred, or the precise words that were used. On this paragraph, see *Zhonghua renmin gongheguo waijiao shi*, p. 19; *Jianguo yilai Mao Zedong wengao*, vol. 1, p. 211; Shi Zhe, *Zai lishi juren shenbian*, p. 438; Wei Shiyan, "Diyi ci fangjian Sulian," p. 305; *Wang Dongxing riji*, pp. 171–72; Goncharov, Lewis, and Xue, *Uncertain Partners*, p. 92, with source references.

182. Shi Zhe, *Zai lishi juren shenbian*, p. 439.

183. *Khrushchev Remembers: The Last Testament*, p. 240.

184. On this paragraph, see Shi Zhe, *Zai lishi juren shenbian*, p. 439; Shi and Chen, "With Mao and Stalin," p. 83; *Wang Dongxing riji*, p. 172; Wang Junyan, *Zhongguo waijiao yanyi*, pp. 23–24, with no indication of sources. On the public reactions in China and the West, see Goncharov, Lewis, and Xue, *Uncertain Partners*, p. 93.

185. *Pravda*, January 2, 1950. The interview was published on the same day by the Xinhua News Agency and appeared on January 3, 1950, in *Renmin ribao*; see *Jianguo yilai Mao Zedong wengao*, vol. 1, pp. 206–7. On this paragraph, see *Zhonghua renmin gongheguo waijiao shi*, p. 20; Goncharov, Lewis, and Xue, *Uncertain Partners*, p. 92.

186. Kovalev's report for Stalin, January 2, 1950, on his talk with Mao Zedong of January 1, 1950, AP RF, f. 3, op. 65, d. 533, ll. 61–62, reproduced from a copy of the original, cited in Ledovskii, "Vizit Mao Tszeduna v Sovetskii Soiuz," pt. 1, pp. 28–29.

187. Kovalev's report for Stalin, January 2, 1950, on his talk with Mao Zedong of January 1, AP RF, f. 3, op. 65, d. 533, ll. 61–62, reproduced from a copy of the original, cited in Ledovskii, "Vizit Mao Tszeduna v Sovetskii Soiuz," pt. 1, pp. 29–30.

188. Zhu Zhongli, *Mao Zedong Wang Jiaxiang*, p. 188. On this paragraph, see *Zhonghua renmin gongheguo waijiao shi*, p. 20; Goncharov, Lewis, and Xue, *Uncertain Partners*, pp. 92–93, 106.

189. On this and the following sentence, see Mao Zedong's talk with Iudin, March 31, 1956, Soviet record of the talk, in *PDV* 5 (1994), p. 106. On the "interview," see Goncharov, Lewis, and Xue, *Uncertain Partners*, pp. 92–93, although this source was not available to them at the time.

190. Mao Zedong's talk with Roshchin, January 1, 1950, Roshchin's report of the talk, AVP RF, f. 100, op. 43, p. 302, d. 10, l. 1.

191. Telegram from Zhou Enlai to Mao Zedong, December 31, 1949, AP RF, f. 45, op. 1, d. 334, ll. 2, 12, cited in Ledovskii, "Vizit Mao Tszeduna v Sovetskii Soiuz," pt. 1, pp. 31, 38.

192. See also Vladislav Zubok and Constantine Pleshakov, *Inside the Kremlin's Cold War*, p. 60.

193. Cf. Chen Jian, *China's Road*, p. 89.

194. Cf. Vladislav Zubok, "'To Hell with Yalta!'" p. 35; Goncharov, Lewis, and Xue, *Uncertain Partners*, pp. 98, 101; Melvin P. Leffler, *A Preponderance of Power*, pp. 336–38. For further details on Truman's press conference of January 5, 1950, see this chapter.

195. Telegram from Mao Zedong to Zhou Enlai, January 2, 1949, in *Jianguo yilai Mao Zedong wengao*, vol. 1, pp. 211–12. On this talk, cf. Shi Zhe, *Zai lishi juren shenbian*, pp. 439–40. Shi Zhe apparently based his account on the telegram of January 2, 1949.

196. Telegram from Mao Zedong to Zhou Enlai, January 2, 1949, in *Jianguo yilai Mao Zedong wengao*, vol. 1, pp. 211–12.

197. Ibid., p. 212.

198. Ibid. Shi Zhe confirmed the most important part of the conversation, but his report was clearly based on Mao's telegram. See Shi Zhe, *Zai lishi juren shenbian*, pp. 439–40.

199. Telegram from Mao Zedong to Zhou Enlai, January 2, 1950, in *Jianguo yilai Mao Zedong wengao*, vol. 1, p. 212.

200. Cf. this chapter above.

201. Shi Zhe, *Zai lishi juren shenbian*, p. 440; *Wang Dongxing riji*, p. 177.

202. *Wang Dongxing riji*, p. 177.

203. Mao Zedong to Kovalev, January 12, 1950, AP RF, f. 3, op. 65, d. 533, l. 73, cited in Ledovskii, "Vizit Mao Tszeduna v Sovetskii Soiuz," pt. 1, p. 47.

204. On this paragraph, see *Izvestiia*, January 17 and 18, 1950, in each case on p. 1; *Pravda*, January 17 and 18, 1950, p. 1; *Renmin ribao*, January 18, 1950, p. 1; *Wang Dongxing riji*, pp. 179–80; Shi Zhe, *Zai lishi juren shenbian*, pp. 442–44; AP RF, f. 3, op. 65, d. 533, l. 73.

205. On this paragraph, see *Pravda*, January 18, 1950, p. 1; *Izvestiia*, January 18, 1950, p. 1.

206. *Pravda*, January 4, 1950, p. 3. For further details, see chapter 3 above.

207. On the details in this paragraph, see ibid.

208. On this and the following two sentences, see "O funktsiiach dikatatury proletariata, vypolniaemykh rezhimom Narodnoi demokratii," *Pravda*, January 5, 1950, p. 3, col. 6.

209. On this paragraph, see *Pravda*, January 1–3, 5, 6, 5–15, 18, 19, 25, and 26, 1950.

210. *Pravda*, January 5–9, 1950.

211. Specifically, on December 16, 18, and 25, and on December 27 and 31, 1949. Cf. *Pravda*, October 2, 1949–January 4, 1950.

212. On this paragraph, see Kovalev, "Dialog Stalina s Mao Tszedunom," pt. 1, pp. 90–91, 92–93 n. 11; Bo, *Ruogan zhongda juece*, vol. 1, p. 41; *Khrushchev Remembers: The Last Testament*, pp. 243–44; Shi Zhe and Chen Jian, *With Mao and Stalin*, p. 52; Goncharov, Lewis, and Xue, *Uncertain Partners*, pp. 97, 319 n. 111.

213. On this paragraph, see *Khrushchev Remembers: The Last Testament*, pp. 243–44; Kovalev, "Dialog Stalina s Mao Tszedunom," pt. 1, p. 91.

214. On this and the following four sentences, see Shi Zhe, *Zai lishi juren shenbian*, pp. 393, 438; Shi Zhe, "Gao Gang ceng jianyi jiang Dongbei huagui Sulian ma?" p. 85; Li Haiwen, "Mao Zedong Liu Shaoqi," p. 96; Li Haiwen, "A Distortion of History," p. 64. Shi Zhe's statements about Stalin's distancing himself from Kovalev, which are paraphrased here, include a few contradictions that cannot be explained. On the whole, one gets the impression that Shi Zhe was not informed at the time about the background and contents of the materials Stalin handed over. Kovalev's statements, on the other hand, seem reliable.

215. Kovalev, "Dialog Stalina s Mao Tszedunom," pt. 2, p. 90.

216. Bo, *Ruogan zhongda juece*, vol. 1, p. 41.6

217. Mao Zedong's talk with Iudin, March 31, 1956, Soviet record of the talk, in *PDV* 5 (1994), p. 106.

218. For further details, see chapter 3 above.

219. On this paragraph, see Liu Shaoqi to O. Chechetkina, January 27, 1950, in B.T. Kulik, "Kitaiskaia Narodnaia Respublika v period stanovleniia (1949–1952 gg.)," pt. 2, p. 82.
220. On this paragraph, see ibid., pp. 82–83.
221. On the Wang Ming case, see chapter 1 above.
222. On this paragraph, see the convincing argument in Goncharov, Lewis, and Xue, *Uncertain Partners*, pp. 98, 100. Cf. Ronald L. McGlothlen, *Controlling the Waves*, p. 108; Qiang Zhai, *The Dragon, the Lion, and the Eagle*, p. 44.
223. For further details, see this chapter above.
224. On this paragraph, see Kovalev, "Dialog Stalina s Mao Tszedunom," pt. 2, p. 83.
225. Telegram from Mao Zedong to Liu Shaoqi, January 13, 1950, in *Jianguo yilai Mao Zedong wengao*, vol. 1, p. 235. On this paragraph, see Goncharov, Lewis, and Xue, *Uncertain Partners*, pp. 98–99; Qiang, *Dragon*, p. 44.
226. *Pravda*, January 6, 1950, p. 4; January 9, p. 3; January 13, p. 4; January 19, p. 4. France was also occasionally accused of supporting the KMT government; see *Pravda*, January 8, 1950, p. 3; January 13, p. 4. Great Britain was accused of secretly "activating the fight against the people's liberation movements in the countries of Asia and the Pacific" in the context of the Commonwealth; *Pravda*, January 8, 1950, p. 4.
227. For further details, see chapter 3 above.
228. Goncharov, Lewis, and Xue, *Uncertain Partners*, pp. 99–100, with source references. The reference here (n. 123) suggesting that Moscow approved a trip to Moscow by the air force commander, Liu Yalou, in mid-January 1950, is inaccurate. According to a more recent Chinese source, Liu had already received Stalin's approval for Soviet assistance in the development of a Chinese air force during the Moscow visit in August 1949. See *Zhonggong zhongyang zai Xiangshan*, p. 454. For further details, see chapter 3 above.
229. Goncharov, Lewis, and Xue, *Uncertain Partners*, p. 100, with source reference.
230. The People's Republic of China had already made a similar demand to UN Secretary-General Trygve Lie on November 18, 1949. See Bevin Alexander, *The Strange Connection*, p. 94.
231. On this paragraph, see Vyshinskii's record of the talk, AVP RF, f. 0100, op. 43, p. 302, d. 8, ll. 2–3. The contents of Mao Zedong's report on this section of the talk are consistent with Vyshinskii's version; see telegram from Mao Zedong to Zhou Enlai and the CC of the CCP, January 7, 1950, in *Jianguo yilai Mao Zedong wengao*, vol. 1, pp. 219–20; for an English-language version, see Goncharov, Lewis, and Xue, *Uncertain Partners*, p. 246. Out of consideration for the hostile relationship between Moscow and Belgrade, Mao instructed Liu and Zhou to send the telegram to all members of the Security Council with the exception of Yugoslavia; see *Jianguo yilai Mao Zedong wengao*, vol. 1, p. 221.
232. Telegram from Mao Zedong to Liu Shaoqi and Zhou Enlai, January 7, 1950, *Jianguo yilai Mao Zedong wengao*, vol. 1, p. 221; final version of the text in ibid. (note, p. 222) and in *Pravda*, January 13, 1950, p. 4.
233. On this paragraph, see *Pravda*, January 13, 1950, and January 14, 1950, in each case on p. 4.
234. The talk is reproduced in this and the following paragraph on the basis of Vyshinskii's minutes of the talk, January 13, 1950, AP RF, f. 3, op. 65, d. 364, ll. 94–96, cited in Ledovskii, "Vizit Mao Tszeduna v Sovetskii Soiuz," pt. 1, p. 47–50. Mao's report on the talk in his telegram to Liu Shaoqi, January 13, 1950, was kept considerably shorter, but it corresponds in the main points with Vyshinskii's description. Mao also emphasized that the USSR was supporting the Chinese demand for the expulsion of the KMT representative, while the United States, Great Britain, and the majority of the Security Council members opposed it. See *Jianguo yilai Mao Zedong wengao*, vol. 1, p. 235; an English-language version is in Goncharov, Lewis, and Xue, *Uncertain Partners*, p. 250.
235. Telegram from Mao Zedong to Liu Shaoqi, January 13, 1950, in *Jianguo yilai Mao Zedong wengao*, vol. 1, p. 235.
236. Telegram from Mao Zedong to Liu Shaoqi, January 18, 1950, and text of the cable to the UN in ibid., pp. 242–43; an English-language version appears in Goncharov, Lewis, and Xue, *Uncertain Partners*, pp. 252–53.
237. On this paragraph, see *Pravda*, January 15, 17, 19, and 20, 1950, in each case on p. 4. For transparent propaganda purposes, *Pravda* claimed in its reports that Yugoslavia had also voted for the Soviet proposal. See *Pravda*, January 15, 1950, p. 4; McGlothlen, *Controlling the Waves*, pp. 155–56.
238. Alexander, *Strange Connection*, p. 95; Goncharov, Lewis, and Xue, *Uncertain Partners*, pp. 100–101, with source references.
239. On this paragraph, see Alexander, *Strange Connection*, pp. 93–94; Leffler, *Preponderance of Power*, p. 338.
240. Dean Acheson, "Crisis in Asia—an Examination of U.S. Foreign Policy," cited in Goncharov, Lewis, and Xue, *Uncertain Partners*, pp. 254–55.

241. Cf. ibid., p. 101.

242. Memorandum by the Undersecretary of State to the Executive Secretary of the National Security Council, August 19, 1949, in *FRUS 1949, The Far East: China*, vol. 9, p. 535, cited in John Hansen Feaver, *The Truman Administration and China, 1945–1950*, p. 402; Goncharov, Lewis, and Xue, *Uncertain Partners*, p. 101.

243. Telegram from Mao Zedong to Liu Shaoqi, January 18, 1950, in *Jianguo yilai Mao Zedong wengao*, vol. 1, p. 241.

244. On this paragraph, see telegram from Mao Zedong to Liu Shaoqi, January 13, 1950, in ibid., p. 235; Goncharov, Lewis, and Xue, *Uncertain Partners*, p. 102; Alexander, *The Strange Connection*, p. 95.

245. Vyshinskii's record of a talk with Mao Zedong on January 12, 1950, and January 13, 1950, AP RF, f. 3, op. 65, d. 364, l. 97, cited in Ledovskii, "Vizit Mao Tszeduna v Sovetskii Soiuz," pt. 1, p. 50.

246. Ibid., pp. 50–51.

247. *Renmin ribao*, January 19, 1950, p. 1. Cf. Alexander, *Strange Connection*, p. 95; Leffler, *Preponderance of Power*, p. 340, in which the date is noted inaccurately.

248. Telegram from Mao Zedong to Liu Shaoqi, January 17, 1950, in *Jianguo yilai Mao Zedong wengao*, vol. 1, pp. 238–39.

249. Cf. Chen Jian, *China's Road*, pp. 104–5; Goncharov, Lewis, and Xue, *Uncertain Partners*, pp. 106–7.

250. Leffler, *Preponderance of Power*, p. 340.

251. Cf. Goncharov, Lewis, and Xue, *Uncertain Partners*, p. 106, with source references.

252. See, for example, the caricatures in *Renmin ribao*, January 14, 1950, p. 1; January 16, 1950, p. 6; and January 19, 1950, p. 1.

253. *Renmin ribao*, January 13, 1950, p. 6.

254. According to Shi Zhe, Molotov gave Mao the text of the speech; Shi Zhe, *Zai lishi juren shenbian*, p. 454.

255. According to Shi Zhe, Ulan Bator was also supposed to issue such a statement; ibid.

256. On this paragraph, see Molotov's record of the talk, AVP RF, f. 07, op. 23a, p. 18, d. 234, ll. 1–7, cited in Paul Wingrove, "Mao in Moscow, 1949–50," p. 321; Shi Zhe, *Zai lishi juren shenbian*, pp. 454–55; *Pravda*, January 21, 1950, p. 2; *Izvestiia*, January 21, 1950, p. 2; *Renmin ribao*, January 21, 1950, p. 1. According to Shi Zhe, the Mongolian PR also published a corresponding statement on January 21 in the name of the foreign minister; see Shi Zhe, *Zai lishi juren shenbian*, p. 454. We have not been able to verify this statement.

257. Telegram from Mao Zedong to Liu Shaoqi, also sent to Hu Qiaomu for his information, January 19, 1950, in *Jianguo yilai Mao Zedong wengao*, vol. 1, pp. 245–48; *Renmin ribao*, January 21, 1950, p. 1.

258. *Pravda*, January 21, 1950, p. 2; and *Izvestiia*, January 21, 1950, p. 2.

259. On this paragraph, see Shi Zhe, *Zai lishi juren shenbian*, pp. 455–56.

260. Molotov's record of the talk, AVP RF, f. 07, op. 23a, p. 18, d. 234, ll. 1–7, cited in Wingrove, "Mao in Moscow," p. 321.

261. *Renmin ribao*, January 21, 1950, and January 22, 1950; *Pravda*, January 21, 1950, and January 22, 1950.

262. On this paragraph, see Shi Zhe, *Zai lishi juren shenbian*, p. 457.

263. On this paragraph, see ibid., p. 458; Vladimir Skosyrev, "Stalin i Mao molcha nabliudali, kak tantsuet politbiuro" (interview with Shi Zhe), p. 3; *Khrushchev Remembers*, pp. 289–90.

264. Text of the message in *Izvestiia*, January 31, 1950, p. 2.

265. Telegram from Mao Zedong to Liu Shaoqi, January 17, 1950, in *Jianguo yilai Mao Zedong wengao*, vol. 1, p. 238.

266. On this paragraph, see Chen Jian, *China's Road*, p. 104; Goncharov, Lewis, and Xue, *Uncertain Partners*, p. 107.

267. Shi and Chen, "With Mao and Stalin," p. 52: *Zhonghua renmin gongheguo waijiao shi*, pp. 16–17; *Wang Dongxing riji*, pp. 164–65; Wei Shiyan, "Diyi ci fangjian Sulian," p. 304.

268. Letter from Stalin to Mao Zedong, January 6, 1950, AP RF, f. 45, op. 1, d. 334, l. 16, reproduced from a copy of the original, cited in Ledovskii, "Vizit Mao Tszeduna v Sovetskii Soiuz," pt. 1, p. 39.

269. *Izvestiia*, January 31, 1960, p. 2.

270. *Khrushchev Remembers: The Glasnost Tapes*, p. 155.

271. Goncharov, Lewis, and Xue claim that "at Mao's insistence, he [Stalin] recognized the revolutionary regime of Ho Chi Minh," but fail to provide documentary evidence. See Goncharov, Lewis, and Xue, *Uncertain Partners*, p. 211.

272. Wu Xiuquan, *Huiyi yu huainian*, p. 242.
273. Ibid.; *Khrushchev Remembers: The Glasnost Tapes*, pp. 155–56.
274. *Khrushchev Remembers: The Glasnost Tapes*, pp. 155–56.
275. Ibid., p. 156.
276. On this and the following sentence, see Chen Jian, *China's Road*, pp. 104–5, with source references.
277. On this and the following sentence, see the memorandum of the head of the Department for Eastern Europe in the U.S. State Department, January 31, 1950, in *FRUS 1950*, vol. 6, *East Asia and the Pacific*, p. 710.
278. Acheson's statement of February 1, 1950, in ibid., p. 711.
279. Ibid., pp. 716, 720–22.
280. Acheson's memorandum for the president, February 2, 1950, in ibid., p. 717.
281. Goncharov, Lewis, and Xue, *Uncertain Partners*, p. 108. On Truman's statement of January 5, 1950, see this section above.
282. Cf. Alexander, *Strange Connection*, p. 95; Leffler, *Preponderance of Power*, p. 340.
283. Goncharov, Lewis, and Xue, *Uncertain Partners*, p. 104.
284. Letter from Mao Zedong to Stalin, January 2, 1950, AP RF, f. 45, op. 1, d. 334, l. 15, cited in Ledovskii, "Vizit Mao Tszeduna v Sovetskii Soiuz," pt. 1, p. 30.
285. Letter from Stalin to Mao Zedong, January 6, 1950, AP RF, f. 45, op. 1, d. 334, l. 16, reproduced from a copy of the original, cited in ibid. p. 35.
286. On this and the following sentence, see Vyshinskii's report of a talk with Mao Zedong, January 6, 1950, AVP RF, f. 0100, op. 43, p. 302, d. 8, l. 1; Telegram from Mao Zedong to Zhou Enlai and the CC of the CCP, in *Jianguo yilai Mao Zedong wengao*, vol. 1, p. 219.
287. Vyshinskii's report of a talk with Mao Zedong on January 13, 1950, AP RF, f. 3, op. 65, d. 364, ll. 97–98, cited in Ledovskii, "Vizit Mao Tszeduna v Sovetskii Soiuz," pt. 1, p. 51.
288. *Zhonggong zhongyang zai Xiangshan*, p. 454. For further details, see chapter 3 above. On this paragraph, see the telegram from Liu Shaoqi to Stalin, December 30, 1949, AP RF, f. 45, op. 1, d. 334, l. 3, reproduced from a copy of the original, cited in Ledovskii, "Vizit Mao Tszeduna v Sovetskii Soiuz," pt. 1, p. 30; telegram from Mao Zedong to Stalin, February 2, 1950, AP RF, f. f. 45, op. 1, d. 334, l. 2, cited in ibid., p. 31.
289. Stalin sent Liu Shaoqi a telegram with the same contents on January 8, 1950, AP RF, f. 45, op. 1, d. 334, l. 17, cited in ibid., pp. 46–47. Vyshinskii sent the telegram to the military attaché at the Soviet embassy in Peking, who delivered it to Liu by hand, ibid., p. 47.
290. On this paragraph, see the letter from Stalin to Mao Zedong, January 6, 1950, AP RF, f. 45, op. 1, d. 334, l. 16; Vyshinskii's report on a talk with Mao Zedong on January 6, 1950, AVP, f. 0100, op. 43, p. 302, d. 8, ll. 1–2; telegram from Mao Zedong, January 7, 1950, in: *Jianguo yilai Mao Zedong wengao*, vol. 1, p. 219.
291. For further details on developments in Sinkiang, see this chapter above. On this paragraph, see the telegram from Peng Dehuai to Mao Zedong, December 29, 1949, AP RF, f. 45, op. 1, d. 334, ll. 4–7, reproduced from the copy of the original cited in Ledovskii, "Vizit Mao Tszeduna v Sovetskii Soiuz," pt. 1, pp. 33–36; letter from Mao Zedong to Stalin, January 2, 1950, AP RF, f. 45, op. 1, d. 334, l. 2, cited in ibid., p. 31. On the conditions leading to the telegram and its date, see *Peng Dehuai zhuan*, p. 390.
292. Telegram from Peng Dehuai to Mao Zedong, December 29, 1949 reproduced from a copy of the original, AP RF, f. 45, op. 1, d. 334, ll. 6–7, in Ledovskii, "Vizit Mao Tszeduna v Sovetskii Soiuz," pp. 35–36.
293. "Zapis' besedy Stalina s Mao 16 dekabria 1949 g.," AP RF, f. 45, op. 1, d. 329, l. 13.
294. *Sovetsko-kitaiskie otnosheniia*, pp. 227–28.
295. Letter from Stalin to Mao Zedong, January 6, 1950, AP RF, f. 45, op. 1, d. 334, l. 16, cited in Ledovskii, "Vizit Mao Tszeduna v Sovetskii Soiuz," pt. 1, p. 39.
296. AP RF, f. 45, op. 1, d. 329, l. 37. For further details, see this chapter below.
297. Shi Zhe, *Zai lishi juren shenbian*, p. 441.
298. On this and the rest of this section, see the telegram from Mao Zedong to the CC of the CCP, January 3, 1950, in *Jianguo yilai Mao Zedong wengao*, vol. 1, pp. 211–13; an English-language version is in Goncharov, Lewis, and Xue, *Uncertain Partners*, pp. 243–44.
299. Because they did not know the contents of the December 16, 1949, talk between Stalin and Mao, Goncharov, Lewis, and Xue reached mistaken conclusions; see Goncharov, Lewis, and Xue, *Uncertain Partners*, p. 95.
300. Cf. the somewhat different interpretation of these instructions from Mao in Goncharov, Lewis, and Xue, *Uncertain Partners*, p. 96.

301. Zhou Enlai to Shibaev, the Soviet chargé d'affaires in Peking, January 4, 1950, AVP RF, f. 0100, op. 43, p. 302, d. 10, l. 45.

302. Telegrams from Mao Zedong to the CC of the CCP, January 5, 1950, in *Jianguo yilai Mao Zedong wengao*, vol. 1, pp. 215, 217.

303. On this and the following sentence, see Shi Zhe, *Zai lishi juren shenbian*, p. 444; Wu Xiuquan, *Huiyi yu huainian*, p. 236. On January 18, 1950, for example, Zhou Enlai called Mao Zedong by telephone from Sverdlovsk; see Mao's telegram to Liu Shaoqi, January 18, 1950, in *Jianguo yilai Mao Zedong wengao*, vol. 1, p. 242.

304. For further details, see this chapter above.

305. Ibid.

306. On this paragraph, see Vyshinskii's report of a talk with Mao Zedong on January 6, 1950, AVP RF, f. 0100, op. 43, d. 8, ll. 1–4.

307. On this paragraph, see ibid., ll. 4–5. Fedorenko, who clearly drew up the record of this talk, again confused the Sino-Soviet treaty (*dogovor*) of August 14, 1945, with the Yalta agreement (*soglashenie*) of February 11, 1945, as he had done before in the talk between Stalin and Mao on December 16 apparently out of ignorance of the complicated subject matter. See this chapter above.

308. See this chapter above.

309. Cf. Zubok, "'To Hell with Yalta!'" p. 26.

310. On this and the following two paragraphs, see Kovalev's report to Stalin on his January 6, 1950, talk with Mao, dated January 7, 1950, AP RF, f. 3, op. 65, d. 533, ll. 63–64, cited in Ledovskii, "Vizit Mao Tszeduna v Sovetskii Soiuz," pt. 1, pp. 45–46.

311. On this, see the same section cited in n. 298.

312. Kovalev's report to Stalin, cited in Goncharov, Lewis, and Xue, *Uncertain Partners*, pp. 247–48. The report is dated January 1950 here. It was drawn up between January 10 and January 20.

313. AVP RF, f. 07, op. 23a, p. 18, d. 235, ll. 1–115 passim.

314. Ibid., l. 42.

315. Ibid., ll. 12–14. The draft bears a signature that cannot be identified fully (Shesterikov?).

316. Ibid., ll. 15–19. The draft bears the signatures of Fedorenko, Roshchin, Buev, Kurdiukov, and Gribanov, ibid., l. 15.

317. Ibid., ll. 20–25. Kurdiukov and Gribanov signed the draft.

318. Ibid., ll. 26–29.

319. MID is the abbreviation for Ministerstvo Inostrannykh Del (Ministry of Foreign Affairs).

320. AVP RF, f. 07, op. 23a, p. 18, d. 235, ll. 31–34.

321. Ibid., ll. 1–6. The draft of a "clause in the signing protocol" was also included; ibid., l. 7.

322. Ibid., ll. 42–50. Copies went to Politburo members Beriia, Malenkov, Kaganovich, and Bulganin. English text in the appendix, document 7.

323. AVP RF, f. 07, op. 23a, p. 18, d. 235, ll. 1–4. The marginalia refer to specific articles in the four treaties mentioned.

324. The Chinese and Russian versions were both equally legally binding. Chinese version in *Treaties Between the Republic of China and Foreign States*, pp. 505–7; Russian version in *Sovetsko-kitaiskie otnosheniia*, pp. 219–20, English version in *Treaties Between the Republic of China and Foreign States*, pp. 505–7.

325. English-language versions of most of these treaties can be found in the *United Nations Treaty Series*.

326. *Sbornik deistvuiushchikh dogovorov, soglashenii i konventsii, zakliuchennykh SSSR s inostrannymi gosudarstvami*, vol. 13, pp. 15–22. English versions in *United Nations Treaty Series*, vol. 48, 1950: treaty with Romania, pp. 196, 198, 200; treaty with Hungary, pp. 170, 172, 174; treaty with Bulgaria, pp. 142, 144, 146.

327. Text of the treaty in *Sovetsko-mongol'skie otnosheniia, 1921–1966*, pp. 156–58. English version in *United Nations Treaty Series*, vol. 48, 1950, pp. 184, 186.

328. Cf. Boris Meissner, *Das Ostpaktsystem*, pt. 1, pp. I–III. Italics added by the author.

329. Italics added by the author.

330. One of the leading German experts on Eastern European constitutional and international law also assumed that by this they "probably wanted to demonstrate a particularly close bond between the partners." See Dietrich Frenzke, "Bündnis- und Freundschaftsverträge der UdSSR mit asiatischen Staaten," p. 136.

331. "Zapis' besedy Stalina s Mao 16 dekabria 1949 g.," AP RF, f. 45, op. 1, d. 329, l. 2.

332. Telegrams from Mao Zedong to the CC of the CCP, January 2, 1950, and January 3, 1950, in *Jianguo yilai Mao Zedong wengao*, vol. 1, pp. 211, 213.

333. Cf. Frenzke, "Bündnis- und Freundschaftsverträge der UdSSR," pp. 135–36.

334. Alexander Uschakow and Dietrich Frenzke, *Der Warschauer Pakt und seine bilateralen Bündnisverträge*, pp. 395–99, 401–5.

335. *Sovetsko-mongol'skie otnosheniia*, pp. 156–58; Meissner, *Das Ostpaktsystem*, pt. 2, pp. 39–40, 44–46; Uschakow and Frenzke, *Der Warschauer Pakt*, pp. 382–85.

336. Uschakow and Frenzke, *Der Warschauer Pakt*, pp. 390–94.

337. Cf. this chapter below.

338. On this, see Uschakow and Frenzke, *Der Warschauer Pakt*, pp. 223–46 passim.

339. AVP RF, f. 07, op. 23a, p. 18, d. 235, l. 16–19.

340. Draft of January 10, 1950, ibid., ll. 20–22.

341. AVP RF, f. 07, op. 23a, p. 18, d. 235, l. 26.

342. Ibid., l. 13.

343. Cf. *Sbornik deistvuiushchikh dogovorov*, vol. 13, pp. 16, 20–21.

344. Ibid., p. 18.

345. Text of the treaty in *Sovetsko-mongol'skie otnosheniia*, pp. 156–58.

346. For further details, see this section above.

347. On this paragraph, see Uschakow and Frenzke, *Der Warschauer Pakt*, pp. 174–78, esp. p. 177.

348. AVP RF, f. 07, op. 23a, p. 18, d. 235, l. 13. The three mistakes in the first draft of the text of Article 3 were apparently the result of carelessness.

349. Ibid., l. 2.

350. The Soviet-Bulgarian treaty of alliance of March 18, 1948, in *Sbornik deistvuiushchikh dogovorov*, vol. 13, p. 16.

351. AVP RF, f. 07, op. 23a, p. 18, d. 235, ll. 18, 20, 24.

352. Both the Russian and Chinese versions of the three agreements were equally binding legally. The Chinese version is in *Treaties Between the Republic of China and Foreign States*, pp. 512–20; the Russian version is in *Sovetsko-kitaiskie otnosheniia*, pp. 198–203. An English-language version is in *Treaties Between the Republic of China and Foreign States*, pp. 512–20. For further details on the Yalta agreement, the August 14, 1945 treaty, and the three other agreements, see chapter 2 above.

353. For further details, see this chapter above.

354. AVP RF, f. 07, op. 23a, p. 18, d. 235, ll. 13–14. English text in the appendix, document 1.

355. Ibid., l. 18. English text in the appendix, document 2.

356. Ibid., l. 15.

357. Ibid., ll. 18, 20. A line is drawn through Article 7 in the consultants' draft; ibid., l. 18.

358. Ibid., l. 24. English text in the appendix, document 3.

359. Ibid., l. 20.

360. The protocol established among other things that China leased half of the harbor facilities and equipment free of charge to the USSR. Text in *Treaties Between the Republic of China and Foreign States*, p. 520. The Soviet side did not publish the protocol at that time; see AVP RF, f. 07, op. 23a, p. 18, d. 235, l. 122. It is therefore not included in a 1959 collection of Soviet treaties; cf. *Sovetsko-kitaiskie otnosheniia*, p. 203.

361. AVP RF, f. 07, op. 23a, p. 18, d. 235, l. 33. English text in the appendix, document 4.

362. Italics added by the author.

363. On this paragraph, see AVP RF, f. 07, op. 23a, p. 18, d. 235, ll. 3–4. The passage "and also the protocol on the agreement on Dairen harbor" was crossed out by hand. English text in the appendix, document 5.

364. Ibid., ll. 5–6. English text in the appendix, document 6.

365. Ibid., ll. l. 49. English text in the appendix, document 8.

366. Ibid., l. 50. English text in the appendix, document 9.

367. Ibid.

368. Cf. Frenzke, "Bündnis- und Freundschaftsverträge der UdSSR," p. 135. The two draft protocols are in AVP RF, f. 07, op. 23a, p. 18, d. 235, ll. 49–50.

369. AVP RF, f. 07, op. 23a, p. 18, d. 235, ll. 117–20.

370. Ibid., l. 7.

371. Handwritten letter from Durdenevskii sent to Vyshinskii, January 13–14, 1950; underlining in the original (rendered here as italics); AVP RF, f. 07, op. 23a, p. 18, d. 235, l. 117.

372. AVP RF, f. 07, op. 23a, p. 18, d. 235, ll. 41–44.

373. Ibid., l. 44. The author has rendered the titles of the documents more concisely than in the original without leaving out anything of significance.

374. Ibid., ll. 42–43.

375. The English texts of the first three documents are included in the appendix; see documents 7, 8, and 9.

376. On this, see this chapter below.

377. Telegram from Mao Zedong to the CC of the CCP, January 2, 1950, in *Jianguo yilai Mao Zedong wengao*, vol. 1, p. 212.

378. Telegram from Mao Zedong to the CC of the CCP, January 5, 1950, in ibid., pp. 215–16.

379. *Zhou Enlai nianpu, 1949–1976*, Vol. 1, p. 21; Shi Zhe, *Zai lishi juren shenbian*, p. 444; *Chinese Communist Who's Who*, vol. 2, p. 122. Soviet reports gave some of the given names of members of the delegation slightly differently, which may have been the result of mistakes due to mishearing or transcription errors; see *Izvestiia*, January 21, 1950, p. 1.

380. For further details, see chapter 3 above and this chapter above.

381. *Wang Dongxing riji*, pp. 176, 196; Shi Zhe, *Zai lishi juren shenbian*, p. 461; *Zhonghua renmin gongheguo waijiao shi*, p. 21. The last-mentioned source merely noted that Saifudin had arrived from Sinkiang.

382. Wolfgang Bartke, *Who's Who in the People's Republic of China*, p. 51; Goncharov, Lewis, and Xue, *Uncertain Partners*, p. 244 n. 12.

383. Säjpidin Äzizi, *Ömür Dastani*, vol. 2, pp. 460–70; Allen S. Whiting and General Sheng Shi-ts'ai, *Sinkiangt*, pp. 142–43.

384. On Saifudin, see Howard L. Boorman and Richard C. Howard, *Biographical Dictionary of Republican China*, vol. 3, pp. 87–89; *Chinese Communist Who's Who*, vol. 2, p. 150; Bartke, *Who's Who*, p. 312.

385. Telegram from Mao Zedong to the CC of the CCP, January 5, 1950, in *Jianguo yilai Mao Zedong wengao*, vol. 1, p. 215.

386. On this paragraph, see *Izvestiia*, January 21, 1950, p. 1; Shi Zhe, *Zai lishi juren shenbian*, p. 444; Wu Xiuquan, *Huiyi yu huainian*, p. 236.

387. On this paragraph, see *Izvestiia*, January 21, 1950, p. 1; *Renmin ribao*, January 22, 1950, p. 1. Wang Dongxing reported inaccurately that Molotov met Zhou. See *Wang Dongxing riji*, p. 191.

388. On this paragraph, see *Pravda*, January 21, 1950, p. 1; *Renmin ribao*, January 22, 1950, p. 1, and January 24, 1950, p. 1. Wu Xiuquan and Roshchin can also be recognized in the photograph that appeared on the front page of *Izvestiia* on January 21, 1950.

389. On this paragraph, see *Pravda*, January 21, 1950, p. 1.

390. *Wang Dongxing riji*, pp. 191–92; Shi Zhe, *Zai lishi juren shenbian*, p. 445.

391. *Zhonghua renmin gongheguo waijiao shi*, p. 21.

392. On this paragraph, see *Pravda*, January 22, 1950, p. 1; *Renmin ribao*, January 23, 1950, p. 1; *Wang Dongxing riji*, p. 192; Mao Zedong's telegram to Liu Shaoqi, January 25, 1950, in *Jianguo yilai Mao Zedong wengao*, vol. 1, p. 251. Wang Dongxing claimed that Mao made a speech at the celebrations. This is not confirmed by any other source. See *Wang Dongxing riji*, p. 192.

393. The following reconstruction of the talk is based, where it is not otherwise noted, on the Soviet minutes: *Zapis' besedy I.V. Stalina s predsedatelem Tsentral'nogo Narodnogo Pravitel'stva Kitaiskoi Narodnoi Respubliki Mao-Tsze-dunom 22 ianvaria 1950 g.* (henceforth: *Zapis' besedy Stalina s Mao 22 ianvaria 1950 g.*), AP RF, f. 45, op. 1, d. 329, ll. 29–38. In addition to this final version, the author had access to the draft Fedorenko and Roshchin drew up on January 23, 1950, and forwarded to Vyshinskii the following day for confirmation: AP RF, f. 45, op. 1, d. 329, ll. 40–49. It included several handwritten corrections by an unknown hand that were almost exclusively stylistic or orthographic in nature, which were all adopted in the final version. We will refer to the four corrections involving content later when we reproduce the relevant portion of the minutes. English version of the talk in *CWIHPB* 6–7 (Winter 1995–1996), pp. 7–9. Conspectus of the talk in Ledovskii, "Peregovory I.V. Stalina s Mao Tszedunom," pp. 37–41. On the duration of and participants in the talk, see "Posetiteli kremlevskogo kabineta," pt. 2, p. 6. Cf. Chen Jian, "Comparing Russian and Chinese Sources," pp. 20–21; Westad, "Unwrapping the Stalin-Mao Talks," pp. 23–24; Zubok, "'To Hell with Yalta!'" pp. 24–27. The reproduction of the January 22, 1950, talk in Goncharov, Lewis, and Xue is to a great extent incomplete and is in part misleading; see Goncharov, Lewis, and Xue, *Uncertain Partners*, pp. 110–13, 120, 126. This is no criticism of the authors. Like all authors writing before 1995, they did not have access to the Soviet minutes of the meeting and had to rely in large part on Shi Zhe's meager and often inaccurate statements.

394. *Pravda*, January 23, 1950.

395. "Zapis' besedy Stalina s Mao 22 ianvaria 1950 g.," AP RF, f. 45, op. 1, d. 329, l. 29.

396. For further details, see this chapter above.

397. On this and the following sentence, see the text of the treaty in *Sovetsko-kitaiskie othosheniia*, pp. 196–98, in particular Articles 1 and 8.

398. For further details, see this chapter above.
399. "Zapis' besedy Stalina s Mao 22 ianvaria 1950 g.," AP RF, f. 45, op. 1, d. 329, ll. 29–30.
400. Ibid., l. 30.
401. Ibid.; italics by the author.
402. Ibid., ll. 30–31.
403. For further details, see this chapter above.
404. "Zapis' besedy Stalina s Mao 22 ianvaria 1950 g.," AP RF, f. 45, op. 1, d. 329, l. 31.
405. For further details, see this chapter above.
406. "Zapis' besedy Stalina s Mao 22 ianvaria 1950 g.," AP RF, f. 45, op. 1, d. 329, l. 31.
407. Telegram from Stalin to Mikoyan (for Mao Zedong), February 5, 1949, cited in Ledovskii, "Sekretnaia missiia A.I. Mikoiana v Kitai," pt. 1, p. 106.
408. Mao Zedong to Iudin, July 22, 1958, Chinese record of the talk, in *Dang de wenxian* 1 (1994), p. 16.
409. Telegram from Mikoyan to Stalin, February 6, 1949, AP RF, f. 39, op. 1, d. 39, ll. 78–79, copy of the original, cited in Ledovskii, "Vizit Mao Tszeduna v Sovetskii Soiuz," pt. 1, p. 13; Cf. also Ledovskii, "Sekretnaia missiia A.I. Mikoiana v Kitai," pt. 2, p. 101.
410. *Zhonghua renmin gongheguo waijiao shi*, p. 13; Shi Zhe, *Zai lishi juren shenbian*, p. 405; Kovalev, "Dialog Stalina s Mao Tszedunom," pt. 2, p. 86.
411. AP RF, f. 45, op. 1, d. 328, l. 49, cited in Ledovskii, "Sekretnaia missiia delegatsii TsK," p. 42.
412. For further details, see this chapter above.
413. "Zapis' besedy Stalina s Mao 22 ianvaria 1950 g.," AP RF, f. 45, op. 1, d. 329, l. 31.
414. Ibid., l. 32.
415. On this sentence, see AP RF, f. 39, op. 1, d. 39, ll. 78–79, cited in Ledovskii, "Sekretnaia missiia A.I. Mikoiana v Kitai," pt. 2, pp. 100–101; *Zhonghua renmin gongheguo waijiao shi*, p. 13; Shi Zhe, *Zai lishi juren shenbian*, p. 405; "Zapis' besedy Stalina s Mao 16 dekabria 1949 g.," AP RF, f. 45, op. 1, d. 239, ll. 10–12. For further details, see chapter 3 above.
416. "Zapis' besedy Stalina s Mao 22 ianvaria 1950 g.," AP RF, f. 45, op. 1, d. 329, l. 32. The question mark and exclamation point after "Yalta conference" are in the Soviet protocol.
417. "Zapis' besedy Stalina s Mao 16 dekabria 1949 g.," AP RF, f. 45, op. 1, d. 329, l. 10.
418. "Zapis' besedy Stalina s Mao 22 ianvaria 1950 g.," AP RF, f. 45, op. 1, d. 329, l. 32.
419. Ibid.
420. Goncharov, Lewis, and Xue made a similar argument. They assumed that Mao was acting tactically on the basis of the Chinese expression *yuqu guyu* (make concessions to achieve a benefit). See Goncharov, Lewis, and Xue, *Uncertain Partners*, p. 112.
421. Cf. Dieter Heinzig, "Verriet Stalin im Jahre 1945 die kommunistische Revolution in China?" p. 9.
422. Zubok and Pleshakov, *Inside the Kremlin's Cold War*, p. 60.
423. Kovalev's report to Stalin, cited in Goncharov, Lewis, and Xue, *Uncertain Partners*, p. 248.
424. Shi Zhe's literal citation has "immediately withdraw [*xianzai chetui*]." See Shi Zhe, *Zai lishi juren shenbian*, p. 445.
425. On this paragraph, see "Zapis' besedy Stalina s Mao 22 ianvaria 1950 g.," AP RF, f. 45, op. 1, d. 329, l. 32. Shi Zhe reproduced Stalin's two alternatives in almost the same way. He cited this passage literally and thus clearly based his account on the Chinese minutes. See Shi Zhe, *Zai lishi juren shenbian*, p. 445.
426. For further details on this paragraph, see this chapter above. See the text of the "commission draft" on the Port Arthur and Dairen question in document 8 of the appendix.
427. "Zapis' besedy Stalina s Mao 22 ianvaria 1950 g.," AP RF, f. 45, op. 1, d. 329, ll. 32–33. Supported only by a sentence that is not indicated as a literal citation in Shi Zhe, Goncharov, Lewis, and Xue assumed that Mao proposed to treat the three issues of the CCR, Port Arthur, and Dairen in a single document on January 22. The three authors drew far-reaching conclusions from this that are not convincing. See Goncharov, Lewis, and Xue, *Uncertain Partners*, pp. 111–13 passim. We can assume that Mao made no such proposal. It does not appear in the Soviet minutes.
428. On this paragraph, see "Zapis' besedy Stalina s Mao 22 ianvaria 1950 g.," AP RF, f. 45, op. 1, d. 329, ll. 33–34. The last sentence in the draft of the protocol reads "Furthermore [*k tomu zhe*], the result is a house with open doors." Ibid., l. 44. On the Dairen issue, Shi said literally "can be arranged by China itself [*keyi you Zhongguo ziji chuli*]." See Shi Zhe, *Zai lishi juren shenbian*, p. 447.
429. See the text of the commission draft on the Port Arthur and Dairen question in document 8 of the appendix.
430. On this paragraph, see "Zapis' besedy Stalina s Mao 22 ianvaria 1950 g.," AP RF, f. 45, op. 1, d. 329, l. 34.

431. Cf. Zubok, "'To Hell with Yalta!'" p. 25; See also this chapter above.
432. "Zapis' besedy Stalina s Mao 22 ianvaria 1950 g.," AP RF, f. 45, op. 1, d. 329, l. 34.
433. Telegram from Mikoyan to Stalin, February 6, 1949, AP RF, f. 39, op. 1, d. 39, ll. 12–13, copy of the original, cited in Ledovskii, "Vizit Mao Tszeduna v Sovetskii Soiuz," pt. 1, pp. 12–13; Ledovskii, "Sekretnaia missiia A.I. Mikoiana v Kitai," pt. 2, pp. 100–101.
434. On this paragraph, see "Zapis' besedy Stalina s Mao 16 dekabria 1949 g.," AP RF, f. 45, op. 1, d. 329, l. 11. See this chapter above.
435. On this paragraph, see Kovalev's report to Stalin, cited in Goncharov, Lewis, Xue, *Uncertain Partners*, pp. 247–48.
436. Article 6 of the agreement of August 14, 1945; see *Treaties Between the Republic of China and Foreign States*, p. 513.
437. Goncharov, Lewis, and Xue, *Uncertain Partners*, pp. 247–48.
438. Ibid., pp. 248–49.
439. "Zapis' besedy Stalina s Mao 22 ianvaria 1950 g.," AP RF, f. 45, op. 1, d. 329, l. 34. Shi Zhe's account of the discussion of the CCR on January 22, 1950, is abbreviated but agrees on content. See Shi Zhe, *Zai lishi juren shenbian*, p. 446.
440. "Zapis' besedy Stalina s Mao 22 ianvaria 1950 g.," AP RF, f. 45, op. 1, d. 329, l. 34.
441. Ibid., l. 34–35.
442. Ibid., l. 35.
443. Ibid.
444. Ibid.
445. *Zhonggong zhongyang zai Xiangshan*, p. 455; *Mao Zedong nianpu*, vol. 3, p. 541. See chapter 3 above.
446. "Zapis' besedy Stalina s Mao 22 ianvaria 1950 g.," AP RF, f. 45, op. 1, d. 329, l. 36.
447. On this and the rest of the paragraph, see ibid.
448. The draft of the protocol originally read "June," but was corrected by hand to read "July." AP RF, f. 45, op. 1, d. 329, l. 46. The agreement was signed on July 30.
449. On this paragraph, see "Zapis' besedy Stalina s Mao 22 ianvaria 1950 g.," AP RF, f. 45, op. 1, d. 329, l. 36.
450. The phrase "comrade Stalin said joking" was added by hand in the draft of the protocol. Ibid., l. 47.
451. The draft of the protocol read "extremely [*ves'ma*] positive results." Ibid.
452. Ibid., ll. 36–37.
453. Ibid., l. 37.
454. *Pravda*, July 31,1949. The circumstances of this event are described in chapter 3 above.
455. "Zapis' besedy Stalina s Mao 22 ianvaria 1950 g.," AP RF, f. 45, op. 1, d. 329, l. 37.
456. For further details, see chapter 3 above and this chapter above.
457. On this paragraph, see "Zapis' besedy Stalina s Mao 22 ianvaria 1950 g.," AP RF, f. 45, op. 1, d. 329, l. 38.
458. Ibid.
459. Goncharov, Lewis, and Xue mistakenly claim that an exchange of opinions on the independence of the Mongolian PR that was reported by Shi Zhe took place on January 22, 1950. See Goncharov, Lewis, and Xue, *Uncertain Partners*, p. 120.
460. Cf. chapter 3 above (Liu's written report to Stalin).
461. On this and the following sentence, see "Zapis' besedy Stalina s Mao 22 ianvaria 1950 g.," AP RF, f. 45, op. 1, d. 329, ll. 31, 38.
462. Documents from both sides demonstrate that they originally planned to sign the agreement in January. In his telegram to the CC of the CCP of January 2, 1950, Mao Zedong gave the "end of the month" as the deadline he was seeking. See *Jianguo yilai Mao Zedong wengao*, vol. 1, p. 212. The Soviet treaty draft of January 22, 1950, also called for signature in January. See AVP RF, f. 07, op. 23a, p. 18, d. 235, l. 48. English text in the appendix, document 7.
463. Talk between Vyshinskii and Zhou Enlai, February 13, 1950, Vyshinskii's record of the conversation, AVP RF, f. 0100, op. 43, p. 302, d. 8, l. 18; talk between Vyshinskii and Wang Jiaxiang on February 13, 1950, Vyshinskii's record of the conversation, ibid., l. 17.
464. Ibid., l. 37.
465. On this and the following sentence, see Mao Zedong to Ambassador Iudin, March 31, 1956, Soviet record of the talk, in *PDV* 5 (1994), p. 106.
466. As an example, see the comment Ambassador Wang Jiaxiang addressed to Gromyko, which sounds

almost subservient: General Liu Yalou would sign the air transport agreement, "if the Soviet side raised no objections." Gromyko replied coldly that it was the Chinese side's business who signed. It would certainly never have occurred to Gromyko to ask whether the Chinese side would have anything against it if the Soviet comrade XY signed. See AVP RF, f. 0100, op. 43, p. 302, d. 8, l. 36.

467. Chinese-Soviet communiqué about the signature of the treaty and agreement on February 14, 1950, in *Izvestiia*, February 15, 1950, p. 1; *Zhonghua renmin gongheguo duiwai guanxi wenjian ji*, vol. 1, p. 75; English-language version in *Sino-Soviet Treaty and Agreements*, pp. 1–3.

468. *Zhonghua renmin gongheguo duiwai guanxi wenjian ji*, vol. 1, p. 75.

469. On this and the rest of the paragraph, see Vyshinskii's letter to Stalin, February 1, 1950, AP RF, f. 3, op. 65, d. 369, l. 6.

470. Ibid., l. 21.

471. *Jianguo yilai Mao Zedong wengao*, vol. 1, p. 251.

472. "Posetiteli kremlevskogo kabineta," pt. 2, pp. 6–7. The session lasted from 10:10 through 11:40 P.M.; ibid.

473. On this, see Wu Xiuquan, *Huiyi yu huainian*, p. 237; *Wang Dongxing riji*, p. 194; Shi Zhe, *Zai lishi juren shenbian*, p. 447; Goncharev, Lewis, and Xue, *Uncertain Partners*, pp. 113, 322 n. 10, with further source references.

474. This draft is not included in *opis' 23a, delo 235*, of the secret file of Vyshinskii's secretariat (file 07) of the AVP RF, where the treaty drafts of both sides were stored.

475. See the text of the commission draft, insofar as it deals with the treaty, in the appendix, document 7. On the creation of the commission draft, see this chapter above.

476. Shi Zhe, *Zai lishi juren shenbian*, p. 447; Wu Xiuquan, *Huiyi yu huainian*, p. 237.

477. *Wang Dongxing riji*, p. 194.

478. Text of the Chinese treaty draft in the appendix, document 10. On this and the following sentence, cf. AVP RF, f. 07, op. 23a, p. 18, d. 235, ll. 35–55, 59–72. According to information from a contemporary witness, a committee session was held on January 26. See *Wang Dongxing riji*, p. 195.

479. AP RF, f. 3, op. 65, d. 369, ll. 1, 7–10; AVP RF, f. 07, op. 23a, p. 18, d. 235, ll. 73–76, 84–88.

480. Telegram from Mao Zedong to Liu Shaoqi, January 25, 1950, *Jianguo yilai Mao Zedong wengao*, vol. 1, p. 251.

481. Texts of these drafts: AVP RF, f. 07, op. 23a, p. 18, d. 235, ll. 73, 77–83. English text of the Soviet draft of an agreement on Port Arthur, Dairen, and the CCR in the appendix, document 11.

482. See this chapter above.

483. On this paragraph, see Vyshinskii's letter to Stalin, February 12, 1950, AP RF, f. 3, op. 65, d. 369, ll. 48–53; talk between Vyshinskii and Zhou Enlai, February 13, 1950, AVP RF, f. 0100, op. 43, p. 302, d. 8, l. 18; talk between Gromyko and Liu Yalou, January 26, 1950, AVP RF, f. 0100, op. 43, p. 302, d. 8, l. 6

484. On this and the following sentence, see Vyshinskii's letter to Stalin, February 1, 1950, AP RF, f. 3, op. 65, d. 369, ll. 1–21.

485. Vyshinskii's letter to Stalin, February 3, 1950, ibid., l. 42; AVP RF, f. 07, op. 23a, p. 18, ll. 97–100.

486. On this paragraph, see Vyshinskii's letter to Stalin, February 3, 1950, AP RF, f. 3, op. 65, d. 369, ll. 42–47.

487. *Wang Dongxing riji*, p. 200.

488. Vyshinskii to Stalin, February 12, 1950, AP RF, f. 3, op. 65, d. 369, ll. 48–53.

489. Talk between Vyshinskii and Zhou Enlai, February 13, 1950, Vyshinskii's record of the discussion, AVP RF, f. 0100, op. 43, p. 302, d. 8, l. 18; talk between Vyshinskii and Wang Jiaxiang, February 13, 1950, ibid., l. 17.

490. Texts in *Zhonghua renmin gongheguo tiaoyue ji*, vol. 1, pp. 110–21, 126–30.

491. Text in *Izvestiia*, February 15, 1950, p. 1; *Renmin ribao*, February 15, 1950, p. 2; *Sovetsko-kitaiskie otnosheniia*, pp. 219–20; *Zhonghua renmin gongheguo tiaoyue ji*, vol. 1, pp. 1–3. English text in the appendix, document 12.

492. Text in *Izvestiia*, February 15, 1950, pp. 1–2; *Renmin ribao*, February 15, 1950, p. 2; *Sovetsko-kitaiskie otnosheniia*, pp. 221–22; *Zhonghua renmin gongheguo tiaoyue ji*, vol. 1, p. 3–5. English text in the appendix, document 13.

493. Text in *Izvestiia*, February, 15, 1950, p. 2; *Renmin ribao*, February 15, 1950, p. 2; *Sovetsko-kitaiskie otnosheniia*, pp. 223–24.; *Zhonghua renmin gongheguo duiwai guanxi wenjian ji*, vol. 1, pp. 79–80. English text in the appendix, document 14.

494. Soviet-Chinese communiqué of February 14, 1950 on the signing of the treaty and agreement on February 14, 1950, in *Izvestiia*, February 15, 1950, p. 1; *Renmin ribao*, February 15, 1950, p. 2; *Zhonghua*

renmin gongheguo duiwai guanxi wenjian ji, vol. 1, pp. 3–5. English text in *Sino-Soviet Treaty and Agreements*, pp. 1–3. This communiqué also mentions an exchange of notes on the resolution of the Soviet government that the Soviet Union would hand over assets acquired from Japanese proprietors without compensation to the People's Republic of China and a further exchange of notes on the resolution of the Soviet government on the surrender without compensation of the buildings of the former Soviet military settlement in Peking. These exchanges of notes are not presented as "agreements" above as they merely constitute the confirmation of unilateral acts by the USSR.

495. Report in *Izvestiia*, March 29, 1950, p. 2; *Renmin ribao*, March 29, 1950, p. 1; *Sovetsko-kitaiskie otnosheniia*, pp. 227–28.

496. Ibid.

497. Report in *Izvestiia*, April 2, 1950, p. 2; *Renmin ribao*, April 2, 1950, p. 1; *Sovetsko-kitaiskie otnosheniia*, pp. 228–29.

498. Report in *Izvestiia*, April 21, 1950, p. 3; *Renmin ribao*, April 21, 1950, p. 1; *Sovetsko-kitaiskie otnosheniia*, p. 229. The text of the agreement was published in 1957 in a Chinese collection of treaties; see *Zhonghua renmin gongheguo tiaoyue ji*, vol. 1, pp. 47–49.

499. Report in *Izvestiia*, April 21, 1950; *Renmin ribao*, April 21, 1950; *Sovetsko-kitaiskie otnosheniia*, p. 229. The protocol is described as an "agreement" here. The text of the protocol was published (without a list of goods) in 1957 in a Chinese collection of treaties; see *Zhonghua renmin gongheguo tiaoyue ji*, vol. 1, pp. 50–51.

500. Report in *Izvestiia*, April 21, 1950; *Renmin ribao*, April 21, 1950; *Sovetsko-kitaiskie otnosheniia*, p. 229.

501. The protocol was published in 1957 in a Chinese collection of treaties; see *Zhonghua renmin gongheguo tiaoyue ji*, vol. 1, pp. 51–59.

502. Letter from Molotov, Mikoyan, and Vyshinskii to Stalin, February 22, 1950, AVP RF, f. 07, op. 23a, p. 18, d. 235, l. 42.

503. The term *soglashenie* can mean either "arrangement" or "agreement."

504. By "democratic countries," Zhou meant states under Communist rule.

505. Chicherin's speech to the Fifth All-Russian Soviet Congress on July 4, 1918, in G.V. Chicherin, *Stat'i i rechi po voprosam mezhdunarodnoi politiki*, pp. 58–59. On this, cf. Dieter Heinzig, *Sowjetische Militärberater bei der Kuomintang, 1923–1927*, p. 23.

506. On this paragraph, Vyshinskii's record of the conversation, AVP RF, f. 0100, op. 43, p. 302, d. 8, ll. 18–19.

507. On this paragraph, see the text of the telegram in *Jianguo yilai Mao Zedong wengao*, vol. 1, pp. 262–63; cf. an English-language version of the text and commentary in Goncharov, Lewis, and Xue, *Uncertain Partners*, pp. 124–25, 258; *Renmin ribao*, February 15, 1950, p. 3.

508. The text of the treaty in the commission draft is included as document 7 in the appendix. On the commission draft, see this chapter above.

509. AVP RF, f. 07, op. 23a, p. 18, d. 235, l. 51–55. English text included as document 10 in the appendix.

510. See chapter 4 above.

511. Ibid.

512. Shi Chzhe, "Soprovozhdaia predsedatelia Mao," p. 146. On this subject, cf. Shi Zhe, *Zai lishi juren shenbian*, p. 447, and Goncharov, Lewis, and Xue, *Uncertain Partners*, p. 116, with further source references. Shi Zhe's claim is occasionally accepted uncritically in the Western literature. See, for example, Goncharov, Lewis, and Xue, *Uncertain Partners*, p. 116; Murray, *Western versus Chinese Realism*, p. 463.

513. Cited in Goncharov, Lewis, and Xue, *Uncertain Partners*, p. 116.

514. See chapter 4 above.

515. See chapter 4 above ("Preparations for the Negotiations").

516. See Goncharov, Lewis, and Xue, *Uncertain Partners*, p. 118.

517. See ibid., pp. 118–19.

518. Italics by the author.

519. On this and the following sentence, see the four drafts with handwritten corrections stored in Vyshinskii's secret file, AVP RF, f. 07, op. 23a, p. 18, d. 235, ll. 56–72. Molotov, Mikoyan, and Vyshinskii forwarded the draft with their proposed corrections to Stalin on January 28; copies went to Politburo members Malenkov, Beriia, Kaganovich, and Bulganin. Ibid., l. 64.

520. The final version presented to the Chinese side that forms the basis for the following analysis is in ibid., ll. 74–76, 85–88.

521. Vyshinskii noted on the text on which the corrections leading to the final version were written by hand, "On January 28 with comrade Stalin." See ibid., l. 65.

522. Ibid., ll. 73, 84.

523. Cf. Goncharov, Lewis, and Xue, *Uncertain Partners*, pp. 118–19.

524. On this paragraph, see *Sovetsko-kitaiskie otnosheniia*, p. 22; *Zhonghua renmin gongheguo tiaoyue ji*, vol. 1, p. 2.

525. Vyshinskii to Stalin, February 1, 1950, AP RF, f. 3, op. 65, d. 369, l. 20.

526. See document 12 in the appendix.

527. Letter from Vyshinskii to Stalin, February 3, 1950, AP RF, f. 3, op. 65, d. 369, l. 42.

528. Kulik, "Kitaiskaia Narodnaia Respublika," pt. 2, p. 78, with no source references. Kulik maintained—not very convincingly—that the agreement granted both sides equal rights; see B.T. Kulik, *Sovetsko-kitaiskii raskol*, pp. 14–15.

529. *Wang Dongxing riji*, p. 201.

530. Shi Zhe, *Zai lishi juren shenbian*, p. 446; *Wang Dongxing riji*, p. 202; Liu Jiecheng, *Mao Zedong yu Si-da-lin*, pp. 297–98; Shen Zhihua, *Chaoxian zhanzheng jiemi*, p.77. It is not known on which day this talk was held. Shi Zhe gives the impression that the exchange had taken place during the talk on principles on January 22; see Shi Zhe, *Zai lishi juren shenbian*, p. 446. This claim has since been refuted by the Soviet protocol, which is now accessible; see this chapter above. Another archival source allows us to demonstrate that the discussion took place before February 10; see AP RF, f. 3, op. 65, d. 369, l. 48.

531. Telegram from Liu Shaoqi to Mao Zedong, January 2, 1950, cited in *Liu Shaoqi nianpu, 1878–1969*, vol. 2, p. 237.

532. The Soviet draft originally read "agreed to the following."

533. In Russian, "*Dal'nevostochnyi krai.*"

534. This refers to Kazakhstan, Uzbekistan, Kirghizia, Tajikistan, and Turkmenistan.

535. Attachment to the letter from Vyshinskii to Stalin, February 12, 1950, AP RF, f. 3, op. 65, d. 369, l. 53. An identically worded draft can be found in AVP RF, f. 07, op. 23a, p. 18, d. 235, l. 135. Both texts include the underlined passages with the note "The underlined words are corrections by Zhou Enlai."

536. On this paragraph, see the letter from Vyshinskii to Stalin, February 12, 1950, AP RF, f. 3, op. 65, d. 369, ll. 48–49. Cf. also Goncharov, Lewis, Xue, *Uncertain Partners*, p. 121.

537. For further details, see this section below.

538. On Acheson's remarks, see this chapter above.

539. On this paragraph, see the text of the telegram in *Jianguo yilai Mao Zedong wengao*, vol. 1, p. 262; cf. Goncharov, Lewis, and Xue, *Uncertain Partners*, pp. 124–25, 258.

540. Zhou Enlai to Fedorenko, March 14, 1953, AVP RF, f. 0100, op. 46, d. 10, p. 361, ll. 68–71, cited in Kulik, "Kitaiskaia Narodnaia Respublika," pt. 2, pp. 77–78.

541. On this paragraph, see *Khrushchev Remembers: The Last Testament*, p. 284; Wingrove, "Mao in Moscow," p. 329, on the basis of documents in the AVP RF; *Zhonghua renmin gongheguo waijiao shi*, p. 39.

542. *Zhonghua renmin gongheguo waijiao shi*, p. 78, with no source reference.

543. Kulik, "Kitaiskaia Narodnaia Respublika," pt. 2, p. 78.

544. Text of the two drafts of January 22, 1950, in the appendix, documents 8 and 9. On this paragraph, see this chapter above.

545. On this paragraph, see this chapter above.

546. *Wang Dongxing riji*, p. 195; telegram from Mao Zedong to Liu Shaoqi, January 25, 1950, in *Jianguo yilai Mao Zedong wengao*, vol. 1, p. 251.

547. Text in AVP RF, f. 07, op. 23a, p. 18, d. 235, ll. 77–79. Vyshinskii presented the text to Stalin, Molotov, Malenkov, Beriia, Mikoyan, Kaganovich, and Bulganin. Ibid., l. 79. English text in the appendix, document 11.

548. AVP RF, f. 07, op. 23a, p. 18, d. 235, ll. 73, 77–79.

549. Cf. the text of the draft in the appendix, document 9.

550. See this chapter above.

551. See details in Heinzig, *Sowjetische Militärberater*, pp. 25–27.

552. Text of the January 22 and January 29 drafts in the appendix, documents 8 and 11.

553. Cf. this chapter above.

554. On this and the following sentence, see *Sovetsko-kitaiskie otnosheniia*, p. 201.

555. On this paragraph, see Vyshinskii's letter to Stalin, February 1, 1950, AP RF, f. 3, op. 65, d. 369, ll. 1–3, 11–14. The italicized passages were added by Zhou.

556. Ibid., l. 3.

557. Ibid., l. 12.

558. Cf. the text of the February 14, 1950, agreement in *Izvestiia*, February 15, 1950, p. 2. English text in the appendix, document 13.

559. Wu Xiuquan, "'Zhong-Su youhao tongmeng huzhu tiaoyue' qianding de jingguo," p. 14.

560. AP RF, f. 3, op. 65, d. 369, l. 11.

561. Secret protocol, August 14, 1945, cited in Bruce A. Elleman, "The End of Extraterritoriality in China," p. 81 with reference to archival materials.

562. For further details, see this chapter below.

563. On the history of Soviet extraterritorial rights in China, see Elleman, "End of Extraterritoriality," pp. 65–89; Bruce A. Elleman, *Diplomacy and Deception*, pp. 177–94.

564. Specifically in the talk on principles, January 22, 1950, see this chapter above.

565. See this chapter above.

566. Letter from Vyshinskii to Stalin, February 1, 1950, AP RF, f. 3, op. 65, d. 369, ll. 21–22.

567. On this and the rest of the paragraph, see Vyshinskii's letter to Stalin, February 3, 1950, AP RF, f. 3, op. 65, d. 369, l. 45.

568. See this chapter above.

569. On the aforementioned demands by both sides to modify the agreement, see the text signed on February 14, 1950, in the appendix, document 13.

570. *Zhonghua renmin gongheguo waijiao shi*, p. 39.

571. Soviet-Chinese communiqué, September 15, 1952. See *Izvestiia*, September 16, 1952; *Sovetsko-kitaiskie otnosheniia*, pp. 277–78.

572. On this and the following two sentences, see talk between Stalin and Zhou Enlai, August 20, 1952, AP RF, f. 45, op. 1, d. 329, ll. 55–56. See also the talk between Stalin and Zhou Enlai, September 3, 1952, ibid., l. 76.

573. On this and the following sentence, see *Izvestiia*, September 16, 1952; *Sovetsko-kitaiskie otnosheniia*, pp. 278–79.

574. *Izvestiia*, October 12, 1954; *Sovetsko-kitaiskie otnosheniia*, pp. 302–3.

575. On this paragraph, *Zhonghua renmin gongheguo waijiao shi*, p. 39; *Khrushchev Remembers: The Last Testament*, p. 284. For further details on the agreement on the withdrawal from Port Arthur, see Helmut Handzik, *Politische Bedingungen sowjetischer Truppenabzüge, Drei Fallstudien: Die fünfziger Jahre*, pp. 48–50.

576. *Sovetsko-kitaiskie otnosheniia*, p. 200.

577. AVP RF, f. 07, op. 23a, p. 18, d. 235, ll. 73, 80.

578. Ibid., l. 80. See also AP RF, f. 3, op. 65, d. 369, l. 22.

579. Goncharov, Lewis, and Xue inaccurately assume that no customs inspections were planned in the 1945 agreement, either. See Goncharov, Lewis, and Xue, *Uncertain Partners*, p. 126.

580. On this paragraph, see Vyshinskii's letter to Stalin, February 1, 1950, AP RF, f. 3, op. 65, d. 369, ll. 3, 15; Vyshinskii's letter to Stalin, February 3, 1950, ibid., l. 42.

581. Vyshinskii's letter to Stalin (apparently in February 1950), cited in Wingrove, "Mao in Moscow," p. 325, based on a document in the AVP RF. Wingrove does not reproduce the wording of the change Vyshinskii suggested.

582. Ibid., p. 326.

583. Ibid.

584. On this paragraph, see ibid. Goncharov, Lewis, and Xue also assume (with no source reference) that Stalin rejected Zhou's demand. See Goncharov, Lewis, and Xue, *Uncertain Partners*, pp. 126–27.

585. For further details, see chapter 3 above.

586. For further details, see this chapter above.

587. Letter from Molotov, Mikoyan, and Vyshinskii to Stalin, January 22, 1950, AVP RF, f. 07, op. 23a, p. 18, d. 235, l. 42.

588. Ibid., ll. 73, 81–82.

589. On this paragraph, see Vyshinskii's letter to Stalin, February 1, 1950, AP RF, f. 3, op. 65, d. 369, ll. 4, 17–19. The new wording for Articles 1, 3, and 5 that the Chinese proposed was based on the Russian version of Zhou Enlai's draft; see ibid., ll. 17–19. Vyshinskii reproduced the Chinese suggestion for Articles 1 and 5 slightly differently in his letter to Stalin; see ibid., l. 4.

590. *Sovetsko-kitaiskie otnosheniia*, pp. 223–24. English text in the appendix, document 14.

591. On this and the following sentence, see *Khrushchev Remembers*, ed. E. Crankshaw, p. 463.

592. AVP RF, f. 07, op. 23a, p. 18, d. 235, ll. 73, 83.

593. Ibid., l. 83. Goncharov, Lewis, and Xue do not mention this secret protocol. See Goncharov, Lewis, and Xue, *Uncertain Partners*, pp. 121–27.

594. On this paragraph, see Vyshinskii's letter to Stalin, February 1, 1950, AP RF, f. 3, op. 65, d. 369, ll. 5–6.

595. On this paragraph, see Vyshinskii's letter to Stalin, February 3, 1950, AP RF, f. 3, op. 65, d. 369, ll. 43–44.

596. On this paragraph, see the telegram from Mao Zedong to Liu Shaoqi and Hu Qiaomu, February 14, 1950, in *Jianguo yilai Mao Zedong wengao*, vol. 1, p. 262; cf. Goncharov, Lewis, and Xue, *Uncertain Partners*, p. 258.

597. See further above in this section.

598. On the details in this paragraph, see chapter 3 above.

599. Letter from Vsevolod D. Durdenevskii to Vyshinskii, January 16, 1950, AVP RF, f. 07, op. 23a, p. 18, d. 235, l. 123.

600. On this and the following sentence, see "Protokol o rezul'tatakh provedeniia plebistsita po voprosu o nezavizimosti Mongol'skoi Narodnoi Respubliki," ibid., ll. 129–33.

601. See ibid., ll. 124–27.

602. S. Luzianin, "Ialtinskaia konferentsiia i problemy mezhdunarodno-pravovogo oformleniia MNR nakanune i v gody vtoroi mirovoi voiny," p. 59. Luzianin bases his account here, as well as his description of the prehistory of the plebiscite, on previously unpublished Russian archival materials; ibid., pp. 56–60.

603. On this paragraph, see *Zhonghua renmin gongheguo duiwai guanxi wenjian ji*, vol. 1, pp. 13–14.

604. On this paragraph, see Shi Zhe, *Zai lishi juren shenbian*, p. 450. Cf. Wu Xiuquan, *Huiyi yu huainian*, pp. 240–41. Goncharov, Lewis, and Xue put this talk off to January 22. The corresponding Soviet protocol makes no reference to this. The discussion must have taken place before January 31, however, as on that day Zhou presented the Soviet side with a draft of an exchange of notes in which, among other things, the independence of the Mongolian People's Republic was recognized.

605. Article 55 of the Joint Program of the Political Consultative Conference of the Chinese People, September 29, 1949, in *Zhonghua renmin gongheguo duiwai guanxi wenjian ji*, vol. 1, p. 1; *Zhonggong zhongyang wenjian xuanji*, vol. 12, p. 595.

606. Shi Zhe, *Zai lishi juren shenbian*, p. 450.

607. On this and the following sentence, see Vyshinskii's letter to Stalin, February 1, 1950, AP RF, f. 3, op. 65, d. 369, ll. 1, 21.

608. *Izvestiia*, February 15, 1950, p. 1; *Renmin ribao*, February 15, 1950, p. 2; *Zhonghua renmin gongheguo diuwai guanxi wenjian ji*, vol. 1, p. 75.

609. On this and the following sentence, Mao Zedong to Japanese guests, July 10, 1964. See *Mao Zedong tongzhi shi dangdai zui weida de Ma-ke-si Lie-ning zhuyi zhe*, p. 540; Dennis J. Doolin, *Territorial Claims in the Sino-Soviet Conflict*, p. 43; *Khrushchev Remembers: The Last Testament*, pp. 285–86. On the Mongolian question in Sino-Soviet relations, see chapter 2 above; cf. Dieter Heinzig, *Der sowjetisch-chinesische Grenzkonflikt*, pp. 18–20, 22.

610. On this and the following sentence, see Vyshinskii's letter to Stalin, February 3, 1950, AP RF, f. 3, op. 65, d. 369, ll. 46–47; Shi Zhe, *Zai lishi juren shenbian*, p. 452; *Renmin ribao*, February 15, 1950, p. 2; Wu Xiuquan, *Huiyi yu huainian*, p. 244; *Wang Dongxing riji*, p. 200; *Zhou Enlai waijiao huodung dashi ji, 19491975*, p. 16.

611. Shi Zhe, *Zai lishi juren shenbian*, p. 453.

612. Vyshinkii's letter to Stalin, February 3, 1950, AP RF, f. 3, op. 65, d. 369, ll. 46–47.

613. On this paragraph, see Zhu Zhongli, *Mao Zedong Wang Jiaxiang*, p. 182.

614. On this paragraph, see Fedorenko, "Stalin i Mao Tszedun," pt. 2, pp. 83–85; *Bol'shaia Sovetskaia Entsiklopediia*, 3d ed., vol. 6, p. 598. Wang Jiaxiang's wife notes, however, that Chen Boda had become carried away as he sat in the czar's box watching the performance in Mao's place. As the curtain fell, he had applauded wildly and had thus offended national dignity. Mao and Wang Jiaxiang also shared this opinion. It is possible that Chen was only clapping politely and that Wang used the incident to lower Mao's opinion of Chen. It is striking that Wang's wife's memoirs constantly present Chen in a negative light. See Zhu Zhongli, *Mao Zedong Wang Jiaxiang*, pp. 182–83.

615. Wu Xiuquan, *Huiyi yu huainian*, p. 244.

616. On this and the following two sentences, see Fedorenko, "Stalin i Mao Tszedun," pt. 2, pp. 83–85.

617. For further details, see this chapter above.

618. Vyshinkii's letter to Stalin, February 3, 1950, AP RF, f. 3, op. 65, d. 369, l. 46.

619. On this and the following sentence, see letter from Vyshinskii to Stalin, February 12, 1950, AP RF, f. 3, op. 65, d. 369, l. 51.

620. *Renmin ribao*, February 15, 1950, p. 2; Mao to Zhu Zhongli (Wang Jiaxiang's wife), cited in Zhu Zhongli, *Mao Zedong Wang Jiaxiang*, pp. 183–84. Zhu mistakenly indicates that Mao went to the performance with Stalin and that the audience met them with enthusiastic ovations. She is apparently confusing this with the cultural program on December 22, 1949. See ibid., p. 183; cf. this chapter above.

621. On this paragraph, see photograph of the treaty signing in *Izvestiia*, February 15, 1950; Shi Zhe, *Zai lishi juren shenbian*, pp. 461–63; Wu Xiuquan, *Huiyi yu huainian*, pp. 238–39.

622. *Izvestiia*, February 15, 1950, p. 2.

623. Ibid., p. 1; Shi Zhe, Zai lishi juren shenbian, p. 463.

624. On this and the two following sentences, see Vyshinskii's letter to Stalin, February 12, 1950, AP RF, f. 3, op. 65, d. 369, l. 51.

625. On this and the following sentence, see Zhu Zhongli, *Mao Zedong Wang Jiaxiang*, pp. 190–91.

626. On this and the following sentence, see Shi Zhe, *Zai lishi juren shenbian*, pp. 463–64.

627. On this and the rest of the paragraph, see Fedorenko, "Stalin i Mao Tszedun," pt. 2, p. 86.

628. Zhu Zhongli, *Mao Zedong Wang Jiaxiang*, p. 191.

629. On this paragraph, see Shi Zhe, *Zai lishi juren shenbian*, pp. 464–66; Fedorenko, "Stalin i Mao Tszedun," pt. 2, pp. 88–89; Wu Xiuquan, *Huiyi yu huainian*, p. 242; *Wang Dongxing riji*, p. 206; *Zhu Zhongli, Mao Zedong Wang Jiaxiang*, p. 191. Fedorenko passes over both speeches without comment and instead reports on a short toast by Stalin in which he thanked history for having "given us such an outstanding Marxist-Leninist and such a fearless Communist as Mao Zedong."

630. *Izvestiia*, February 16, 1950, p. 6.

631. On this paragraph, *Pravda*, February 15, 1950, pp. 1–2; *Pravda*, February 16, 1950, p. 1; *Izvestiia*, February 15, 1950, pp. 1–2; *Izvestiia*, February 16, p. 1; *Renmin ribao*, February 15, 1950, pp. 1–3; *Wang Dongxing riji*, p. 207.

632. On this paragraph, see Wu Xiuquan, *Huiyi yu huainian*, p. 242; Shi Zhe, *Zai lishi juren shenbian*, p. 466. There is an anecdotal description of the banquet in Zhu Zhongli, *Mao Zedong Wang Jiaxiang*, pp. 192–94.

633. Chinese text in *Zhonghua renmin gongheguo duiwai guanxi wenjian ji*, vol. 1, pp. 82–83; Russian text in *Izvestiia*, February 18, 1950, p. 1.

634. On this paragraph, see *Izvestiia*, February 18, 1950, p. 1; Shi Zhe, *Zai lishi juren shenbian*, p. 468; *Wang Dongxing riji*, p. 207.

635. On this paragraph, see *Izvestiia*, February 18, 1950, p. 1 (Ye Zilong is mistakenly transcribed Ye Jielong here); Shi Zhe, *Zai lishi juren shenbian*, p. 469; Wu Xiuquan, *Huiyi yu huainian*, p. 245; *Wang Dongxing riji*, p. 209.

636. On this paragraph, see Shi Zhe, *Zai lishi juren shenbian*, pp. 469–71; *Wang Dongxing riji*, pp. 207–10; Wu Xiuquan, *Huiyi yu huainian*, p. 245.

637. On this paragraph, see *Renmin ribao*, March 5, 1950, p. 1; *Izvestiia*, March 7, 1950, p. 3; *Zhou Enlai nianpu, 1949–1976*, vol. 1, p. 26; Shi Zhe, *Zai lishi juren shenbian*, p. 472; *Wang Dongxing riji*, pp. 210–11; Wu Xiuquan, *Huiyi yu huainian*, p. 245.

638. On this paragraph, see *Zhou Enlai nianpu, 1949–1976*, vol. 1, p. 25; Zhou Enlai to Vyshinskii, February 13, 1950, Vyshinskii's record, AVP RF, f. 0100, op. 43, p. 302, d. 8, l. 18. See also Wang Jiaxiang to Vyshinskii, February 13, 1950, Vyshinskii's record, ibid., l. 17.

639. On this paragraph, see Wu Xiuquan, *Huiyi yu huainian*, p. 246.

640. Gromyko to Azizov (Saitudin), February 27, 1950, AVP RF, f. 0100, op. 43, p. 302, d. 8, l. 29.

641. For further details on this paragraph, see chapter 3 above. It is not entirely clear whether a formal or informal agreement had been reached at that time. The text is not yet accessible.

642. Letter from Molotov, Mikoyan, and Vyshinskii to Stalin, January 22, 1950, AVP RF, f. 07, op. 23a, p. 18, d. 235, l. 42.

643. Vyshinskii's letter to Stalin, February 12, 1950, AP RF, f. 3, op. 65, d. 369, ll. 48, 50.

644. On this paragraph, see Vyshinskii's letter to Stalin, February 12, 1950, AP RF, f. 3, op. 65, d. 369, ll. 50–51.

645. Murray, *Western versus Chinese Realism*, pp. 461–62; Wingrove, "Mao in Moscow," pp. 330–31. Both sources are based on documents in the AVP RF.

646. Gromyko's record of the talk, March 3, 1950, AVP RF, f. 0100, op. 43, p. 302, d. 8, ll. 31–33.

647. Talk between Gromyko and Wang Jiaxiang, March 8, 1950, Gromyko's record of the talk, AVP RF, f. 0100, op. 43, p. 302, d. 8, l. 37.

648. Ibid., ll. 39–40.

649. Text of the draft in ibid., ll. 51–53. English text of the draft in the appendix, document 15. On this

paragraph, see talk between Gromyko and Wang Jiaxiang, March 22, 1950, Gromyko's record, March 23, 1950, ibid., ll. 49–50.

650. Ibid., ll. 55–56; Murray, *Western versus Chinese Realism*, p. 462, with source references.

651. On this paragraph, see talks between Gromyko and Wang Jiaxiang, March 23, 1950, Gromyko's record, March 24, 1950, AVP RF, f. 0100, op. 43, p. 302, d. 8, ll. 55–56; Murray, *Western versus Chinese Realism*, p. 462, with source references.

652. For further details, see chapter 3 above.

653. Elleman, "End of Extraterritoriality," p. 81.

654. On this and the following, see Goncharov, Lewis, and Xue, *Uncertain Partners*, pp. 25–26.

655. On this and the following sentence, see Shi Zhe, *Zai lishi juren shenbian*, p. 446.

656. Ibid., p. 415. For further details, see chapter 3 above.

657. Shi Zhe, *Zai lishi juren shenbian*, p. 446.

658. See text of the treaty in *Sovetsko-kitaiskie otnosheniia*, pp. 82–85; here, p. 85,

659. The text of the protocol is included as an "appendix" to the trade agreement of June 16, 1939, with the relevant clause (Section 1, Paragraph 8) in a Taiwanese treaty collection from the year 1958; see *Treaties Between the Republic of China and Foreign States*, pp. 494–97; see esp. pp. 495–96. It is missing from the official Soviet document collection on Sino-Soviet relations, as is the aforementioned treaty of February 22, 1929—however, it is not included in the cited Taiwanese collection, either.

660. On this paragraph, see Elleman, "End of Extraterritoriality," pp. 71–81; Elleman, *Diplomacy and Deception*, pp. 177–94.

661. Text in AVP RF, f. 07, op. 23a, d. 221, ll. 21–24. Vyshinskii forwarded the draft to Stalin on March 16, 1950, and the Council of Ministers of the USSR adopted it with a few corrections on March 21, 1950; ibid., l. 19.

662. Note from the Soviet embassy in Peking to the Foreign Ministry of the People's Republic of China, July 18, 1960, Russian text in SAPMO-B-Arch, J IV/2/202/280, unpaginated; English translation in *CWIHPB* 8–9 (Winter 1996–97), pp. 249–50. See also the letter from Khrushchev, July 18, 1960, in which he made a copy of the July 18, 1960, note available to the East German CP leadership; SAPMO-B-Arch, J IV/2/202/280, unpaginated. Commentary on the note is provided by Chen Jian, "A Crucial Step toward the Breakdown of the Sino-Soviet Alliance," pp. 246, 249.

663. Wu Xiuquan, *Huiyi yu huainian*, pp. 247–48.

664. *Izvestiia*, March 29, 1950, p. 2; *Renmin ribao*, March 29, 1950, p. 1; *Sovetsko-kitaiskie otnosheniia*, pp. 227–28.

665. On this paragraph, see ibid.

666. Kulik, "Kitaiskaia Narodnaia Respublika," pt. 2, p. 78, with no source reference.

667. Shi Zhe, *Zai lishi juren shenbian*, p. 446.

668. On this paragraph, see "Memuary Nikity Sergeevicha Khrushcheva," pp. 68, 75. An English-language version of the cited passages, which often differs so much from the original (because of its inaccuracies, abbreviations, additions by the publisher, and mistranslations), that one can no longer call it a translation, can be found in *Khrushchev Remembers*, ed. E. Crankshaw, p. 463, and in *Khrushchev Remembers: The Last Testament*, pp. 240–41.

669. Wu Xiuquan, *Huiyi yu huainian*, p. 248.

670. Telegram from Peng Dehuai to Mao Zedong, December 29, 1949, which Mao forwarded to Stalin on January 2, 1950, AP RF, f. 45, op. 1, d. 334, ll. 2, 7, copy of the original, cited in Ledovskii, "Vizit Mao Tszeduna v Sovetskii Soiuz," pt. 1, pp. 31, 36; *Peng Dehuai zhuan*, p. 390. For further details, see this chapter above.

671. On this and the following sentence, see *Liu Shaoqi nianpu, 1898–1969*, vol. 2, p. 237.

672. See this chapter above.

673. Letter from Stalin to Mao Zedong, January 6, 1950, AP RF, f. 45, op. 1, d. 334, l. 16, cited in Ledovskii, "Vizit Mao Tszeduna v Sovetskii Soiuz," pt. 1, p. 39.

674. On this and the rest of the paragraph, see Vyshinskii's letter to Stalin, February 3, 1950, AP RF, f. 3, op. 65, d. 369, ll. 44–45.

675. *Khrushchev Remembers: The Glasnost Tapes*, p. 143.

676. "Memuary Nikity Sergeevicha Khrushcheva," p. 75.

677. According to an unnamed high-ranking Soviet diplomat in 1992, cited in Goncharov, Lewis, and Xue, *Uncertain Partners*, p. 125.

678. On this and the following sentence, see AVP RF, f. 07, op. 23a, p. 18, d. 235, ll. 42–44.

679. On this paragraph, see AVP RF, f. 0100, op. 43, p. 302, d. 8, ll. 29–30. See also AP RF, f. 3, op. 65, d. 369, ll. 43–44.

680. On this paragraph, see Gromyko's talk with Wang Jiaxiang, March 2, 1950, AVP RF, f. 0100, op. 43, p. 302, d. 8, ll. 33–34; Gromyko's talk with Saifudin's, March 4, 1950, ibid., ll. 35, 40; Gromyko's talk with Wang Jiaxiang, March 8, 1950, ibid., l. 40.

681. On this paragraph, see Gromyko's talk with Wang Jiaxiang, March 10, 1950, ibid., ll. 41–43.

682. On this paragraph, see Gromyko's talk with Wang Jiaxiang, March 16, 1950, ibid., ll. 44–46; Gromyko's talk with Wang Jiaxiang, March 18, 1950, ibid., l. 47.

683. On this paragraph, see Gromyko's talks with Wang Jiaxiang, March 16, 18, 22, and 23, 1950, ibid., ll. 45, 48, 50, 55–56. The communiqué was published in *Izvestiia*, March 29, 1950, p. 2; *Renmin ribao*, p. 1; *Sovetsko-kitaiskie otnosheniia*, pp. 227–28.

684. On this paragraph, see the draft by Liu Shaoqi of an internal Party directive, March 30, 1950, in *Liu Shaoqi nianpu, 1898–1969*, vol. 2, p. 246.

685. See further above in this section.

686. Kulik, "Kitaiskaia Narodnaia Respublika," pt. 2, pp. 77–78.

687. Text of the communiqué in *Sovetsko-kitaiskie otnosheniia*, p. 303.

688. "Memuary Nikity Sergeevicha Khrushcheva," p. 75.

689. Liu Shaoqi's letter to Stalin, July 6, 1949, AP RF, f. 45, op. 1, d. 328, l. 54, from a copy of the original, in Ledovskii, "Sekretnaia missiia delegatsii TsK," p. 47. See also chapter 3 above.

690. Talk between Stalin and Mao, December 16, 1949, see chapter 4 above. Talk between Mao and Kovalev, December 20, 1949; see chapter 4 above.

691. AVP RF, f. 0100, op. 43, p. 302, d. 8, l. 7.

692. On this, see this chapter above (section "The Agreement of March 27, 1950, on the Establishment of Two Mixed Joint-Stock Companies").

693. On this paragraph, see *Izvestiia*, April 2, 1950, p. 2; *Renmin ribao*, April 2, 1950, p. 1; *Sovetsko-kitaiskie otnosheniia*, pp. 228–29.

694. AVP RF, f. 07, op. 23a, p. 18, d. 235, ll. 42–44.

695. For further details, see chapter 3 above.

696. On this paragraph, see AVP RF, f. 0100, op. 43, p. 302, d. 8, ll. 6–7, 9. Soviet archival documents transcribed the name of the Chinese general incorrectly as "Liu Yaolou"; see, for example, ibid., ll. 6–9.

697. On this paragraph, see ibid., l. 9.

698. On this paragraph, see the session on February 8, 1950, ibid., ll. 10–14.

699. Ibid., l. 15.

700. Vyshinskii's letter to Stalin, February 12, 1950, AP RF, f. 3, op. 65, d. 369, l. 50.

701. Zhou Enlai's talk with Vyshinskii, February 13, 1950, AVP RF, f. 0100, op. 43, p. 302, d. 8, l. 18; Vyshinskii's talk with Wang Jiaxiang, February 13, 1950, ibid., l. 17.

702. On this and the following sentence, see talk between Gromyko and Liu Yalou, February 14, 1950, ibid., ll. 20–21.

703. On this paragraph, see talk between Gromyko and Liu Yalou, February 18, 1950, ibid., ll. 22–23.

704. On this and the following sentence, see the talk between Gromyko and Wang Jiaxiang, March 16, 1950, ibid., l. 45.

705. Talk between Gromyko and Wang Jiaxiang, February 21, 1950, ibid., l. 24.

706. On this and the following sentence, talk between Gromyko and Wang Jiaxiang, March 18, 1950, ibid., ll. 47–48.

707. On this paragraph, see *Izvestiia*, March 29, 1950, and April 2, 1950; *Renmin ribao*, March 29, 1950, and April 2, 1950; *Sovetsko-kitaiskie otnosheniia*, pp. 223–24, 228–29.

708. *Izvestiia*, April 21, 1950, p. 3.

709. *Pravda*, July 31, 1949.

710. For further details, see chapter 3 above.

711. On this paragraph, see AVP RF, f. 0100, op. 43, p. 302, d. 8, l. 26; AVP RF, f. 07, op. 23a, p. 18, d. 235, l. 42; AP RF, f. 3, op. 65, d. 369, ll. 45–46.

712. Vyshinskii's record of the talk on the February 11, 1950, session is located in AP RF, f. 3, op. 65, d. 369, l. 49.

713. On this and the following sentence, see AP RF, f. 45, op. 1, d. 329, l. 13.

714. On this and the following sentence, see ibid., l. 36.

715. On this and the following sentence, see AVP RF, f. 07, op. 23a, p. 18, d. 237, ll. 2, 241.

716. AVP RF, f. 07, op. 23a, d. 236, ll. 32, 126, cited in Kulik, "Kitaiskaia Narodnaia Respublika," pt. 2, p. 75.

717. AVP RF, f. 07, op. 23a, p. 18, d. 237, l. 126.

718. On this paragraph, see ibid., ll. 1–120. Molotov, Mikoyan, and Vyshinskii presented a corresponding list to Stalin on February 13, 1950, ibid., ll. 121–80, 185–240. See also O.B. Borisov and B.T. Koloskov, *Sovetsko-kitaiskie otnosheniia, 1945–1970*, pp. 47, 49–50.

719. On this and the following sentence, see Vyshinskii's record of the talk on, February 12, 1950, AP RF, f. 3, op. 65, d. 369, ll. 49–50.

720. Talk between Vyshinskii and Zhou Enlai, February 13, 1950, AVP RF, f. 0100, op. 43, p. 302, d. 8, l. 18. Wang Jiaxiang made a similar statement to Vyshinskii on the same day; see ibid., l. 17.

721. On this paragraph, see AVP RF, f. 07, op. 23a, p. 18, d. 237, ll. 1–2, 241.

722. Ibid., l. 242.

723. On this paragraph, see ibid., ll. 241–49.

724. On this paragraph, see Wu Xiuquan, *Huiyi yu huainian*, pp. 246–47.

725. Text of the February 13, 1950, Soviet draft in AVP RF, f. 07, op. 23a, p. 18, d. 237, ll. 181–83; official text of the agreement in *Zhonghua renmin gongheguo tiaoyue ji*, vol. 1, pp. 47–49.

726. For further details, see further above in this section.

727. See further above in this section.

728. Chao Kang and Feng-Hwa Mah, "A Study of the Rouble-Yuan Exchange Rate," p. 193.

729. On this paragraph, see Zhou Enlai's March 2, 1950, draft of a telegram from Mao Zedong addressed to Li Fuchun, Wang Jiaxiang, and Liu Yalou, cited in *Zhou Enlai nianpu, 1949–1976*, vol. 1, p. 26; Zhou Enlai's speech to cadres in the Northeast Bureau of the CCP in Shenyang, March 3, 1950, ibid.

730. On this paragraph, see *Zhonghua renmin gongheguo tiaoyue ji*, vol. 1, p. 50.

731. There was only a brief report on this protocol at the time. See *Izvestiia*, April 21, 1950; *Renmin ribao*, April 21, 1950; *Sovetsko-kitaskie otnosheniia*, p. 229. The text of the protocol was published in 1957, without lists of goods, in a Chinese collection of treaties; see *Zhonghua renmin gongheguo tiaoyue ji*, vol. 1, pp. 50–51.

732. The protocol was published in 1957 in a Chinese collection of treaties; see *Zhonghua renmin gongheguo tiaoyue ji*, vol. 1, pp. 51–59.

733. The author had access only to a Soviet draft of this protocol. See AVP RF, f. 07, op. 23a, p. 18, d. 237, l. 184.

734. On this paragraph, see Borisov and Koloskov, *Sovetsko-kitaiskie otnosheniia*, p. 52.

735. On this paragraph, see Wu Xiuquan, *Huiyi yu huainian*, p. 249.

736. On this, see chapter 3 above.

737. On this and the following two sentences, see Mao Zedong to Ambassador Iudin, October 29, 1957, Iudin's report of the talk, October 30, 1957; see "Telegramma posla SSSR P.F. Iudina," AP RF, f. 3, op. 65, d. 589, ll. 173–79, p. 112.

738. On this and the following paragraph, see Shi Zhe, *Zai lishi juren shenbian*, pp. 209–10. See also Goncharov, Lewis, and Xue, *Uncertain Partners*, p. 127.

Notes to Chapter 5

1. For more on Chinese nationalism as the motivation for the Chinese revolution in imperial, republican, and Communist China, see Jonathan Unger, ed., *Chinese Nationalism*; in particular, see James Townsend, "Chinese Nationalism," pp. 1–30.

2. For references, see Brian Joseph Murray, *Western versus Chinese Realism*, pp. 137–38.

3. "Memuary Nikity Sergeevicha Khrushcheva," p. 73. According to two Japanese authors, Khrushchev said in the secret speech to the Twentieth Party Congress on February 25, 1956, in which he settled accounts with Stalin, that the Soviet dictator had treated Mao like a beggar. The version of the speech published in the United States in 1956 to which the two authors refer is not genuine. See Nagai Yonosuke and Iriye Akira, *The Origins of the Cold War in Asia*, pp. 213, 222. The text of the speech published in the USSR under Gorbachev does not contain any such observation by Khrushchev. Cf. "O kul'te lichnosti i ego posledstviiach: Doklad Pervogo sekretaria TsK KPSS tov. Khrushcheva N.S. XX s"ezdu Kommunisticheskoi partii Sovetskogo Soiuza," pp. 128–70.

4. Mao Zedong, September 24, 1962, at the Tenth Plenary Session of the Tenth Party Congress, cited in *Mao Zedong tongzhi shi dangdai zui weida de Ma-ke-si Lie-ning zhuyi zhe*, p. 432.

5. Mao Zedong to a meeting of provincial and city secretaries of the CCP, January 1957, cited in ibid., pp. 163–64.

6. Mao Zedong at the Chengdu Conference, March 10, 1958, cited in ibid., pp. 163–64. It is unclear what Mao meant by "border questions."

7. On this and the following three sentences, see *Zhonghua renmin gongheguo waijiao shi*, p. 19. Zhou is apparently referring to the second half of December 1949 here.

8. On this and the rest of the paragraph, see "Memuary Nikity Sergeevicha Khrushcheva," pp. 73–74.

9. Khrushchev is apparently referring here to Mao's outburst to Kovalev and Fedorenko and to his threat that he would leave in late December. See chapter 4 above.

10. Mao Zedong to Iudin, July 22, 1958, Chinese record of the talk, in *Dang de wenxian* 1 (1994), pp. 16–17.

11. Ibid., p. 17. It is unclear to what specific situation this complaint is referring.

12. Mao Zedong to Iudin, March 31, 1956, Soviet record of the talk, in *PDV* 5 (1994), pp. 106–7. Cf. Mao's statements to the Yugoslav ambassador in September 1956, in *Mao Zedong waijiao wenxuan*, p. 254.

13. On this and the following sentence, see Mao Zedong, "On the Ten Great Relations [*Lun shi da guanxi*]," speech delivered at a Politburo session, April 25, 1956, in *Mao Zedong xuanji*, vol. 5, p. 286; Mao Zedong in a speech to the Tenth Plenary Session of the CC of the Eighth CCP Party Congress, September 24, 1962, in *Mao Zedong tongzhi*, p. 432. Some earlier versions of the speech of April 25, 1956, do not include this passage. See John K. Leung and Michael Y.M. Kao, *The Writings of Mao Zedong, 1949–1976*, vol. 2, pp. 43–44, 59–61.

14. Telegram from Stalin to Ambassador Terentii Shtykov, January 30, 1950, AVP RF (no retrieval number). The author holds a copy. On this and the following paragraph, see Evgueni Bajanov, "Assessing the Politics of the Korean War, 1949–61," p. 87.

15. On this and the rest of the paragraph, Mao Zedong in a talk with Iudin on March 31, 1956, reproduced in a telegram from Iudin to Moscow, April 20, 1956, AVP RF (no retrieval number). The author holds a copy of the original. This passage was omitted from a reproduction of the telegram published in 1994 in Moscow with no indication of the omission—evidence that Soviet-style censorship continues to falsify documents in Russia today. See Dieter Heinzig, "Stalin, Mao, Kim and Korean War Origins, 1950," p. 240.

16. On this and the following two sentences, see Ivan Arkhipov, "1970s . . . ," pp. 40–45.

17. M.S. Kapitsa, *Sovetsko-kitaiskie otnosheniia*, pp. 349, 372.

18. On this paragraph, see *Khrushchev Remembers*, ed. E. Crankshaw, pp. 462, 466–67; "Memuary Nikity Sergeevicha Khrushcheva," p. 80.

19. On this, see Zhou Enlai's internal report on his talks with the Soviet leadership, January 1957, in Shi Zhongquan, *Zhou Enlai de zhuoyue fengxian*, pp. 302–3.

20. Stuart Schram, *Mao Tse-tung*, p. 268.

21. Zhou Wenqi and Chu Liangru, eds., *Gongchan guoji, Sulian he Zhongguo gongchandang guanxi biannianshi, 1919–1991*, p. 513. On this paragraph, see ibid., pp. 512–14; Zhang Shu Guang and Chen Jian, "The Emerging Disputes Between Beijing and Moscow," pp. 154–59; Vladislav Zubok, "The Mao-Khrushchev Conversations, 31 July–3 August 1958 and 2 October 1959," pp. 244–60.

22. The author thanks a former official in the Soviet CC apparatus who worked on China and does not wish to be identified for providing the following information in this paragraph. The interview took place in September 1993 in Moscow.

23. On this, see Jonathan D. Pollack, *Perception and Action in Chinese Foreign Policy*, passim. Cf. a different view in Zubok, "Mao-Khrushchev Conversations," p. 246.

24. *Renmin ribao*, August 15, 1963; Zhou and Chu, *Gongchan guoji*, p. 517; Arkhipov, "1970s . . . ," p. 44; Alfred D. Low, *The Sino-Soviet Dispute*, pp. 98–99. In August 1955, the Soviet Party leadership had approved a request from Peking for assistance in the training of professionals in the peaceful use of atomic energy; see TsKhSD, f. 4, op. 9, d. 1347, ll. 121–23.

25. On this paragraph, see O.B. Borisov and B.T. Koloskov, *Sovetsko-kitaiskie otnosheniia, 1945–1970*, pp. 210–16; *Long Live Leninism*, passim; Donald S. Zagoria, *The Sino-Soviet Conflict, 1956–1961*, pp. 319–27.

26. Khrushchev used the expressions *staraia kalosha* and *rvanaia tuflia.* (See n. 28.)

27. The original quote read "*Budda, kotoryi svoi teorii iz nosa vykovyrivaet.*" (See n. 28.)

28. The author received the information here and in the rest of the paragraph from Soviet contemporary witnesses, one of whom attended the Bucharest meeting, while the other assessed the protocols of the Bucharest talks as an adviser to the CC of the CPSU. They wish to remain anonymous. The interviews were conducted in September 1993 in Moscow. On this, cf. also L. Delyusin, "1950s and 1960s . . . ," p. 56.

29. Note from the Soviet embassy in Peking to the Ministry of Foreign Affairs of the People's Republic of China, July 18, 1960, SAPMO-B-Arch, J IV/2/202/280, unpaginated (Russian text); an English translation appeared in *CWIHPB* 8–9 (Winter 1996–97), pp. 249–50.

30. The author received the information reproduced here and in the following sentence from an adviser on China active at the time in the CC apparatus, who wishes to remain anonymous. The interview was conducted in September 1993 in Moscow. The Chinese term for "bald-headed idiot" was *guangtu de shagua.*

31. Arkhipov, "1970s . . . ," p. 47.

32. On this, see Peter J. Opitz, *Chinas Aussenpolitik*, pp. 47–56.

33. The author received this information from an adviser to the Soviet CC who worked on China at the time and wishes to remain anonymous. The interview was conducted in March 1995 in Moscow.

34. This and the rest of the paragraph were reported by a Soviet diplomat, who wishes to remain anonymous. He was taking care of the Chinese delegation and was present at the outburst. The interview was conducted in September 1993 in Moscow.

35. On this and the following sentence, see the speech by Zhou Enlai on August 23, 1968, reproduced in *Peking Review* 34 (1968), suppl., pp. iii–iv.

36. Harry Gelman, *The Soviet Far East Buildup and Soviet Risk-Taking Against China*, pp. 13–14, 19–20; text of the treaty of alliance in *Sovetsko-mongol'skie otnosheniia, 1921–1966*, pp. 320–22.

37. On this, see A.I. Elizavetin, "Peregovory Kosygina i Chzhou Enlaia v Pekinskom aeroportu," pt. 1, pp. 39–63, and pt. 2, 1993, pp. 107–19.

38. On the details in this paragraph, see Dieter Heinzig, *Der sowjetisch-chinesische Grenzkonflikt*, passim.

39. On this and following sentence, see Lowell Dittmer, *Sino-Soviet Normalization and Its International Implications, 1945–1990*, pp. 195–216.

40. Ibid., pp. 217–30.

41. For further details, see Dieter Heinzig, "China as a Regional and Global Challenge to the Soviet Union," pp. 41–61.

Notes to Appendix

1. Article 7 of the draft of a Treaty of Friendship, Cooperation, and Mutual Assistance between the Union of Soviet Socialist Republics and the People's Republic of China of January 5, 1950; AVP RF, f. 07, op. 23a, p. 18, d. 235, ll. 13–14. Original in Russian.

2. Article 7 of the draft of a Treaty of Friendship, Cooperation, and Mutual Assistance between the Union of Soviet Socialist Republics and the People's Republic of China of January 9, 1950; AVP RF, f. 07, op. 23a, p. 18, d. 235, l. 18. Original in Russian.

3. Article 7 of the draft of a Treaty of Friendship, Cooperation, and Mutual Assistance between the Union of Soviet Socialist Republics and the People's Republic of China of January 10, 1950; AVP RF, f. 07, op. 23a, p. 18, d. 235, l. 24. Original in Russian.

4. Article 7 of the draft of a Treaty of Friendship, Cooperation, and Mutual Assistance between the Union of Soviet Socialist Republics and the People's Republic of China of January 12, 1950; AVP RF, f. 07, op. 23a, p. 18, d. 235, l. 33. Original in Russian.

5. Article 7 of the draft of a Treaty of Friendship, Alliance, and Mutual Assistance between the Union of Soviet Socialist Republics and the People's Republic of China of January 16, 1950; AVP RF, f. 07, op. 23a, p. 18, d. 235, ll. 3–4. Original in Russian.

6. Ibid., 1.5. Original in Russian.

7. Ibid., 1.6. Original in Russian.

8. Ibid., ll. 45–48. Original in Russian.

9. Ibid., 1. 49. Original in Russian.

10. Ibid., l. 50. Original in Russian.

11. Ibid., ll. 38–40. Translated from Russian. The Chinese original is not yet accessible.

12. Ibid., ll. 77–79. Original in Russian.

13. *Izvestiia*, February 15, 1950, p. 1. Chinese text in *Zhonghua renmin gongheguo duiwai guanxi wenjian ji*, vol. 1, pp. 75–77, and *Zhonghua renmin gongheguo tiaoyue ji*, vol. 1, pp. 1–3. The English version presented here follows the text published in *United Nations Treaty Series*, vol. 226 (1956), no. 3103, pp. 12, 14, 16.

14. *Izvestiia*, February 15, 1950, p. 2. Chinese text in *Zhonghua renmin gongheguo duiwai guanxi wenjian ji*, vol. 1, pp. 77–79, and *Zhonghua renmin gongheguo tiaoyue ji*, vol. 1, pp. 3–5. The English

version presented here follows the text published in *United Nations Treaty Series*, vol. 226 (1956), no. 3105, pp. 38, 40, 42.

15. *Izvestiia*, February 15, 1950, p. 2. Chinese text in *Zhonghua renmin gongheguo duiwai guanxi wenjian ji*, vol. 1, pp. 79–80, and *Zhonghua renmin gongheguo tiaoyue ji*, vol. 1, pp. 45–46. The English version presented here follows the text published in *United Nations Treaty Series*, vol. 226 (1956), no. 3104, pp. 26, 28.

16. In Russian, the word is *razorennost',* which means "impoverishment" or "destruction."

17. AVP RF, f. 0100, op. 43, p. 302, d. 8, ll. 51–53. Translated from Russian. The Chinese original is not yet accessible.

18. Henceforth (to simplify): CPG of the PRC.

19. Henceforth (to simplify): USSR.

Bibliography

We have used the following bibliographic abbreviations in the notes and bibliography:

AFMROC	Archives of the Foreign Ministry of the Republic of China (*Waijiaobu dang'an zixun chu*), Taipei
AP RF	Archiv Prezidenta Rossiiskoi Federatsii (Archives of the President of the Russian Federation), Moscow
AVP RF	Arkhiv Vneshnei Politiki Rossiiskoi Federatsii (Archives of Foreign Policy of the Russian Federation), Moscow
FRUS	U.S. Department of State, *Foreign Relations of the United States*, Washington, DC
NiNI	*Novaia i Noveishaia Istoriia* (Modern and Contemporary History), Moscow
PDV	*Problemy Dal'nego Vostoka* (Problems of the Far East), Moscow
RTsKhIDNI	Rossiiskii Tsentr Khraneniia i Izucheniia Dokumentov Noveishei Istorii (Russian Center for Storage and Investigation of Documents of Contemporary History), Moscow; in 2000, renamed Rossiiskii Gosudarstvennyi Arkhiv Sotsial'no-politicheskoi Istorii (GASPI; State Archives of Sociopolitical History)
SAPMO-B-Arch	Stiftung Archiv der Parteien and Massenorganisationen der ehemaligen DDR im Bundesarchiv (Foundation Archives of the Parties and Mass Organizations of the Former GDR in the Federal Archives), Berlin
CQ	*The China Quarterly*, London
CHC	*The Cambridge History of China*
CWIHPB	*Cold War International History Project Bulletin*, Washington, DC
TsGAOR	Tsentral'nyi Gosudarstvennyi Arkhiv Oktiabrskoi Revoliutsii (Central State Archives of the October Revolution), Moscow
TsKhSD	Tsentr Khraneniia Sovremennoi Dokumentatsii (Storage Center for Contemporary Documentation), Moscow

Acheson, Dean. "Crisis in Asia—An Examination of U.S. Foreign Policy." *Department of State Bulletin* (January 23, 1950), pp. 111–18.

———. *Present at the Creation: My Years in the State Department.* New York: Norton, 1969.

Alexander, Bevin. *The Strange Connection: Intervention in China, 1944–1972.* Westport, CT: Greenwood, 1992.

Arkad'ev, A.A. "Politika Iaponii v otnoshenii SSSR posle nachala vtoroi mirovoi voiny (1939–1941)." In *SSSR—Iaponii: K 50-letiiu ustanovleniia sovetsko-iaponskikh diplomaticheskikh otnoshenii (1925–1975).* Moscow: Nauka, 1978.

Arkhipov, Ivan. "1970s . . ." In *USSR—China in the Changing World: Soviet Sinologists on the History and Prospects of Soviet-Chinese Relations*, pp. 39–51. Moscow: Novosti Press Agency, 1989.

Bagby, Wesley M. *The Eagle-Dragon Alliance: America's Relations with China in World War II.* Newark: University of Delaware Press, 1992.

Bajanov, Evgueni, "Assessing the Politics of the Korean War, 1949–61." *CWIHPB* 6–7 (Winter 1995–1996), pp. 54, 87–91.

Barker, Elisabeth. *The British Between the Superpowers, 1945–50.* London: Macmillan, 1983.

Bartke, Wolfgang. *Biographical Dictionary and Analysis of China's Party Leadership, 1922–1988.* Munich: Saur, 1990.

———. *Who's Who in the People's Republic of China.* Brighton, England: Harvesters, 1981.

Bazhanov, E.P. *Kitai i vneshnii mir.* Moscow: Mezhdunarodnye otnosheniia, 1990.

———. "Sovetsko-kitaiskie otnosheniia: uroki proshlogo i sovremennost'." *NiNI* 2 (1989), pp. 3–25.

Benton, Gregor, and Allan Hunter, eds. *Wild Lily, Prairie Fire: China's Road to Democracy, Yan'an to Tian'anmen, 1942–1944.* Princeton, NJ: Princeton University Press, 1995.

Bi Xiantian. "Zhanhou Dongbei wenti yu Zhong-Su guanxi zouxiang." *Jindaishi yanjiu* 1 (1996), pp. 53–76.

Bol'shaia Sovetskaia Entsiklopediia. 3rd ed. 30 vols. Moscow: Izdatel'stvo Sovetskoi Entsiklopediia, 1971.

Boorman, Howard L., and Richard C. Howard, eds. *Biographical Dictionary of Republican China.* Vols. 1–4. New York: Columbia University Press, 1967–71, 1979.

Borisov, O. *Sovetskii Soiuz i mandzhurskaia revoliutsionnaia baza, 1945–1949.* 2nd ed., Moscow: Mysl', 1977.

Borisov, O.B., and B.T. Koloskov. *Sovetsko-kitaiskie otnosheniia, 1945–1970: Kratkii ocherk.* 2nd ed. Moscow: Mysl', 1971.

Borodin, B.A. *Pomoshch' SSSR kitaiskomu narodu v antiiaponskoi voine, 1937–1941.* Moscow: Mysl', 1965.

Bo Yibo. *Ruogan zhongda juece yu shibian de huigu.* Vol. 1. Peking: Zhonggong zhongyang dangxiao chubanshe, 1991.

Brandt, Conrad. *Stalin's Failure in China.* New York: W.W. Norton, 1966.

Brandt, Conrad, Benjamin Schwartz, and John K. Fairbank. *A Documentary History of Chinese Communism.* New York: Atheneum, 1966.

Braun, Otto. *A Comintern Agent in China.* London: C. Hurst, 1982.

A Brief History of the Chinese National Revolutionary Forces. Comp. Hu Pu-yu. Taipei: Chung Wu, 1971.

Byron, John, and Robert Pack. *The Claws of the Dragon: Kang Sheng—The Evil Genius Behind Mao—and His Legacy of Terror in the People's Republic of China.* New York: Simon & Schuster, 1992.

Carr, E.H. *Twilight of the Comintern, 1930–1935.* New York: Pantheon Books, 1982.

Chalidze, V., ed. *Nikita Khrushchev: Vospominaniia. Izbrannye otryvki.* New York: Chalidze Publications, 1979.

Chang, Gordon H. *Friends and Enemies: The United States, China, and the Soviet Union, 1948–1972.* Stanford, CA: Stanford University Press, 1990.

Chang Kuo-t'ao. Introduction to *Collected Works of Liu Shao-ch'i Before 1944.* Hong Kong: Union Research Institute, 1969.

———. *The Rise of the Chinese Communist Party.* Vol. 1, *1921–1927.* Lawrence: University Press of Kansas, 1971.

———. *The Rise of the Chinese Communist Party.* Vol. 2, *1928–1938.* Lawrence: University Press of Kansas, 1972.

Chao Kang, and Feng-Hwa Ma. "A Study of the Rouble-Yuan Exchange Rate." *CQ* (January–March 1964), pp. 192–204.

Ch'en, Jerome. "The Chinese Communist Movement to 1927." In *CHC*, vol. 12. *Republican China, 1912–1949*, pt. 1, ed. John K. Fairbank and Denis Twitchett, pp. 505–26. Cambridge, England: Cambridge University, 1983.

———. "The Communist Movement, 1927–1937." In Fairbank and Feuerwerker, eds., *CHC*, vol. 13, *Republican China, 1912–1949*, pt. 2, pp. 168–229.

———. "Resolution of the Tsunyi Conference." *CQ* 40 (October–December 1969), pp. 1–38.

Chen Dunde. *Mao Zedong yu Jiang Jieshi, 1949–1976.* Peking: Ba yi chubanshe, 1993.

Cheng, Tien-fong. *A History of Sino-Russian Relations.* Washington, DC: Public Affairs, 1957.

Chen Jian. *China's Road to the Korean War: The Making of the Sino-American Confrontation.* New York: Columbia University Press, 1994.

———. "Comparing Russian and Chinese Sources: A New Point of Departure for Cold War History." *CWIHPB* 6–7 (Winter 1995–96), pp. 20–21.

———. "A Crucial Step Toward the Breakdown of the Sino-Soviet Alliance: The Withdrawal of Soviet Experts from China in July 1960." *CWIGHPB* (Winter 1996–97), pp. 246, 249.

———. "The Myth of America's 'Lost Chance' in China: A Chinese Perspective in the Light of New Evidence." *Diplomatic History* 1 (Winter 1997), pp. 77–86.

———. "The Sino-Soviet Alliance and China's Entry into the Korean War." *The Cold War International History Project.* Working Paper no. 1. Washington, DC: June 1992.

———. "The Ward Case and the Emergence of Sino-American Confrontation, 1948–1950." *Australian Journal of Chinese Affairs*, July 30, 1993, pp. 149–70.

Cherepanov, A.I. *Zapiski voennogo sovetnika v Kitae.* 2nd ed. Moscow: Nauka, 1976.

Chiang Kai-shek. *Soviet Russia in China: A Summing-Up at Seventy.* New York: Farrar, Straus & Cudahy, 1957.

Chicherin, G.V. *Stat'i i rechi po voprosam mezhdunarodnoi politiki.* Moscow: Izdatel'stvo sotsial'no-ekonomicheskoi literatury, 1961.

The China White Paper: August 1949. Originally issued as *United States Relations with China: With Special Reference to the Period 1944–1949.* Department of State Publication 3537, Far Eastern Series 30. Stanford, CA: Stanford University Press, 1967.

Chinese Communist Who's Who. 2 vols. Taipei: Institute of International Relations, 1970–71.

Die chinesische Frage auf dem 8. Plenum der Exekutive der Kommunistischen Internationale. 1928. Reprint, Milan: Feltrinelli, 1967.

Chuikov, V. I. *Missiia v Kitae: Zapiski voennogo sovetnika.* Moscow: Nauka, 1981.

Clubb, O. Edmund. *China and Russia: The "Great Game."* New York: Columbia University Press, 1971.

Cochran, Thomas B., Robert S. Norris, and Oleg A. Bukharin. *Making the Russian Bomb: From Stalin to Yeltsin.* Boulder, CO: Westview, 1995.

Cohen, Warren, I. *America's Response to China: An Interpretative History of Sino-American Relations.* New York: John Wiley & Sons, 1971.

———. "Introduction: Was There a 'Lost Chance' in China?" *Diplomatic History* 1 (Winter 1997), pp. 71–75.

Compton, Boyd. *Mao's China: Party Reform Documents, 1942–1944.* Seattle: University of Washington Press, 1952.

The Conference of Berlin (The Potsdam Conference), 1945. Vol. 2. *FRUS.* Diplomatic Papers. Washington, DC: U.S. Government Printing Office, 1960.

The Conferences at Malta and Yalta, 1945. FRUS. Diplomatic Papers. Washington, DC: U.S. Government Printing Office, 1955.

Cong Yan'an dao Beijing. Peking: Zhongyang wenxian chubanshe, 1993.

Dai Qing. *Wang Shiwei and White Lilies: Rectification and Purges in the Chinese Communist Party, 1942–1944.* Armonk, NY: M.E. Sharpe, 1994.

Dangdai zhongguo waijiao. Peking: n.p., 1990.

Dedijer, Vladimir. *Tito Speaks: His Self-Portrait and Struggle with Stalin.* London: Weidenfeld & Nicolson, 1953.

Deliusin, L. "SSSR—KNR: ostavliaia proshloe v proshlom." *Documenty i fakty* 5 (1990), pp. 4–5.

Delyusin, L. "1950s and 1960s . . ." In *USSR—China in the Changing World: Soviet Sinologists on the History and Prospects of Soviet-Chinese Relations*, pp. 52–65. Moscow: Novosti Press Agency, 1989.

Deng Xiaoping wenxuan. Vol. 1. 2nd ed. Peking: Renmin chubanshe, 1994.

Dittmer, Lowell. *Sino-Soviet Normalization and Its International Implications, 1945–1990.* Seattle: University of Washington Press, 1992.

Djilas, Milovan. *Conversations with Stalin.* New York: Harcourt, Brace & World, 1962.

Domes, Jürgen. *Vertagte Revolution: Die Politik der Kuomintang in China, 1923–1937.* Berlin: De Gruyter, 1969.

Doolin, Dennis J. *Territorial Claims in the Sino-Soviet Conflict: Documents and Analysis.* Hoover Institution Studies 7. Stanford, CA: Stanford University Press, 1965.

Dupuy, Trevor N. *The Military History of the Chinese Civil War.* New York: F. Watts, 1969.

Eastman, Lloyd E. *Seeds of Destruction: Nationalist China in War and Revolution.* Stanford, CA: Stanford University Press, 1984.

———. "Nationalist China During the Nanking Decade, 1927–1937." In Fairbank and Feuerwerker, eds., *CHC*, vol. 13, *Republican China, 1912–1949*, pt. 2, pp. 116–67.

Elizavetin, A.I. "Peregovory Kosygina i Chzhou Enlaia v Pekinskom aeoportu." Parts 1–2. *PDV* 5 (1992), pp. 39–63; 1 (1993), pp. 107–19.

Elleman, Bruce A. *Diplomacy and Deception: The Secret History of Sino-Soviet Diplomatic Relations, 1917–1927.* Armonk, NY: M.E. Sharpe, 1997.

———. "The End of Extraterritoriality in China: The Case of the Soviet Union, 1917–1960." *Republican China* 21, no. 2 (April 1996), pp. 65–89.

Esherick, Joseph W., ed. *Lost Chance in China: The World War II Despatches of John S. Service.* New York: Random House, 1974.

Fairbank, John K., and Albert Feuerwerker, eds. *CHC.* Vol. 13, *Republican China, 1912–1949*, pt. 2. Cambridge, England: Cambridge University Press, 1986.

Feaver, John Hansen. *The Truman Administration and China, 1945–1950: The Policy of Restrained Intervention.* Ph.D. diss. Oklahoma University, 1980. Ann Arbor, MI: University Microfilms International, 1983.

Fedorenko, N.T. "Stalin i Mao Tszedun." Parts 1–2. *NiNI* 5 (1992), pp. 98–113; 6 (1992), pp. 83–95.
———. "Nochnye besedy." *Pravda*, October 23, 1988.
———. *"Stalin i Mao: besedy v Moskve."* *PDV* 1 (1989), pp. 149–64.
Filosofskaia Entsiklopediia. Vol. 5. Moscow: Sovetskaia Entsiklopediia, 1970.
Fischer, Louis. *Russia's Road from Peace to War: Soviet Foreign Relations, 1917–1941.* New York: Harper & Row, 1969.
Frenzke, Dietrich. "Bündnis- und Freundschaftsverträge der UdSSR mit asiatischen Staaten." In *Die Sowjetunion und Asien in den achtziger Jahren: Ziele und Grenzen sowjetischer Politik zwischen Indischem Ozean und Pazifik*, ed. Joachim Glaubitz and Dieter Heinzig, pp. 133–55. Baden-Baden: Nomos, 1988.
FRUS 1944. Vol. 6, *China.* Washington, DC: U.S. Government Printing Office, 1967.
FRUS 1945. Vol. 7, *The Far East: China.* Washington, DC: U.S. Government Printing Office, 1969.
FRUS 1946. Vol. 9, *The Far East: China.* Washington, DC: U.S. Government Printing Office, 1972.
FRUS 1947. Vol. 7, *The Far East: China.* Washington, DC: U.S. Government Printing Office, 1972.
FRUS 1948. Vol. 7, *The Far East: China.* Washington, DC: U.S. Government Printing Office, 1973.
FRUS 1949. Vol. 8, *The Far East: China.* Washington, DC: U.S. Government Printing Office, 1978.
FRUS 1949. Vol. 9, *The Far East: China.* Washington, DC: U.S. Government Printing Office, 1975.
FRUS 1950. Vol. 6, *East Asia and the Pacific.* Washington, DC: U.S. Government Printing Office, 1976.
Furuya Keiji. *Chiang Kai-shek: His Life and Times.* Abr. English ed. by Chun-ming Chang. New York: St. John's University Press, 1981.
Garthoff, Raymond L. "The Soviet Intervention in Manchuria, 1945–46." In *Sino-Soviet Military Relations*, ed. Raymond L. Garthoff. New York: Praeger, 1966.
Garver, John W. "Chiang Kai-shek's Quest for Soviet Entry into the Sino-Japanese War." *CQ* 102 (1987), pp. 295–316.
———. *Chinese-Soviet Relations, 1937–1945.* New York: Oxford University Press, 1988.
———. "Comment: Mao, the Comintern and the Second United Front." *CQ* 129 (1992), pp. 171–79.
———. *Foreign Relations of the People's Republic of China.* Englewood Cliffs, NJ: Prentice-Hall, 1992.
———. "Little Chance." *Diplomatic History* 1 (Winter 1997), pp. 87–94.
———. "The Origins of the Second United Front: The Comintern and the Chinese Communist Party." *CQ* 113 (1988), pp. 29–59.
Gelman, Harry. *The Soviet Far East Buildup and Soviet Risk-Taking Against China.* Rand Publications Series, no. R-2943–AF. Santa Monica, CA: Rand, 1982.
Gibanskii, L. Ia. "Kominform v deistvii, 1947–1948 gg.: Po arkhivnym documentam." Parts 1–2. *NiNI* 1 (1996), pp. 149–70; 2 (1996), pp. 157–72.
Gittings, John. *Survey of the Sino-Soviet Dispute: A Commentary and Extracts from the Recent Polemics, 1963–1967.* London: Oxford University Press, 1968.
Glaubitz, Joachim, and Dieter Heinzig, ed. *Die Sowjetunion und Asien in den achtziger Jahren: Osteuropa und der internationale Kommunismus.* Baden-Baden: Nomos, 1988.
Glunin, V.I. *Kommunisticheskaia Partiia Kitaia nakanune i vo vremia natsional'noi revoliutsii (1921–1927 gg.).* 2 vols. Moscow: Institut Dal'nego Vostoka AN SSSR (Dlia sluzhebnogo pol'zovaniia), 1975.
Go Khen'iui et al., eds. *VKP(b), Komintern i natsional'no-revoliutsionnoe dvizhenie v Kitae. Dokumenty.* Vol. 1, *1920–1925.* Moscow: RTsKhIDNI, Institut Dal'nego Vostoka RAN, Vostochno-aziatskii Seminar Svobodnogo Universiteta Berlina, 1994.
Goldstein, Steven M. "Sino-American Relations, 1948–1950." In *Sino-American Relations, 1945–1955: A Joint Reassessment of a Critical Decade*, eds. Harry Harding, and Yuan Ming, pp. 120–42. Wilmington, DE: Scholarly Resources, 1989.
Goncharov, Sergei N., John W. Lewis, and Xue Litai. *Uncertain Partners: Stalin, Mao and the Korean War.* Stanford, CA: Stanford University Press, 1993.
Gongchan guoji he Zhongguo geming guanxi jishim, 1919–1943. Peking: Zhongguo shehui kexue chubanshe (xian guonei faxing), 1983.
Gongchan guoji yu Zhongguo geming guanxi lunwen ji. Shanghai: n.p., 1985.
Gongchan guoji yu Zhongguo geming ziliao xuanji (1925–1927). Peking: Renmin chubanshe (neibu faxing), 1985.
Grigor'ev, A.M. "Pavel Mif (1901–1938)." In *Vidnye sovetskie kommunisty—uchastniki kitaiskoi revoliutsii*, pp. 88–101. Moscow: Nauka, 1970.
Gwertzman, Bernard. "Literary Detective on Trail of Soviet 'Diary.'" *International Herald Tribune*, January 13, 1976.

Handzik, Helmut. *Politische Bedingungen sowjetischer Truppenabzüge, Drei Fallstudien: Asien 1925, 1946 und 1948.* SWP—S 332 Fo.Pl.I3b/86. Ebenhausen, Germany: Stiftung Wissenschaft und Politik, 1986.

———. *Politische Bedingungen sowjetischer Truppenabzüge. Drei Fallstudien: Die fünfziger Jahre.* SWP—S352 Fo.Pl.2.2/90. Ebenhausen: Stiftung Wissenschaft und Politik, 1990.

Harriman, W. Averell, and Elie Abel. *Special Envoy to Churchill and Stalin, 1941–1946.* New York: Random House, 1975.

Head, William Pace. *America's China Sojourn.* Ph.D. diss., Florida State University, 1980. Ann Arbor, MI: University Microfilms International, 1983.

He Di. "The Evolution of the Chinese Communist Party's Policy Toward the United States, 1944–1949." In *Sino-American Relations, 1945–1955: A Joint Reassessment of a Critical Decade*, ed. Harry Harding, and Yuan Ming, pp. 31–50. Wilmington, DE: Scholarly Resources, 1989.

———. "The Last Campaign to Unify China: The CCP's Unmaterialized Plan to Liberate Taiwan, 1949–1950." *Chinese Historians* 1 (Spring 1992), pp. 1–16.

Heinzig, Dieter. "China as a Regional and Global Challenge to the Soviet Union." *Sino-Soviet Affairs* 3 (1987), pp. 41–61.

———. "From the 'Middle of the Road' to 'Yibian Dao': Changes in CCP Foreign Policy in 1949." In *Notions et perceptions du changement en Chine: Mémoires de l'Institut des Hautes Études Chinoises*, vol. 36, pp. 59–72. Paris: Collège de France, Institut des Hautes Études Chinoises, 1994.

———. "Otto Braun and the Tsunyi Conference." *CQ* 42 (1970), pp. 131–35.

———. "The Otto Braun Memoirs and Mao's Rise to Power." *CQ* 46 (1991), pp. 274–88.

———. *Der sowjetisch-chinesische Grenzkonflikt: Ursachen—Ablauf—Perspektiven.* Berichte, no. 17 (1979). Cologne: Bundesinstituts für Ostwissenschaftliche und Internationale Studien, 1979.

———. *Sowjetische Militärberater bei der Kuomintang, 1923–1927.* Baden-Baden: Nomos, 1978.

———. "Stalin, Mao, Kim and Korean War Origins, 1950: A Russian Documentary Discrepancy." *CWIHPB* 8–9 (Winter 1996–97), p. 240.

———. "Verriet Stalin im Jahre 1945 die kommunistische Revolution in China?" *Asien* 56 (July 1995), pp. 6–15.

History of the Sino-Japanese War (1937–1945). Comp. Hsu Long-hsuen and Chang Ming-kai. Taipei: Chung Wu, 1971.

Hsiao Tso-liang. *Power Relations Within the Chinese Communist Movement, 1930–1934.* Seattle: University of Washington Press, 1961.

Huang Hua. "Nanjing jiefang chuqi wo tong Si-tu-lei-deng de jici jiechu." In *Xin Zhongguo waijiao fengyun*, pp. 23–32. Peking: Shijie zhishi chubanshe, 1990.

Huang Xiurong. *Gongchan guoji yu Zhongguo geming guanxi shi.* 2 vols. Peking: Zhonggong zhongyang dangxiao chubanshe, 1989.

Huang Zheng, ed. *Liu Shaoqi yi sheng.* Peking: Zhongyang wenxian chubanshe, 1995.

Hu Chi-hsi. "Hua Fu, the Fifth Encirclement Campaign and the Tsunyi Conference." *CQ* 43 (1970), pp. 31–46.

Hunt, Michael H. *The Genesis of Chinese Communist Foreign Policy.* New York: Columbia University Press, 1996.

Huntington, Samuel P. *The Clash of Civilizations and the Remaking of World Order.* New York: Simon & Schuster, 1996.

Hu Qiaomu. *Huiyi Mao Zedong.* Peking: Xinhua chubanshe, 1994.

Iriye Akira. *The Cold War in Asia: A Historical Introduction.* Englewood Cliffs, NJ: Prentice-Hall, 1974.

Isaacs, Harold R. *The Tragedy of the Chinese Revolution.* 2nd rev. ed. New York: Atheneum, 1961.

Istoriia Vsesoiuznoi Kommunisticheskoi Partii (bolshevikov): Kratkii kurs. Moscow: Gosudarstvennoe izdatel'stvo politichesko literatury, 1938.

Iur'ev, M.F. *Vooruzhennye sily KPK v osvoboditel'noi bor'be kitaiskogo naroda: 20–40-e gody.* Moscow: Nauka, 1983.

Jiang Jingguo. *Fengyu zhong de ningjing.* Taipei: Shidian chubanshe, 1974.

———. *Fuzhong zhi yuan.* Taipei: Yushi shudian chubanshe, 1960.

Jianguo yilai Mao Zedong wengao. 6th ed., vol. 1. Peking: Zhongyang wenxian chubanshe, 1992.

Johnson, Chalmers A. *Peasant Nationalism and Communist Power.* Stanford, CA: Stanford University Press, 1962.

Kaliagin, A. Ia. *Po neznakomym dorogam: Zapiski voennogo sovetnika v Kitae.* 2nd enl. ed. Moscow: Nauka, 1979.

Kampen, Thomas. *Mao Zedong, Zhou Enlai and the Evolution of the Chinese Communist Leadership.* Copenhagen: NIAS, 2000.

———. "The Zunyi Conference and Further Steps in Mao's Rise to Power." *CQ* 17 (1989), pp. 118–34.
Kapitsa, M.S. "Na raznykh paralleliakh." *Aziia i Afrika segodnia* 5 (1995), pp. 12–18.
———. *Na raznykh paralleliakh: Zapiski diplomata.* Moscow: Kniga I biznes, 1996.
———. *Sovetsko-kitaiskie otnosheniia.* Moscow: Gosudarstvennoe izdatel'stvo politicheskoi literatury, 1958.
Khazanov, A. "Informatsiia iz pervykh ruk." *Aziia i Afrika segodnia* 5 (1997).
Khrushchev, Nikita. *Khrushchev Remembers.* With an introduction, commentary, and notes by Edward Crankshaw. London: Deutsch, 1971.
———. *Khrushchev Remembers: The Glasnost Tapes.* Boston: Little, Brown, 1990.
———. *Khrushchev Remembers: The Last Testament.* Boston: Little, Brown, 1974.
Kimball, Warren F., ed. *Churchill and Roosevelt: The Complete Correspondence.* 3 vols. Princeton, NJ: Princeton University, 1984.
Kindermann, Gottfried-Karl. *Der Ferne Osten in der Weltpolitik des industriellen Zeitalters.* Munich: Deutscher Taschenbuch Verlag, 1970.
Klein, Donald W., and Anne B. Clark. *Biographic Dictionary of Chinese Communism.* 2 vols. Cambridge, MA: Harvard University Press, 1971.
Klochko, Mikhail A. *Soviet Scientist in Red China.* New York: Hollis & Carter, 1964.
Kovalev, I.V. "Dialog Stalina s Mao Tszedunom." Parts 1–2. *PDV* 6 (1991), pp. 83–93; 1–3 (1992), pp. 77–91.
Kuhfus, Peter M. "Die 'Wladimirow-Tagebücher' in neuem Licht." *Osteuropa* 1 (1991), pp. 69–77.
———. "'Ein Held, der mit der Zeit geht': Über eine verbotene russische Mao-Edition und den chinesisch-sowjetischen Dialog." In *Chinesische Intellektuelle im 20. Jahrhundert: Zwischen Tradition und Moderne*, ed. Karl-Heinz Pohl; Gudrun Wacker; and Liu Huiru, pp. 209–331. Hamburg: Institut für Asienkunde, 1993.
Kukushkin, K.V. "Komintern i edinyi natsional'nyi antiiaponskii front v Kitae (1935–1943)." In *Komintern i vostok*, pp. 350–79. Moscow: Nauka, 1969.
Kulik, B.T. "Kitaiskaia Narodnaia Respublika v period stanoveleniia (1949–1952 gg.) (Po materialam Arkhiva vneshnei politiki RF)." Parts 1–2. *PDV* 5 (1994), pp. 111–25; 6 (1994), pp. 73–83.
———. *Sovetsko-kitaiskii raskol: prichiny i posledstviia.* Moscow: Institut Dal'nego Vostoka RAN, 2000.
Kuo, Warren. *Analytical History of the Chinese Communist Party.* 4 vols. Taipei: Institute for International Relations, 1966–71.
Kuo Heng-yü. *Die Komintern und die Chinesische Revolution: Die Einheitsfront zwischen der KP Chinas und der Kuomintang, 1924–1927.* Paderborn, Germany: Ferdinand Schöningh, 1979.
———. *Maos Weg zur Macht und die Komintern.* Paderborn, Germany: Ferdinand Schöningh, 1975.
Ledovskii, A.M. *Delo Gao Gan—Zhao Shushi.* Moscow: Institut Dal'nego Vostoka AN SSSR (*Dlia sluzhebnogo pol'zovaniia*), 1990.
———. *Kitaiskaia politika SShA i sovetskaia diplomatiia, 1942–1954.* Moscow: Nauka, 1985.
———. "Na diplomaticheskoi rabote v Kitae v 1942–1952 gg." *NiNI* 6 (1993), pp. 102–32.
———. Nankinskie zapiski. Unpublished manuscript. Dieter Heinzig archives, Erftstadt, Germany.
———. "Peregovory I.V. Stalina s Mao Tszedunom v dekabre 1949–fevrale 1950 g.: Novye archivnye documenty." *NiNI* 1 (1997), pp. 23–47.
———. "Peregovory I.V. Stalina s Sun Tszywenem o sovetsko-kitaiskom dogovore, iiun'-avgust 1945 g." Unpublished manuscript. Dieter Heinzig archives, Erftstadt, Germany.
———. "Sekretnaia missiia A.I. Mikoiana v Kitai (ianvar'-fevral' 1949 g.)." Parts 1–2. *PDV* 2 (1995), pp. 96–111; 3 (1995), pp. 94–105.
———. "Sekretnaia missia delegatsii TsK Kommunisticheskoi Partii Kitaia v Moskvu v iiune–avguste 1949 g." Unpublished manuscript. Dieter Heinzig archives, Erftstadt, Germany.
———. *SSSR i Stalin v sud'bakh Kitaia. Dokumenty i svidetel'stva uchastnika sobytii 1937–52.* Moscow: Pamiatniki istoricheskoi mysli, 1999.
———. *SSSR, SShA i narodnaia revoliutsiia v Kitae.* Moscow: Nauka, 1979.
———. "Stalin i Chan Kaishi: Sekretnaia missiia syna Chan Kajshii v Moskvu. Dekabr' 1945–ianvar' 1946 g." *NiNI* 4 (1996), pp. 100–29.
———. "Stenogrammy peregovorov I.V. Stalina s Chzhou En'laem v avguste-sentiabre 1952 g." *NiNI* 2 (1997), pp. 69–86.
———. "Vizit Mao Tszeduna v Sovetskii Soiuz v dekabre 1949–fevral' 1950 g.: Nachalo otnoshenii mezhdu Kitaiskoi Narodnoi Respublikoi i Sovetskim Soiuzom." Parts 1–2. Unpublished manuscript. Dieter Heinzig archives, Erftstadt, Germany.
———. "Vizit v Moskvu delegatsii Kommunisticheskoi partii Kitaia v iiune-avguste 1949 g." Part 1–2. *PDV* 4 (1996), pp. 66–83; 5 (1996), pp. 84–94.

———. "The World Reaction to the Soviet-Chinese Treaty" In *The 1945 Sino-Soviet Treaty of Friendship and Alliance*, ed. Tai Wan-chin, pp. 153–59. Taipei: Tamkang University, 1995.
———. "Zapiski diplomata." *PDV* 1 (1991), pp. 108–18.
Leffler, Melvyn P. *A Preponderance of Power: National Security, the Truman Administration, and the Cold War.* Stanford, CA: Stanford University Press, 1992.
Leong Sow-Theng. *Sino-Soviet Diplomatic Relations, 1917–1926.* Canberra: Australian National University Press, 1976.
Leonhard, Wolfgang. *Child of the Revolution.* London: Collins, 1957.
Leung, John K., and Michael Y.M. Kao. *The Writings of Mao Zedong, 1949–1976.* Vol. 2. Armonk, NY: M.E. Sharpe, 1992.
Levine, Steven I. *Anvil of Victory: The Communist Revolution in Manchuria, 1945–1948.* New York: Columbia University Press, 1987.
———. "Soviet-American Rivalry in Manchuria and the Cold War." In *Dimensions of China's Foreign Relations*, ed. Hsueh Chün-tu. New York: Praeger, 1977.
Liang Chin-tung. *General Stilwell in China, 1942–1944: The Full Story.* Asia in the Modern World Series, no. 12. Jamaica, NY: St. John's University Press, 1972.
———. "The Sino-Soviet Treaty." In *Nationalist China During the Sino-Japanese War, 1937–1945. (Dui Ri kangzhan zhi Zhongguo guomin zhengfu)*, ed. Paul K.T. Si. Hicksville, NY: Exposition Press, 1977.
Lie-duo-fu-si-ji, An. "Mi-gao-yang yu Mao Zedong de mimi tanpan (1949 nian 1–2 yue) (zhong)." *Dang de wenxian* 1 (1996), pp. 90–96.
Li Haiwen. "A Distortion of History: An Interview with Shi Zhe about Kovalev's Memoirs." *Chinese Historians* 2 (Fall 1992), p. 64.
———. "Mao Zedong Liu Shaoqi fang Su ji Zhong-Su liang dang guanxi zhong de yixie wenti. Shi Zhe fang tan lu." *Renwu* 2 (1993), pp. 90–96.
———. "Mao Zedong yu Si-da-lin jiaowang." *Liaowang* 52 (1992), pp. 11–13.
Lin Jun. *Zhong Su guanxi, 1689–1989.* Harbin: n.p., 1989.
Litten, Freddy. *Otto Brauns frühes Wirken in China (1932–1935).* Arbeiten aus dem Osteuropa-Institut München. Working Papers, no. 124. Munich: Osteuropa-Institut, 1988.
Liu, F.F. *A Military History of Modern China.* Princeton, NJ: Princeton University, 1956.
Liu Jiecheng. *Mao Zedong yu Si-da-lin.* Peking: Zhonggong zhongyang dangxiao chubanshe, 1993.
———. *Mao Zedong yu Si-da-lin huiwu jishi.* Peking: Zhonggong zhongyang dangxiao chubanshe, 1997.
Liu Shao-chi. *On the Party.* 3rd. ed. Peking: Foreign Language Press, 1951.
Liu Shaoqi nianpu, 1898–1969. 2 vols. Peking: Zhongyang wenxian chubanshe, 1996.
Liu Shaoqi xuanji. Tokyo: n.p., 1967.
Liu Xiao. *Chushi Sulian ba nian.* Peking: Zhonggong dangshi ziliao chubanshe, 1986.
Liu Zhiqing. "Shi lun 'Zhong-Su youhao tongmen tiaoyue' di qianding." *Huozhong shifan daxue yanjiusheng xuebao* 2 (1988), pp. 69–74.
Li Yueran. *Waijiao wutai shang de xin Zhongguo lingxiu: Li Yueran huyi.* Peking: Jiefangjun chubanshe, 1989.
Li Zhisui. *Mao Zedong siren yisheng.* Taipei: Shibao wenhua, 1994.
———. *The Private Life of Chairman Mao.* London: Chatto & Windus, 1994.
Long Live Leninism. Peking: Foreign Languages Press, 1960.
Low, Alfred D. *The Sino-Soviet Dispute: An Analysis of the Polemics.* Rutherford, NY: Farleigh Dickinson University Press, 1976.
Luzianin, S. "Ialtinskaia konferentsiia i problemy mezhdunarodno-pravovogo oformleniia MNR nakanune i v gody vtoroi mirovoi voiny." *PDV* 6 (1995), pp. 52–60.
Mackerras, Colin, and Robert Chan. *Modern China: A Chronology from 1842 to the Present.* London: Thames & Hudson, 1982.
Mao Zedong. *Selected Works of Mao Tse-tung.* 5 vols. Peking: Foreign Languages Press, 1961–77.
Mao Zedong ji. Ed. Minoru Takeuchi. 10 vol. Tokyo: Suo suo sha, 1970.
Mao Zedong nianpu, 1893–1949. 3 vols. Peking: Zhonggong wenxian chubanshe, 1993.
Mao Zedong tongzhi shi dangdai zui weida de Ma-ke-si Lie-ning zhuyi zhe. N.p., 1969.
Mao Zedong waijiao wenxuan. Peking: n.p., 1994.
Mao Zedong xuanji. Vols. 2–5. Peking: Renmin chubanshe, 1953, 1960, 1977.
McGlothlen, Ronald L. *Controlling the Waves: Dean Acheson and U.S. Foreign Policy in Asia.* New York: W.W. Norton, 1993.
McLane, Charles B. *Soviet Policy and the Chinese Communists, 1931–1946.* New York: Columbia University Press, 1958.

Meissner, Boris. *Das Ostpaktsystem: Dokumentensammlung.* Part 1, *Das sowjetische Paktsystem*; Part 2, *Das sowjetische Paktsystem in Asien.* Hamburg: Forschungsstelle für Völlkerrecht und Ausländisches Öffentliches Recht der Universität Hamburg, 1951.

Meliksetov, Arlen. "1940s. . ." In *USSR-China in the Changing World: Soviet Sinologists on the History and Prospects of Soviet-Chinese Relations*, pp. 23–38. Moscow: Novosti Press Agency, 1989.

"Memuary Nikity Sergeevicha Khrushcheva." *Voprosy istorii* 11–12 (1992), pp. 65–85.

Military Campaigns in China: 1924–1950. Taipei: Office of Military History, 1966.

Murray, Brian Joseph. *Western versus Chinese Realism: Soviet-American Diplomacy and the Chinese Civil War, 1945–1950.* Ph.D. diss., Columbia University. Ann Arbor, MI: Dissertation Services, 1995.

Nagai Yonosuke, and Iriye Akira. *The Origins of the Cold War in Asia.* New York: Columbia University, 1977.

Na kitaiskoi zemle: Vospominaniia sovetskikh dobrovol'tsev, 1925–1945. Moscow: Nauka, 1974.

Nash drug Kitai: Slovar'-spravochnik. Moscow: Gospolitizdat, 1959.

Niu Jun. "The Origins of the Sino-Soviet Alliance." In Westad, ed., *Brothers in Arms*, pp. 47–89.

Nollau, Günther. *International Communism and World Revolution: History and Methods.* London: Hollis & Carter, 1959.

North, Robert C. *Moscow and the Chinese Communists.* 2nd ed. Stanford, CA: Stanford University, 1963.

O'Ballance, Edgar. *The Red Army of China.* London: Faber & Faber, 1962.

"O kul'te lichnosti i ego posledstviiach: Doklad Pervogo sekretaria TsK KPSS tov. Khrushcheva N.S. XX s"ezdu Kommunisticheskoi partii Sovetskogo Soiuza." *Izvestiia Ts KPSS* 3 (1989), pp. 128–70.

Opitz, Peter J. *Chinas Aussenpolitik: Ideologische Prinzipien—strategische Konzepte.* Zurich: Edition Interfrom, 1977.

Ovchinnikov, Yuri, ed. and trans. "Comintern-CCP Relations: The Anti-Japanese United Front in China; New Documents from the Comintern Archives." Parts 1–2. *Chinese Law and Government*, no. 1 (January–February 1997); no. 2 (March–April 1997).

Peng Dehuai zhuan. Peking: Dangdai Zhongguo chubanshe, 1993.

Peng Zhen. "Dongbei jiefang zhanzheng de tou jiu yue." *Zhonggong dangshi fengyun lu.* Peking: 1990.

Pepper, Suzanne. "The KMT-CCP Conflict, 1945–1949." In Fairbank and Feuerwerker, eds. *CHC*, vol. 13, pp. 723–88. *Republican China, 1912—1949*, pt. 2, pp. 723–88.

Pollack, Jonathan D. *Perception and Action in Chinese Foreign Policy: The Quemoi Decision*. 2 vols. Ann Arbor, MI: Xerox University Microfilms, 1976.

"Posetiteli kremlevskogo kabineta I.V. Stalina." Parts 1–2. *Istoricheskii Arkhiv* 5–6 (1996), pp. 3–61; 1 (1997), pp. 3–39.

Qiang Zhai. *The Dragon, the Lion, and the Eagle: British-American-Chinese Relations, 1949–1958.* Kent, Ohio: Kent State University, 1994.

Qiu Jing. "Mao Zedong yu Si-da-lin de huiwu." *Dang de wenxian* 2 (1996), pp. 51–55.

Rea, Kenneth W., and John C. Brewer, eds. *The Forgotten Ambassador: The Reports of John Leighton Stuart, 1946–1949.* Boulder, CO: Westview, 1981.

Reardon-Anderson, James. *Yenan and the Great Powers: The Origins of Chinese Communist Foreign Policy, 1944–1946.* New York: Columbia University Press, 1980.

Rue, John E. *Mao Tse-tung in Opposition.* Stanford, CA: Stanford University Press, 1966.

Russko-kitaiskie otnosheniia, 1698–1916. Ofitsial'nye dokumenty. Moscow: Izdatel'stvo rostochnoi literatury, 1958.

Saich, Tony, ed. *The Origins of the First United Front in China: The Role of Sneevliet (alias Maring).* 2 vols. Leiden: Brill, 1991.

———. *The Rise to Power of the Chinese Communist Party.* Armonk. NY: M.E. Sharpe, 1996.

Säjpidin Äzizi. *Ömür Dastani.* Vol. 2. Peking: n.p., 1990.

Sbornik deistvuiushchikh dogovorov, soglashenii i konventsii, zakliuchennykh SSSR s inostrannymi gosudarstvami. Vol. 13. Moscow: Mezhdunarodnye otnosheniia, 1965.

Schram, Stuart R. *Mao Tse-tung.* Harmondsworth, England: Penguin Books, 1966.

———. *Mao Zedong: A Preliminary Reassessment.* Hong Kong: n.p., 1983.

———. *The Political Thought of Mao Tse-tung.* Rev. and enl. ed. New York: Praeger, 1969.

Service, John S. *The Amerasia Papers: Some Problems in the History of U.S.-China Relations.* University of California Center for Chinese Studies. China Research Monographs, no. 7. Berkeley: University of California Press, 1971.

Sheng, Michael M. "America's Lost Chance in China? A Reappraisal of Chinese Communist Policy Toward the United States Before 1945." *Australian Journal of Chinese Affairs* 29 (1993), pp. 135–57.

———. *Battling Western Imperialism: Mao, Stalin, and the United States.* Princeton, NJ: Princeton University Press, 1997.
———. *Maoist Dualism and Chinese Communist Foreign Relations, 1935–1949.* Ph.D. diss., York University. Ann Arbor, MI: UMI Dissertation Services, 1994.
———. "Mao, Stalin, and the Formation of the Anti-Japanese United Front: 1935–1937." *CQ* 129 (1992), pp. 150–70.
———. "Response: Mao and Stalin: Adversaries or Comrades." *CQ* 129 (1992), pp. 180–83.
———. "The Triumph of Internationalism: The CCP-Moscow Relations Before 1949." *Diplomatic History* 1 (Winter 1997), pp. 95–104.
Shen Zhihua. *Chaoxian zhanzheng jiemi.* Hong Kong: n.p., 1995.
Shi Chzhe. "Soprovozhdaia predsedatelia Mao." *PDV* 1 (1989), p. 146.
Shi Yinhong. "Cong Ya-er-ta mimi xieding dao 1945 nan Zhong Su youhao tongmeng tiaoyue." *Jinxiandai guoji guanxi* 12 (1986), pp. 61–70.
Shi Yuanhua. *Zhonghua minguo waijiao shi.* Shanghai: Shanghai renmin chubanshe, 1994.
Shi Zhe. "Gao Gang ceng jianyi jiang Dongbei huagui Sulian ma?" *Yan Huang chunqiu* 4 (1993), pp. 84–85.
———. "Peitong Mao zhuxi fang Su." *Renwu* 5 (1988), pp. 5–21.
———. "Peitong Mao zhuxi fang Su." In *Feng yu gu: Shi Zhe huiyi lu*, ed. Shi Qiulang, pp. 47–72. Peking: Hongqi chubanshe, 1992.
———. "With Mao and Stalin: The Reminiscences of Mao's Interpreter." Pt. 2. *Chinese Historians* 1 (Spring 1993), pp. 67–90.
———. *Zai lishi juren shenbian: Shi Zhe huiyi lu.* Peking: Zhongyang wenxian chubanshe, 1991.
Shi Zhe and Chen Jian. "With Mao and Stalin: Shi Zhe's Memoirs." Unpublished draft. Dieter Heinzig archives, Erftstadt, Germany.
Shi Zhongquan. *Zhou Enlai de zhuoyue fengxian.* Peking: Zhonggong zhongyuan dangxiao chubanshe, 1993.
Sino-Soviet Treaty and Agreements (signed in Moscow on February 14, 1950). Peking: Foreign Language Press, 1951.
Skosyrev, Vladimir. "Stalin i Mao molcha nabliudali, kak tantsuet politbiuro." *Nezavisimaia gazeta*, July 5, 1994.
Sladkovskii, M.I. *Istoriia torgovo-ekonomicheskikh otnoshenii SSSR s Kitaem* (1917–1974). Moscow: Nauka, 1977.
Snow, Edgar. *The Other Side of the River: Red China Today.* London: Gollancz, 1963.
———. *Red Star over China.* New York: Grove, 1961.
Sovetsko-kitaiskie otnosheniia, 1917–1957: Sbornik dokumentov. Moscow: Izdatel'stvo vostochnoi literatury, 1959.
Sovetsko-mongol'skie otnosheniia: Sbornik dokumentov. Moscow: Nauka, 1966.
Stalin, I.V. *Sochineniia.* Vol. 10. Moscow: Gosudarstvennoe izdatel'tvo politicheskoi literatury, 1954.
Stolberg, Eva-Maria. *Stalin und die chinesischen Kommunisten, 1945–1953: Eine Studie zur Entstehungsgeschichte der sowjetisch-chinesischen Allianz vor dem Hintergrund des Kalten Krieges.* Stuttgart: Franz Steiner, 1997.
Sto sorok besed s Molotovym: Iz dnevnika F. Chueva. Moscow: Terra, 1991.
Tai Wan-chin. "The Conclusion of the 1945 Sino-Soviet Treaty of Friendship and Alliance: Joseph Stalin's Negotiating Style and Skills." In *The 1945 Sino-Soviet Treaty of Friendship and Alliance*, ed. Tai Wan-chin, pp. 105–33. Taipei: Tamkang University,1995.
Tao Wenzhao, Yang Kuisong, and Wang Jianlang. *Kang Ri zhanzheng shiqi zhong Zhongguo duiwai guanxi.* Peking: Zhonggong dangshi chubanshe, 1995.
Teiwes, Frederick C. *The Formation of the Maoist Leadership, from the Return of Wang Ming to the Seventh Party Congress.* SOAS University of London, Research Notes and Studies Series, no. 10. London: Contemporary China Institute, School of Oriental and African Studies, 1994.
———. *Politics at Mao's Court: Gao Gang and Party Factionalism in the Early 1950s.* Armonk, NY: M.E. Sharpe, 1990.
"Telegramma posla SSSR P.F. Iudina Pervomu sekretariu TsK KPSS N.S. Khrushchevu 30 oktiabria 1957 goda." *Istochnik* 4 (1996), pp. 110–14.
Thornton, Richard C. *The Comintern and the Chinese Communists, 1928–1931.* Seattle: University of Washington Press, 1969.
———. "The Emergence of a New Comintern Strategy for China: 1928." In *The Comintern: Historical Highlights. Essays, Recollections, Documents*, ed. Milohrad M. Drachkovitch and Branko Lazitch. New York: Praeger; London: Pall Mall, 1966.

———. *China: A Political History, 1917–1980.* Boulder, CO: Westview, 1982.
———. *Odd Man Out: Truman, Stalin, Mao, and the Origins of the Korean War.* Washington, DC: Brassey's, 2000.
Tikhvinskii, S.L. *Kitai v moei zhizni.* Moscow: Nauka, 1992.
———. "O 'sekretnom demarshe' Chzhou En'laia i neofitsial'nykh peregovorakh KPK s amerikantsami v iiune 1949 g." *PDV* 3 (1994), pp. 133–38.
———. "Perepiska I.V. Stalina s Mao Tszedunom v ianvare 1949 g." *NiNI* 4–5 (1994), pp. 132–40.
Titarenko, M.L., et al., eds. *VKP(b), Komintern i natsional'no-revoliutsionnoe dvizhenie v. Kitae: Dokumenty.* Vol. 2, *1926–1927.* Parts 1–2. Moscow: RTsKhIDNI, Institut Dal'nego Vostoka RAN, Vostochno-aziatskii Seminar Svobodnogo Universiteta Berlina, 1996.
Townsend, James. "Chinese Nationalism." In Unger, ed., *Chinese Nationalism*, pp. 1–30.
Treaties Between the Republic of China and Foreign States (1927–1957). Zhong wai tiaoyue jibian (Zhonghua minguo shiliu nian zhi sishiliu nian). Waijiaobu bian. Taipei: Ministry of Foreign Affairs, 1958.
Truman, Harry S. *Year of Decisions: 1945.* Vol. 1 of *The Memoirs of Harry S. Truman.* London: Hodder & Stoughton, 1956.
Tsou, Tang. *America's Failure in China, 1941–1950.* 2 vols. Chicago: University of Chicago Press, 1967.
Tuchman, Barbara. *Sand Against the Wind: Stilwell and the American Experience in China, 1911–45.* London: Macmillan, 1971.
Tucker, Nancy Bernkopf. *Patterns in the Dust: Chinese-American Relations and the Recognition Controversy, 1949–1950.* New York: Columbia University Press, 1983.
Ulam, Adam B. *Expansion and Coexistence: The History of Soviet Foreign Policy, 1917–67.* London: Secker & Warburg, 1968.
Unger, Jonathan, ed. *Chinese Nationalism.* Armonk, NY: M.E. Sharpe, 1996.
United Nations Treaty Series, vol. 48. New York: United Nations, 1950.
———, vol. 226. New York: United Nations, 1956.
Uschakow, Alexander, and Dietrich Frenzke. *Der Warschauer Pakt und seine bilateralen Bündnisverträge: Analyse und Texte.* Berlin: Berlin Verlag Arno Spitz, 1987.
Van Min. *Polveka KPK i predatel'stvo Mao Tsze-duna.* Moscow: Izdatel'stvo politicheskoi literatury, 1975.
Van Slyke, Lyman P., ed. *The Chinese Communist Movement: A Report of the United States War Department, July 1945.* Stanford, CA: Stanford University Press, 1968.
Vasold, Manfred. "Versäumte Gelegenheiten? Die amerikanische Chinapolitik im Jahre 1949." *Vierteljahreshefte für Zeitgeschichte* 2 (1983), pp. 242–71.
Vladimirov, O. *Sovetsko-kitaiskie otnosheniia v sorokovykh-vos'midesiatykh godach.* Moscow: Mezhdunarodnye otnosheniia, 1984. (This book was removed from Soviet bookshelves shortly after publication apparently because its aggressively anti-Chinese tone no longer conformed to the new policy of seeking rapprochement with Peking.)
Vladimirov, O., and V. Riazantsev. *Stranitsy politicheskoi biografii Mao Tsze-duna.* Moscow: Izdatel'stvo politicheskoi literatury, 1973.
Vladimirov, P.P. *Osobyi raion Kitaia, 1942–1945.* Moscow: Izdatel'stvo Agenstva pechati Novosti, 1973.
Vlasov, Iurii. "Povest' ob ottse." Parts 1–2. *PDV* 6 (1990), pp. 124–34; 1 (1991), pp. 171–81.
———. "Pravda ob ottse." *Argumenty i fakty* 18 (1990), p. 4.
Vlasov, Peter. *The Vladimirov Diaries: Yenan, China, 1942–1945.* Garden City, NY: Doubleday, 1975.
Vlasov, Yuri. "The Story of My Father." Parts 1–2. *Far Eastern Affairs* (Moscow) no. 1 (1991), pp. 189–207; no. 2 (1991), pp. 193–208.
Voennaia pomoshch' SSSR v osvoboditel'noi bor'be kitaiskogo naroda. Moscow: Voennoe izdatel'stvo Ministerstva Oborony SSSR, 1975.
Volkogonov, Dmitri. *Stalin: Triumph and Tragedy.* Rocklin, CA: Prima, 1992.
Wang Chien-min. *Zhongguo gongchandang shi gao.* Vol. 2. Taipei: n.p., 1965.
Wang Dongxing. *Wang Dongxing riji.* Peking: Zhongguo shehui kexue chubanshe, 1993.
Wang Fangming. "'Shishi jiu shi, duli sikao'—huiyi Mao zhuxi 1957 nian yici qinqie jianghua." *Renmin ribao*, January 2, 1979.
Wang Jin. "Lüe lun xin Zhongguo chengli chuqi de dui Su 'yi bian dao' zhengce." *Zhonggong dangshi yanjiu, zengkan* (1990), pp. 108–12.
Wang Junyan. *Zhongguo waijiao yanyi: Xin Zhongguo shiqi.* Peking: n.p., 1995.
Wan Min. *Polveka KPK i predatel'stvo Mao Tszeduna.* Moscow: Izdatel'stvo politicheskoi literatury, 1975.
Wassmund, Hans. *Grundzüge der Weltpolitik: Daten und Tendenzen von 1945 bix zur Gegenwart.* Munich: Beck, 1982.

Wei, Henry. *China and Soviet Russia.* New York: Nostrand, 1956.
Wei Shiyan. "Diyi ci fangjian Sulian." In *Mao Zedong de zuji*, pp. 302–8. Peking: Zhonggong dangshi chubanshe, 1993.
Westad, Odd Arne. *Cold War and Revolution: Soviet-American Rivalry and the Origins of the Civil War.* New York: Columbia University Press, 1993.
———. "Losses, Chances, and Myths: The United States and the Creation of the Sino-Soviet Military Alliance, 1945–1950." *Diplomatic History* 1 (Winter 1997), pp. 105–15.
———. "Unwrapping the Stalin-Mao Talks: Setting the Record Straight." *CWIHPB* 6–7 (Winter 1996–97), pp. 23–24.
———, ed. *Brothers in Arms: The Rise and Fall of the Soviet Alliance, 1945–1963.* Washington, DC: Woodrow Wilson Center; Stanford, CA: Stanford University Press, 1998.
White, Theodore H. *In Search of History: A Personal Adventure.* New York: Warner Books, 1979.
Whiting, Allen S. *China Crosses the Yalu: The Decision to Enter the Korean War.* New York: Macmillan, 1960.
———. *Soviet Policies in China, 1917–1924.* Stanford, CA: Stanford University Press, 1953.
Whiting, Allen S., and General Sheng Shi-ts'ai. *Sinkiang: Pawn or Pivot.* East Lansing: Michigan State University Press, 1958.
Wilbur, Martin C. *The Nationalist Revolution in China, 1923–1928.* London: Cambridge University Press, 1983.
Wilbur, Martin C., and Julie Lien-ying How. *Missionaries of Revolution: Soviet Advisers and Nationalist China, 1920–1927.* Cambridge, MA: Harvard University Press, 1989.
Wilson, Dick. *Zhou Enlai: A Biography.* New York: Viking, 1984.
Wingrove, Paul. "Mao in Moscow, 1949–50: Some New Archival Evidence." *Journal of Communist Studies and Transition* 4 (December 1995), pp. 309–34.
Wu, Aitchen K. *China and the Soviet Union: A Study of Sino-Soviet Relations.* London: Methuen, 1950.
Wu Tien-wei. *The Sian Incident: A Pivotal Point in Modern Chinese History.* University of Michigan Center for Chinese Studies. Michigan Papers in Chinese Studies, no. 26. Ann Arbor: University of Michigan Press, 1976.
Wu Xiuquan. *Huiyi yu huainian.* Peking: Zhongggong zhongyang dangxiao chubanshe, 1991.
———. "'Zhong-Su youhao tongmeng huzhu tiaoyue' qianding de jingguo." In *Xin Zhongguo waijiao fengyun*, pp. 10–14. Peking: Shijie zhishi chubanshe, 1990.
Wylie, Raymond F. *The Emergence of Maoism: Mao Tse-tung, Ch'en Po-ta, and the Search for Chinese Theory, 1935–1945.* Stanford, CA: Stanford University Press, 1980.
Xiang Qing. "Guanyu Si-da-lin ceng fou quanzu jiefang dajun guo jiang zhi wojian." *Dang de wenxian* 6 (1989).
Xiang Qing, You Zhifu, and Liu Dexi, eds. *Sulian yu Zhongguo geming.* Peking: Zhongyang bianyi chubanshe, 1994.
Xia Zhenduo. "Sulian hongjun jinzhu 10 nian." In *Zhonggong dangshi ziliao*, 43 (1992), pp. 125–53.
Xu Yan. *Sulian chubing Zhongguo Dongbei jishi.* Hong Kong: Tiandi tushu youxian gongsi, 1993.
Yang, Benjamin. "The Zunyi Conference, or One Step in Mao's Rise to Power: A Survey of Historical Studies of the Chinese Communist Party." *CQ* 106 (1986), pp. 235–71.
Yang Kuisong. "Hua-de shijian yu xin Zhongguo dui Mei zhengce de queding." *Lishi yanjiu* 5 (1994), pp. 104–17.
———. "The Soviet Factor and the CCP's Policy Toward the United States in the 1940s." *Chinese Historians* 1 (Spring 1992), pp. 17–34.
———. "Zhongguo gongchandang duoquan Dongbei zhanlüe yanbien yu Sulian." *Zhonggong dangshi yanjiu, zengkan* (1990), pp. 60–71.
———. *Zhongjian didai de geming: Zhongguo geming de celüe zai guoji beijing xia de yanbian.* Peking: Zhonggong zhongyang dangxiao chubanshe, 1992.
Yang Yunruo. *Gongchan guoji he zhongguo geming guanxi jishi (1919–1943).* Peking: Zhongguo shehui kexue chubanshe (xian guonei faxing), 1983.
Yang Yunruo and Yang Kuisong. *Gongchan guoji he Zhongguo geming.* Shanghai: Shanghai renmin chubanshe, 1988.
Young, Arthur N. *China and the Helping Hand, 1937–1945.* Cambridge, MA: Harvard University Press, 1963.
Yuan Ming. "The Failure of Conception: America's China Policy, 1949–50." In *Sino-American Relations, 1945–1955: A Joint Reassessment of a Critical Decade*, ed. Harry Harding and Yuan Ming, pp. 143–54. Wilmington, DE: Scholarly Resources, 1989.

Yu Zhan and Zhang Guangyou. "Guanyu Si-da-lin ceng fou quanzu wo guo Changjiang di tantao." In *Xin Zhongguo waijiao fengyun*, pp. 15–21. Peking: Shijie zhishi chubanshe, 1990.

Yu Zhanbang. "Mao Zedong yu Zhang Zhizhong de yi ci zhongyao tanhua." *Zhonggong dangshi ziliao* 48 (1993), pp. 148–55.

Zagoria, Donald S. *The Sino-Soviet Conflict, 1956–1961.* Princeton, NJ: Princeton University Press, 1962.

Zaitsev, Valentin. "Liubimaia mnogotirazhka prezidenta Taivania." *Zhurnalist* 2 (1996), pp. 42–45.

"Zapis' besedy I. V. Stalina s predsedatelem Tsentral'nogo Narodnogo Pravitel'stva Kitaiskoi Narodnoi Respubliki Mao-Tsze-dunom 22 ianvaria 1950 g." AP RF, f. 45, op. 1, d. 329, ll. 29–38.

"Zapis' besedy tovarishcha Stalina I.V. s predsedatelem Tsentral'nogo Narodnogo Pravitel'stva Kitaiskoi Narodnoi Respubliki Mao Tsze-dunom 16 dekabria 1949 g." AP RF, f. 45, op. 1, d. 239, ll. 9–17.

Zeng Kelin. *Rongma shengya de huiyi.* Peking: Renmin jiefangjun chubanshe, 1992.

———. "Xinhuo." In *Geming huilu.* Vol. 7. Shenyang: 1983.

Zhai Qiang. *The Dragon, the Lion, and the Eagle. Chinese/British/American Relations, 1949–1958.* Kent, OH: Kent State University Press, 1994.

Zhang Shu Guang and Chen Jian. "The Emerging Disputes Between Bejing and Moscow: Ten Newly Available Documents, 1956–1958." *CWIHPB* 6–7 (Winter 1995–96), pp. 148–63.

Zhonggong dangshi dashi nianbao shuoming. Peking: n.p., 1983.

Zhonggong zhongyang wenjian xuanji. 18 vols. Peking: Zhonggong zhongyang dangxiao chubanshe, 1992.

Zhonggong zhongyang zai Xiangshan. Peking: Zhonggong dangshi chubanshe, 1993.

Zhongguo gongchandang chuangye sanshi nian (1919–1949). Peking: n.p., 1991.

Zhongguo gongchandang dangzhang. 4th ed. N.p.: Huadong xinhua shudian chuban, 1949.

Zhonghua renmin gongheguo duiwai guanxi wenjian ji. Vol. 1: *1949–1951.* Peking: Shijie zhishi chubanshe, 1957.

Zhonghua renmin gongheguo shilu. Vol. 1, pt. 1. Changchun: Jilin renmin chubanshe, 1994.

Zhonghua renmin gongheguo tiaoyue ji. Vol. 1: *1949–1951.* Peking: Falü chubanshe, 1957.

Zhonghua renmin gongheguo waijiao shi, 1949–1956. Peking: Shijie zhishi chubanshe, 1994.

Zhou Enlai nianpu, 1898–1949. Peking: Zhongyang wenxian chubanshe, 1990.

Zhou Enlai nianpu, 1949–1976. Vol. 1. Peking : Zhongyang wenxian chubanshe, 1997.

Zhou Enlai waijiao huodung dashi ji, 1949–1975. Peking: Shijie zhishi chubanshe, 1993.

Zhou Enlai zhuan, 1898–1949. Peking: Zhongyang wenxian chubanshe, 1989.

Zhou Wenqi. "Guanyu Si-da-lin ceng fou quanzu wo jun daguo Changjiang wenti de jiutao." Appeared in an internal publication of the Central Party School (*zhongyang dangxiao*), title unknown, pp. 9–10. Peking, 1992.

Zhou Wenqi and Chu Liangru, eds. *Gongchan guoji, Sulian he Zhongguo gongchandang guanxi biannianshi, 1919–1991.* Wuhan: Hubei renmin chubanshe (xian guonei faxing), 1993.

Zhu De zhuan. Peking: Renmin chubanshe, Zhongyang wenxian chubanshe, 1993.

Zhu Yuanshi. "Liu Shaoqi 1949 nian mimi fang Su." *Dang de wenxian* 3 (1991), pp. 74–80.

Zhu Zhongli. *Mao Zedong Wang Jiaxiang zai wo de shenghuo zhong.* Peking: Zhonggong zhongyang dangxiao chubanshe, 1995.

Ziemann, Han-jung. *Die Beziehungen Sinkiangs (Ostturkestan) zu China und der UdSSR, 1917–1945.* Bochum, Germany: Studienverlag Brockmeyer, 1984.

Zubok, Vladislav. "The Mao-Khrushchev Conversations, 31 July–3 August 1958 and 2 October 1959." *CWIHPB* 12–13 (Fall–Winter 2001), pp. 244–72.

———. "'To Hell with Yalta!'—Stalin Opts for a New Status Quo." *CWIHPB* 6–7 (Winter 1995–96), pp. 24–27.

Zubok, Vladislav, and Constantine Pleshakov. *Inside the Kremlin's Cold War: From Stalin to Khrushchev.* Cambridge, MA: Harvard University Press, 1996.

Index

Dieter Heinzig completed a doctorate in East European history, Chinese history, and public law at Bochum University, Germany. Until his retirement in 1997, he did research at the Federal Institute for East European and International Studies in Cologne. He was deputy director and head of the Asia Department of the Federal Institute.